Structural Dynamics with X-ray and Electron Scattering

Theoretical and Computational Chemistry Series

Editor-in-chief:
Jonathan Hirst, *University of Nottingham, Nottingham, UK*

Advisory board:
Dongqing Wei, *Shanghai Jiao Tong University, China*
Jeremy Smith, *Oakridge National Laboratory, USA*

Titles in the series:
1: Knowledge-based Expert Systems in Chemistry: Not Counting on Computers
2: Non-Covalent Interactions: Theory and Experiment
3: Single-Ion Solvation: Experimental and Theoretical Approaches to Elusive Thermodynamic Quantities
4: Computational Nanoscience
5: Computational Quantum Chemistry: Molecular Structure and Properties *in Silico*
6: Reaction Rate Constant Computations: Theories and Applications
7: Theory of Molecular Collisions
8: *In Silico* Medicinal Chemistry: Computational Methods to Support Drug Design
9: Simulating Enzyme Reactivity: Computational Methods in Enzyme Catalysis
10: Computational Biophysics of Membrane Proteins
11: Cold Chemistry: Molecular Scattering and Reactivity Near Absolute Zero
12: Theoretical Chemistry for Electronic Excited States
13: Attosecond Molecular Dynamics
14: Self-organized Motion: Physicochemical Design based on Nonlinear Dynamics
15: Knowledge-based Expert Systems in Chemistry: Artificial Intelligence in Decision Making
16: London Dispersion Forces in Molecules, Solids and Nano-structures: An Introduction to Physical Models and Computational Methods
17: Machine Learning in Chemistry: The Impact of Artificial Intelligence
18: Tunnelling in Molecules: Nuclear Quantum Effects from Bio to Physical Chemistry
19: Understanding Hydrogen Bonds: Theoretical and Experimental Views
20: Computational Techniques for Analytical Chemistry and Bioanalysis
21: Effects of Electric Fields on Structure and Reactivity: New Horizons in Chemistry
22: Multiscale Dynamics Simulations: Nano and Nano-bio Systems in Complex Environments
23: Exploration on Quantum Chemical Potential Energy Surfaces: Towards the Discovery of New Chemistry
24: Cheminformatics and Bioinformatics at the Interface with Systems Biology: Bridging Chemistry and Medicine
25: Structural Dynamics with X-ray and Electron Scattering

How to obtain future titles on publication:
A standing order plan is available for this series. A standing order will bring delivery of each new volume immediately on publication.

For further information please contact:
Book Sales Department, Royal Society of Chemistry, Thomas Graham House, Science Park, Milton Road, Cambridge, CB4 0WF, UK
Telephone: +44 (0)1223 420066, Fax: +44 (0)1223 420247
Email: booksales@rsc.org
Visit our website at books.rsc.org

Structural Dynamics with X-ray and Electron Scattering

Edited by

Kasra Amini
Max-Born-Institut, Germany
Email: kasra.amini@mbi-berlin.de

Arnaud Rouzée
Max-Born-Institut, Germany
Email: arnaud.rouzee@mbi-berlin.de

and

Marc J. J. Vrakking
Max-Born-Institut, Germany
Email: marc.vrakking@mbi-berlin.de

ROYAL SOCIETY
OF **CHEMISTRY**

Theoretical and Computational Chemistry Series No. 25

Print ISBN: 978-1-83767-114-4
PDF ISBN: 978-1-83767-156-4
EPUB ISBN: 978-1-83767-158-8
Print ISSN: 2041-3181
Electronic ISSN: 2041-319X

A catalogue record for this book is available from the British Library

The Royal Society of Chemistry is a charity, registered in England and Wales, Number 207890, and a company incorporated in England by Royal Charter (Registered No. RC000524), registered office: Burlington House, Piccadilly, London W1J 0BA, UK, Telephone: +44 (0) 20 7437 8656.

For further information see our web site at www.rsc.org

Printed in the United Kingdom by CPI Group (UK) Ltd, Croydon, CR0 4YY, UK

Preface

With the advent of ultrafast lasers,[1-3] one of the most insightful methods to investigate photoinduced processes in molecules and condensed matter is the implementation of pump–probe experiments,[4,5] which traces back to the seminal work of Abraham and Lemoine at the end of the 19th century.[6] In a pump–probe experiment, a first short laser pulse (typically referred to as the "pump") initiates a dynamical process that is subsequently interrogated after a variable time delay using a second short laser pulse (typically referred to as the "probe"). The probe laser maps the time-evolving, optical pump-induced dynamics of the system under investigation onto a suitable experimental observable.[7] This mapping is performed similarly to stroboscopic photographic measurements that enable an observer to capture a picture of rapidly moving objects using multiple flashes of light.[8] Observation of the dependence of this observable on the pump–probe time delay encodes the dynamical response of the system. It provides time-domain information that complements and frequently supersedes the frequency-domain information that can be obtained through more traditional spectroscopy approaches.

The development of Ti:Sapphire and Kerr lens mode-locking technology (for reviews, see ref. 3 and 9) spurred the rapid proliferation of ultrafast laser techniques, for example, in the extension towards both shorter and longer wavelengths employing non-linear wave-mixing techniques such as by optical parametric amplification.[10,11] In particular, the application of ultrafast lasers and ultrafast pump–probe spectroscopy experienced tremendous expansion, culminating in the Nobel Prize in Chemistry that was awarded in 1999 to Prof. Ahmed Zewail of Caltech for the development of the field of femtochemistry.[12] Although this reward may be regarded as clear recognition of the impact of pump–probe spectroscopy, the implementation and interpretation of these experiments are not without challenges.

Theoretical and Computational Chemistry Series No. 25
Structural Dynamics with X-ray and Electron Scattering
Edited by Kasra Amini, Arnaud Rouzée and Marc J. J. Vrakking
© The Royal Society of Chemistry 2024
Published by the Royal Society of Chemistry, www.rsc.org

The response of matter to incident light is frequently very complex and usually depends on both the electronic response and the motion of the nuclei in a manner that cannot easily be separated.[13] As a consequence, understanding such experiments required the development of sophisticated computational methods[14] that enable the tracking of the dynamics of the system with many coupled degrees of freedom. For this reason, significant efforts are now being pursued to develop experimental techniques that either focus on the purely electronic response of a system to an incident light pulse or methods that primarily study structural changes during nuclear dynamics.

The typical timescale for purely electronic motion is in the attosecond domain. Nowadays, the electronic response of matter to incident light, and the early stages of the coupling of electronic to nuclear degrees of freedom, have become critical points of interest in attosecond science.[15–18] In this field, pump–probe spectroscopy is implemented, making use of attosecond laser pulses that are created by the process of high-harmonic generation (HHG),[19,20] or, more recently, by special modes of operation of X-ray free electron lasers.[21,22] In contrast, nuclear motion is revealed in unprecedented detail using several new experimental techniques based on ultrafast X-ray[23,24] or electron diffraction[25,26] and microscopy,[27] which nowadays can be performed with a time resolution in the femtosecond domain.[28–38] Although X-ray and electron diffraction are primarily based on the scattering of incident X-rays or electrons from the electrons in a sample, the localisation of the electrons around the nuclei permits these methods to track the electron density distributions and nuclear motion in the sample with sub-atomic (sub-Ångström) spatial resolution.

This book aims to provide the interested reader with the latest recent developments in the structural dynamics of gas-phase molecules and condensed matter physics, particularly in connection with the recent emergence of ultrafast time-resolved experimental methods based on X-ray and electron diffraction. Figure 1 provides a schematic overview of our book structure, highlighting the areas of overlap and differences between gas-phase molecular dynamics and condensed matter physics, and the X-ray and electron scattering methods employed to study them. At the core of the combined works is the capability to provide a complete picture of the coupled electron, nuclear and spin ultrafast dynamics in photoexcited samples.

The book aims to both present the current state-of-the-art and to make this material accessible to students and researchers who have no prior experience in this research field. For this reason, the book is divided into two parts (see the black border in Figure 1). The book starts with three introductory chapters that provide introduction to ultrafast molecular spectroscopy in the gas phase (Chapter 1, M. Gühr), ultrafast spectroscopy in condensed matter (Chapter 2, M. Beye) and the theory of time-dependent scattering processes in ultrafast X-ray and electron scattering on molecules (Chapter 3, M. Simmermacher *et al.*). Note that introductions to the theory of time-dependent scattering processes in condensed matter are contained in Chapter 4 by Hauf *et al.* (phonon dynamics and charge density waves studied

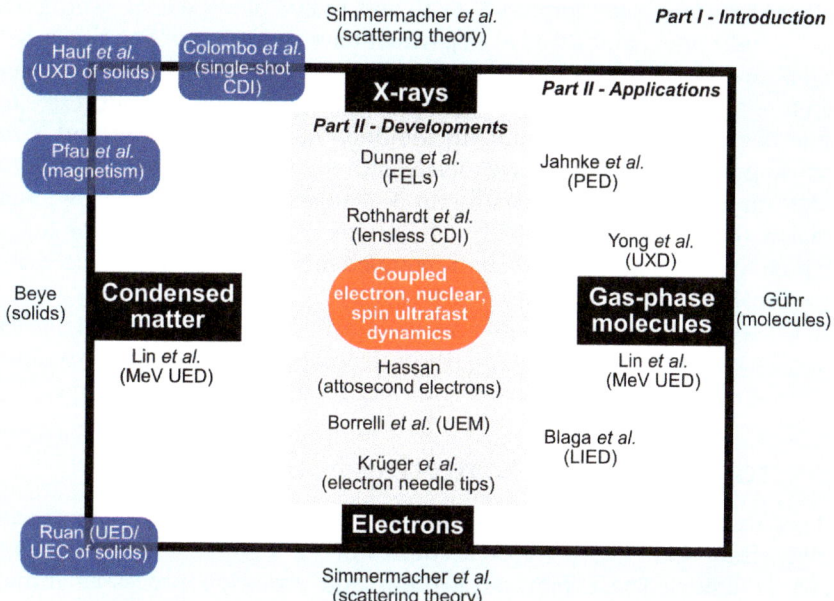

Figure 1 Schematic view of this book, illustrating the role of the different chapters within the scope of the book. The book is split into two parts: (i) introduction, and (ii) applications and developments. The chapter contributions are arranged around the four pillars of the book (X-ray and electron methods employed in the study of gas-phase molecules and condensed matter), the combination of which can provide details of the coupled electronic, nuclear and spin ultrafast dynamics of samples, wherever appropriate.

by X-ray scattering), Chapter 7 by Pfau *et al.* (magnetisation and electron-spin dynamics studied by extreme ultraviolet and X-ray methods), and Chapter 11 by Ruan (lattice dynamics studied by ultrafast electron diffraction and crystallography). The latter chapter by Ruan also provides a thorough overview of the history, formalisms, technological developments and applications of ultrafast electron diffraction and microscopy. Moreover, Chapter 5 by Colombo *et al.* provides a systematic guide to the data analysis and interpretation of single-shot coherent diffraction imaging (CDI) results (*e.g.*, details on solving the inverse scattering problem, phase retrieval algorithms, and correctly interpreting CDI data).

The second half of the book contains expert contributions by leading researchers that have pioneered the use of ultrafast electron and X-ray diffraction and scattering methods in gas-phase photochemistry and condensed matter research (*e.g.*, see Chapters 4, 5, 7–9, and 12 by Hauf *et al.*, Colombo *et al.*, Pfau *et al.*, Dunne *et al.*, Yong *et al.*, and Lin *et al.*). These chapters are underpinned and supplemented by several chapters describing the state-of-the-art ultrafast electron and X-ray sources (*e.g.*, see Chapters 4, 11, 12, 15 and 16 by Hauf *et al.*, Ruan, Lin *et al.*, Borrelli *et al.*, and Krüger *et al.*)

and methods (*e.g.*, see Chapters 5, 6, 10 and 13 by Colombo *et al.*, Rothhardt *et al.*, Jahnke *et al.*, and Blaga *et al.*), which have recently reached the attosecond domain[21,36–41] (Chapter 14, M. T. Hassan). In the years to come, we may therefore anticipate a synthesis of (i) attosecond methods that address electronic degrees of freedom, and (ii) methods based on electron and X-ray scattering, which until now have mostly, although not exclusively, been used to determine time-dependent structural changes. In this manner, we may envision the methods described in this book forming the basis for future research that will be able to probe the coupled motion of electrons, nuclei and spins on all relevant timescales (see Figure 1).

In the remainder of this chapter, we will further discuss and investigate the synergies and differences between (i) spectroscopy and scattering/diffraction methods, and (ii) X-ray diffraction and electron diffraction/microscopy.

Spectroscopy *Versus* Scattering and Diffraction

In spectroscopy experiments, information on a system of interest is typically derived from wavelength-resolved studies of the light that the system absorbs or emits. These light absorption and emission processes induce transitions between the quantum states of the system, which are described by quantum mechanics. Spectroscopy experiments sparked the development of quantum mechanics. The quantisation of energy was first introduced by Planck in 1901[42] in order to explain the spectrum of electromagnetic radiation emitted by a black body in thermal equilibrium. This was extended a few years later by Einstein[43] through the photoelectric effect, where he postulated that light can be considered to consist of energy quanta. These initial studies, together with the observation of discrete spectral lines in the absorption spectrum of the hydrogen atom, led to the first development of the atomic model by Rutherford, Bohr and Sommerfeld, in which an atom was described as a dense nucleus of positive charge surrounded by lower-mass electrons orbiting with distinct, quantised energy states given by their angular momentum.[44] This first formulation of quantum mechanics was based on the idea that the observed spectral emission lines of the hydrogen atom corresponded to transitions of electrons between orbits. This initial description, together with the "théorie des quanta" by de Broglie in 1924,[45,46] provided the stepping stones for the development of modern quantum mechanics by Heisenberg, Max Born, Jordan and Schrödinger, to cite a few. Schrödinger described the quantum "eigenstate" of an electron in an atom as a wave function, $\Psi(r)$,[47] with the squared modulus, $|\Psi(r)|^2$, providing the probability distribution of finding the electron at a particular position, r, or with a given momentum, at a given time. In this description, the emission lines of the hydrogen atoms were simply given by the energy difference between two eigenstates of the system (through the solution of the time-independent Schrödinger equation).

The eigenstate picture in quantum mechanics does not explain why quantum systems can display dynamical behaviour. This behaviour can be understood from the superposition principle of quantum mechanics, which states that if a system has several eigenstates (*e.g.*, the different Rydberg states of an atom or the different rotational or vibrational states of a molecule), then a coherent superposition of these states is also a possible state of the system.[48] This is exploited in pump–probe experiments, where the pump laser populates a coherent superposition of eigenstates (a "wavepacket"), which is mapped using the probe laser onto a suitable observable. The observed time dependence in the pump–probe experiment derives from the different rates at which eigenstates of different energy in the coherent superposition accumulate phase in the time-interval between the pump and the probe laser interaction, and can be modelled by solving the time-dependent Schrödinger equation. When large numbers of quantum states are coherently excited in an experiment, the observed dynamics may somewhat resemble that of a classical system that evolves under the same equations of motion.

Following the first demonstrations of these concepts in experiments by – especially – Zewail and co-workers,[12] pump–probe techniques have been extended to systems of increasing complexity, and more sophisticated observables have been introduced.[5,49] For example, in pump–probe experiments involving ionisation, measurements of the kinetic energy and angular distribution of photoelectrons have been introduced, which can, for example, very sensitively capture situations where the nature of the electronic state changes during the dynamics (see ref. 50–53 for gas-phase studies and ref. 54 for a recent review of angle-resolved photoemission studies of quantum materials, including a time-resolved investigation). Nevertheless, the interpretation of these experiments is seldom trivial. The outcome of a pump–probe experiment can be reliably predicted when sufficient information about the system is available ahead of time. For example, the nuclear dynamics of a molecular pump–probe experiment using time-resolved ("transient") absorption or time-resolved photoelectron spectroscopy as its observable can be reliably predicted if the relevant potential energy surfaces are available ahead of time, but, in contrast, it is very difficult to infer information on the molecular structure (and on nuclear wavepackets), which is needed to obtain a mechanistic understanding of the dynamics, directly from these experimental observables. For many pump–probe experiments that have appeared in the literature, this means that they either could have been predicted based on information available prior to the experiment or the experimental work needed to be accompanied by extensive theoretical work to draw significant conclusions.

The modern ultrafast structural methods that are the main topic of this book seek to find a way out of this conundrum. They do so by replacing the reliance on spectroscopy with a reliance on scattering/diffraction. Scattering is a process that occurs whenever light or an elementary quantum particle, such as an electron, collides with another particle, thereby changing its direction and (possibly) its energy. Within quantum mechanics, the result of the scattering is an outgoing wave that typically covers a substantial angular range.

Diffraction is based on the fact that scattering of a particle from multiple, nearby scattering centres (for example, the different atoms that together form a molecule) leads to very pronounced interferences in the scattered waves. A particularly vivid example is Bragg diffraction, where efficient specular reflection of X-rays from a crystal is observed when the wavelength λ of the X-rays and the angle of incidence θ satisfy the equation $n\lambda = 2d \sin \theta$, where d is the separation between the crystal layers.[55] The beauty of this measurement is that it provides the lattice spacing d without needing to know anything about the composition of the crystal or the quantum states of its constituent atoms.

With the emergence of modern synchrotrons at many locations around the world, X-ray crystallography has become one of the most important and widely used scientific techniques, with applications that range from resolving the essential building blocks of life (*e.g.*, the structures of proteins such as myoglobin and haemoglobin,[56,57] and the double-helix structure of DNA[58]) to the determination of the structure of the Severe Acute Respiratory Syndrome Coronavirus 2 (SARS-CoV-2) that causes COVID-19,[59,60] and numerous novel drugs developed by the pharmaceutical industry.[61] However, these X-ray diffraction experiments at synchrotrons are often unsuitable for ultrafast pump–probe experiments, for two reasons. The first reason is that the pulse durations at synchrotrons are typically a few tens of picoseconds (*i.e.*, about a thousand times longer than the typical timescale for atomic motion). This problem can be overcome by implementing slicing schemes,[62] which, however, leads to a significant reduction of the available X-ray flux that is prohibitive for many applications. A second reason is that X-ray-induced damage will occur in many samples of interest before a sufficient X-ray fluence has illuminated the sample. These problems are circumvented by some of the ultrashort X-ray sources introduced and exploited in this book.

On the one hand, modern ultrafast X-ray sources such as the Linac Coherent Light Source (LCLS) at Stanford[63] (see Chapter 8, M. Dunne *et al.*) concentrate all the X-ray photons in a pulse that is typically a few tens of femtoseconds long, and in some cases, even of sub-femtosecond duration.[21,22] Such pulses are perfect for capturing snapshots of a molecule that is undergoing a photoinduced structural transformation (see Chapter 9, H. Yong *et al.*) or of condensed matter that is undergoing a photoinduced phase transition (see Chapter 8, M. Dunne *et al.*). On the other hand, these ultrashort and ultrabright X-ray pulses will scatter and diffract off the sample within a time interval that is so short that the atoms do not significantly change their position during the X-ray pulse. Although extensive X-ray-induced ionisation processes inevitably lead to the destruction of the sample during the experiment, the "diffract-before-destroy" approach (see Chapter 5, A. Colombo *et al.*) allows structural determination on timescales before significant X-ray-induced structural changes take place, and has led to the development of serial crystallography protocols[64–66] at X-ray free electron lasers where structures are determined from nano-crystals[64] that are too small to be used at a synchrotron.

It should be noted that X-ray free electron lasers like the LCLS,[63] SACLA[67] and the EuXFEL[68] not only offer unique novel opportunities when they are applied to time-resolved X-ray diffraction but also permit several further advances that are significant in the context of the topics considered within this book. First, using X-rays facilitates the interpretation of pump–probe spectroscopy, given that X-ray absorption occurs in particular by core rather than valence electrons. This implies that ultrashort X-ray pulses can be used to configure pump–probe experiments where the use of X-rays that are near-resonant with an atomic edge creates conditions where the detection is both site- and element-specific.[69–72] This aspect is discussed in the introductory chapters on ultrafast molecular spectroscopy in the gas phase (Chapter 1, M. Gühr) and ultrafast spectroscopy in condensed matter (Chapter 2, M. Beye). In addition, site- and element-specific experiments can be performed in combination with X-ray dichroism studies to investigate magnetic phenomena by offering direct access to the magnetic moments of specific atom elements within a material. This aspect is extensively discussed in Chapter 7 by Pfau *et al.*, which presents pioneering X-ray resonant scattering and holography experiments on magnetisation dynamics in thin films and multilayers on the femtosecond to nanosecond time scale.

Besides producing ultrashort X-ray pulse durations, the X-ray free electron lasers can also produce extremely high focused intensities, leading to a situation where atoms and molecules can be stripped of most of their electrons.[73,74] In the case of molecules, this results in a violent Coulomb explosion, where the ionic fragments are ejected with momenta that reflect the original position of the atom within the molecule.[74] Using detectors that can measure both the position- and time-of-impact of all fragment ions with high precision,[75] the 3D momenta of all fragment ions can be measured in coincidence, and reconstruction of the original molecular structure prior to X-ray ionisation has been demonstrated.[76] In the near future, using a high repetition rate that will be available at X-ray free electron lasers like LCLS-II (see Chapter 8, M. Dunne *et al.*), it will be possible to configure pump–probe experiments where Coulomb explosion imaging (CEI) will be used to monitor pump-laser induced structural changes in molecules as a function of time.

A further application of X-ray free electron lasers described in this book (see Chapter 10, T. Jahnke and D. Rolles) uses the so-called photoelectron holography method. This is the first example of an electron diffraction experiment induced by X-rays. Here, under suitably chosen experimental conditions, the ionisation of a molecule by an X-ray laser pulse leads to the formation of high-energy electrons with a de Broglie wavelength ($\lambda = \pi(2/E_{kin})^{1/2}$, where E_{kin} is the kinetic energy of the outgoing photoelectron) that is small compared to the internuclear distances that exist within the molecule. As a result, a hologram will be formed, in which pathways through which the photoelectron leaves the molecule without scattering can interfere with pathways where the photoelectron scatters off one or more of the neighbouring atoms as they leave the molecule.[77,78]

The method has been implemented to determine static molecular structures[79] and follow photoinduced structural changes.[80]

The opportunities offered by modern X-ray free electron lasers are extremely exciting, but the development and use of these machines requires massive financial investment. For this reason, it is understandable that alternative approaches are also being pursued. An attractive example is reported in Chapter 4 by Hauf *et al.*, who used a Cu K_α source[81] that is based on irradiation of a Cu tape with intense mid-infrared (MIR, 5.0 μm) laser radiation from an optical parametric chirped pulse amplifier (OPCPA).[82] As described in the chapter, transient electronic charge densities can be extracted from measured diffraction patterns, which reveal phonon dynamics in polar crystals and phonon-driven charge dynamics in molecular crystals and ferroelectrics. Another approach reported in Chapter 6 by Rothhardt *et al.* is to use the high-order harmonic generation process, which, as previously discussed, allows for the generation of attosecond extreme ultraviolet and soft X-ray pulses from the irradiation of (typically) an atomic gas target with an intense visible or near-infrared (NIR) laser pulse.[19,20] The photon energy achieved by HHG is much lower than at a free electron laser. However, table-top HHG setups are quite compact, and the XUV/soft X-ray pulses are fully coherent and inherently synchronised with the laser, making them ideal sources for time-resolved diffractive imaging studies. The large bandwidth of ultrashort XUV pulses obtained by HHG introduces new challenges for understanding and analysing ultrafast coherent diffractive imaging studies. In Chapter 6, Rothhardt *et al.* reviews recent techniques, such as lensless coherent diffractive imaging,[83] Fourier transform holography (FTH),[84] and ptychography[85] based on broadband HHG sources. Initial experimental studies aiming for nanoscale imaging of ultrafast dynamics in condensed matter are also included, such as imaging nanoscale heat transport and ultrafast spin dynamics in a nickel antenna and in a Co/Pd multilayer sample, respectively.

From X-ray Diffraction to Electron Diffraction and Microscopy

In addition to laboratory-scale hard X-ray sources, a second major alternative to X-ray free electron lasers is the substitution of X-ray diffraction with electron diffraction. An electron with a 150 eV kinetic energy already has a de Broglie wavelength of as little as 1 Ångström (which, as a laser wavelength, would imply the use of 12.4 keV photons!). In fact, electron bunches with de Broglie wavelengths in the 0.01–0.1 Ångström range can readily be obtained using a photocathode source and a compact accelerator. Moreover, the scattering probability for a few tens of keV electrons is about five orders of magnitude higher than that of X-rays.[86]

However, using an ultrashort electron pulse comes with its own challenges. First, electrons repel each other, and that means that unlike a laser beam, the pulse duration and energy distribution of a high-density electron bunch rapidly deteriorate and broaden during propagation.[87] Accordingly, the use of photocathode-based electron sources with bunches containing large numbers of electrons is typically accompanied by electron acceleration to as high kinetic energies as possible ($\sim$ 100 keV in the case of a lab-scale, non-relativistic, electron photocathode (see Chapter 11, Ruan), and 2−4 MeV in the case of the SLAC MeV-UED instrument (see Chapter 12, Lin *et al.*). In addition, the electron bunch spreading due to space charge can be counteracted by either limiting the number of electrons in the bunch, using very compact setups, and by the implementation of electron pulse compression stages based, for instance, on a radio frequency (RF) compression cavity,[88–90] single-cycle THz fields[34] or DC-field strategies and dipole magnets such as mirror compressors.[91] In these approaches, a temporal "refocusing" of the electron pulse at the sample position is achieved by decelerating the faster electrons and accelerating the slower ones. In this manner, relativistic electron bunches with 10^5 electrons at 7 MeV on target with a duration of sub-10 fs RMS have been accomplished[92] using a bunching cavity, while in the non-relativistic regime, Baum *et al.* demonstrated a pulse duration of 28 fs RMS at 25 keV in a single-electron pulse by employing a microwave RF compression field.[93] We note that in this latter study, temporal spreading due to Coulomb repulsion is minimised by using a single-electron pulse, while the microwave field is used to compensate for quantum effects during propagation (*i.e.*, the temporal broadening of the wavepacket due to the finite energy bandwidth of the electron pulse). For bunches containing a large number of electrons, RF cavities generally allow for compression down to $\sim$ 100 fs. RF cavities are particularly efficient for compensating space-charge-induced dispersion effects. However, the synchronisation of the laser cavity leads to an additional timing jitter and drift, limiting the total temporal resolution of the instrument. Phase-locked synchronisation schemes are therefore required to achieve the best temporal resolution.[94] A second challenge arises from the fact that, although their kinetic energy may be near-relativistic, the velocity of a non-relativistic electron bunch never reaches the speed of light. Accordingly, the velocity mismatch between the optical-pump pulse and non-relativistic electron-probe pulse limits the total temporal resolution when propagating through a sample. This problem can be minimised by working with thin samples or employing laser beams with a tilted wavefront so that the wavefront propagates at the speed of the electron bunch.[95]

The continuous technological development over the last 20 years has led to a dramatic improvement in the overall time-resolution achieved in ultrafast electron diffraction and microscopy experiments, which has allowed the possibility to atomically and temporally resolve a large variety of ultrafast dynamical processes, especially in chemistry and material sciences (Chapter 11, Ruan and Chapter 12, Lin *et al.*). Non-relativistic UED has been initially used to probe the ultrafast dynamics of the 1,2-diiodotetra-fluoroethane ($C_2F_4I_2$) chemical reaction and the ring-opening dynamics of 1,3-cyclohexadiene (CHD) with picosecond time resolution.[25] Since then, non-relativistic UED has been exploited to image the molecular motions in the organic salt $(EDO-TTF)_2PF_6$ undergoing a photoinduced insulator-to-metal phase transition,[96] the lattice and charge density reorganisations occurring during the semiconductor-to-metal transition in vanadium dioxide (VO_2),[97] the laser-induced melting dynamics of aluminium[98] and bismuth,[99] and the laser-induced alignment and bond breaking dynamics in several gas-phase molecular systems,[100,101] to cite a few. Ultrafast photochemical dynamics in isolated molecules and solution, including coherent vibrational wavepacket dynamics, molecular dissociation, ring-opening and isomerisation dynamics, have been investigated using the SLAC MeV-UED instrument.[102–108] MEV-UED studies of material sciences could also temporally and spatially resolve charge density waves in crystals.[109,110]

New schemes are now being developed with an emphasis on further improving the capability of ultrafast electron diffraction and microscopy to image ultrafast dynamics down to the attosecond timescale, allowing for mapping ultrafast electronic dynamics in molecules and solids. Laser-based approaches such as the optical gating approach in which a laser beam (CW or pulsed) interacts with an electron pulse inside a nanostructure or a membrane have been proposed and established to generate a train of attosecond electron pulses,[38,111,112] and significant efforts are now being pursued to generate a single-isolated attosecond electron pulse using an attosecond light-field synthesizer.[113,114] This new development is introduced in Chapter 14 by Hassan. Femtosecond-scale electron pulses can now be generated from a CW electron source by traversing a continuous electron beam through an RF cavity, causing the transverse deflection of electrons. The deflected electrons are then chopped by a downstream aperture, generating ~ 100 fs electron pulses. This new technique is discussed in Chapter 15 by Borrelli *et al.* Ultrafast needle tip-based electron sources are the source of choice for producing transversely coherent electrons, which play a crucial role in high-resolution electron microscopy measurements. Nevertheless, significant progress is still being made with tip-based electron sources that, for example, recently demonstrated the generation of correlated two-electron states with sub-Poissonian statistics, further pushing the boundaries of quantum electron optics. This enables electron heralding, where one electron can be employed to measure a target sample while the second electron can then reference the first electron's existence. Needle sources also provide possibilities to generate attosecond electron pulses through dielectric acceleration on photonic chips.[115,116] These topics are covered in Chapter 16 by Krüger *et al.*

In the landscape of ultrafast electron imaging techniques, an alternative and imaginative way of performing electron diffraction experiments has recently emerged. It uses electrons extracted from the same molecule they are supposed to interrogate. This can be achieved using an ultrashort X-ray laser pulse to ionise the sample of interest, as in the abovementioned photoelectron holography method (Chapter 10, T. Jahnke and D. Rolles) or through a strong laser field. In the latter case, the technique is called Laser-Induced Electron Diffraction[117] (LIED, see Chapter 13, Blaga and DiMauro). In these experiments, molecules under investigation are ionised using strong-field ionisation, typically using a laser pulse in the NIR or MIR wavelength range. The oscillating laser field accelerates part of the outgoing electron wavepacket before being driven back and rescattered on its parent ion. The portion of the electron wavepacket that has rescattered will then be accelerated again by the laser field before detection. Information about the molecular structure is then encoded in the diffraction pattern contained in the kinetic and angle-resolved photoelectron momentum distribution.[118–132] As for any diffractive imaging experiment, the returning electron wavepacket at the time of recollision should have a de Broglie wavelength that is small enough to be sensitive to the molecule's structure. The use of NIR/MIR laser pulses ensures that the ponderomotive acceleration of the ionised electron reaches kinetic energies of a few hundred eV and, therefore, an Ångström-scale de Broglie wavelength, allowing experiments to be performed with a spatial resolution of a few picometres.[129] A drawback of using NIR/MIR laser pulses is that the ionisation and rescattering cross-section dramatically decreases with the laser wavelength. However, this can be remedied by measuring rescattering data with a high repetition rate and high average power femtosecond laser system containing sufficient statistics within reasonable acquisition times.[133] A very attractive property of LIED is that ionisation and recollision occur within an optical cycle of the laser field (*i.e.*, on a timescale of a few fs for NIR wavelengths), allowing for time-resolved studies with few-fs to attosecond time resolution. So far, LIED has been used to image relatively small gas-phase molecules,[129,130,132] and many ongoing works exist in extending the method to probe complex molecular systems and dynamics.[131,134–136]

The recent development of experimental methods and technological improvement of ultrashort X-ray and electron sources over the last two decades has been very impressive, providing scientists with new ways to investigate the fundamental properties of matter. The ultimate goal to resolve all aspects of photoinduced dynamics in gas-phase and condensed matter with attosecond and atomic spatiotemporal resolution is within reach. In turn, our ability to map the dynamical function of more complex systems is expected to substantially impact the development of novel technologies for light harvesting applications, combustion science, photosynthesis and catalysis, and for elaborating quantum materials with specific properties. By providing the reader with a journey through this emerging field, we hope to give a glimpse of all the new, amazing and exciting discoveries ahead of us and provide inspirational encouragement to advance the field further.

Kasra Amini
Arnaud Rouzée
Marc J. J. Vrakking

References

1. A. J. DeMaria, D. A. Stetser and H. Heynau, *Appl. Phys. Lett.*, 2004, **8**, 174–176.
2. C. V. Shank, *et al.*, *Appl. Phys. Lett.*, 1982, **40**, 761–763.
3. U. Keller, *Nature*, 2003, **424**, 831–838.
4. R. G. W. Norrish and G. Porter, *Nature*, 1949, **164**, 658–658.
5. M. Maiuri, M. Garavelli and G. Cerullo, *J. Am. Chem. Soc.*, 2020, **142**, 3–15.
6. H. Abraham and J. Lemoine, *C. R. Acad. Sci. Hebd. Seances Acad. Sci. D*, 1899, **129**, 206–208.
7. T. Elsaesser, *Chem. Rev.*, 2017, **117**, 10621–10622.
8. H. E. Edgerton and J. R. Killian, *Stroboscopic Revolution in Photography*, 1979.
9. S. Backus, *et al.*, *Rev. Sci. Instrum.*, 1998, **69**, 1207–1223.
10. C. Manzoni and G. Cerullo, *J. Opt.*, 2016, **18**, 103501.
11. H. Fattahi, *et al.*, *Optica*, 2014, **1**, 45–63.
12. A. H. Zewail, *J. Phys. Chem. A*, 2000, **104**, 5660–5694.
13. T. R. Nelson, *et al.*, *Chem. Rev.*, 2020, **120**, 2215–2287.
14. J. C. Tully, *J. Chem. Phys.*, 2012, **137**, 22A301.
15. F. Krausz and M. Ivanov, *Rev. Mod. Phys.*, 2009, **81**, 163–234.
16. U. Thumm *et al.*, in *Photonics*, John Wiley & Sons, Ltd, 2015, pp. 387–441.
17. M. Nisoli, *et al.*, *Chem. Rev.*, 2017, **117**, 10760–10825.
18. S. Ghimire, *et al.*, *J. Phys. B: At., Mol. Opt. Phys.*, 2014, **47**, 204030.
19. P. Agostini and L. F. DiMauro, *Rep. Prog. Phys.*, 2004, **67**, 813.
20. A. L'Huillier, in *Attosecond and XUV Physics*, John Wiley & Sons, Ltd, 2014, pp. 321–338.
21. J. Duris, *et al.*, *Nat. Photonics*, 2020, **14**, 30–36.
22. N. Huang, *et al.*, *Innovation*, 2021, **2**, 100097.
23. H. Ihee, *et al.*, *Science*, 2005, **309**, 1223–1227.
24. H. N. Chapman, *et al.*, *Nat. Phys.*, 2006, **2**, 839–843.
25. H. Ihee, *et al.*, *Science*, 2001, **291**, 458–462.
26. A. H. Zewail, *Annu. Rev. Phys. Chem.*, 2006, **57**, 65–103.
27. A. H. Zewail, *Science*, 2010, **328**, 187–193.
28. H. Ki, *et al.*, *Annu. Rev. Phys. Chem.*, 2017, **68**, 473–497.
29. M. Bargheer, *et al.*, *Chem. Phys. Chem.*, 2006, 7, 783–792.
30. M. Buzzi, M. Först and A. Cavalleri, *Philos. Trans. R. Soc., A*, 2019, **377**, 20170478.
31. M. P. Minitti, *et al.*, *Phys. Rev. Lett.*, 2015, **114**, 255501.
32. B. Stankus, *et al.*, *J. Phys. B: At., Mol. Opt. Phys.*, 2020, **53**, 234004.
33. G. Sciaini and R. D. Miller, *Rep. Prog. Phys.*, 2011, **74**, 096101.
34. C. Kealhofer, *et al.*, *Science*, 2016, **352**, 429–433.
35. A. Feist, *et al.*, *Ultramicroscopy*, 2017, **176**, 63–73.
36. P. Baum and A. H. Zewail, *Proc. Natl. Acad. Sci. U. S. A.*, 2007, **104**, 18409–18414.
37. M. T. Hassan, *et al.*, *Nat. Photonics*, 2017, **11**, 425–430.

38. K. E. Priebe, *et al.*, *Nat. Photonics*, 2017, **11**, 793–797.
39. H. Y. Kim, *et al.*, *Nature*, 2023, **613**, 662–666.
40. N. Hartmann, *et al.*, *Nat. Photonics*, 2018, **12**, 215–220.
41. R. N. Coffee, *et al.*, *Philos. Trans. R. Soc., A*, 2019, **377**, 20180386.
42. M. Planck, *Ann. Phys.*, 1901, **309**, 353–363.
43. A. Einstein, *Ann. Phys.*, 1905, **17**, 13244.
44. N. Bohr, *Philos. Mag.*, 1913, **26**, 476–502.
45. L. de Broglie, PhD thesis, Migration-université en cours d'affectation, 1924.
46. L. de Broglie, *Ann. Phys.*, 1925, **10**, 22–128.
47. E. Schrödinger, *Ann. Phys.*, 1926, **13**, 437–490.
48. C. Cohen-Tannoudji, B. Diu and F. Laloë, *Quantum mechanics*, John Wiley & Sons, New York, 1977.
49. J. Lloyd-Hughes, *et al.*, *J. Phys.: Condens. Matter*, 2021, **33**, 353001.
50. D. M. Neumark, *Annu. Rev. Phys. Chem.*, 2001, **52**, 255–277.
51. A. Stolow, A. E. Bragg and D. M. Neumark, *Chem. Rev.*, 2004, **104**, 1719–1758.
52. A. Stolow and J. G. Underwood, *Adv. Chem. Phys.*, 2008, **139**, 497–583.
53. M. S. Schuurman and V. Blanchet, *Phys. Chem. Chem. Phys.*, 2022, **24**, 20012–20024.
54. J. A. Sobota, Y. He and Z.-X. Shen, *Rev. Mod. Phys.*, 2021, **93**, 025006.
55. W. H. Bragg and W. L. Bragg, *Proc. R. Soc. London, Ser. A*, 1913, **88**, 428–438.
56. M. F. Perutz, Nobel Foundation, 1962.
57. J. C. Kendrew, Nobel Foundation, 1962.
58. J. D. Watson and F. H. C. Crick, *Nature*, 1953, **171**, 737–738.
59. W. Dai, *et al.*, *Science*, 2020, **368**, 1331–1335.
60. L. Zhang, *et al.*, *Science*, 2020, **368**, 409–412.
61. A. L. Carvalho, J. Trincão and M. J. Romão, in *Ligand-Macromolecular Interactions in Drug Discovery: Methods and Protocols*, ed. A. C. A. Roque, Humana Press, Totowa, NJ, 2010, pp. 31–56.
62. R. W. Schoenlein, *et al.*, *Science*, 2000, **287**, 2237–2240.
63. P. Emma, *et al.*, *Nat. Photonics*, 2010, **4**, 641–647.
64. H. N. Chapman, *et al.*, *Nature*, 2011, **470**, 73–77.
65. S. Boutet, *et al.*, *Science*, 2012, **337**, 362–364.
66. T. R. M. Barends, *et al.*, *Nat. Rev. Methods Primers*, 2022, **2**, 1–24.
67. T. Ishikawa, *et al.*, *Nat. Photonics*, 2012, **6**, 540–544.
68. W. Decking, *et al.*, *Nat. Photonics*, 2020, **14**, 391–397.
69. B. K. McFarland, *et al.*, *Nat. Commun.*, 2014, **5**, 4235.
70. P. Wernet, *et al.*, *Nature*, 2015, **520**, 78–81.
71. T. J. A. Wolf, *et al.*, *Nat. Commun.*, 2017, **8**, 29.
72. D. Mayer, *et al.*, *Nat. Commun.*, 2022, **13**, 198.
73. L. Young, *et al.*, *Nature*, 2010, **466**, 56–61.
74. A. Rudenko, *et al.*, *Nature*, 2017, **546**, 129–132.
75. J. Ullrich, *et al.*, *Rep. Prog. Phys.*, 2003, **66**, 1463.
76. R. Boll, *et al.*, *Nat. Phys.*, 2022, **18**, 423–428.
77. A. Landers, *et al.*, *Phys. Rev. Lett.*, 2001, **87**, 013002.

78. F. Krasniqi, *et al.*, *Phys. Rev. A: At., Mol., Opt. Phys.*, 2010, **81**, 033411.

79. R. Boll, *et al.*, *Phys. Rev. A: At., Mol., Opt. Phys.*, 2013, **88**, 061402.

80. G. Kastirke, *et al.*, *Phys. Rev. X*, 2020, **10**, 021052.

81. A. Koç, *et al.*, *Opt. Lett.*, 2021, **46**, 210–213.

82. L. von Grafenstein, *et al.*, *Opt. Lett.*, 2017, **42**, 3796–3799.

83. R. L. Sandberg, *et al.*, *Phys. Rev. Lett.*, 2007, **99**, 098103.

84. R. L. Sandberg, *et al.*, *Opt. Lett.*, 2009, **34**, 1618–1620.

85. L. Loetgering, S. Witte and J. Rothhardt, *Opt. Express*, 2022, **30**, 4133–4164.

86. L. Ma, *et al.*, *Struct. Dyn.*, 2020, 7, 034102.

87. B. J. Siwick, *et al.*, *J. Appl. Phys.*, 2002, **92**, 1643–1648.

88. T. van Oudheusden, *et al.*, *Phys. Rev. Lett.*, 2010, **105**, 264801.

89. M. Gao, *et al.*, *Opt. Express*, 2012, **20**, 12048–12058.

90. O. Zandi, *et al.*, *Struct. Dyn.*, 2017, **4**, 044022.

91. M. Mankos, *et al.*, *Ultramicroscopy*, 2017, **183**, 77–83.

92. J. Maxson, *et al.*, *Phys. Rev. Lett.*, 2017, **118**, 154802.

93. A. Gliserin, *et al.*, *Nat. Commun.*, 2015, **6**, 8723.

94. M. R. Otto, *et al.*, *Struct. Dyn.*, 2017, **4**, 051101.

95. P. Baum and A. H. Zewail, *Proc. Natl. Acad. Sci. U. S. A.*, 2006, **103**, 16105–16110.

96. M. Gao, *et al.*, *Nature*, 2013, **496**, 343–346.

97. V. R. Morrison, *et al.*, *Science*, 2014, **346**, 445–448.

98. B. J. Siwick, *et al.*, *Science*, 2003, **302**, 1382–1385.

99. G. Sciaini, *et al.*, *Nature*, 2009, **458**, 56–59.

100. C. J. Hensley, J. Yang and M. Centurion, *Phys. Rev. Lett.*, 2012, **109**, 133202.

101. J. Yang, *et al.*, *Nat. Commun.*, 2015, **6**, 8172.

102. J. Yang, *et al.*, *Nat. Commun.*, 2016, 7, 11232.

103. J. Yang, *et al.*, *Science*, 2018, **361**, 64–67.

104. J. Yang, *et al.*, *Science*, 2020, **368**, 885–889.

105. T. J. A. Wolf, *et al.*, *Nat. Chem.*, 2019, **11**, 504–509.

106. E. G. Champenois, *et al.*, *Science*, 2021, **374**, 178–182.

107. J. Cao, X. Wang and D. Zhong, *Science*, 2021, **374**, 34–35.

108. M.-F. Lin, *et al.*, *Science*, 2021, **374**, 92–95.

109. A. Zong, *et al.*, *Sci. Adv.*, 2018, **4**, eaau5501.

110. A. Kogar, *et al.*, *Nat. Phys.*, 2020, **16**, 159–163.

111. Y. Morimoto and P. Baum, *Nat. Phys.*, 2018, **14**, 252–256.

112. A. Ryabov, *et al.*, *Sci. Adv.*, 2020, **6**, eabb1393.

113. H. Alqattan, *et al.*, *APL Photonics*, 2022, 7, 041301.

114. D. Hui, *et al.*, *Nat. Photonics*, 2022, **16**, 33.

115. N. Schönenberger, *et al.*, *Phys. Rev. Lett.*, 2019, **123**, 264803.

116. D. S. Black, *et al.*, *Phys. Rev. Lett.*, 2019, **123**, 264802.

117. T. Zuo, A. D. Bandrauk and P. B. Corkum, *Chem. Phys. Lett.*, 1996, **259**, 313–320.

118. M. Meckel, *et al.*, *Science*, 2008, **320**, 1478–1482.

119. D. Ray, *et al.*, *Phys. Rev. Lett.*, 2008, **100**, 143002.

120. Z. Chen, A.-T. Le, T. Morishita and C. D. Lin, *Phys. Rev. A: At., Mol., Opt. Phys.*, 2009, **79**, 033409.
121. C. D. Lin, *et al.*, *J. Phys. B: At., Mol. Opt. Phys.*, 2010, **43**, 122001.
122. C. I. Blaga, *et al.*, *Nature*, 2012, **483**, 194–197.
123. M. G. Pullen, *et al.*, *Nat. Commun.*, 2015, **6**, 1–6.
124. B. Wolter, *et al.*, *Science*, 2016, **354**, 308–312.
125. M. G. Pullen, *et al.*, *Nat. Commun.*, 2016, **7**, 11922.
126. Y. Ito, *et al.*, *Struct. Dyn.*, 2016, **3**, 034303.
127. F. Schell, *et al.*, *Sci. Adv.*, 2018, **4**, eaap8148.
128. E. T. Karamatskos, *et al.*, *J. Chem. Phys.*, 2019, **150**, 244301.
129. K. Amini and J. Biegert, in *Advances In Atomic, Molecular, and Optical Physics*, ed. L. F. Dimauro, H. Perrin and S. F. Yelin, Academic Press, 2020, vol. 69, pp. 163–231.
130. B. Belsa, *et al.*, *Struct. Dyn.*, 2021, **8**, 014301.
131. X. Liu, *et al.*, *Commun. Chem.*, 2021, **4**, 1–7.
132. A. Sanchez, *et al.*, *Nat. Commun.*, 2021, **12**, 1520.
133. B. Wolter, *et al.*, *Phys. Rev. X*, 2015, **5**, 021034.
134. K. Amini, *et al.*, *Proc. Natl. Acad. Sci. U. S. A.*, 2019, **116**, 8173–8177.
135. H. Fuest, *et al.*, *Phys. Rev. Lett.*, 2019, **122**, 053002.
136. F. Brausse, *et al.*, *Phys. Rev. Lett.*, 2020, **125**, 123001.

Contents

Theoretical and Computational Chemistry Series No. 25
Structural Dynamics with X-ray and Electron Scattering
Edited by Kasra Amini, Arnaud Rouzée and Marc J. J. Vrakking
© The Royal Society of Chemistry 2024
Published by the Royal Society of Chemistry, www.rsc.org

CHAPTER 1

Ultrafast Molecular Spectroscopy in the Gas Phase

M. GÜHR[a,b]

[a] Deutsches Elektronen-Synchrotron DESY, Notkestr. 85, 22607 Hamburg, Germany; [b] Institute of Physical Chemistry, University of Hamburg, Martin-Luther- King-Platz 6, 20146 Hamburg, Germany
Email: markus.guehr@desy.de

1.1 Introduction

All properties of molecules, with their emerging complexity for chemistry and biology, result from the quantum laws of interaction between positively charged nuclei and negatively charged electrons. Formulating the equations of motion for molecular constituents is simple; the simulation and the experimental measurement of molecular dynamics provide formidable challenges at the very edge of technical possibilities.

The masses and forces determine the timescale for the motion of atoms within molecules. Two unscreened nitrogen nuclei at an Ångstrom (Å) distance repel each other with a force of about 1 μN. Screening of electrons stabilizes the repulsion and the N≡N triple bond in molecular nitrogen (N_2) is described as a spring with a constant of 0.2 μNÅ^{-1}. With a reduced mass of 7 amu, the motion between two nitrogen nuclei in N_2 can be described using a harmonic oscillator with a period of ~ 15 femtoseconds (fs). Heavier nuclei will undergo slightly slower dynamics; nevertheless, the governing timescale for any change in the molecular geometry is the few to hundreds of femtosecond timescale. Therefore, any experimental scheme to investigate the geometry changes of molecules needs to measure observables on the femtosecond timescale.

Theoretical and Computational Chemistry Series No. 25
Structural Dynamics with X-ray and Electron Scattering
Edited by Kasra Amini, Arnaud Rouzée and Marc J. J. Vrakking
© The Royal Society of Chemistry 2024
Published by the Royal Society of Chemistry, www.rsc.org

The electrons, in contrast, are much lighter particles. The forces on electrons and nuclei are Coulomb forces, which are the same for both classes of particles. Thus, the electrons usually move on a much faster, attosecond timescale, providing an even more challenging task for delivering experimental observables. The difference between the timescales of electronic and nuclear motion forms the ground for the Born–Oppenheimer approximation, which presents the framework for many of our interpretations of molecular dynamics. Essentially, it allows thinking about molecular geometry changes to be determined by gradients on a potential, formed by the nuclear repulsion and the electrons instantaneously adapting to any change in the molecular geometry. In the electronic ground state, this approximation describes the dynamics very well. The paradigm is entirely different for photoexcited molecular dynamics, where the Born–Oppenheimer approximation cannot fully describe the molecular dynamics. The timescale for the electronic dynamics is inversely-proportional to the energy spacing ΔE of electronic potentials. For molecular geometries in which Born–Oppenheimer potentials become energetically close, their corresponding electronic dynamics is slow, and the very concept of electronic–nuclear separation does not hold anymore. Consequently, the molecule can switch to another electronic potential, allowing it to take a very different geometrical path compared to the original potential.

In recent years, molecular science has discovered many relevant photo-induced molecular processes where the mechanism described earlier plays an important role.[1–4] Examples include bacterial light-harvesting and vision,[5] excited-state photoisomerization,[6–8] and the ultraviolet photoprotection of nucleobases.[9–11] The latter is used as a central example for ultrafast molecular dynamics and probing in this chapter. Increasing our fundamental understanding of photoinduced molecular dynamics thus directly influences our capability to predict and design specially tailored molecule-light interactions in the context of green chemistry[12,13] and alternative energy.[14]

We face a complex problem in which changes in the electronic state and nuclear geometry can be coupled on a femtosecond timescale. This highly entangled electronic–nuclear motion also yields observables influenced by electronic and geometric degrees of freedom. For decades, ultrafast molecular scientists have been seeking methods to increase the sensitivity to either the electronic or the nuclear geometry part of the dynamics. We are currently witnessing a rapid change in the experimental possibilities to disentangle molecular electronic and nuclear dynamics. Due to the fast development of ultrashort and bright X-ray pulses from free-electron lasers, we can now probe electronic changes with high element and local specificity within a molecule, providing accurate statements on the valence electronic dynamics. The diffraction of ultrashort X-ray and electron pulses delivers precise nuclear geometry changes in photoexcited molecules.

At the same time, the simulation of the electron–nuclear coupling can now include quantum effects in the propagation and coupling. Directly comparing calculated spectroscopic and diffraction observables from quantum dynamical simulations with experimental observables has become possible. Investigating molecules in the gas phase plays a special role in this context because the simulation of single, small chromophores can be accomplished with very high accuracy. A direct comparison of gas-phase experimental and simulation results thus yields essential information on the validity of simulation methods. Gas-phase molecular experiments allow for quantum manipulation techniques requiring long coherence times, such as molecular alignment. In addition, gas-phase experimental observables include charge particle detection with powerful coincidence methods.

This chapter introduces photoexcited molecular dynamics and ultrafast spectroscopy methods which so far have provided most of our knowledge of molecular dynamics. The following section will present the basic concept of a potential energy surface based on the Born–Oppenheimer approximation and highlight the situations in which it cannot be applied. Section 1.3 introduces the most prominent methods to probe molecular dynamics in the gas phase *via* ionization. Valence ionization, photoion, and photoelectron spectroscopy have built most of our current understanding of ultrafast molecular processes in the gas phase. Due to the availability of ultrashort extreme ultraviolet (XUV) and soft X-ray pulses, core ionization has started showing new observables. We are only beginning to understand the differences that the observables generated by core ionization provide as compared to valence ionization probes. This differentiation will also be made in Section 1.4, where several exemplary time-resolved X-ray absorption and photoelectron spectroscopy measurements are presented, demonstrating the capability of X-ray-based probing methods to measure ultrafast dynamics in the electronic degrees of freedom with increased sensitivity. The chapter ends with a short discussion of the novel and complementary information that time-resolved diffraction experiments deliver, with their sensitivity mainly to the nuclear geometry changes of the molecule. Throughout the remainder of this chapter, the dynamics of nucleobases will be used to exemplify the insight that can be gained with ultrafast spectroscopic methods. These systems have attracted the attention of theoretical and experimental scientists because of their relevance to life and their richness in phenomena, such as internal conversion and intersystem crossing.

1.2 Concepts for Molecular Dynamics

The following section introduces the basic theoretical concepts of molecular dynamics. It starts with the full Schrödinger equation of molecules and the separation of electronic and nuclear motion in the framework of the Born–Oppenheimer approximation (BOA). Based on the BOA, we discuss vibrational and rotational wavepackets. The breakdown of the BOA leads to changes in the electronic state and coupled electronic–nuclear dynamics.

1.2.1 Molecular Hamiltonian and the Born–Oppenheimer Approximation

All complexity we observe in the molecular world relies on the interaction between charged atomic nuclei and electrons within molecules. These small particles obey the laws of quantum mechanics; thus, we have to describe the system within a Schrödinger equation $H\xi = W\xi$, with wavefunction ξ, energies W and a Hamiltonian operator H. Using atomic units $\left(\hbar = \dfrac{1}{4\pi\varepsilon_0} = e = m_e = 1 \right)$, one can write the full Hamiltonian H for a molecule consisting of M nuclei and N electrons as

$$H = -\sum_{j=1}^{N} \frac{1}{2}\nabla_j^2 - \sum_{j=1}^{N}\sum_{a=1}^{M} \frac{Z_a}{r_{ja}} + \sum_{j=1}^{N}\sum_{k>j}^{N} \frac{1}{r_{jk}} + \sum_{a=1}^{M}\sum_{b>a}^{M} \frac{Z_a Z_b}{R_{ab}} - \sum_{a=1}^{M} \frac{1}{2M_a}\nabla_a^2,$$

(1.1a)

which can be combined to

$$H = H_e + \sum_{a=1}^{M}\sum_{b>a}^{M} \frac{Z_a Z_b}{R_{ab}} - \sum_{a=1}^{M} \frac{1}{2M_a}\nabla_a^2 = H_e + \sum_{a=1}^{M}\sum_{b>a}^{M} \frac{Z_a Z_b}{R_{ab}} + T.$$ (1.1b)

Here, the operators ∇_j^2 and ∇_a^2 are the Laplace operators for electron and nuclear motion, respectively, M_a and Z_a are the mass and charge of the ath nucleus, respectively, while r_{ja} is the distance between the jth electron and ath nucleus, r_{jk} is the distance between the two electrons j and k, and R_{ab} is the distance between nuclei a and b. The first three terms in eqn (1.1a), electronic kinetic energy, electron–nuclear Coulomb interaction, and electron–electron Coulomb interaction, form the so-called electronic Hamilton operator H_e. The fourth term represents the nuclear repulsion, and the last term is the nuclear kinetic energy operator T.

For the total wavefunction of the system $\xi(r_1, r_2 \ldots r_N, R_1, R_2 \ldots R_M) = \xi(r, R)$ we choose an Ansatz

$$\xi(r, R) = \sum_{j=1}^{N} \phi_j(r; R)\Phi_j(R),$$ (1.2)

with $\phi_j(r; R)$ being the solution of the electronic Schrödinger equation

$$H_e(r, R)\phi_j(r; R) = W_j(R)\phi_j(r; R).$$ (1.3)

Here, W_j is the energy of the electronic state j. The semicolon in the electronic wavefunction $\phi_j(r; R)$ indicates that this function only depends parametrically on the nuclear coordinates R. The purely nuclear wavefunction for the jth electronic state is given as $\Phi_j(R)$.

The task is now to solve the stationary Schrödinger equation $H\xi(r, R) = W\xi(r, R)$, with new eigenvalues W. In the following, we adopt an argument from the book by Szabo and Ostlund, stating that the light electrons move faster than the nuclei, and one might approximate the electronic coordinates in H by their 'averaged values, averaged over the electronic wave function' (see page 44 of ref. 15). Thus, all dependence on the electronic coordinates falls out of the Schrödinger equation to give

$$\left(T + E_j(R) + \sum_{a=1}^{M} \sum_{b>a}^{M} \frac{Z_a Z_b}{R_{ab}}\right)\Phi_j(R) = \left(T + W_j^{\text{tot}}(R)\right)\Phi_j(R) = W_j\Phi_j(R). \quad (1.4)$$

Eqn (1.4) is the Schrödinger equation of the nuclear wavefunctions considering the averaged molecular motion in the jth electronic state. The influence of the electrons in their jth state is described as a potential $W_j^{\text{tot}}(R)$, which is also called the Born–Oppenheimer potential energy surface or potential surface (PES). The expression for the full Schrödinger equation of the molecular system can be found, for example, in ref. 16–18. In contrast to eqn (1.4), the full Schrödinger equation also contains coupling terms, which intertwine the nuclear wavefunction on this jth electronic state with all the other electronic states. The terms read as

$$\frac{1}{2M_a}\langle\phi_j(r;R)|\nabla_a^2|\phi_i(r;R)\rangle\Phi_i(R), \quad (1.5)$$

and

$$\frac{1}{2M_a}\langle\phi_j(r;R)|\nabla_a|\phi_i(r;R)\rangle\nabla_a\Phi_i(R). \quad (1.6)$$

The BOA neglects these coupling elements. Historically more accurate, the adiabatic approximation means neglecting the elements with $i \neq j$, while all elements, even those with $i = j$, are neglected in the BOA. Formal justifications for the BOA can be found in ref. 3 and 19. We use a more intuitive reasoning. The nuclear gradients applied on electronic wavefunctions in eqn (1.5) and (1.6) describe the changes in the electronic wavefunction with a change in the nuclear position. The electronic Hamiltonian contains the electronic gradients on electronic wavefunctions, which represent the changes in that same electronic wavefunction on the electron coordinate. Both are of the same order of magnitude. Thus, the pre-factor $\frac{1}{2M_a}$ diminishes the coupling in eqn (1.5) and (1.6) compared to the terms of the electronic Hamiltonian.

A simpler, classical argument might appear more intuitive. The light electrons move fast in the vicinity of the heavy and slow nuclei. The electrons can adapt 'instantaneously', on the timescale of the nuclei, to any change in the nuclear geometry. Therefore, the electrons do not possess any 'memory' of past nuclear motion. This argument has the same meaning as neglecting the nuclear gradient matrix elements given above or assuming that the electronic wavefunction only depends parametrically on the nuclear coordinates, as given by the semicolon in $\phi_i(r; R)$.

We now turn to the potential energy surfaces (PESs) resulting from the BOA. The lowest PES, $W_0^{tot}(R)$, is called the electronic ground state. Its shape determines the molecular geometry, vibrational modes, and chemical reactivity of the molecule except in photochemistry involving also energetically higher-lying PESs. The excited PES $W_{j>0}^{tot}(R)$ is important for the interaction of the molecule with light. The population of an excited electronic state is achieved if a non-zero dipole transition moment from the electronic ground PES exists. The shape of the excited state determines the fate of the photoexcited molecule, including its dissociation, fluorescence, internal conversion, or intersystem crossing.

To facilitate the connection to later sections in this chapter, we discuss the excited-state processes in a time-dependent quantum mechanical picture. An excellent introduction to this topic is presented in the book by Tannor;[20] here, we only summarize some key points. The solution of the Schrödinger equation (eqn (1.4)) provides a set of nuclear eigenfunctions and eigenenergies, one set for each electronic PES. These nuclear eigenfunctions of the jth electronic state that are indexed according to a vibrational quantum number ν, $\Phi_{\nu,j}(R)$, can be superimposed to form a nuclear vibrational wavepacket given by

$$\Psi(R, t) = \sum_{\nu} c_{\nu} \Phi_{\nu, j}(R) \exp\left(-i \frac{W_j^{\nu}}{h} t\right), \qquad (1.7)$$

with the phases according to the energies of the nuclear wavefuntions W_j^{ν}.

A wavepacket is excited with an ultrashort, and thus spectrally broad, laser pulse. In the case of only one ground-state vibrational wavefunction being populated, the coefficients c_{ν} of the individual wavefunctions are the product of Franck–Condon factors, the electronic dipole matrix element of states 0 and j (in the framework of the Condon approximation), and the laser's complex amplitude. For several populated ground-state vibrational wavefunctions, one repeats the creation of a wavepacket, and then takes the sum of the absolute square of the different wavepackets. This process is called 'incoherent addition' as phase information is not used in the addition process. For strongly repulsive excited states, one can determine the shape of the nuclear wavepacket by projecting the shape of the lower energy nuclear wavefunction on the upper potentials, as worked out by Schincke.[21]

Changes in the nuclear coordinates of M atoms in a molecule can occur across 3M-6 dimensions (3M-5 dimensions for linear molecules). This is because the 3M individual atomic translational degrees of freedom of every atom are diminished by three centre-of-mass translational degrees of freedom and three (two for linear molecules) rotational degrees of freedom.

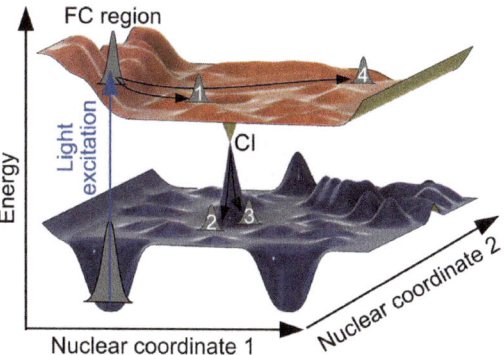

Figure 1.1   Overview of a vibrational wavepacket moving on potential energy surfaces, PESs. The optical excitation launches a wavepacket in the Franck–Condon region. On the excited PES (red surface), the wavepacket moves into the direction of conical intersection, CI (path 1), or a dissociative direction (path 4). The CI connects the two PESs, and the excited-state wavepacket undergoes internal conversion into the ground state. At the ground state, the wavepacket may bifurcate towards the geometry of the initial excitation (path 2) or an isomerized geometry (path 3).

The propagation of this nuclear wavepacket on the 3*M*-6 dimensional PES (see Figure 1.1) follows the shape of the PES, which determines how the molecule's geometry changes. Launched at the Frank–Condon geometry by an ultrafast laser pulse, the wavepacket will move along the steepest gradient. At the same time, it will also broaden due to the potential's anharmonicity.[22] If that wavepacket flows along an energetic valley corresponding to fragmentation, the molecule will dissociate with high probability on an ultrafast timescale (see path 4 in Figure 1.1). If the wavepacket stays trapped in the excited potential (not shown in Figure 1.1), it will fluoresce back to the ground state on a nanosecond scale. However, the molecule can also transition to a different electronic state through non-radiative means by intersystem crossing (ISC) and internal conversion (IC). We will now investigate the basic principles of ISC and IC.

1.2.2   The Breakdown of the BOA, Avoided Crossings and Conical Intersections

When the wavepacket encounters an area with close-lying PESs (path 1 in Figure 1.1), the energy spacing $|E_j(R) - E_i(R)|$ between two, or generally even more PES, becomes small. Here, the paradigm of fast electrons and slow nuclei underlying the BOA must be revisited. The inverse of the energy spacing of eigenstates gives the quantum timescale for atomic and electronic motion. Thus, electronic motion becomes slow as PESs approach, and the very meaning of a PES on which nuclear wavepackets move becomes invalid. Formally arguing, the matrix element of the nuclear gradient on an electronic wavefunction is $\langle \phi_j(r; R) | \nabla_a | \phi_i(r; R) \rangle \sim 1/|E_j(R) - E_i(R)|$. Thus, the matrix elements that have been neglected in the BOA become crucial in areas where electronic states become close in energy. Consequently, the different electronic states mix, and nuclear wavepackets will, in part (or even entirely), leave their original electronic state j into state i.

The so-called conical intersection (CI) is the prototypical topology for fast radiationless transitions on photoexcited molecular states. Despite already being conceptually described in 1927 by Teller,[23] it was considered as a rare topological PES artefact. In the 1990s, CIs were identified as one of the critical mechanisms driving photoexcited molecular dynamics.[2,4,17] The CIs are involved in light harvesting and vision,[5] excited-state photoisomerization of important molecular markers,[6-8] and the ultraviolet photoprotection of nucleobases.[9-11]

In a particular sub-space containing two PESs, a CI appears with the shape of a cone between two PESs (see Figure 1.1), which acts as a "funnel" that allows the transfer of a portion of the nuclear wavepacket from one PES to another. Mathematically, a CI can be described as a linear splitting of degeneracy between two PESs from one point into the directions

$$h_{ij} = \frac{1}{|E_j(\vec{R}) - E_i(\vec{R})|} \langle \phi_i(r; R) | \nabla_a H_{el} | \phi_j(r; R) \rangle \text{ and } g_{ij} = \nabla_a (\langle \phi_i(r; R) | H_{el} | \phi_i(r; R) \rangle - \langle \phi_j(r; R) | H_{el} | \phi_j(r; R) \rangle).$$ While the two-dimensional space spanned by the h_{ij} and g_{ij} vectors is a sub-space for the linear degeneracy, the CIs are embedded in higher subspaces or so-called 'seams' of $3M$-6-2 dimensions.[6,24] The CI's effectiveness in redistributing the molecular population is determined by its position in the nuclear coordinate space and its energy.[25] For example, a CI is inactive if it is located in a region of geometry space that is not sampled by the molecular vibrational wavepacket. In addition, it is generally not the lowest-lying CI (so-called 'minimum energy CI') along a seam that determines the nonradiative transition rate, but the region along the seam first encountered by the wavepacket. The velocity of the wavepacket also determines the transition rate, described by a Landau–Zener mechanism.[26]

1.2.3   Molecular Electronic Absorption Spectra

Visible and ultraviolet (UV) light generally interact with the electronic degrees of freedom, inducing molecular population in the Franck–Condon geometry in an electronically excited state (see Figure 1.1). The structure of the absorption spectrum from the electronic ground to the excited state is strongly related to the ultrafast dynamics of the excited state, especially for dilute gas-phase targets with long coherence times. The dipole matrix element determines the strength of this interaction. Vibrational and rotational excitation accompany the electronic absorption; their respective amplitudes are separated out of the dipole matrix element (see textbooks such as ref. 27–29).

The formal connection between time-dependent molecular motion and frequency-domain spectroscopy was elaborated by Heller.[30] It states that the absorption spectrum is the Fourier transform of the so-called autocorrelation function:

$$\varepsilon(\omega) = C\omega \int_{-\infty}^{\infty} e^{i(\omega + E_0)t} \langle \phi | \phi(t) \rangle \mathrm{d}t \qquad (1.8)$$

Here, C is a constant and E_0 is the energy of the nuclear ground state in the electronic ground state, from which the absorption process is assumed to start. The nuclear wavefunction on an excited PES is given by ϕ, launched in the Franck–Condon region. It is obtained by multiplication of the ground-state nuclear wavefunction with the electronic transition moment. In the Condon approximation, the latter is assumed to be constant. The excited-state nuclear wavefunction is, in general, not static and will propagate under the influence of the gradient in the excited PES, resulting in $\phi(t)$. The autocorrelation of the excited-state vibrational wavepacket is given as $\langle \phi | \phi(t) \rangle$. In the case where the excited-state Born–Oppenheimer potential is guiding the $\phi(t)$ permanently away from the Franck–Condon region, the autocorrelation function $\langle \phi | \phi(t) \rangle$ will show an amplitude at $t = 0$ and decay to zero afterwards. This decay in time will give rise to a broad, unstructured absorption spectrum $\varepsilon(\omega)$, typical for dissociative excited states. If the excited state is bound, the $\phi(t)$ will periodically revisit the Franck–Condon region, and the autocorrelation function $\langle \phi | \phi(t) \rangle$ will show a periodic modulation, resulting in a periodic structure in the absorption spectrum. The frequency differences in the absorption spectrum are inversely proportional to the periodicity of the autocorrelation function. Thus, we can already obtain important information on the excited-state shape and the ultrafast molecular processes by analyzing the absorption spectrum. However, the limiting factor is that all these statements are limited to the Franck–Condon region. One does not obtain any details on the molecular dynamics outside the Franck–Condon geometry.

1.2.4 Vibrational and Rotational Wavepackets

We can generally distinguish two different types of excited-state surfaces. In bound PESs, the nuclear wavepacket will return, to some extent, to the Franck–Condon region leading to an oscillatory motion of the wavepacket. In contrast, the molecular PES can also be dissociative, or a strong non-radiative transition can drive the nuclear wavepacket to other PES(s). In this case, the nuclear wavepacket will not return to the Franck–Condon region.

Prototypical examples of bound motion are found in many diatomic molecules. Early studies on the halogens and di-alkali molecules[31–36] showed textbook-style vibrational recurrences. Visible or near-infrared ultrafast laser pulses were used for wavepacket excitation, showing periodic modulations with a few 100 fs periods. The prototypical examples of dissociative wavepackets are methyl iodide (CH_3I) and its close relative trifluoroiodomethane (CF_3I). Launching the molecule on the A band using a UV pulse leads to the dissociation of the C–I bond within a short (~ 100 fs)

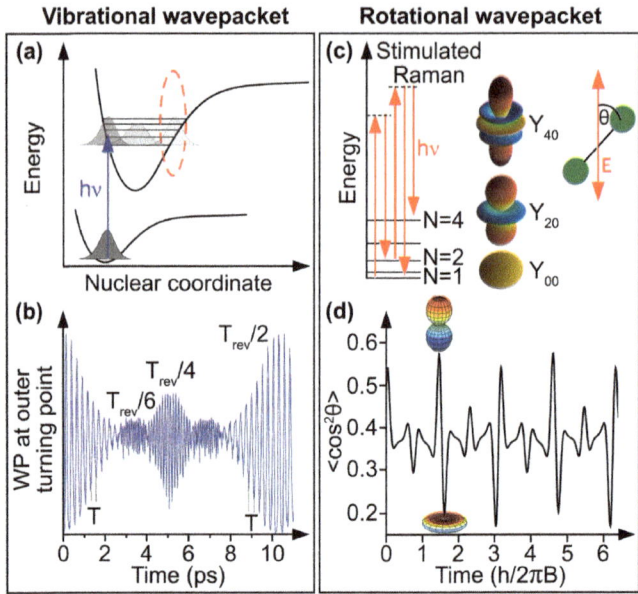

Vibrational wavepacket   **Rotational wavepacket**

Figure 1.2   Vibrational (a and b) and rotational (c and d) wavepacket dynamics.
(a) A short light pulse creates a coherent superposition of several
rotational–vibrational states on the excited potential energy surface,
PES. We concentrate on the vibrational part and show the propagation
of the vibrational wavepacket on the excited PES to the outer turning
point. (b) The wavepacket at the outer turning point. In anharmonic
potentials, the vibrational wavepacket undergoes periodic dispersion
and revivals. (c) The creation of a rotational wavepacket of a diatomic
molecule using non-resonant short pulses occurs via stimulated non-
resonant Raman transitions, coupling rotational states in the electronic
ground state. (d) The rotational wavepacket modulates the molecular
alignment $\langle \cos^2 \theta \rangle$ in time. Here, θ is the angle between the laser
polarization and the molecular axis, as indicated in (c).

timescale since the generated nuclear wavepacket on the excited state evolves
towards the C–I nuclear coordinate that is non-bonding (*i.e.*, large C–I bond
length).[37–40]

Now let us consider a vibrational wavepacket simulation on electronically
excited Br_2, specifically the wavepacket population at the outer turning point
wof the excited PES (see Figure 1.2a and b). Initially, one identifies the wavepacket oscillating between inner and outer turning points, as in a harmonic
oscillator potential. The vibrational states for a harmonic oscillator are
evenly spaced in terms of energy by ω_e, leading to the oscillation of a wavepacket with a vibrational period $T \sim 1/\omega_e$. A typical anharmonic potential
(see Figure 1.2a) provides a minor correction to this even spacing, reducing
the differences among consecutive eigenstates for higher energies in the
PES. This reduces the modulation contrast, as shown in Figure 1.2b, over
the first few oscillations due to wavepacket dispersion. The wavepacket

shows modulations with revivals spaced by fractions of the revival time $T_{rev} = 2\pi/\omega_e x_e$, where x_e is the anharmonicity.[41,42] In general, the effect of wavepacket dispersion sets a limit to temporal and spatial resolution in any experiment with molecules. One might try to set appropriate chirps to overcome the dispersive effects by focusing the wavepacket at particular times in the wavepacket evolution for very simple systems.[22] The scheme, however, works only partially in systems with more than one vibrational dimension and different anharmonicities in these dimensions. The consequences of the wavepacket effects for diffraction experiments have been discussed in detail in ref. 43.

The nuclear wavefunction in the excited states has both vibrational eigenvalues and rotational eigenvalues. Thus, besides vibrational wave-packet dynamics, the excited state also exhibits rotational wavepacket features, as in the example of I_2.[31,32] The rotational structure can be well-separated from the vibrational structure, as the timescales for vibration and rotation are separated by about three orders of magnitude. For instance, the B state vibrational period in I_2 is on the order of 300 fs, while the rotational period is on the order of 600 ps.

A slight deviation from discussing electronic excited-state phenomena is taken here to briefly consider rotational wavepackets in molecular ground states. For gas-phase spectroscopy, the method of ground-state rotational wavepackets has proved powerful in manipulating the angular degrees of freedom and fixing the molecular axis in space for a short time. Ref. 44 gives a thorough review of the subject. The technique essentially uses the interaction of the molecular polarizability tensor α with a strong but non-resonant AC electric field E from a laser. Let us consider the inter-action of a strong, non-resonant laser field with a diatomic molecule (see Figure 1.2c). The laser induces a dipole $p = \alpha E$, which interacts again with the laser's electric field. In the case of a linearly polarized laser field, the dipole minimizes its energy by rotating to align parallel with the electric field vector of the laser field. This dipole is along the internuclear direction for diatomics, as this has the highest polarizability. The non-resonant, short laser pulse induces stimulated Raman transitions among the rotational eigenstates in a quantum picture and creates a rotational wavepacket. Here, we only show the transitions originating from the lowest state Y_{00}. We show the expectation value of the squared cosine of the angle between the molecular axis and laser polarization in Figure 1.2d. The molecular alignment revives at regular intervals (see Figure 1.2d). The particular advantage of this method to align molecules is that the revivals modulate the molecular angular shape even in the absence of an electric field if the rotational coherences are not perturbed, which is not the case for dilute gas-phase experiments. Thus, additional experimental steps can be performed on aligned molecules in the absence of strong laser fields. Early experiments probed the change in birefringence associated with rotational wavepackets.[45] Ion imaging technology allowed full angular resolution of the rotational wavepackets.[46,47] The alignment method is now used to define the molecular axis in the laboratory frame, such as in the context of high harmonic generation[48] (see Chapter 6), excited-state ultrafast spectroscopy[49] and strong-field ionization.[50]

1.2.5   Internal Conversion and Intersystem Crossing – Dynamics of Nucleobases

Internal conversion (IC) and intersystem crossing (ISC) provide a pathway for a nuclear wavepacket to leave an excited-state PES, which was initially populated after the optical excitation of a molecule, due to couplings to other electronic states. In the case of IC, this happens without changing the molecule's spin, whereas it happens under a spin change in ISC. The standard textbooks on photochemistry provide an in-depth review of these processes.[51,52] This section will discuss these processes in nucleobases, since Section 1.4 will feature some of the ultrafast experiments on nucleobases.

Nucleobases are a crucial class of molecules. They encode genetic information and function in a variety of roles in cell metabolism. However, the nucleobases' high absorption cross-section for UV light poses a significant challenge to their ability to function correctly in crucial biological processes. This is because the absorbed UV light is energetic enough to change chemical bonds within and among molecules. For example, nucleobases form photo-induced lesions within nucleic acids, such as the cyclopyrimidine dimer or the 6-4 lesion.[53] These lesions are found in UV-irradiated human skin tissue and can lead to the formation of skin cancer if not repaired. Nevertheless, one of the key aspects that protect biological life from UV damage is the low quantum yield for the photolesions, avoiding them by ultrafast relaxation processes, including IC and ISC, that ultimately lead to the relaxation of the nuclear wavepacket to the electronic ground state.[9–11,54,55]

In the following, we concentrate on the pyrimidine nucleobases. These molecules are initially excited to $\pi\pi^*$ states by UV light, with a single occupation of the π and π^* orbitals. *Ab initio* quantum chemical studies of isolated thymine point towards two main possibilities for the ultrafast decay from $\pi\pi^*$ towards the electronic ground state, in which the molecule is completely protected from dimer formation. A femtosecond decay from $\pi\pi^*$ directly into the vibrationally hot ground state was suggested as the dominant relaxation channel through an ethene-like conical intersection.[56–58] In addition, an indirect decay from $\pi\pi^*$ to the spectroscopically dark $n\pi^*$ (doubly occupied π, singly occupied n and π^*) was suggested. Its relative weight compared to the first channel and its relaxation rates are a matter of debate.[58,59] The origin of the controversy arises from the reaction barrier at the $\pi\pi^*$ state in the indirect relaxation path to $n\pi^*$. Suppose this reaction barrier is high, as predicted using a particular electronic structure approximation (CAS-SCF).[59,60] In that case, the population is trapped in a local $\pi\pi^*$ minimum, requiring a few picoseconds for thymine to empty this. If, however, the barrier is low, a 100 fs decay into this channel is expected. We will show various experiments on this topic in Section 1.4, which provide partially contradicting results.

However, the case of ISC can also be made using the example of nucleo-bases with the sulfur-substituted thio-nucleobases. Thionated nucleobases are an important class among the modified (noncanonical) nucleobases,[61] because they play an essential role in medication and in a basic and applied research context. Their current and projected future medical use is im-munosuppression[62] and therapy for certain types of cancer.[63] Their research relies on their photoinduced cross-linking capabilities (for an application example, see ref. 64). These sulfur-substituted nucleobases exhibit remark-able differences compared to canonical ones. Photoexcitation of thionu-cleobases produces long-living triplet states, leading to cross-linking[65,66] and the creation of reactive singlet oxygen through a reaction with the triplet-ground state O_2 molecules in the vicinity of the photoexcited thionucleo-base.[67,68] On the one hand, these properties create a higher risk of skin cancer in patients treated with thionucleobases;[69] on the other hand, they might open the path for targeted photoinduced tumor therapy.[70,71]

The efficient creation of long-lived triplet states after UV irradiation is attributed to the photoinduced relaxation involving ultrafast ISC. Various pathways are suggested for the thionated pyrimidine nucleobases. In the case of pyrimidine nucleobases, they again involve $n\pi^*$ states, with n now being a sulfur lone pair orbital (see ref. 72, 73 and references therein).

1.2.6 Challenges in the Simulation and Measurements of Excited-state Dynamics

Many of the time-resolved cases presented in Section 1.4 are either collaborative studies of theorists and experimentalists or experimental studies referring to a particular result from a theoretical publication to aid the interpretation of the experimental data. Therefore, it is worth pointing out the challenges that theory encounters in simulating molecular dynamics.

Even under the Born–Oppenheimer approximation, the PESs need to be calculated. The analytical treatment of the electronic Schrödinger equation can only be accomplished for molecules as simple as H_2^+. The more com-plex systems appearing in this chapter need approximations and numerical tools. Quantum chemists have designed various methods to approximate the PESs, either based on refinements of the Hartree–Fock[15] scheme or density functional theory (DFT).[74] All those methods have their particular strengths and weaknesses, which are presented in the respective books or review papers.[15,18,75]

The propagation of molecules in regions with non-adiabatic mixing is an additional challenge. For a recent in-depth discussion, we refer to ref. 76 and 77. In summary, one of the most used approaches is the Tully surface hopping method, in which an ensemble of classical trajectories represents the nuclear wavepacket. The trajectories follow the gradients on the excited states. Single trajectories can hop from one PES to another depending on the amplitude of the coupling matrix elements, thereby implementing the mixing by classical means.[78-80] The spreading of the nuclear wavefunction on the excited state using non-stationary Gaussian basis functions facilitates a quantum treatment of the mixed electronic–nuclear dynamics.[81-84] Grid-based quantum methods, such as multi-configurational time-dependent Hartree,[85] are the most precise methods for excited-state propagation but require comparably more intense computational resources.

These simulations necessarily contain approximations. It is important to ask if they describe the processes accurately enough to be comparable to the experiment. Only close collaboration among experimentalists and theorists can solve this crucial task. Experiments in the gas phase play a unique role in this context. Gas phase spectroscopy allows the use of prototypical systems that are complex enough to reveal the coupled electron–nuclear dynamics. Yet the molecules are simple enough to perform simulations using a variety of approaches, from very approximate to a very high level of accuracy, without including a solvent. Moreover, we can use sophisticated charge particle probing of molecular dynamics containing significant information about the molecular electronic states and nuclear geometry. Combining the charge particle observables with various triggers for their emission, from multi-photon infrared (IR) to (vacuum-)ultraviolet (VUV) and X-rays, facilitates an understanding that can inform the choice of approximations to be used in simulations, as will be demonstrated in Section 1.4.

1.3   Molecular Ionization

Probing molecular dynamics in the gas phase is mainly accomplished *via* ionization. We present the valence and core ionization of molecules, which populate cationic states. Although these cationic states also show interesting dynamics, we concentrate on information about the neutral molecule that can be extracted from the ionization process. This will be essential for discussing the ultrafast pump–probe experiments presented in the following section.

1.3.1   Valence Ionization, Photoemission, and Fragmentation

The concepts of valence ionization are important as many of the gas-phase ultrafast methods employ valence ionization to probe the ultrafast molecular dynamics in neutral, electronically excited states. The production of photo-ions, photoelectrons, or both in coincidence serves as an observable which depend on the delay between pump and probe pulse, called the pump–probe delay. Valence ionization from the electronic ground state of the molecule can be accomplished using VUV light. There is a rich literature on VUV photoelectron spectroscopy using rare gas plasma emission lines and synchrotron light.[86-89]

Let us consider a scheme describing valence photoelectron spectroscopy and photoion concepts in the example of thymine (see Figure 1.3), where an orbital picture of photoionization with valence orbitals is presented

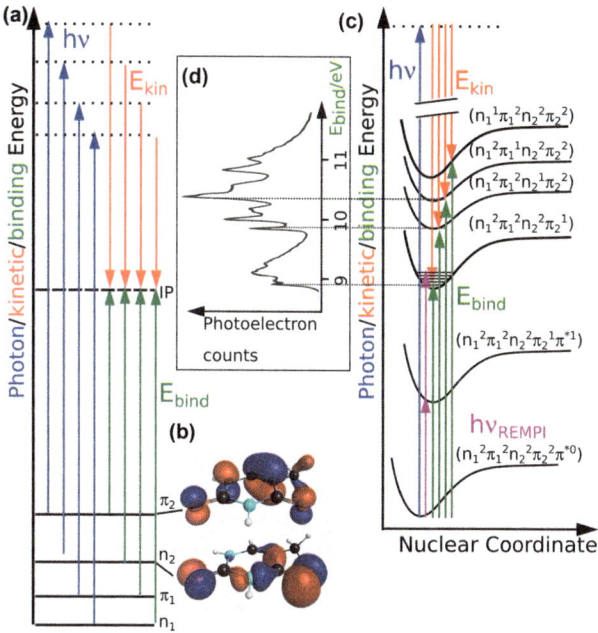

Figure 1.3   Molecular valence ionization. (a) Orbital diagram of valence ionization on the example of thymine, showing the least bound orbitals n_1, π_1, n_2, and π_2. (b) The n-orbitals are lone-pair oxygen orbitals, which are dominated by the oxygen 2p character, as shown in the orbital plot of n_2. The π orbitals are delocalized over the molecular ring, as exemplified in the orbital plot of π_2. The ionization using photons $h\nu$ results in electrons of kinetic energy E_{kin} or binding energy E_{bind}. (c) The multi-electron PES corresponding to the ionization process. Ionization from the ground state with occupation $(n_1^2\pi_1^2n_2^2\pi_2^2)$ results in the highest kinetic energy electrons for creation of the $(n_1^2\pi_1^2n_2^2\pi_2^1)$ cationic state, which corresponds to π_2 ionization in the orbital diagram. (d) Photoelectron spectrum showing peaks according to the cationic state ionization potentials and vibrational structure. Reproduced from ref. 91 with permission from AIP Publishing, Copyright 2015. The photon energy in the reference was varied between 17 and 150 eV, but for this spectrum it is not explicitly given in the reference.

(see Figure 1.3a). Ionization of the ground-state molecule using light of energy $h\nu$ leads to the creation of photoelectrons with specific kinetic energies $E_{kin} = h\nu - E_{bind}$. Here, E_{bind} is the binding energy calculated for a neutral molecule. Koopmans' theorem states that the binding energy of an electron from a particular orbital is its Hatree–Fock orbital energy. We indicate the four least-bound orbitals n_1, π_1, n_2, and π_2 with orbital plots of n_2, and π_2 (see Figure 1.3b) together with the multi-electron picture of photoionization (see Figure 1.3c). Here, the PES concept is used; thus, vibrational excitation can be included in the description of ionization. We start from the ground state with an electronic configuration $(n_1^2\pi_1^2n_2^2\pi_2)$, skipping lower-lying occupied and higher unoccupied orbitals. The different ionization channels are related to the emission of electrons from single orbitals here, leading to the lowest cationic state $(n_1^2\pi_1^2n_2^2\pi_2^1)$ and higher-lying cationic states $(n_1^2\pi_1^2n_2^1\pi_2^2)$, $(n_1^2\pi_1^1n_2^2\pi_2^2)$ and $(n_1^1\pi_1^2n_2^2\pi_2^2)$.[90] What was called binding energy E_{bind} in an orbital model can alternatively be called the ionization energy for the different channels, and they are 8.8 eV, 9.8 eV, 10.3 eV, and 10.8 eV, respectively[91] (see Figure 1.3d). As shown for the cationic ground state, vibrational degrees of freedom can be excited under molecular ionization. This leads to a vibrational structure in the thymine photoelectron spectrum[91] (see Figure 1.3d).

The lowest cationic states of a molecule can be very well described as a neutral electronic configuration with only one orbital unoccupied. From the ground state, this state is reached by removing only one electron without rearranging the orbital occupation. This fact is used in ultrafast spectroscopy for disentangling different excited electronic states *via* the so-called Koopmans' correlations. Any two neutral and singly ionized states are said to be Koopmans' correlated if the transition occurs exclusively by removing one electron without changing the occupation of any other orbital than the ionized one. Koopmans' correlated transitions are stronger compared to non-Koopmans' correlated transitions. For higher-lying cationic states starting at about 14 eV binding energy in thymine,[90] the cationic states are increasingly mixed electronic configurations. In that case, the ionization process cannot be described well using Koopmans' theorem. Apart from the pure kinetic energy, the angular characteristics of the emitted electrons also provide important information about the cationic state after valence ionization.[89,92]

As a technically simpler alternative to photoelectron spectroscopy, the excitation of cationic molecular states can also be detected *via* the ions created. In general, photon energies close to the lowest molecular ionization energy create the parent ion, the unfragmented ionized molecule. The molecule tends to fragment into ionic and neutral species with increasing photon energy. For example, the appearance energies of certain neutral fragments for thymine have been previously reported in ref. 93. From 10.7 eV photon energy towards higher photon energies, the very stable neutral fragment HNCO (isocyanic acid) is created. Between 10.7 eV and 14.4 eV, HCN and CO are found with increasing energy. Subsequently, two HNCO fragments split off. The onset of the fragmentation channels and the fragmentation pathways can be very accurately determined using threshold photoelectron–photoion (PEPICO) coincidences,[94] as fragmentation resulting from only one single (or several but energetically degenerate) cationic states is isolated. The PEPICO technique has been widely applied in the investigation of valence ionization[92,95] and combustion product formation.[96–98] Describing the processes leading to fragmentation in the cationic manifold is quite complex. These states form a very dense band of overlapping PESs for high excitation energies. The population in these PESs then rapidly relaxes to lower-energy cationic states with high vibrational energy.

An alternative ionization scheme to the above-discussed single-photon ionization is multi-photon ionization (MPI). In MPI, multiple, in part resonant, transitions are utilized to accomplish molecular ionization.[99] This nonlinear process is driven using nanosecond and shorter laser pulses, provided they have sufficient intensity. In case one of the steps towards ionization is resonant with a molecular transition, the process is called resonance-enhanced multiphoton photoionization (REMPI), and it can lead to the population of the lowest cationic state (see Figure 1.3c). The scheme explores the ro-vibrational levels of these resonant states.[100] The combination with additional infrared or UV lasers allows for exploring vibrational and electronic excitations by depleting the ground electronic or vibrational state population for multiphoton-ionization.[101] The advancement in REMPI with shorter, femtosecond pulses has led to the development of the ultrafast pump–probe method with ion and photoelectron detection. At the temporal overlap of pump and probe pulses, the process is identical to REMPI. Dynamical information about the resonant state is obtained by introducing a time delay between the two pulses.

Additional information on the molecular ionization process is contained in the angular and momentum distributions of the ions. Velocity map imaging (VMI) is a very powerful technique that can measure these distributions for ions and electrons.[102–104] Ionic fragments can possess a momentum that is non-zero because the cationic molecule (or a neutral intermediate in the case of REMPI) dissociates with recoil kinetic energy into two or more fragments. The fragments expand around the laser-molecule interaction region to form a Newton sphere. The ions in this sphere are projected onto a position-sensitive detector by applying a specially shaped extraction field that preserves the initial three-dimensional momentum distributions of generated fragments towards the detector.

Strong-field interactions allow for Coulomb-explosion imaging, a scheme that derives information from the momentum distribution of an exploding molecule that is multiply charged. To achieve this, coincidence (or covariance) methods are typically combined with strong-field ionization. The method is not restricted to valence ionization; it is a popular tool in conjunction with XUV, X-ray, ion, and electron excitation of a gas-phase molecule. The coincident detection of ions allows for reconstruction of the molecule's orientation in the laboratory frame where only one molecule is ionized in the interaction region by a single excitation pulse. The momentum of the ions is determined by accelerating the ions onto a position-sensitive detector, measuring the (x, y) two-dimensional hit position on the detector plane and the time-of-flight for every fragment.[105–107] A beautiful example for the power of this method is the determination of the chirality ('handedness') of individual molecules by laser-induced Coulomb-explosion imaging.[108] The chiral bromochlorofluoromethane was ionized up to a +5 charge state, and the momenta of five separate ion fragments were determined. Visualizing the momenta as vectors in a single plot then allows to reconstruct the ionized enantiomer on a shot-to-shot basis.

1.3.2   Core Ionization, Photoemission, and Fragmentation

The interaction of light with core electrons happens mostly in the X-ray domain. Ultrafast gas-phase spectroscopy with X-rays has been mainly performed with soft X-rays, as many suitable sample systems with high vapour pressure contain lighter elements. Nevertheless, hard X-ray spectroscopy can also be applied to study small halogenated molecules or neutral and small metal compounds such as iron pentacarbonyl.

The definition of a core electron is not entirely set, but a few conditions apply. Core electrons are deeper bound than valence electrons. In contrast to valence electrons, they do not strongly contribute to molecular bonds apart from screening the nucleus. Their binding energies are element-specific and provide a fingerprint for elements. Furthermore, the core electrons are strongly localized due to their high binding energies. These properties provide any probe technique involving core electrons with a high element and site-selective specificity.

Famous examples of core-level qualities in the VUV domain (which we define here arbitrarily from 6 to 100 eV photon energy) are 3p levels of 3d transition metals and shallow core levels of halogens. The next deepest typical core levels in the XUV up to 300 eV are sulfur 2s, 2p (around 230 and 170 eV, respectively), and the carbon 1s level at 280 eV. The nitrogen and oxygen 1s levels at 410 and 540 eV follow in the soft X-ray range. Spectroscopy on those core levels is sufficient to address important problems in organic molecules. In addition to the significantly large core ionization energy changes, there are much more subtle eV-scale effects, called chemical or site-specific shifts, as explained below.[109–111] The core-electron–light interaction gives rise to a couple of possible observables. In the case of a resonant core–valence excitation, the absorption in the gas phase is most conveniently detected from ion or Auger–Meitner electron yield. Non-resonant core-ionization can be monitored either by photoelectron spectroscopy, ion spectroscopy, Auger–Meitner spectroscopy, or with a small yield in the soft X-ray domain, by photon emission.

An overview of resonant soft X-ray absorption and soft X-ray photoelectron spectroscopy can be given regarding the single-electron orbital and multi-electron PES pictures shown in Figure 1.4. The method of X-ray absorption spectroscopy is discussed in great detail in the book by Stöhr.[112] For X-ray molecular photoelectron spectroscopy, the reader is referred to ref. 109 and 113.

Resonant X-ray absorption is defined as a transition from a core orbital to an unoccupied valence orbital (see Figure 1.4a). The transitions from the different localized core orbitals differ by more than 100 eVs. For instance, the neighbouring elements nitrogen and oxygen have their 1s-valence transitions at 410 and 530 eV, respectively. The orbital picture gives intuition about the energetic position of the core–valence resonances as exemplified using the O1s-π^* absorption in thymine. The strength of the resonances is determined by the dipole matrix element between the molecular ground

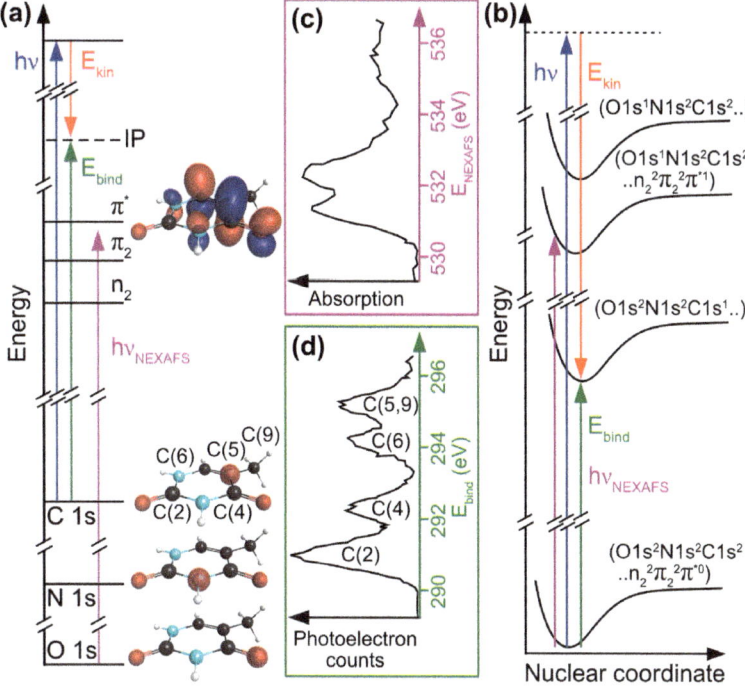

Figure 1.4 Core ionization and core–valence absorption schemes using the example of thymine. (a) The orbital picture shows some of the core 1s levels bound by ~ 540, 410 and 280 eV for the O, N and C 1s orbitals, respectively. The core orbitals are shown as a red shell around the atom; they let the C(5) and one of the N atoms appear in red. Ionization with soft X-ray photons with energy $h\nu$ leads to photoelectrons with kinetic energy E_{kin}, shown for photoelectrons from the C(5)1s orbital. Alternatively, soft X-ray radiation can induce core–valence transitions leading to near-edge X-ray fine structure (NEXAFS), exemplified by an O1s-π^* absorption. The n_1 and π_1 orbitals are not shown. (b) Multi-electron NEXAFS and photoemission picture. The O1s-π^* absorption is a transition between the $\left(\text{O1s}^2\text{N1s}^2\text{C1s}^2\ldots n_2^2\pi_2^2\pi^{*0}\right)$ ground state and $\left(\text{O1s}^1\text{N1s}^2\text{C1s}^2\ldots n_2^2\pi_2^2\pi^{*1}\right)$ core–valence excited state. (c) The corresponding absorption spectrum with the split O1s-π^* resonance due to two close-lying $\left(\text{O1s}^1\text{N1s}^2\text{C1s}^2\ldots n_2^2\pi_2^2\pi^{*1}\right)$ states. (d) The photoelectron spectrum resulting from the five carbon atoms in thymine. The multi-electron state after carbon 1s core-ionization is $(\text{O1s}^2\text{N1s}^2\text{C1s}^1\ldots)$. The site-specific chemical shift separates the spectrum into four discernable lines. Reproduced from ref. 114 with permission from Elsevier, Copyright 2008.

state and the core–valence ionized state. In a multi-electron picture, this corresponds to a transition between the $\left(\text{O1s}^2\text{N1s}^2\text{C1s}^2\ldots n_2^2\pi_2^2\pi^{*0}\right)$ ground state and $\left(\text{O1s}^1\text{N1s}^2\text{C1s}^2\ldots n_2^2\pi_2^2\pi^{*1}\right)$ core–valence excited state. Let us consider the near-edge X-ray absorption fine structure (NEXAFS) spectrum of thymine (see Figure 1.4c) with the data taken from ref. 114. The O1s-π^* resonance is split into two lines arising from transitions from the two different oxygen atoms to two slightly shifted π^* orbitals (not shown in the figure). As the photon energy increases, the absorption includes molecular Rydberg states, building a dense band of states towards the ionization potential (IP), not shown in the spectrum. Beyond the IP, the absorption ionizes the molecule directly, creating a photoelectron. Absorption below the IP thus contains information about the unoccupied molecular orbitals of the molecule and the onset of direct ionization. Optical transitions change the valence structure of the molecule, promoting an electron into an initially unoccupied π^* orbital and leaving a hole in a π_2 or n_2 orbital. The hole in an initially occupied valence orbital opens a new absorption channel energetically below the core-LUMO transition, as will be shown later in Section 1.4. The multi-electron picture of NEXAFS also allows us to explain the vibrational features in the spectra. However, in the case of core–valence transitions, those are less pronounced than valence spectroscopy because of lifetime broadening, as discussed below. Beyond the IP, the X-ray interaction leaves the molecules in an $\left(O1s^1 \ldots n_2^2 \pi_2^2 \pi^{*0}\right)$ state and creates a photoelectron.

The orbital picture (see Figure 1.4a) is also intuitive for core-level photoelectron or X-ray photoelectron spectroscopy (XPS), as we approximate the kinetic energy of the core electron again by the difference of photon and 1s orbital binding energy. Here, we demonstrate core-level photoelectron spectroscopy on the example of the carbon core levels. In the multi-electron picture, the carbon 1s ionization leads to an $(O1s^2 N1s^2 C1s^1 \ldots)$ PES. We use the thymine carbon 1s photoelectron spectrum (see Figure 1.4d) to discuss the concept of the chemical shift, which manifests itself as an eV-scale splitting of features in the NEXAFS and XPS spectra.[109–111] Electrons of a particular element but in different chemical environments will possess a shift in their ionization energy, leading to a situation where two atoms of the same element can differ in core binding energy by several eVs. For example, the carbon atom C(2) in thymine (see inset of Figure 1.4a) has a portion of its valence charge pulled away by the more electronegative oxygen and nitrogen atoms in its vicinity. In this chemical environment, the screening of the nuclear potential in the C(2) atom is reduced, thus shifting the C(2) binding energy to higher values. In contrast, carbon atoms C(5) and C(9) are only surrounded by hydrogen or carbon atoms, having equal electronegativity to the C(5) and C(9), resulting in no charge abstraction and, thus, no core shift.[114] The simple model described here is called the 'potential model'.[110] It can successfully describe trends in the spectra on a qualitative level. However, a complete calculation of the shifts also contains other effects, such as the reorganisation of the electronic structure upon core-ionization, called the core-hole screening.[111] The core-level shift mechanisms have been intensively studied and simulated in the core spectra of ethyl-trifluoroacetate.[115] Chemical shifts also appear in resonant absorption spectra, which is exemplified by the comparison of NEXAFS spectra with XPS spectra of thymine in ref. 114.

The transitions into the continuum are shaped by the core-continuum matrix element determining the absorption strength and angular emission characteristics. The final continuum states are not just free atom continuum states with simple spherical symmetry, but the molecule provides a complex potential energy landscape in which the photoelectron scatters, thus involving highly mixed angular momentum contributions.[116,117]

The absorption and photoelectron spectra are limited in resolution by the core hole decay process, typically taking a couple of femtoseconds in the soft X-ray domain. The core hole decay mostly occurs *via* the Auger–Meitner process, which for the 1s core hole has a lifetime of a few femtoseconds.[118] In this process, one electron fills the core hole while another electron leaves the atom with the excess energy. The decay *via* X-ray emission is governed by the Einstein rate constant for spontaneous emission. This process is much slower than Auger–Meitner decay in the soft X-ray range, thus delivering a yield that is about three orders of magnitude lower than the Auger–Meitner yield. As the X-ray energy increases, spontaneous X-ray emission dominates due to the cubic scaling of Einstein-rates with the frequency. Although Auger–Meitner electron emission has been employed in time-resolved experiments,[119–122] its detailed discussion will not be presented here due to limited space. The reader is referred to ref. 123–125 for a detailed discussion of the Auger–Meitner emission process.

The interaction of soft X-ray light with core levels creates ions due to Auger–Meitner decay. For resonant absorption, they are mostly singly charged; for nonresonant absorption, they are mostly doubly charged. With increasing core-electron binding energies, higher charge states are generated by Auger–Meitner decay cascades. The core-ionization step is strongly localized at a particular site within the molecule. The fragmentation of the molecule results from a complex interplay of nuclear dynamics on the core-ionized (core excited in resonance) states, Auger–Meitner decay into dicationic states (cationic states in resonance), and the continued nuclear dynamics and electronic transitions in this dense band of states. Thus, the fragmentation pattern is difficult to predict and has little dependence on the excitation conditions. In fact, for most cases, including thymine and uracil,[126,127] very little site-selectivity is observed in the fragmentation channels. In some cases, however, the relative abundance of fragments is strongly modulated by the core-ionization site, for instance, by fast dissociative dynamics due to the resonant excitation into an antibonding valence orbital.[128,129]

Coincidence methods combined with soft X-rays can measure photoelectron angular distributions in the molecular frame using non-aligned molecular targets. The ion and electron momenta are recorded by multiparticle coincidence for one molecule at a time interacting with the X-ray beam from a synchrotron. Coincident ion-momentum detection allows fixing the molecular axes in the laboratory frame. The electron emission direction can then be referenced to the molecular frame. The photoelectron distribution is sensitive to the molecule's structure as the emitted electron can scatter in the potential formed by the nuclei and electrons. This observation gave rise to the terminology of 'illuminating molecules from within'[130] (see Chapter 10). For molecules undergoing an X-ray-induced dissociation, this method allowed to monitor the nuclear dynamics on core-ionized states without explicitly implementing a pump–probe method.[131] This is made possible by sorting the events according to kinetic energy release in the ions, indicating the internuclear position of the molecule at the time of Auger–Meitner decay.

X-ray emission has so far not been used for probing time-resolved experiments in the gas phase. Due to the relatively poor efficiency and small angular acceptance of gratings under grazing incidence, the signal levels are relatively low compared to charge particle detection.[132,133] Synchrotron experiments without time resolution have been performed for high-pressure gases and reveal highly structured emission spectra.[134–136]

1.4 Ultrafast Methods for Probing Molecular Dynamics

This central section of the current chapter discusses exemplary ultrafast experiments of gas-phase molecular dynamics. We divided the section according to the strategy presented in the previous one. Valence probing *via* photoion, photoelectron, and high field methods is presented first. We then switch to recent results from core hole probing molecular dynamics. Methods and understanding of gas-phase molecular dynamics up to now have been highly coupled to the developments of short-pulse light sources. Generally, the gas-phase molecular science community has continuously helped in developing new light sources and invented experimental methods for their utilization. We will thus first present a methodological section on the most prominent light sources for ultrafast molecular dynamics experiments in the gas phase.

An overview of ultrafast laser technology can be found in various textbooks and reviews.[137,138] For early femtosecond experiments, the colliding-pulse dye lasers were the source of choice.[139] The invention of the chirped pulse amplifier[140] together with the implementation of titanium-doped sapphire technology, enabled the generation of mJ-level laser pulses,[141,142] marking a revolutionary breakthrough in ultrafast science. These systems allowed for relatively simple conversion into the IR, visible, and UV spectral ranges using optical parametric amplifiers.[143–148] At the same time, the high peak powers allowed for strong-field probing schemes (see Chapter 13) as well as strong-field high harmonic generation[149–151] (see Chapter 6) conversion into the extreme ultraviolet,[152] and finally, the soft X-ray range.[153]

The second line of ultrafast sources was developed with accelerator-based technology. The combination of lasers and synchrotron light sources lead to the first breakthrough reaching femtosecond pulses in the X-ray domain using the pulse slicing technique.[154] The beginning of the new millennium changed the landscape for ultrafast spectroscopy following the development of the first short-wavelength free-electron lasers (FELs; see Chapter 8). The evolution began with the free-electron laser in Hamburg (FLASH) at DESY, which spanned the XUV to the water window spectral range, using a superconducting accelerator coupled to undulators which deliver an effective pulse rate of a few kHz.[155] The LCLS,[156] SACLA,[157] SWISS-FEL,[158] and PAL-FEL,[159] which are all based on warm copper-cavity accelerators, extended the photon energy range and delivered femtosecond pulses of soft- and hard-X-rays at repetition rates of around 100 Hz. All the abovementioned FEL sources start from noise with a self-amplified spontaneous emission (SASE) process, resulting in randomly structured spectral pulse shapes. The FERMI FEL is also based on warm accelerator technology. Still, in contrast to the other FELs mentioned above, FERMI is seeded, which results in close to Fourier-limited and stable spectral pulse shape in the XUV region[160] but with a repetition rate of 50 Hz. A new generation of FELs aims to extend the repetition of FELs significantly beyond the 1 kHz level. For example, the European XFEL operates in the hard- and soft X-ray region in the SASE mode at an effective rate of 27 kHz.[161] The LCLS II and the SHINE FELs will soon deliver soft- and hard X-ray pulses at up to MHz repetition rates.

1.4.1   Ultrafast Valence Probe Spectroscopy

In this sub-section, molecular dynamics studied *via* probing of valence electrons will be discussed. The valence electron probe of excited-state dynamics can be accomplished by ionization either with single-photon transitions using UV or VUV pulses or through multi-photon transitions using pulses of longer wavelengths. In gas-phase ultrafast spectroscopy, probing *via* ionization has a clear technical advantage because charged particle detection methods are very well-developed for these targets.

Resonant valence transitions are a comparably rare probe method for gas phase targets. However, transient absorption spectroscopy in transmission is the most prominent ultrafast technique for solvated molecules.[162] Here, an ultrashort excitation pulse launches the molecule in its excited state, and a second, delayed, typically broadband, pulse probes the excited-state absorption, stimulated emission, and ground-state bleach signals in the spectrally resolved continuum. In gas-phase samples, the achievable column density is typically lower, depending on the sample pressure curve and the onset of pyrolysis, sometimes by ten orders of magnitude. Nevertheless, transient absorption experiments have been performed by means of transmission measurements in gas cells. In VUV spectroscopy based on harmonic sources, the technique is also employed in core-level probing,[163] as elaborated in Section 1.4.2.

Valence probing *via* molecular ionization and the detection of molecular ions has been used on many different molecules in the gas phase. Fuss and co-workers have measured a large range of molecular excited-state dynamics; some examples include cyclohexadiene,[164] and transition-metal carbonyl complexes.[165] These studies utilize multi-photon infrared transitions to ionize the molecules from the electronically excited neutral states. The decay of parent ions and ionic fragments is used to deduce information about the neutral molecular states, such as molecular dissociation, electronic relaxation, and vibrational energy re-distribution. For instance, the appearance of a fragment channel can indicate molecular dissociation in the neutral states. However, in many cases, the attribution of dissociation on the neutral channels is not self-evident.

The strong-field ionization domain for probing has been investigated thoroughly by Weinacht, Matsika *et al.* using nucleobases as an example.[166–168] The close experimental–theoretical collaboration allowed the attribution of fragments to final-ionic states. The highest occupied molecular orbitals can be related to measured ion yields by varying the relative polarization of the UV pump and the strong-field probe pulse. The delay-dependent behaviour in the parent and the most intense fragment ions in the mass spectrum of uracil in these strong-field measurements reveal short ($\sim$100 fs) and long time scale (a few picoseconds) decay pathways.[167]

The interpretation of photoion data in terms of electronic and nuclear geometry changes is quite challenging. Both the electronic relaxation and nuclear geometry changes influence the probe transition probability and thus shape the delay dependence of parent and fragment ions. The comparison of experimental time constants to simulations can exclude specific molecular pathways. In addition, different fragment channels can be correlated to excited cationic states using VUV spectroscopy at synchrotrons.[94] Thus, the relative dynamics in parent and fragmentation channels can be an additional parameter in interpreting molecular dynamics.[169] For example, the ion momentum's interpretative power has been demonstrated in vibrational wavepacket dynamics of simple diatomics[170,171] and the prototypal C–I bond dissociation in CH_3I.[37,172] In alignment experiments, the angular information extracted from the ionic fragments is a powerful tool for assessing the degree of alignment.[46,173]

Time-resolved valence photoelectron spectroscopy provides additional experimental information compared to ion spectroscopy, such as the kinetic energy and angular distribution of photoelectrons. We refer to ref. 174–178. The highest possible electron kinetic energy is $E_{kin} = h\nu_{pump} + h\nu_{probe} - E_{bind}$, where $h\nu_{pump}$ and $h\nu_{probe}$ are the pump and probe photon energies, respectively, and E_{bind} is the binding energy for the creation of a particular cationic state.

Let us consider a valence photoelectron experiment (see Figure 1.5a). We assume an electronic ground state configuration $n_2^2\pi_2^2\pi^{*0}$ and a UV excited $\pi\pi^*$ state with orbital occupation $n_2^2\pi_2^1\pi^{*1}$. Probing the neutral $\pi\pi^*$ state with orbital occupation $n_2^2\pi_2^1\pi^{*1}$ is possible *via* a transition to the $n_2^2\pi_2^1\pi^{*0}$ cationic state. The latter is the probe transition with the lowest possible photon energy. If the molecule undergoes an internal conversion to a neutral $n\pi^*$ state with orbital occupation $n_2^1\pi_2^2\pi^{*1}$, the lowest cationic state for probing would be $n_2^1\pi_2^2\pi^{*0}$. In both cases, we assume that the dominant ionization is the removal of only one electron without rearrangement of the orbital occupation for other electrons. This assumption is certainly valid for the lowest cationic states, and their energy can be calculated relatively accurately

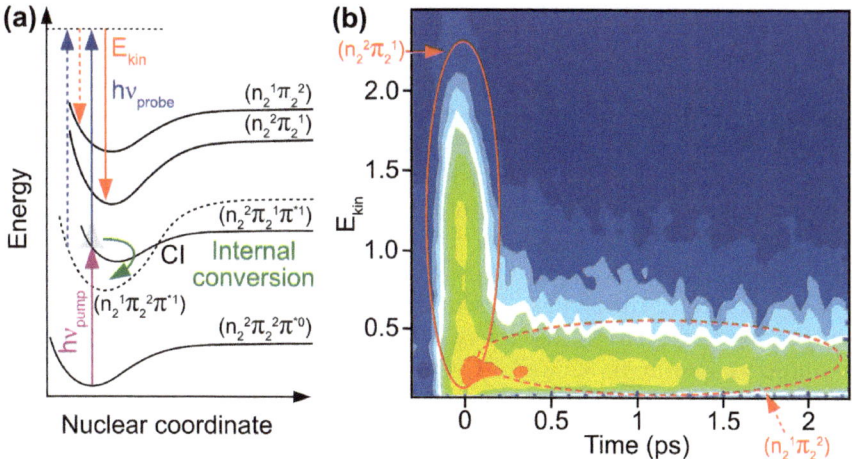

Figure 1.5 Time-resolved valence photoelectron spectroscopy on the example of thymine, data from ref. 179. (a) Pump–probe scheme. An excitation pulse of photon energy $h\nu_{\text{pump}}$ prepares the molecule on the $n_2^2\pi_2^1\pi^{*1}$ state. Probing *via* a time-delayed pulse $h\nu_{\text{probe}}$ creates a photoelectron with the lowest cationic state $n_2^2\pi_2^1$. After internal conversion through the CI into the $n_2^1\pi_2^2\pi^{*1}$ state, the higher lying $n_2^1\pi_2^2$ cationic state will be populated in the probe step, resulting in lower energy photoelectrons. The energetic orbital ordering follows the cationic channels. The $n_2^1\pi_2^2\pi^{*1}$ is energetically below the $n_2^2\pi_2^1\pi^{*1}$ state, which the simple orbital energy picture cannot explain. (b) The valence photoelectrons spectrum shows a fast decay of the photoelectron kinetic energy, with the interpretation of the respective dicationic channels overlapped. This has been interpreted as an ultrafast, sub-100 fs internal conversion from $n_2^2\pi_2^1\pi^{*1}$ to $n_2^1\pi_2^2\pi^{*1}$ in ref. 179. The interpretation of the very same dataset is fundamentally different in ref. 59, stating an ultrafast geometrical change in the state followed by picosecond internal conversions.

using Koopmans' theorem (see Section 1.3.1). However, for higher cationic states, correlation effects play an increasingly important role, and one cannot assume Koopmans' correlations to hold anymore. The different cationic states reached *via* Koopmans' correlations from the $\pi\pi^*$ and $n\pi^*$ states have different ionization energies and, thus, different electron kinetic energy in the photoelectron spectra.

Ullrich *et al.* performed the first time-resolved valence photoelectron experiments on nucleobases,[179,180] specifically with thymine as an example (see Figure 1.5b). They found three different time constants in the decay of the photoelectron spectra: a fast sub-50 fs component, an intermediate 500–800 fs component, and a longer few-picosecond time constant. Related to these time constants, Ullrich *et al.* found changes in the photoelectron kinetic energy spectra. A decrease in the photoelectron kinetic energy is associated with the observed very fast decay. The data were interpreted using the abovementioned Koopmans' correlations. The quickly decaying spectrum could be explained by the vertical ionization energy of the cationic $n_2^2\pi_2^1\pi^{*0}$ state (vertical referring to the Franck–Condon region) corrected by the excess vibrational energy of the molecule in the $\pi\pi^*$ state. The lower kinetic energy spectrum associated with the 500–800 fs processes could be consistently explained by the vertical IP of the cationic $n_2^1\pi_2^2\pi^{*0}$ state corrected by the excess vibrational energy of the molecule in the $n\pi^*$ state. The conclusion, therefore, was an ultrafast, sub-50 fs $\pi\pi^*$–$n\pi^*$ internal conversion. Using this explanation, any reaction barrier in the $\pi\pi^*$ state path (see Section 1.2.5) leading to other electronically excited states would be entirely negligible. The change in ionization energy with molecular geometry changes was neglected in this interpretation, which gave rise to a new joint simulation–experiment study.[59] The paper included a simulation of the expected geometry changes of the molecule after preparation at the Franck–Condon point of the $\pi\pi^*$ state. The simulations found an increasing energetic gap between the $\pi\pi^*$ state and ionic states for the predicted geometry change path from the Franck–Condon region to a predicted minimum on the $\pi\pi^*$ state. In addition, the simulations found a decreasing Franck–Condon factor for probing along this path. These simulations could explain the decreasing kinetic energy and overall signal strength by ultrafast changes in the molecular geometry without internal conversion on the sub-picosecond timescale. The few picosecond timescales from the experiments were then attributed to internal conversion, in contrast to the earlier interpretation of the IC as the fastest sub-50 fs process.[179] In addition, the study found a barrier between the Franck–Condon point and accessible CIs, which indicated the appearance of internal conversion in the longer time constants of the experiment. The appearance of the new $n_2^1\pi_2^2\pi^{*0}$ state photoelectrons could be explained by a geometry-dependent increase in the $n_2^1\pi_2^2\pi^{*0}$ ionization channel from the $\pi\pi^*$ state. We will return to this topic in the next sub-section.

Combining the measurement of the photoelectron kinetic energies with their angular distribution provides additional observables for interpreting molecular dynamics.[181] Probe transitions from the neutral excited states to different Koopmans' correlated continua possess different photoelectron angular distributions (PADs). However, the particular difficulty is that molecules are usually randomly oriented in the laboratory frame, which smears out the single-molecule PAD. This problem can be overcome through two different solutions. In the first case, only a single molecule is present in the interaction region of the pump and probe pulses, and the generated photoelectrons are measured in coincidence with molecular ions. This type of coincidence technique, combining electron kinetic energy with PADs, has found an application in the time-resolved investigation of ultrafast molecular dissociation.[182,183] This technique can only be applied as long as the axial-recoil approximation holds, which means that the emission of each ion proceeds in a linear trajectory from the molecular centre. The second solution employs the strong-field nonresonant molecular alignment technique discussed in Section 1.2.4. The molecules are aligned and undergo rotational wavepacket dynamics in the electronic ground state. At a rotational revival, the molecules are optically excited into the neutral state of interest and then ionized by a probe pulse. This alignment approach was applied to carbon disulfide (CS_2), significantly increasing the signal-to-noise ratio of the measured PAD compared to a randomly oriented and aligned isotropic CS_2 molecular ensemble.[49]

Using VUV light to realise the ionizing probe step has an important advantage in monitoring molecular dynamics. As mentioned above, the gap between the neutral states and Koopmans' correlated cationic states might increase along the molecular path on the excited states, as has been simulated for the case of nucleobases.[59,184] The total energy deposited in the neutral states under investigation remains constant in isolated molecules and is determined by the ground state's vibrational energy and the pump pulse's photon energy. Due to electronic relaxation, however, energy flows out of the electronically stored energy into vibrational energy. This leads to different IPs, as mentioned above. Within the vibrational degrees of freedom, energy also undergoes a dynamic exchange. In the language of PESs, the molecular population explores the energetically-allowed area on the high dimensional potential energy surface. Some of these explored vibrational degrees of freedom have negligible Franck–Condon overlap for the probe transition to cationic states. As the energy in the degrees of freedom with good Franck–Condon overlap decreases and flows into degrees of freedom with negligible Franck–Condon overlap or increased IP, the 'apparent' internal energy measured using the ionizing probe pulse decreases, and the probe pulse quantum energy might not be sufficient anymore for ionization, as pointed out in ref. 176 and 184.

The problem of an increasing energy gap can also be overcome using multi-photon probe transitions. However, they provide difficulties in interpreting the dynamics as intermediate resonances along the molecular relaxation path actively shape the delay-dependent ion or electron signal.[185]

Ultrafast time-resolved VUV-probe photoelectron spectroscopy has successfully tracked the transient dissociation dynamics of simple diatomic molecules.[186,187] It has also been recently extended to the study of more complex molecules.[188–191] In these cases, the VUV pulses were created by four-wave-mixing or strong-field high harmonic generation. Ultrafast VUV probing combined with photoion kinetic energy probing has been applied to resolve the dynamics of small molecules such as ethylene[192] and water.[193] It is also noteworthy to mention the use of VUV light as a pump pulse instead of a probe, specifically the VUV-pump IR/VIS-probe work by Calegari *et al.* and Mikosch *et al.* in the study of molecular dynamics on cationic states generated by the VUV pump pulse.[194,195]

1.4.2 Ultrafast Core Level Spectroscopy

After discussing the rich information that can be obtained from valence electron probing in time-resolved absorption, photoion, and photoelectron spectroscopy, we now turn to probing *via* the core levels. The differences between valence- and core-level probing have already been discussed in Section 1.3. In short, core-level probing provides a very different, localized point-of-view of molecular dynamics. Highly element-specific information on the molecular dynamics can be derived from core-level probing. In addition, we have introduced the terminology of the chemical shift. Thus, core-level probes can deliver element-specific information, and different sites in the molecule occupied by the same element can also be distinguished.

We first concentrate on studies in which ultrafast VUV-to-soft-X-ray absorption is used for core-level probing of optically induced molecular dynamics. High-harmonic generation (HHG) sources provide a structured, broad continuum of photon energies in the form of femtosecond and attosecond pulses,[196] and they are therefore suitable for time-resolved absorption studies. The first such experiments were performed between 20 eV and 70 eV photon energy, where the sources deliver a relatively high flux. The shallowest core levels of halogens could be observed in experiments on dissociating CH_2Br_2[197] and CH_3I.[198]

With increasing flux at higher photon energies, the experimental capabilities of HHG sources were extended to studies at the carbon 1s and sulfur 2p levels. Wörner and co-workers employed a HHG source to investigate the dynamics of CF_4^+ and SF_6^+ at the C(1s) and S(2p) core levels; both molecular ions were produced *via* a preceding strong-field ionization step.[199] The NEXAFS spectra of the ionized species exhibited strong changes in the measured transient signal, reflecting the ionization-induced dissociation of CF_4^+ and SF_6^+ into CF_3^+ and SF_5^+, respectively. The change in symmetry of the dissociated species gave rise to splitting in the unoccupied orbitals, leading to the splitting of the X-ray absorption lines. Leone and co-workers used NEXAFS spectroscopy at the carbon K edge to investigate the UV-induced ring-opening of 1,3-cyclohexadiene (CHD).[200] The spectroscopic signatures identified the electronic structure and its dynamics around the pericyclic minimum, a point in the excited-state landscape at which the population either returns to the ring-closed ground state of CHD or travels further to form ring-opened 1,3-5-hexatriene. Moreover, a time constant of 60 fs for the creation and 110 fs for the decay of the electronic structure at this point was extracted.[200] These interpretations of the experimental data were aided by simulations of the NEXAFS spectra, identifying a 282 eV absorption feature indicative of the electronic structure at the pericyclic minimum. The ring-opening of CHD will be discussed again in the context of time-resolved diffraction experiments in Section 1.5 (and Chapters 9 and 12).

In line with the special emphasis on nucleobases, we present an oxygen K-edge absorption study of thymine in more detail. Section 1.2.5 describes that the participation and time constant of the $n\pi^*$ PES in the molecular relaxation process are controversial for thymine and uracil. The original interpretation and reinterpretation of the time-resolved valence photoelectron data in Section 1.4.1 demonstrated the challenges in

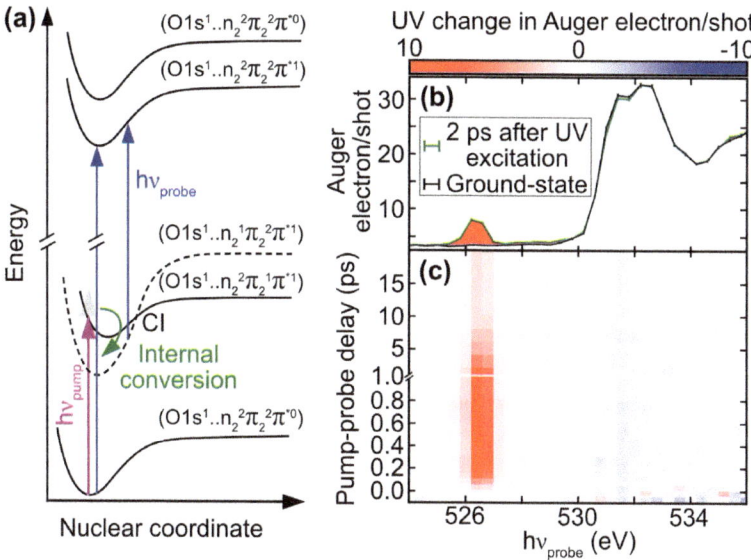

Figure 1.6 Time-resolved X-ray absorption spectroscopy of thymine at the oxygen K-edge. (a) The same excitation scheme as shown in Figure 1.5. Probing now occurs *via* probe transitions to the core–valence excited $(O1s^1 \cdots n_2^2 \pi_2^2 \pi^{*1})$ state. (b) NEXAFS spectrum of the thymine ground state (black) exhibits the O1s-π* transitions are shown in Figure 1.4. The molecular excitation with $h\nu_{\text{pump}}$ changes this spectrum, giving rise to a new band at 526 eV (see red shaded), indicating a strong 1s-n transition and thus the internal conversion from ππ* to nπ*. (c) The false colour representation of the difference signal showing the new band (red) with a rise of 60 fs and a drop of several picoseconds. Panels (b and c) reproduced from ref. 201, https://doi.org/10.1038/s41467-017-00069-7, under the terms of the CC BY 4.0 license http://creativecommons.org/licenses/by/4.0/.

attributing the electronic structure based on valence electron data only. This triggered us to perform a novel experiment in which the electronic state of the molecule could be probed with high fidelity through NEXAFS spectroscopy at the oxygen K-edge (see Figure 1.6a). As in the ultrafast valence-probe experiments, the thymine molecule is prepared using an excitation pulse $h\nu_{\text{pump}}$ on the pulse $h\nu_{\text{pump}}$ at the ππ* $(n_2^2 \pi_2^2 \pi^{*1})$ state. An nπ* $(n_2^1 \pi_2^2 \pi^{*1})$ state possesses a half-empty oxygen n-orbital and is essentially an oxygen 2p orbital. The hole in the n-orbital gives rise to an entirely new and strong absorption line in the NEXAFS spectrum corresponding to an O1s–O2p atomic absorption. In the multi-electron picture, the O1s-π* resonance is depicted as $(O1s^2 \ldots n_2^2 \pi_2^2 \pi^{*0})$ to $(O1s^1 \ldots n_2^2 \pi_2^2 \pi^{*1})$. The O1s-n $\left((O1s^2 \ldots n_2^2 \pi_2^2 \pi^{*1}) \rightarrow (O1s^1 \ldots n_2^2 \pi_2^2 \pi^{*1})\right)$ transition has the same final state as the O1s-π* resonance. The energy is thus lower than the O1s-π* resonance by $h\nu_{\text{pump}}$. The O1s-n resonance is strong because O1s–O2p is an atomic transition. This demonstrates that the strong localization of core holes can detect electronic features related to heteroatoms in organic photochemistry.

The black curve in Figure 1.6b shows the NEXAFS spectrum of thymine in its ground state. In this experiment, the molecule's absorption is measured as the yield of Auger–Meitner-electrons resulting from the oxygen core-hole decay. The two absorption maxima between $h\nu_{pump} = 530$ and 534 eV belong to electronic transitions from the O1s orbitals of the two different oxygen atoms to unoccupied molecular π^* orbitals, as explained in the context of Figure 1.4. The UV excitation promotes an electron from a π bonding orbital to a π^* anti-bonding orbital. The NEXAFS spectrum taken 2 ps after UV excitation (green line in Figure 1.6b) exhibits a new maximum at $h\nu_{pump} = 526$ eV (see red shaded area in Figure 1.6b), which is due to the $\left(\text{O1s}^2 \ldots n_2^1 \pi_2^2 \pi^{*1}\right) \rightarrow \left(\text{O1s}^1 \ldots n_2^2 \pi_2^2 \pi^{*1}\right)$ resonance. The NEXAFS spectrum thus directly reflects the $\pi\pi^* \rightarrow n\pi^*$ IC of thymine.[201]

The difference-spectrum of the molecule as a function of $h\nu_{prove}$ and pump–probe delay (see Figure 1.6c) exhibits more details of the molecular dynamics. The red features indicate more absorption, and thus an increased Auger–Meitner yield due to excitation with UV light $h\nu_{pump}$, whereas the blue features indicate a UV-induced bleach of the absorption. One can identify the onset of the feature at $h\nu_{prove} = 526$ eV shortly after excitation with a (60 ± 30) fs delay. The selectivity of this resonance to the $n\pi^*$ state thus confirms that the $\pi\pi^* \rightarrow n\pi^*$ IC is accomplished very quickly. This result confirms the initial interpretation of the valence photoelectron spectra.[179] The simulation-guided interpretation of the experiments in ref. 59 and 179 attributed a much longer time-constant on the picosecond scale to the IC, in conflict with the time-resolved NEXAFS results.

The above-discussed thymine example shows the new point of view that core-level absorption studies provide in investigating IC involving states with lone pair orbitals. Core-level absorption studies are particularly applicable to a large class of compounds that possess heteroatoms possessing lone-pair orbitals. The new absorption line's dominant and spectrally unspoiled appearance makes this a sensitive tool for the study of electronic effects within the molecule. Geometry changes are a second-order cause for the appearance of this absorption line, as verified through experiments[201] and calculations.[201,202]

An important molecule in this class of $\pi\pi^* \rightarrow n\pi^*$ IC is azobenzene. Simulations of this molecule exhibit a new absorption line in the NEXAFS spectrum of azobenzene when populated in the $n\pi^*$ state.[202,203] The strength of the absorption line does not change substantially as the molecule undergoes geometrical changes as long as it is in the $n\pi^*$ state. However, the precise photon energy of the line in the pre-edge region depends on the geometry in an intuitive way.[203]

First implementations of time-resolved core-level photoelectron spectroscopy have been established at the least bound halogen core levels, in analogy to the early developments of core-level resonant absorption. The prototypical dissociative system CH_3I was investigated using 4d photoelectrons emitted from the iodine atom.[38] The molecular and the atomic 4d photoelectron contribution could be identified, and the appearance of the atomic photoelectron line could be correlated with the dissociative time constant after UV excitation.

Simulations of the time-dependent carbon 1s photoemission of CH_3I show rich details obtained with eV level resolution.[204] In the Franck–Condon range of the excited state, a main 1s photoelectron peak is accompanied by a 10 eV split correlation satellite feature. Upon C–I dissociation, these lines move close and swap their relative strength. The lower photoelectron line belongs to a core-hole creation accompanied by a charge redistribution from iodine to the C1s ionized CH_3, leading to a dissociation limit with a neutral CH_3 radical and an iodine cation. The UV-induced dissociation separates the two fragments before the X-ray interaction. The subsequent X-ray photoemission occurs at an extended C–I bond, changing the photoionization cross-sections. The aforementioned core-ionization channel accompanied by charge redistribution shows a reduction in the cross-section because charge cannot efficiently transfer over the longer C–I bond distance. In contrast, the Franck–Condon satellite peak, which does not correspond to a charge redistribution, exhibits an increase in cross-section with increasing internuclear distance. The core-hole screening effects in the final state are also investigated in the iron-3d time-resolved photoelectron spectroscopy of dissociating iron pentacarbonyl.[205]

A recent experimental XPS study on the thionucleobase 2-thiouracil, 2-tUra (see Figure 1.7) extended the concept of the chemical shift to the excited state of molecules.[206] Here, the UV-excited molecular dynamics is investigated using XPS of the sulfur 2p core level. The excitation is analogous to the case of thymine. A UV pulse excites a singlet $\pi\pi^*$ state, which undergoes ultrafast IC to a singlet $n\pi^*$ state as the first step of electronic relaxation before undergoing an ISC to triplet states in this type of molecule.

The photoelectron spectrum of the excited (see the orange line in Figure 1.7b) and unexcited (blue line) 2-tUra molecules show a difference (green line) the latter is investigated as a function of UV-X-ray time delay (false colour plot in Figure 1.7c). The spin–orbit splitting, which is 1.2 eV for the sulfur core level of 2-tUra,[207] cannot be identified here. The FEL experiment at FLASH2 was performed with the non-monochromatized beam resulting in ~ 4 eV spectral resolution. One can identify a shift of the excited-state photoelectron spectrum to lower kinetic energies, meaning that the binding energy of the sulfur 2p electron increases with UV excitation. This shift of the excited-state core-level ionization potential depends strongly on the local charge on the probed sulfur atom. Ref. 206 demonstrates a linear correlation between charge and ionization energy. This kind of dependency is completely analogous to the chemical shift concept in the ground state[110] and the term 'excited-state chemical shift' was thus chosen.[206] Interestingly, the charge on the sulfur, and thus the excited-state chemical shift, depends mainly on the electronic state of the molecule and only second-order on the molecular geometry. Thus, as in the case of X-ray absorption, time-resolved XPS is mostly sensitive to the dynamically evolving electronic structure of the molecule.

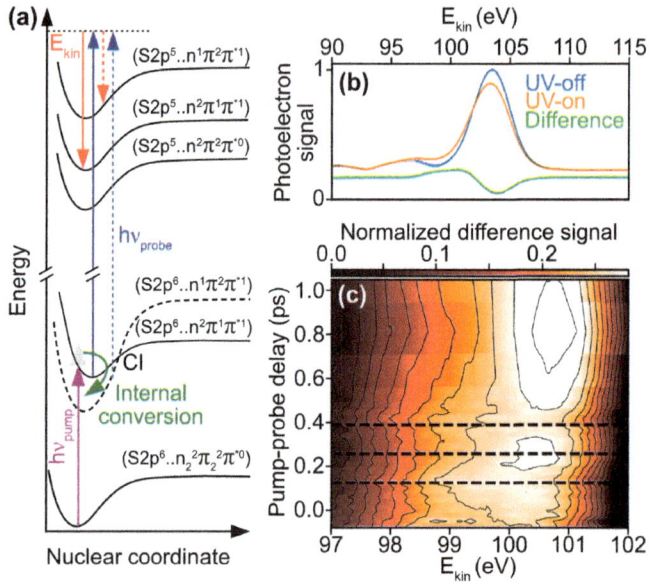

Figure 1.7 Time-resolved X-ray photoelectron spectroscopy (XPS) on 2-thiouracil (2-tUra). (a) UV excitation with a photon energy $h\nu_{pump}$ launches the molecule at a $(n^2\pi^1\pi^{*1})$ state. A predicted $\pi\pi^* \to n\pi^*$ IC to a $(n^1\pi^2\pi^{*1})$ state as in the nucleobase thymine is the first step in electronic relaxation. The molecule is probed *via* a short pulse of photon energy $h\nu_{pump}$ creating photoelectrons *via* the cationic channels $(S2p^5\cdots n^2\pi^1\pi^{*1})$ and $(S2p^5\cdots n^1\pi^2\pi^{*1})$ for the $(n^2\pi^1\pi^{*1})$ and $(n^1\pi^2\pi^{*1})$ states, respectively. (b) XPS spectrum of 2-tUra in the ground state (blue). The spin–orbit splitting is not visible within the experimental resolution. UV excitation shifts the XPS towards lower kinetic energies (orange), creating a typical differential line shape difference spectrum (green). (c) Zoom into the difference spectrum, normalized on the maximum in the E_{kin} range 98–102 eV. The false colour representation shows an oscillation period of 250 fs within the first 600 fs after UV excitation. The black dotted lines mark the times of extremes in the shift. Panels (b and c) reproduced from ref. 206, https://doi.org/10.1038/s41467-021-27908-y, under the terms of the CC BY 4.0 license http://creativecommons.org/licenses/by/4.0/.

The strongest shifts of about 4eV in the ionization potential of the sulfur 2p electrons are predicted for $n\pi^*$ states, as the strongly localized sulfur lone pair orbital n is missing an electron. All other states are predicted to have lower ionization potentials. The positive part of the difference spectrum was investigated further (see Figure 1.7c), normalized to the maximum of the difference signal. With adapted colour scale and contour lines, one identifies oscillations in the photoelectron spectrum. The photoelectron signal oscillates with a period of 250 fs (see dashed lines). The pattern can be followed up to a 0.6 ps delay; beyond 0.6 ps, the experimental delay resolution is insufficient. Simulations of the molecular dynamics show an oscillating population exchange among the different electronic states after molecular excitation.[208,209] Comparing the experimental oscillations with the theoretically predicted $^1n\pi^*$ electronic population shows a good agreement. This example demonstrates the power of time-resolved XPS for identifying the electronic dynamics of excited molecules.

Time-resolved Auger–Meitner spectroscopy allows the deduction of local bond changes around the probed atom. The principle was demonstrated in thymine[119] and 2-thiouracil.[121] For thymine, the elongation of the C–O bond upon UV absorption could be observed *via* time-resolved Auger–Meitner spectroscopy at the oxygen K-edge. For 2-thiouracil, the time-resolved Auger–Meitner spectroscopy was performed at the sulfur $L_{2,3}$ edges. Here, the observed bond dynamics of expanding and shrinking distance are confirmed with the predicted trajectories from calculations.[209]

The fragmentation after Auger–Meitner decay is also used in time-resolved experiments on the exemplary case of dissociative CH_3I. In work by Erk *et al.*,[210] the C–I bond dissociation was initiated by a multiphoton-IR transition, and followed by a time-delayed X-ray pulse that created an iodine M-shell core hole that undergoes an Auger–Meitner decay. The observed delay-dependent change in the ion channels can be translated into a change in the interatomic distance during the photodissociation process using classical Newtonian mechanics. No highly charged iodine cations are found for early delays as the positive charge, initially localized on the iodine, spreads over the whole molecule. However, highly charged iodine fragments are observed for longer delays and, thus, larger bond distances due to the loss in electronic overlap. Amini *et al.*[211] induced the dissociation of CH_3I and more complex 2,6-difluoroiodobenzene following the absorption of one UV photon, and the XUV probe pulse site selectively ionized the iodine core levels. The conclusion of charge transfer at small bond lengths in this single-photon UV-excitation study confirmed the results of Erk *et al.*'s multi-photon IR-excitation study.

1.4.3 Diffraction (with X-rays and Electrons)

So far in this chapter, the time-resolved spectroscopic observables have been interpreted in statements about molecular dynamics. We have shown that changes in both the nuclear geometry and the electronic structure can lead to a modulation of the spectroscopic observables. Using only the experimental results, there is often an ambivalence in attributing an observed time-constant to either a change in geometry or a change in electronic state, as the example of the nucleobase elaborated above shows. Combining experiments with theory, in particular by including a simulation of the observable, is an often fruitful approach. Exploring new observables related to core electrons can provide a different perspective on molecular dynamics, which according to the first core-electron experiments, shows a strong sensitivity to the electronic part of the dynamics.

Many contributions in this book focus on molecular ultrafast diffraction studies with X-rays and electrons. From the perspective of this chapter, diffraction is a complementary technique to spectroscopic observables. While the recently elaborated methods of ultrafast probing *via* core levels show an increased sensitivity to molecular electronic changes, the diffraction methods show an increased sensitivity to changes in nuclear geometry. The word sensitivity is not to be taken as an exclusive statement; the observables in both methods also show some modulation in the measured signal (*e.g.*, inelastic electron scattering; see Chapter 3) due to the other degree of freedom (*e.g.*, electronic dynamics).

Within the past couple of years, the rapid success in ultrafast X-ray (see Chapter 9) and electron (see Chapters 10–13) diffraction, referred to as UXD and UED, respectively, has changed the field of ultrafast gas-phase molecular dynamics profoundly. We contrast a few cases mentioned in the context above with their ultrafast diffraction measurements. We start with very simple vibrational and rotational wavepackets presented in Section 1.2.4 as the first examples of ultrafast measurements. The iodine molecule I_2 was a prototype for developing and interpreting vibrational wavepackets. The earlier experiments on I_2 showed that the PES for the vibrational wavepackets could be reconstructed by scanning the excitation of the wavepacket energetically within that potential. Thus, it was also possible to attribute a bond distance to the wavepacket after the potential construction.[31,32] On the other hand, diffraction is directly sensitive to the internuclear distance as the bond distance modulates the interference pattern of the X-ray or electron waves scattered at the atomic centres. Even for an isotropic ensemble of molecules, the information on the bond distance is encoded in the molecular scattering intensity, I_{mol}, pattern as a function of momentum transfer, s, according to (see Chapter 3)

$$I_{mol}(s) = \sum_{i,j=1}^{M} |f_i||f_j|\cos\left(\Delta\eta_{ij}\right)\sin\left(sr_{ij}\right)/sr_{ij}, \tag{1.9}$$

where f_i is the scattering amplitude of the *i*th atom, Δn_{ij} is the phase difference of the scattered wave from the *i*th and *j*th atom, and r_{ij} is the distance between the *i*th and *j*th atoms. The equation uses the independent atom model, neglecting any effects due to molecular bonding.[212] Thus, diffraction measures the distances between pairs of atoms within molecules, and for diatomic molecules, this is the complete information needed to track their nuclear dynamics. One of the earliest applications of relativistic, femtosecond electron pulses available at SLAC[213] for gas-phase diffraction was the direct determination of the interatomic distance $r_{ij} = R$ in a vibrational wavepacket of I_2.[214] The experiment showed the feasibility of the UED method in gas-phase nuclear dynamics and also delivered information on changes in the wavepacket shape during propagation on the molecular excited-state potential.

In analogy to these vibrational proof-of-principle experiments, studies on rotational wavepackets and rotational alignment have been performed with

UXD and UED.[215,216] In the example from ref. 217, molecular alignment was achieved using a long, nanosecond pulse and probed with femtosecond UXD. In ref. 217, a femtosecond pulse was used for inducing alignment in a molecule, and the subsequently generated rotational wavepacket was detected *via* UED.

Two of the molecular samples studied with ultrafast spectroscopic experiments discussed in the previous sections have also been investigated recently by UXD and UED. The CF_3I molecule, a close relative of the prototypical CH_3I system exhibiting C–I photodissociation dynamics, was investigated *via* UED after UV excitation to the A-band dissociative states and after two-photon excitation into Rydberg and charge pair states.[218] The two signals are disentangled by using the scattering angle characteristic of the diffraction signal. The two-photon channel in CH_3I induces molecular geometry changes that sample a CI between Rydberg and ion-pair states. The scattering signal orthogonal to the laser polarization is transformed into a pair correlation function, and the CI can be identified. The one-photon transition excites molecules with the C–I axis parallel to the excitation polarization; the two-photon transition is orthogonal to that axis. The ultrafast dissociation of CF_3I *via* A-band states could be observed as a vanishing signal in the pair distribution function for the C–I and F–I ground state distances. The F–I feature showed delayed dynamics because the fluorine atoms react slightly later than the C–I bond due to the dissociative character of the C–I bond in the A-band states.

This UED work complements time-resolved photoelectron spectroscopy studies on CH_3I.[38,204] The diffraction experiment is sensitive to the distance between scattering atomic centres in an evolving transient molecular structure and delivers the closest view of the changing geometry. In comparison, the XPS line shift during the molecular dissociation presents the 'electron view' of the UV-induced dissociation process. In the experimental study by Brauße *et al.*,[38] a molecular feature was identified in the iodine 4d line that transforms into an atomic feature. The time resolution was not sufficient to follow the dissociation with intermediate steps. In the simulations by Inhester *et al.*,[204] exact details about the change in bonding with increasing C–I distance can be identified, as mentioned in the previous section.

The prototypical photoinduced ring opening of 1,3-cyclohexadiene (CHD) has also been investigated using time-resolved X-ray (see Chapter 9) and electron (see Chapter 12) diffraction techniques, apart from the electronically sensitive soft X-ray absorption spectroscopy mentioned above.[200] The first time-resolved CHD diffraction study by Minitti *et al.*[219,220] was performed using hard X-rays at the LCLS FEL, covering a momentum transfer range of 4.5 Å^{-1}, corresponding to a spatial resolution of $\frac{2\pi}{4.5}$Å $= 1.4$ Å. On comparing the time-resolved diffraction signal to the simulated diffraction signal from trajectory studies, the authors extracted the most dominant paths reproducing the experimental data. A second time-resolved diffraction study on CHD was performed with UED at SLAC's ultrafast MeV electron facility, covering a higher momentum range of 10 $Å^{-1}$.[221] This study revealed a real-space transformation of the diffraction signals. The ring-opening process can be followed by an increase in the carbon-pair distances after ring opening and the coherent oscillations in the measured signal due to the formation of the vibrationally-hot 1,3,5-hexatriene photoproduct. Time-resolved carbon K-edge NEXAFS spectroscopy[200] of the same ring-opening CHD reaction delivers complementary information by probing the unoccupied orbitals. As in the case of thymine, the existence of an excitation-induced pre-edge feature completely depends on the electronic structure. Geometrical changes shift the photon energy of the pre-edge feature. Similar to the case of thymine, a passage through conical intersections leads to changes in the electronic structure. A combination of diffraction and NEXAFS studies provides a detailed view of both the electronic and nuclear dynamics occurring during molecular ring-opening. The diffraction experiments do not identify the pericyclic point; they do not show any special sensitivity to this region of the PESs. The time-resolved NEXAFS measurements seem to pronounce the pericyclic point due to its particular electronic structure compared to regions sampled before and after.

Ultrafast diffraction experiments also contain traces of the electronic structure, just like ultrafast core-level experiments also show some sensitivity to changes in the molecular geometry. Hard X-rays scatter at the molecular electrons as Thompson scattering prefers light electrons, while nuclei are too heavy to oscillate in the X-ray field.[222] Transitioning from heavier to lighter atoms, the X-ray scattering signal from the electrons gets weaker, but also, more of the electrons actively participate in bonding, and thus one gains sensitivity on the valence electron structure. The participation of valence electrons in gas-phase X-ray diffraction has been observed upon the valence electron excitation into Rydberg states[223] and shows up as corrections in the diffraction intensities of different excited states[224] (see Chapter 9). In contrast to X-rays, electrons interact *via* Rutherford scattering on all charged particles. The direct trace of valence electron dynamics in ultrafast electron diffraction has been observed for an $n\pi^*$ to ground state IC of pyridine.[225] Yang *et al.* show a strong experimental feature between 0.3 and 0.7 $Å^{-1}$, which fits the expected $n\pi^*$ dynamics. The inelastically scattered signal at small momentum transfer contains a negative contribution, scaling with electron correlation. With two electrons in the very compact lone pair n orbital, the electronic ground state has a higher dynamic electron correlation level than the $n\pi^*$ state with one electron in the n orbital. Thus, with the population of the $n\pi^*$ state, the scattering signal at low momentum transfer increases. The ~ 2 ps internal conversion into the ground state leads to a decrease in this signal.

1.5   Future of Molecular Dynamics Probing

We are fortunate to perform molecular science in a time of rapid change in the invention and utilization of new ultrafast sources of light and electrons.

The development of VUV and soft X-ray FELs has offered unique possibilities for exploring ultrafast molecular dynamics over the past years. At the same time, strong-field HHG sources offer increasing photon energies for time-resolved core-level studies at much smaller laser laboratories than large FEL facilities. Infrared-driven HHG sources already cover the sulfur L-and carbon K-edges with rapid extension to the nitrogen and oxygen K-edges.[199,200,226,227] Harmonic sources provide intrinsically broadband, attosecond train pulses. Using these HHG sources to explore molecular excited-state dynamics with transient femtosecond and attosecond absorption spectroscopy will provide new insight into many different molecular systems in the gas phase and thin liquid sheets.

At the same time, soft X-ray FELs become more controlled in their temporal and spectral properties. First, attosecond pulses in the soft X-ray domain have been demonstrated and utilized.[228] Overcoming SASE spectral and temporal fluctuations using seed lasers makes the FEL sources more suitable for spectroscopic purposes and allows for the utilization of experimental schemes that cannot work with single-shot read-out techniques. In particular, the attosecond and seeding efforts pave the way for controlled nonlinear spectroscopy schemes in the X-ray domain.[229–235]

One of the parameters of FELs in the context of molecular dynamics probing is the increasing repetition rate based on superconducting (SC) accelerator technology. Repetition rates up to a MHz will be realized soon, providing several breakthroughs in molecular dynamics studies. First, this will make the exploration of very dilute samples possible. For example, time-resolved spectroscopy performed today at rather small molecular systems with a high vapour pressure can be extended to complex, large molecules (which often have very low vapour pressures) by means of electrospray ionization. Second, these MHz sources will have better feedback systems and will be much more stable in their intensity, temporal and spectral properties. This, in turn, simplifies the data analysis, making it much more similar to small tabletop experiments and, thus, much more user-friendly.

Time-resolved gas-phase diffraction experiments at FELs and relativistic electron sources are a growing research field and have already, in a few years, provided a revolutionary insight into the structural dynamics of a range of selected samples and compounds, as will be described in many chapters that follow in this book. The future developments in X-ray diffraction are coupled with the FEL developments given above, with their advantages for more dilute systems and better control over the experiment. An increase in the repetition rate is highly likely and advantageous for electron sources. The photoinjector used at an FEL is the source for ultrafast diffraction experiments with relativistic MeV electrons. Future developments in non-relativistic UED employing <200 keV electrons (see Chapters 14 and 16) also hold the promise to study ultrafast structural dynamics with sub-100 fs time resolution at high repetition rates.[236–239]

Acknowledgements

I am highly grateful for many discussions and experiments with the members of the nucleobase photoprotection collaboration, the authors of ref. 119 and 201, and the thionucleobase collaboration, the authors of ref. 206. I thank Brian McFarland, Thomas Wolf, Dennis Mayer, Fabiano Lever and David Picconi for working intensely together on ultrafast X-ray Auger–Meitner, X-ray absorption, and X-ray photoemission spectroscopy.

References

1. M. A. Robb, F. Bernardi and M. Olivucci, *Pure Appl. Chem.*, 1995, **67**, 783–789.
2. D. Yarkony, *Rev. Mod. Phys.*, 1996, **68**, 985–1013.
3. *Conical intersections electronic structure, dynamics & spectroscopy*, ed. W. Domcke, D. Yarkony and H. Köppel, World Scientific, River Edge, N.J., London, 2004.
4. W. Domcke, D. R. Yarkony and H. Köppel, *Conical intersections: theory, computation and experiment*, World Scientific Publishing Company, Singapore, 2011.
5. D. Polli, P. Altoè, O. Weingart, K. M. Spillane, C. Manzoni, D. Brida, G. Tomasello, G. Orlandi, P. Kukura, R. A. Mathies, M. Garavelli and G. Cerullo, *Nature*, 2010, **467**, 440–443.
6. B. G. Levine and T. J. Martínez, *Annu. Rev. Phys. Chem.*, 2007, **58**, 613–634.
7. G. Cui, Z. Lan and W. Thiel, *J. Am. Chem. Soc.*, 2012, **134**, 1662–1672.
8. A. Toniolo, S. Olsen, L. Manohar and T. J. Martínez, *Faraday Discuss.*, 2004, **127**, 149–163.
9. C. E. Crespo-Hernández, B. Cohen, P. M. Hare and B. Kohler, *Chem. Rev.*, 2004, **104**, 1977–2020.
10. C. T. Middleton, K. de La Harpe, C. Su, Y. K. Law, C. E. Crespo-Hernández and B. Kohler, *Annu. Rev. Phys. Chem.*, 2009, **60**, 217–239.
11. W. J. Schreier, P. Gilch and W. Zinth, *Annu. Rev. Phys. Chem.*, 2015, **66**, 497–519.
12. P. T. Anastas and J. C. Warner, *Green chemistry: theory and practice*, Oxford University Press, Oxford [England], New York, 1998.
13. A. Albini and M. Fagnoni, *Green Chem.*, 2004, **6**, 1.
14. *Molecular Devices for Solar Energy Conversion and Storage*, ed. G. Boschloo, A. Hagfeldt and H. Tian, Springer, Singapore, 2018.
15. A. Szabo, *Modern Quantum Chemistry: Introduction to Advanced Electronic Structure Theory*, Dover Publications, Mineola, New York, 2012.
16. W. Kolos and L. Wolniewicz, *J. Chem. Phys.*, 1964, **41**, 3663–3673.
17. S. Matsika and P. Krause, *Annu. Rev. Phys. Chem.*, 2011, **62**, 621–643.
18. F. Jensen, *Introduction to computational chemistry*, Wiley, Chichester, UK, Hoboken, NJ, 3rd edn, 2017.
19. M. Born and R. Oppenheimer, *Ann. Phys.*, 1927, **389**, 457–484.
20. D. J. Tannor, *Introduction to quantum mechanics: a time-dependent perspective*, University Science Books, Sausalito, Calif, 2007.

21. R. Schinke, *Photo dissociation dynamics*, Cambridge University Press, Cambridge, 1993.
22. M. Gühr, H. Ibrahim and N. Schwentner, *Phys. Chem. Chem. Phys.*, 2004, **6**, 5353.
23. E. Teller, *J. Phys. Chem.*, 1937, **41**, 109–116.
24. G. J. Atchity, S. S. Xantheas and K. Ruedenberg, *J. Chem. Phys.*, 1991, **95**, 1862–1876.
25. M. S. Schuurman and A. Stolow, *Annu. Rev. Phys. Chem.*, 2018, **69**, 427–450.
26. J. P. Malhado and J. T. Hynes, *Chem. Phys.*, 2008, **347**, 39–45.
27. H. Haken and H. C. Wolf, *Physics and elements of quantum chemistry: introduction to experiments and theory*, 2004.
28. P. W. Atkins and R. Friedman, *Molecular quantum mechanics*, Oxford University Press, Oxford, New York, 5th edn, 2011.
29. I. V. Hertel, C.-P. Schulz and I. V. Hertel, *Molecules and photons – spectroscopy and collisions*, Springer, Heidelberg Berlin, 2015.
30. E. J. Heller, *Acc. Chem. Res.*, 1981, **14**, 368–375.
31. M. Gruebele, G. Roberts, M. Dantus, R. M. Bowman and A. H. Zewail, *Chem. Phys. Lett.*, 1990, **166**, 459–469.
32. M. Gruebele and A. H. Zewail, *J. Chem. Phys.*, 1993, **98**, 883–902.
33. N. F. Scherer, R. J. Carlson, A. Matro, M. Du, A. J. Ruggiero, V. Romero-Rochin, J. A. Cina, G. R. Fleming and S. A. Rice, *J. Chem. Phys.*, 1991, **95**, 1487–1511.
34. I. Fischer, D. M. Villeneuve, M. J. J. Vrakking and A. Stolow, *J. Chem. Phys.*, 1995, **102**, 5566–5569.
35. M. J. J. Vrakking, D. M. Villeneuve and A. Stolow, *Phys. Rev. A: At., Mol., Opt. Phys.*, 1996, **54**, R37–R40.
36. T. Baumert, M. Grosser, R. Thalweiser and G. Gerber, *Phys. Rev. Lett.*, 1991, **67**, 3753–3756.
37. W. G. Roeterdink and M. H. M. Janssen, *Chem. Phys. Lett.*, 2001, **345**, 72–80.
38. F. Brauße, G. Goldsztejn, K. Amini, R. Boll, S. Bari, C. Bomme, M. Brouard, M. Burt, B. C. de Miranda, S. Düsterer, B. Erk, M. Géléoc, R. Geneaux, A. S. Gentleman, R. Guillemin, I. Ismail, P. Johnsson, L. Journel, T. Kierspel, H. Köckert, J. Küpper, P. Lablanquie, J. Lahl, J. W. L. Lee, S. R. Mackenzie, S. Maclot, B. Manschwetus, A. S. Mereshchenko, T. Mullins, P. K. Olshin, J. Palaudoux, S. Patchkovskii, F. Penent, M. N. Piancastelli, D. Rompotis, T. Ruchon, A. Rudenko, E. Savelyev, N. Schirmel, S. Techert, O. Travnikova, S. Trippel, J. G. Underwood, C. Vallance, J. Wiese, M. Simon, D. M. P. Holland, T. Marchenko, A. Rouzée and D. Rolles, *Phys. Rev. A*, 2018, **97**, 043429.
39. Z. Yang, K. Schnorr, A. Bhattacherjee, P.-L. Lefebvre, M. Epshtein, T. Xue, J. F. Stanton and S. R. Leone, *J. Am. Chem. Soc.*, 2018, **140**, 13360–13366.
40. M. L. Murillo-Sánchez, J. González-Vázquez, M. E. Corrales, R. de Nalda, E. Martínez-Núñez, A. García-Vela and L. Bañares, *J. Chem. Phys.*, 2020, **152**, 014304.

41. I. Averbukh and N. F. Perelman, *Phys. Lett. A*, 1989, **139**, 449–453.
42. T. Lohmüller, V. Engel, J. A. Beswick and C. Meier, *J. Chem. Phys.*, 2004, **120**, 10442–10449.
43. A. Kirrander and P. Weber, *Appl. Sci.*, 2017, **7**, 534.
44. H. Stapelfeldt and T. Seideman, *Rev. Mod. Phys.*, 2003, **75**, 543–557.
45. J. P. Heritage, T. K. Gustavson and C. H. Lin, *Phys. Rev. Lett.*, 1975, **34**, 1299–1302.
46. F. Rosca-Pruna and M. J. J. Vrakking, *Phys. Rev. Lett.*, 2001, **87**, 153902–153904.
47. M. D. Poulsen, E. Péronne, H. Stapelfeldt, C. Z. Bisgaard, S. S. Viftrup, E. Hamilton and T. Seideman, *J. Chem. Phys.*, 2004, **121**, 783–791.
48. B. K. McFarland, J. P. Farrell, P. H. Bucksbaum and M. Gühr, *Science*, 2008, **322**, 1232.
49. C. Z. Bisgaard, O. J. Clarkin, G. Wu, A. M. D. Lee, O. Gessner, C. C. Hayden and A. Stolow, *Science*, 2009, **323**, 1464–1468.
50. D. Pavičić, K. F. Lee, D. M. Rayner, P. B. Corkum and D. M. Villeneuve, *Phys. Rev. Lett.*, 2007, **98**, 243001.
51. N. J. Turro, *Principles of molecular photochemistry: an introduction*, University Science Books, Sausalito, Calif, 2009.
52. M. Klessinger and J. Michl, *Excited states and photochemistry of organic molecules*, VCH, New York, 1995.
53. A. C. Kneuttinger, G. Kashiwazaki, S. Prill, K. Heil, M. Müller and T. Carell, *Photochem. Photobiol.*, 2014, **90**, 1–14.
54. R. Improta, F. Santoro and L. Blancafort, *Chem. Rev.*, 2016, **116**, 3540–3593.
55. M. Barbatti, A. C. Borin and S. Ullrich, Photoinduced Phenomena in Nucleic Acids I: *Nucleobases in the gas phase and in solvents*, Springer, Cham, 2015.
56. S. Perun, A. L. Sobolewski and W. Domcke, *J. Phys. Chem. A*, 2006, **110**, 13238–13244.
57. M. Merchán, R. González-Luque, T. Climent, L. Serrano-Andrés, E. Rodríguez, M. Reguero and D. Peláez, *J. Phys. Chem. B*, 2006, **110**, 26471–26476.
58. D. Asturiol, B. Lasorne, M. A. Robb and L. Blancafort, *J. Phys. Chem. A*, 2009, **113**, 10211–10218.
59. H. R. Hudock, B. G. Levine, A. L. Thompson, H. Satzger, D. Townsend, N. Gador, S. Ullrich, A. Stolow and T. J. Martínez, *J. Phys. Chem. A*, 2007, **111**, 8500–8508.
60. J. J. Szymczak, M. Barbatti, J. T. Soo Hoo, J. A. Adkins, T. L. Windus, D. Nachtigallová and H. Lischka, *J. Phys. Chem. A*, 2009, **113**, 12686–12693.
61. T. Carell, C. Brandmayr, A. Hienzsch, M. Müller, D. Pearson, V. Reiter, I. Thoma, P. Thumbs and M. Wagner, *Angew. Chem., Int. Ed.*, 2012, **51**, 7110–7131.
62. P. F. Swann, T. R. Waters, D. C. Moulton, Y.-Z. Xu, Q. Zheng, M. Edwards and R. Mace, *Science*, 1996, **273**, 1109–1111.

63. T. S. Gee, K.-P. Yu and B. D. Clarkson, *Cancer*, 1969, **23**, 1019–1032.
64. K. Kramer, T. Sachsenberg, B. M. Beckmann, S. Qamar, K.-L. Boon, M. W. Hentze, O. Kohlbacher and H. Urlaub, *Nat. Methods*, 2014, **11**, 1064–1070.
65. X. Zhang, G. Jeffs, X. Ren, P. O'Donovan, B. Montaner, C. M. Perrett, P. Karran and Y.-Z. Xu, *DNA Repair*, 2007, **6**, 344–354.
66. R. Brem, I. Daehn and P. Karran, *DNA Repair*, 2011, **10**, 869–876.
67. H. Kuramochi, T. Kobayashi, T. Suzuki and T. Ichimura, *J. Phys. Chem. B*, 2010, **114**, 8782–8789.
68. M. Pollum, L. Martínez-Fernández and C. E. Crespo-Hernández, in *Photoinduced Phenomena in Nucleic Acids I*, ed. M. Barbatti, A. C. Borin and S. Ullrich, Springer International Publishing, Cham, 2014, vol. 355, pp. 245–327.
69. S. Euvrard, J. Kanitakis and A. Claudy, *N. Engl. J. Med.*, 2003, **348**, 1681–1691.
70. A. Massey, Y.-Z. Xu and P. Karran, *Curr. Biol.*, 2001, **11**, 1142–1146.
71. O. Reelfs, P. Karran and A. R. Young, *Photochem. Photobiol. Sci.*, 2012, **11**, 148–154.
72. S. Arslancan, L. Martínez-Fernández and I. Corral, *Molecules*, 2017, **22**, 998.
73. B. Ashwood, M. Pollum and C. E. Crespo-Hernández, *Photochem. Photobiol.*, 2019, **95**, 33–58.
74. E. Runge and E. K. U. Gross, *Phys. Rev. Lett.*, 1984, **52**, 997–1000.
75. J. Westermayr and P. Marquetand, *Chem. Rev.*, 2021, **121**, 9873–9926.
76. B. F. E. Curchod and T. J. Martínez, *Chem. Rev.*, 2018, **118**, 3305–3336.
77. L. M. Ibele and B. F. E. Curchod, *Phys. Chem. Chem. Phys.*, 2020, **22**, 15183–15196.
78. J. C. Tully and R. K. Preston, *J. Chem. Phys.*, 1971, **55**, 562–572.
79. J. C. Tully, *J. Chem. Phys.*, 1990, **93**, 1061–1071.
80. J. C. Tully, *J. Chem. Phys.*, 2012, **137**, 22A301.
81. G. A. Worth, M. A. Robb and I. Burghardt, *Faraday Discuss.*, 2004, **127**, 307.
82. B. Lasorne, M. J. Bearpark, M. A. Robb and G. A. Worth, *Chem. Phys. Lett.*, 2006, **432**, 604–609.
83. D. V. Makhov, C. Symonds, S. Fernandez-Alberti and D. V. Shalashilin, *Chem. Phys.*, 2017, **493**, 200–218.
84. M. Ben-Nun, J. Quenneville and T. J. Martínez, *J. Phys. Chem. A*, 2000, **104**, 5161–5175.
85. *Multidimensional quantum dynamics: MCTDH theory and applications*, ed. H.-D. Meyer, F. Gatti and G. Worth, Wiley-VCH, Weinheim, Chichester, 2009.
86. D. W. Turner, *Molecular photoelectron spectroscopy: a handbook of He 584 Å spectra*, Wiley Interscience, London, New York, 1970.
87. J. H. D. Eland, *Photoelectron spectroscopy: an introduction to ultraviolet photoelectron spectroscopy in the gas phase*, Butterworths, London, 1974.
88. J. Berkowitz, *Photoabsorption, photoionization, and photoelectron spectroscopy*, Academic Press, New York, 1979.

89. *VUV and soft X-ray photoionization*, ed. U. Becker and D. A. Shirley, Plenum Press, New York, 1996.
90. A. B. Trofimov, J. Schirmer, V. B. Kobychev, A. W. Potts, D. M. P. Holland and L. Karlsson, *J. Phys. B*, 2006, **39**, 305–329.
91. K. D. Fulfer, D. Hardy, A. A. Aguilar and E. D. Poliakoff, *J. Chem. Phys.*, 2015, **142**, 224310.
92. J. H. D. Eland, *J. Chem. Phys.*, 1979, **70**, 2926–2933.
93. H.-W. Jochims, M. Schwell, H. Baumgärtel and S. Leach, *Chem. Phys.*, 2005, **314**, 263–282.
94. B. Sztáray, A. Bodi and T. Baer, *J. Mass Spectrom.*, 2010, **45**, 1233–1245.
95. J. H. D. Eland, *Rev. Sci. Instrum.*, 1978, **49**, 1688–1690.
96. P. Oßwald, P. Hemberger, T. Bierkandt, E. Akyildiz, M. Köhler, A. Bodi, T. Gerber and T. Kasper, *Rev. Sci. Instrum.*, 2014, **85**, 025101.
97. D. Krüger, P. Oßwald, M. Köhler, P. Hemberger, T. Bierkandt, Y. Karakaya and T. Kasper, *Combust. Flame*, 2018, **191**, 343–352.
98. M. B. Prendergast, B. B. Kirk, J. D. Savee, D. L. Osborn, C. A. Taatjes, P. Hemberger, S. J. Blanksby, G. da Silva and A. J. Trevitt, *Phys. Chem. Chem. Phys.*, 2019, **21**, 17939–17949.
99. V. S. Letokhov, *Laser Photoionization Spectroscopy*, Elsevier, 1987, pp. 117–149.
100. M. N. R. Ashfold and J. D. Howe, *Annu. Rev. Phys. Chem.*, 1994, **45**, 57–82.
101. E. Nir, Ch Plützer, K. Kleinermanns and M. de Vries, *Eur. Phys. J. D*, 2002, **20**, 317–329.
102. D. W. Chandler and P. L. Houston, *J. Chem. Phys.*, 1987, **87**, 1445–1447.
103. A. Eppink and D. H. Parker, *Rev. Sci. Instrum.*, 1997, **68**, 3477–3484.
104. M. N. R. Ashfold, N. H. Nahler, A. J. Orr-Ewing, O. P. J. Vieuxmaire, R. L. Toomes, T. N. Kitsopoulos, I. A. Garcia, D. A. Chestakov, S.-M. Wu and D. H. Parker, *Phys. Chem. Chem. Phys.*, 2006, **8**, 26–53.
105. J. Ullrich, R. Moshammer, R. Dörner, O. Jagutzki, V. Mergel, H. Schmidt-Böcking and L. Spielberger, *J. Phys. B*, 1997, **30**, 2917–2974.
106. R. Dörner, V. Mergel, O. Jagutzki, L. Spielberger, J. Ullrich, R. Moshammer and H. Schmidt-Böcking, *Phys. Rep.*, 2000, **330**, 95–192.
107. J. Ullrich, R. Moshammer, A. Dorn, R. Dorner, L. P. H. Schmidt and H. Schmidt-Böcking, *Rep. Prog. Phys.*, 2003, **66**, 1463–1545.
108. M. Pitzer, M. Kunitski, A. S. Johnson, T. Jahnke, H. Sann, F. Sturm, L. P. H. Schmidt, H. Schmidt-Bocking, R. Dorner, J. Stohner, J. Kiedrowski, M. Reggelin, S. Marquardt, A. Schiesser, R. Berger and M. S. Schoffler, *Science*, 2013, **341**, 1096–1100.
109. K. Siegbahn, *ESCA applied to free molecules*, North-Holland Pub. Co, Amsterdam, 1969.
110. U. Gelius, *Phys. Scr.*, 1974, **9**, 133–147.
111. N. Mårtensson and A. Nilsson, *J. Electr. Spectr. Rel. Phen.*, 1995, **75**, 209–223.
112. J. Stöhr, *NEXAFS spectroscopy*, Springer, Berlin, New York, 1st edn, corr. print., 1996.
113. S. Hüfner, *Photoelectron spectroscopy: principles and applications*, Springer, Berlin, New York, 3rd edn, 2003.

114. O. Plekan, V. Feyer, R. Richter, M. Coreno, M. de Simone, K. C. Prince, A. B. Trofimov, E. V. Gromov, I. L. Zaytseva and J. Schirmer, *Chem. Phys.*, 2008, **347**, 360–375.

115. O. Travnikova, K. J. Børve, M. Patanen, J. Söderström, C. Miron, L. J. Sæthre, N. Mårtensson and S. Svensson, *J. Electron Spectrosc. Relat. Phenom.*, 2012, **185**, 191–197.

116. G. Fronzoni, M. Stener and P. Decleva, *Phys. Chem. Chem. Phys.*, 1999, **1**, 1405–1414.

117. A. P. P. Natalense and R. R. Lucchese, *J. Chem. Phys.*, 1999, **111**, 5344–5348.

118. W. Bambynek, B. Crasemann, R. Fink, H. Freund, H. Mark, C. Swift, R. Price and P. Rao, *Rev. Mod. Phys.*, 1972, **44**, 716–813.

119. B. K. McFarland, J. P. Farrell, S. Miyabe, F. Tarantelli, A. Aguilar, N. Berrah, C. Bostedt, J. D. Bozek, P. H. Bucksbaum, J. C. Castagna, R. N. Coffee, J. P. Cryan, L. Fang, R. Feifel, K. J. Gaffney, J. M. Glownia, T. J. Martinez, M. Mucke, B. Murphy, A. Natan, T. Osipov, V. S. Petrović, S. Schorb, Th. Schultz, L. S. Spector, M. Swiggers, I. Tenney, S. Wang, J. L. White, W. White and M. Gühr, *Nat. Comm.*, 2014, **5**, 4235.

120. T. Wolf, F. Holzmeier, I. Wagner, N. Berrah, C. Bostedt, J. Bozek, P. Bucksbaum, R. Coffee, J. Cryan, J. Farrell, R. Feifel, T. Martinez, B. McFarland, M. Mucke, S. Nandi, F. Tarantelli, I. Fischer and M. Gühr, *Appl. Sci.*, 2017, 7, 681.

121. F. Lever, D. Mayer, D. Picconi, J. Metje, S. Alisauskas, F. Calegari, S. Düsterer, C. Ehlert, R. Feifel, M. Niebuhr, B. Manschwetus, M. Kuhlmann, T. Mazza, M. S. Robinson, R. J. Squibb, A. Trabattoni, M. Wallner, P. Saalfrank, T. J. A. Wolf and M. Gühr, *J. Phys. B At. Mol. Opt. Phys.*, 2020, **54**, 014002.

122. T. J. A. Wolf, A. C. Paul, S. D. Folkestad, R. H. Myhre, J. P. Cryan, N. Berrah, P. H. Bucksbaum, S. Coriani, G. Coslovich, R. Feifel, T. J. Martinez, S. P. Moeller, M. Mucke, R. Obaid, O. Plekan, R. J. Squibb, H. Koch and M. Gühr, *Faraday Discuss.*, 2021, **228**, 555.

123. W. E. Moddeman, T. A. Carlson, M. O. Krause, B. P. Pullen, W. E. Bull and G. K. Schweitz, *J. Chem. Phys.*, 1971, **55**, 2317.

124. W. Mehlhorn, Atomic Auger spectroscopy: Historical perspective and recent highlights, X-ray and inner-shell processes: 18th International Conference, *AIP Conf. Proc.*, Chicago, Illinois, USA, 2000, vol. 506, pp. 33–56.

125. B. Lohmann, *Angle and Spin Resolved Auger Emission: Theory and Applications to Atoms and Molecules*, Springer, 2008.

126. E. Itälä, D. T. Ha, K. Kooser, M. A. Huels, E. Rachlew, E. Nõmmiste, U. Joost and E. Kukk, *J. Electron Spectrosc. Relat. Phenom.*, 2011, **184**, 119–124.

127. E. Itälä, D. T. Ha, K. Kooser, E. Nõmmiste, U. Joost and E. Kukk, *Int. J. Mass Spectrom.*, 2011, **306**, 82–90.

128. L. Schwob, S. Dörner, K. Atak, K. Schubert, M. Timm, C. Bülow, V. Zamudio-Bayer, B. von Issendorff, J. T. Lau, S. Techert and S. Bari, *J. Phys. Chem. Lett.*, 2020, **11**, 1215–1221.

129. M. Gerlach, F. Fantuzzi, L. Wohlfart, K. Kopp, B. Engels, J. Bozek, C. Nicolas, D. Mayer, M. Gühr, F. Holzmeier and I. Fischer, *J. Chem. Phys.*, 2021, **154**, 114302.

130. A. Landers, T. Weber, I. Ali, A. Cassimi, M. Hattass, O. Jagutzki, A. Nauert, T. Osipov, A. Staudte, M. H. Prior, H. Schmidt-Böcking, C. L. Cocke and R. Dörner, *Phys. Rev. Lett.*, 2001, **87**, 013002.

131. G. Kastirke, M. S. Schöffler, M. Weller, J. Rist, R. Boll, N. Anders, T. M. Baumann, S. Eckart, B. Erk, A. De Fanis, K. Fehre, A. Gatton, S. Grundmann, P. Grychtol, A. Hartung, M. Hofmann, M. Ilchen, C. Janke, M. Kircher, M. Kunitski, X. Li, T. Mazza, N. Melzer, J. Montano, V. Music, G. Nalin, Y. Ovcharenko, A. Pier, N. Rennhack, D. E. Rivas, R. Dörner, D. Rolles, A. Rudenko, P. Schmidt, J. Siebert, N. Strenger, D. Trabert, I. Vela-Perez, R. Wagner, T. Weber, J. B. Williams, P. Ziolkowski, L. P. H. Schmidt, A. Czasch, F. Trinter, M. Meyer, K. Ueda, P. V. Demekhin and T. Jahnke, *Phys. Rev. X*, 2020, **10**, 021052.

132. J. Nordgren, G. Bray, S. Cramm, R. Nyholm, J.-E. Rubensson and N. Wassdahl, *Rev. Sci. Instrum.*, 1989, **60**, 1690.

133. J. Nordgren, in *New directions in research with third-generation soft X-ray synchrotron radiation sources*, ed. A. S. Schlachter and F. J. Wuilleumier, Kluwer Academic, Dordrecht, Boston, 1994.

134. J. Nordgren, P. Glans, K. Gunnelin, J. Guo, P. Skytt, C. Såthe and N. Wassdahl, *Appl. Phys. Mater. Sci. Process*, 1997, **65**, 97–105.

135. A. Pietzsch, Y.-P. Sun, F. Hennies, Z. Rinkevicius, H. O. Karlsson, T. Schmitt, V. N. Strocov, J. Andersson, B. Kennedy, J. Schlappa, A. Föhlisch, J.-E. Rubensson and F. Gel'mukhanov, *Phys. Rev. Lett.*, 2011, **106**, 153004.

136. J. Nordgren and J.-E. Rubensson, *J. Electron Spectrosc. Relat. Phenom.*, 2013, **188**, 3–9.

137. C. Rulliere, *Femtosecond Laser Pulses: Principles and Experiments*, Springer, 2005.

138. U. Keller, *Ultrafast Lasers: a comprehensive introduction to fundamental principles with practical applications*, Spinger Nature, Cham, 2021.

139. R. L. Fork, B. I. Greene and C. V. Shank, *Appl. Phys. Lett.*, 1981, **38**, 671–672.

140. D. Strickland and G. Mourou, *Opt. Commun.*, 1985, **56**, 219–221.

141. J. Squier, D. Harter, F. Salin and G. Mourou, *Opt. Lett.*, 1991, **16**, 324.

142. J. D. Kmetec, J. J. Macklin and J. F. Young, *Opt. Lett.*, 1991, **16**, 1001.

143. J. A. Giordmaine and R. C. Miller, *Phys. Rev. Lett.*, 1965, **14**, 973–976.

144. R. Baumgartner and R. Byer, *IEEE J. Quantum Electron.*, 1979, **15**, 432–444.

145. T. Wilhelm, J. Piel and E. Riedle, *Opt. Lett.*, 1997, **22**, 1494.

146. G. Cerullo, M. Nisoli and S. De Silvestri, *Appl. Phys. Lett.*, 1997, **71**, 3616–3618.

147. A. Shirakawa and T. Kobayashi, *Appl. Phys. Lett.*, 1998, **72**, 147–149.

148. C. Manzoni and G. Cerullo, *J. Opt.*, 2016, **18**, 103501.

149. K. J. Schafer, B. Yang, L. F. DiMauro and K. C. Kulander, *Phys. Rev. Lett.*, 1993, **70**, 1599–1602.

150. J. L. Krause, K. J. Schafer and K. C. Kulander, *Phys. Rev. Lett.*, 1992, **68**, 3535–3538.
151. P. B. Corkum, *Phys. Rev. Lett.*, 1993, **71**, 1994–1997.
152. J. J. Macklin, J. D. Kmetec and C. L. Gordon, *Phys. Rev. Lett.*, 1993, **70**, 766–769.
153. T. Popmintchev, M. C. Chen, A. Bahabad, M. Gerrity, P. Sidorenko, O. Cohen, I. P. Christov, M. M. Murnane and H. C. Kapteyn, *Proc. Natl. Acad. Sci. U. S. A.*, 2009, **103**, 13279.
154. R. W. Schoenlein, *Science*, 2000, **287**, 2237–2240.
155. W. Ackermann, G. Asova, V. Ayvazyan, A. Azima, N. Baboi, J. Baehr, V. Balandin, B. Beutner, A. Brandt, A. Bolzmann, R. Brinkmann, O. I. Brovko, M. Castellano, P. Castro, L. Catani, E. Chiadroni, S. Choroba, A. Cianchi, J. T. Costello, D. Cubaynes, J. Dardis, W. Decking, H. Delsim-Hashemi, A. Delserieys, G. Di Pirro, M. Dohlus, S. Duesterer, A. Eckhardt, H. T. Edwards, B. Faatz, J. Feldhaus, K. Floettmann, J. Frisch, L. Froehlich, T. Garvey, U. Gensch, C. Gerth, M. Goerler, N. Golubeva, H.-J. Grabosch, M. Grecki, O. Grimm, K. Hacker, U. Hahn, J. H. Han, K. Honkavaara, T. Hott, M. Huening, Y. Ivanisenko, E. Jaeschke, W. Jalmuzna, T. Jezynski, R. Kammering, V. Katalev, K. Kavanagh, E. T. Kennedy, S. Khodyachykh, K. Klose, V. Kocharyan, M. Koerfer, M. Kollewe, W. Koprek, S. Korepanov, D. Kostin, M. Krassilnikov, G. Kube, M. Kuhlmann, C. L. S. Lewis, L. Lilje, T. Limberg, D. Lipka, F. Loehl, H. Luna, M. Luong, M. Martins, M. Meyer, P. Michelato, V. Miltchev, W. D. Moeller, L. Monaco, W. F. O. Mueller, A. Napieralski, O. Napoly, P. Nicolosi, D. Noelle, T. Nunez, A. Oppelt, C. Pagani, R. Paparella, N. Pchalek, J. Pedregosa-Gutierrez, B. Petersen, B. Petrosyan, G. Petrosyan, L. Petrosyan, J. Pflueger, E. Ploenjes, L. Poletto, K. Pozniak, E. Prat, D. Proch, P. Pucyk, P. Radcliffe, H. Redlin, K. Rehlich, M. Richter, M. Roehrs, J. Roensch, R. Romaniuk, M. Ross, J. Rossbach, V. Rybnikov, M. Sachwitz, E. L. Saldin, W. Sandner, H. Schlarb, B. Schmidt, M. Schmitz, P. Schmueser, J. R. Schneider, E. A. Schneidmiller, S. Schnepp, S. Schreiber, M. Seidel, D. Sertore, A. V. Shabunov, C. Simon, S. Simrock, E. Sombrowski, A. A. Sorokin, P. Spanknebel, R. Spesyvtsev, L. Staykov, B. Steffen, F. Stephan, F. Stulle, H. Thom, K. Tiedtke, M. Tischer, S. Toleikis, R. Treusch, D. Trines, I. Tsakov, E. Vogel, T. Weiland, H. Weise, M. Wellhoeffer, M. Wendt, I. Will, A. Winter, K. Wittenburg, W. Wurth, P. Yeates, M. V. Yurkov, I. Zagorodnov and K. Zapfe, *Nat. Photonics*, 2007, **1**, 336–342.
156. P. Emma, R. Akre, J. Arthur, R. Bionta, C. Bostedt, J. Bozek, A. Brachmann, P. Bucksbaum, R. Coffee, F.-J. Decker, Y. Ding, D. Dowell, S. Edstrom, A. Fisher, J. Frisch, S. Gilevich, J. Hastings, G. Hays, P. Hering, Z. Huang, R. Iverson, H. Loos, M. Messerschmidt, A. Miahnahri, S. Moeller, H.-D. Nuhn, G. Pile, D. Ratner, J. Rzepiela, D. Schultz, T. Smith, P. Stefan, H. Tompkins, J. Turner, J. Welch, W. White, J. Wu, G. Yocky and J. Galayda, *Nat. Photonics*, 2010, **4**, 641–647.

157. T. Ishikawa, H. Aoyagi, T. Asaka, Y. Asano, N. Azumi, T. Bizen, H. Ego, K. Fukami, T. Fukui, Y. Furukawa, S. Goto, H. Hanaki, T. Hara, T. Hasegawa, T. Hatsui, A. Higashiya, T. Hirono, N. Hosoda, M. Ishii, T. Inagaki, Y. Inubushi, T. Itoga, Y. Joti, M. Kago, T. Kameshima, H. Kimura, Y. Kirihara, A. Kiyomichi, T. Kobayashi, C. Kondo, T. Kudo, H. Maesaka, X. M. Maréchal, T. Masuda, S. Matsubara, T. Matsumoto, T. Matsushita, S. Matsui, M. Nagasono, N. Nariyama, H. Ohashi, T. Ohata, T. Ohshima, S. Ono, Y. Otake, C. Saji, T. Sakurai, T. Sato, K. Sawada, T. Seike, K. Shirasawa, T. Sugimoto, S. Suzuki, S. Takahashi, H. Takebe, K. Takeshita, K. Tamasaku, H. Tanaka, R. Tanaka, T. Tanaka, T. Togashi, K. Togawa, A. Tokuhisa, H. Tomizawa, K. Tono, S. Wu, M. Yabashi, M. Yamaga, A. Yamashita, K. Yanagida, C. Zhang, T. Shintake, H. Kitamura and N. Kumagai, *Nat. Photonics*, 2012, **6**, 540–544.

158. E. Prat, R. Abela, M. Aiba, A. Alarcon, J. Alex, Y. Arbelo, C. Arrell, V. Arsov, C. Bacellar, C. Beard, P. Beaud, S. Bettoni, R. Biffiger, M. Bopp, H.-H. Braun, M. Calvi, A. Cassar, T. Celcer, M. Chergui, P. Chevtsov, C. Cirelli, A. Citterio, P. Craievich, M. C. Divall, A. Dax, M. Dehler, Y. Deng, A. Dietrich, P. Dijkstal, R. Dinapoli, S. Dordevic, S. Ebner, D. Engeler, C. Erny, V. Esposito, E. Ferrari, U. Flechsig, R. Follath, F. Frei, R. Ganter, T. Garvey, Z. Geng, A. Gobbo, C. Gough, A. Hauff, C. P. Hauri, N. Hiller, S. Hunziker, M. Huppert, G. Ingold, R. Ischebeck, M. Janousch, P. J. M. Johnson, S. L. Johnson, P. Juranić, M. Jurcevic, M. Kaiser, R. Kalt, B. Keil, D. Kiselev, C. Kittel, G. Knopp, W. Koprek, M. Laznovsky, H. T. Lemke, D. L. Sancho, F. Löhl, A. Malyzhenkov, G. F. Mancini, R. Mankowsky, F. Marcellini, G. Marinkovic, I. Martiel, F. Märki, C. J. Milne, A. Mozzanica, K. Nass, G. L. Orlandi, C. O. Loch, M. Paraliev, B. Patterson, L. Patthey, B. Pedrini, M. Pedrozzi, C. Pradervand, P. Radi, J.-Y. Raguin, S. Redford, J. Rehanek, S. Reiche, L. Rivkin, A. Romann, L. Sala, M. Sander, T. Schietinger, T. Schilcher, V. Schlott, T. Schmidt, M. Seidel, M. Stadler, L. Stingelin, C. Svetina, D. M. Treyer, A. Trisorio, C. Vicario, D. Voulot, A. Wrulich, S. Zerdane and E. Zimoch, *Nat. Photonics*, 2020, **14**, 748–754.

159. I. Nam, C.-K. Min, B. Oh, G. Kim, D. Na, Y. J. Suh, H. Yang, M. H. Cho, C. Kim, M.-J. Kim, C. H. Shim, J. H. Ko, H. Heo, J. Park, J. Kim, S. Park, G. Park, S. Kim, S. H. Chun, H. Hyun, J. H. Lee, K. S. Kim, I. Eom, S. Rah, D. Shu, K.-J. Kim, S. Terentyev, V. Blank, Y. Shvyd'ko, S. J. Lee and H.-S. Kang, *Nat. Photonics*, 2021, **15**, 435–441.

160. E. Allaria, R. Appio, L. Badano, W. A. Barletta, S. Bassanese, S. G. Biedron, A. Borga, E. Busetto, D. Castronovo, P. Cinquegrana, S. Cleva, D. Cocco, M. Cornacchia, P. Craievich, I. Cudin, G. D'Auria, M. Dal Forno, M. B. Danailov, R. De Monte, G. De Ninno, P. Delgiusto, A. Demidovich, S. Di Mitri, B. Diviacco, A. Fabris, R. Fabris, W. Fawley, M. Ferianis, E. Ferrari, S. Ferry, L. Froehlich, P. Furlan, G. Gaio, F. Gelmetti, L. Giannessi, M. Giannini, R. Gobessi, R. Ivanov, E. Karantzoulis, M. Lonza, A. Lutman, B. Mahieu, M. Milloch, S. V.

Milton, M. Musardo, I. Nikolov, S. Noe, F. Parmigiani, G. Penco, M. Petronio, L. Pivetta, M. Predonzani, F. Rossi, L. Rumiz, A. Salom, C. Scafuri, C. Serpico, P. Sigalotti, S. Spampinati, C. Spezzani, M. Svandrlik, C. Svetina, S. Tazzari, M. Trovo, R. Umer, A. Vascotto, M. Veronese, R. Visintini, M. Zaccaria, D. Zangrando and M. Zangrando, *Nat. Photonics*, 2012, **6**, 699–704.

161. W. Decking, S. Abeghyan, P. Abramian, A. Abramsky, A. Aguirre, C. Albrecht, P. Alou, M. Altarelli, P. Altmann, K. Amyan, V. Anashin, E. Apostolov, K. Appel, D. Auguste, V. Ayvazyan, S. Baark, F. Babies, N. Baboi, P. Bak, V. Balandin, R. Baldinger, B. Baranasic, S. Barbanotti, O. Belikov, V. Belokurov, L. Belova, V. Belyakov, S. Berry, M. Bertucci, B. Beutner, A. Block, M. Blöcher, T. Böckmann, C. Bohm, M. Böhnert, V. Bondar, E. Bondarchuk, M. Bonezzi, P. Borowiec, C. Bösch, U. Bösenberg, A. Bosotti, R. Böspflug, M. Bousonville, E. Boyd, Y. Bozhko, A. Brand, J. Branlard, S. Briechle, F. Brinker, S. Brinker, R. Brinkmann, S. Brockhauser, O. Brovko, H. Brück, A. Brüdgam, L. Butkowski, T. Büttner, J. Calero, E. Castro-Carballo, G. Cattalanotto, J. Charrier, J. Chen, A. Cherepenko, V. Cheskidov, M. Chiodini, A. Chong, S. Choroba, M. Chorowski, D. Churanov, W. Cichalewski, M. Clausen, W. Clement, C. Cloué, J. A. Cobos, N. Coppola, S. Cunis, K. Czuba, M. Czwalinna, B. D'Almagne, J. Dammann, H. Danared, A. de Zubiaurre Wagner, A. Delfs, T. Delfs, F. Dietrich, T. Dietrich, M. Dohlus, M. Dommach, A. Donat, X. Dong, N. Doynikov, M. Dressel, M. Duda, P. Duda, H. Eckoldt, W. Ehsan, J. Eidam, F. Eints, C. Engling, U. Englisch, A. Ermakov, K. Escherich, J. Eschke, E. Saldin, M. Faesing, A. Fallou, M. Felber, M. Fenner, B. Fernandes, J. M. Fernández, S. Feuker, K. Filippakopoulos, K. Floettmann, V. Fogel, M. Fontaine, A. Francés, I. F. Martin, W. Freund, T. Freyermuth, M. Friedland, L. Fröhlich, M. Fusetti, J. Fydrych, A. Gallas, O. García, L. Garcia-Tabares, G. Geloni, N. Gerasimova, C. Gerth, P. Geßler, V. Gharibyan, M. Gloor, J. Głowinkowski, A. Goessel, Z. Gołębiewski, N. Golubeva, W. Grabowski, W. Graeff, A. Grebentsov, M. Grecki, T. Grevsmuehl, M. Gross, U. Grosse-Wortmann, J. Grünert, S. Grunewald, P. Grzegory, G. Feng, H. Guler, G. Gusev, J. L. Gutierrez, L. Hagge, M. Hamberg, R. Hanneken, E. Harms, I. Hartl, A. Hauberg, S. Hauf, J. Hauschildt, J. Hauser, J. Havlicek, A. Hedqvist, N. Heidbrook, F. Hellberg, D. Henning, O. Hensler, T. Hermann, A. Hidvégi, M. Hierholzer, H. Hintz, F. Hoffmann, M. Hoffmann, M. Hoffmann, Y. Holler, M. Hüning, A. Ignatenko, M. Ilchen, A. Iluk, J. Iversen, J. Iversen, M. Izquierdo, L. Jachmann, N. Jardon, U. Jastrow, K. Jensch, J. Jensen, M. Jeżabek, M. Jidda, H. Jin, N. Johansson, R. Jonas, W. Kaabi, D. Kaefer, R. Kammering, H. Kapitza, S. Karabekyan, S. Karstensen, K. Kasprzak, V. Katalev, D. Keese, B. Keil, M. Kholopov, M. Killenberger, B. Kitaev, Y. Klimchenko, R. Klos, L. Knebel, A. Koch, M. Koepke, S. Köhler, W. Köhler, N. Kohlstrunk, Z. Konopkova, A. Konstantinov, W. Kook, W. Koprek, M. Körfer, O. Korth, A. Kosarev, K. Kosiński,

D. Kostin, Y. Kot, A. Kotarba, T. Kozak, V. Kozak, R. Kramert, M. Krasilnikov, A. Krasnov, B. Krause, L. Kravchuk, O. Krebs, R. Kretschmer, J. Kreutzkamp, O. Kröplin, K. Krzysik, G. Kube, H. Kuehn, N. Kujala, V. Kulikov, V. Kuzminych, D. La Civita, M. Lacroix, T. Lamb, A. Lancetov, M. Larsson, D. Le Pinvidic, S. Lederer, T. Lensch, D. Lenz, A. Leuschner, F. Levenhagen, Y. Li, J. Liebing, L. Lilje, T. Limberg, D. Lipka, B. List, J. Liu, S. Liu, B. Lorbeer, J. Lorkiewicz, H. H. Lu, F. Ludwig, K. Machau, W. Maciocha, C. Madec, C. Magueur, C. Maiano, I. Maksimova, K. Malcher, T. Maltezopoulos, E. Mamoshkina, B. Manschwetus, F. Marcellini, G. Marinkovic, T. Martinez, H. Martirosyan, W. Maschmann, M. Maslov, A. Matheisen, U. Mavric, J. Meißner, K. Meissner, M. Messerschmidt, N. Meyners, G. Michalski, P. Michelato, N. Mildner, M. Moe, F. Moglia, C. Mohr, S. Mohr, W. Möller, M. Mommerz, L. Monaco, C. Montiel, M. Moretti, I. Morozov, P. Morozov, D. Mross, J. Mueller, C. Müller, J. Müller, K. Müller, J. Munilla, A. Münnich, V. Muratov, O. Napoly, B. Näser, N. Nefedov, R. Neumann, R. Neumann, N. Ngada, D. Noelle, F. Obier, I. Okunev, J. A. Oliver, M. Omet, A. Oppelt, A. Ottmar, M. Oublaid, C. Pagani, R. Paparella, V. Paramonov, C. Peitzmann, J. Penning, A. Perus, F. Peters, B. Petersen, A. Petrov, I. Petrov, S. Pfeiffer, J. Pflüger, S. Philipp, Y. Pienaud, P. Pierini, S. Pivovarov, M. Planas, E. Pławski, M. Pohl, J. Polinski, V. Popov, S. Prat, J. Prenting, G. Priebe, H. Pryschelski, K. Przygoda, E. Pyata, B. Racky, A. Rathjen, W. Ratuschni, S. Regnaud-Campderros, K. Rehlich, D. Reschke, C. Robson, J. Roever, M. Roggli, J. Rothenburg, E. Rusiński, R. Rybaniec, H. Sahling, M. Salmani, L. Samoylova, D. Sanzone, F. Saretzki, O. Sawlanski, J. Schaffran, H. Schlarb, M. Schlösser, V. Schlott, C. Schmidt, F. Schmidt-Foehre, M. Schmitz, M. Schmökel, T. Schnautz, E. Schneidmiller, M. Scholz, B. Schöneburg, J. Schultze, C. Schulz, A. Schwarz, J. Sekutowicz, D. Sellmann, E. Semenov, S. Serkez, D. Sertore, N. Shehzad, P. Shemarykin, L. Shi, M. Sienkiewicz, D. Sikora, M. Sikorski, A. Silenzi, C. Simon, W. Singer, X. Singer, H. Sinn, K. Sinram, N. Skvorodnev, P. Smirnow, T. Sommer, A. Sorokin, M. Stadler, M. Steckel, B. Steffen, N. Steinhau-Kühl, F. Stephan, M. Stodulski, M. Stolper, A. Sulimov, R. Susen, J. Świerblewski, C. Sydlo, E. Syresin, V. Sytchev, J. Szuba, N. Tesch, J. Thie, A. Thiebault, K. Tiedtke, D. Tischhauser, J. Tolkiehn, S. Tomin, F. Tonisch, F. Toral, I. Torbin, A. Trapp, D. Treyer, G. Trowitzsch, T. Trublet, T. Tschentscher, F. Ullrich, M. Vannoni, P. Varela, G. Varghese, G. Vashchenko, M. Vasic, C. Vazquez-Velez, A. Verguet, S. Vilcins-Czvitkovits, R. Villanueva, B. Visentin, M. Viti, E. Vogel, E. Volobuev, R. Wagner, N. Walker, T. Wamsat, H. Weddig, G. Weichert, H. Weise, R. Wenndorf, M. Werner, R. Wichmann, C. Wiebers, M. Wiencek, T. Wilksen, I. Will, L. Winkelmann, M. Winkowski, K. Wittenburg, A. Witzig, P. Wlk, T. Wohlenberg, M. Wojciechowski, F. Wolff-Fabris, G. Wrochna, K. Wrona, M. Yakopov, B. Yang, F. Yang, M. Yurkov, I. Zagorodnov, P. Zalden, A. Zavadtsev, D. Zavadtsev, A. Zhirnov,

A. Zhukov, V. Ziemann, A. Zolotov, N. Zolotukhina, F. Zummack and D. Zybin, *Nat. Photonics*, 2020, **14**, 391–397.

162. A. Rosspeintner, B. Lang and E. Vauthey, *Annu. Rev. Phys. Chem.*, 2013, **64**, 247–271.

163. Z.-H. Loh, M. Khalil, R. E. Correa and S. R. Leone, *Rev. Sci. Instrum.*, 2008, **79**, 073101.

164. W. Fuß, W. E. Schmid and S. A. Trushin, *J. Chem. Phys.*, 2000, **112**, 8347.

165. S. A. Trushin, W. Fuss, W. E. Schmid and K. L. Kompa, *J. Phys. Chem. A*, 1998, **102**, 4129–4137.

166. M. Kotur, T. C. Weinacht, C. Zhou and S. Matsika, *Phys. Rev. X*, 2011, **1**, 021010.

167. M. Kotur, C. Zhou, S. Matsika, S. Patchkovskii, M. Spanner and T. C. Weinacht, *Phys. Rev. Lett.*, 2012, **109**, 203007.

168. S. Matsika, M. Spanner, M. Kotur and T. C. Weinacht, *J. Phys. Chem. A*, 2013, **117**, 12796–12801.

169. M. S. Robinson, M. Niebuhr, F. Lever, D. Mayer, J. Metje and M. Gühr, *Chem. – Eur. J.*, 2021, **27**, 11418–11427.

170. Th. Ergler, B. Feuerstein, A. Rudenko, K. Zrost, C. D. Schroeter, R. Moshammer and J. Ullrich, *Phys. Rev. Lett.*, 2006, **97**, 103004.

171. Y. H. Jiang, A. Rudenko, J. F. Perez-Torres, O. Herrwerth, L. Foucar, M. Kurka, K. U. Kuehnel, M. Toppin, E. Plesiat, F. Morales, F. Martin, M. Lezius, M. F. Kling, T. Jahnke, R. Doerner, J. L. Sanz-Vicario, J. van Tilborg, A. Belkacem, M. Schulz, K. Ueda, T. J. M. Zouros, S. Duesterer, R. Treusch, C. D. Schroeter, R. Moshammer and J. Ullrich, *Phys. Rev. A*, 2010, **81**, 051402(R).

172. R. de Nalda, J. Durá, A. García-Vela, J. G. Izquierdo, J. González-Vázquez and L. Bañares, *J. Chem. Phys.*, 2008, **128**, 244309.

173. E. T. Karamatskos, S. Raabe, T. Mullins, A. Trabattoni, P. Stammer, G. Goldsztejn, R. R. Johansen, K. Długołecki, H. Stapelfeldt, M. J. J. Vrakking, S. Trippel, A. Rouzée and J. Küpper, *Nat. Commun.*, 2019, **10**, 3364.

174. A. Stolow, *Annu. Rev. Phys. Chem.*, 2003, **54**, 89–119.

175. A. Stolow, A. E. Bragg and D. M. Neumark, *Chem. Rev.*, 2004, **104**, 1719–1758.

176. I. V. Hertel and W. Radloff, *Rep. Prog. Phys.*, 2006, **69**, 1897–2003.

177. T. Suzuki, *Int. Rev. Phys. Chem.*, 2012, **31**, 265–318.

178. N. Kotsina and D. Townsend, *Phys. Chem. Chem. Phys.*, 2021, **23**, 10736–10755.

179. S. Ullrich, T. Schultz, M. Z. Zgierski and A. Stolow, *Phys. Chem. Chem. Phys.*, 2004, **6**, 2796.

180. S. Ullrich, T. Schultz, M. Z. Zgierski and A. Stolow, *J. Am. Chem. Soc.*, 2004, **126**, 2262–2263.

181. T. Suzuki, *Annu. Rev. Phys. Chem.*, 2006, **57**, 555–592.

182. J. A. Davies, J. E. LeClaire, R. E. Continetti and C. C. Hayden, *J. Chem. Phys.*, 1999, **111**, 1–4.

183. O. Gessner, A. M. D. Lee, J. P. Shaffer, H. Reisler, S. V. Levchenko, A. I. Krylov, J. G. Underwood, H. Shi, A. L. L. East, D. M. Wardlaw, E. T. H. Chrysostom, C. C. Hayden and A. Stolow, *Science*, 2006, **311**, 219–222.
184. M. Barbatti and S. Ullrich, *Phys. Chem. Chem. Phys.*, 2011, **13**, 15492.
185. M. Koch, T. J. A. Wolf and M. Gühr, *Phys. Rev. A*, 2015, **91**, 031403(R).
186. L. Nugent-Glandorf, M. Scheer, D. Samuels, A. Mulhisen, E. Grant, X. Yang, V. Bierbaum and S. Leone, *Phys. Rev. Lett.*, 2001, **87**, 193002.
187. P. Wernet, M. Odelius, K. Godehusen, J. Gaudin, O. Schwarzkopf and W. Eberhardt, *Phys. Rev. Lett.*, 2009, **103**, 013001.
188. T. J. A. Wolf, R. M. Parrish, R. H. Myhre, T. J. Martínez, H. Koch and M. Gühr, *J. Phys. Chem. A*, 2019, **123**, 6897–6903.
189. S. L. Horton, Y. Liu, R. Forbes, V. Makhija, R. Lausten, A. Stolow, P. Hockett, P. Marquetand, T. Rozgonyi and T. Weinacht, *J. Chem. Phys.*, 2019, **150**, 174201.
190. Y. Liu, P. Chakraborty, S. Matsika and T. Weinacht, *J. Chem. Phys.*, 2020, **153**, 074301.
191. A. S. Chatterley, F. Lackner, D. M. Neumark, S. R. Leone and O. Gessner, *Phys. Chem. Chem. Phys.*, 2016, **18**, 14644–14653.
192. T. K. Allison, H. Tao, W. J. Glover, T. W. Wright, A. M. Stooke, C. Khurmi, J. van Tilborg, Y. Liu, R. W. Falcone, T. J. Martínez and A. Belkacem, *J. Chem. Phys.*, 2012, **136**, 124317.
193. A. Baumann, S. Bazzi, D. Rompotis, O. Schepp, A. Azima, M. Wieland, D. Popova-Gorelova, O. Vendrell, R. Santra and M. Drescher, *Phys. Rev. A*, 2017, **96**, 013428.
194. F. Calegari, D. Ayuso, A. Trabattoni, L. Belshaw, S. De Camillis, S. Anumula, F. Frassetto, L. Poletto, A. Palacios, P. Decleva, J. B. Greenwood, F. Martin and M. Nisoli, *Science*, 2014, **346**, 336–339.
195. M. C. E. Galbraith, S. Scheit, N. V. Golubev, G. Reitsma, N. Zhavoronkov, V. Despré, F. Lépine, A. I. Kuleff, M. J. J. Vrakking, O. Kornilov, H. Köppel and J. Mikosch, *Nat. Commun.*, 2017, **8**, 1018.
196. P. B. Corkum and F. Krausz, *Nat. Phys.*, 2007, **3**, 381–387.
197. Z.-H. Loh and S. R. Leone, *J. Chem. Phys.*, 2008, **128**, 204302.
198. A. R. Attar, A. Bhattacherjee and S. R. Leone, *J. Phys. Chem. Lett.*, 2015, **6**, 5072–5077.
199. Y. Pertot, C. Schmidt, M. Matthews, A. Chauvet, M. Huppert, V. Svoboda, A. von Conta, A. Tehlar, D. Baykusheva, J.-P. Wolf and H. J. Wörner, *Science*, 2017, **355**, 264–267.
200. A. R. Attar, A. Bhattacherjee, C. D. Pemmaraju, K. Schnorr, K. D. Closser, D. Prendergast and S. R. Leone, *Science*, 2017, **356**, 54–59.
201. T. J. A. Wolf, R. H. Myhre, J. P. Cryan, S. Coriani, R. J. Squibb, A. Battistoni, N. Berrah, C. Bostedt, P. Bucksbaum, G. Coslovich, R. Feifel, K. J. Gaffney, J. Grilj, T. J. Martinez, S. Miyabe, S. P. Moeller, M. Mucke, A. Natan, R. Obaid, T. Osipov, O. Plekan, S. Wang, H. Koch and M. Gühr, *Nat. Comm*, 2017, **8**, 29.
202. C. Ehlert, M. Gühr and P. Saalfrank, *J. Chem. Phys.*, 2018, **149**, 144112.

203. F. Segatta, A. Nenov, S. Orlandi, A. Arcioni, S. Mukamel and M. Garavelli, *Faraday Discuss.*, 2020, **221**, 245–264.

204. L. Inhester, Z. Li, X. Zhu, N. Medvedev and T. J. A. Wolf, *J. Phys. Chem. Lett.*, 2019, **10**, 6536–6544.

205. T. Leitner, I. Josefsson, T. Mazza, P. S. Miedema, H. Schröder, M. Beye, K. Kunnus, S. Schreck, S. Düsterer, A. Föhlisch, M. Meyer, M. Odelius and P. Wernet, *J. Chem. Phys.*, 2018, **149**, 044307.

206. D. Mayer, F. Lever, D. Picconi, J. Metje, S. Alisauskas, F. Calegari, S. Düsterer, C. Ehlert, R. Feifel, M. Niebuhr, B. Manschwetus, M. Kuhlmann, T. Mazza, M. S. Robinson, R. J. Squibb, A. Trabattoni, M. Wallner, P. Saalfrank, T. J. A. Wolf and M. Gühr, *Nat. Commun.*, 2022, **13**, 198.

207. B. M. Giuliano, V. Feyer, K. C. Prince, M. Coreno, L. Evangelisti, S. Melandri and W. Caminati, *J. Phys. Chem. A*, 2010, **114**, 12725–12730.

208. S. Mai, P. Marquetand and L. González, *J. Phys. Chem. Lett.*, 2016, 7, 1978–1983.

209. S. Mai, F. Plasser, M. Pabst, F. Neese, A. Köhn and L. González, *J. Chem. Phys.*, 2017, **147**, 184109.

210. B. Erk, R. Boll, S. Trippel, D. Anielski, L. Foucar, B. Rudek, S. W. Epp, R. Coffee, S. Carron, S. Schorb, K. R. Ferguson, M. Swiggers, J. D. Bozek, M. Simon, T. Marchenko, J. Küpper, I. Schlichting, J. Ullrich, C. Bostedt, D. Rolles and A. Rudenko, *Science*, 2014, **345**, 288–291.

211. K. Amini, E. Savelyev, F. Brauße, N. Berrah, C. Bomme, M. Brouard, M. Burt, L. Christensen, S. Düsterer, B. Erk, H. Höppner, T. Kierspel, F. Krecinic, A. Lauer, J. W. L. Lee, M. Müller, E. Müller, T. Mullins, H. Redlin, N. Schirmel, J. Thøgersen, S. Techert, S. Toleikis, R. Treusch, S. Trippel, A. Ulmer, C. Vallance, J. Wiese, P. Johnsson, J. Küpper, A. Rudenko, A. Rouzée, H. Stapelfeldt, D. Rolles and R. Boll, *Struct. Dyn.*, 2018, **5**, 014301.

212. R. Srinivasan, V. A. Lobastov, C.-Y. Ruan and A. H. Zewail, *Helv. Chim. Acta*, 2003, **86**, 1761–1799.

213. S. P. Weathersby, G. Brown, M. Centurion, T. F. Chase, R. Coffee, J. Corbett, J. P. Eichner, J. C. Frisch, A. R. Fry, M. Gühr, N. Hartmann, C. Hast, R. Hettel, R. K. Jobe, E. N. Jongewaard, J. R. Lewandowski, R. K. Li, A. M. Lindenberg, I. Makasyuk, J. E. May, D. McCormick, M. N. Nguyen, A. H. Reid, X. Shen, K. Sokolowski-Tinten, T. Vecchione, S. L. Vetter, J. Wu, J. Yang, H. A. Dürr and X. J. Wang, *Rev. Sci. Instrum.*, 2015, **86**, 073702.

214. J. Yang, M. Guehr, X. Shen, R. Li, T. Vecchione, R. Coffee, J. Corbett, A. Fry, N. Hartmann, C. Hast, K. Hegazy, K. Jobe, I. Makasyuk, J. Robinson, M. S. Robinson, S. Vetter, S. Weathersby, C. Yoneda, X. Wang and M. Centurion, *Phys. Rev. Lett.*, 2016, **117**, 173002.

215. J. Küpper, S. Stern, L. Holmegaard, F. Filsinger, A. Rouzée, A. Rudenko, P. Johnsson, A. V. Martin, M. Adolph, A. Aquila, S. Bajt, A. Barty, C. Bostedt, J. Bozek, C. Caleman, R. Coffee, N. Coppola, T. Delmas, S. Epp, B. Erk, L. Foucar, T. Gorkhover, L. Gumprecht, A. Hartmann,

R. Hartmann, G. Hauser, P. Holl, A. Hömke, N. Kimmel, F. Krasniqi, K.-U. Kühnel, J. Maurer, M. Messerschmidt, R. Moshammer, C. Reich, B. Rudek, R. Santra, I. Schlichting, C. Schmidt, S. Schorb, J. Schulz, H. Soltau, J. C. H. Spence, D. Starodub, L. Strüder, J. Thøgersen, M. J. J. Vrakking, G. Weidenspointner, T. A. White, C. Wunderer, G. Meijer, J. Ullrich, H. Stapelfeldt, D. Rolles and H. N. Chapman, *Phys. Rev. Lett.*, 2014, **112**, 083002.

216. J. Yang, M. Guehr, T. Vecchione, M. S. Robinson, R. Li, N. Hartmann, X. Shen, R. Coffee, J. Corbett, A. Fry, K. Gaffney, T. Gorkhover, C. Hast, K. Jobe, I. Makasyuk, A. Reid, J. Robinson, S. Vetter, F. Wang, S. Weathersby, C. Yoneda, X. Wang and M. Centurion, *Faraday Discuss.*, 2016, **194**, 563–581.

217. J. Yang, M. Guehr, T. Vecchione, M. S. Robinson, R. Li, N. Hartmann, X. Shen, R. Coffee, J. Corbett, A. Fry, K. Gaffney, T. Gorkhover, C. Hast, K. Jobe, I. Makasyuk, A. Reid, J. Robinson, S. Vetter, F. Wang, S. Weathersby, C. Yoneda, M. Centurion and X. Wang, *Nat. Comm.*, 2016, **7**, 11232.

218. J. Yang, X. Zhu, T. J. A. Wolf, Z. Li, J. P. F. Nunes, R. Coffee, J. P. Cryan, M. Gühr, K. Hegazy, T. F. Heinz, K. Jobe, R. Li, X. Shen, T. Veccione, S. Weathersby, K. J. Wilkin, C. Yoneda, Q. Zheng, T. J. Martinez, M. Centurion and X. Wang, *Science*, 2018, **361**, 64–67.

219. M. P. Minitti, J. M. Budarz, A. Kirrander, J. Robinson, T. J. Lane, D. Ratner, K. Saita, T. Northey, B. Stankus, V. Cofer-Shabica, J. Hastings and P. M. Weber, *Faraday Discuss.*, 2014, **171**, 81–91.

220. M. P. Minitti, J. M. Budarz, A. Kirrander, J. S. Robinson, D. Ratner, T. J. Lane, D. Zhu, J. M. Glownia, M. Kozina, H. T. Lemke, M. Sikorski, Y. Feng, S. Nelson, K. Saita, B. Stankus, T. Northey, J. B. Hastings and P. M. Weber, *Phys. Rev. Lett.*, 2015, **114**, 255501.

221. T. J. A. Wolf, D. M. Sanchez, J. Yang, R. M. Parrish, J. P. F. Nunes, M. Centurion, R. Coffee, J. P. Cryan, M. Gühr, K. Hegazy, A. Kirrander, R. K. Li, J. Ruddock, X. Shen, T. Vecchione, S. P. Weathersby, P. M. Weber, K. Wilkin, H. Yong, Q. Zheng, X. J. Wang, M. P. Minitti and T. J. Martínez, *Nat. Chem.*, 2019, **11**, 504–509.

222. R. Santra, *J. Phys. B: At., Mol. Opt. Phys.*, 2009, **42**, 169801.

223. H. Yong, N. Zotev, J. M. Ruddock, B. Stankus, M. Simmermacher, A. M. Carrascosa, W. Du, N. Goff, Y. Chang, D. Bellshaw, M. Liang, S. Carbajo, J. E. Koglin, J. S. Robinson, S. Boutet, M. P. Minitti, A. Kirrander and P. M. Weber, *Nat. Commun.*, 2020, **11**, 2157.

224. B. Stankus, H. Yong, N. Zotev, J. M. Ruddock, D. Bellshaw, T. J. Lane, M. Liang, S. Boutet, S. Carbajo, J. S. Robinson, W. Du, N. Goff, Y. Chang, J. E. Koglin, M. P. Minitti, A. Kirrander and P. M. Weber, *Nat. Chem.*, 2019, **11**, 716–721.

225. J. Yang, X. Zhu, J. P. F. Nunes, J. K. Yu, R. M. Parrish, T. J. A. Wolf, M. Centurion, M. Gühr, R. Li, Y. Liu, B. Moore, M. Niebuhr, S. Park, X. Shen, S. Weathersby, T. Weinacht, T. J. Martinez and X. J. Wang, *Science*, 2020, **368**, 885–889.

226. F. Silva, S. M. Teichmann, S. L. Cousin, M. Hemmer and J. Biegert, *Nat. Commun.*, 2015, **6**, 6611.
227. D. Popmintchev, B. R. Galloway, M.-C. Chen, F. Dollar, C. A. Mancuso, A. Hankla, L. Miaja-Avila, G. O'Neil, J. M. Shaw, G. Fan, S. Ališauskas, G. Andriukaitis, T. Balčiunas, O. D. Mücke, A. Pugzlys, A. Baltuška, H. C. Kapteyn, T. Popmintchev and M. M. Murnane, *Phys. Rev. Lett.*, 2018, **120**, 093002.
228. J. Duris, S. Li, T. Driver, E. G. Champenois, J. P. MacArthur, A. A. Lutman, Z. Zhang, P. Rosenberger, J. W. Aldrich, R. Coffee, G. Coslovich, F.-J. Decker, J. M. Glownia, G. Hartmann, W. Helml, A. Kamalov, J. Knurr, J. Krzywinski, M.-F. Lin, J. P. Marangos, M. Nantel, A. Natan, J. T. O'Neal, N. Shivaram, P. Walter, A. L. Wang, J. J. Welch, T. J. A. Wolf, J. Z. Xu, M. F. Kling, P. H. Bucksbaum, A. Zholents, Z. Huang, J. P. Cryan and A. Marinelli, *Nat. Photonics*, 2020, **14**, 30–36.
229. N. Rohringer and R. Santra, *Phys. Rev. A: At., Mol., Opt. Phys.*, 2007, **76**, 033416.
230. S. Mukamel, D. Abramavicius, L. Yang, W. Zhuang, I. V. Schweigert and D. V. Voronine, *Acc. Chem. Res.*, 2009, **42**, 553–562.
231. C. Weninger, M. Purvis, D. Ryan, R. London, J. Bozek, C. Bostedt, A. Graf, G. Brown, J. Rocca and N. Rohringer, *Phys. Rev. Lett.*, 2013, **111**, 233902.
232. V. Kimberg, A. Sanchez-Gonzalez, L. Mercadier, C. Weninger, A. Lutman, D. Ratner, R. Coffee, M. Bucher, M. Mucke, M. Agåker, C. Såthe, C. Bostedt, J. Nordgren, J. E. Rubensson and N. Rohringer, *Faraday Discuss.*, 2016, **194**, 305–324.
233. S. Mukamel, D. Healion, Y. Zhang and J. D. Biggs, *Annu. Rev. Phys. Chem.*, 2013, **64**, 101–127.
234. L. Young, K. Ueda, M. Gühr, P. H. Bucksbaum, M. Simon, S. Mukamel, N. Rohringer, K. C. Prince, C. Masciovecchio, M. Meyer, A. Rudenko, D. Rolles, C. Bostedt, M. Fuchs, D. A. Reis, R. Santra, H. Kapteyn, M. Murnane, H. Ibrahim, F. Légaré, M. Vrakking, M. Isinger, D. Kroon, M. Gisselbrecht, A. L'Huillier, H. J. Wörner and S. R. Leone, *J. Phys. B: At., Mol. Opt. Phys.*, 2018, **51**, 032003.
235. J. T. O'Neal, E. G. Champenois, S. Oberli, R. Obaid, A. Al-Haddad, J. Barnard, N. Berrah, R. Coffee, J. Duris, G. Galinis, D. Garratt, J. M. Glownia, D. Haxton, P. Ho, S. Li, X. Li, J. MacArthur, J. P. Marangos, A. Natan, N. Shivaram, D. S. Slaughter, P. Walter, S. Wandel, L. Young, C. Bostedt, P. H. Bucksbaum, A. Picón, A. Marinelli and J. P. Cryan, *Phys. Rev. Lett.*, 2020, **125**, 073203.
236. T. van Oudheusden, P. L. E. M. Pasmans, S. B. van der Geer, M. J. de Loos, M. J. van der Wiel and O. J. Luiten, *Phys. Rev. Lett.*, 2010, **105**, 264801.
237. M. R. Otto, L. P. René de Cotret, M. J. Stern and B. J. Siwick, *Struct. Dyn.*, 2017, **4**, 051101.
238. Y. Xiong, K. J. Wilkin and M. Centurion, *Phys. Rev. Res.*, 2020, **2**, 043064.
239. D. Filippetto, P. Musumeci, R. K. Li, B. J. Siwick, M. R. Otto, M. Centurion and J. P. F. Nunes, *Rev. Mod. Phys.*, 2022, **94**, 045004.

CHAPTER 2

Ultrafast Spectroscopy in Solid Matter

M. BEYE

Deutsches Elektronen-Synchrotron DESY, Notkestr. 85,
22607 Hamburg, Germany
Email: martin.beye@desy.de

2.1 Introduction

In this chapter, a fundamental basis is laid in solid-state physics, crucial for later discussions of ultrafast dynamics in solids (see Section 2.2). While only a short summary of these concepts is given in this chapter, the reader is referred to standard textbooks on solid-state physics[1–5] for further details. Describing these well-known concepts from a different viewpoint may also interest the advanced reader. For example, we emphasise the importance of solid-state phase transitions (see Section 2.3), which play a crucial role in modern applications,[6–9] where a change in the phase of materials can alter the function of devices. Optical lasers can trigger such phase transitions, providing access to the material's ultrafast dynamics through pump–probe measurements. Performing ultrafast dynamical studies gives a detailed understanding of the inner workings of solids and provides access to un-usual, potentially metastable states that do not exist in the material's equilibrium phase.

Given that some of these studies employ X-ray spectroscopic methods, we will discuss the applicability and information content of these probing techniques (see Section 2.4). Moreover, several selected examples of time-resolved X-ray spectroscopy experiments on solids are briefly highlighted (see Section 2.5), and the interested reader is referred to the relevant

Theoretical and Computational Chemistry Series No. 25
Structural Dynamics with X-ray and Electron Scattering
Edited by Kasra Amini, Arnaud Rouzée and Marc J. J. Vrakking

literature for further details. This chapter concludes with a short discussion of future developments in the application of X-ray spectroscopy making full use of the coherence properties of modern ultrashort pulsed X-ray sources (see Section 2.6). New spectroscopy tools can enable a better understanding of the complex properties of matter and can lead to improved tailoring of materials to the requirements of modern applications.

2.2 Fundamental Concepts in Solid-state Physics

The solid state of matter is characterised by a dense arrangement of atoms, which maintain their initial spatial structure even in the absence of external forces, in contrast to the gas or liquid phases. At typical solid densities, even small volumes of solid matter (on the order of cubic millimetres) contain many atoms (on the order of 10^{26}), all bound in a specific structural arrangement that minimises the total energy.

A ubiquitous challenge in solid-state physics is the appropriate connection of the macroscopic state to the microscopic driving forces. Macroscopic properties can easily be measured, for example, the material's density, compressibility, and shear modulus, as well as electrical and thermal conductivity and its magnetic susceptibility. The challenge lies in the microscopic explanation of how this abundance of atoms interacts in the quantum world to yield their corresponding macroscopic behaviour. Quantum theory requires approximations in order to make appropriate calculations tractable. Experimental input is necessary to continuously benchmark computational results, while often new and surprising experimental observations also require novel concept developments within new theoretical frameworks.

Since symmetry is one of the most powerful concepts in physics, it is imperative to analyse the symmetry properties of solids first. Slightly counterintuitively, solids are characterised by strong symmetry breaking,[6-9] while gas and liquid phases often possess global continuous symmetries, and their observations do not get altered when the observer is translated in space by small or large amounts (the sample is homogeneous) or rotated in space by small or large angles (the sample is isotropic). The situation is entirely different in solids. A general translation or rotation yields a completely different environment for the observer. Nevertheless, a small set of symmetries remains, and it is these that characterise the solid state. The determining symmetry of a crystalline solid is a set of discrete translations that lead the observer back to an environment identical to its original configuration. Although only infinitely-sized solids strictly hold this translational symmetry, the consideration is still approximately correct for finite-sized solids because of the magnitude difference between the necessary translations on the sub-nanometre scale and the millimetre sizes of small solids.

In crystals, one can identify a limited set of the smallest translations that transform the solid back onto itself. The part of the crystal that repeats itself after such translations yields the definition of the unit cell (see Figure 2.1a), which is not unique. The unit cell can be thought to be constructed from

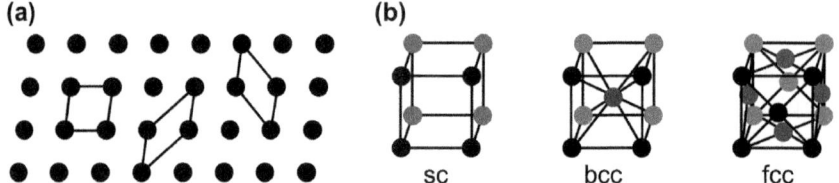

Figure 2.1   (a) A crystal shows a regular arrangement of atoms. The smallest
repeating structural unit is called the unit cell. The choice of a unit
cell is not unique, and three examples of choosing a unit cell in a regular
two-dimensional lattice are shown. (b) For simple materials in solid-state
physics, the cubic symmetry often occurs. Conventional unit cells can be
derived from the simple cubic ("sc") cell, where an additional atom is
centred in the body of the cube ("bcc", body-centred cubic) or atoms are
located on the faces of the cube ("fcc", face-centred cubic).[1-5]

three basis vectors in three dimensions: a, b, and c. The set of translations T
that lead the infinite crystal back onto itself is then given by:

$$T = m \times a + n \times b + o \times c, \quad \text{with } m, n, o \in \mathbb{Z} \text{ integer numbers.} \qquad (2.1)$$

The set of symmetry operations that can be performed on a crystal in three
dimensions defines its membership in one of the 230 space groups. The set
of translational symmetries in a crystal that can fill the three-dimensional
space can be classified further into 14 different Bravais lattices. In elemental
solids or crystals of simple elemental composition with not too many different atoms in a unit cell, the structure of the crystal is often relatively
simple with large symmetries, such as, for example, all sides of the unit cell
being equally long (cubic, see Figure 2.1b), or two sides being equal and all
angles equating to 90° (tetragonal).[1-4]

The translational symmetry operation gives rise to a periodicity in the
structure that can be described in the Fourier space of periodic functions.
The three-dimensional Fourier transformation of the solid lattice yields the
reciprocal lattice, and it holds analogous symmetries and properties to the
real-space representation of the solid structure. The reciprocal lattice can
be directly derived from the real-space basis vectors and yields a set of
reciprocal lattice vectors G:

$$G = h \times \hat{a} + k \times \hat{b} + l \times \hat{c}$$

with h, k, $l \in \mathbb{Z}$ integers and the reciprocal lattice basis   (2.2a)

$$\hat{a} = \frac{2\pi b \times c}{V}, \quad \hat{b} = \frac{2\pi c \times a}{V}, \quad \hat{c} = \frac{2\pi a \times b}{V}, \quad V = a \cdot (b \times c) \qquad (2.2b)$$

Since a crystal is fully described by specifying the interior of the unit cell, it is
sufficient to specify quantities in the interior of the first Brillouin zone,
which is the reciprocal-space analogue of the unit cell.[1-5]

In reality, the nature of the symmetries in a crystal gives significant convenience to the reciprocal-space description of quantum states in the crystal, as opposed to the situation in atomic gases. In an atom, the nucleus defines the spatial coordinate of the atom and a description in spherical coordinates is best adapted to the symmetry of the forces between electrons and the nucleus. Therefore, the ansatz to use spherical harmonics to derive appropriate quantum numbers is well-suited in this case.

In a translational periodic crystal instead, the reciprocal lattice with the crystal momentum k as a vector in this space is the best symmetry-adapted description of the system. An evaluation of energies of quantum states in the reciprocal space will yield simple eigenstates of the Hamiltonian, which have the same symmetry. A successful but very simplified description of the solid is the nearly-free electron picture, where only the symmetry character of the crystal remains, but no binding energies nor forces.[1-5] In the nearly-free electron picture, electron wavefunctions can be described following Bloch's theorem. The theorem states that in a crystal with a periodic potential, the electron eigenstates of the Hamiltonian can be factorised into a plane wave part ($e^{ik \cdot r}$) and a function $u_k^\alpha(r)$ that repeats periodically between unit cells (α is an additional index to characterise the state in addition to the crystal momentum k, Ψ is the wavefunction, and r is the spatial coordinate):

$$\Psi_k^\alpha(r) = e^{ik \cdot r} u_k^\alpha(r) \tag{2.3}$$

The best description of electrons in a crystal is as modified plane waves, characterised by the momentum vector k. The natural connection between lattice symmetry and periodic functions is the reason why solid-state physics puts great effort into the reciprocal space description of crystals. The determining quantity of electronic states in a crystal is, therefore, the three-dimensional band structure (*i.e.*, the energy of quantum states numbered by the crystal momentum k), which is a three-dimensional vector. It is worthwhile to note that displayed band structures often only show one-dimensional cuts along high-symmetry directions of the full three-dimensional structure.[1-5]

Another powerful concept in physics is separating complex systems into smaller, largely independent sub-systems that are first described within themselves and are only later coupled with their interaction as a small perturbation. This approach is at the heart of perturbation theory, which is a powerful tool to make the overwhelming number of degrees of freedom tractable in describing the solid state. The sub-systems (often also called the degrees of freedom of the solid) that are often considered independently are, for example:[1-5]

- *The lattice system:* the periodic arrangement of the nuclei (see Chapters 4, 11 and 12 for more details),
- *The electronic distribution:* the localisation or delocalisation of electrons around specific nuclei (in the simplest case, in an ionic crystal, an electron is fully transferred from the cation to the anion). In metal

oxides, partial charge transfer between oxygen and the metal occurs. Charges can also be periodically arranged independently from the lattice symmetry in incommensurate charge-density waves[10] (see Chapter 4),

- *The spin system:* the alignment of unpaired electronic spins to a quantisation axis (in the spin system, well-known macroscopic effects manifest themselves microscopically: ferromagnetism is due to the uniaxial parallel alignment of spins. Antiferromagnetic order also describes a parallel arrangement of spins but with alternating directions.[1–5] See Chapter 7 for more details),
- *The orbital system:* the survival and preferential occupation of specific atomic orbitals largely localised on specific nuclei (this situation is frequently encountered with the atomic 3d, 4d or 4f orbitals in transition metal or rare earth compounds, where specific orbitals gain or lose energy depending on the local fields in the crystal and then become preferentially occupied or empty[11]).

These degrees of freedom can often be decoupled and solved independently, and portions of the solid's functionality can be based on each degree of freedom. For example, the magnetic character of a solid (see Chapter 7) is solely based on the spin sub-system, and often this sub-system can be treated independently from the others, at least in equilibrium. The coupling between these sub-systems and their description is a matter of current research and will be discussed in further detail below.

The different degrees of freedom also have their corresponding excitation spectrum, which can often be numbered using the crystal momentum. The different excitations have different characters and names:

- *Lattice excitations, i.e., phonons:* the quantised deviation of atomic positions from the translationally symmetric crystal lattice following excitation of only the lattice structure (see Chapters 4, 11 and 12; this is analogous to vibrational motion in molecular physics). Due to the mass of the nuclei, phonon crystal momenta are large at comparably low energies with slow time constants. The phonon spectrum largely determines the heat capacity and transport properties of a solid.[1–5]
- *Electronic excitations:* excitations of the electronic system determine the optical properties of a solid (colour, reflectivity, optical absorbance, *etc.*) and charge transport properties (electrical conductivity). Excitation energies can be considerable, and electron dynamics are the fastest processes in a solid. Different electronic excitations have different names and encompass, for example, the bound electron–hole pairs called excitons.[1–5]
- *Magnetic excitations:* excitations of the spin system can lead to a change in the magnetisation state of a material. Spin waves are called magnons (see Chapter 7). The spin system can interact with other degrees of freedom, and the relevant time and energy scales in the spin system are typically intermediate in between electron and phonon excitation time

and energy scales.[2-5] Skyrmions are complex local non-equilibrium arrangements of spins (see Chapter 7).

Often, the most interesting physics occurs when the different sub-systems interact. At positions in *k*-space, where excitations of the non-interacting sub-systems would have a similar energy, new excitations with coupled character and modified energy appear and with newly assigned names, for example, polarons or polaritons, as a result of the coupling between the lattice and electrons.[1-5]

The ground state of the solid system is thus determined by the intricate interplay of different sub-systems. For example, the distance between nuclei has a strong influence on the balance between the potential and kinetic energy of electrons and thus their overall state, while it also directly influences the spin–spin interaction strengths (as well as the effective sign of the spin–spin interaction).[1-5] Although large energy scales are at play and the different states of each sub-system often have vastly different energies, the system's total energy is minimised. It can occur that although changing the atomic positions is energetically very unfavourable for the lattice system by several electronvolts (eV) per atom, this energy cost is counterbalanced by an energy gain of similar size in the electronic or spin systems. The ground state of the solid is thus determined by a subtly balanced minimisation of the different possible combinations of states of each sub-system.

Let us now consider the influence of entropy and temperature.[12] It becomes apparent that small changes in temperature, pressure, chemical compositions, external fields or the like of solids can alter the balance of the minimised energies. A completely new combination of states of the sub-systems can lead to a slightly lower total energy, although each sub-system may have a vastly different state and energy. The system can be very susceptible to significant changes in each sub-system in response to a small change in environmental conditions. A new ground state can appear as a consequence. The new ground state can have different macroscopic characteristics and, thus, different application functionality. When the new ground state is sufficiently different from the original state, the transformation is called a phase transition, which will be the topic of the next section.[6-9]

2.3 Phase Transitions in Solids

In addition to the well-known phase transitions where a liquid solidifies or boils, one can also encounter phase transitions between different solid phases (see Figure 2.2a and b). Since they are often not as disruptive as the former, some phase transitions only become apparent upon closer inspection. In any case, the physical characteristics of the solid can dramatically change, for example, a metal can become an insulator. Upon metal–insulator transition (MIT), many macroscopic properties of the sample can change besides the electrical conductivity (*e.g.*, the thermal conductivity and optical properties[10]).

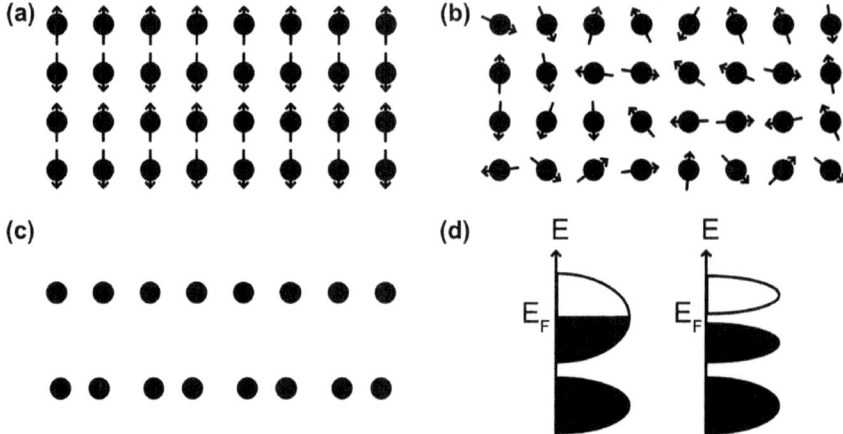

Figure 2.2   (a) The different degrees of freedom in a solid can lead to ordering
phenomena beyond the lattice structure. The panel shows the antiferromagnetic ordering of local atomic spins. Spins align parallel to
each other in one direction, while they are strictly antiparallel along
another direction. (b) A solid–solid phase transition with spin-sym-
metry breaking. An antiferromagnet can lose the spin order, although
the lattice order stays intact. The system remains solid, but the spins
orient randomly. (c) The Peierls mechanism describes how the total
energy of a system can be lowered by altering the distance between
neighbouring atoms.[15] A regular chain of atoms can spontaneously
order into pairs of atoms, lowering the total energy of the system.
(d) The Peierls[15] and the Mott[16] mechanisms explain the appearance of
insulators, although a half-filled band crosses the Fermi level (E_F).
Usually, this should lead to a system with metallic behaviour. Through
the Peierls or Mott mechanisms, the band splits up directly at the Fermi
level and a gap forms. The system turns insulating.

The MIT is one example of a solid–solid phase transition with great application potential, especially in information technology, where the phase
state of the material could encode and store or route and switch bits of information. A system can spontaneously undergo a phase transition if the free
energy F (*i.e.*, the combination of internal energy U and entropy S at a given
temperature T) of the new phase is lower, given by

$$\Delta F = F_{new} - F_{old} < 0, \tag{2.4a}$$

and

$$F = U - TS. \tag{2.4b}$$

In order to classify the various kinds of observed phase transitions, in the
early 20th century, Paul Ehrenfest considered the first derivative of the free
energy at the phase transition point. Ehrenfest observed discontinuities for
some phase transitions, which he called first-order phase transitions, while

other phase transitions only exhibited discontinuities in the second derivative of the free energy, termed second-order phase transitions.[13] Later, examples were found that did not fit into this classification scheme and the definitions were slightly altered. For example, first-order phase transitions are presently understood to encompass all transitions involving latent heat transfer. Typically, heat is transferred into or out of the system during the phase transition at a fixed temperature. Often, first-order phase transitions are also accompanied by inhomogeneity and hysteresis, while second-order phase transitions, also called continuous phase transitions, are frequently encountered when analysing solid–solid phase transitions, but they regularly do not involve latent heat, nor hysteresis or inhomogeneity.[6–9]

In the strictest sense, phase transitions only occur between the ground states for a given set of thermodynamic variables. Metastability is not accounted for, although transitions into long-lived metastable phases occur often and have interesting, interconnected properties as well.[6–9] Disordered states can be metastable and need sufficient atomic mobility to form an ordered state. For example, mobilities are low at sufficiently low temperatures, and atoms move slowly enough so that metastable phases can exist for comparably long times. To complicate matters, the exact nature of a metastable phase may depend on how the state has been populated. This starkly contrasts the ground state, which is well-defined by the current thermodynamic parameters, independent of how this state has been reached. Metastable phases, pathways to produce them, and their properties are an active field of research and hold great promise for tuning the material's properties for future applications.[14]

Generally, phase transitions break symmetries (see Figure 2.2a and b).[6–9] As described above, the transition into the solid phase breaks translational symmetry and the isotropy of the liquid phase. This lower symmetry state has less entropy and is thus only favoured if the energy gain is sufficient at low temperatures. The entropy term in the free energy (multiplied with the temperature) is why states of lower symmetry are found at lower temperatures. A spontaneous symmetry breaking, like condensation into a regular lattice, can occur at lower temperatures if the gain in binding energy outweighs the loss in entropy at the phase transition temperature. Also, here, it is not small changes in entropy and energy that find a balance. The changes in entropy and internal energy at the melting point are substantial, but the change in internal energy perfectly balances the change in entropy. This change in entropy and internal energy defines the phase transition temperature.

Peierls has shown that spontaneous symmetry breaking of a regular lattice can lower the internal energy.[15] This symmetry breaking can be visualised directly in a linear chain of atoms with binding energies that do not linearly depend on distance (see Figure 2.2c). When an atom is moved away from its equilibrium position, the gain in internal energy on one side of the atom is more significant than the loss on the other. Moving every second atom away from the equilibrium position by the same amount lowers the total internal energy with minimal effect on the other parameters of the chain, but most prominently, the symmetry is spontaneously broken, and the unit cell is doubled. A periodic lattice distortion is introduced into the system. In reality, the free energy minimum depends on temperature and the symmetry is broken only if the gain in internal energy outweighs the reduced entropy. Nevertheless, the susceptibility towards Peierls distortions is a general phenomenon encountered in many solids.

Furthermore, the Peierls mechanism can also describe metal–insulator transitions: if the linear chain has an electronic band half-filled with electrons, the system is metallic since there are sufficient empty states around the highest occupied states called the Fermi level. The Peierls distortion now equally splits this band into two, while the gain in internal energy causes the unoccupied part of the band to increase in energy and the occupied part of the band to reduce in energy. A gap opens between the occupied and unoccupied parts of the band, and the system becomes insulating (see Figure 2.2d). In summary, the Peierls mechanism describes a very general coupling between the electronic and lattice systems, where the spontaneous introduction of a lattice distortion lowers the energy and might be accompanied by a transition from a metal to an insulator.

Mott devised an alternative model of an MIT without involving the coupling between the lattice and electrons.[16] He proposed a model solely considering correlations among electrons. It describes how a system with a half-filled band that should be metallic would become insulating instead. His proposal is based on a system where the half-filled band is derived from atomically localised states. Each state should thus be filled with one electron and would give room for another electron with opposite spin. The Mott transition occurs when the repelling interaction between two electrons occupying the same site becomes more significant than the loss in kinetic energy by localising the charges on a specific atom. The electrons can no longer move from one atom to another and remain localised. The half-filled band splits into two, and the system becomes insulating (see Figure 2.2d). This mechanism does not involve the lattice and is purely based on electron correlation effects (*i.e.*, the interaction energy of two electrons occupying the same atomic site). The correlation energies are notoriously hard to calculate, and measurement through experiments is necessary.

In real systems, there is often no clear distinction between the Mott or Peierls mechanisms since localisation-driven phases are also susceptible to Peierls distortions. At the same time, the localisation in a distorted Peierls chain could also be explained by large correlation energies. It is therefore an interesting scientific question for experimentalists to find clues to explain the underlying mechanism of a phase transition. One of the main questions is whether the system would remain insulating even when the lattice distortion is (transiently) removed. Another interesting question is the timescale the system requires to become metallic when heated into the metallic phase. It is generally expected that the lattice reacts slower than an electronic mechanism, and studying the metal–insulator characteristics on an ultrafast timescale would shed light on the dominant mechanism that drives the insulating behaviour in the first place.[17]

In order to study such phenomena, the sample needs to be brought across the phase transition point on a timescale that allows to disentangle the lattice from electron contributions to the phase transition dynamics, requiring measurements with temporal resolutions of tens of femtoseconds. Ultrashort laser pulses can induce phase transitions, and most prominently, titanium sapphire-based 800 nm laser technology has often been used in the past for this purpose.[17] The absorption of optical light can lead to non-specific excitations of electrons from various occupied band states to unoccupied band states. Due to the large width of the valence bands in solids, many carriers can be excited into a multitude of states. Only in rare cases in solids can specific resonances be excited with a negligible probability of causing additional (unwanted) excitations in the system. When restricted to 800 nm pulses, the opportunities for resonant excitations essentially vanish. The absent specificity and non-resonant nature of optical light-induced electronic excitations in solids introduce an abundance of entropy to the solid. In other words, the light pulse acts merely as a heating source of the solid.[18]

The interaction of optical light with the sub-systems in a solid is entirely dominated by the absorption of light in the electronic system and the excitation of electrons into empty states (see Figure 2.3). Although the photo-absorption initially creates an unusual, non-thermal electron distribution

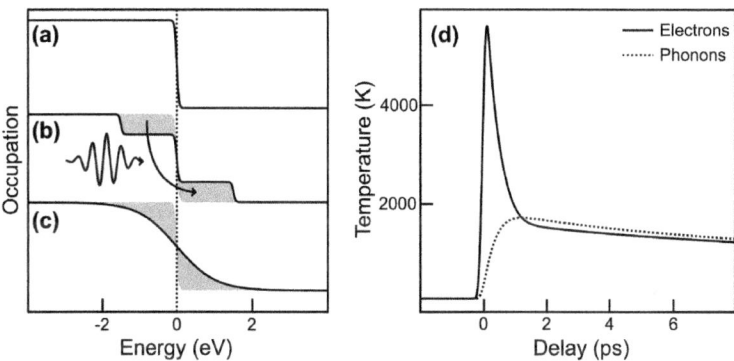

Figure 2.3 Processes after ultrafast optical excitation of a solid. (a–c) Electron distributions are shown. (a) Before laser excitation, the electron occupation follows a Fermi distribution for a low temperature. (b) After excitation with an 800 nm laser pulse (1.55 eV photon energy), electrons are transferred from below to above the Fermi level (at the energy origin, marked as a dashed line). A Fermi distribution cannot describe the electron occupation distribution, which is called non-thermal. (c) Within femtoseconds, electron–electron scattering redistributes energy among the electrons and an equilibrated Fermi distribution is assumed, albeit at an elevated temperature (*e.g.*, here shown for 3000 K). (d) The temperature evolution of the electronic system after laser excitation in a typical metal assuming thermal equilibrium according to the two-temperature model. While the electronic system reaches high temperatures quickly, the phonon system reacts slower and heats to lower equilibrium temperatures because of the different heat capacities.

(see Figure 2.3b), scattering processes among the valence electrons are extremely fast and occur within femtoseconds.[19] On this timescale, energy is exchanged between valence electrons, and the original non-thermal electron distribution develops towards a thermal distribution and assumes a Fermi occupation distribution (see Figure 2.3c). Before the internal equilibration of electrons towards the Fermi distribution, one cannot assign a temperature to the electronic system. After internal equilibration, the electron distribution according to the Fermi function allows us to define an electronic temperature independent of the other sub-systems. In many analyses of experiments, the initial non-thermal state is beyond the temporal resolution of the experiment and is often ignored.

Coupling between the electronic sub-system and the other sub-systems in the solid typically occurs on longer timescales.[20] Therefore, even several tens of femtoseconds after the initial excitation, the solid can populate an exotic state with a hot but already internally thermalised electron system (temperatures of several 1000 K are easily reached), while the other sub-systems, especially the lattice, are still cold. The main reason for this decoupling is the considerable mass difference between electrons and nuclei (see Figure 2.3d).

Typically, the coupling between electrons and phonons becomes relevant on a timescale of about 100 fs. Energy is transferred from the electrons to lattice vibrations. Phonon states are preferentially excited that strongly couple to the excited electrons. This again creates a non-thermal phonon occupation that needs to thermalise through further energy exchanges, either *via* an internal coupling or again mediated by electrons. The timescales for phonon excitation through the hot electron system and the timescales for phonon thermalisation (*i.e.*, the internal equilibration to a thermal (Boltzmann) distribution of the phonons) largely overlap. In most analyses of experiments, the non-thermal phonon state is thus neglected, and the phonon system gets characterised by its transient temperature. Only recently, studies have begun distinguishing non-thermal phonon occupations and grasping their relevance for the observed dynamics.[21]

When the hot electrons couple to the lattice, the vastly different heat capacities play an essential role. The comparably low heat capacity of the electronic system (*i.e.*, a small energy deposition increases the temperature significantly) easily yields high electronic temperatures of several 1000 K at modest optical excitations. In contrast, the much higher heat capacity of the phonon system translates to much lower temperature increases after the energy is transferred away from the electronic system and equilibrated with the lattice (see Figure 2.3d).[22] This enables the experimental opportunity to transiently create very high electron temperatures far beyond all phase transition temperatures. While after equilibration of the electronic system with the lattice, the total system of electrons and nuclei finds itself again below the phase transition temperatures. This decoupling of the electron and lattice systems in transient non-equilibrium states holds vast opportunities to study the inner workings of complex materials.

In general, the relevant thermal constants of the electronic and lattice systems are pretty well-known and tabulated. It is therefore straightforward to formulate the coupled differential equations for the electron and lattice temperatures which are based on the optical pulse as a source of energy input to the electronic system, the heat capacity of the electrons, the coupling of electrons to the lattice through its heat capacity and heat transfer into the sample in the lattice. This so-called two-temperature model can then make predictions of the thermal evolutions of the electron and lattice, the respective peak temperatures and the equilibration time scales.[22] Although it always assumes thermal distributions and leaves out many details of the optical excitation of electrons and the mechanism of how the electrons transfer energy to the lattice, this model is rather powerful as a first approach to the general phenomena in solids after optical excitation. Furthermore, the two-temperature model lends itself to extension with further sub-systems possessing their own heat capacity and coupling constants. For example, considering magnetisation dynamics after optical excitation, a three-temperature model that further includes the thermal evolution of the spin system can be employed. Even for surface chemical reactions, the temperature of atoms or molecules on the surface can be included (still excluding non-thermal effects in such considerations).[23]

In a more atomistic view, the electronic excitation in the solid leads to the occupation of formerly unoccupied states. These states can have antibonding character and, thus, in the presence of anharmonic interatomic potentials, have a different equilibrium atomic distance than the original state. The electronic system then exerts a force on the atoms to assume a new equilibrium position. This force triggers lattice vibrations and excites phonons, which is the primary mechanism behind electron–phonon coupling. If the electronic excitation is sufficiently strong, the atoms can gain enough momentum to break the bonding lattice arrangement. The lattice melts into a disordered state. If this occurs directly without reaching an internal thermal equilibrium of the phonons, this melting occurs non-thermally and may lead to a state different from the thermally molten state of the material. Furthermore, non-thermal melting can proceed on picosecond timescales, which is much faster than thermal melting since it does not proceed through the thermalised phonon system.[24]

Before we turn to some examples of ultrafast studies of solids after femtosecond optical excitation, we will introduce a set of methods that use the energy dependence of X-ray photons to study solids, namely, X-ray spectroscopic methods.

2.4 X-ray Spectroscopy on Solids

X-ray spectroscopy tools can directly probe the electronic structure of solids. X-rays with photon energies above several tens of electronvolts allow interaction with electrons tightly bound to the nuclei, the core electrons. The binding energies of core electrons are particular for each element in the periodic table and differ even between neighbouring elements, often by a large fraction (on the order of 10%) of the actual energy due to the different number of charges in the nucleus.[25] In the soft X-ray regime (100–2000 eV), core level binding energies often differ by many tens of electronvolts. Since photon energies can be separated or tuned at modern X-ray facilities by fractions of an electronvolt, different elements, even in complex compound materials, can be specifically studied with X-ray spectroscopy techniques. All elements of the periodic table have important core-level binding energies in the soft X-ray range (see Figure 2.4 for the first 28 elements).

Small variations in binding energies are due to the different chemical environments in which the same element exists in the compound. Therefore, identification of, for example, oxygen atoms with different bonding partners (and therefore different chemical environments) in one compound becomes tractable.[26] The variations in core-level binding energies due to the chemical environment result from different electron densities around the probed nuclei depending on the electronegativity of the neighbouring atoms. Very electronegative atoms attract electrons and thus reduce the electron density on neighbouring atoms. The change in electron density leads to different electric potentials around the core levels, yielding the so-called chemical shifts in the core level binding energies of up to several electronvolts.

X-ray photons can resonantly excite different core levels by choosing the X-ray photon energy such that the related signals are resonantly enhanced by many orders of magnitude. This resonant enhancement allows isolation of responses from the active elements even on top of the background of many inactive species. A prominent example that should be mentioned in this context is the case of high-temperature superconductors.[27] Here, the highest transition temperatures are reached in the cuprate class of compounds, where the superconducting mechanism occurs in the copper oxide planes of multi-element compounds as complex as $Bi_2Sr_2Ca_2Cu_3O_{10}$. Tuning the photon energies to the copper resonance allows to distinguish signatures of mechanisms directly responsible for or related to the superconductivity, while the many electrons that form a passive background do not contribute to the signal.

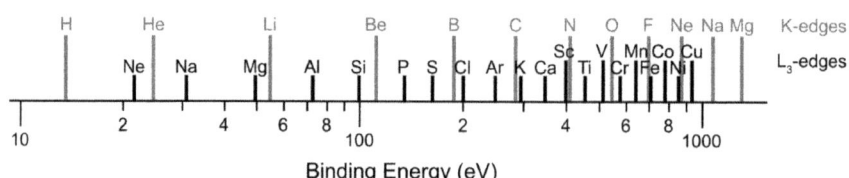

Figure 2.4   Core level binding energies in the soft X-ray region of the first 28 elements of the periodic table.[25] The top row displays the K-edges of the elements in grey (*i.e.*, the 1s core level binding energies), while the row below shows the L-edge binding energies (here displayed as the L_3-edge energy, the binding energy of the $2p_{3/2}$ levels).

In addition to the element specificity and chemical state selectivity, as illustrated above, X-ray interactions are subject to stringent selection rules that give access to atomic symmetry properties that usually vanish when considering band-like states in solids. For example, the angular momentum characteristics of delocalised states are not a good quantum number anymore in extended solids. The interaction of extended solids with X-rays projects the band states onto the atomic core states that do not hybridise with neighbouring atoms and thus retain all symmetry properties of the atomic orbitals. The different orbitals can even be selected by appropriately tuning the X-ray energy. Dipole selection rules have to be strictly obeyed in the soft X-ray range. For example, tuning the X-ray photon energy to a p-symmetric core state provides a probe of dipole transitions into s- and d-symmetric band states according to dipole selection rules $(\Delta l = 1)$. By addressing core states of different symmetry, one can thus construct a complete picture of the valence states, including their symmetry properties.[11] In addition, using polarised X-rays, one can even access directional properties of the valence states: if the X-rays are polarised along the vertical axis and tuned to an s-symmetric core state, for example, transitions into vertically extended p-like states are selected. If the sample is oriented relative to the X-ray polarisation, the different directions of valence states can be selectively studied.[11]

In a prominent application of all these features, Nyberg *et al.*[28] used soft X-ray spectroscopy and the selection rules to study the local molecular orbital energies of a molecule adsorbed on a surface. Adsorbing glycine molecules on a Cu(110) surface aligns the molecules along the rows of copper atoms. The measured soft X-ray spectrum contains signals that are site-selective and orbital-selective. Site-selectivity is achieved by selecting the X-ray energy that corresponds to a specific core level energy of an element of interest (*e.g.*, nitrogen, oxygen, and carbon). The orbital-selectivity arises from using different polarisations of the incoming X-ray photons in combination with the appropriate rotation of the sample. This allows us to capture a complete picture of which orbitals are localised at which atoms and what their orientation is, with a sketch of molecular geometry (see Figure 2.5).

In general, it is essential to note that any spectroscopic technique, in fact, probes essentially the final state of the transition.[11] In a strict sense, this is true when the initial state of the spectroscopic transition is the ground state and can be defined at the origin of an energy axis. For example, the measured energy is required in absorption spectroscopy to create a specific final state, including all multi-electron reactions arising from the absorption event. In X-ray absorption spectroscopy, the measured absorption energies correspond to the energy necessary to create a core-excited final state from the ground state. The final state includes the response of all electrons to the sudden change in charge distribution. Since the core electron that was formerly localised close to the nucleus is transferred to a much more spatially diffuse valence state, this electron does not screen the nucleus anymore,

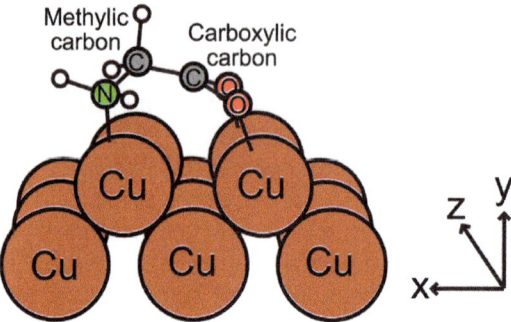

Figure 2.5   A glycine molecule adsorbed on a copper(110) surface.[28] This specific
surface of copper terminates in rows of Cu atoms. The glycine mol-
ecules adsorb in a specific geometry, as shown with the nitrogen atom
binding to one row of copper atoms, while the oxygens bind to the
neighbouring row. With core-level spectroscopy, the different orbitals
located at the different elements of the molecule, carbon, nitrogen and
oxygen, can be addressed separately. The chemical shift of the carb-
oxylic and the methylic carbon atom environments allows us to study
them specifically as well. Since the molecules are oriented with respect
to the surface, the polarisation of the X-rays allows us to study orbitals
specifically along the x-, y- and z-directions to obtain a complete
picture.

leading to the sudden appearance of an unscreened charge. The local electron density around the core hole has important implications on the multi-
electron screening response, influencing the observed chemical shift of the
core electron binding energy. The system will seek to reach its lowest energy
configuration. A higher local electron density near the created core hole can
yield a more efficient screening. Thus, a core-level binding energy shifted to
lower energies typically reflects a higher local charge density.

Furthermore, this so-called final state rule allows the various X-ray spectroscopy methods to be classified according to the different nature of the
final states. However, their classification can significantly vary by the
measured quantity and the instrumentation required to perform the measurement. As discussed below, the following will overview some of the most
important (soft) X-ray spectroscopy methods performed at modern X-ray
facilities (see Figure 2.6). All spectroscopy methods can employ the above-
discussed properties of X-rays: (i) the element selectivity and chemical state
specificity, and (ii) the stringent selection rules concerning symmetry of the
orbitals and directionality relative to the X-ray polarisation. Time-resolved
analogues of these X-ray methods exist. Core-hole lifetimes have a typical
value of several femtoseconds,[29] and this is sufficiently short for capturing
well-resolved snapshots of transiently evolving electronic structures, even
when using the decay of core holes as a spectroscopic probe.

X-ray photoelectron spectroscopy (XPS) is one of the spectroscopy methods
that lead to a core-ionised state (*i.e.*, a state where the core electron is
promoted to a free continuum state) at sufficiently high photon energies

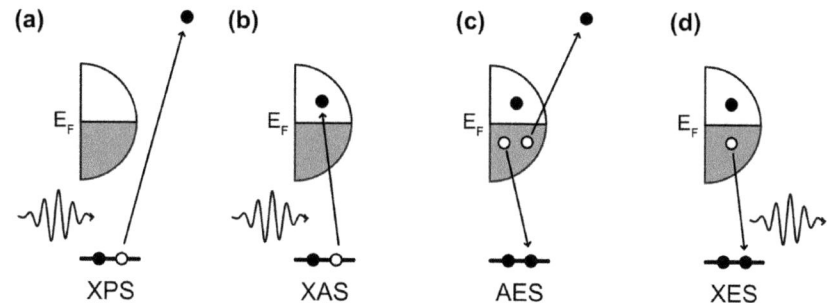

Figure 2.6 Overview of some of the important X-ray spectroscopy tools.[11] The electronic configuration in the final state of the spectroscopy is shown in the valence states, occupied up to the Fermi level E_F, and in the core levels, here usually occupied with two electrons (black) in the ground state, such as a 1s core level. Holes are shown as white circles. (a) X-ray photoelectron spectroscopy (XPS). Here, a core electron is emitted into vacuum, and its energy is analysed to determine the core level binding energy. (b) X-ray absorption spectroscopy (XAS). Here, the core electron is excited into an empty valence state and thus measures the (symmetry-resolved) unoccupied density of states. The XAS process is fundamental for near-edge X-ray absorption fine structure (NEXAFS) measurements and X-ray absorption near-edge spectroscopy (XANES). (c and d) The following panels show the decay processes of the core holes (here, after resonant excitation). (c) Auger electron spectroscopy (AES). Here, an Auger–Meitner electron is emitted that is energy-analysed. This process is the majority decay channel for soft X-ray excitations. (d) X-ray emission spectroscopy (XES). Here, an electron fills the core hole, and a photon is emitted with a characteristic energy. The analogous process also lies behind resonant inelastic X-ray scattering (RIXS).

(see Figure 2.6a).[30] In XPS, the kinetic energy of the emitted electron (E_{kin}) is measured with an electron energy analyser. Since the X-ray photon energy (E_γ) is well-known and set appropriately and work functions (Φ) are tabulated or can be independently measured, the binding energy (E_B) of the core level of interest can be determined using

$$E_B = E_\gamma - E_{kin} - \Phi. \tag{2.5}$$

As discussed earlier, the core-level binding energies relate directly to the chemical element, the nature of the core level and the chemical environment. Applying XPS for time-resolved studies, shifts in the core-level binding energies give direct access to the change in the chemical environment.

For the measurement of the kinetic energy of electrons, photoelectron analysers are required, which have multiple uses. For example, modern photoelectron analysers can also measure the direction (*i.e.*, the transversal momentum) of the emitted electrons. This yields information on the emission angles and, thus, on the momentum of the electrons inside the solid. This is known as angle-resolved photoemission spectroscopy (ARPES) and gives direct access to band structures in the solid (albeit in the presence of an electron hole in the framework of "the final state rule", as discussed earlier).[31] This is one of the most powerful spectroscopic techniques that can also be combined with different X-ray photon energies: here, tuning the photon energy to an elemental resonance can enhance the signal of electrons from this element. Higher photon energies can increase the inelastic mean free path of the emitted electrons inside the solid (*i.e.*, the length that energetically undisturbed electrons travel without losing energy).[32] Selecting a specific photon energy allows us to choose the average probing depth of the method. In general, photoelectron spectroscopy is a very surface-sensitive method with typical inelastic mean free paths of several nanometres (*i.e.*, several atomic layers only) for electrons possessing several hundred electronvolt kinetic energy. This makes electron spectroscopy an interesting tool to study dynamics on or near the surface directly but is less attractive for dynamics deep inside the bulk that do not extend sufficiently close to the surface.

The recent appearance of modern electron analysers with very good momentum resolution and high detection efficiency led to the revival of X-ray photoelectron diffraction (XPD) studies in solids where the diffraction pattern of electrons from solids is measured.[33] Compared with extensive simulations, such measurements can yield information about the structural distributions of the elements inside the solid. The substantial scattering cross-section of the electrons leads to multiple scattering events per electron, and together with the multitude of photoelectron emission sites in the solid, the analysis of such studies is not easy but is tractable when compared to results from modern simulation tools. Analysing dynamic changes in these patterns is desirable and can yield interesting results.

X-ray absorption spectroscopy (XAS) is another method that generates a core hole but promotes the electron resonantly into unoccupied valence states (see Figure 2.6b).[11,34] The structure of the spectra then represents the structure of the unoccupied valence states. The photon energies used in XAS are typically lower than those in XPS. Here, the measured XAS data contains information on the probability of exciting the electron from the core level into this unoccupied (valence) state, which is, in a simplified picture, given by the (dipole-operator mediated) projection of the core state onto the unoccupied state. This overlap of core and valence orbitals determines the cross-section to absorb a photon. The most straightforward approach to measure this absorption cross-section is to analyse the fraction of photons transmitted through a sample as a function of X-ray photon energy. This direct measurement approach, although, is problematic since the absorption cross-sections are very large, especially for soft X-rays around the core-level resonances. Only very thin samples on the order of 100 nm thickness transmit a sizable number of photons. Generally, a typical solid material often cannot be prepared in such a thin, well-defined state. However, it is still possible to indirectly measure X-ray photon absorption cross-sections in thicker solid samples through yield methods as explained in the following.

It is generally assumed that the X-ray photon absorption event creates a core hole that decays with a largely energy-independent probability into secondary particles that can be measured. In the soft X-ray region, the most likely core hole decay pathway is through the Auger–Meitner decay, where a free electron is emitted.[11] Measuring the total number of emitted electrons, the total electron yield, thus gives an electron current that is largely proportional to the X-ray absorption cross-section. Different decay channels can be measured and distinguished in the partial electron yield method with a coarse energy filtering of the emitted electrons. With some setups, the emitted electron current is directly measured, while it is much easier to measure the electron current that flows from the Earth's ground potential into the sample to compensate for the emitted charges. This drain current can easily be measured when the sample is not electrically grounded, but the ground connection is established *via* a pico-amperemeter. As with all electron-based measurements, the escape depth of electrons is limited, and the measurement results are dominated by core decays close to the surface. The dynamics are generally different in the bulk of the material, whereby the electron yield measurements may not give a good signal.

Fluorescent decay is an additional decay channel, which is more sensitive to the bulk structure and dynamics but accounts for less than one percent of the core hole decays in the soft X-ray range. Here, photodiodes can be used to measure the photons emitted from the sample through fluorescent decays. Care needs to be taken to prevent the saturation of diodes by background light (*e.g.*, from the laser pulse or laboratory environment) during the experiment. These diodes are often sensitive to emitted electrons that may dominate the signal. Precautions need to be taken if a pure photon signal is needed. Due to the additional dipole transition for the fluorescent emission, additional selection rules are at play for the fluorescent yield. In several cases, the fluorescent yield does not directly measure the absorption cross-section but a slightly modified quantity in intensity.[35] Selecting the energy of the emitted photons then leads to a partial fluorescence yield measurement that is background-free, which is interesting for very dilute samples, but the signal strength is usually very low.

In general, X-ray absorption spectroscopy (XAS) can be split into several sub-categories depending on the energy range employed and the community, such as NEXAFS (near-edge absorption fine structure) or XANES (X-ray absorption near-edge spectroscopy) closely around resonances, or even EXAFS (extended X-ray absorption fine structure) where an extended range up to several hundred electronvolts above resonances is studied.[34,36] In this range above resonance, the original X-ray photon absorption cross-section is modified by the multi-scattering events of the photoemitted electron on the surrounding atoms. The spectral structures analysed in EXAFS thus yield geometric structural information on the neighbouring environment around the resonantly selected atom.[36]

The decays of core excitations can also be used directly to learn more about the sample. Measuring the Auger–Meitner electrons with an electron energy analyser yields the energetic structure of the involved states. This method is called Auger electron spectroscopy (AES, see Figure 2.6c).[37] The method is less frequently used since three different states are involved in the Auger–Meitner decays: (i) the core hole state that decays in the first place, (ii) the state where the filling electron comes from, and (iii) the state where the emitted electron comes from. The measured spectra represent a convolution of all three.

Recently, a strong interest has emerged in analysing the specific energies of the photons emitted in fluorescent decay (see Figure 2.6d).[38] If the excitation is close to resonance, the photon-in–photon-out process must be considered a scattering event. The energy difference between the incoming photons and the emitted photon can be constant, and the energy is used to excite a specific final state of this process. Such a resonant inelastic X-ray scattering (RIXS) process can create final states with low-energy excitations, which can be directly probed. These encompass charge excitations as well as phonon or magnon excitations. In addition to an optical Raman process, the intermediate core excited state yields more information due to the X-ray's important role in resonantly enhancing the signal of interest, thanks to its element specificity and symmetry selectivity. Furthermore, the X-ray photons carry substantial momenta in contrast to optical photons. Depending on the angular arrangement of the experimental setup (X-ray source, sample, and photon spectrometer), a net momentum is transferred to the sample in the photon-in–photon-out process, and excitations from nearly across the entire first Brillouin zone become selectively accessible. In addition, some spectrometers can also analyse the polarisation of the outgoing radiation. This can be used to separate the contribution to the scattering signal coming from charge excitations with conserved polarisation from those contributions that flip the polarisation, for example, as a result of magnetic excitations. With ever-improving energy resolutions, RIXS has evolved into a fruitful tool for studying low-energy excitations in materials with high specificity.[38]

In the strict sense, the RIXS process is characterised by a well-defined final state and a constant energy difference between incoming and outgoing photons regardless of small changes in the incoming photon energy. The measured signal always shows additional contributions of photons emitted at constant energy called fluorescence photons. Fluorescence becomes particularly strong above resonances, especially in systems where the core excitation is well-screened (*i.e.*, where it couples strongly to the environment) such that any additional incoming energy can be absorbed by the environment, leading to photon emission at constant energy. This process is dominant in metals or adsorbates on metals, and is utilised in X-ray emission spectroscopy (XES) or fluorescence spectroscopy.[39,40] In contrast to the RIXS signals, XES does not exhibit strong resonance effects and, in particular, does not allow for momentum-selective studies since additional momentum can be transferred to the environment.

Nevertheless, XES still probes the occupied electronic structure with the element specificity and symmetry selectivity of X-ray spectroscopy.[39,40]

With the combination of XES and XAS, the occupied and unoccupied valence electronic structure becomes directly accessible with element specificity. When both methods are performed at similar photon energies using the same core level, it is appealing to display the obtained states on a common energy scale, for example, in order to extract the energy width of band gaps between occupied and unoccupied states. This is not straightforward: XES probes a charged final state possessing a hole in the probed valence state, while XAS probes a charge neutral state with a hole in the core state. Screening energies associated with core holes are usually larger than those associated with valence holes. In order to quantitatively extract the energy width of band gaps, the difference in screening energies needs to be accounted for.[41,42]

2.5 Applications of Time-resolved Spectroscopy on Solids

In this section, three examples of time-resolved spectroscopy on solids will be discussed from three very different processes with diverse applications to showcase the breadth of the applicability of the discussed techniques. First, a surface chemical reaction studied in real-time while forming a chemical bond will be discussed.[43] This is followed by details of the non-thermal melting of silicon, where the electronic structure analysis using X-rays revealed an interesting first-order phase transition between two different liquid phases along the melting pathway.[44] Finally, X-ray spectroscopy of a correlated transition metal oxide model compound, nickel oxide, undergoing a photoinduced reaction, will be discussed together with details of excitations induced by an optical laser with a photon energy smaller than the energy width of the band gap of the material.[45]

In the first example,[43] the oxidation of carbon monoxide (CO) adsorbed on a metal surface will be discussed (see also Figure 2.7). This reaction plays a

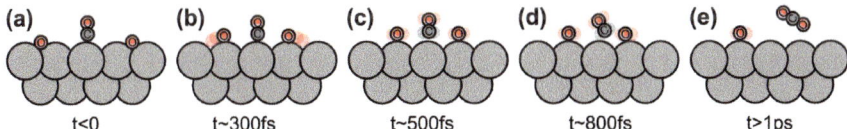

(a)	(b)	(c)	(d)	(e)
t<0	t~300fs	t~500fs	t~800fs	t>1ps

Figure 2.7 Summary of the CO oxidation reaction mechanism on a metal surface as determined from ultrafast X-ray absorption and emission spectroscopy.[43] (a) Before the pump laser arrives, the carbon monoxide (CO) molecule sits on top of a metal atom while the oxygen atoms of molecular oxygen (O_2) are adsorbed atomically on hollow sites. (b) At 300 fs after photoexcitation, the oxygen atoms are activated and move towards the bridge site. (c) The CO molecule is also activated on a 500 fs timescale. (d) After approximately 800 fs, the CO molecules and oxygen atoms begin interacting with one another such that a new C=O bond is formed to generate the carbon dioxide (CO_2) product molecule. (e) The CO_2 molecule desorbs from the ruthenium surface on a timescale of >1 ps.

vital role in cleaning exhaust gases emitted from every modern fuel-driven car which contains a platinum catalyst where highly poisonous CO is oxidised to non-toxic carbon dioxide (CO_2). The element selectivity of X-ray spectroscopy here is very useful since it allows us to distinguish the signals arising from the reacting molecules while being insensitive to the large, passive background of the metal surface. The small number of molecules in the low-density gas phase does not visibly contribute to the signals.

Performing XAS and XES measurements at the carbon and oxygen K-edges allows us to track in real-time the evolving electronic structure site selectively. The X-ray spectra were recorded with ultrashort X-ray pulses from a free-electron laser at selected delays after optical excitation. The intuition that connected the observed spectral changes with different configurations of the molecules was corroborated by molecular dynamics simulations combined with simulated X-ray spectra. The following dynamic picture emerges. Before the reaction starts, gas-phase CO molecules adsorb on a ruthenium metal surface (see Figure 2.7a). Looking at the atomic configuration of the surface, one can identify three different adsorption sites: (i) a CO molecule can sit on top of a single ruthenium atom that is quite far from other ruthenium atoms, (ii) or the molecule can achieve a higher coordination number by bridging two neighbouring ruthenium sites, (iii) or the molecule can be strongly coordinated in the so-called hollow site in between three ruthenium atoms. While the CO molecule has a strong internal bond between the atoms, the adsorption to the surface is weak, and it sits upright at the top site, with the carbon atom chemisorbed to the metal and the oxygen atom pointing away from the surface. When gas-phase oxygen molecules impinge on the surface, they can get rather strongly bound to the surface at the hollow sites. Bonding with the surface, in turn, weakens and subsequently dissociates the O–O molecular bond to generate two oxygen atoms. At room temperature, the CO molecules adsorbed on ruthenium do not undergo a reaction with the oxygen atoms, and no CO_2 is formed.[46] It has been shown that ultrafast heating of the electronic system of the metal with an optical laser pulse can excite the oxygen atoms out of the strongly bound hollow site, weakening the atom–surface bond and moving the oxygens to the weaker bound bridge sites. This excitation occurs during the first 300 fs after optical excitation (see Figure 2.7b).[47] Within the first 500 fs after excitation, vibrational modes of the CO molecules become excited and become more mobile on the surface (see Figure 2.7c). If now a CO molecule and an oxygen atom come close enough to form a bond, typically within less than a picosecond after excitation (see Figure 2.7d), the OC=O bond can become sufficiently stable such that a stable CO_2 molecule desorbs and is emitted into the gas phase (see Figure 2.7e). In a large fraction of cases, even if the reactants collide with each other, no product is formed and the OC=O bond that formed transiently breaks apart again.[43]

As a second example, a study on the non-thermal melting of a silicon crystal will be discussed (see Figure 2.8).[44] In this study, X-ray emission spectra after non-resonant excitation of silicon were recorded using pulses

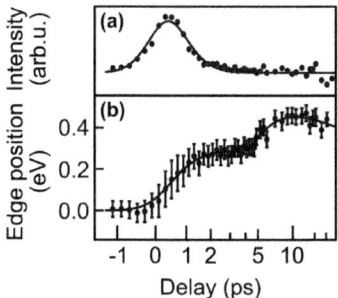

Figure 2.8   Non-thermal melting of crystalline silicon.[44] Data were acquired with ultrafast X-ray emission spectroscopy at the silicon L-edge after optical laser excitation. (a) The signal extracted from hot electrons above the band gap. The hot electrons appear instantaneously with the optical laser excitation while they decay within the first picosecond. (b) The electronic structure changes during the phase transitions. This is visible in the edge position of the occupied valence band. The valence band edge shifts within 1 ps to an intermediate state, which can be identified with an unusual transient state that is liquid but with the same density as the solid. It takes ∼5 ps for the gap to fully close during a second step-like evolution. After 5 ps, the equilibrium liquid state is reached.

from a free-electron laser. In this case, non-resonant excitation had the advantage that the full photon flux of the laser could be used without monochromatisation, and the signal levels were thus enhanced without negatively affecting the recorded signals. The emission spectra showed the symmetry-resolved occupied part of the electronic structure of silicon as a function of time. Non-thermal melting of the sample was induced with intense optical laser pulses. Before laser excitation (*i.e.*, at negative pump–probe delays), the spectra from the sample appear undisturbed and closely resemble the analogues acquired at other X-ray facilities.[48] Directly after the laser excitation, new signals appeared at energies higher than previously observed. This signature can be related to hot electrons that occupy states above the bandgap and yield a sizable new signal at higher than previously observed energies (see Figure 2.8a). Within the first picosecond after excitation, the band edge below the band gap shifts upwards until the gap is partially filled. Several picoseconds later, the band edge shifts again, and the band gap becomes completely filled (see Figure 2.8b). The measured data were then compared to calculations for different states of silicon which predicted that the ultrafast and non-thermal excitation of silicon destroyed the regular crystalline structure on a very short timescale to yield a transient half-metallic state that has a comparable density to the solid but lacks the highly symmetric structure of the latter. This half-metallic state can be identified with the transient signature in the spectra, where the band gap is only partially filled.[49]

The equilibrium liquid phase of silicon has a higher density than the solid, also found in water and ice. This surprising densification upon melting is often a consequence of a rather space-consuming solid crystalline network (in the present case with a tetragonal four-fold coordination) that can melt into a less space-consuming, higher coordinated disordered liquid phase. In order to equilibrate the non-thermally molten phase, many atoms must move macroscopically to realise the density increase. This densification occurs on a slower timescale and necessarily proceeds *via* the formation of empty voids in between patches of equilibrated liquid silicon. The equilibrium liquid state of silicon is calculated to be metallic, and the second, now complete closing of the band gap in the electronic structure, can be related to the equilibration and densification of the structurally disordered sample. The observed time delay and step-like behaviour of the transition into the equilibrium liquid state suggest that the two observed liquid phases – the transient low-density liquid immediately after laser excitation and the equilibrated higher-density liquid phase observed after more than 5 ps – are actually connected *via* a first-order phase transition and can thus only proceed after the latent heat of the phase transition is transferred from the laser-excited electronic system to the lattice. This interesting observation supports models of liquid–liquid phase transitions in tetrahedrally coordinating materials.[44]

In the last example, details of the photoinduced dynamics will be given following the non-resonant sub-gap optical excitation of a nickel oxide thin film studied with X-ray spectroscopy at the nickel L-edge and oxygen K-edge.[45] The main results from this study are summarised (see Figure 2.9).

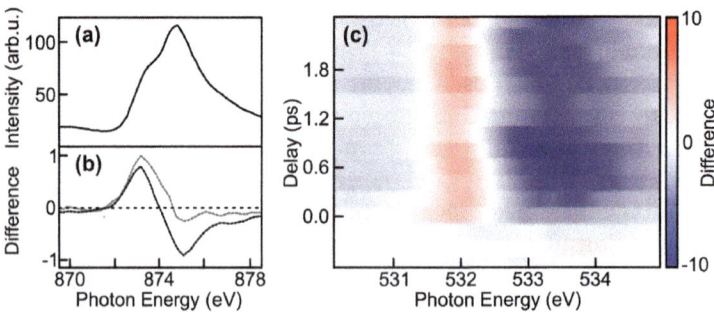

Figure 2.9 Ultrafast dynamics in NiO observed with time-resolved X-ray reflectivity measurements[45] at the (a and b) nickel L_2-edge and (c) oxygen K-edge. (a) Spectrum characteristic for linear dichroism effects due to antiferromagnetic alignment in the sample. The double structure evolves upon changes in relative polarisation or as a function of spin temperature. (b) Difference spectrum at the nickel L_2-edge following excitation with the optical laser. The grey dashed line summarises the "early" delay of <400 fs, while the black dashed line is the difference at longer pump–probe delays. Here, the long delay differences can be directly related to a heated spin system, while the short time delay difference has no resemblance to an equilibrium spectrum. (c) Difference spectrum measured at the oxygen K-edge. Blue colours show a loss of intensity, while red colours display an intensity increase. The red band around 532 eV indicates a shift of the band edge to lower energies together with a weak oscillation due to the excited magnon mode as a function of delay.

In this study, the reflectivity of the sample was analysed for the largest signal-to-noise ratio. Through the Fresnel equations, the reflectivity is directly related to the real and imaginary parts of the refractive index, which are coupled to each other through the Kramers–Kronig transformation. This allows reflectivity data to be related to absorption spectra (*i.e.*, the imaginary part of the refractive index). Spectral features usually agree in terms of energy. The spectral intensity dependence of reflectivity spectra and absorption spectra can show similar behaviour when the reflectivity is measured at rather glancing angles. In time-resolved studies, the analysis is even easier since the focus is on dynamical changes in the spectra in contrast to the ground state spectrum.

Nickel oxide samples at room temperature exhibit an antiferromagnetic ordering of the electron spins. The spin order, the occupation of 3d levels, and the crystal lattice charges lead to a specific and well-studied spectral structure in the nickel L_2-edge around 874 eV (see Figure 2.9a).[50] This spectral structure can be directly related to antiferromagnetism even when probed with linearly polarised X-rays in X-ray magnetic linear dichroism (XMLD; see Chapter 7 for more details). The sample was excited with 800 nm optical laser pulses. The photon energy at 800 nm (1.55 eV) is within the band gap of the material (around 4 eV), and the sample is relatively transparent for this wavelength (since excitation from the valence to the conduction band is not possible at 800 nm). Nevertheless, some absorption occurs and appears to induce further defect states with an energy inside the bandgap.

The observed spectral changes at the nickel L-edge suggest that a modification of the spin system may have occurred after photoexcitation (see Figure 2.9b). On a sub-ps timescale, the changes indicate a non-thermal distribution of electrons and spins in the system since the transient spectra cannot be related to reference data measured at different temperatures in thermal equilibrium. Only on a longer timescale do the transient spectra directly resemble reference spectra at elevated spin temperature. The observed temperature increase agrees with the absorbed energy of the optical pump pulse and the low absorption cross-section of the sub-gap optical wavelength. The spectral changes observed at the leading edge of the oxygen K-edge (around 532 eV) can be directly related to the band gap in the material.[51] The leading edge shows a general shift to lower energies and displays an oscillation with a period of around 1 THz (see Figure 2.9c). This frequency is characteristic of the eigenfrequency of a prominent magnon mode in nickel oxide, and it has been proposed that this magnon mode directly modifies the electronic structure of the sample through correlation effects.[52] The magnon excitation also results in an average spin canting away from the parallel ground state orientation, which yields an average energetic lowering of the unoccupied band above the band gap. Corroborated by calculations, the observations in the spectra can therefore be related to the emission of magnons *via* the sub-gap optical excitation. This study demonstrates the capability of time-resolved X-ray spectroscopy to reveal a wealth of information about the non-thermal and thermal states of the sub-systems in the material. It also reveals details about the complex couplings between different excitation modes and transient electronic structure modification when other sub-systems are excited.[45]

In summary, these three examples demonstrate the substantial information content that can be obtained with X-ray-based techniques.

2.6 Novel X-ray Techniques

Future applications of X-ray spectroscopy to time-resolved studies using free electron lasers (FELs; see Chapter 8) and laboratory-based high harmonic generation (HHG; see Chapter 6) sources hold great promise. These sources generate intense and ultrashort X-ray pulses that are transversally fully coherent. Although most FEL pulses have poor longitudinal coherence, some FELs can directly generate fully longitudinally coherent pulses through the so-called "seeding", or the use of monochromators can enhance the coherence after the pulse generation.[53]

With these X-ray pulse properties, which go well beyond what was routinely available before, new breakthrough developments in X-ray spectroscopy techniques are anticipated, some of which will be discussed below.[54–56] A summary of some underlying processes in these emerging techniques is sketched (see Figure 2.10).

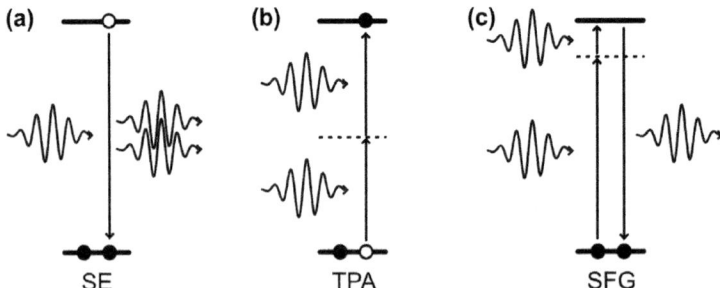

Figure 2.10 Schematic overview of several non-linear processes observed with X-rays.[54–56] (a) Stimulated emission (SE) schematic. An incoming X-ray photon interacts with a core-excited atom that causes the emission of an additional photon with the same properties as the incoming photon. After the interaction, two photons leave the sample. (b) Two-photon absorption (TPA) schematic. An atom is excited by the simultaneous absorption of two photons which goes through a virtual state (dashed). Since selection rules for such a transition are different, other states become accessible than in single-photon absorption. (c) Sum-frequency generation (SFG) process schematic. The displayed process only involves one high-energy photon and one low-energy photon. A third photon is emitted at the sum of the energy of both incoming photons. The displayed process is symmetry-forbidden in centrosymmetric samples and is thus highly surface sensitive where centrosymmetry is intrinsically broken. Since non-linear cross-sections are generally small in the X-ray range, resonant enhancement potentially of intermediate dark states can increase signal levels.

Stimulated emission is the most fundamental effect that forms the basis of multi-dimensional and wave-mixing spectroscopy (see Figure 2.10a). In stimulated emission, the interaction of an atom with a first X-ray photon creates a core-excited state. Through the interaction with a second incoming X-ray photon (the stimulating photon), the core-excited state decays and an additional X-ray photon is emitted that is identical in energy, phase and direction to the stimulating photon. In recent experiments, this process has been observed from gaseous, liquid and solid samples, where the stimulating photon was either a spontaneously emitted fluorescence photon from the sample (this process is then called amplified spontaneous emission, ASE) or an external photon from the X-ray source.[57–59]

The attractiveness of stimulated emission lies in the possibility of enhancing the yield for photon core decays at the expense of Auger–Meitner decays, increasing the measured signal level for spectroscopy methods such as RIXS or XES. The fluorescence probability is between 0.1% and 1% in the soft X-ray range, so a signal increase of two-to-three orders of magnitude is possible. In addition, when triggered externally, stimulated emission can form a highly directional beam. This is advantageous for efficient detection since the angular acceptance of typical soft X-ray gratings is usually limited to between 10^{-4} and 10^{-6} of the full solid angle. Spontaneous emission is emitted rather isotropically such that only a tiny fraction of the emitted photons can be detected. The directional beam from stimulated emission may be completely collected even with a standard soft X-ray grating, so the detection efficiency is increased by orders of magnitude.

Searching for stimulated X-ray effects, experimentalists also observed that the high core-excitation densities yield a wealth of concurrent decays on the femtosecond timescale, which generates all kinds of additional excitations in the sample that may have destructive effects on the sample and may mask the signals of interest. During the femtosecond core-hole lifetime, spontaneous decays occur, such as Auger–Meitner processes which are the primary decay channel in the soft X-ray range. In this process, electronic excitations and energetic electrons deep inside the sample can be generated on the timescale of a few femtoseconds. The energetic electrons may even create a multitude of additional excitations through ultrafast electron–electron scattering at neighbouring sites. In order to observe stimulated transitions between core and valence states, the presence of a core hole and an electron in the valence state is necessary. The creation of secondary excitations proceeds first by the decay (and thus the disappearance of the core hole) and secondly by creating a multitude of holes in valence states such that the conditions for stimulated emission quickly vanish already within a few femtoseconds.[60] The only possible way to observe stimulated X-ray emission is to use pulses substantially shorter than the core-hole lifetime (*e.g.*, a pulse duration of about less than a femtosecond). While such pulses are hardly available, the large energy bandwidth associated with sub-fs pulses may limit the achievable spectroscopic information that can be extracted from such processes.

An intense, single-colour pulse can further induce two-photon absorption (see Figure 2.10b).[61,62] The simultaneous absorption of two photons occurs at the same resonance as with standard single-photon absorption, but the selection rules are different since the angular momentum of the two photons can be combined parallel ($\Delta l = \pm 2$) or antiparallel ($\Delta l = 0$) to each other. Such a spectroscopic technique thus can probe states of opposite symmetry character than in single-photon absorption. This holds excellent promise, especially for studies on spectroscopically important elements such as carbon, nitrogen and oxygen. For example, to perform element-selective X-ray spectroscopy on these elements, only the 1s core level can be resonantly excited. Single photon dipole transitions thus can probe only the p-symmetric valence density of states at these elements. With two-photon absorption, s- and d-symmetric valence states become uniquely accessible for element-specific X-ray spectroscopy.

Splitting a single-colour pulse that impinges from two different directions onto the sample at temporal overlap creates a grating-like interference pattern on the sample.[63,64] The excitations created from this grating exhibit a local periodic density modulation. This periodic density modulation may create local excitations that induce a periodic variation of the refractive index for an additional probe pulse. The probe pulse will be diffracted from this grating into a new direction, and a background-free signal can be recorded in this new direction. By varying the delay between the pulse that created the grating and the probe pulse, the dephasing and rephasing of the grating can be probed in real-time, characteristic of the velocity of the excitations in the sample and their diffusion. Such transient gratings in the X-ray range thus give access to the dispersion of many excitations in samples at wave vectors that are not accessible with optical methods.

Further opportunities arise when pulses of different wavelengths overlap in time and space at the sample (see Figure 2.10c).[65,66] Here, the wavelength mixing of these pulses can yield new information on the coherent interaction between different excited states. The direct sum and difference frequency generation is subject to selection rules prohibiting such signals in centrosymmetric systems. The surface of a solid breaks the symmetry, and such signals can be generated only in a region close to the surface. The interaction of more photons is subject to different selection rules, and novel selectivity can arise and be combined with the specificity of X-rays.

While FELs strive to produce multiple X-ray wavelengths simultaneously, it is also interesting to mix laser-induced optical excitations with element-selective X-ray excitations to study the evolution of these couplings in space and time. By choosing a right combination of wavelengths in the suitable experimental geometries, background-free signals could be generated with high efficiency and significant information content on the coupling between different local and delocalised states driving the dynamics and functionality of the system. The first steps have been taken into developing experimental protocols for such studies, and theoretical developments have been made.

Acknowledgements

The work with its focus points that led to the writing of this chapter has been influenced by discussions and exchanges with many excellent scientists at various stages of their and my life. I am honoured to have had the chance to work with outstanding people throughout my career, and I am indebted to many of them. Especially, the stimulating environments at the workplaces at the University of Hamburg, at the Helmholtz-Zentrum Berlin, at SLAC at Stanford University, and the Deutsche Elektronen-Synchrotron DESY with the FEL FLASH and all involved colleagues, also behind the scenes, have contributed significantly to the ability to write this chapter.

References

1. C. Kittel, *Introduction to Solid State Physics*, Wiley, New York, 8th edn, 2005.
2. N. W. Ashcroft and N. D. Mermin, *Solid State Physics*, Saunders College Publishing, New York, 1st edn, 1976.
3. P. Phillips, *Advanced Solid State Physics*, Cambridge University Press, Cambridge, 2nd edn, 2012.
4. S. H. Simon, *The Oxford Solid State Basics*, Oxford University Press, Oxford, 1st edn, 2013.
5. P. M. Chaikin and T. C. Lubensky, *Principles of condensed matter physics*, Cambridge University Press, Cambridge, 1st edn, 1995.
6. R. V. Solé, *Phase Transitions*, Princeton University Press, Princeton, 1st edn, 2011.
7. I. Herbut, *A Modern Approach to Critical Phenomena*, Cambridge University Press, Cambridge, 1st edn, 2007.
8. U. C. Täuber, *Critical Dynamics: A Field Theory Approach to Equilibrium and Non-Equilibrium Scaling Behavior*, Cambridge University Press, Cambridge, 1st edn, 2014.
9. N. Goldenfeld, *Lectures on Phase Transitions and the Renormalization Group*, Westview Press, Boulder, 1st edn, 1992.
10. E. Dagotto, *Nanoscale Phase Separation and Colossal Magnetoresistance*, Springer, Berlin, 1st edn, 2003.
11. F. de Groot and A. Kotani, *Core Level Spectroscopy of Solids*, CRC Press, Boca Raton, 1st edn, 2008.
12. C. Kittel and H. Kroemer, *Thermal Physics*, W. H. Freeman and Company, New York, 2nd edn, 1980.
13. P. Ehrenfest, *Proc. Royal Acad. Amsterdam*, 1933, **36**, 153.
14. P. H. Poole, F. Sciortino, U. Essmann and H. E. Stanley, *Nature*, 1992, **360**, 324.
15. R. E. Peierls, presented in part at cours de l'école d'été de physique théorique, Les Houches, 1953.
16. N. F. Mott, *Proc. Phys. Soc., London, Sect. A*, 1949, **62**, 416.

17. S. Hellmann, T. Rohwer, M. Kalläne, K. Hanff, C. Sohrt, A. Stange, A. Carr, M. M. Murnane, H. C. Kapteyn, L. Kipp, M. Bauer and K. Rossnagel, *Nat. Commun.*, 2012, **3**, 1069.
18. P. Beaud, A. Caviezel, S. O. Mariager, L. Rettig, G. Ingold, C. Dornes, S.-W. Huang, J. A. Johnson, M. Radovic, T. Huber, T. Kubacka, A. Ferrer, H. T. Lemke, M. Chollet, D. Zhu, J. M. Glownia, M. Sikorski, A. Robert, H. Wadati, M. Nakamura, M. Kawasaki, Y. Tokura, S. L. Johnson and U. Staub, *Nat. Mater.*, 2014, **13**, 923.
19. H.-S. Rhie, H. A. Dürr and W. Eberhardt, *Phys. Rev. Lett.*, 2003, **90**, 247201.
20. N. Thielemann-Kühn, D. Schick, N. Pontius, C. Trabant, R. Mitzner, K. Holldack, H. Zabel, A. Föhlisch and C. Schüßler-Langeheine, *Phys. Rev. Lett.*, 2017, **119**, 197202.
21. F. Murphy-Armando, É. D. Murray, I. Savić, M. Trigo, D. A. Reis and S. Fahy, *Appl. Phys. Lett.*, 2023, **122**, 012202.
22. P. B. Allen, *Phys. Rev. Lett.*, 1987, **59**, 1460.
23. M. Brandbyge, P. Hedegård, T. F. Heinz, J. A. Misewich and D. M. Newns, *Phys. Rev. B: Condens. Matter Mater. Phys.*, 1995, **52**, 6042.
24. A. Rousse, C. Rischel, S. Fourmaux, I. Uschmann, S. Sebban, G. Grillon, P. Balcou, E. Förster, J. P. Geindre, P. Audebert, J. C. Gauthier and D. Hulin, *Nature*, 2001, **410**, 65.
25. J. A. Bearden and A. F. Burr, *Rev. Mod. Phys.*, 1967, **39**, 125.
26. K. Siegbahn, *Philos. Trans. R. Soc., A*, 1970, **268**, 33.
27. E. Dagotto, *Rev. Mod. Phys.*, 1994, **66**, 763.
28. M. Nyberg, J. Hasselström, O. Karis, N. Wassdahl, M. Weinelt, A. Nilsson and L. G. M. Pettersson, *J. Chem. Phys.*, 2000, **112**, 5420.
29. C. Nicolas and C. Miron, *J. Electron Spectrosc. Relat. Phenom.*, 2012, **185**, 267.
30. S. Hüfner, *Photoelectron Spectroscopy*, Springer, Berlin, 3rd edn, 2013.
31. A. Damascelli, *Phys. Scr.*, 2004, **T109**, 61.
32. J. Woicik, *Hard X-ray Photoelectron Spectroscopy (HAXPES)*, Springer, Berlin, 1st edn, 2016.
33. O. Fedchenko, A. Winkelmann, K. Medjanik, S. Babenkov, D. Vasilyev, S. Chernov, C. Schlueter, A. Gloskovskii, Y. Matveyev, W. Drube, B. Schönhense, H. J. Elmers and G. Schönhense, *New J. Phys.*, 2019, **21**, 113031.
34. J. Stöhr, *NEXAFS Spectroscopy*, Springer, Berlin, 1st edn, 1992.
35. F. de Groot, M. A. Arrio, P. Sainctavit, C. Cartier and C. T. Chen, *Solid State Commun.*, 1994, **92**, 991.
36. J. J. Rehr and R. C. Albers, *Rev. Mod. Phys.*, 2000, **72**, 621.
37. C. C. Chang, *Surf. Sci.*, 1971, **25**, 53.
38. L. J. P. Ament, M. van Veenendaal, T. P. Devereaux, J. P. Hill and J. van den Brink, *Rev. Mod. Phys.*, 2011, **83**, 705.
39. J. Nordgren, G. Bray, S. Cramm, R. Nyholm, J.-E. Rubensson and N. Wassdahl, *Rev. Sci. Instrum.*, 1989, **60**, 1690.
40. J. Nordgren and J.-E. Rubensson, *J. Electron Spectrosc. Relat. Phenom.*, 2013, **188**, 3.
41. M. Beye, F. Hennies, M. Deppe, E. Suljoti, M. Nagasono, W. Wurth and A. Föhlisch, *New J. Phys.*, 2010, **12**, 043011.

42. P. S. Miedema, M. Beye, R. Könnecke, G. Schiwietz and A. Föhlisch, *J. Electron Spectrosc. Relat. Phenom.*, 2014, **197**, 37.
43. H. Öström, H. Öberg, H. Xin, J. LaRue, M. Beye, M. Dell'Angela, J. Gladh, M. L. Ng, J. A. Sellberg, S. Kaya, G. Mercurio, D. Nordlund, M. Hantschmann, F. Hieke, D. Kühn, W. F. Schlotter, G. L. Dakovski, J. J. Turner, M. P. Minitti, A. Mitra, S. P. Moeller, A. Föhlisch, M. Wolf, W. Wurth, M. Persson, J. K. Nørskov, F. Abild-Pedersen, H. Ogasawara, L. G. M. Pettersson and A. Nilsson, *Science*, 2015, **347**, 978.
44. M. Beye, F. Sorgenfrei, W. F. Schlotter, W. Wurth and A. Föhlisch, *Proc. Natl. Acad. Sci. U. S. A.*, 2010, **107**, 16772.
45. X. Wang, R. Y. Engel, I. Vaskivskyi, D. Turenne, V. Shokeen, A. Yaroslavtsev, O. Grånäs, R. Knut, J. O. Schunck, S. Dziarzhytski, G. Brenner, R.-P. Wang, M. Kuhlmann, F. Kuschewski, W. Bronsch, C. Schüßler-Langeheine, A. Styervoyedov, S. S. P. Parkin, F. Parmigiani, O. Eriksson, M. Beye and H. A. Dürr, *Faraday Discuss.*, 2022, **237**, 300.
46. M. Bonn, S. Funk, C. Hess, D. N. Denzler, C. Stampfl, M. Scheffler, M. Wolf and G. Ertl, *Science*, 1999, **285**, 1042.
47. M. Dell'Angela, T. Anniyev, M. Beye, R. Coffee, A. Föhlisch, J. Gladh, T. Katayama, S. Kaya, O. Krupin, J. LaRue, A. Møgelhøj, D. Nordlund, J. K. Nørskov, H. Öberg, H. Ogasawara, H. Öström, L. G. M. Pettersson, W. F. Schlotter, J. A. Sellberg, F. Sorgenfrei, J. J. Turner, M. Wolf, W. Wurth and A. Nilsson, *Science*, 2013, **339**, 1302.
48. S. Eisebitt, J. Lüning, J.-E. Rubensson, A. Settels, P. H. Dederichs, W. Eberhardt, S. N. Patitsas and T. Tiedje, *J. Electron Spectrosc. Relat. Phenom.*, 1998, **93**, 245.
49. P. Ganesh and M. Widom, *Phys. Rev. Lett.*, 2009, **102**, 075701.
50. D. Alders, L. H. Tjeng, F. C. Voogt, T. Hibma, G. A. Sawatzky, C. T. Chen, J. Vogel, M. Sacchi and S. Iacobucci, *Phys. Rev. B: Condens. Matter Mater. Phys.*, 1998, **57**, 11623.
51. O. Grånäs, I. Vaskivskyi, X. Wang, P. Thunström, S. Ghimire, R. Knut, J. Söderström, L. Kjellsson, D. Turenne, R. Y. Engel, M. Beye, J. Lu, D. J. Higley, A. H. Reid, W. Schlotter, G. Coslovich, M. Hoffmann, G. Kolesov, C. Schüßler-Langeheine, A. Styervoyedov, N. Tancogne-Dejean, M. A. Sentef, D. A. Reis, A. Rubio, S. S. P. Parkin, O. Karis, J.-E. Rubensson, O. Eriksson and H. A. Dürr, *Phys. Rev. Res.*, 2022, **4**, L032030.
52. C. Tzschaschel, K. Otani, R. Iida, T. Shimura, H. Ueda, S. Günther, M. Fiebig and T. Satoh, *Phys. Rev. B*, 2017, **95**, 174407.
53. E. Allaria, R. Appio, L. Badano, W. A. Barletta, S. Bassanese, S. G. Biedron, A. Borga, E. Busetto, D. Castronovo, P. Cinquegrana, S. Cleva, D. Cocco, M. Cornacchia, P. Craievich, I. Cudin, G. D'Auria, M. Dal Forno, M. B. Danailov, R. De Monte, G. De Ninno, P. Delgiusto, A. Demidovich, S. Di Mitri, B. Diviacco, A. Fabris, R. Fabris, W. Fawley, M. Ferianis, E. Ferrari, S. Ferry, L. Froehlich, P. Furlan, G. Gaio, F. Gelmetti, L. Giannessi, M. Giannini, R. Gobessi, R. Ivanov, E. Karantzoulis, M. Lonza, A. Lutman, B. Mahieu, M. Milloch, S. V. Milton, M. Musardo, I. Nikolov, S. Noe, F. Parmigiani, G. Penco, M. Petronio, L. Pivetta, M. Predonzani, F. Rossi, L. Rumiz, A. Salom, C. Scafuri, C. Serpico, P. Sigalotti, S. Spampinati,

C. Spezzani, M. Svandrlik, C. Svetina, S. Tazzari, M. Trovo, R. Umer, A. Vascotto, M. Veronese, R. Visintini, M. Zaccaria, D. Zangrando and M. Zangrando, *Nat. Photonics*, 2012, **6**, 699.
54. N. Rohringer, *Philos. Trans. R. Soc., A*, 2019, **377**, 20170471.
55. A. Zong, B. R. Nebgen, S.-C. Lin, J. A. Spies and M. Zuerch, *Nat. Rev. Mater.*, 2023, **8**, 224.
56. S. R. Leone and D. M. Neumark, *Science*, 2023, **379**, 536.
57. N. Rohringer, D. Ryan, R. A. London, M. Purvis, F. Albert, J. Dunn, J. D. Bozek, C. Bostedt, A. Graf, R. Hill, S. P. Hau-Riege and J. J. Rocca, *Nature*, 2012, **481**, 488.
58. M. Beye, S. Schreck, F. Sorgenfrei, C. Trabant, N. Pontius, C. Schüßler-Langeheine, W. Wurth and A. Föhlisch, *Nature*, 2013, **501**, 191.
59. H. Yoneda, Y. Inubushi, K. Nagamine, Y. Michine, H. Ohashi, H. Yumoto, K. Yamauchi, H. Mimura, H. Kitamura, T. Katayama, T. Ishikawa and M. Yabashi, *Nature*, 2015, **524**, 446.
60. S. Schreck, M. Beye, J. A. Sellberg, T. McQueen, H. Laksmono, B. Kennedy, S. Eckert, D. Schlesinger, D. Nordlund, H. Ogasawara, R. G. Sierra, V. H. Segtnan, K. Kubicek, W. F. Schlotter, G. L. Dakovski, S. P. Moeller, U. Bergmann, S. Techert, L. G. M. Pettersson, P. Wernet, M. J. Bogan, Y. Harada, A. Nilsson and A. Föhlisch, *Phys. Rev. Lett.*, 2014, **113**, 153002.
61. K. Tamasaku, E. Shigemasa, Y. Inubushi, T. Katayama, K. Sawada, H. Yumoto, H. Ohashi, H. Mimura, M. Yabashi, K. Yamauchi and T. Ishikawa, *Nat. Photonics*, 2014, **8**, 313.
62. K. Tamasaku, E. Shigemasa, Y. Inubushi, I. Inoue, T. Osaka, T. Katayama, M. Yabashi, A. Koide, T. Yokoyama and T. Ishikawa, *Phys. Rev. Lett.*, 2018, **121**, 083901.
63. F. Bencivenga, R. Cucini, F. Capotondi, A. Battistoni, R. Mincigrucci, E. Giangrisostomi, A. Gessini, M. Manfredda, I. P. Nikolov, E. Pedersoli, E. Principi, C. Svetina, P. Parisse, F. Casolari, M. B. Danailov, M. Kiskinova and C. Masciovecchio, *Nature*, 2015, **520**, 205.
64. J. R. Rouxel, D. Fainozzi, R. Mankowsky, B. Rösner, G. Seniutinas, R. Mincigrucci, S. Catalini, L. Foglia, R. Cucini, F. Döring, A. Kubec, F. Koch, F. Bencivenga, A. Al Haddad, A. Gessini, A. A. Maznev, C. Cirelli, S. Gerber, B. Pedrini, G. F. Mancini, E. Razzoli, M. Burian, H. Ueda, G. Pamfilidis, E. Ferrari, Y. Deng, A. Mozzanica, P. J. M. Johnson, D. Ozerov, M. G. Izzo, C. Bottari, C. Arrell, E. J. Divall, S. Zerdane, M. Sander, G. Knopp, P. Beaud, H. T. Lemke, C. J. Milne, C. David, R. Torre, M. Chergui, K. A. Nelson, C. Masciovecchio, U. Staub, L. Patthey and C. Svetina, *Nat. Photonics*, 2021, **15**, 499.
65. J. D. Gaynor, A. P. Fidler, Y.-C. Lin, H.-T. Chang, M. Zuerch, D. M. Neumark and S. R. Leone, *Phys. Rev. B*, 2021, **103**, 245140.
66. H. Rottke, R. Y. Engel, D. Schick, J. O. Schunck, P. S. Miedema, M. C. Borchert, M. Kuhlmann, N. Ekanayake, S. Dziarzhytski, G. Brenner, U. Eichmann, C. von Korff Schmising, M. Beye and S. Eisebitt, *Sci. Adv.*, 2022, **8**, eabn5127.

CHAPTER 3

Theory of Time-dependent Scattering

M. SIMMERMACHER,*[a] P. M. WEBER[b] AND A. KIRRANDER*[a]

[a] Physical and Theoretical Chemistry Laboratory, Department of Chemistry, University of Oxford, South Parks Road, Oxford OX1 3QZ, UK; [b] Department of Chemistry, Brown University, Providence, RI 02912, USA
*Emails: mats.simmermacher@chem.ox.ac.uk;
adam.kirrander@chem.ox.ac.uk

3.1 Introduction

The last two decades have seen a proliferation of ultrafast pump–probe scattering experiments that exploit short pulses of X-ray photons or high-energy electrons to probe dynamics in excited targets. This has yielded remarkable insights into time-dependent structural changes and beyond. In this chapter, we present a unified framework for time-resolved non-resonant X-ray and electron scattering.[†] Such a framework permits the analysis and simulation of current and future ultrafast scattering experiments, establishes a rigorous and fundamental understanding of the experimental observables, and provides the necessary foundation to describe experiments that may probe electron dynamics as well. Rather than following the standard Lippmann–Schwinger route of static scattering, our derivations are based on physically equivalent,[1] but explicitly time-dependent perturbation theory that directly accounts for the non-stationary nature of ultrafast experiments. This applies to both X-ray and electron scattering, placing

---

[†] The more common term for ultrafast electron scattering is ultrafast electron diffraction. Here, we use the term scattering for both X-rays and electrons throughout.

Theoretical and Computational Chemistry Series No. 25
Structural Dynamics with X-ray and Electron Scattering
Edited by Kasra Amini, Arnaud Rouzée and Marc J. J. Vrakking
© The Royal Society of Chemistry 2024
Published by the Royal Society of Chemistry, www.rsc.org

them on an equal footing. The text explores the expressions for isolated gas-phase molecules in detail.

The chapter is structured such that the main results in the form of differential scattering cross-sections are presented first, followed by their derivations, which for X-ray scattering take the starting point in non-relativistic quantum electrodynamics. The reader may choose to skip these derivations. The section on electron scattering mirrors that on X-ray scattering. It is self-contained, but elements common with X-ray scattering are described more briefly and distinguishing features are emphasized. The chapter finishes with an overview of how scattering cross-sections may be calculated and how rotational averaging and molecular alignment can be taken into account.

3.2 Non-resonant X-ray Scattering

3.2.1 X-ray Scattering by a General Target

The observable that is measured in ultrafast non-resonant X-ray scattering (see Chapter 9) is the differential X-ray scattering cross-section, $d\sigma/d\Omega$, at pump–probe delay time τ. Here, $d\sigma/d\Omega$ is defined by the number of photons scattered into a solid angle Ω. For a general target of charged particles such as an atom or a molecule, the differential scattering cross-section is given as[2-5]

$$\frac{d\sigma}{d\Omega} = \frac{1}{2\pi}\left(\frac{d\sigma}{d\Omega}\right)_{Th} \int_0^\infty w(\omega_s)\frac{\omega_s}{\omega_0}\int\int_{-\infty}^{+\infty} I(t-\tau)e^{i(\omega_0-\omega_s)\delta}$$
$$\times C(\delta)\mathcal{L}(q,t,\delta)dt d\delta d\omega_s, \tag{3.1}$$

where $(d\sigma/d\Omega)_{Th}$ is the Thomson scattering cross-section, later defined in eqn (3.19), that describes the scattering of a photon by a free electron. Moreover, ω_0 is the mean angular frequency of the incident photon and ω_s the angular frequency of the scattered photon. The window, or detector response, function $w(\omega_s)$ accounts for the probability that a scattered photon with ω_s is measured by the detector. $I(t-\tau)$ and $C(\delta)$ are the photon intensity and the linear coherence function of the incident X-ray pulse at the position of the target at times t and δ, respectively, and τ is the pump–probe delay time at which the X-ray pulse is centered. The integrals over t and δ share the same limits of integration. The quantity $\mathcal{L}(q,t,\delta)$ is the X-ray scattering probability at point q in reciprocal space at times t and δ:

$$\mathcal{L}(q,t,\delta) = \langle\Psi(t)|e^{i\hat{H}_0\delta/2\hbar}\hat{L}^\dagger(q)e^{-i\hat{H}_0\delta/\hbar}\hat{L}(q)e^{i\hat{H}_0\delta/2\hbar}|\Psi(t)\rangle. \tag{3.2}$$

The matrix element in eqn (3.2) implies integration over all internal coordinates of the target's constituent particles. The target is described by its time-dependent state $|\Psi(t)\rangle$ that is prepared by the pump pulse which precedes the scattering probe. The exponential time-evolution operators contain the imaginary unit ι, the Hamiltonian of the unperturbed target $\hat{H}_0$, and the reduced Planck constant, $\hbar = h/(2\pi)$. Finally, $\hat{L}(q)$ is the one-electron X-ray scattering operator, defined as

$$\hat{L}(\boldsymbol{q}) = \sum_{n}^{N_e} e^{\iota \boldsymbol{q} \cdot \boldsymbol{r}_n}. \tag{3.3}$$

Here, the momentum transfer vector $\boldsymbol{q}$ is defined as the vector difference between the wave vectors of the incident and scattered photons, $\boldsymbol{q} = \boldsymbol{k}_0 - \boldsymbol{k}_s$.[‡] The sum over n runs over all N_e electrons of the target with coordinates $\boldsymbol{r}_n$ (*cf.* the discussion below eqn (3.16)).

Reading eqn (3.2) from right to left, the X-ray scattering probability can be understood as follows: the target in state $|\Psi(t)\rangle$ at time t is propagated back in time to $t - \delta/2$, interacts with the X-ray field to yield the first amplitude, and is then propagated forward in time to $t + \delta/2$ to interact a second time, yielding the second amplitude. Both amplitudes interfere to constitute the overall X-ray scattering probability. The state is finally propagated back to t in order to match $\langle \Psi(t)|$, closing the bracket.

It is important to note the difference between t and δ: while t is simply the real time in which the target evolves and the X-ray pulse moves through space, as described by $|\Psi(t)\rangle$ and $I(t - \tau)$, respectively, δ relates to the fact that X-ray amplitudes at two different interaction times can interfere. The presence of δ in $\mathcal{L}(q, t, \delta)$ is thus a direct consequence of the finite duration and coherence time of the X-ray pulse, as also reflected by the presence of the linear coherence function $C(\delta)$ in eqn (3.1).

Before eqn (3.1) will be simplified further for the specific case of a molecule in the gas phase, its derivation from first principles will be outlined in the following subsection. (The reader who is not interested in the derivation may continue directly with Section 3.2.2.)

3.2.1.1 Derivation of the Scattering Cross-section

Often, X-ray scattering is described in terms of the first Born approximation to the Lippmann–Schwinger equation.[1] Here, we will apply first-order time-dependent perturbation theory instead, which is equivalent to the first Born approximation but allows a more straightforward and perhaps more intuitive description of time-resolved scattering by a target in a non-stationary state. In this framework, the differential scattering cross-section is given by the overlap of the first-order correction to the target's wave function,[2]

$$\frac{d\sigma}{d\Omega} = \int_0^\infty w(\omega_s) \rho(\omega_s) \lim_{t \to \infty} \langle \Psi^{(1)}(t) | \Psi^{(1)}(t) \rangle d\omega_s. \tag{3.4}$$

[‡] The norm of the wave vector, $k = |\boldsymbol{k}|$, relates to the angular frequency of the photon, ω, *via* $k = \omega/c$, where c is the speed of light.

The limit of the overlap matrix element as $t \to \infty$ ensures that the interaction with the entire X-ray pulse is taken into account. Similarly, the integration over the angular frequency ω_s accounts for the detection of all scattered photons for which $w(\omega_s) > 0$. At each value of ω_s, the respective contribution to the scattering signal is weighted by the density of states of the scattered photons, given by

$$\rho(\omega_s) = \frac{\omega_s^2 V}{8\pi^3 c^3},\tag{3.5}$$

where c is the speed of light and V the quantization volume. V is an artificial quantity used to quantize and normalize the states of the photons and will cancel out later. Finally, eqn (3.4) contains the first-order time-dependent wave function that couples the photon states and the target,[1,6,7]

$$\left|\Psi^{(1)}(t)\right\rangle = -\frac{\imath}{\hbar}\int_{-\infty}^{t} \hat{U}(t,t')\hat{O}(t')\left|\Psi(t')\right\rangle \mathrm{d}t'.\tag{3.6}$$

Here, $|\Psi(t')\rangle$ is the field-free non-stationary state of the target that we already encountered in the matrix element in eqn (3.2). This state is perturbed at time t' by the operator $\hat{O}(t')$ which will be defined below. The time-evolution operator $\hat{U}(t,t') = \exp[-\imath \hat{H}_0(t - t')/\hbar]$ then propagates the perturbed state from time t' to t. In non-resonant X-ray scattering, the perturbation operator $\hat{O}(t')$ couples the incident and the scattered X-ray photons in states $|\psi_{uk_0}\rangle$ and $|\psi_{vk_s}\rangle$, respectively, *via* the interaction Hamiltonian $\hat{H}_{\mathrm{int}}(t)$ mediated by the target, and is given as

$$\hat{O}(t) = \left\langle \psi_{vk_s}\left|\hat{H}_{\mathrm{int}}(t)\right|\psi_{uk_0}\right\rangle.\tag{3.7}$$

The indices u and v in eqn (3.7) refer to the polarization of the incident and scattered photons. Within minimal coupling and Coulomb gauge, the interaction Hamiltonian is[6,8,9]

$$\hat{H}_{\mathrm{int}}(t) = \sum_{\alpha}\left(\frac{q_\alpha^2}{2m_\alpha}\hat{A}^2(\boldsymbol{r}_\alpha, t) + \frac{q_\alpha}{m_\alpha}\hat{\boldsymbol{p}}_\alpha\cdot\hat{\boldsymbol{A}}(\boldsymbol{r}_\alpha, t)\right).\tag{3.8}$$

The sum over α in eqn (3.8) runs over all constituent particles of the target. The quantities q_α, m_α, and $\boldsymbol{r}_\alpha$ denote a particle's charge, mass, and position, respectively, and $\hat{\boldsymbol{p}}_\alpha$ is its momentum operator.

$\hat{\boldsymbol{A}}(\boldsymbol{r}_\alpha, t)$ is the quantized vector potential operator of the incident X-ray field at position $\boldsymbol{r}_\alpha$ and time t. Expanding the vector potential in terms of plane waves yields

$$\hat{\boldsymbol{A}}(\boldsymbol{r}, t) = \sum_u \sum_k \sqrt{\frac{\hbar}{2\varepsilon_0 V\omega_k}}\left(\boldsymbol{\epsilon}_u\hat{a}_{uk}\mathrm{e}^{\imath(\boldsymbol{k}\cdot\boldsymbol{r} - \omega_k t)} + \boldsymbol{\epsilon}_u^*\hat{a}_{uk}^\dagger\mathrm{e}^{-\imath(\boldsymbol{k}\cdot\boldsymbol{r} - \omega_k t)}\right),\tag{3.9}$$

with ε_0 being the vacuum permittivity and $\boldsymbol{\epsilon}_u$, $\boldsymbol{k}$, and ω_k being the polarization vector, the wave vector, and the angular frequency of a photon, respectively. The operators $\hat{a}_{uk}^{\dagger}$ and $\hat{a}_{uk}$ are bosonic creation and annihiliation operators in second quantization. As their names suggest, they either create or annihilate a photon in a field mode characterized by $\boldsymbol{\epsilon}_u$ and $\boldsymbol{k}$. Making use of their commutation relation,

$$\hat{a}_{u'k'}\hat{a}_{uk}^{\dagger} = \hat{a}_{uk}^{\dagger}\hat{a}_{u'k'} + \delta_{uu'}\delta_{kk'}, \tag{3.10}$$

it is straightforward to solve the matrix element in eqn (3.7). The $\delta_{uu'}$ and $\delta_{kk'}$ are Kronecker deltas that equal unity for $u = u'$ or $k = k'$ and zero otherwise.

Considering that each elementary scattering event leads to a single X-ray photon that is scattered into a particular direction with a specific energy and polarization, the state of the scattered photon in eqn (3.7), $|\psi_{vk_s}\rangle$, can be conveniently described using a single-photon number state,

$$|\psi_{vk_s}\rangle = |vk_s\rangle = \hat{a}_{vk_s}^{\dagger}|\text{vac}\rangle, \tag{3.11}$$

with polarization and wave vectors $\boldsymbol{\epsilon}_v$ and $\boldsymbol{k}_s$, respectively. Here, $|\text{vac}\rangle$ is the vacuum state with no photons. It is excited by the operator $\hat{a}_{vk_s}^{\dagger}$ which creates a photon characterized by the field mode $|vk_s\rangle$.

For the incident X-ray photon, the state $|\psi_{uk_0}\rangle$ has to resemble a time-dependent pulse propagating through space. For that, a single field mode as in eqn (3.11) is not sufficient, since it is completely delocalized and independent of time. Instead, $|\psi_{uk_0}\rangle$ can be expressed as a singe-photon multimode wave packet,[2,8,9]

$$|\psi_{uk_0}\rangle = \sum_{k} c_{k-k_0}|uk\rangle = \sum_{k} c_{k-k_0}\hat{a}_{uk}^{\dagger}|\text{vac}\rangle, \tag{3.12}$$

where the sum runs over different norms of the wave vector, $k = |\boldsymbol{k}|$, and c_{k-k_0} are the expansion coefficients for given values of k peaked at $k = k_0$.[§] It is assumed that all field modes have the same polarization, $\boldsymbol{\epsilon}_u$, and the same direction of propagation, $\boldsymbol{k}/k$. This corresponds to the assumption of a linearly polarized pulse. Evaluating the expectation value of the light intensity operator, it can be shown that the electric field amplitude $\mathcal{E}(\boldsymbol{r}, t)$ at point $\boldsymbol{r}$ and time t for the wave packet in eqn (3.12) is[8,10]

$$\mathcal{E}(\boldsymbol{r}, t) = \sqrt{\frac{\hbar\omega_0}{2\varepsilon_0 V}}\, h(\boldsymbol{r}, t)\mathrm{e}^{-\imath\omega_0 t}. \tag{3.13}$$

[§] The single-photon multimode wave packet, eqn (3.12), involves a discrete sum over field modes rather than a continuous integral because the states are quantized in a box of finite volume V, reflected by the appearance of V in the density of states in eqn (3.5).

Here, the fraction $\sqrt{\hbar\omega_0/(2\varepsilon_0 V)}$ corresponds to the amplitude of the classical electric field, ω_0 is the carrier frequency of the pulse, and $h(r,t)$ is its envelope function. For high photon energies and a relatively small bandwidth, *i.e.*, for $\omega_k/\omega_0 \approx 1$, the envelope function is defined as

$$h(\mathbf{r}, t) = \sum_k c_{k-k_0} e^{\imath((\mathbf{k}-\mathbf{k}_0)\cdot\mathbf{r} - (\omega_k - \omega_0)t)}. \qquad (3.14)$$

Considering furthermore that even a 1 fs short pulse spreads over 300 nm, while atoms and molecules extend only over Ångströms or a few nanometres at most, the electric field envelope has almost the same value at any point in the volume of the target at a given time t. Hence, it can be assumed that, in the volume of the target, $h(r,t)$ is independent of r and only depends on t. For a transform-limited pulse, the envelope can be approximated by a Gaussian function centred at time τ with standard deviation σ,

$$h(\mathbf{r}, t) \approx h_{\mathrm{p}}(t - \tau) = N_{\mathrm{h}} e^{-\frac{(t-\tau)^2}{4\sigma^2}}, \qquad (3.15)$$

where N_{h} is a factor that ensures normalization of the envelope function. Replacing $h(r,t)$ with $h_{\mathrm{p}}(t - \tau)$ in eqn (3.13), the electric field amplitude becomes $\mathcal{E}(r, t) \approx \mathcal{E}(t - \tau)$.

Turning back to the perturbation that couples the states of the incident and the scattered photons to the state of the target, eqn (3.7), insertion of the photon states from eqn (3.11) and (3.12) yields

$$\hat{O}(t) = \sum_\alpha \sum_k c_{k-k_0} \frac{q_\alpha^2}{2m_\alpha} \left\langle vk_{\mathrm{s}} \middle| \hat{A}^2(r_\alpha, t) \middle| uk \right\rangle$$

$$= \boldsymbol{\epsilon}_u \cdot \boldsymbol{\epsilon}_v^* \frac{\mathcal{E}(t - \tau)}{\omega_0} \sqrt{\frac{2\hbar}{\varepsilon_0 V \omega_{\mathrm{s}}}} e^{\imath\omega_{\mathrm{s}} t} \sum_\alpha \frac{q_\alpha^2}{2m_\alpha} e^{\imath(k_0 - k_{\mathrm{s}})\cdot r_\alpha}. \qquad (3.16)$$

Here, only the $\hat{A}^2$ term of the interaction Hamiltonian, eqn (3.8), contributes. Matrix elements of the $\hat{p}\cdot\hat{A}$ term vanish for the initial and final states defined in eqn (3.11) and (3.12) due to an uneven number of creation and annihilation operators. The $\hat{p}\cdot\hat{A}$ term can only contribute to scattering *via* second-order in time-dependent perturbation theory. It becomes important when resonant scattering or scattering at low mean-photon energies such as Raman scattering is considered. For hard X-rays far from absorption edges, however, the $\hat{A}^2$ term in first-order time-dependent perturbation theory dominates and the second-order contribution from the $\hat{p}\cdot\hat{A}$ term is negligible.

According to the interaction Hamiltonian in eqn (3.8), the sum over α runs over all constituent particles of the target. Each particle's contribution to the scattering amplitude *via* eqn (3.16) is weighted by the ratio of its squared charge q_α^2 and its mass m_α. Since the mass of an atomic nucleus is at least three orders of magnitude larger than the mass of an electron,[¶] nuclei scatter X-rays significantly weaker than electrons. Scattering from the nuclei of carbon-12, oxygen-16, and silicon-28, for example, is eight orders of magnitude weaker than scattering from the atoms' electrons. For a target that is composed of both electrons and nuclei, X-ray scattering by the nuclei is thus negligible and the sum over all charged particles can be reduced to a sum over N_e electrons. Hence, the perturbation operator becomes

$$\hat{O}(t) = \boldsymbol{\epsilon}_u \cdot \boldsymbol{\epsilon}_v^* \frac{\mathcal{E}(t-\tau)}{\omega_0} \frac{e^2}{m_e} \sqrt{\frac{\hbar}{2\varepsilon_0 V \omega_s}} e^{\iota \omega_s t} \hat{L}(\boldsymbol{q}), \tag{3.17}$$

where the momentum transfer vector $\boldsymbol{q}$ and the one-electron scattering operator $\hat{L}(\boldsymbol{q})$ that appeared in eqn (3.3) are introduced.

Insertion of eqn (3.17) into the first-order wave function in eqn (3.6) and subsequent insertion of $|\Psi^{(1)}(t)\rangle$ into eqn (3.4) now yields the differential scattering cross-section as

$$\begin{aligned}
\frac{d\sigma}{d\Omega} = {} & \frac{1}{2\pi} \left(\frac{d\sigma}{d\Omega}\right)_{\text{Th}} \int_0^\infty w(\omega_s) \frac{\omega_s}{\omega_0} \int\int_{-\infty}^{+\infty} e^{\iota(\omega_0 - \omega_s)(t'' - t')} \\
& \times \frac{c}{V} h_{\text{p}}(t' - \tau) h_{\text{p}}(t'' - \tau) \\
& \times \left\langle \Psi(t'') \left| \hat{L}^\dagger(\boldsymbol{q}) e^{-\iota \hat{H}_0 (t'' - t')/\hbar} \hat{L}(\boldsymbol{q}) \right| \Psi(t') \right\rangle dt' dt'' d\omega_s,
\end{aligned} \tag{3.18}$$

where the differential Thomson scattering cross-section, $(d\sigma/d\Omega)_{\text{Th}}$, which describes the non-relativistic scattering by a free electron, is given by[6,8,12,13]

$$\left(\frac{d\sigma}{d\Omega}\right)_{\text{Th}} = r_e^2 P = \frac{e^4 P}{16\pi^2 \varepsilon_0^2 m_e^2 c^4}. \tag{3.19}$$

Here, r_e is the classical electron radius and $P = |\boldsymbol{\epsilon}_u \cdot \boldsymbol{\epsilon}_v^*|^2$ is the polarization factor (*i.e.*, the absolute squared dot product of the polarization vectors of the incident and scattered photons, $\boldsymbol{\epsilon}_u$ and $\boldsymbol{\epsilon}_v^*$, respectively). For the linearly polarized light available at free-electron laser and synchrotron sources, the polarization factor is

$$P = \sin^2 \phi_u + \cos^2 \phi_u \cos^2 \theta_s, \tag{3.20}$$

where θ_s is the scattering angle and ϕ_u the azimuthal angle relative to $\boldsymbol{\epsilon}_u$ within the plane orthogonal to $\boldsymbol{k}_0$.[12,14] If the incident beam is perpendicular to the detector, ϕ_u equals the azimuthal angle on the detector.

[¶] The proton–electron mass ratio is approximately 1836.[11]

Eqn (3.18) can be further simplified through substitution of the two time variables t' and t'' with $t - \delta/2$ and $t + \delta/2$, respectively. This leads to

$$
\frac{d\sigma}{d\Omega} = \frac{1}{2\pi}\left(\frac{d\sigma}{d\Omega}\right)_{\mathrm{Th}} \int_0^\infty w(\omega_s)\frac{\omega_s}{\omega_0}\int\int_{-\infty}^{+\infty} e^{\iota(\omega_0 - \omega_s)\delta}
$$

$$
\times \frac{c}{V}h_{\mathrm{p}}(t - \delta/2 - \tau)h_{\mathrm{p}}(t + \delta/2 - \tau) \tag{3.21}
$$

$$
\times \left\langle \Psi(t + \delta/2)\left|\hat{L}^\dagger(\boldsymbol{q})e^{-\iota\hat{H}_0\delta/\hbar}\hat{L}(\boldsymbol{q})\right|\Psi(t - \delta/2)\right\rangle
$$

$$
\times \, dt\,d\delta\,d\omega_{\mathrm{s}}.
$$

Making use of the fact that the envelope function $h_{\mathrm{p}}(t \pm \delta/2 - \tau)$ is approximated by a Gaussian function, eqn (3.15), the two time variables t and δ are separable and the product of the two envelope functions in eqn (3.21) becomes

$$
h_{\mathrm{p}}(t - \delta/2 - \tau)h_{\mathrm{p}}(t + \delta/2 - \tau) = |N_{\mathrm{h}}|^2\, e^{-\frac{(t-\tau)^2}{2\sigma^2}}e^{-\frac{\delta^2}{8\sigma^2}}. \tag{3.22}
$$

With that, the photon intensity,

$$
I(t - \tau) = \frac{c}{V}\left|h_{\mathrm{p}}(t - \tau)\right|^2 = \frac{c}{V}|N_{\mathrm{h}}|^2\, e^{-\frac{(t-\tau)^2}{2\sigma^2}}, \tag{3.23}
$$

and the linear coherence function,

$$
C(\delta) = \sqrt{h_{\mathrm{p}}(\delta)} = e^{-\frac{\delta^2}{8\sigma^2}}, \tag{3.24}
$$

of the incident Gaussian-shaped X-ray pulse can be identified in the second line of eqn (3.21) and the expression for the differential X-ray scattering cross-section in eqn (3.1) is finally obtained.

3.2.2   Scattering by Gas-phase Molecules

After the expression for the differential X-ray scattering cross-section for a general target composed of electrons and atomic nuclei, eqn (3.1), has been derived, it can be further simplified specifically for scattering by a molecule in the gas phase.[5,15,16] To do so, the time-dependent state $|\Psi(t)\rangle$ of the molecule can be expressed using the Born–Huang expansion,

$$
|\Psi(t)\rangle = \sum_{k=1}^N |\varphi_k(\bar{\boldsymbol{R}})\rangle|\chi_k(t)\rangle, \tag{3.25}
$$

where $|\varphi_k(\bar{\boldsymbol{R}})\rangle$ is an adiabatic eigenstate of the electronic Hamiltonian that parametrically depends on the set of nuclear coordinates $\bar{\boldsymbol{R}}$. The sum over k

includes all N electronic eigenstates that have non-zero population at any time during the dynamics induced by the pump pulse. $|\chi_k(t)\rangle$ is the nuclear wave packet of the respective electronic state and includes the phase factor of the electronic state. The nuclear wave packets can be expanded as a time-dependent superposition of rovibrational states $|\chi_{\bar{k}}\rangle$,

$$|\chi_k(t)\rangle = \sum_{\nu_k} \sum_{J_k} a_{\bar{k}}(t)|\chi_{\bar{k}}\rangle, \tag{3.26}$$

with $\bar{k}$ defining the combined set of electronic, vibrational, and rotational quantum numbers, $\bar{k} = \{k, \nu_k, J_k\}$.

Substitution of eqn (3.25) and (3.26) for $|\Psi(t)\rangle$ and insertion of the corresponding identity operator,

$$\hat{1} = \sum_{\bar{k}}^{\infty} |\varphi_k(\boldsymbol{R})\rangle |\chi_{\bar{k}}\rangle \langle \chi_{\bar{k}}| \langle \varphi_k(\boldsymbol{R})|, \tag{3.27}$$

after the exponential time-evolution operators in eqn (3.1) lead to,

$$\frac{\mathrm{d}\sigma}{\mathrm{d}\Omega} = \left(\frac{\mathrm{d}\sigma}{\mathrm{d}\Omega}\right)_{\mathrm{Th}} \sum_{i,j}^{N} \sum_{\nu_i, J_i}^{\infty} \sum_{\nu_j, J_j}^{\infty} \sum_{\bar{f}}^{\infty} \mathcal{N}_{\bar{f}\bar{i}j} \int_{-\infty}^{+\infty} I(t-\tau) a_{\bar{j}}^*(t) a_{\bar{i}}(t) \mathrm{d}t \tag{3.28}$$

$$\times \langle \chi_{\bar{j}}|L_{\bar{f}j}^*(\tilde{\boldsymbol{q}},\bar{\boldsymbol{R}})|\chi_{\bar{f}}\rangle \langle \chi_{\bar{f}}|L_{\bar{f}i}(\tilde{\boldsymbol{q}},\bar{\boldsymbol{R}})|\chi_{\bar{i}}\rangle.$$

where the appearance of different state indices i and j accounts for the possibility that the pump pulse may have excited the molecule into a coherent superposition of eigenstates. In the derivation of eqn (3.28), several approximations are made. First, it is assumed that the action of the time-evolution operator upon the eigenstates is adiabatic,[5,16]

$$\mathrm{e}^{-\imath\hat{H}_0\delta/\hbar}|\varphi_k(\boldsymbol{R})\rangle|\chi_{\bar{k}}\rangle \approx \mathrm{e}^{-\imath E_{\bar{k}}\delta/\hbar}|\varphi_k(\boldsymbol{R})\rangle|\chi_{\bar{k}}\rangle, \tag{3.29}$$

meaning that the nuclear kinetic energy part of the Hamiltonian $\hat{H}_0$ does not couple to other electronic states. This approximation may seem crude since non-adiabatic effects often play a crucial role in the dynamics of electronically excited molecules. However, the approximation in eqn (3.29) only concerns the propagation in time δ. The molecular wave packet $|\Psi(t)\rangle$ can still evolve non-adiabatically in time t. In fact, the propagation in δ mainly affects the spectral density of the pulse, given as

$$F\left(\omega_{\mathrm{s}} + \omega_{\bar{f}\bar{i}j} - \omega_0\right) = \frac{1}{2\pi} \int_{-\infty}^{+\infty} C(\delta) \mathrm{e}^{-\imath\left(\omega_{\mathrm{s}} + \omega_{\bar{f}\bar{i}j} - \omega_0\right)\delta} \mathrm{d}\delta, \tag{3.30}$$

where $\omega_{\mathrm{s}} + \omega_{\bar{f}\bar{i}j}$ is the angular frequency of a photon prior to being scattered and $\omega_{\bar{f}\bar{i}j} = (E_{\bar{f}} - [E_{\bar{i}} + E_{\bar{j}}]/2)/\hbar$ is the average angular frequency of the two

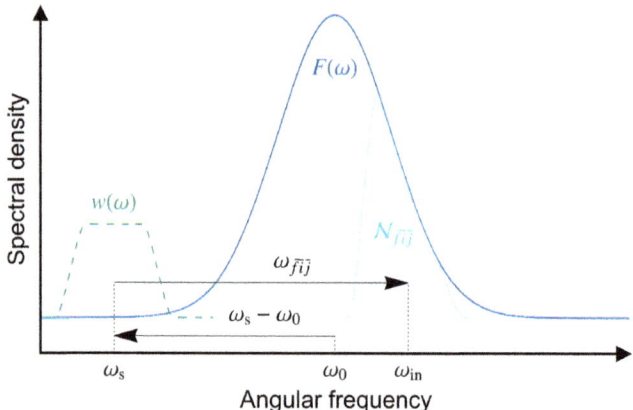

Figure 3.1   Illustration of eqn (3.30) and (3.31). Shown are the spectral density $F(\omega)$ of a transform-limited, Gaussian-shaped X-ray pulse, its mean angular frequency ω_0, the angular frequency of a scattered photon ω_s, the average angular frequency of two rovibronic transitions $\omega_{\bar{f}\bar{i}\bar{j}}$, and the angular frequency of the incident photon prior to scattering $\omega_{in} = \omega_s + \omega_{\bar{f}\bar{i}\bar{j}}$. Furthermore, an example of a window (or detector-response) function $w(\omega)$, here centred at ω_s, is shown in dashed lines. The corresponding spectral density of detectable photons that fall into the window after a shift by $\omega_{\bar{f}\bar{i}\bar{j}}$ (filled curve) is sketched. The integral of this curve, $\mathscr{N}_{\bar{f}\bar{i}\bar{j}}$, accounts for the number of detectable photons.

rovibronic transitions, $\bar{f} \leftarrow \bar{i}$ and $\bar{f} \leftarrow \bar{j}$. Each of these transitions corresponds to one of the two scattering amplitudes in eqn (3.28) (*i.e.*, $L_{\bar{f}\bar{i}}(\tilde{\boldsymbol{q}},\bar{R})$ and $L_{\bar{f}\bar{j}}^{*}(\tilde{\boldsymbol{q}},\bar{R})$). The spectral density in eqn (3.30) is the Fourier transform of the linear coherence function $C(\delta)$ from time δ into frequency space. For a transform-limited, Gaussian-shaped pulse defined in eqn (3.23) and (3.24), $F(\omega_s + \omega_{\bar{f}\bar{i}\bar{j}} - \omega_0)$ is peaked at zero where the angular frequency of an incident photon is equal to the mean angular frequency, $\omega_s + \omega_{\bar{f}\bar{i}\bar{j}} = \omega_0$. Such a spectral density and ω_s, ω_0, and $\omega_{\bar{f}\bar{i}\bar{j}}$ are illustrated in Figure 3.1.

Second, it is assumed that the angular frequencies of the scattered photons are not very different from those of the incident photons such that the fraction ω_s/ω_0 in eqn (3.1) is approximately unity and the momentum transfer vector is independent of ω_s. The latter is made explicit by the tilde on top of $\tilde{q}$ and allows the scattering amplitudes to be moved out of the integral over ω_s. This significant simplification is known as the Waller–Hartree approximation.[17] For the scattering of hard X-rays by molecules with rovibronic transition energies mostly on the order of a few eV, the Waller–Hartree approximation usually works very well.

The remaining integral over ω_s,

$$\mathscr{N}_{\bar{f}\bar{i}\bar{j}} = \int_0^{\infty} w(\omega_s) F\left(\omega_s + \omega_{\bar{f}\bar{i}\bar{j}} - \omega_0\right) d\omega_s, \qquad (3.31)$$

now only involves the spectral density from eqn (3.30) and the window function $w(\omega_s)$ that accounts for the probability of a photon with angular frequency

ω_s being measured by the detector. The factor $\mathcal{N}_{\bar{f}\bar{i}\bar{j}}$ weights the contribution from the states with rovibronic quantum numbers $\bar{f}$, $\bar{i}$, and $\bar{j}$, thereby accounting for the number of detectable photons with angular frequencies ω_s. A window function and its effect upon the spectral density are illustrated in Figure 3.1 as well. Due to the weighting by $\mathcal{N}_{\bar{f}\bar{i}\bar{j}}$, the differential scattering cross-section in eqn (3.28) can be used to describe X-ray scattering experiments with high energy resolution that may partially distinguish between different rovibrational transitions induced by inelastic scattering.

Moreover, eqn (3.28) involves X-ray scattering amplitudes, also referred to as form factors or one-electron scattering matrix elements,

$$L_{fi}(\tilde{q}, \bar{R}) = \langle \varphi_f(\bar{R})|\hat{L}(\tilde{q})|\varphi_i(\bar{R})\rangle = \langle \varphi_i(\bar{R})|\hat{L}^{\dagger}(\tilde{q})|\varphi_f(\bar{R})\rangle^*, \tag{3.32}$$

which are expectation values of the one-electron scattering operator, $\hat{L}(\tilde{q})$, introduced in eqn (3.3), and of the electronic states $|\varphi_f(\bar{R})\rangle$ and $|\varphi_i(\bar{R})\rangle$. The scattering matrix elements can be written as the Fourier transform of the one-electron (transition) density,

$$\rho_{fi}(r, \bar{R}) = \langle \varphi_f(\bar{R})|\hat{\rho}(r)|\varphi_i(\bar{R})\rangle, \tag{3.33}$$

from real-space into reciprocal-space,

$$L_{fi}(\tilde{q}, \bar{R}) = \int_{-\infty}^{+\infty} e^{i\tilde{q}\cdot r} \rho_{fi}(r, \bar{R}) dr. \tag{3.34}$$

Here,

$$\hat{\rho}(r) = \sum_{n}^{N_e} \delta(r - r_n)$$

is the one-electron density operator where $\delta(r - r_n)$ is the Dirac delta function that sifts out the single electronic coordinate r.[18]

For atoms in their electronic ground states, the rotationally-averaged elastic elements $L_{ii}(\tilde{q})$ are the well-known atomic form factors for X-ray scattering, $f_A^x(\tilde{q})$.[13] Here, $\tilde{q}$ is the norm of the momentum transfer vector $\tilde{q}$. These form factors are used within the independent atom model, which is discussed in Section 3.4.2. The evaluation of one-electron scattering matrix elements is briefly discussed in Section 3.4.

3.2.2.1 Intermediate Energy Resolution

Making use of the window function $w(\omega_s)$ in eqn (3.31), different regimes of detector response (or energy-resolution) can be distinguished.[5,19] An evaluation of the differential scattering cross-section *via* the infinite sum of rovibrational eigenstates, $|\chi_{\bar{f}}\rangle$, is necessary only if the detector can discriminate between photons with energies that differ on the order of ro-vibrational transitions. Presently, experiments do not provide such a high energy resolution. In the following, we will discuss the case of an intermediate energy resolution where the detector is largely insensitive to rovibrational but still sensitive to electronic transition energies. The discussion of this case is not only physically instructive but also lays the foundation for experiments that may provide additional information by probing scattering matrix elements in eqn (3.32) more selectively. To account for the intermediate energy resolution, the window function $w(\omega_s)$ can be chosen such that the weights $\mathcal{N}_{\bar{f}ij}$ in eqn (3.31) are the same for all rotational and vibrational quantum numbers and only depend on the electronic states, $\mathcal{N}_{\bar{f}ij} \approx \mathcal{N}_{fij}$. Under this condition, the identity in the subspace of nuclear coordinates,

$$\hat{\mathbf{1}}_{\bar{R}} = \sum_{\nu_k, J_k}^{\infty} |\chi_{\bar{k}}\rangle \langle \chi_{\bar{k}}|, \tag{3.35}$$

can be applied to remove the sum over rovibrational states $|\chi_{\bar{f}}\rangle$ in eqn (3.28). The differential scattering cross-section then simplifies to

$$\frac{d\sigma}{d\Omega} = \left(\frac{d\sigma}{d\Omega}\right)_{\text{Th}} \sum_{i,j}^{N} \sum_{f}^{\infty} \mathcal{N}_{fij} \int_{-\infty}^{+\infty} I(t-\tau)\langle \chi_j(t)|L_{fj}^*(\tilde{\boldsymbol{q}},\bar{\boldsymbol{R}})L_{fi}(\tilde{\boldsymbol{q}},\bar{\boldsymbol{R}})|\chi_i(t)\rangle dt, \tag{3.36}$$

where eqn (3.26) has been used to reintroduce the nuclear wave packets. The differential scattering cross-section in eqn (3.36) now depends solely on the electronic quantum numbers i, j, and f and not on rovibrational quantum numbers. The two electronic states with indices i and j are initially populated and the state with index f is a final state that can be excited by an inelastically scattered X-ray photon. Consequently, three physically distinct components of the scattering signal[5,20,21] can be identified:

(1) **Elastic X-ray scattering**: when no electronic transition takes place, no energy is transferred between the photon and the molecule, and the scattering is electronically elastic. This is the case for all terms with $i=j=f$ in eqn (3.36). The elastic X-ray scattering signal is given by

$$\frac{d\sigma_{\text{el}}}{d\Omega} = \left(\frac{d\sigma}{d\Omega}\right)_{\text{Th}} \mathcal{N} \sum_{i}^{N} \int_{-\infty}^{+\infty} I(t-\tau)\langle \chi_i(t)||L_{ii}(\tilde{\boldsymbol{q}},\bar{\boldsymbol{R}})|^2|\chi_i(t)\rangle dt. \tag{3.37}$$

Here, $\mathcal{N}$ no longer depends on the electronic states.

(2) **Inelastic X-ray scattering**: when an electronic transition, $f \leftarrow i$, takes place, energy is transferred between the photon and the molecule, and the scattering is electronically inelastic. This is the case for all terms

with $i = j \neq f$ in eqn (3.36). The inelastic X-ray scattering signal is given by

$$\frac{\mathrm{d}\sigma_{in}}{\mathrm{d}\Omega} = \left(\frac{\mathrm{d}\sigma}{\mathrm{d}\Omega}\right)_{Th} \sum_i^N \sum_{f \neq i}^{\infty} \mathscr{N}_{fi} \int_{-\infty}^{+\infty} I(t - \tau)\langle\chi_i(t)||L_{fi}(\tilde{\boldsymbol{q}}, \bar{\boldsymbol{R}})|^2|\chi_i(t)\rangle\mathrm{d}t. \quad (3.38)$$

Since a photon cannot couple two states with different spin multiplicities, transitions that imply a change of spin multiplicity are forbidden. The scattering operator can furthermore only lead to single excitations or single ionisations of the initial state.[72] Other than that, no selection rules exist for inelastic X-ray scattering.

(3) **Coherent mixed X-ray scattering**: when scattering amplitudes of two or more electronic states interfere coherently, the scattering is coherent mixed.[4,5,16,19,21–27] This is the case for all terms with $i \neq j$ in eqn (3.36). The coherent mixed X-ray scattering signal is given by

$$\frac{\mathrm{d}\sigma_{cm}}{\mathrm{d}\Omega} = 2\left(\frac{\mathrm{d}\sigma}{\mathrm{d}\Omega}\right)_{Th} \sum_i^{N-1} \sum_{j>i}^{N} \sum_f^{\infty} \mathscr{N}_{fij} \int_{-\infty}^{+\infty} I(t - \tau)$$

$$\times \mathrm{Re}\left[\langle\chi_j(t)|L_{fj}^*(\tilde{\boldsymbol{q}}, \bar{\boldsymbol{R}})L_{fi}(\tilde{\boldsymbol{q}}, \bar{\boldsymbol{R}})|\chi_i(t)\rangle\right]\mathrm{d}t. \quad (3.39)$$

This component may originate, for instance, from a molecule being coherently excited by the pump pulse or by passing through a conical intersection.[71] It shows a rapid beating on the timescale of the electronic motion and therefore requires femtosecond or even sub-femtosecond temporal resolution to be measurable.

The three components (see Figure 3.2 for an illustration) add up incoherently,

$$\frac{\mathrm{d}\sigma}{\mathrm{d}\Omega} = \frac{\mathrm{d}\sigma_{el}}{\mathrm{d}\Omega} + \frac{\mathrm{d}\sigma_{in}}{\mathrm{d}\Omega} + \frac{\mathrm{d}\sigma_{cm}}{\mathrm{d}\Omega}, \quad (3.40)$$

and are always measured together.

3.2.2.2 No Energy Resolution

We now turn to scattering in the limit of no energy resolution, which is currently the most common scenario. In most time-resolved X-ray scattering experiments, the detector is also insensitive to energy differences on the order of electronic transitions. Hence, the weights in eqn (3.31) are effectively the same for all rotational, vibrational, and electronic quantum numbers, $\mathscr{N}_{\overline{fij}} \approx \mathscr{N}$. Then, the identity in the subspace of electronic coordinates,

$$\hat{1}_r = \sum_k^{\infty} |\varphi_k(\bar{\boldsymbol{R}})\rangle\langle\varphi_k(\bar{\boldsymbol{R}})|, \quad (3.41)$$

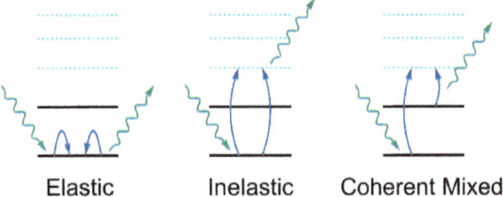

Elastic   Inelastic   Coherent Mixed

Figure 3.2   Illustration of the three components of the X-ray scattering signal in eqn (3.37)–(3.39). The horizontal lines represent electronic states of a molecule and the curly arrows represent incident and scattered photons (green). The curved arrows depict interfering scattering amplitudes (blue) and their corresponding electronic transitions. Elastic scattering, $d\sigma_{el}/d\Omega$, leaves the electronic state of the molecule unchanged while inelastic scattering, $d\sigma_{in}/d\Omega$, involves excitation (or deexcitation). In contrast to these two components, coherent mixed scattering, $d\sigma_{cm}/d\Omega$, implies an interference of scattering amplitudes of two coherently populated initial states. While an inelastically scattered photon loses (or gains) the energy of the corresponding electronic transition, $\Delta E = E_f - E_i$, the energy of a photon that undergoes coherent mixed scattering is shifted by the average transition energy, $\Delta E = E_f - [E_i + E_j]/2$. Here, E_f, E_i, and E_j are the energies of the final and initial states, respectively. In experiments, coherent mixed scattering usually vanishes due to insufficient temporal resolution or symmetry.

can be applied to remove the sum over the electronic eigenstates $|\varphi_f(\bar{R})\rangle$ as well and the differential scattering cross-section in eqn (3.36) further simplifies to

$$\frac{d\sigma}{d\Omega} = \left(\frac{d\sigma}{d\Omega}\right)_{\text{Th}} \sum_{i,j}^{N} \int_{-\infty}^{+\infty} \mathcal{I}(t-\tau)\langle\chi_j(t)|\Lambda_{ji}(\tilde{\boldsymbol{q}},\bar{\boldsymbol{R}})|\chi_i(t)\rangle dt, \qquad (3.42)$$

where $\mathcal{I}(t-\tau) = \mathcal{N}I(t-\tau)$ is the detectable fraction of the photon intensity and $\Lambda_{ji}(\tilde{\boldsymbol{q}},\bar{\boldsymbol{R}})$ is the total X-ray scattering probability of the molecule with nuclear geometry $\bar{\boldsymbol{R}}$ in electronic states $|\varphi_i(\bar{\boldsymbol{R}})\rangle$ and $|\varphi_j(\bar{\boldsymbol{R}})\rangle$, given by

$$\Lambda_{ji}(\tilde{\boldsymbol{q}}, \bar{\boldsymbol{R}}) = \langle\varphi_j(\bar{\boldsymbol{R}})|\hat{L}^\dagger(\tilde{\boldsymbol{q}})\hat{L}(\tilde{\boldsymbol{q}})|\varphi_i(\bar{\boldsymbol{R}})\rangle$$

$$= \sum_{m,n}^{N_e} \langle\varphi_j(\bar{\boldsymbol{R}})|e^{i\tilde{\boldsymbol{q}}\cdot(\boldsymbol{r}_n - \boldsymbol{r}_m)}|\varphi_i(\bar{\boldsymbol{R}})\rangle. \qquad (3.43)$$

The matrix element in eqn (3.43) contains the product of two one-electron scattering operators, $\hat{L}^\dagger(\tilde{\boldsymbol{q}})\hat{L}(\tilde{\boldsymbol{q}})$. The scattering probability $\Lambda_{ji}(\tilde{\boldsymbol{q}},\bar{\boldsymbol{R}})$ can be split into its one- and two-electron parts,

$$\Lambda_{ji}(\tilde{\boldsymbol{q}},\bar{\boldsymbol{R}}) = N_e\delta_{ij} + \Lambda_{ji}^{2e}(\tilde{\boldsymbol{q}},\bar{\boldsymbol{R}}). \qquad (3.44)$$

The first term on the right-hand side, $N_e\delta_{ij}$, is the one-electron part that can be derived setting $m = n$ in the second line of eqn (3.43). Then, the coordinates r_m and r_n refer to the same electron and the matrix element reduces to N_e times the overlap integral $\langle \varphi_j(\bar{R})|\varphi_i(\bar{R})\rangle$. For orthonormal states, this overlap integral is equal to Kronecker delta δ_{ij}, which is unity when $i=j$ and zero when $i \neq j$. Hence, the one-electron part of the total scattering probability is just the number of electrons N_e and only contributes if both electronic states are the same. The second term on the right-hand side, $\Lambda_{ji}^{2e}(\bar{q},\bar{R})$, is the two-electron scattering matrix element that follows from $m \neq n$ in eqn (3.43). It can be written as the two-fold Fourier transform of the two-electron (transition) density,

$$\rho_{ji}(r_1, r_2, \bar{R}) = \langle \varphi_j(\bar{R})|\hat{\rho}(r_1, r_2)|\varphi_i(\bar{R})\rangle, \tag{3.45}$$

from real into reciprocal space,

$$\Lambda_{ji}^{2e}(\bar{q},\bar{R}) = 2 \int \int_{-\infty}^{+\infty} e^{i\bar{q}\cdot(r_2 - r_1)} \rho_{ji}(r_1, r_2, \bar{R})dr_1 dr_2. \tag{3.46}$$

Here,

$$\hat{\rho}(r_1, r_2) = \frac{1}{2} \sum_m^{N_e} \sum_{n \neq m}^{N_e} \delta(r_1 - r_m)\delta(r_2 - r_n)$$

is the two-electron density operator that sifts out the two coordinates r_1 and r_2 *via* the Dirac delta functions $\delta(r_1 - r_m)$ and $\delta(r_2 - r_n)$.[18]

For an individual atom, the diagonal element $\Lambda_{ii}(\bar{q})$ of the electronic ground state $|\varphi_i\rangle$ equals the sum of its absolute squared atomic form factor and its incoherent scattering function, $\Lambda_{ii}(\bar{q}) = |f_A^x(\bar{q})|^2 + S_A(\bar{q})$.[13] The evaluation of two-electron scattering matrix elements for both atoms and molecules based on electronic structure theory is discussed in Section 3.4.1.

Analogous to eqn (3.36) before, three physically distinct components can be identified in the differential scattering cross-section in eqn (3.42):

(1) **Background X-ray scattering**: the one-electron part of eqn (3.44) with $i=j$ leads to a structureless, time-independent background,

$$\frac{d\sigma_{bg}}{d\Omega} = \left(\frac{d\sigma}{d\Omega}\right)_{Th} \mathcal{I}_\tau N_e, \tag{3.47}$$

that is proportional to the total number of detectable X-ray photons, $\mathcal{I}_\tau = \int \mathcal{I}(t-\tau)dt$, and to the number of electrons, N_e.

(2) **Excess X-ray scattering**: the two-electron part of eqn (3.44) with $i=j$ leads to a time- and structure-dependent component that Waller and Hartee have termed "excess scattering",[17]

$$\frac{d\sigma_{ex}}{d\Omega} = \left(\frac{d\sigma}{d\Omega}\right)_{Th} \sum_i^N \int_{-\infty}^{+\infty} \mathcal{I}(t-\tau)\langle \chi_i(t)|\Lambda_{ii}^{2e}(\bar{q},\bar{R})|\chi_i(t)\rangle dt. \tag{3.48}$$

(3) **Coherent mixed X-ray scattering**: the two-electron part of eqn (3.42) with $i \neq j$ leads to coherent mixed scattering,

$$\frac{d\sigma_{cm}}{d\Omega} = 2\left(\frac{d\sigma}{d\Omega}\right)_{Th} \sum_{i}^{N-1} \sum_{j>i}^{N} \int_{-\infty}^{+\infty} \mathcal{I}(t-\tau)$$

$$\times \mathrm{Re}\left[\langle\chi_j(t)|\Lambda_{ji}^{2e}(\tilde{\boldsymbol{q}},\bar{\boldsymbol{R}})|\chi_i(t)\rangle\right]dt, \tag{3.49}$$

as already discussed around eqn (3.39). Since the product of one-electron scattering operators, $\hat{L}^\dagger(\tilde{\boldsymbol{q}})\hat{L}(\tilde{\boldsymbol{q}})$, in eqn (3.43) is symmetric under inversion of the electronic coordinates, this component vanishes in the limit of no energy resolution for two electronic states with different inversion symmetries.[3,28] The detection of coherent mixed scattering between states that transform differently under inversion therefore requires not only femtosecond or sub-femtosecond pulses but also detectors with at least intermediate energy resolution.

Again, the three components add up incoherently,

$$\frac{d\sigma}{d\Omega} = \frac{d\sigma_{bg}}{d\Omega} + \frac{d\sigma_{ex}}{d\Omega} + \frac{d\sigma_{cm}}{d\Omega}, \tag{3.50}$$

and are always measured together. The sum of the background and excess scattering is often called total scattering. It is equal to the sum of the elastic and inelastic components when all scattered photons are detected irrespective of their energies. Eqn (3.47) and (3.48) thus offer a convenient way to circumvent the infinite sum over final states $|\varphi_f(\bar{\boldsymbol{R}})\rangle$ in eqn (3.38) that would otherwise render a converged calculation of the full inelastic component impossible.[72] Instead, it can be expressed simply as the difference between the total and elastic scattering from eqn (3.37),

$$\frac{d\sigma_{in}}{d\Omega} = \frac{d\sigma_{bg}}{d\Omega} + \frac{d\sigma_{ex}}{d\Omega} - \frac{d\sigma_{el}}{d\Omega}. \tag{3.51}$$

The incoherent scattering functions of atoms in their electronic ground states, $S(\tilde{q})$, (see Section 3.4.2) can be calculated accordingly,

$$S(\tilde{q}) = N_e + \Lambda_{ii}^{2e}(\tilde{q}) - |L_{ii}(\tilde{q})|^2, \tag{3.52}$$

where index i refers to the electronic ground state of the atom and the elements $\Lambda_{ii}^{2e}(\tilde{q})$ and $L_{ii}(\tilde{q})$ are rotationally averaged.

We note again that, in the limit of no energy resolution and within the Waller–Hartree approximation, the form of the linear coherence function, $C(\delta)$, and that of the spectral density, $F(\omega_s + \omega_{\tilde{f}ij} - \omega_0)$, defined in eqn (3.30), do not matter. The scattering signal is not affected by $C(\delta)$ and

$F(\omega_s + \omega_{f\bar{i}j} - \omega_0)$ and only depends on the detectable fraction of the photon intensity, $\mathcal{I}(t-t)$, introduced in eqn (3.42). Without energy-resolving detectors, none of the components of the ultrafast scattering signal is sensitive to the degree of coherence of the X-ray pulse.

3.3 High-energy Electron Scattering

3.3.1 Electron Scattering by a General Target

Analogous to ultrafast X-ray scattering discussed in the previous section, the observable that is measured in ultrafast electron scattering is the differential electron scattering cross-section, $d\sigma/d\Omega$, at pump–probe delay time τ. Here, $d\sigma/d\Omega$ is defined by the number of electrons scattered into a solid angle Ω. For a general target of charged particles such as an atom or a molecule, the differential scattering cross-section is given as

$$\frac{d\sigma}{d\Omega} = \frac{1}{2\pi} \int_0^\infty \left(\frac{d\sigma}{d\Omega}\right)_{\text{Ru}} w(\omega_s) \frac{\omega_s}{\omega_0} \int \int_{-\infty}^{+\infty} I(t-\tau) e^{i(\omega_0 - \omega_s)\delta}$$

$$\times C(\delta) \mathcal{Z}(\boldsymbol{s}, t, \delta) dt d\delta d\omega_s, \tag{3.53}$$

where $(d\sigma/d\Omega)_{\text{Ru}}$ is the Rutherford scattering cross-section, later defined in eqn (3.68), that describes scattering of an electron by a free and singly-charged particle. Moreover, ω_0 is the mean angular de Broglie frequency of the incident electron and ω_s is the angular de Broglie frequency of the scattered electron.[||] The window function $w(\omega_s)$ accounts for the probability that a scattered electron with ω_s is measured by the detector. $I(t-\tau)$ and $C(\delta)$ are the electron intensity and the linear coherence function of the incident electron pulse at the location of the target at times t and δ, respectively. Notably, $C(\delta)$ reflects the longitudinal coherence of the electron pulse and the generally low transverse coherence of electron pulses does not concern us here. The quantity $\mathcal{Z}(\boldsymbol{s}, t, \delta)$ is the electron scattering probability at point $\boldsymbol{s}$ in reciprocal space and at times t and δ, and is defined as

$$\mathcal{Z}(\boldsymbol{s}, t, \delta) = \langle \Psi(t) | e^{i\hat{H}_0 \delta/2\hbar} \hat{Z}^\dagger(\boldsymbol{s}) e^{-i\hat{H}_0 \delta/\hbar} \hat{Z}(\boldsymbol{s}) e^{i\hat{H}_0 \delta/2\hbar} | \Psi(t) \rangle. \tag{3.54}$$

As before, the matrix element in eqn (3.54) implies integration over all internal coordinates of the target's constituent particles, with the target described by the time-dependent state $|\Psi(t)\rangle$ and $\hat{H}_0$ being the Hamiltonian

---

[||] The angular de Broglie frequency is $\omega = 2\pi\nu/\lambda$, with ν being the velocity of the electron and $\lambda = h/p$ being the de Broglie wavelength, where h is the Planck constant and $p = \gamma m_e \nu$ is the relativistic momentum. The latter involves the electron's rest mass m_e, the Lorentz factor $\gamma = 1/\sqrt{1 - \nu^2/c^2}$, and the speed of light c.

of the unperturbed target. Finally, $\hat{Z}(s)$ is the one-particle electron scattering operator, given as

$$\hat{Z}(s) = \hat{L}(s) - \sum_{A}^{N_n} Z_A e^{is \cdot R_A}, \qquad (3.55)$$

which contains the one-electron X-ray scattering operator $\hat{L}(s)$ defined in eqn (3.3). Here, the momentum transfer vector s is defined as the vector difference between the de Broglie wave vectors of the incident and scattered electrons, $s = k_0 - k_s$.** The second term on the right-hand side of eqn (3.55) furthermore refers to scattering by atomic nuclei. The sum over A runs over all N_n nuclei of the target with atomic numbers Z_A and coordinates R_A. The bracket in eqn (3.54) can be read as a sequence of time-propagations in δ interspersed by two interactions between the electron and the target, analogous to the description of eqn (3.2) in Section 3.2.1. As before, it is important to note the difference between the times t and δ, with t being the realtime and δ allowing for scattering amplitudes at different interaction times within the pulse to interfere.

The expression for the differential electron scattering cross-section given in eqn (3.53) is thus remarkably similar to the expression for the scattering of X-rays in eqn (3.1), even though the scattered particles and the physical nature of their interaction with the target are different. Apart from the fact that the particle intensity $I(t - \tau)$ and linear coherence function $C(\delta)$ refer to an electron pulse in the former and to an X-ray pulse in the latter case, the expressions only differ in two points: (i) the electron scattering probability $\mathcal{Z}(s, t, \delta)$ also includes scattering by atomic nuclei and (ii) the electron scattering signal is weighted by the differential Rutherford scattering cross-section, $(d\sigma/d\Omega)_{Ru}$, while the X-ray scattering signal is proportional to the differential Thomson scattering cross-section, $(d\sigma/d\Omega)_{Th}$. The close similarity between eqn (3.53) and (3.1) has the convenient consequence that both electron and X-ray scattering can be expressed in terms of the same scattering matrix elements, as will become more obvious later.

Before eqn (3.53) will be simplified further for a molecule in the gas phase, its derivation from first principles will be outlined in the following subsection. (The reader who is not interested in the derivation may continue directly with Section 3.3.2.)

3.3.1.1   Derivation of the Scattering Cross-section

As in the case of ultrafast X-ray scattering earlier, first-order time-dependent perturbation theory is used to explicitly account for time-resolved scattering

---

** In ultrafast electron scattering, it is customary to use s rather than q, which is the common notation for the momentum transfer vector in X-ray scattering.

by a target in a non-stationary state. Again, the differential scattering cross-section is given by the overlap of the first-order correction to the target's wave function, eqn (3.4). The only difference with respect to X-ray scattering occurs in the ansatz for the perturbation operator $\hat{O}(t)$ that couples the incident and scattered electrons to the target: the states of the electrons as well as the interaction Hamiltonian in eqn (3.8) have to be different. While the X-ray photons are coupled to the target *via* the square of the vector potential as shown in eqn (3.16), the interaction between the incident and scattered electrons and the charged particles of the target is electrostatic.[29] Hence, the interaction Hamiltonian is

$$\hat{H}_{\text{int}} = \frac{e^2}{4\pi\varepsilon_0} \left(-\sum_A^{N_n} Z_A |\boldsymbol{r} - \boldsymbol{R}_A|^{-1} + \sum_n^{N_e} |\boldsymbol{r} - \boldsymbol{r}_n|^{-1} \right), \qquad (3.56)$$

where the first term on the right-hand side refers to the Coulomb attraction of the incident electron by the scattering nuclei and the second term refers to its repulsion by the scattering electrons.

Analogous to the single-photon number state in eqn (3.11), the state of the scattered electron can be described by a single plane-wave state,

$$|\psi_{k_s}\rangle = |k_s\rangle = \frac{1}{\sqrt{V}} e^{i(k_s \cdot r - \omega_s t)}, \qquad (3.57)$$

with de Broglie wave vector $\boldsymbol{k}_s$. The factor $1/\sqrt{V}$ ensures normalization since the plane-wave state is quantized in a box of finite volume V. Note that the state $|\psi_{k_s}\rangle$ and its interaction are straightforwardly written in first quantization, meaning that no creation and annihilation operators are employed. Also, the "polarization" of the electron, its spin, is not considered, since spin does not enter the expression for the Coulomb interaction, which only depends on the particles' charges and their distances.

As in eqn (3.12), the state of the incident electron has to resemble a time-dependent pulse that propagates through space. For that, a coherent superposition of different plane-wave states, a wave packet, is chosen,

$$|\psi_{k_0}\rangle = \sum_k c_{k-k_0} |k\rangle = \frac{1}{\sqrt{V}} \sum_k c_{k-k_0} e^{i(k \cdot r - \omega_k t)}. \qquad (3.58)$$

Here, the sum runs over the norm of the de Broglie wave vector of the respective plane-wave states, $k = |\boldsymbol{k}|$,[††] and c_{k-k_0} are the expansion coefficients at given values of k peaked at $k = k_0$. It is assumed that all states have the same direction of propagation, $\boldsymbol{k}/k$.

[††] With $k = 2\pi/\lambda$, where λ is the de Broglie wavelength defined in footnote || on page 101.

With eqn (3.56)–(3.58), the perturbation operator that couples the incident and scattered electrons to the target becomes

$$
\hat{O}(t) = \sum_k c_{k-k_0} \langle \boldsymbol{k}_s | \hat{H}_{\text{int}} | \boldsymbol{k} \rangle
$$

$$
= \frac{e^2}{4\pi\varepsilon_0 V} e^{i\omega_s t} \left(-\sum_A^{N_n} Z_A \mathcal{V}_A(\boldsymbol{r}, t) + \sum_n^{N_e} \mathcal{V}_n(\boldsymbol{r}, t) \right), \tag{3.59}
$$

where $\mathcal{V}_\alpha(\boldsymbol{r}, t)$ contains the Fourier-transformed Coulomb potential of a scattering particle with index α, given by

$$
\mathcal{V}_\alpha(\boldsymbol{r}, t) = \sum_k c_{k-k_0} e^{-i\omega_k t} \int e^{i(\boldsymbol{k}-\boldsymbol{k}_s)\cdot\boldsymbol{r}} |\boldsymbol{r} - \boldsymbol{r}_\alpha|^{-1} d\boldsymbol{r}
$$

$$
= \sum_k c_{k-k_0} e^{-i\omega_k t} \frac{4\pi}{|\boldsymbol{k}-\boldsymbol{k}_s|^2} e^{i(\boldsymbol{k}-\boldsymbol{k}_s)\cdot\boldsymbol{r}_\alpha} \tag{3.60}
$$

$$
\approx \frac{4\pi}{s^2} e^{i\boldsymbol{s}\cdot\boldsymbol{r}_\alpha} h(\boldsymbol{r}_\alpha, t) e^{-i\omega_0 t}.
$$

Here, the integral over $\boldsymbol{r}$ is solved according to Bethe[30] and the momentum transfer vector $\boldsymbol{s}$ and its norm s are introduced. Moving from the second to the third line of eqn (3.60), it is assumed that the distribution of k around k_0 is narrow such that $|\boldsymbol{k}-\boldsymbol{k}_s| \approx |\boldsymbol{k}_0 - \boldsymbol{k}_s| = s$ is a valid approximation. It holds as long as the bandwidth of the pulse is small in comparison to its mean-electron energy. Moreover, $h(\boldsymbol{r}_\alpha, t)$ is the envelope function of the electron pulse at position $\boldsymbol{r}_\alpha$ and time t, defined as

$$
h(\boldsymbol{r}, t) = \sum_k c_{k-k_0} e^{i((\boldsymbol{k}-\boldsymbol{k}_0)\cdot\boldsymbol{r} - (\omega_k - \omega_0)t)}, \tag{3.61}
$$

and is analogous to the envelope function of the X-ray pulse in eqn (3.14).

Considering that even a 10 fs short electron pulse with a mean kinetic energy of only 10 keV spreads over roughly 580 nm, while atoms and molecules extend over Ångströms or a few nanometres at most, the envelope function has almost the same value at any point in the volume of the target at a given time t. This applies even more to MeV pulses that spread over micrometres in 10 fs. Hence, it can be assumed that, in the volume of the target, $h(\boldsymbol{r},t)$ is independent of $\boldsymbol{r}$ and only depends on t. Usually, it can be approximated by a Gaussian function $h_p(t-\tau)$ centred at time τ as defined in eqn (3.15). Thus, the perturbation operator in eqn (3.59) becomes

$$
\hat{O}(t) = \frac{e^2}{\varepsilon_0 V s^2} h_p(t - \tau) e^{-i(\omega_0 - \omega_s)t} \hat{Z}(\boldsymbol{s}), \tag{3.62}
$$

where $\hat{Z}(s)$ is the one-particle electron scattering operator from eqn (3.55). Insertion of the perturbation operator into the first-order wave function in eqn (3.6) and subsequent insertion of $|\Psi^{(1)}(t)\rangle$ and of the density of final states of the scattered electrons,

$$\rho(\omega_s) = \frac{\omega_s V m_e}{8\pi^3 \hbar c}, \tag{3.63}$$

into eqn (3.4), now yields the differential electron scattering cross-section,

$$\frac{d\sigma}{d\Omega} = \frac{e^4 m_e}{8\pi^3 \hbar^3 c\varepsilon_0^2 V} \int_0^\infty \frac{w(\omega_s)}{s^4} \omega_s \int\int_{-\infty}^{+\infty} e^{i(\omega_0 - \omega_s)(t'' - t')}$$

$$\times h_p(t' - \tau) h_p(t'' - \tau) \tag{3.64}$$

$$\times \langle \Psi(t'')|\hat{Z}^\dagger(s) e^{-i\hat{H}_0(t'' - t')/\hbar} \hat{Z}(s)|\Psi(t')\rangle \, dt' dt'' d\omega_s.$$

As before, eqn (3.64) can be simplified by substitution of $t - \delta/2$ and $t + \delta/2$ for the two time variables t' and t'', respectively. This leads to

$$\frac{d\sigma}{d\Omega} = \frac{e^4 m_e}{8\pi^3 \hbar^3 c\varepsilon_0^2 V} \int_0^\infty \frac{w(\omega_s)}{s^4} \omega_s \int\int_{-\infty}^{+\infty} e^{i(\omega_0 - \omega_s)\delta}$$

$$\times h_p(t - \delta/2 - \tau) h_p(t + \delta/2 - \tau) \tag{3.65}$$

$$\times \langle \Psi(t + \delta/2)|\hat{Z}^\dagger(s) e^{-i\hat{H}_0\delta/\hbar} \hat{Z}(s)|\Psi(t - \delta/2)\rangle$$

$$\times dt d\delta d\omega_s.$$

Making use of the fact that the envelope function $h_p(t \pm \delta/2 - \tau)$ approximately resembles a Gaussian function, the time variables t and δ are separable according to eqn (3.22). Hence, the electron intensity,

$$I(t - \tau) = \frac{\hbar\omega_0}{m_e c V} |h_p(t - \tau)|^2, \tag{3.66}$$

and the linear coherence function $C(\delta)$, eqn (3.24), of the incident Gaussian-shaped electron pulse can be identified in the second line of eqn (3.65). The remaining physical constants equal the Bohr radius,

$$a_0 = \frac{4\pi\varepsilon_0 \hbar^2}{e^2 m_e}, \tag{3.67}$$

which appears in the differential Rutherford scattering cross-section of a free and singly-charged particle,[‡‡][1,6,13,29]

$$\left(\frac{d\sigma}{d\Omega}\right)_{Ru} = \frac{4}{a_0^2}\frac{1}{s^4}. \tag{3.68}$$

With eqn (3.67) and (3.68), the expression for the differential electron scattering cross-section in eqn (3.53) is finally obtained.

3.3.2 Electron Scattering by Gas-phase Molecules

After the differential electron scattering cross-section for a general target made up from electrons and atomic nuclei, eqn (3.53), has been derived, the expression can be simplified specifically for scattering by a molecule in the gas phase. As in the case of ultrafast X-ray scattering before, such a simplification is achievable by substitution of the Born–Huang expansion, eqn (3.25), for the time-dependent state $|\Psi(t)\rangle$ and insertion of the identity, eqn (3.27), behind each of the exponential time-evolution operators in eqn (3.54). The effect of these operators upon the eigenstates is approximated using eqn (3.29) and the Waller–Hartree approximation[17] is applied again to set $\omega_s/\omega_0 \approx 1$ and $s \approx \tilde{s}$, meaning that the momentum transfer vector no longer depends on ω_s. With eqn (3.26), (3.30), and (3.31), the differential electron scattering cross-section thus becomes

$$\frac{d\sigma}{d\Omega} = \left(\frac{d\sigma}{d\Omega}\right)_{Ru} \sum_{i,j}^{N} \sum_{\nu_i,J_i}^{\infty} \sum_{\nu_j,J_j}^{\infty} \sum_{\tilde{f}}^{\infty} \mathscr{N}_{\tilde{f}ij} \int_{-\infty}^{+\infty} I(t-\tau)a_j^*(t)$$

$$\times a_i(t)dt\langle\chi_{\tilde{j}}|Z_{fj}^*(\tilde{s},\bar{R})|\chi_{\tilde{f}}\rangle\langle\chi_{\tilde{f}}|Z_{fi}(\tilde{s},\bar{R})|\chi_{\tilde{i}}\rangle. \tag{3.69}$$

This expression can be used to describe electron scattering experiments with high energy resolution that may partially distinguish between different rovibrational transitions induced by inelastic scattering. The weight $\mathscr{N}_{\tilde{f}ij}$ that appears in eqn (3.69) is defined in eqn eqn (3.31).

The differential electron scattering cross-section in eqn (3.69) contains electron scattering amplitudes, also referred to as form factors,

$$Z_{fi}(\tilde{s},\bar{R}) = L_{fi}(\tilde{s},\bar{R}) - \delta_{fi}\sum_A^{N_n} Z_A e^{i\tilde{s}\cdot R_A}, \tag{3.70}$$

[‡‡] At $s=0$, the Rutherford cross-section becomes singular. This is a consequence of the long-range character of the Coulomb potential and reflects that the asymptotic form of the incident and scattered plane-wave states, eqn (3.57) and (3.58), breaks down for electron scattering in forward direction.[31] In practice, the smallest value of momentum transfer, $s_{min}>0$, is often defined and the scattering signal is evaluated at values $s \geq s_{min}$ only. Since the scattering is not measured in forward direction in experiments anyway, this limitation usually does not constitute a problem. For scattering by neutral atoms and molecules, the screening of the nuclear Coulomb potential by the electrons furthermore ensures that the elastic differential scattering cross-section becomes finite even at $s=0$.

which contain the one-electron scattering matrix elements $L_{fi}(\tilde{\boldsymbol{s}},\bar{\boldsymbol{R}})$ introduced in eqn (3.32) as well as a term that describes scattering by nuclei. This nuclear term only contributes if the amplitude is electronically elastic (*i.e.*, if $f=i$, as follows from the Kronecker delta δ_{fi}). For all electronically inelastic amplitudes with $f \neq i$, the electron scattering and X-ray scattering amplitudes are identical, $Z_{fi}(\tilde{\boldsymbol{s}},\bar{\boldsymbol{R}}) = L_{fi}(\tilde{\boldsymbol{s}},\bar{\boldsymbol{R}})$. Intuitively, the incident electron has to interact with the electrons of the molecule, not with its nuclei, to induce an electronic transition, $f \leftarrow i$.

Apart from the additional nuclear term in eqn (3.70) and the proportionality to $(d\sigma/d\Omega)_{Ru}$, the expression for the time-resolved scattering of electrons by molecules in the gas phase, eqn (3.69), is identical to the corresponding expression for the scattering of X-rays, eqn (3.28). Both techniques thus provide, in principle, the same information about the molecule. The differential electron scattering cross-section, however, decays significantly faster with increasing $\tilde{s}$ due to the scaling by $1/\tilde{s}^4$ of the differential Rutherford cross-section, which attenuates local maxima of the scattering signal. Since the position and shape of these maxima are crucial for the determination of molecular structures, X-ray scattering tends to offer better contrast than electron scattering. The attenuation also means that X-ray scattering provides a stronger signal at large values of $\tilde{q}$ than electron scattering at large values of $\tilde{s}$.[32] Nevertheless, electron scattering has at least three potential advantages over X-ray scattering. First, the scaling by $1/\tilde{s}^4$ also implies that the scattering signal is enhanced at small values of $\tilde{s}$ and might therefore be more sensitive to changes in that range of momentum transfer, as demonstrated recently.[33] Second, the fact that electron scattering probes the position of the nuclei not just indirectly *via* electrons but directly *via* the scattering by the nuclei themselves aids the determination of atomic positions, in particular for lighter elements such as hydrogen.[34] Third, electron scattering is sensitive to the molecular structure over a wider range of momentum transfer.[32] At large values of $\tilde{q}$, X-ray scattering is predominantly inelastic, and thus mostly incoherent, while the electron scattering signal becomes increasingly elastic with increasing $\tilde{s}$ and exhibits structural interferences where X-ray scattering does not (see also Figure 3.3 below). On the condition that high-quality data can be obtained at large values of momentum transfer despite the unfavourable $1/\tilde{s}^4$ scaling, electron scattering could therefore provide more detailed information about structural dynamics. This is, again, a consequence of the scattering by the nuclei.

3.3.2.1 Intermediate Energy Resolution

Making use of the window function $w(\omega_s)$ in eqn (3.31), different regimes of detector response, or energy-resolution, can be distinguished as in X-ray scattering. We consider now scattering in the limit of intermediate energy resolution defined in Section 3.2.2.1. The window function can therefore

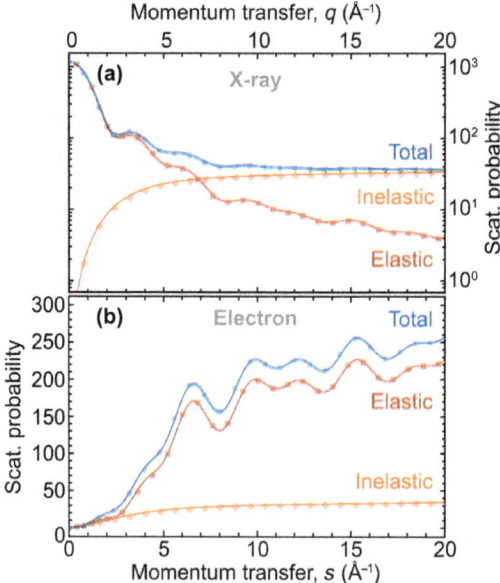

Figure 3.3 Isotropic total, elastic, and inelastic scattering probabilities of CHF_3 in its electronic ground state as a function of momentum transfer, $\tilde{q}$ or $\tilde{s}$. (a) X-Ray scattering probabilities in units of the differential Thomson scattering cross-section. (b) Electron scattering probabilities in units of the differential Rutherford scattering cross-section. (a and b) The solid lines show the probabilities calculated with the IAM, while the markers refer to accurate reference data computed[47] from the electronic wave function obtained with CASSCF(8,8)/cc-pVTZ implemented in MOL-PRO.[69,70] The underlying molecular geometry is optimized with CCSD(T)/cc-pVQZ.

be chosen such that the weights $\mathcal{N}_{\overline{fij}}$ in eqn (3.31) are the same for all rotational and vibrational quantum numbers and only depend on the electronic states, $\mathcal{N}_{\overline{fij}} \approx \mathcal{N}_{fij}$. With the identity in the subspace of nuclear coordinates, eqn (3.35), the sum over the rovibrational states $|\chi_{\bar{f}}\rangle$ in eqn (3.69) can be removed and the differential scattering cross-section simplifies to

$$\frac{d\sigma}{d\Omega} = \left(\frac{d\sigma}{d\Omega}\right)_{Ru} \sum_{i,j}^{N} \sum_{f}^{\infty} \mathcal{N}_{fij} \int_{-\infty}^{+\infty} I(t-\tau) \langle \chi_j(t) | Z_{fj}^{\star}(\tilde{s}, \bar{R}) Z_{fi}(\tilde{s}, \bar{R}) | \chi_i(t) \rangle dt, \quad (3.71)$$

where eqn (3.26) has been used to reintroduce the nuclear wave packets. The differential scattering cross-section in eqn (3.71) depends solely on the electronic quantum numbers i, j, and f and not on rovibrational quantum numbers now. The two electronic states with indices i and j are initially populated, the state with index f is a final state that can be excited by an inelastically scattered electron. Consequently, three physically distinct components of the scattering signal can be identified in eqn (3.71):

(1) **Elastic electron scattering**: when no electronic transition takes place, no energy is transferred between the electron and the molecule, and the scattering is electronically elastic. This is the case for all terms with $i=j=f$. The elastic electron scattering signal is given by

$$\frac{d\sigma_{el}}{d\Omega} = \left(\frac{d\sigma}{d\Omega}\right)_{Ru} \mathcal{N} \sum_{i}^{N} \int_{-\infty}^{+\infty} I(t-\tau)\langle\chi_i(t)||Z_{ii}(\tilde{\mathbf{s}},\bar{\mathbf{R}})|^2|\chi_i(t)\rangle\,dt. \tag{3.72}$$

Here, the elastic electron scattering probability,

$$|Z_{ii}(\tilde{\mathbf{s}},\bar{\mathbf{R}})|^2 = |L_{ii}(\tilde{\mathbf{s}},\bar{\mathbf{R}})|^2 + \sum_{A}^{N_n} Z_A^2$$

$$+2\sum_{A}^{N_n-1}\sum_{B>A}^{N_n} Z_A Z_B \cos[\tilde{\mathbf{s}}\cdot(\mathbf{R}_B - \mathbf{R}_A)] \tag{3.73}$$

$$-2\sum_{A}^{N_n} Z_A \text{Re}\left[L_{ii}(\tilde{\mathbf{s}},\bar{\mathbf{R}})e^{-i\tilde{\mathbf{s}}\cdot\mathbf{R}_A}\right],$$

contains terms that describe scattering by the molecule's electrons only (first line) and its nuclei only (first and second line), and an interference between electronic and nuclear scattering amplitudes (third line).

(2) **Inelastic electron scattering**: when an electronic transition, $f \leftarrow i$, takes place, energy is transferred between the electron and the molecule, and the scattering is electronically inelastic. This is the case for all terms with $i=j\neq f$. The inelastic electron scattering signal is given by

$$\frac{d\sigma_{in}}{d\Omega} = \left(\frac{d\sigma}{d\Omega}\right)_{Ru} \sum_{i}^{N}\sum_{f\neq i}^{\infty} \mathcal{N}_{fi} \int_{-\infty}^{+\infty} I(t-\tau)\langle\chi_i(t)||L_{fi}(\tilde{\mathbf{s}},\bar{\mathbf{R}})|^2|\chi_i(t)\rangle\,dt. \tag{3.74}$$

As discussed below eqn (3.70) before, the inelastic component does not involve any term that refers to scattering by the nuclei. It is purely electronic. Apart from different prefactors, inelastic electron scattering and X-ray scattering, eqn (3.74) and (3.38), are thus identical. As in X-ray scattering, transitions that imply a change of spin multiplicity are forbidden, because the Coulomb interaction between the incident electron and the electrons of the molecule cannot couple two states with different spin multiplicities.[§§] The scattering operator can furthermore only lead to single excitations or single ionisations of the initial state.[72] Other than that, no selection rules exist for inelastic electron scattering.

(3) **Coherent mixed electron scattering:** when scattering amplitudes of two or more electronic states interfere coherently, the scattering is coherent mixed. This is the case for all terms with $i \neq j$. The coherent mixed electron scattering signal is given by

$$\frac{\mathrm{d}\sigma_{\mathrm{cm}}}{\mathrm{d}\Omega} = 2\left(\frac{\mathrm{d}\sigma}{\mathrm{d}\Omega}\right)_{\mathrm{Ru}} \sum_{i}^{N-1}\sum_{j>i}^{N}\sum_{f}^{\infty} \mathcal{N}_{fij} \int_{-\infty}^{+\infty} I(t-\tau)$$

$$\times \mathrm{Re}[\langle\chi_j(t)|Z_{fj}^*(\tilde{\mathbf{s}},\bar{\mathbf{R}})Z_{fi}(\tilde{\mathbf{s}},\bar{\mathbf{R}})|\chi_i(t)\rangle]\mathrm{d}t. \tag{3.75}$$

This component is analogous to the coherent mixed signal proposed in X-ray scattering, eqn (3.39). It shows the same rapid beating and cannot be resolved with pulses that are long compared to the timescale of electronic motion. Interestingly, the coherent mixed component in ultrafast electron scattering is not purely electronic but also contains a cross-term between scattering by the molecule's electrons and nuclei,

$$Z_{fj}^*(\tilde{\mathbf{s}},\bar{\mathbf{R}})Z_{fi}(\tilde{\mathbf{s}},\bar{\mathbf{R}}) = L_{fj}^*(\tilde{\mathbf{s}},\bar{\mathbf{R}})L_{fi}(\tilde{\mathbf{s}},\bar{\mathbf{R}}) - \delta_{fj}\sum_{A}^{N_{\mathrm{n}}}Z_A L_{fi}(\tilde{\mathbf{s}},\bar{\mathbf{R}})\mathrm{e}^{-i\tilde{\mathbf{s}}\cdot\mathbf{R}_A}$$

$$-\delta_{fi}\sum_{A}^{N_{\mathrm{n}}}Z_A L_{fj}^*(\tilde{\mathbf{s}},\bar{\mathbf{R}})\mathrm{e}^{i\tilde{\mathbf{s}}\cdot\mathbf{R}_A}. \tag{3.76}$$

In this regard, it is different from coherent mixed scattering of X-rays, eqn (3.39), that only involves the electronic term.

The three components add up incoherently as in eqn (3.40) and are always measured together. Figure 3.2 that was introduced to illustrate the components of ultrafast X-ray scattering can be applied to the respective components of ultrafast electron scattering as well.

3.3.2.2 No Energy Resolution

Now, the most common case of a detector that is also insensitive to energy differences on the order of electronic transitions is considered. The weights

---

[§§] The incident electron can couple states with different spin multiplicities *via* magnetic interaction. This magnetic interaction is, however, significantly weaker than the dominant Coulomb interaction and thus not considered here.

in eqn (3.31) are thus effectively the same for all rotational, vibrational, and electronic quantum numbers, $\mathscr{N}_{\bar{f}ij} \approx \mathscr{N}$. With the identity in the subspace of electronic coordinates, eqn (3.41), the sum over the electronic eigenstates $|\varphi_{\bar{f}}(\bar{R})\rangle$ in eqn (3.71) can be removed and the differential scattering cross-section further simplifies to

$$\frac{d\sigma}{d\Omega} = \left(\frac{d\sigma}{d\Omega}\right)_{\mathrm{Ru}} \sum_{i,j}^{N} \int_{-\infty}^{+\infty} \mathcal{I}(t-\tau)\langle\chi_j(t)|\Xi_{ji}(\tilde{\boldsymbol{s}},\bar{\boldsymbol{R}})|\chi_i(t)\rangle \, dt, \qquad (3.77)$$

where $\mathcal{I}(t-\tau) = \mathscr{N}I(t-\tau)$ is the detectable fraction of the electron intensity and $\Xi_{ji}(\tilde{\boldsymbol{s}},\bar{\boldsymbol{R}})$ is the total electron scattering probability of the molecule with nuclear geometry $\bar{\boldsymbol{R}}$ in electronic states $|\varphi_i(\bar{R})\rangle$ and $|\varphi_j(\bar{R})\rangle$,

$$\begin{aligned}
\Xi_{ji}(\tilde{\boldsymbol{s}},\bar{\boldsymbol{R}}) = \delta_{ij}&\left(N_e + \sum_{A}^{N_n} Z_A^2 + 2 \sum_{A}^{N_n-1}\sum_{B>A}^{N_n} Z_A Z_B \cos[\tilde{\boldsymbol{s}}\cdot\boldsymbol{R}_{AB}]\right)\\
&+ \Lambda_{ji}^{2e}(\tilde{\boldsymbol{s}},\bar{\boldsymbol{R}}) - 2\sum_{A}^{N_n} Z_A \mathrm{Re}\left[L_{ji}(\tilde{\boldsymbol{s}},\bar{\boldsymbol{R}})e^{-i\tilde{\boldsymbol{s}}\cdot\boldsymbol{R}_A}\right],
\end{aligned} \qquad (3.78)$$

that contains one- and two-electron scattering matrix elements from X-ray scattering, $L_{ji}(\tilde{\boldsymbol{s}},\bar{\boldsymbol{R}})$ and $\Lambda_{ji}^{2e}(\tilde{\boldsymbol{s}},\bar{\boldsymbol{R}})$, given in eqn (3.34) and (3.46). Here, $\boldsymbol{R}_{AB} = \boldsymbol{R}_B - \boldsymbol{R}_A$ is the distance vector between two nuclei of the molecule. The two-electron scattering matrix element refers to scattering by the molecule's electrons only while the one-electron scattering matrix element contributes to a cross-term that describes the interference between electronic and nuclear scattering amplitudes. The remaining terms in the first line of eqn (3.78) are non-zero only for diagonal elements, $\Xi_{ii}(\tilde{\boldsymbol{s}},\bar{\boldsymbol{R}})$. They account for incoherent scattering by the electrons as well as for scattering by the nuclei.

In analogy to the total X-ray scattering probability, eqn (3.44), the total electron scattering probability in eqn (3.78) can be split into an incoherent one-particle and a coherent two-particle part,

$$\Xi_{ji}(\tilde{\boldsymbol{s}},\bar{\boldsymbol{R}}) = \delta_{ij}\left(N_e + \sum_{A}^{N_n} Z_A^2 \right) + \Xi_{ji}^{2\mathrm{p}}(\tilde{\boldsymbol{s}},\bar{\boldsymbol{R}}), \qquad (3.79)$$

where the two-particle part is,

$$\begin{aligned}
\Xi_{ji}^{2\mathrm{p}}(\tilde{\boldsymbol{s}},\bar{\boldsymbol{R}}) = 2\delta_{ij}&\sum_{A}^{N_n-1}\sum_{B>A}^{N_n} Z_A Z_B \cos[\tilde{\boldsymbol{s}}\cdot\boldsymbol{R}_{AB}]\\
&+ \Lambda_{ji}^{2e}(\tilde{\boldsymbol{s}},\bar{\boldsymbol{R}}) - 2\sum_{A}^{N_n} Z_A \mathrm{Re}\left[L_{ji}(\tilde{\boldsymbol{s}},\bar{\boldsymbol{R}})e^{-i\tilde{\boldsymbol{s}}\cdot\boldsymbol{R}_A}\right].
\end{aligned} \qquad (3.80)$$

Again, three physically distinct components can be identified in the differential scattering cross-section in eqn (3.77):

(1) **Background electron scattering**: the one-particle part of eqn (3.79) leads to a time-independent, structureless background,

$$\frac{\mathrm{d}\sigma_{\mathrm{bg}}}{\mathrm{d}\Omega} = \left(\frac{\mathrm{d}\sigma}{\mathrm{d}\Omega}\right)_{\mathrm{Ru}} \mathcal{I}_\tau \left(N_{\mathrm{e}} + \sum_{A}^{N_{\mathrm{n}}} Z_A^2\right), \tag{3.81}$$

that is proportional to the total number of detectable electrons from the pulse, $\mathcal{I}_\tau = \int \mathcal{I}(t-\tau)\mathrm{d}t$. It contains the number of electrons, N_{e}, and the sum of the squared atomic numbers, Z_A^2.

(2) **Excess electron scattering**: the two-particle part of eqn (3.79) with $i=j$ leads to a time- and structure-dependent component analogous to the excess scattering in ultrafast X-ray scattering, eqn (3.48),

$$\frac{\mathrm{d}\sigma_{\mathrm{ex}}}{\mathrm{d}\Omega} = \left(\frac{\mathrm{d}\sigma}{\mathrm{d}\Omega}\right)_{\mathrm{Ru}} \sum_{i}^{N} \int_{-\infty}^{+\infty} \mathcal{I}(t-\tau)\langle\chi_i(t)|\Xi_{ii}^{2\mathrm{P}}(\tilde{\boldsymbol{s}},\bar{\boldsymbol{R}})|\chi_i(t)\rangle\mathrm{d}t. \tag{3.82}$$

(3) **Coherent mixed electron scattering**: the two-particle part of eqn (3.79) with $i \neq j$ leads to coherent mixed scattering,

$$\frac{\mathrm{d}\sigma_{\mathrm{cm}}}{\mathrm{d}\Omega} = 2\left(\frac{\mathrm{d}\sigma}{\mathrm{d}\Omega}\right)_{\mathrm{Ru}} \sum_{i}^{N-1}\sum_{j>i}^{N} \int_{-\infty}^{+\infty} \mathcal{I}(t-\tau)\,\mathrm{Re}\left[\left\langle\chi_j(t)\middle|\Xi_{ji}^{2\mathrm{P}}(\tilde{\boldsymbol{s}},\bar{\boldsymbol{R}})|\chi_i(t)\right\rangle\right]\mathrm{d}t$$
$$\tag{3.83}$$

as already discussed around eqn (3.75). For two electronic states with different inversion symmetries, both terms that contribute to $\Xi_{ji}^{2\mathrm{P}}(\tilde{\boldsymbol{s}},\bar{\boldsymbol{R}})$ vanish.

3.3.3 Relativistic Effects

If the velocity of the incident electrons approaches the speed of light, one may need to account for relativistic effects. This usually applies to MeV pulses (see Chapter 12) with velocities of more than 90% of the speed of light.[¶] Even though a fully relativistic description of ultrafast electron scattering would require a theoretical framework rooted in relativistic quantum electrodynamics,[‖] the non-relativistic theory discussed in this chapter can still be applied to all but the most extreme experiments with GeV electrons.

[¶] The total relativistic energy of an electron is $E = \gamma m_{\mathrm{e}}c^2$, where m_{e} is the electron's rest mass, $\gamma = 1/\sqrt{1 - v^2/c^2}$ the Lorentz factor, c the speed of light, and v the velocity of the electron. The velocity as a function of E is thus $v = c\sqrt{1 - (m_{\mathrm{e}}c^2/E)^2}$.

[‖] Accounting for effects such as magnetic spin–spin interaction, spin–orbit interaction, and radiative coupling of the electrons to quantized electromagnetic field modes.

For mean electron energies on the order of one MeV, it is usually sufficient to apply the relativistic value of the electron's mass and de Broglie wavelength (see footnote ‖ on page 101).[13] This also implies that the rest mass of the electron m_e present in eqn (3.67) has to be replaced with γm_e where γ is the Lorentz factor. The differential scattering cross-sections must therefore be multiplied with γ^2. In addition to this, but only if the Waller–Hartree approximation breaks down, the scaling by $1/s^4$ contained within the Rutherford scattering cross-section, $(d\sigma/d\Omega)_{Ru}$, in eqn (3.53) has to be modified to $1/(s^2 - w^2)^2$ with $w = m_e c / \hbar(\gamma_0 - \gamma_s)$ to account for relativistic retardation.[29] Here, γ_0 and γ_s are the Lorentz factors of the incident and scattered electrons, respectively. Considering that the Waller–Hartree approximation works increasingly well for increasing initial velocities and is essentially exact when MeV electrons are employed, the modification of $1/s^4$ is usually negligible and only the multiplication of the cross-section with γ^2 is required.*** The Lorentz factor, however, is independent of the momentum transfer and thus leaves the scattering patterns unaltered.

We note that, even though multiplication with γ^2 leads to an increase in the differential scattering cross-section with the incident electron's velocity at any given value of $\tilde{s}$, the magnitude of the total observable signal does not increase with the energy of the electrons.[35] The effect of the Lorentz factor upon the total integrated scattering cross-section, σ_t, is compensated by the shift of the scattering signal towards smaller scattering angles θ_s that takes place when the velocity of the electrons increases. This shift implies that the area into which most electrons are scattered drops and hence σ_t would quickly approach zero if it were not for γ. Taking both effects into account, the total cross-section is almost constant in the MeV range.

For mean-electron energies of several MeV, relativistic corrections to the electron scattering amplitudes, eqn (3.70), and probabilities, eqn (3.78), may become necessary as well. The expressions for the differential scattering cross-sections, eqn (3.71) and (3.77), however, remain the same apart from the aforementioned multiplication with γ^2. This aligns with the common practice in MeV experiments to utilise relativistically corrected atomic form factors in the analysis.[36] It is only in the extremely relativistic regime of GeV pulses beyond the scope of this chapter that the derived expressions for the differential electron scattering cross-sections themselves become inadequate.[29]

3.4 Calculation of Scattering Cross-sections

3.4.1 *Ab Initio* Electronic Structure Theory

The simulation of time-resolved differential scattering cross-sections given by eqn (3.36) and (3.42) for X-rays and by eqn (3.71) and (3.77) for electrons requires that one- and two-electron scattering matrix elements are evaluated. Following

---

***Within the Waller–Hartree approximation, the relativistic momentum transfer is given as $\tilde{s} = 2\sqrt{E^2 - (m_e c^2)^2}/(c\hbar)\sin(\theta_s/2)$, where E is the total mean-electron energy and θ_s is the scattering angle.

eqn (3.34) and (3.46), these matrix elements can be obtained as Fourier transforms of a molecule's one- and two-electron (transition) densities. These densities can, in turn, be computed from electronic structure codes either *via* density functional theory (DFT) or wave function-based methods such as configuration interaction (CI) or multi-configurational self-consistent field (MCSCF).[18] To take one example, the commonly used atomic form factors, which are rotationally averaged elastic one-electron scattering matrix elements of single atoms in their electronic ground states, are calculated from *ab initio* Hartree–Fock (HF) densities.[13] Several methods for performing the Fourier transform either numerically[37-40] or analytically[41-47] have been published. These methods not only differ in their specific approaches for evaluating the Fourier integral but also in whether they account explicitly for rotational averaging (see Section 3.4.3), provide only one- or also two-electron scattering matrix elements, and, similarly, calculate only diagonal elastic and total or also off-diagonal inelastic and coherent mixed components.

3.4.2 Independent Atom Model

The computationally demanding evaluation of accurate scattering matrix elements by means of *ab initio* electronic structure theory briefly discussed in the previous subsection is not always necessary. Many experiments can be simulated with sufficient accuracy by employing a much simpler and cheaper method termed the independent atom model (IAM). The IAM was originally devised by Debye,[48] Heisenberg,[49] Morse,[50] and Bewilogua[51] and has found widespread use in the analysis of scattering experiments and X-ray crystallography. The underlying idea of the IAM is that the scattering probability of a molecule can be approximated using tabulated scattering amplitudes of the molecule's constituent atoms.[52] It yields scattering signals that are in good agreement with experimental data if the target molecules are in their electronic ground states and their composition is not dominated by light elements, hydrogen in particular. Indeed, the limitations of IAM were first noted for accurate X-ray crystallography data collected from organic molecular crystals, where the spherical approximation for the hydrogen atoms was found to be inadequate.[53] In general, the IAM cannot account for more subtle electronic effects such as chemical bonding, electronic excitation, or electron dynamics probed by the coherent mixed component. In the following, we introduce the independent atom model for perfectly aligned molecules and then turn to a discussion of rotationally-averaged scattering probabilities in the next subsection.

Within the IAM, the elastic X-ray scattering amplitude of a molecule is approximated by a coherent sum of atomic form factors for X-ray scattering, $f_A^x(\tilde{q})$, which are isotropic scattering amplitudes, eqn (3.34), of atoms in their electronic ground states.[13] The approximate elastic scattering amplitude is

$$L(\tilde{q}, \bar{R}) = \sum_A^{N_n} f_A^x(\tilde{q}) e^{i\tilde{q} \cdot R_A}. \qquad (3.84)$$

The sum over A runs over all N_n atoms of the molecule. The exponential phase factors, $\exp[\imath \tilde{q} \cdot R_A]$, account for the position of the atoms, R_A, and thereby for the phase relation between the scattering amplitudes of different atoms. Note that $L(\tilde{q},\bar{R})$ carries no indices, in contrast to the exact one-electron scattering matrix element in eqn (3.34), because the IAM does not allow a distinction between different electronic states. With the amplitude in eqn (3.84), the elastic X-ray scattering probability that appears in eqn (3.37) is given by

$$|L(\tilde{q},\bar{R})|^2 = \sum_A^{N_n} \sum_B^{N_n} f_A^x(\tilde{q}) f_B^x(\tilde{q}) e^{\imath \tilde{q} \cdot R_{AB}}, \tag{3.85}$$

where $R_{AB} = R_B - R_A$ is the distance vector between two atoms of the molecule. Inserting eqn (3.84) into eqn (3.70), the corresponding elastic electron scattering amplitude within IAM is

$$Z(\tilde{s},\bar{R}) = L(\tilde{s},\bar{R}) - \sum_A^{N_n} Z_A e^{\imath \tilde{s} \cdot R_A} = \sum_A^{N_n} f_A^e(\tilde{s}) e^{\imath \tilde{s} \cdot R_A}, \tag{3.86}$$

where the atomic form factors for electron scattering,

$$f_A^e(\tilde{s}) = f_A^x(\tilde{s}) - Z_A, \tag{3.87}$$

are introduced. Apart from an arbitrary factor of -1 and the scaling by $1/\tilde{s}^2$, this definition of $f_A^e(\tilde{s})$ is identical to the commonly used Mott–Bethe formula.[30,54] With the amplitude in eqn (3.86), the elastic electron scattering probability that appears in eqn (3.72) can be approximated as

$$|Z(\tilde{s},\bar{R})|^2 = \sum_A^{N_n} \sum_B^{N_n} f_A^e(\tilde{s}) f_B^e(\tilde{s}) e^{\imath \tilde{s} \cdot R_{AB}}, \tag{3.88}$$

which is identical to eqn (3.85) for X-ray scattering apart from the different atomic form factors.

While the IAM cannot describe individual inelastic transitions, the total inelastic component can be approximated by an incoherent sum of incoherent scattering functions, $S_A(\tilde{q})$.[13,49,50] These functions are obtained by subtracting the elastic from the total X-ray scattering probabilities of the individual atoms as is done in eqn (3.52). The total X-ray scattering probability that appears in the diagonal part of eqn (3.42) is thus,

$$\Lambda(\tilde{q},\bar{R}) = |L(\tilde{q},\bar{R})|^2 + \sum_A^{N_n} S_A(\tilde{q}). \tag{3.89}$$

Similarly, the total electron scattering probability that appears in the diagonal part of eqn (3.77) becomes,

$$\Xi(\tilde{s},\bar{R}) = |Z(\tilde{s},\bar{R})|^2 + \sum_A^{N_n} S_A(\tilde{s}). \tag{3.90}$$

In Figure 3.3, scattering probabilities of ground state fluoroform (CHF_3) calculated using the IAM (solid lines) are compared to accurate data computed from the molecule's electronic wave function (markers). The probabilities are rotationally averaged and fully isotropic (see Section 3.4.3 below). The IAM shows an excellent agreement with the reference data both for X-ray and electron scattering and for the total scattering probabilities as well as for their elastic and inelastic components. This excellent agreement can be attributed to three reasons mainly: (i) the carbon and fluorine atoms are already relatively electron-rich and their electron densities are not altered much by the chemical bonding; (ii) the accurate scattering probabilities are calculated for the molecule's electronic ground state; and (iii) absolute scattering probabilities are displayed. Finally, the rotational averaging obscures some detail. If the comparison is made for aligned molecules, then further, albeit small, differences will be noted. The overall good agreement between the IAM and the reference data illustrates why the independent atom model is a valuable and widely applied tool for the analysis and simulation of scattering experiments.

3.4.3 Rotational Averaging

While the previous subsection introduced the independent atom model for perfectly aligned molecules, experimental scattering signals measured in the gas phase arise from molecules that are partially aligned at most.[73,74] The scattering probabilities thus have to be rotationally averaged to allow a direct comparison with experimental data. The expressions for the differential scattering cross-sections, eqn (3.36) and (3.42) for X-ray scattering and eqn (3.71) and (3.77) for electron scattering, account for that by bracketing the scattering probabilities with the time-dependent nuclear wave packets. The expressions for the elastic X-ray and electron scattering signals, eqn (3.37) and (3.72), for example, contain the following matrix elements,

$$\langle \chi_i(t) | |L_{ii}(\tilde{\boldsymbol{q}},\bar{\boldsymbol{R}})|^2 | \chi_i(t) \rangle = \int |\chi_i(\bar{\boldsymbol{R}},t)|^2 |L_{ii}(\tilde{\boldsymbol{q}},\bar{\boldsymbol{R}})|^2 \mathrm{d}\bar{\boldsymbol{R}},$$
$$\langle \chi_i(t) | |Z_{ii}(\tilde{\boldsymbol{s}},\bar{\boldsymbol{R}})|^2 | \chi_i(t) \rangle = \int |\chi_i(\bar{\boldsymbol{R}},t)|^2 |Z_{ii}(\tilde{\boldsymbol{s}},\bar{\boldsymbol{R}})|^2 \mathrm{d}\bar{\boldsymbol{R}},$$

$$(3.91)$$

where $\chi_i(\bar{\boldsymbol{R}},t)$ is the nuclear wave packet of an electronic state with index i. The set of internal nuclear coordinates, $\bar{\boldsymbol{R}}$, includes the vibrational and rotational coordinates of the molecule. The integration over $\bar{\boldsymbol{R}}$ therefore implies rotational averaging of the elastic scattering probabilities with respect to the molecule's rotational density. The latter can be described explicitly by separating the rotational and vibrational motion. For that, the nuclear wave packet in eqn (3.91) can be expressed using a product ansatz,

$$|\chi_i(t)\rangle = |\chi_i^{\mathrm{rot}}(t)\rangle |\chi_i^{\mathrm{vib}}(t)\rangle.$$

$$(3.92)$$

Here, $|\chi_i^{\mathrm{rot}}(t)\rangle$ and $|\chi_i^{\mathrm{vib}}(t)\rangle$ are rotational and vibrational wave packets that depend solely on the molecule's rotational and vibrational degrees of freedom, respectively. To solve the resulting integrals analytically, the rotational wave packet can be expanded in terms of Wigner D-matrices,[55]

$$\langle \alpha, \beta, \gamma | \chi_i^{\text{rot}}(t) \rangle = \sum_J \sum_{M=-J}^{J} \sum_{K=-J}^{J} c_{M,K}^{i,J}(t) D_{M,K}^J(\alpha, \beta, \gamma), \qquad (3.93)$$

where $\{\alpha, \beta, \gamma\}$ are the three Euler angles, the rotational coordinates in the molecular frame of reference, $\{J, M, K\}$ are the rotational quantum numbers, and $c_{M,K}^{i,J}(t)$ are the time- and phase-dependent expansion coefficients. We note that, for a linear molecule with $K = 0$, the Wigner D-matrices reduce to spherical harmonics,

$$D_{M,0}^J(\alpha, \beta, \gamma) = \sqrt{\frac{4\pi}{2J+1}} Y_J^{M*}(\beta, \alpha). \qquad (3.94)$$

The expansion in eqn (3.93) is exact and can, in principle, include any Wigner D-matrix. To keep matters simple, we will now consider a specific case of alignment that is still applicable to all types of molecules but relies on three Wigner D-matrices only. At the expense of generality, this restriction will enable us to derive a few closed-form expressions that are not only instructive but also permit the analysis and simulation of many ultrafast scattering experiments in the gas phase.[3,56–58]

To begin with, a molecule is usually in a thermal ensemble in its electronic ground state before the arrival of the pump pulse. Although many different rotational eigenstates may be populated, the resulting rotational density is completely isotropic and independent of time. Disregarding random phases of individual eigenstates, the initial rotational wave packet in the electronic ground state (GS) is therefore

$$\langle \alpha, \beta, \gamma | \chi_{\text{GS}}^{\text{rot}}(t \ll t_0) \rangle = \frac{1}{2\sqrt{2\pi}} = \frac{1}{2\sqrt{2\pi}} D_{0,0}^0(\alpha, \beta, \gamma), \qquad (3.95)$$

with $t \ll t_0$ referring to a time well before the arrival of the pump pulse centred at t_0. The coefficient $1/(2\sqrt{2\pi})$ ensures normalization under integration over the Euler angles and $1 = D_{0,0}^0(\alpha, \beta, \gamma)$ is used. Once the pump pulse interacts with the molecule, the rotational density becomes partially anisotropic and a degree of alignment is induced. For linearly polarized light and resonant one-photon excitation in the visible or UV range of the electromagnetic spectrum (*i.e.*, within the weak-field limit and the electric dipole approximation), the rotational wave packet of the electronically excited state will be proportional to $\cos \beta$. This follows directly from the interaction Hamiltonian,

$$\hat{H}_{\text{int}}(t) = -\boldsymbol{\mu}_{ji} \cdot \boldsymbol{\epsilon} \mathcal{E}(t) = -\mu_{ji} \mathcal{E}(t) \cos \beta. \qquad (3.96)$$

Here, $\boldsymbol{\mu}_{ji}$ is a transition dipole moment vector, μ_{ji} is its norm, $\boldsymbol{\epsilon}$ and $\mathcal{E}(t)$ are the polarization vector and the electric field amplitude of the pump pulse, respectively, and β is the angle between $\boldsymbol{\mu}_{fi}$ and $\boldsymbol{\epsilon}$. Hence, the rotational wave packet in the electronically excited state (ES) is

$$\langle \alpha, \beta, \gamma | \chi_{\text{ES}}^{\text{rot}}(t \gg t_0) \rangle = \frac{a_1 \sqrt{3}}{2\sqrt{2\pi}} \cos \beta = \frac{a_1 \sqrt{3}}{2\sqrt{2\pi}} D_{0,0}^1(\alpha, \beta, \gamma), \qquad (3.97)$$

where $t \gg t_0$ implies a time after the pump pulse has interacted with the molecule, $\cos \beta = D_{0,0}^1(\alpha,\beta,\gamma)$, and a_1 is a coefficient proportional to the square root of the excitation fraction, ζ, such that $|a_1|^2 \approx \zeta$. The factor $\sqrt{3}/(2\sqrt{2\pi})$ ensures normalization of the Wigner D-matrix. We note that the wave packet in eqn (3.97) does not incorporate the effect of rotational dephasing, which, over time, reduces the degree of alignment. Rotational dephasing would effectively decrease the ratio of the anisotropic to the isotropic part of the rotational density. We will comment on that again once we have obtained the rotationally-averaged scattering probability for the excited state.

Moreover, the pump pulse depletes the rotational wave packet in the electronic ground state and transfers amplitude back from the excited to the ground state *via* stimulated emission. The latter is a two-photon process and, within the perturbative weak-field limit, significantly weaker than the one-photon driven depletion of the ground state. While the direct contribution of this two-photon process to the scattering signal is negligible, the effect of its interference with the remaining ground state amplitude may not be.[59] If the probabilities of the one- and two-photon processes scale with ζ and ζ^2, respectively, the interference between the transferred and depleted ground state amplitudes scales roughly with $\zeta\sqrt{1-\zeta} \approx \zeta$. The two-photon process can therefore indirectly affect the scattering signal on the same order of magnitude as the one-photon absorption and should be considered. At times $t \gg t_0$, the rotational wave packet in the electronic ground state (GS) is thus

$$\langle \alpha, \beta, \gamma | \chi_{\text{GS}}^{\text{rot}}(t \gg t_0) \rangle = \frac{a_0}{2\sqrt{2\pi}} + \frac{a_2\sqrt{5}}{2\sqrt{2\pi}} \cos^2 \beta$$

$$= \frac{1}{2\sqrt{2\pi}} \left(a_0 + \frac{a_2\sqrt{5}}{3} \right) D_{0,0}^0(\alpha, \beta, \gamma) \qquad (3.98)$$

$$+ \frac{a_2\sqrt{5}}{3\sqrt{2\pi}} D_{0,0}^2(\alpha, \beta, \gamma),$$

where the coefficients a_0 and a_2 account for the depletion of the initial wave packet, eqn (3.95), and the strength of the two-photon process, meaning that $|a_0|^2 \approx 1 - \zeta$ and $|a_2|^2 \approx \zeta^2$. Since the two-photon process involves the interaction Hamiltonian in eqn (3.96) twice, the transferred amplitude has to be proportional to

$$\cos^2\beta = \left|D^1_{0,0}(\alpha,\beta,\gamma)\right|^2 = \frac{1}{3}D^0_{0,0}(\alpha,\beta,\gamma) + \frac{2}{3}D^2_{0,0}(\alpha,\beta,\gamma). \tag{3.99}$$

The factors $1/(2\sqrt{2}\pi)$ and $\sqrt{5}/(2\sqrt{2}\pi)$ in eqn (3.98) ensure normalization. Again, we note that the wave packet in eqn (3.98) does not incorporate the effect of rotational dephasing, which would cause the rotational density of the ground state to become more isotropic over time.

With the two rotational wave packets, eqn (3.97) and (3.98), at hand, expressions for the rotational average of the scattering probabilities can be derived. Here, we will illustrate the procedure by application to the independent atom model, but several approaches exist that analytically average accurate scattering probabilities obtained from *ab initio* electronic structure theory.[40,41,43,47] Starting from eqn (3.85) and (3.88) and realizing that only the exponentials $\exp[\imath\tilde{q}\cdot R_{AB}]$ and $\exp[\imath\tilde{s}\cdot R_{AB}]$ therein depend on the Euler angles, rotational averaging requires the evaluation of the following integrals:[†††]

$$\left\langle\chi^{rot}_{GS}\left|e^{\imath\tilde{q}\cdot R_{AB}}\right|\chi^{rot}_{GS}\right\rangle \approx \frac{1}{8\pi^2}\left(|a_0|^2 + b_{0,2}\right)\left\langle D^0_{0,0}\left|e^{\imath\tilde{q}\cdot R_{AB}}\right|D^0_{0,0}\right\rangle$$
$$+ \frac{1}{4\pi^2}b_{0,2}\left\langle D^0_{0,0}\left|e^{\imath\tilde{q}\cdot R_{AB}}\right|D^2_{0,0}\right\rangle, \tag{3.100}$$

$$\left\langle\chi^{rot}_{ES}\left|e^{\imath\tilde{q}\cdot R_{AB}}\right|\chi^{rot}_{ES}\right\rangle = \frac{3}{8\pi^2}|a_1|^2\left\langle D^1_{0,0}\left|e^{\imath\tilde{q}\cdot R_{AB}}\right|D^1_{0,0}\right\rangle. \tag{3.101}$$

In eqn (3.100), all terms proportional to $|a_2|^2$, which relate solely to the amplitude gained by the two-photon process, are neglected. This approximation is valid as long as the excitation fraction is small (*i.e.*, not much larger than 10%). The remaining contribution from the interference is weighted by the coefficient $b_{0,2} = \sqrt{5}/3\mathrm{Re}[a^*_0 a_2]$. Now, the matrix elements on the right-hand sides of eqn (3.100) and (3.101) can be solved by expanding the exponential in terms of spherical harmonics, $Y^{k*}_l(\theta_{\tilde{q}},\phi_{\tilde{q}})$ and $Y^m_l(\theta_{AB},\phi_{AB})$, giving[55]

$$e^{\imath\tilde{q}\cdot R_{AB}} = 4\pi\sum_{l=0}^{\infty}\sum_{m=-l}^{l}\sum_{k=-l}^{l}\imath^l j_l(\tilde{q}R_{AB})D^l_{m,k}(\alpha,\beta,\gamma)$$
$$\times Y^{k*}_l(\theta_{\tilde{q}},\phi_{\tilde{q}})Y^m_l(\theta_{AB},\phi_{AB}). \tag{3.102}$$

Here, $j_l(\tilde{q}R_{AB})$ is a spherical Bessel function, $\{\tilde{q},\theta_{\tilde{q}},\phi_{\tilde{q}}\}$ are the spherical coordinates of the momentum transfer vector $\tilde{q}$ in the laboratory frame, and

---

[†††] Evidently, the integrals for $\exp[\imath\tilde{s}\cdot R_{AB}]$ are identical.

$\{R_{AB}, \theta_{AB}, \phi_{AB}\}$ are the spherical coordinates of the distance vector $\boldsymbol{R}_{AB}$ in the molecular frame. The Wigner D-matrices $D^l_{m,k}(\alpha, \beta, \gamma)$ transform from one frame into the other. With eqn (3.102), the integrals over the Euler angles contained in eqn (3.100) and (3.101) become

$$\left\langle D^0_{0,0} \middle| D^l_{m,k} \middle| D^0_{0,0} \right\rangle = 8\pi^2 \delta_{l,0} \delta_{m,0} \delta_{k,0}, \tag{3.103}$$

$$\left\langle D^0_{0,0} \middle| D^l_{m,k} \middle| D^2_{0,0} \right\rangle = \frac{8\pi^2}{5} \delta_{l,2} \delta_{m,0} \delta_{k,0}, \tag{3.104}$$

$$\left\langle D^1_{0,0} \middle| D^l_{m,k} \middle| D^1_{0,0} \right\rangle = \frac{8\pi^2}{3} \delta_{l,0} \delta_{m,0} \delta_{k,0} + \frac{16\pi^2}{15} \delta_{l,2} \delta_{m,0} \delta_{k,0}, \tag{3.105}$$

where the orthogonality of the Wigner D-matrices,[55]

$$\int_0^{2\pi} \int_0^{\pi} \sin\beta \int_0^{2\pi} \left(D^J_{M,K}(\alpha, \beta, \gamma)\right)^* D^l_{m,k}(\alpha, \beta, \gamma) \mathrm{d}\alpha \mathrm{d}\beta \mathrm{d}\gamma = \frac{8\pi^2}{2l+1} \delta_{lJ} \delta_{mM} \delta_{kK}, \tag{3.106}$$

and eqn (3.99) are used. With eqn (3.103)–(3.105), the two matrix elements in eqn (3.100) and (3.101) now become

$$\begin{aligned}\left\langle \chi^{\mathrm{rot}}_{\mathrm{GS}} \middle| e^{i\tilde{\boldsymbol{q}} \cdot \boldsymbol{R}_{AB}} \middle| \chi^{\mathrm{rot}}_{\mathrm{GS}} \right\rangle &= \left(|a_0|^2 + b_{0,2}\right) j_0(\tilde{q} R_{AB}) \\ &\quad - 2 b_{0,2} \, j_2(\tilde{q} R_{AB}) P_2(\cos\theta_{\tilde{q}}) P_2(\cos\theta_{AB}),\end{aligned} \tag{3.107}$$

$$\begin{aligned}\left\langle \chi^{\mathrm{rot}}_{\mathrm{ES}} \middle| e^{i\tilde{\boldsymbol{q}} \cdot \boldsymbol{R}_{AB}} \middle| \chi^{\mathrm{rot}}_{\mathrm{ES}} \right\rangle &= |a_1|^2 j_0(\tilde{q} R_{AB}) \\ &\quad - 2 |a_1|^2 j_2(\tilde{q} R_{AB}) P_2(\cos\theta_{\tilde{q}}) P_2(\cos\theta_{AB}).\end{aligned} \tag{3.108}$$

The last lines of eqn (3.107) and (3.108) contain second-order Legendre polynomials, $P_2(\cos\theta)$, that derive from the spherical harmonics in eqn (3.102),

$$Y^0_2(\theta, \phi) = \sqrt{\frac{5}{4\pi}} P_2(\cos\theta). \tag{3.109}$$

With the matrix elements in eqn (3.107) and (3.108) and the elastic scattering probability in eqn (3.85), the rotationally-averaged elastic X-ray scattering probabilities within the IAM are

$$\begin{aligned}\left\langle \chi^{\mathrm{rot}}_{\mathrm{GS}} \middle| |L(\tilde{\boldsymbol{q}}, \bar{\boldsymbol{R}})|^2 \middle| \chi^{\mathrm{rot}}_{\mathrm{GS}} \right\rangle &= \left(|a_0|^2 + b_{0,2}\right) \mathcal{S}^x_0(\tilde{q}, \bar{\boldsymbol{R}}_{\mathrm{vib}}) \\ &\quad - b_{0,2} \mathcal{S}^x_2(\tilde{q}, \bar{\boldsymbol{R}}_{\mathrm{vib}}) P_2(\cos\theta_{\tilde{q}}),\end{aligned} \tag{3.110}$$

for the electronic ground state and

$$\left\langle \chi^{\mathrm{rot}}_{\mathrm{ES}} \middle| |L(\tilde{\boldsymbol{q}}, \bar{\boldsymbol{R}})|^2 \middle| \chi^{\mathrm{rot}}_{\mathrm{ES}} \right\rangle = |a_1|^2 \left(\mathcal{S}^x_0(\tilde{q}, \bar{\boldsymbol{R}}_{\mathrm{vib}}) - \mathcal{S}^x_2(\tilde{q}, \bar{\boldsymbol{R}}_{\mathrm{vib}}) P_2(\cos\theta_{\tilde{q}}) \right), \tag{3.111}$$

for the electronically excited state. Here, $\bar{R}_{\mathrm{vib}}$ refers to the vibrational degrees of freedom and $\mathcal{S}_0^x(\tilde{q}, \bar{R}_{\mathrm{vib}})$ describes the isotropic Debye scattering of X-rays,[‡‡‡]

$$\mathcal{S}_0^x(\tilde{q}, \bar{R}_{\mathrm{vib}}) = \sum_A^{N_n} \sum_B^{N_n} f_A^x(\tilde{q}) f_B^x(\tilde{q}) j_0(\tilde{q} R_{AB}), \qquad (3.112)$$

and $\mathcal{S}_2^x(\tilde{q}, \bar{R}_{\mathrm{vib}})$ is the radial part of anisotropic scattering,

$$\mathcal{S}_2^x(\tilde{q}, \bar{R}_{\mathrm{vib}}) = 2 \sum_A^{N_n} \sum_B^{N_n} f_A^x(\tilde{q}) f_B^x(\tilde{q}) j_2(\tilde{q} R_{AB}) P_2(\cos \theta_{AB}). \qquad (3.113)$$

The expressions for the rotationally-averaged elastic electron scattering probabilities that can be derived from eqn (3.88) are almost identical to eqn (3.110)–(3.113) for X-ray scattering. The only difference is that the atomic form factors for X-ray scattering, $f_A^x(\tilde{q})$, in eqn (3.112) and (3.113) have to be replaced with the atomic form factors for electron scattering, $f_A^e(\tilde{s})$, introduced in eqn (3.87).

The aforementioned effect of rotational dephasing can now be accounted for heuristically by decreasing the relative contribution of anisotropic scattering, $\mathcal{S}_0^x(\tilde{q})$, in eqn (3.110) and (3.111). Thereby, the scattering signal becomes more isotropic over time.

Finally, the rotationally-averaged total scattering probabilities that appear in eqn (3.42) and (3.77) can be obtained by adding the incoherent scattering functions, $S_A(\tilde{q})$, from eqn (3.89) to the isotropic Debye scattering, $\mathcal{S}_0^x(\tilde{q}, \bar{R}_{\mathrm{vib}})$, in eqn (3.112). Within the IAM, inelastic scattering is assumed to be incoherent and thus purely isotropic.

3.4.4 Vibrational Dynamics

With the rotationally-averaged scattering probabilities at hand, we are now left with the task of bracketing them with the vibrational wave packets, $|\chi_i^{\mathrm{vib}}(t)\rangle$, introduced in eqn (3.92). These wave packets describe how a photochemical reaction proceeds and can be obtained from quantum dynamics simulations using methods such as the split operator,[60] Multi-Configuration Time-Dependent Hartree (MCTDH),[61] variational Multi-Configuration Gaussian (vMCG),[62] Multi-Configurational Ehrenfest (MCE),[63] or Ab Initio Multiple Spawning (AIMS).[64] One widely applied code that can be used to run MCTDH and vMCG is QUANTICS.[65,66] Alternatively, surface hopping,[67] a popular mixed quantum-classical method, can be used to describe the vibrational motion in terms of classical trajectories rather than quantum-mechanical wave packets. The effect of the pump pulse can be accounted for either explicitly *via* an appropriate interaction Hamiltonian,

---

[‡‡‡] Note that $j_0(\tilde{q} R_{AB}) = \mathrm{sinc}(\tilde{q} R_{AB})$.

for example eqn (3.96), or by projecting the Franck–Condon wave packet from the electronic ground to the desired excited state. Finally, the matrix elements of the scattering probabilities and the vibrational wave packets can be integrated numerically or, if surface hopping is used, by averaging over the available trajectories for any given point in time. Some examples of quantum molecular dynamics simulations in the context of scattering can be found in ref. 15, 34 and 68.

Acknowledgements

AK acknowledges funding from the Engineering and Physical Sciences Research Council (EP/V049240/2, EP/V006819/2, EP/X026698/1, and EP/X026973/1), the Leverhulme Trust (RPG-2020-208), and the Swedish Collegium for Advanced Study supported by the Erling-Persson Family Foundation and the Knut and Alice Wallenberg Foundation. PMW acknowledges funding from the U.S. Department of Energy, Office of Science, Basic Energy Sciences, under awards DE-SC0017995 and DE-SC0020276. The authors are grateful to Prof. Martin Centurion for feedback on the manuscript and to Dr Andrés Moreno Carrascosa for his computational support.

References

1. J. J. Sakurai, *Modern Quantum Mechanics*, Cambridge University Press, Cambridge, 3rd edn, 2021.
2. N. E. Henriksen and K. B. Møller, *J. Phys. Chem. B*, 2008, **112**, 558–567.
3. U. Lorenz, K. B. Møller and N. E. Henriksen, *Phys. Rev. A: At., Mol., Opt. Phys.*, 2010, **81**, 023422.
4. G. Dixit, O. Vendrell and R. Santra, *Proc. Natl. Acad. Sci. U. S. A.*, 2012, **109**, 11636–11640.
5. M. Simmermacher, A. Moreno Carrascosa, N. E. Henriksen, K. B. Møller and A. Kirrander, *J. Chem. Phys.*, 2019, **151**, 174302.
6. J. J. Sakurai, *Advanced Quantum Mechanics*, Addison-Wesley, Reading, Mass, 1st edn, 1967.
7. S. P. A. Sauer, *Molecular Electromagnetism: A Computational Chemistry Approach*, Oxford University Press, Oxford, 1st edn, 2011.
8. R. Loudon, *The Quantum Theory of Light*, Oxford University Press, Oxford, 3rd edn, 2000.
9. C. Cohen-Tannoudji, J. Dupont-Roc and G. Grynberg, *Atom-Photon Interactions: Basic Processes and Applications*, Wiley-VCH, Weinheim, 1st edn, 2004.
10. C. Cohen-Tannoudji, J. Dupont-Roc and G. Grynberg, *Photons and Atoms: Introduction to Quantum Electrodynamics*, Wiley, New York, 1st edn, 1997.
11. E. Tiesinga, P. J. Mohr, D. B. Newell and B. N. Taylor, *Rev. Mod. Phys.*, 2021, **93**, 025010.
12. J. Als-Nielsen and D. McMorrow, *Elements of Modern X-ray Physics*, Wiley, Hoboken, 2nd edn, 2011.

13. E. Prince, *International Tables for Crystallography, Volume C: Mathematical, Physical and Chemical Tables*, Wiley, New York, 3rd edn, 2004.

14. B. Stankus, H. Yong, J. Ruddock, L. Ma, A. Moreno Carrascosa, N. Goff, S. Boutet, X. Xu, N. Zotev, A. Kirrander, M. P. Minitti and P. M. Weber, *J. Phys. B: At. Mol. Opt. Phys.*, 2020, **53**, 234004.

15. A. Kirrander, K. Saita and D. V. Shalashilin, *J. Chem. Theory Comput.*, 2016, **12**, 957–967.

16. M. Simmermacher, N. E. Henriksen, K. B. Møller, A. Moreno Carrascosa and A. Kirrander, *Phys. Rev. Lett.*, 2019, **122**, 073003.

17. I. Waller and D. R. Hartree, *Proc. R. Soc. London, Ser. A*, 1929, **124**, 119–142.

18. T. Helgaker, P. Jørgensen and J. Olsen, *Molecular Electronic-Structure Theory*, Wiley, Chichester, 1st edn, 2000.

19. G. Dixit, J. M. Slowik and R. Santra, *Phys. Rev. A: At., Mol., Opt. Phys.*, 2014, **89**, 043409.

20. J. Cao and K. R. Wilson, *J. Phys. Chem. A*, 1998, **102**, 9523–9530.

21. M. Kowalewski, K. Bennett and S. Mukamel, *Struct. Dyn.*, 2017, **4**, 054101.

22. G. Dixit and R. Santra, *J. Chem. Phys.*, 2013, **138**, 134311.

23. M. Simmermacher, N. E. Henriksen and K. B. Møller, *Phys. Chem. Chem. Phys.*, 2017, **19**, 19740–19749.

24. K. Bennett, M. Kowalewski, J. R. Rouxel and S. Mukamel, *Proc. Natl. Acad. Sci. U. S. A.*, 2018, **115**, 6538–6547.

25. G. Hermann, V. Pohl, G. Dixit and J. C. Tremblay, *Phys. Rev. Lett.*, 2020, **124**, 013002.

26. H. Yong, S. M. Cavaletto and S. Mukamel, *J. Phys. Chem. Lett.*, 2021, **12**, 9800–9806.

27. S. Giri, J. C. Tremblay and G. Dixit, *Phys. Rev. A*, 2021, **104**, 053115.

28. F. Allum, K. Amini, M. Ashfold, D. Bansal, R. J. F. Berger, M. Centurion, G. Dixit, D. Durham, E. Fasshauer, J. P. Figueira Nunes, I. Fischer, G. Grell, M. Ivanov, A. Kirrander, O. Kornilov, C. Kuttner, K. Lopata, L. Ma, V. Makhija, A. Maxwell, A. Moreno Carrascosa, A. Natan, D. Neumark, S. Pratt, A. Röder, D. Rolles, J. M. Rost, T. Sekikawa, M. Simmermacher, A. Stolow, E. Titov, J. C. Tremblay, P. M. Weber, H. Yong and L. Young, *Faraday Discuss.*, 2021, **228**, 161–190.

29. M. Inokuti, *Rev. Mod. Phys.*, 1971, **43**, 297–347.

30. H. Bethe, *Ann. Phys.*, 1930, **397**, 325–400.

31. V. G. Baryshevskii, I. D. Feranchuk and P. B. Kats, *Phys. Rev. A: At., Mol., Opt. Phys.*, 2004, **70**, 052701.

32. L. Ma, H. Yong, J. D. Geiser, A. Moreno Carrascosa, N. Goff and P. M. Weber, *Struct. Dyn.*, 2020, 7, 034102.

33. J. Yang, X. Zhu, J. P. F. Nunes, J. K. Yu, R. M. Parrish, T. J. A. Wolf, M. Centurion, M. Gühr, R. Li, Y. Liu, B. Moore, M. Niebuhr, S. Park, X. Shen, S. Weathersby, T. Weinacht, T. J. Martinez and X. Wang, *Science*, 2020, **368**, 885–889.

34. M. Stefanou, K. Saita, D. V. Shalashilin and A. Kirrander, *Chem. Phys. Lett.*, 2017, **683**, 300–305.

35. F. M. Rudakov, J. B. Hastings, D. H. Dowell, J. F. Schmerge and P. M. Weber, *AIP Conf. Proc.*, 2006, **845**, 1287–1292.

36. F. Salvat, A. Jablonski and C. J. Powell, *Comput. Phys. Commun.*, 2005, **165**, 157–190.

37. G. Hermann, V. Pohl, J. C. Tremblay, B. Paulus, H.-C. Hege and A. Schild, *J. Comput. Chem.*, 2016, **37**, 1511–1520.

38. V. Pohl, G. Hermann and J. C. Tremblay, *J. Comput. Chem.*, 2017, **38**, 1515–1527.

39. G. Hermann, V. Pohl and J. C. Tremblay, *J. Comput. Chem.*, 2017, **38**, 2378–2387.

40. R. M. Parrish and T. J. Martínez, *J. Chem. Theory Comput.*, 2019, **15**, 1523–1537.

41. J. Wang and V. H. Smith Jr, *Int. J. Quantum Chem.*, 1994, **52**, 1145–1151.

42. A. Debnarova, S. Techert and S. Schmatz, *J. Chem. Phys.*, 2006, **125**, 224101.

43. D. L. Crittenden and Y. A. Bernard, *J. Chem. Phys.*, 2009, **131**, 054110.

44. T. Northey, N. Zotev and A. Kirrander, *J. Chem. Theory Comput.*, 2014, **10**, 4911–4920.

45. A. M. Carrascosa and A. Kirrander, *Phys. Chem. Chem. Phys.*, 2017, **19**, 19545–19553.

46. A. Moreno Carrascosa, H. Yong, D. L. Crittenden, P. M. Weber and A. Kirrander, *J. Chem. Theory Comput.*, 2019, **15**, 2836–2846.

47. N. Zotev, A. Moreno Carrascosa, M. Simmermacher and A. Kirrander, *J. Chem. Theory Comput.*, 2020, **16**, 2594–2605.

48. P. Debye, *Phys. Z.*, 1930, **31**, 419–428.

49. W. Heisenberg, *Phys. Z.*, 1931, **32**, 737–740.

50. P. M. Morse, *Phys. Z.*, 1932, **33**, 443–445.

51. L. Bewilogua, *Phys. Z.*, 1932, **33**, 688–692.

52. J. H. Hubbell, W. J. Veigele, E. A. Briggs, R. T. Brown, D. T. Cromer and R. J. Howerton, *J. Phys. Chem. Ref. Data*, 1975, **4**, 471–538.

53. R. F. Stewart, E. R. Davidson and W. T. Simpson, *J. Chem. Phys.*, 1965, **42**, 3175.

54. N. F. Mott and W. L. Bragg, *Proc. R. Soc. London, Ser. A*, 1930, **127**, 658–665.

55. D. A. Varshalovich, A. N. Moskalev and V. K. Khersonskii, *Quantum Theory of Angular Momentum*, World Scientific, Singapore, 1st edn, 1989.

56. J. C. Williamson and A. H. Zewail, *J. Phys. Chem.*, 1994, **98**, 2766–2781.

57. J. S. Baskin and A. H. Zewail, *ChemPhysChem*, 2005, **6**, 2261–2276.

58. J. S. Baskin and A. H. Zewail, *ChemPhysChem*, 2006, **7**, 1562–1574.

59. U. Lorenz, K. B. Møller and N. E. Henriksen, *New J. Phys.*, 2010, **12**, 113022.

60. M. Feit, J. Fleck and A. Steiger, *J. Comp. Phys.*, 1982, **47**, 412–433.

61. M. Bonfanti, G. A. Worth and I. Burghardt, *Quantum Chemistry and Dynamics of Excited States: Methods and Applications*, John Wiley and Sons, 1st edn, 2020, ch. 12, pp. 383–411.

62. G. A. Worth and B. Lasorne, *Quantum Chemistry and Dynamics of Excited States: Methods and Applications*, John Wiley and Sons, 1st edn, 2020, ch. 13, pp. 413–433.

63. A. Kirrander and M. Vacher, *Quantum Chemistry and Dynamics of Excited States: Methods and Applications*, John Wiley and Sons, 1st edn, 2020, ch. 15, pp. 469–497.

64. B. F. E. Curchod, *Quantum Chemistry and Dynamics of Excited States: Methods and Applications*, John Wiley and Sons, 1st edn, 2020, ch. 14, pp. 435–467.

65. G. A. Worth, K. Giri, G. W. Richings, I. Burghardt, M. H. Beck, A. Jäckle and H.-D. Meyer, *The QUANTICS Package, Version 1.1*, 2015.

66. G. A. Worth, *Comput. Phys. Commun.*, 2020, **248**, 107040.

67. S. Mai, P. Marquetand and L. González, *Quantum Chemistry and Dynamics of Excited States: Methods and Applications*, John Wiley and Sons, 1st edn, 2020, ch. 16, pp. 499–530.

68. A. Kirrander and P. M. Weber, *Appl. Sci.*, 2017, 7, 534.

69. H.-J. Werner, P. J. Knowles, G. Knizia, F. R. Manby and M. Schütz, *Wiley Interdiscip. Rev.: Comput. Mol. Sci.*, 2012, **2**, 242–253.

70. H.-J. Werner, P. J. Knowles, G. Knizia, F. R. Manby and M. Schütz, *MOLPRO, version 2015.1, a package of ab initio programs*, 2015, see http://www.molpro.net.

71. S. P. Neville, A. Stolow and M. S. Schuurman, *J. Phys. B: At. Mol. Opt. Phys.*, 2022, **55**, 044004.

72. K. M. Ziems, M. Simmermacher, S. Gräfe and A. Kirrander, *J. Chem. Phys.*, 2023, **159**, 044108.

73. S. Ryu, R. M. Stratt and P. M. Weber, *J. Phys. Chem. A*, 2003, **107**, 6622–6629.

74. S. Ryu, R. M. Stratt, K. K. Baeck and P. M. Weber, *J. Phys. Chem. A*, 2004, **108**, 1189–1199.

CHAPTER 4

Femtosecond Diffraction with Laser-driven Hard X-ray Sources: Nuclear Motions and Transient Charge Densities

C. HAUF,[†] M. WOERNER AND T. ELSAESSER*

Max-Born-Institut, Berlin, 12489, Germany
*Email: elsasser@mbi-berlin.de

4.1 Introduction

X-ray diffraction has provided a wealth of information on the equilibrium structures of crystalline matter. Diffraction patterns consisting of up to millions of diffraction peaks have been recorded with high-brightness hard X-ray sources such as third- and fourth-generation synchrotrons. Their analysis has given precise information on atomic arrangements and electronic charge distributions in increasingly complex structures, including large proteins and other biomolecular systems. Charge density analysis has revealed fine details of chemical bonding on length scales corresponding to a small fraction of a covalent bond length.

In recent years, the study of transient nonequilibrium structures with femtosecond electron and X-ray diffraction methods has been established as a new area of condensed-matter research.[1-6] The analysis of structure-transforming processes on atomic length and timescales, *i.e.*, in the

---

[†]Current address: Research Neutron Source Heinz Maier-Leibnitz (FRM II), Technical University of Munich, Garching, 85748, Germany.

Theoretical and Computational Chemistry Series No. 25
Structural Dynamics with X-ray and Electron Scattering
Edited by Kasra Amini, Arnaud Rouzée and Marc J. J. Vrakking
© The Royal Society of Chemistry 2024
Published by the Royal Society of Chemistry, www.rsc.org

(sub-)angstrom length and atto- to femtosecond time domain, holds strong potential for understanding the relevant interactions and driving forces behind them. Such knowledge is not only essential for clarifying structure–function relationships at the atomic level but is also relevant for steering functional properties through tailored light–matter interactions.

Most experiments in femtosecond X-ray diffraction are based on a pump–probe approach in which an optical pump pulse initiates a structure-changing process and a hard X-ray probe pulse interacts with the sample at a fixed time delay to generate a momentary X-ray diffraction pattern.[1,2] The sequence of such patterns as a function of delay time allows the reconstruction of the time evolution of the structure and the underlying processes. The time resolution is determined by the duration of pump and probe pulses, the fluctuations in timing between them (temporal jitter), and the group velocity difference between the optical and the X-ray pulse in the sample. Consequently, the generation of ultrashort hard X-ray pulses synchronized with an optical femtosecond laser system is the main prerequisite for this type of study.

Laser-driven sources enable hard X-ray generation in a laboratory frame and offer a time resolution of pump–probe experiments on the order of 100 fs.[7–11] This experimental approach has resulted in a series of pioneering X-ray diffraction studies of structural dynamics in crystalline solids and layered materials. In this way, the foundation has been laid for the field of time-resolved structure research with ultrafast diffraction methods. While early experiments have been performed at a pulse repetition rate below 50 Hz, state-of-the art laser-driven sources provide a stable hard X-ray output at kilohertz repetition rates.[9,12] The latter technology has enabled the implementation of highly sensitive diffraction and scattering schemes avoiding sample destruction and, thus, allowing for observing subtle ultrafast changes in the structure of crystalline matter.

Early work on X-ray diffraction with laser-driven sources has focused on mapping irreversible melting processes, acoustic phonon and strain propagation and/or coherent lattice dynamics such as optical phonon wavepackets or molecular vibrations.[13–17] The implementation of powder diffraction has opened the field of time-resolved charge density analysis by deriving transient electron density maps from powder diffraction patterns.[18–20] In this way, relocations of valence charge become accessible. Such progress has lead to new insight into charge dynamics in polar crystals, including ferroelectric materials. The results have also allowed for exploring the connection between changes of microscopic electron density and macroscopic electric polarizations. A topic of strong current interest is the interplay of lattice and electron motions, for example, charge relocations driven by so-called soft-mode phonon excitations.

Beginning around the year 2000, femtosecond hard X-ray pulses have been generated in accelerator-based sources such as so-called slicing beamlines at synchrotrons and free electron lasers (FELs).[21–26] FELs offer an un-precedented hard X-ray flux on the order of 10^{12} photons per pulse at repetition rates of up to now some 100 Hz. Their extremely high brightness has allowed for the development of novel X-ray diffraction methods such as serial crystallography[27] and resonant X-ray diffraction.[28] Moreover, FELs will pave the way towards X-ray nonlinear optics, *i.e.*, highly nonlinear light–matter interactions at hard X-ray wavelengths.[29,30] Nevertheless, there is a broad range of open scientific questions that can be addressed with laser-driven hard X-ray sources in the laboratory frame, in part offering a higher experimental sensitivity than FEL experiments and avoiding their severe beamtime limitations. In femtosecond X-ray powder diffraction experiments operated at a 1 kHz repetition rate, one can measure transient intensity changes of Bragg reflections as small as $\Delta I_{hkl}/I_{hkl} = 0.001$, with signal-to-noise ratios as large as 5 after up to 50 experiment days.[31] So far, the higher sensitivity of tabletop femtosecond X-ray experiments is predicated on a higher repetition rate and longer data collecting periods compared to those of typical FEL experiments performed in the past.

This chapter gives an overview of recent and current research based on femtosecond X-ray diffraction with laser-based laboratory sources. Experimental aspects, methods for transient charge density analysis, and prototypical results are discussed in a tutorial way. The text is organized as follows: Section 4.2 introduces basic concepts, experimental techniques, and methods of data analysis. Section 4.3 covers the physics of optical phonon excitation, soft modes, electron–phonon coupling, and addresses the connection between microscopic charge densities and macroscopic electric polarizations. Section 4.4 presents results on optically driven phonon dynamics in polar crystals and nanostructures, while Section 4.5 addresses phonon-driven charge dynamics as observed in recent experiments. Conclusions are given in the short final Section 4.6.

4.2   Methods

4.2.1   Laser-driven Sources of Ultrashort Hard X-ray Pulses

Ultrafast X-ray science requires sources for soft and hard X-ray pulses with attosecond to femtosecond durations. In addition to a sufficient photon flux, synchronization of the X-ray pulses with optical pulses is mandatory for implementing pump–probe schemes. There are two major types of attosecond to femtosecond X-ray sources, accelerator-based large-scale infrastructures such as FELs and laser-driven sources covering a spectral range from the extended ultraviolet up to hard X-ray photon energies. The work discussed in this chapter is based on hard X-ray sources driven by femtosecond lasers or pulses from laser-pumped optical parametric chirped pulse amplification (OPCPA) systems at a kilohertz repetition rate.[9,12,32] In the following, the physics of this generation method is presented, including a summary of the current state of technology.

A schematic of a table-top laser-driven pump–probe setup is shown in Figure 4.1a. The optical driver system provides sub-100 fs pulses of millijoule

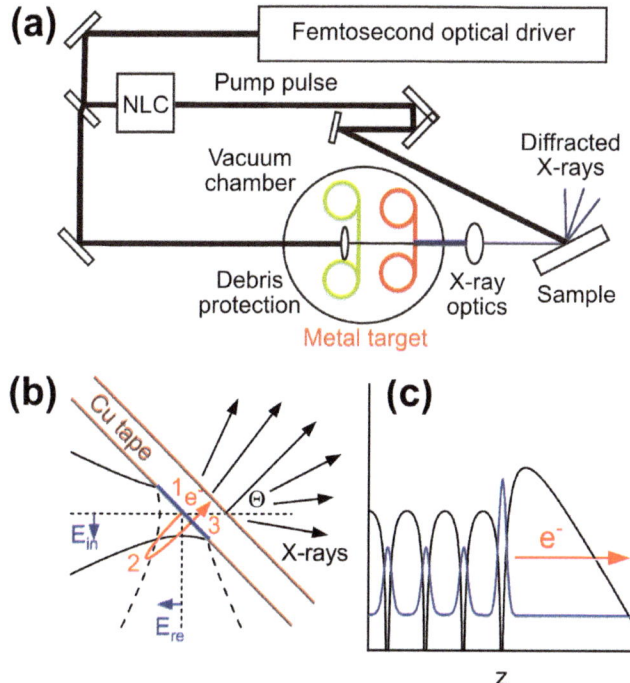

Figure 4.1 (a) Schematic of an experimental setup for femtosecond X-ray diffraction with a laser-driven table-top hard X-ray source. (b) Interaction geometry at the metallic tape target. The numbers refer to the different steps of X-ray generation. (c) Schematic of electron extraction from the target through tunneling induced by the optical field.

energy at a kilohertz repetition rate. Pulses with a centre wavelength around 800 nm are typically generated with amplified Ti:sapphire lasers, while OPCPA systems provide pulses at longer mid-infrared wavelengths up to 5 μm. A major fraction of the driver output is used for X-ray generation, a weaker replica split off with a beamsplitter serves as pump pulse. The wavelength of the pump pulses is often changed by nonlinear optical frequency conversion (NLC), in many cases by generating harmonics of the fundamental wavelengths. The pump pulses are focused onto the sample and induce a structure-changing elementary excitation.

The X-ray driver pulse is focused at a spot size on the order of some micrometers on a thin metal tape target placed in a vacuum chamber (pressure 10^{-4} mbar) which moves continuously with a speed of several cm per second in order to provide a fresh target for each incoming laser pulse. The interaction of the driver with the target generates debris in the form of evaporated metal which is taken up by moving plastic tapes in front of and behind the target. The generated X-rays are collected in the forward direction by an X-ray multilayer optics and focused onto the sample to generate a diffraction pattern. Practically all experiments performed with this scheme have made use of Bragg diffraction from single crystals, *i.e.*, mapped individual diffraction peaks, or from crystalline powder, resulting in Debye–Scherrer ring patterns. The diffracted X-rays are typically recorded with area detectors (not shown in Figure 4.1a).

A major advantage of this setup consists in the comparably short optical path lengths on the order of at most a few metres, *i.e.*, its compactness, and in the negligible timing jitter between the optical pump and the X-ray probe pulses. Both pulses are derived from the same optical driver and are, thus, synchronized on the few-femtosecond timescale of an optical half cycle of the driving laser pulse. In contrast, pump–probe studies with accelerator based X-ray sources and separate optical lasers require highly sophisticated synchronization schemes for reaching a comparable femtosecond time resolution.

The physical processes relevant for X-ray generation at the metal target are discussed next.[33] Figure 4.1b shows the interaction geometry with a thin metallic tape target made from, for example, copper (Cu). The incoming and the reflected p-polarized optical excitation pulse overlap in front of the target, generating a strong optical field component $E_\perp$ perpendicular to the front surface of the metal. For peak intensities of the driver pulses between 10^{16} and 10^{18} W cm^{-2}, $E_\perp$ reaches a value between 2×10^{11} and 2×10^{12} V m^{-1} or 200 and 2000 V nm^{-1}. At such high field strength, Cu atoms in the front layers of the target are field-ionized, *i.e.*, electrons are released by tunneling from bound atomic states into continuum states in vacuum (see Figure 4.1c, and step (1) in Figure 4.1b). Electron tunneling displays a quantum yield close to one for $E_\perp > 1000$ V nm^{-1}.

Electrons in vacuum are accelerated in the strong optical field, a process occurring on the few-femtosecond timescale of a half-cycle during which $E_\perp$ first points away from and then towards the target (see step (2) in Figure 4.1b).[34] This so-called vacuum heating results in relativistic electron kinetic energies of up to several hundreds of keV. The acquired kinetic energy is proportional to $|E_\perp|^2 \lambda^2$, where λ is the centre wavelength of the driving field. The dependence on λ arises because a longer optical cycle $T = \lambda/c$ (*c*: velocity of light) corresponds to a longer acceleration period and a higher kinetic energy attained. This fact makes femtosecond driver pulses with centre wavelengths in the mid-infrared particularly attractive. A driving field centred at $\lambda = 800$ nm has a half cycle of $T/2 = 1.35$ fs, while the half cycle of a $\lambda = 5$ μm pulse lasts for $T/2 = 8.33$ fs.

The spatial trajectory of an accelerated electron depends on the instant in time of the ionization event within the optical cycle of the driving field. For electrons released close to the onset of a half cycle, $E_\perp$ points away from the target for the full acceleration period and the electron moves to large distances away from the target. In contrast, electrons generated at phases between $\pi/2$ and π in a sinusoidal driving field experience a reversal in the sign of $E_\perp$ and are accelerated back towards the target and acquire a high kinetic energy on the order of some 100 keV. A detailed discussion and simulations of phase-dependent electron trajectories have been presented in ref. 33.

On top of the external driving field, positively charged Cu ions in the target material generate an attractive electric space-charge field acting on the electrons as well. This field makes all electrons eventually return to the target.

The returning accelerated electrons enter the metal target, interact with unexcited Cu atoms, and generate both characteristic X-ray emission and Bremsstrahlung. Both types of emission are incoherent and cover the full solid angle [step (3) in Figure 4.1b]. The characteristic emission has a femtosecond time structure which is determined by the duration of the driver pulses and the propagation length of electrons in the target. For a target thickness of approximately 20 µm, the duration of the characteristic X-ray pulses has a value of some 100 fs (see Section 4.2.2). A theoretical model allowing for a quantitative simulation of femtosecond X-ray generation has been presented in ref. 33.

In the present scheme, plasma effects originating from the presence of both vacuum electrons and positive Cu ions in vacuum and in the target play a minor role in X-ray generation. During a 100 fs period, the positive ions move by less than 5 nm and, thus, plasma expansion is negligible. Nevertheless, the positive Cu ions represent a space charge affecting the spatial profile of the driving field $E_\perp$. To limit such effects, driver pulses with a high temporal contrast and a suppression of optical pre-pulses, which generate ions before the arrival of the main driver pulse, are important. In summary, the generation scheme presented here is based on tunneling ionization and vacuum heating, in contrast to X-ray plasma sources driven by picosecond or nanosecond optical pulses.

Femtosecond hard X-ray sources working at kilohertz repetition rates have been implemented with both near- and mid-infrared optical drivers.[8,9,12,32,35,36] While amplified Ti:sapphire driver lasers providing sub-50 fs pulses around 800 nm represent an established technology, recent work has demonstrated the strong potential of mid-infrared drivers based on the OPCPA concept.[32,37] The OPCPA sources typically consist of a multi-stage arrangement of nonlinear optical frequency converters in which their second-order optical nonlinearity is exploited for parametric frequency conversion.[38] A picosecond pump pulse centred at a wavelength λ_1 provides the energy for parametric amplification of femtosecond pulses at longer wavelengths $\lambda_{2,3}$ with $1/\lambda_1 = 1/\lambda_2 + 1/\lambda_3$. This concept has allowed for the generation of sub-100 fs mid-infrared pulses of millijoule pulse energy. The OPCPA setup providing 5 µm idler pulses discussed below is described in full detail in ref. 38.

In the following, the state-of-the-art and the parameters of the generated X-ray pulses are discussed. X-ray emission spectra are summarized in Figure 4.2, where the normalized emitted intensity is plotted as a function of photon energy. Data were measured for a Cu tape target (thickness 20 µm) driven by 800 nm pulses with a duration of 35 fs (blue line in Figure 4.2a) and 80 fs pulses at 5 µm (red line); the optical pulse energy on the target was 2–4 mJ. The spectrum from a Mo target (Figure 4.2b) was also excited with 800 nm pulses. The spectra consist of narrow characteristic

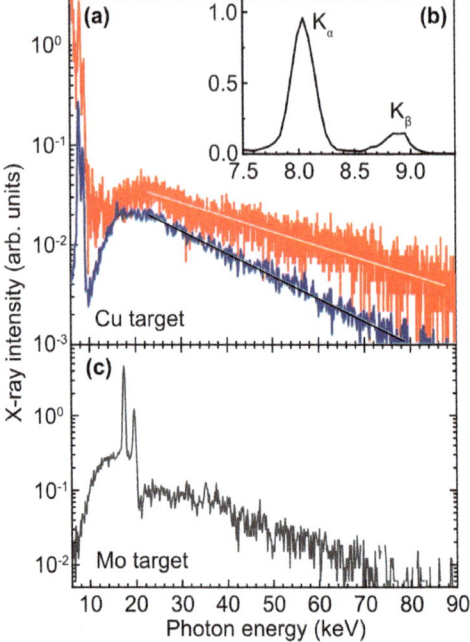

Figure 4.2   (a) X-ray emission spectra of a Cu tape target driven by 800 nm pulses (blue line) and 5 μm pulses (red line). The narrow peaks around 8 keV are characteristic emission, the broad bands represent Bremsstrahlung. (b) Emission spectrum in the range of the Cu Kα and Kβ components (driver wavelength 800 nm). The width of the characteristic emission peaks is determined by the energy resolution of the single-photon X-ray detector. (c) Emission spectrum of a Mo tape target driven by 800 nm pulses, the Kα emission occurs around 17 keV.

Kα and Kβ emissions and a broad band of Bremsstrahlung which extends to photon energies on the order of 100 keV. The spectral width of characteristic emission in Figure 4.2 reflects the energy resolution of the detector. The much smaller intrinsic bandwidth is determined by the few-femtosecond lifetime of the Cu core hole generated by electron impact and has a value of 2 eV. The band shape of Bremsstrahlung is governed by the energy distribution of accelerated electrons. Assuming an exponential high-energy tail (see white and black lines in Figure 4.2a), *i.e.*, a Boltzmann distribution,[39] one extracts a slope $(kT_x)^{-1} = 29$ keV for the 5 μm driver and some 20 keV for the 800 nm driver (T_x: X-ray photon temperature, k: Boltzmann's constant). The higher T_x for the 5 μm driver is due to the higher kinetic energy of electrons, which originates from the longer acceleration period compared to the 800 nm driver.

The generated hard X-ray flux depends strongly on the peak intensity $I \propto |E_\perp|^2$ of the driving optical pulses. Results for the Cu Kα source are presented for centre wavelengths of the driver of 5 μm and 800 nm (Figure 4.3). The experimental results (symbols) display a very steep, nearly exponential rise with

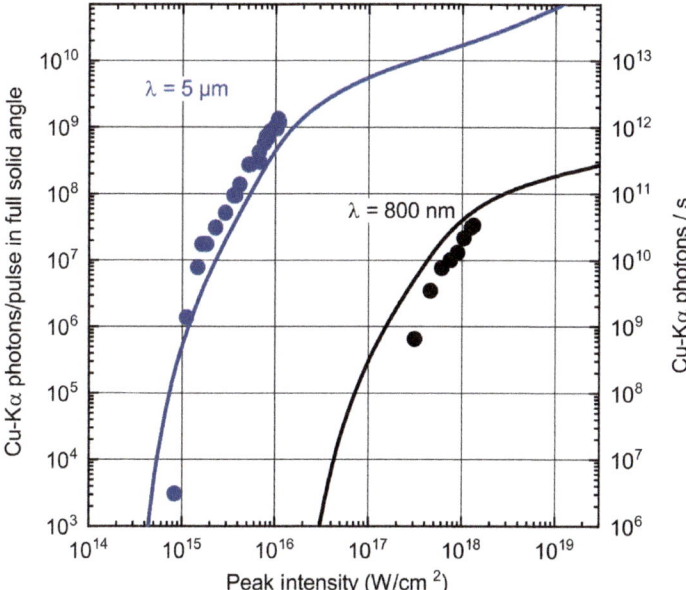

Figure 4.3 Spatially integrated Cu Kα photon flux into the full solid angle as a function of peak intensity of the femtosecond driver pulses centred at 5 μm (blue symbols and line) and at 800 nm (black symbols and line). Symbols represent experimental results while the solid line represents the results of calculations based on the model reported in ref. 33. The maximum flux generated with 5 μm excitation is 30 times higher than for excitation with 800 nm pulses of comparable energy.[32]

an onset of saturation at the highest optical peak intensities. This behaviour is well reproduced by theoretical calculations (solid lines) based on the approach of ref. 33. The results for the 5 μm driver exhibit a maximum flux which is 30 times higher than that from the source driven by 800 nm pulses of comparable energy and reaches a value 1.5×10^9 X-ray photons per pulse or 1.5×10^{12} X-ray photons per second. Moreover, there is a strong re-scaling of the driver intensity when switching from 800 nm to 5 μm pulses, fully in line with the $(E_\perp)^2 \lambda^2 \propto I \lambda^2$ dependence of the X-ray yield with driver intensity I and wavelength λ. The higher kinetic energy achieved by vacuum electrons in the 5 μm optical field, due to the longer optical period, leads to an increase in both the yield of Kα emission and the spectral width of Bremsstrahlung (see Figure 4.2a).

A detailed analysis of the long-term-stability and the pulse-to-pulse fluctuations of the Cu Kα source has been presented in ref. 40. In brief, the X-ray flux varies by ±10% on a timescale of 20 minutes. The shot-to-shot fluctuations are on the order of 40%. The Fourier spectrum of the shot-to-shot fluctuations displays a $1/f$ behaviour for frequencies f between 0.1 and 10 Hz, close to the behaviour of a photon source with pink noise. The duration of the generated Cu Kα pulses is on the order of 100 fs and has been

Table 4.1 Properties of femtosecond laser-driven X-ray sources working at a 1 kHz repetition rate. The spatial resolution and the MEM resolution are estimated from retrieved crystalline structures and electron density maps (MEM: maximum entropy method).

Target material	Cu	Cu	Mo
Driver wavelength (µm)	5	0.8	0.8
Photon energy (keV)	8.05	8.05	17.48
Total Kα flux ($\times 10^{12}$ photons per s)	1.5	0.05	0.0078
Pulse duration	120 fs	~100 fs	—
Source diameter (µm)	34	10	12.6
Spatial resolution (nm)	0.09	0.09	0.043
MEM resolution (nm)	0.07	0.07	0.037

determined by measuring the temporal cross correlation with 800 nm pulses. Such experiments are discussed in Section 4.2.2. A summary of the main source parameters is presented in Table 4.1.

The achievable spatial resolution is mostly limited by the employed wavelength of the femtosecond X-ray pulses, and geometrical restrictions on where the detector can be placed, resulting in an upper limit for the attainable scattering angle. For a typical $2\theta_{hkl}^{max} < 60°$, the spatial resolution is given in the last two rows of Table 4.1, and typically up to some 40 Bragg reflections can be monitored, depending on the crystal structure studied.

4.2.2 Methods of Ultrafast X-ray Diffraction

Pump–probe schemes, with a femtosecond optical excitation pulse and a synchronized hard X-ray probe pulse to be diffracted from the excited sample, are the most common approach in ultrafast X-ray diffraction. Characteristic X-ray pulses from laser-driven sources have a small bandwidth below 5 eV and, thus, mainly Bragg diffraction methods have been applied to study crystalline materials. In the weak scattering or kinematic diffraction limit, maxima of diffracted intensity occur at angular positions defined by the Bragg condition

$$\mathbf{k}' - \mathbf{k} = \mathbf{q} = \mathbf{G}, \quad 2d_{hkl}\sin(\Theta) = m\lambda \qquad (4.1)$$

where $\mathbf{k}$ and $\mathbf{k}'$ are the wavevectors of the incoming and diffracted X-ray beams, respectively, $\mathbf{q}$ is the scattering vector, $\mathbf{G} = (h,k,l)$, where the lattice plane index (hkl) is a reciprocal lattice vector of the crystal lattice under study, d_{hkl} is the separation of neighbouring lattice planes (hkl), 2Θ is the diffraction angle enclosed by $\mathbf{k}$ and $\mathbf{k}'$, λ is the X-ray wavelength, and $m = 1, 2,\ldots$ is an integer number.[41] For single crystals, $\mathbf{G}$ has a well-defined orientation in space and the diffraction pattern consists of individual Bragg peaks. In powders of crystallites, $\mathbf{G}$ has a random orientation, resulting in a diffraction pattern consisting of concentric Debye–Scherrer rings. The absolute value of the scattering vector is given by $|\mathbf{q}| = 4\pi\sin(\Theta)/\lambda$.

In crystallography, the prefactor 4π is typically omitted and $|\mathbf{q}| = \sin(\Theta)/\lambda$. In the following, we use the latter expression.

X-rays interact predominantly with electrons in the crystal. The intensity of a Bragg peak or ring I_{hkl} is proportional to $|F_{hkl}|^2$ with the structure factor F_{hkl} given by the spatial Fourier transform of the electron density $\rho(\mathbf{r})$:

$$F_{hkl} = \int_{UC} d^3 r \rho(\mathbf{r}) \exp(i\mathbf{G} \cdot \mathbf{r}) \tag{4.2}$$

Here, the integral runs over the unit cell (UC). The electron density $\rho(\mathbf{r})$ of crystals consisting of heavier atoms is dominated by their core electrons and, thus, the diffraction pattern directly reflects the atomic positions. For crystals containing light atoms, valence electrons make a substantial contribution to $\rho(\mathbf{r})$ and, consequently, contribute appreciably to the diffraction of X-rays. The kinematic limit of X-ray diffraction breaks down if multiple scattering events in the crystalline sample affect the diffracted intensity. Such contributions are taken into account in dynamic diffraction as, for example, described by the Darwin theory.[41]

Thermally activated motions of the crystal lattice affect the observed X-ray diffraction pattern. The resulting reduction in intensity I_{hkl} of the Bragg peaks or Debye–Scherrer rings is described by the Debye–Waller factor which, for a lattice with a single type of atoms undergoing an average elongation u, is given by $I_{hkl}/I_0 = \exp[-(1/3)|\mathbf{G}|^2\langle u^2 \rangle]$. The kinetic energy of a harmonic oscillator of mass M and frequency ω in thermal equilibrium at a temperature T is given by $1/2 M\omega^2 \langle u^2 \rangle = (3/2)kT$, resulting in $I_{hkl}/I_0 = \exp[-kT|\mathbf{G}|^2/(M\omega^2)]$ (I_0: intensity diffracted in the absence of thermal motion and k: Boltzmann's constant). A more detailed discussion of the Debye–Waller factor and the related phenomenon of thermal diffuse scattering has been given in ref. 42.

There are different types of photoinduced structural dynamics that can be mapped directly through time-resolved X-ray diffraction. Changes of the interlayer distance d in the crystal lattice are connected with a change in size of a unit cell (Figure 4.4a), and give rise to an angular shift of the related diffraction peaks or, in the case of powder diffraction, to a change in diameter of the Debye–Scherrer rings. In the case of an increase in interlayer distance, the corresponding Bragg peak shifts to a smaller diffraction angle. Observation of such changes requires many unit cells to change their dimension. After optical excitation of a small fraction of unit cells, the limit in which most ultrafast diffraction experiments are performed, the characteristic picosecond timescale on which angular shifts occur is set by acoustic phonon propagation, spreading the initial local structure change over many unit cells. In the femtosecond time domain, the lattice geometry appears to be 'frozen', *i.e.*, unchanged compared to equilibrium even in the presence of acoustic phonon propagation.

A change in the structure factor F_{hkl} results in a change in intensity of the diffraction peak or ring (*hkl*) (see Figure 4.4b). Photoinduced changes in

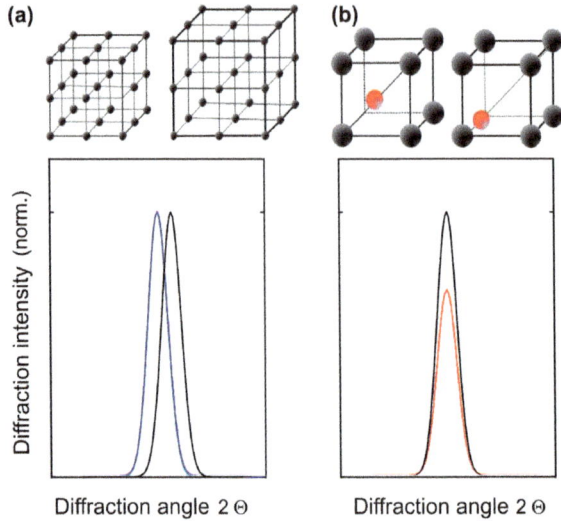

Figure 4.4 (a) Schematic of the angular shift of a Bragg diffraction peak (black line) upon an increase in lattice constant (blue line), as exemplified for a cubic lattice (upper panel). (b) Exemplary decrease (an increase is also possible) of diffracted intensity on a transient Bragg peak (red line) upon a change of the structure factor, as caused by, for example, a local change of the unit cell geometry (upper panel).

charge density within the excited unit cells, optical phonon or other local excitations can cause such a change upon which the angular position of the diffraction peak remains unchanged. The angle-integrated intensity change of a particular Bragg peak or Debye–Scherrer ring is given by

$$\frac{\Delta I_{hkl}(t)}{I^0_{hkl}} = \frac{I_{hkl}(t) - I^0_{hkl}}{I^0_{hkl}} = \frac{|F_{hkl}(t)|^2 - |F^0_{hkl}|^2}{|F^0_{hkl}|^2} \tag{4.3}$$

where $I_{hkl}(t)$ and $F_{hkl}(t)$ are the intensity and the structure factor at delay time t, and I^0_{hkl} and F^0_{hkl} are the intensity and structure factor before excitation.[5] In systems with a known equilibrium structure, *i.e.*, known structure factor F^0_{hkl} and electron density $\rho^0(\mathbf{r})$, the quantity $|F_{hkl}(t)|^2$ is derived from the measured intensity change. The transient electron density $\rho(\mathbf{r},t)$ is then determined by inverting the Fourier integral in eqn (4.2). The latter step requires a solution of the so-called phasing problem by analytical or numerical methods as $F_{hkl}(t)$ is a complex quantity. A detailed discussion of this issue for time-resolved powder diffraction is given in Section 4.2.3.

Ultrafast X-ray diffraction makes use of different interaction and detection geometries as shown in Figure 4.5. In Figure 4.5a, a schematic of a pump–probe experiment with single crystals is shown. The optical pump pulse and the pulse driving the X-ray source are derived from the output of the driver laser with negligible timing jitter. The characteristic X-ray emission from

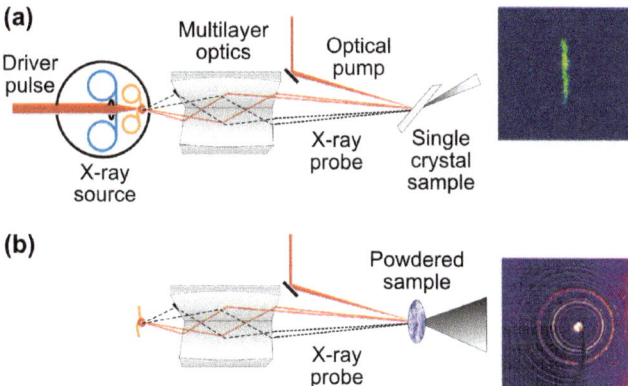

Figure 4.5 (a) Schematic of an optical-pump/X-ray-probe setup for diffraction from single crystals, including a typical diffraction peak. (b) Interaction geometry for femtosecond X-ray power diffraction and diffraction pattern.

the X-ray source is collected in the forward direction through multilayer optics and focused onto the sample. These optics are wavelength-selective and separate the Kα component from the Bremsstrahlung background. The incoming X-ray flux on the sample has a value of up to several 10^8 photons per s with spot sizes between 50 and several hundred micrometres, depending on the particular X-ray optics.[43] Diffraction patterns are recorded with single-element or two-dimensional detectors with a sensitivity down to the single photon level. With a single-crystal sample, different Bragg peaks are recorded sequentially by tilting the crystal and/or moving the detector. The powder diffraction scheme (Figure 4.5b) implements a two-dimensional detector, allowing for the simultaneous recording of many Debye–Scherrer rings.[18] This method has been applied in all femtosecond charge density studies performed so far.

A systematic experimental analysis has identified the main sources of fluctuations and noise in the time-resolved diffraction experiments, most of which originate from the target spooling mechanics.[40] Optimization of the setup and data recording protocol has allowed for measurements close to the shot-noise limit set by the detected X-ray photon flux. Due to the high stability of the experimental setup, data can be reliably recorded over periods of more than 24 hours. In femtosecond powder diffraction experiments, intensity changes of Debye–Scherrer rings as small as $\Delta I_{hkl}/I^0_{hkl} \approx 10^{-3}$ have been measured, a sensitivity not reached so far with accelerator based sources.[20]

An important issue is the time resolution of the experiment, which is defined by the cross-correlation of the optical pump and the X-ray probe pulses. The challenge to measure this cross-correlation has been addressed in experiments with nonresonant excitation of charge-transfer materials far below their fundamental bandgap. In this scheme, the electric field of the pump induces a relocation of electrons without creating real carriers of finite lifetime. As a result, the rise and decay of the electron transfer in this virtual state is entirely determined by the duration of the pump pulse, *i.e.*, the response is quasi-instantaneous. The changes in the spatial electron distribution manifest as changes in the structure factor and, thus, intensity of the diffraction pattern generated with the hard X-ray probe pulse. In the pump–probe experiment, the pump-induced charge dynamics are convoluted with the temporal envelope of the X-ray probe pulses, directly giving the pump–probe cross-correlation.

Results of such a measurement with a polycrystalline sample of $LiBH_4$ are summarized in Figure 4.6.[44] The orthorhombic unit cell of this material contains four pairs of Li^+ and BH_4^- ions with a separation of 2.5 Å. The electric field of the pump pulse induces charge transfer from the electron-rich BH_4^- to the neighbouring positive Li^+ ion, as is evident from the differential electron density map $\Delta\rho(t,\mathbf{r}) = \rho(t,\mathbf{r}) - \rho_0(\mathbf{r})$ recorded at zero pump–probe delay in the $Y = 0.25$ plane of the unit cell (Figure 4.6a). Integrating the differential electron density over the respective volume of the two ionic entities gives the total charge change Δq on each unit. The total charge change is plotted as a function of pump–probe delay for the Li^+ and the BH_4^- ion in Figure 4.6b and c. The amplitudes of the charge increase on Li^+ and the charge decrease on BH_4^- are the same, demonstrating that charge transfer in essentially localized in the ion pair. The time traces of a total width of 120 fs (FWHM) were deconvoluted with

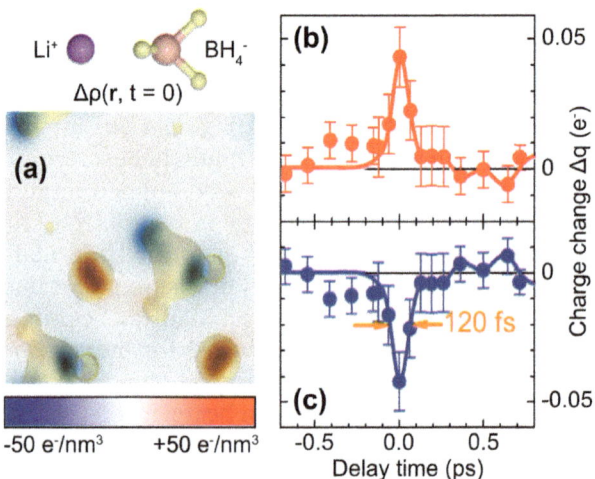

Figure 4.6 (a) Upper panel: Schematic of a Li^+ and a BH_4^- ion pair in $LiBH_4$. Lower panel: Differential charge density map $\Delta\rho(\mathbf{r},t) = \rho(\mathbf{r},t) - \rho_0(\mathbf{r})$ measured at zero delay $t = 0$ between pump and probe pulses ($\rho(\mathbf{r},t)$, $\rho_0(\mathbf{r})$: transient and equilibrium electron density). (b and c) Charge change Δq integrated over the volume of the Li^+ (red symbols) and the BH_4^- ion (blue symbols). The 120 fs width of the transients is determined by the 50 fs duration of the optical and the 100 fs duration of the X-ray pulses.

a pump pulse duration of 50 fs. This procedure gives a duration of the hard X-ray probe pulses of approximately 100 fs, in agreement with theoretical simulations of the X-ray generation process.[33]

The duration of the X-ray pulses depends on the target geometry and thickness. In the transmission geometry presented here, the interaction length is well defined, namely, given by the tape target thickness of 20 μm and supporting the generation of 100 fs X-ray pulses. For thicker targets, in particular those for X-ray generation in reflection, there is a spread of penetration depth with electron energy, resulting in a temporal smearing of X-ray generation and a concomitant temporal pulse broadening, leading to hard X-ray pulse durations on the order of several picoseconds.

4.2.3 Methods for Reconstructing and Analysing Transient Charge Densities

Changes in the scattering vector $\mathbf{q} = \mathbf{k}' - \mathbf{k}$, as revealed by a shift in the position of a single crystal Bragg reflection or the Debye–Scherrer ring pattern from a powder sample, can be directly connected to the change of the unit cell parameters of a crystalline material. Such changes might be due to heating of the material, a phase change upon photoexcitation or a strain wave travelling through the crystal, initiated by an ultrafast stress generation. The presence or absence of systematically 'forbidden' reflections is directly connected to changes in the crystallographic symmetry of the unit cell.

Under certain conditions, for example, for resonant excitation of a single phonon mode of the crystal, the time-resolved change in intensity $\Delta I_{hkl}(t)/I_{hkl}^0$ (eqn (4.3)) of a very small number of reflections (potentially only one) may be sufficient to develop a simple, yet accurate model of some aspects of lattice dynamics within the crystalline material. In general, however, it is necessary to infer the time-dependent electron density distribution $\rho(\mathbf{r},t)$ from the time-dependent intensity changes $\Delta I_{hkl}(t)/I_{hkl}^0$ on a larger number of reflections to reveal electron dynamics and structure changes on atomic length and timescales.

The observable $\Delta I_{hkl}(t)/I_{hkl}^0$ is related to the transient X-ray structure factors $F_{hkl}(t)$ according to eqn (4.3). The conceptually most straightforward approach to obtain $\rho(\mathbf{r},t)$ would be the Fourier series

$$\rho(\mathbf{r}, t) = \frac{1}{V} \sum_{\mathbf{q}} F_{hkl}(t) e^{-2\pi i \mathbf{q} \cdot \mathbf{r}} \tag{4.4}$$

with the volume of the unit cell V and the scattering vector $\mathbf{q}$. In most ultrafast diffraction experiments, however, Debye–Scherrer rings (or single crystal Bragg reflections) up to some maximum scattering vector $\mathbf{q}_{max}$ are detected, where $\mathbf{q}_{max}$ is limited by the X-ray wavelength and detector positioning. On the one hand, this results in a limited spatial resolution of the resulting electron density distribution which has to be accounted for. On the other hand, the reconstruction of $\rho(\mathbf{r},t)$ *via* this route will typically be unreliable since the abrupt end of the Fourier series generates artefacts such as the so-called Gibbs phenomenon, the appearance of ring-like charge structures around the atoms, or even completely unphysical results such as negative values of $\rho(\mathbf{r},t)$ in certain regions in space.[45]

In stationary X-ray crystallography, this problem is usually treated using a model-based approach. First an initial solution to the so called phasing problem is found *via* one of the several methods available. Next, the electron density distribution within the unit cell is modelled *via* atom centred functions, which allows the calculation of a set of structure factors $F^0_{hkl,\text{calc}}$. A frequently applied approach is the Hansen–Coppens-model which describes the intra-atomic and interatomic charge distribution through a multipole expansion of atomic charge densities.[46,47] The parameters within this expansion and the positions of the atoms within the unit cell are then optimized by minimizing the difference between $F^0_{hkl,\text{calc}}$ and the experimentally observed F^0_{hkl} in a least-squares refinement. To reliably converge, however, this method typically requires ten times more observed reflections than refined parameters. Even highly symmetric small molecular crystal structures can yield dozens, if not hundreds, of parameters describing the atomic co-ordinates and thermal motion parameters. As a result, such model-based approaches are usually not feasible for ultrafast diffraction experiments because of the comparably small number of Bragg reflections or Debye–Scherrer rings.

An alternative way to address the challenges of retrieving transient electron densities is the so called maximum entropy method (MEM), which is briefly described in the following. For a more complete overview, the reader is referred to ref. 46. Originally used as an information-theory-based technique to deal with data sets with low signal-to-noise ratios, the MEM has since then found widespread use in various scientific fields.[48,49] Starting in 1982, the MEM has been applied to retrieve charge density maps from steady-state diffraction data without the use of an underlying model.[50] This can be particularly useful when dealing with aperiodic structures such as quasi-crystals or incommensurately modulated structures.

Within the framework of this method, the electron density $\rho(\mathbf{r})$ is numerically reconstructed on a three-dimensional grid of sub-volumes, so called voxels (see Figure 4.7a). Summing over the number of electrons n_i in each individual subvolume for all m voxels constituting the entire unit cell yields the total number of electrons within the unit cell N, which is kept fixed throughout the entire optimization procedure. The information entropy S being maximized in an iterative fashion is defined in analogy to the thermodynamic entropy as

$$S = -\sum_{i=1}^{m} n_i \log(n_i/p_i), \tag{4.5}$$

where p_i is the corresponding value of the prior density $\rho_p(\mathbf{r})$ in the individual voxel. Without other constraints, the maximization of S

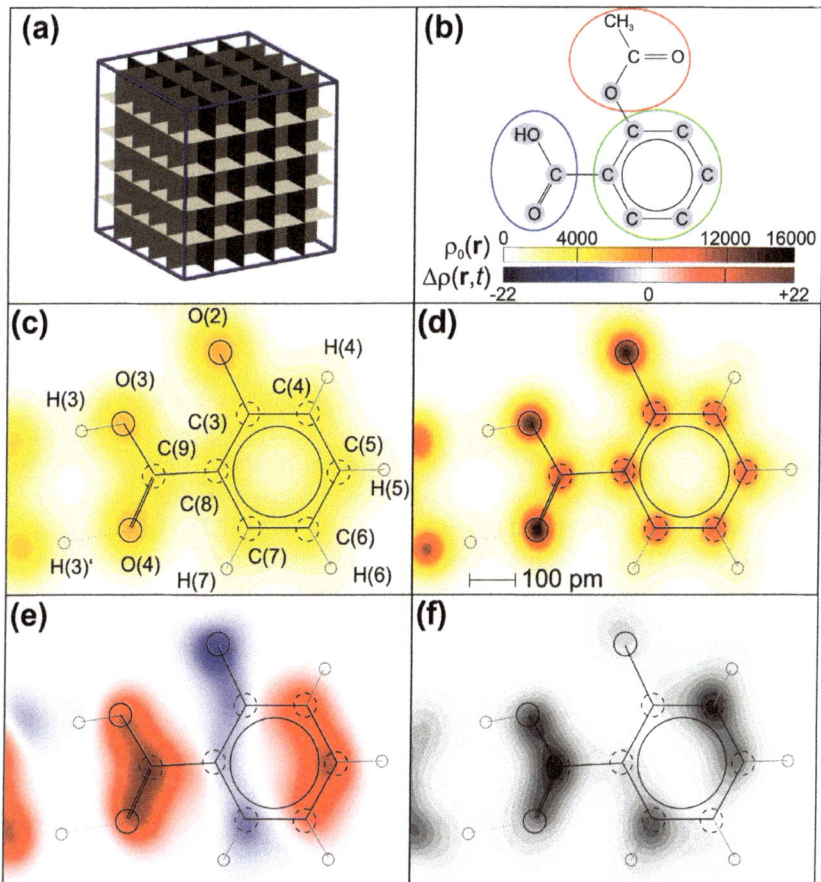

Figure 4.7 (a) Illustration of the division of a unit cell into subvolumes, the so-called voxels. (b) Molecular structure of an aspirin molecule ($C_9H_8O_4$). (c) Charge density map of aspirin in the electronic ground state derived from diffraction data with a maximum scattering vector $|\mathbf{q}_{max}| = 0.27$ Å^{-1}. (d) Same as (c) but for $|\mathbf{q}_{max}| = 0.6$ Å^{-1}. (e) Differential charge density map $\Delta\rho(\mathbf{r},t) = \rho(\mathbf{r},t) - \rho_0(\mathbf{r})$ for $t = 240$ fs derived from time resolved X-ray diffraction data reported in ref. 53. The maximum scattering vector is the same as in panel (c). Regions exhibiting a charge increase/decrease are coloured in red/blue, respectively. (f) Map of the spatially resolved uncertainty of differential charge density maps as derived from maps measured at negative delay times. The maximum uncertainty has a value of 10 to 15% of the maximum $\Delta\rho(\mathbf{r},t)$ in panel (e).

would obviously result in a flat distribution of $\rho(\mathbf{r})$ where every voxel contains the same number of electrons. Therefore, the additional constraint

$$C_F = -1 + \frac{1}{N_F} \sum_{i=1}^{N_F} \left(\frac{\left| F^i_{obs}(\mathbf{q}) - F^i_{MEM}(\mathbf{q}) \right|}{\sigma^i_{obs}(\mathbf{q})} \right)^2 \tag{4.6}$$

is necessary, which introduces the information provided by all N_F unique reflections obtained from the experiment in the form of the observed structure factors $F_{obs}(\mathbf{q})$ and their corresponding standard errors $\sigma_{obs}(\mathbf{q})$. The calculated structure factors $F_{MEM}(\mathbf{q})$ are derived from the momentary $\rho(\mathbf{r})$ *via* a Fourier summation over all voxels. The electron density $\rho(\mathbf{r})$ is then iteratively optimized to maximize the resulting entropy, simultaneously minimizing the difference between the observed and the calculated structure factors in eqn (4.6). The optimized electron density is the 'the least biased possible' result when taking into account only those structure factors that can be observed up to the maximum experimental resolution, since the MEM algorithm is 'maximally noncommittal with regard to missing information'.[48,49]

While this nonlinear algorithm has a unique global solution maximizing the entropy, appropriate convergence is subject to a suitable choice of the starting prior density $\rho_p(\mathbf{r})$. Ideally, $\rho_p(\mathbf{r})$ is chosen to be as close as possible to the final result with the maximal entropy. Since the ground state crystal structure and its associated electron density is usually well known for most materials studied in ultrafast diffraction experiments, it is an ideal starting point to generate a suitable prior density. In a first step, complex structure factors are derived from the known crystal structure up to a resolution significantly higher than $\mathbf{q}_{max}$ to generate a reliable ground-state density.

Using the MEM directly against a set of reflections with $|\mathbf{q}_{hkl}| > |\mathbf{q}_{max}|$ would result in unreliable predictions since the number of unknown reflections would be much larger than the number of unique reflections N_F from the experiment. Furthermore, the spatial resolution of the resulting ground-state density would be much higher than that attainable in the experiment. These challenges can be addressed by multiplying each structure factor $F(\mathbf{q})$ with a resolution dependent factor $\exp[-\ln(2)(|\mathbf{q}|/|\mathbf{q}_{max}|)^2]$, *i.e.*, by introducing a Gaussian weighting of structure factors.[51] The weighted structure factors are then used to generate a suitable prior density $\rho_0(\mathbf{r})$ *via* an initial run of the MEM with a spatial resolution matching the experimental resolution. In the second step, the resolution adapted ground-state density $\rho_0(\mathbf{r})$ is retrieved based on $\rho_p(\mathbf{r})$ employing a set of only those structure factors with $|\mathbf{q}_{hkl}| < |\mathbf{q}_{max}|$. All charge density maps presented in the following were generated in this way with the help of the BayMEM suite of programs.[52]

An exemplary map of the ground-state charge density is shown in the plane of the C_6-ring of crystalline acetylsalicylic acid ($C_9H_8O_4$, aspirin; Figure 4.7c) together with a sketch of the entire molecule (Figure 4.7b).[53] In this case, $|\mathbf{q}_{max}| = \sin\theta_{max}/\lambda \approx 0.27$ Å^{-1} was used for a realistic value for Cu Kα pulses ($\lambda(Cu_{K\alpha}) = 1.5406$ Å). The map shows all non-hydrogen atoms as individual maxima, whose positions closely match the predicted positions from the known ground-state structure. The hydrogen atoms, however, are essentially invisible, as their small contribution to the total electron density is almost completely overshadowed by that of the carbon and oxygen atoms to which they are bonded. As a result, the hydrogen atoms are not resolved as individual maxima. For a shorter X-ray wavelength, the individual atoms are gradually better resolved. As an example, a map of the ground-state density in the same molecular plane of aspirin at a resolution of $|\mathbf{q}_{max}| \approx 0.60$ Å^{-1} is shown in Figure 4.7d with the same contour spacings as in Figure 4.7c. Such a resolution could, for instance, be obtained with molybdenum as a target material for X-ray generation $(\lambda(Mo_{K\alpha}) = 0.7107$ Å$)$. In this map, the heavy atoms are now visible as even sharper maxima, and the bonded atoms are now clearly separated from each other by regions with density values at least one order of magnitude lower than in the core regions. Hydrogen atoms on the other hand are still not resolved as individual maxima but their small contributions to the total density are now slightly visible, for instance, in the carboxyl group of the molecule.

The time resolved electron densities $\rho(\mathbf{r},t)$ are then finally obtained by employing the experimentally observed transient $\Delta I_{hkl}(t)/I^0_{hkl}$ and structure factors derived from them in the MEM analysis. This procedure allows for calculating the differential charge density maps $\Delta\rho(\mathbf{r},t) = \rho(\mathbf{r},t) - \rho_0(\mathbf{r})$ for different delay times t with the equilibrium ground-state density $\rho_0(\mathbf{r})$, which bring out the changes in charge distribution most clearly. Differential maps of aspirin after electronic excitation are derived from the femtosecond experiments reported in ref. 53. The maximal uncertainty of differential charge density shown in Figure 4.7f is on the order of 10 to 15% of the maximum $\Delta\rho(\mathbf{r},t)$ in Figure 4.7e. It should be stressed that the most accurate charge density map $\rho_0(\mathbf{r})$ is required as a reference quantity to determine transient charge densities as deviations from the equilibrium charge distribution.

4.3 Transient Phonon Excitations and Charge Relocations in Polar Crystalline Materials

Phonons are quasi-particles describing quantized excitations of the normal modes of a crystal lattice with a dispersion relation $E_{ph}(\mathbf{q}_{ph})$, where $\mathbf{q}_{ph}$ is the phonon wavevector. For a crystal consisting of unit cells with a perfect long-range order, it is sufficient to consider $E_{ph}(\mathbf{q}_{ph})$ in the first Brillouin zone of q-space. There are three acoustic phonon branches and $3(p-1)$ optical branches for a number of p atoms in the basis of the unit cell. The energy of acoustic phonons vanishes at the Γ point $(\mathbf{q}_{ph} = 0)$ while optical phonons display a nonzero energy at $\mathbf{q}_{ph} = 0$ and can be excited by light with a wavevector $k \ll \pi/a$, where π/a with the real-space lattice constant a is a measure for the extension of the first Brillouin zone.

Time-dependent lattice excitations, their impact on and interaction with electronic charges of the crystal define an important topic of ultrafast X-ray diffraction. Phonon dynamics are made visible in space and time *via* their impact on the spatial distribution of electronic charge and its deviation from thermal equilibrium. In the most elementary case, transient atomic positions are derived from X-ray diffraction patterns dominated by inner-shell electrons attached to the lattice atoms or ions. For valence electrons shared by different atoms, this approximation breaks down and lattice motions are connected with relocations of valence charge. In such a case, a transient lattice excitation has a partly electronic character, a behaviour highly relevant in polar materials and ferroelectrics. Atomic motions in a crystal lattice span a broad range of timescales, from femtosecond oscillation periods of optical phonons up to hundreds of picoseconds propagation time of acoustic phonons.

4.3.1 Phonon Populations and Coherent Phonon Wavepackets

The quantum mechanical description of phonons requires a solution of the many-body Schrödinger equation of the crystal. In the adiabatic approximation, one calculates phonon quantum states in the vibrational potential determined by the ground-state electronic wavefunction. This treatment gives a series of eigenstates for each phonon mode in a generic anharmonic potential (Figure 4.8a).

There are two basic types of excitations relevant in the present context:

(i) Excitation of a nonequilibrium phonon population in an eigenstate, such as, the levels $|1\rangle$, $|2\rangle$, and $|3\rangle$ (green, blue and cyan lines in Figure 4.8a) are connected with a change in the vibrational wavefunction that modifies the occupation probabilities, *i.e.*, diagonal elements in the corresponding density matrix. The larger spatial extension of the vibrational wavefunction in the excited state typically leads to a reduction in the intensity of X-ray Bragg peaks, a behaviour which can be described as an excitation-induced change of the Debye–Waller factor (see Section 4.2.2). In the charge density map of the crystal, one observes a charge-conserving broadening of electron-density peaks at atomic positions. Nonequilibrium phonon excitations mainly decay by radiationless processes of population relaxation mediated through anharmonic couplings to lattice modes at lower frequencies. Typical optical phonon lifetimes are in the few-picosecond regime. It is important to note that excitation of an eigenstate results in a change of an eigenfunction and is not connected with periodic spatial elongations of lattice atoms. The latter require the generation of a phonon wavepacket which is discussed next.

(ii) Excitation of a superposition of phonon states with a well-defined mutual quantum phase generates a quantum-coherent phonon wavepacket. This nonstationary excitation is oscillatory and connected

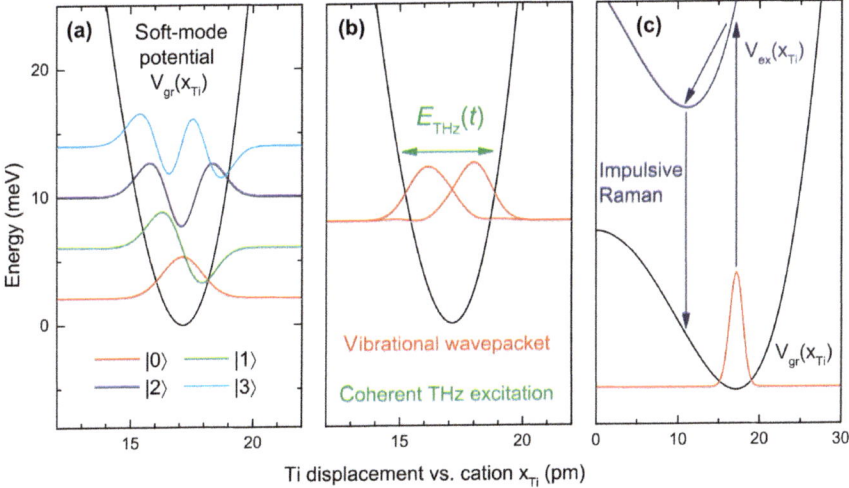

Figure 4.8   Vibrational energy-eigenstates and wavepackets in crystals. (a) The soft-mode potential (black line) of ferroelectric PbTiO₃[78] in the electronic ground state is shown along the soft-mode coordinate x_{Ti} together with the vibrational wavefunctions of the 3 lowest eigenstates $|0\rangle$ (red), $|1\rangle$ (green), $|2\rangle$ (blue), and $|3\rangle$ (cyan). (b) Oscillating vibrational wavepacket excited by coherent THz excitation. (c) Wavepacket dynamics in impulsive resonance Raman scattering. Blue curve: displaced potential surface of the electronically excited state. A first interaction with the driving pulse promotes the vibrational wavepacket to the excited potential surface. After a short propagation on the latter, a second interaction with the driving pulse projects a displaced wavepacket back to the potential surface of the electronic ground state.

with periodic real-space elongations of lattice atoms (Figure 4.8b). The oscillation frequency is determined by the energy spacings of the contributing quantum states which are identical for a harmonic oscillator. Interaction of the coherent wavepacket with other excitations, for example, the thermally excited phonon bath and/or population relaxation, leads to a decay of the quantum phase and, thus, the periodic lattice motions. In most solids, such decoherence processes occur in the subpico- to few-picosecond time range.

Coherent phonon oscillations lead to a periodic modulation of intensity in a subset of Bragg peaks. The experimentally observed oscillation frequency depends strongly on the symmetry properties of the transient nuclear displacements in the unit cell connected to the respective phonon. In case the phonon mode possesses the full symmetry of the unit cell, one directly observes the oscillatory atomic motion in the transient charge density map with the frequency of the phonon. A different behaviour arises for phonons belonging to irreducible representations of lower symmetry of the crystal's space group. Those phonons occur as a coherent superposition of the atomic motion and its space-inverted counterpart at each contributing atom in the unit cell, resulting in the second harmonic of the phonon frequency in the transient charge density map. Typically, an oscillating deformation of the atom's electron density between a spherical and cigar-like shape arises. A simultaneous excitation of several coherent phonon wavepackets in different low-symmetry modes leads to oscillations at the difference and sum frequencies of all contributing modes in the charge density map. Phonon decoherence results in an intensity reduction of Bragg peaks, analogous to the change in the Debye–Waller factor by phonon populations.

4.3.2 Excitation of Phonon Wavepackets

There are various mechanisms for generating phonon wavepackets. In ultrafast X-ray diffraction, both direct excitation *via* light–matter interaction or indirect excitation *via* a relaxation process on a timescale shorter than the phonon period have been applied. The following excitation schemes are important:

(i) In the electronic ground state of the crystal, superpositions of polar-optical phonon states can be directly excited *via* dipole-allowed optical transitions. Mid-infrared or terahertz pulses of sufficient spectral width and a duration significantly shorter than the phonon period generate a wavepacket (Figure 4.8b). In the most elementary case, the quantum-mechanical superposition involves essentially the vibrational ground $|0\rangle$ and first excited state $|1\rangle$.

(ii) Impulsive Raman excitation within the bandwidth of an ultrashort optical pulse can induce a coherent phonon motion in the electronic ground state of a crystal (Figure 4.8c). Of particular relevance are pre-resonant or resonant Raman excitation schemes in which the Raman polarizability is enhanced by the dipole moment of an electronic transition to an excited state. In the wavepacket description developed by Pollard *et al.*,[54] the driving electric field induces a vibrational wavepacket which propagates on the excited-state potential energy surface ($V_{ex}(x_{Ti})$; see the blue curve in Figure 4.8c) along the vibrational coordinate x_{Ti} and is projected back to the ground state by the second interaction with the optical field. Such processes occur within the sub-50 fs decoherence time of the electronic transition and generate a displaced phonon wavepacket in the electronic ground state.

(iii) Coherent phonons in electronically excited states have been generated by the displacive excitation mechanism (Figure 4.9a). The change in electronic charge distribution upon excitation, in our scheme a three-photon process, results in the modification of the vibrational potential, shifting the potential minimum along the particular phonon coordinate. This so-called origin shift occurs on a timescale that is fast compared to the phonon oscillation period. Consequently, the phonon wavepacket generated by optical excitation is displaced with respect to the new potential minimum and, thus, starts to move in an oscillatory fashion with a cosine-like phase.

(iv) Relaxation of optically excited hot carriers, for example, electrons in higher conduction band states, generates incoherent phonons through which excess energy is transferred to the crystal lattice. If such relaxation processes occur on a timescale much shorter than the oscillation period of low-frequency phonons, the latter can be impulsively excited *via* anharmonic phonon–phonon coupling (Figure 4.9b). This again leads to a coherent superposition of phonon states or a wavepacket. In contrast to the direct displacive excitation scenario (Figure 4.9a), incoherent phonon excitation is possible even with a negligible shift of the excited state relative to the ground state potential. Electron relaxation (see orange arrows in Figure 4.9b) leads to a temporally delayed onset of the coherent phonon oscillation with rise times in the range of up to a few picoseconds.

There is a detailed theoretical discussion of phonon wavepackets in the literature based on optical line shape theory methods that addresses the

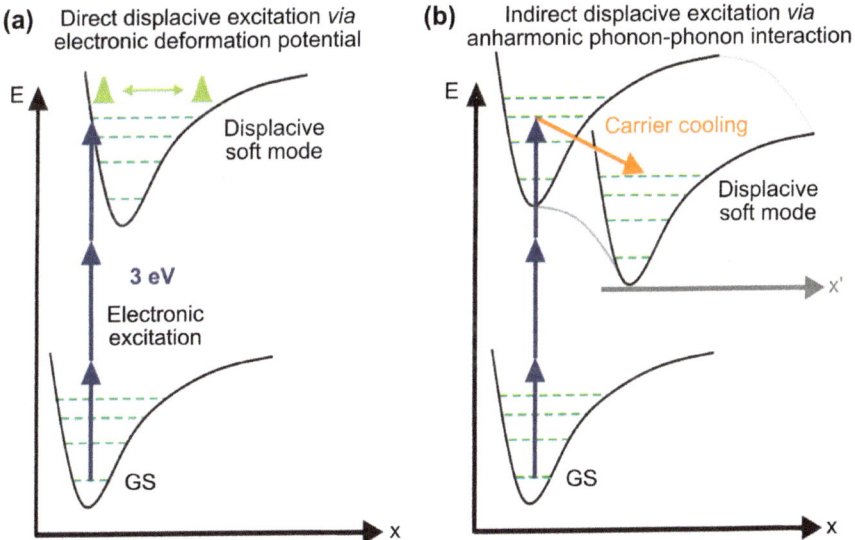

Figure 4.9 Coherent phonons in the electronically excited state of a crystal. (a) Potential-energy surfaces of the electronic ground and excited state along a low-frequency lattice coordinate. Optical excitation *via* three-photon absorption generates a vibrational wavepacket displaced on the potential surface of an electronically excited state, thereby triggering a coherent low-frequency oscillation. (b) For zero displacement of the excited state potential, fast incoherent relaxation (orange arrow) on the multidimensional potential surface of the electronically excited state can generate a displaced wave packet along the x' coordinate after a time delay set by the relaxation kinetics (carrier cooling).

coherence properties and also the phase of the phonon wavepackets in the electronic ground state.[55,56] Wavepacket description in the Raman literature is often limited to the analysis of the observable which is the polarization in the case of spectroscopy. In contrast, the crystallographic observation of vibrational coherence is directly measurable in coordinate space. The excitation mechanisms presented here have been explored in a wide range of femtosecond all-optical experiments and, to lesser extent, in X-ray diffraction experiments, the latter directly mapping the nuclear motions in space and time. Examples will be discussed in the following sections.

4.3.3 Soft Modes – Phonons Involving Electronic Charge Relocations

Phonon excitations of nonmetallic crystals with, at least partly, delocalized valence electrons induce a relocation of electronic charge. Such changes in the spatial electron distribution affect both the dielectric function $\epsilon(\nu)$ and the X-ray structure factors. In this scenario, a phonon represents both a nuclear and electronic degree of freedom and modifies the Coulomb interactions in the crystal. The most elementary approach to describe this phenomenon is the Clausius–Mossotti relation between the macroscopic electric field in the material and the local field experienced by individual dipoles. In ref. 57, it was explicitly shown that the local field acting on an individual, *i.e.*, a spatially distinct dipole, is the macroscopic electric field minus the dipole's own internal field. In this limit, the following relation between the macroscopic dielectric function $\epsilon(\nu)$ and the polarizabilities $\alpha_{el}(\nu)$ and $\alpha_{vib}(\nu)$ of electronic and vibrational dipoles holds in an isotropic material (*e.g.*, cubic structure):

$$3\frac{\epsilon(\nu)-1}{\epsilon(\nu)+2} = N_{el}\alpha_{el}(\nu) + N_{vib}\alpha_{vib}(\nu). \tag{4.7}$$

Here, N_{el} and N_{vib} are the volume densities of electronic and vibrational dipoles, respectively.

Soft modes in ferroelectrics are a prototypical case of hybrid phonons involving an extensive electronic charge transfer upon excitation. The pioneering core–shell model of Cochran[58] is essentially based on eqn (4.7) and explains the emergence of soft modes in ferroelectrics. As an illustrative example, we analyse the soft mode in $SrTiO_3$[59,60] (Figure 4.10). The dielectric function of (epitaxially strained) ferroelectric $SrTiO_3$ is shown around the experimentally observed soft-mode resonance in Figure 4.10a. With eqn (4.7) and the high-frequency dielectric function $\epsilon(\nu \gg 20\ THz) = 5$, one can derive the purely ionic susceptibility (see Figure 4.10b). It is obvious that the local-field correction according to the Clausius–Mossotti relation (4.7) renormalizes the phonon energy strongly, shifting the phonon resonance from 20 THz down to 1 THz. Concomitantly, the character of the soft

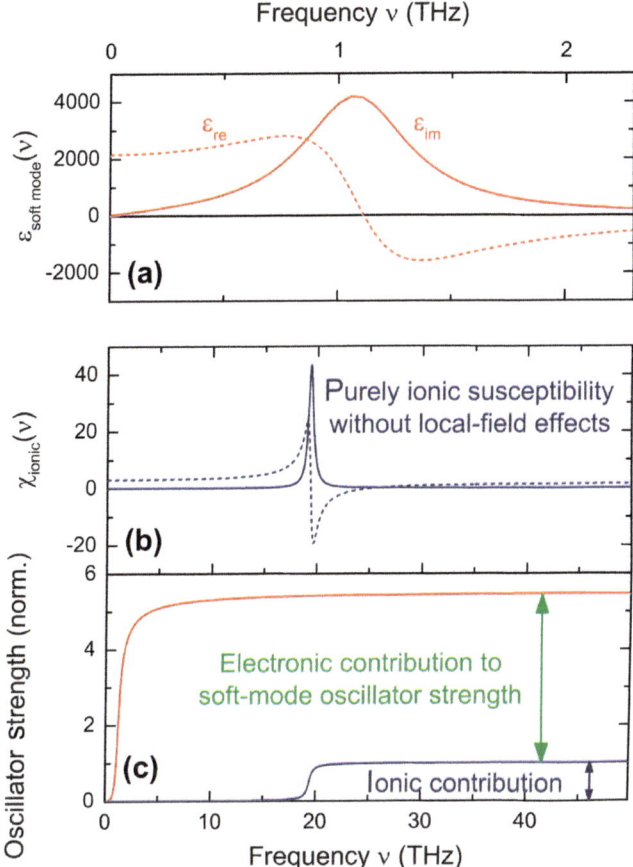

Figure 4.10 Soft modes in perovskites according to ref. 59 and 60. (a) Dielectric function of (epitaxically strained) ferroelectric $SrTiO_3$ around the experimentally observed soft-mode resonance. (b) Ionic susceptibility reconstructed with the help of eqn (4.7) and $\epsilon(\nu \gg 20\ \text{THz}) = 5$. (c) Electronic and ionic contributions to the oscillator strength of the soft mode as calculated from the real part of the frequency-dependent conductivity $\int_0^\nu \text{Re}[\sigma(s)]\,ds$.

mode evolves from a pure lattice motion to a hybrid mode which is dominated by the electronic charge transfer taking place in the Ti–O bonds.[59] A pronounced electronic character of the soft mode is evident from the electronic and ionic contributions to the oscillator strength calculated from the real part of the frequency-dependent conductivity $\int_0^\nu \text{Re}[\sigma(s)]\,ds$ (see Figure 4.10c).

Femtosecond X-ray diffraction experiments have demonstrated the hybrid character of coherent soft-mode wavepackets and revealed the length scale of charge relocations. A prototypical example is discussed in more detail in Section 4.5.2.

4.3.4 Macroscopic Electric Polarization, Microscopic Currents, and Transient Charge Densities

First-principles calculations have established a link between macroscopic electric polarizations and microscopic charge densities. It has been pointed out that polarization differences $\Delta\mathbf{P}_{el}$ between different lattice geometries rather than absolute polarizations are physically meaningful and well-defined quantities.[61] For equilibrium states, such calculations are based on quantum theory and involve geometric quantum phases in the adiabatic limit for calculating electronic polarization differences connected to nuclear displacements within the unit cell of a crystal. Such work has provided quantitative insight in stationary electric polarizations of crystalline ferroelectrics.[61]

A description of non-stationary electronic polarizations such as those induced by, for example, time-dependent soft-mode excitations requires an extension of the existing theory. To address this most relevant issue of structural dynamics, we recently developed a combined theoretical and experimental approach which allows the determination of time-dependent macroscopic polarization changes from transient microscopic charge density maps.[62] The key steps involve deriving the microscopic current density $\mathbf{j}(\mathbf{r},t)$ from experimental transient charge density maps, and calculating the macroscopic polarization change between times t_1 and t_2 [averaged over the unit cell (UC)] from this current density:

$$\Delta\mathbf{P}_{el}(\mathbf{r}, t_1, t_2) = \mathbf{P}_{el}(\mathbf{r}, t_2) - \mathbf{P}_{el}(\mathbf{r}, t_1) = \frac{1}{V_{UC}} \int_{UC} d^3\mathbf{r} \int_{t_1}^{t_2} \mathbf{j}(\mathbf{r}, t)dt. \qquad (4.8)$$

The starting point is the time-dependent change of the electron density $\rho(\mathbf{r},t)$, *i.e.*, the quantity directly accessible by time-resolved X-ray diffraction experiments. The charge density is connected to the microscopic current density $\mathbf{j}(\mathbf{r},t)$ and, in turn, to the velocity field $\mathbf{v}(\mathbf{r},t)$ of the electron ensemble *via* the continuity equation of electronic charge:

$$\frac{\partial\rho(\mathbf{r}, t)}{\partial t} = \dot{\rho}(\mathbf{r}, t) = -\nabla \cdot \mathbf{j}(\mathbf{r}, t) = -\nabla \cdot [\rho(\mathbf{r}, t)\mathbf{v}(\mathbf{r}, t)] \qquad (4.9)$$

At each spatial position $\mathbf{r}$ within the unit cell, the information at hand is $\dot{\rho}(\mathbf{r},t)$ and there are three unknown values for the microscopic current densities $j_x(\mathbf{r},t)$, $j_y(\mathbf{r},t)$, and $j_z(\mathbf{r},t)$. An additional constraint is required for determining the current density vector field $\mathbf{j}(\mathbf{r},t)$ uniquely. To this end, we assume that the entire electron ensemble performs a quasi-adiabatic motion, *i.e.*, the wavefunction of the electronic system stays as close as possible to a certain stationary eigenstate of the electronic Hamiltonian at all times. In appendix A of ref. 62, a 'hydrodynamic' formulation of quantum mechanics has been presented which explicitly shows that the quasi-adiabatic motion of electrons corresponds to the velocity field $\mathbf{v}(\mathbf{r},t) = \mathbf{j}(\mathbf{r},t)/\rho(\mathbf{r},t)$ with the lowest additional kinetic energy of the electrons

$$W_{\text{kin}}(t) = \frac{m_e}{2(-e_0)} \int \mathrm{d}V \rho(\mathbf{r}, t) |\mathbf{v}(\mathbf{r}, t)|^2 = \frac{m_e}{2(-e_0)} \int \mathrm{d}V \frac{|\mathbf{j}(\mathbf{r}, t)|^2}{\rho(\mathbf{r}, t)} \qquad (4.10)$$

relative to the energy of the stationary eigenstate of the electronic Hamiltonian in which the quasi-adiabatic motion occurs (m_e: electron mass and e_0: elementary charge). Thus, the three-dimensional current density $\mathbf{j}(\mathbf{r},t)$, which fulfils the continuity eqn (4.9) and minimizes the kinetic energy of the electrons during their motion (eqn (4.10)), uniquely characterizes the quasi-adiabatic motion of the electron ensemble. The numerical procedure has been described in appendix B of ref. 62.

In contrast to the electron ensemble, the wavefunctions of the individual nuclei are not delocalized within the unit cell. The contribution of nuclear displacements to the macroscopic electric polarization is given by

$$\Delta \mathbf{P}_{\text{nuc}}(t_1, t_2) = \frac{1}{V_{\text{UC}}} \sum_{k}^{\text{nuclei}} e_0 Z_k [\mathbf{R}_k(t_2) - \mathbf{R}_k(t_1)]. \qquad (4.11)$$

The transient spatial position of a nucleus $\mathbf{R}_k(t)$ with charge Z_k can be derived from the transient charge density map by fitting a three-dimensional Gaussian profile to the atom's charge-density peak which is typically dominated by its core electrons. In most cases studied by femtosecond X-ray diffraction so far, the electronic contribution to the macroscopic polarization change in eqn (4.8) dominates by far over the nuclear contribution in eqn (4.11). It is important to note that simplified pictures describing soft modes by moving ions with very large Born effective charges are incorrect, as suspected in ref. 59 for the case of perovskite ferroelectrics.

4.4 Phonon Dynamics in Polar Crystals and Nanostructures

The propagation of optically induced strain and oscillatory phonon motions have been major subjects of femto- to picosecond X-ray diffraction. Compared to all-optical studies of such phenomena, X-ray diffraction maps atomic positions in space and time, thus giving insight into transient structure rather than the transient dielectric function. In this section, we discuss two prototypical cases which illustrate both key experimental aspects and the physics underlying phonon dynamics in polar materials.

4.4.1 Strain Propagation in Bulk LiNbO$_3$

In Section 4.2.2, X-ray diffraction has been discussed in the so-called kinematic limit in which multiple elastic scattering can be safely neglected.

For larger and/or thicker crystals, the integrated intensity measured in reflection $I_0 = \int R(\Theta)d\Theta$ (where Θ is the diffraction angle) can be distinctly different from the predictions of kinematic diffraction theory. In particular, multiple elastic scattering in a thick crystal of small absorption leads to an extinction length much shorter than the X-ray absorption length, thus reducing I_0 significantly. The short extinction length originates from the interference between X-ray waves scattered from unit cells residing at different depths relative to the crystal's surface. An appropriate description of this behaviour requires the application of the dynamic X-ray diffraction theory.[41,63,64]

X-ray diffraction experiments which map the propagation of acoustic strain in thick ideally perfect crystals, consisting of unit cells identical in size and shape, are frequently in the dynamic diffraction limit. To illustrate this behaviour, Figure 4.11a shows the calculated X-ray reflectivity $R_{110}(\Theta)$ on the (110) peak in reflection from an unstrained $LiNbO_3$ crystal with the lattice structure shown in Figure 4.11a. Any deviation from the perfect interference of all contributing lattice planes leads to an increase in $R_{110}(\Theta)$, making it particularly sensitive to lattice distortions by shock waves or strain fronts propagating with the sound velocity inside the crystal. In Figure 4.11b, values of $R_{110}(\Theta)$ are plotted for a partially strained $LiNbO_3$ crystal after propagation of a step-like expansive strain front from the crystal's surface to half the extinction length of the (110) reflection. The partially strained crystal displays two different lattice constants in the probed crystal volume, leading to an integrated reflected intensity $I \approx 1.5 I_0$, *i.e.*, 1.5 times larger than in the unstrained case. For distances of the strain front from the crystal's surface much larger than the X-ray extinction length, the probed volume has a uniform but different lattice constant (Figure 4.11c), resulting in a shifted angular position of the diffraction peak with an integrated reflectivity identical to that of the unstrained crystal.

We now illustrate the relevance of this scenario by discussing a recent femtosecond X-ray diffraction experiment with $LiNbO_3$ single crystals.[65] Optical interband excitation of this ferroelectric material results in an electric shift current induced *via* the bulk photovoltaic effect.[66] The moving carriers interact with the crystal lattice through different types of electron–phonon coupling, the most prominent being the piezoelectric coupling of the current-related polarization to acoustic phonons. In this way, a propagating strain wave is generated. Other mechanisms such as electron–phonon coupling through deformation potentials and anharmonic phonon–phonon couplings play a minor role.

In the experiment, a transient shift current is generated by two-photon absorption of femtosecond pulses centred at 400 nm and the lattice distortion connected with the propagating strain is mapped by diffracting a hard X-ray probe pulse (Cu Kα) in reflection from the excited sample. A schematic of the crystal structure of $LiNbO_3$ (Figure 4.12a) highlights the edge-sharing NbO_6 octahedra viewed along the x-direction. In the ferroelectric phase, congruent $LiNbO_3$ crystallizes in the trigonal crystal system (space group $R3c$, no. 161) with lattice constants of $a = 5.148$ Å and

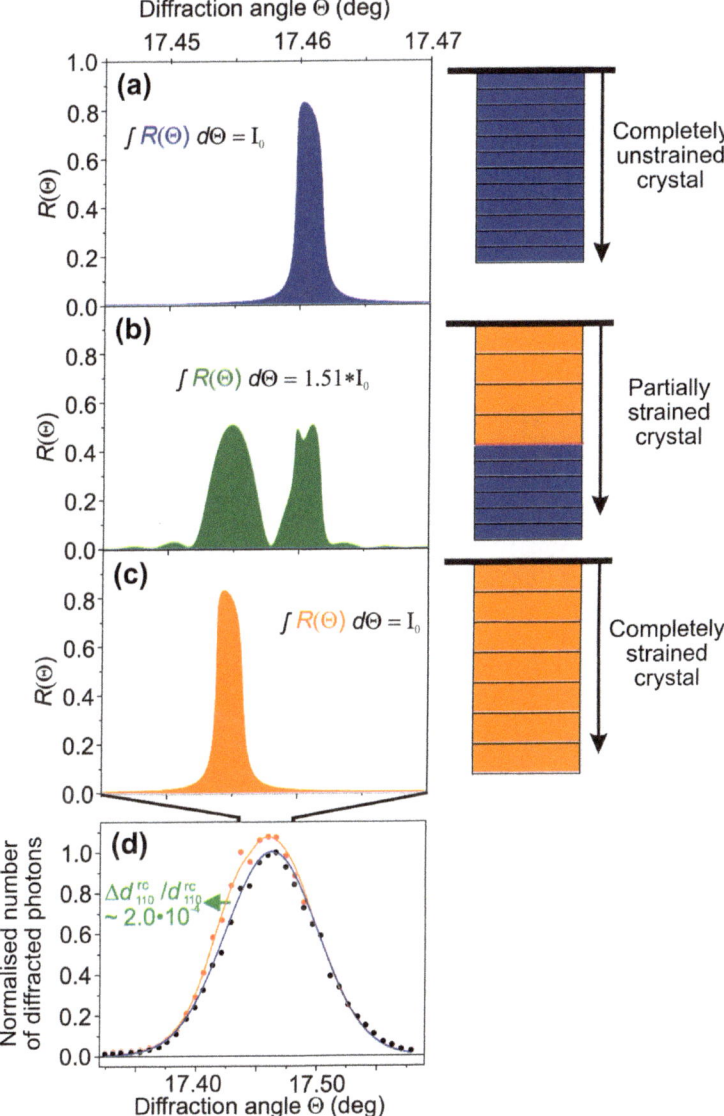

Figure 4.11 (a) X-ray reflectivity $R_{110}(\Theta)$ of an ideally perfect, unstrained LiNbO$_3$ crystal as a function of diffraction angle Θ, as calculated with dynamic X-ray diffraction theory. (b) $R_{110}(\Theta)$ of a partially strained LiNbO$_3$ crystal (*i.e.*, after propagation of a step-like expansive strain front from the crystal's surface to half the extinction length of the (110) reflection). (c) $R_{110}(\Theta)$ of the fully strained LiNbO$_3$ crystal. (d) Rocking curves measured in a femtosecond pump–probe experiment for unstrained (110)-oriented LiNbO$_3$ (black circles) and after a delay of $t = 800$ ps (red circles). The solid lines represent Gaussian lineshape functions taking the angular resolution of the experiment into account. Their widths are predominantly determined by the divergence of the X-ray source.

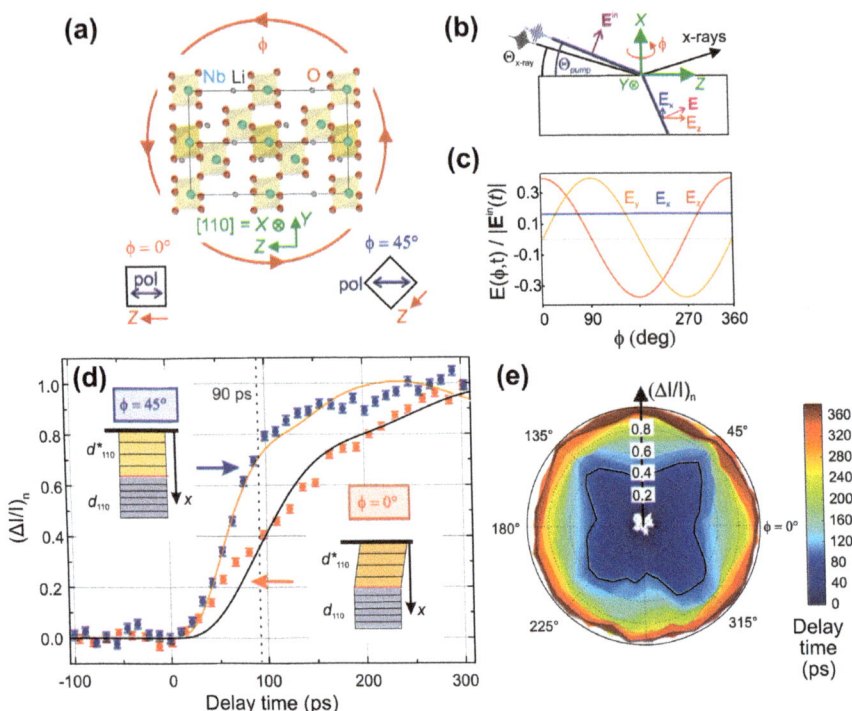

Figure 4.12 (a) Crystal structure of LiNbO$_3$ along the x-direction and definition of the rotation angle ϕ. (b) Pump–probe geometry with electric field components of the pump light inside the crystal. (c) Components of the electric field $\mathbf{E}(\phi, t) = \mathbf{E}(\phi) \cdot E(t)$ as a function of ϕ. (d) Normalized transient X-ray reflectivity $(\Delta I/I)_n$ of the (110) reflection as a function of pump–probe delay for $\phi = 0°$ (red circles) and $45°$ (blue circles). Solid lines: calculated reflectivity transients. Insets: Polarizations of the strain waves for $\phi = 0°$ and $45°$. (e) Polar plot of $(\Delta I/I)_n$ as a function of ϕ. The radial value indicates the percentage of the maximal $(\Delta I/I)_n$ value reached at specific delay times as color-coded. The black line indicates the percentage reached after a delay time of $t = 90$ ps.

$c = 13.863$ Å.[67] The (110) Bragg reflection from a 50 µm thick x-cut single crystal with the (110) lattice planes parallel to the crystal surface was studied in the experiment (diffraction angle $\Theta_{110} = 17.46°$). This geometry with the polar c-axis in the surface plane allows the measurement of the transient X-ray reflectivity for all angles ϕ between the c-axis and the p-polarized electric field of the pump pulse (Figure 4.12b). The pump beam hits the crystal's surface at the Brewster angle, leading to an almost ϕ-independent intensity inside the crystal with electric field components $\mathbf{E}(\phi,t) = \mathbf{E}(\phi) \cdot E(t)$ varying as a function of ϕ (Figure 4.12c).

Time-resolved transients of X-ray reflectivity were measured with pump pulses polarized parallel ($\phi = 0°$, red symbols in Figure 4.12d) and at an angle of $\phi = 45°$ (blue symbols) to the c-axis. The normalized change of diffracted intensity $(\Delta I/I)_n$ is plotted as a function of pump–probe delay t.

Both transients exhibit an increase in diffracted intensity on a timescale of 100 ps and reach a maximum after roughly 300 ps (absolute value $\Delta I/I \sim 15\%$). For $\phi = 45°$, the signal rise is much faster than for $\phi = 0°$. Rocking curves measured at a delay time of 800 ps ($\phi = 45°$, red symbols) are compared to the unexcited crystals (black symbols) in Figure 4.11d. In both cases, the undetermined contributions to the width and shape of the curve by the experimental setup (*e.g.*, divergence and $\Delta E/E$ of the source) are identical, and the observed difference between both curves is solely due to the excitation of the sample. The observed shift to lower diffraction angles in the case of the fully strained sample allows the retrieval of an expansive strain of $\Delta d_{110}/d_{110} = (2.0 \pm 0.5) \times 10^{-4}$.

Time-resolved transients were recorded for the full angular range between $\phi = 0°$ and $360°$ (Figure 4.12e). The radial axis of the polar plot gives the signal relative to its maximum value $(\Delta I/I)_n$ for the particular transient, *i.e.*, the value of ϕ. The coloured areas (legend on the right-hand side) encode the time intervals required to reach a particular value of the normalized signal. The black contour indicates the percentage reached after a delay time of 90 ps. This line has a distinctly non-circular shape, directly reflecting the different rise times of the intensity changes at different ϕ. We observe essentially a four-fold symmetry with a continuous transition between the slow $\phi = (0 + m \cdot 90)°$ and fast rise time $\phi = (45 + m \cdot 90)° \; \forall m \in \mathbf{Z}$.

The X-ray penetration depth into the crystal is limited by extinction to approximately 0.7 µm, requiring a description of the transient data using dynamic diffraction theory (as calculated from the ground state of LiNbO$_3$ using the Darwin formalism described below).[41,63] The much larger optical penetration depth of 15 µm results in a homogeneous excitation of the relevant crystal volume. The mechanical stress originating from the piezoelectric coupling between the shift current and the lattice generates an acoustic strain front which propagates with the sound velocity from the surface into the bulk of the crystal (perpendicular to the surface). The speed of sound is highly anisotropic, as is directly evident from the different rise times of intensity changes measured for different angles ϕ (Figure 4.12e). The time- and diffraction-angle-dependent reflectivity $R_{110}(t,q)$ for the (110) Bragg reflection was calculated from the Darwin formalism by employing the fast matrix transfer method.[42] The reflectivity transients (solid lines in Figure 4.12d) are calculated by comparing the angle-integrated reflectivity of the unstrained crystal ($d_{110} = 2.574$ Å) to that of a partly strained crystal with d_{110}^{*} as a function of the depth penetrated by the strain waves. The interference of X-rays diffracted from strained and unstrained parts of the lattice results in the observed increase in diffracted intensity. The experimental transients are well reproduced with a strain amplitude $\Delta d_{110}/d_{110} = (d_{110}^{*} - d_{110})/d_{110} = 3.3 \times 10^{-4}$ and sound velocities of $v_1 = 6615$ m s^{-1} ($\phi = 45°$) and $v_2 = 4438$ m s^{-1} ($\phi = 0°$), which matches the results obtained by the above analysis of the rocking curves. Details of this analysis have been discussed in ref. 65 which also gives an in-depth discussion of the mechanical stress generated *via* piezoelectric coupling to the shift current. This analysis shows that the electric polarization in the x-direction drives the strain fronts propagating with v_2 while the polarization along y causes the strain to propagate with v_1.

4.4.2 Dynamics of Zone-folded Phonons in Perovskite Superlattices

Crystalline materials of reduced dimensionality display a phonon dispersion different from the bulk. In this context, quasi-two-dimensional (2D) quantum materials and layered structures made from semiconductors and/or ferroelectrics have attracted much interest. The prototypical superlattice (SL) structures consist of a periodic sequence of layers of different composition. The periodicity along their stacking axis is described with the help of the SL period d_{SL} which gives the thickness of the smallest periodically repeated layer arrangement. In the simplest case, this arrangement includes a pair of layers (see the inset of Figure 4.13b). In most SLs, the SL period has values between a few and tens of nanometres, substantially larger than the lattice constants of the materials from which the layers are made.

The periodicity of the SL along its stacking axis results in a substantial change in the dispersion in $\mathbf{q}_{ph}$-space of longitudinal acoustic phonons (see schematic in Figure 4.13b).[68] A bulk acoustic phonon branch is folded back into the so-called mini-Brillouin zone of a width $\pm \pi/d_{SL}$, introducing new SL phonon branches which cross $\mathbf{q}_{ph} = 0$ and develop energy gaps at $\mathbf{q}_{ph} = 0$ and π/d_{SL}. The energy gaps are caused by the (typically small) differences in mass density and sound velocities of the two layer materials. The back-folded phonons can be excited optically and have been mapped by Raman spectroscopy of semiconductor SLs. If several SL phonon states are simultaneously excited in a phase-coherent way, this quantum wavepacket gives rise to nonstationary changes in layer thickness in the SL structure.

The static X-ray Bragg diffraction pattern of a SL consists of a series of SL peaks which reflect the additional periodicity of the structure.[69] The diffraction pattern (black solid line) of a SL consisting of 15 pairs of ferroelectric PbZr$_x$Ti$_{1-x}$O$_3$ (PZT) and metallic SrRuO$_3$ (SRO) layers with a period $d_{SL} = d_{PZT} + d_{SRO} = 4.92$ nm $+ 6.29$ nm $= 11.21$ nm is shown in Figure 4.13c, and the PZT unit cell is shown in Figure 4.13a. The peak intensities of the SL peaks follow envelope functions which are determined by the individual PZT and SRO layers and their respective thickness (blue and orange lines in Figure 4.13c). The simulated diffraction pattern (red solid line) was calculated using dynamic diffraction theory for this SL structure, and is in good agreement with the experiment. The strong peak at $\Theta = 23.3°$ is due to the SrTiO$_3$ (STO) substrate on which the SL was grown.

Coherent dynamics of SL phonons have been investigated in all-optical ultrafast experiments and, to much lesser extent, by time-resolved X-ray diffraction.[17,70,71] The PZT/SRO SL introduced above has been studied through optical pump/X-ray diffraction probe experiments in which structural

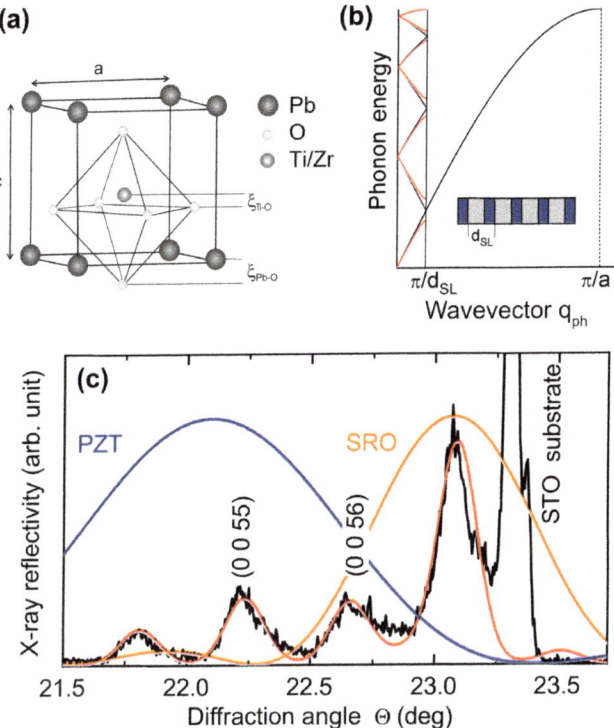

Figure 4.13 (a) Unit cell of PZT. (b) Schematic of phonon dispersion in a super-lattice. The back-folding of an acoustic phonon branch of the bulk Brillouin zone into a mini-Brillouin zone leads to additional phonon branches accessible for optical excitation at $q = 0$. The width of the mini-Brillouin zone is determined by the superlattice period d_{SL}. (c) Static X-ray difraction pattern from a PZT/SRO superlattice consisting of 15 periods ($d_{SL} = 11$ nm). The experimental pattern (black line) displays a series of superlattice peaks which are well reproduced by the simulated pattern (red line). The broad coloured lines are the envelopes of single PZT and SRO layers. The strong peak at $\Theta = 23.3°$ is due to the STO substrate.

dynamics were monitored *via* changes in the stationary SL diffraction pattern (Figure 4.13c). The sample was excited with a 50 fs pulse centred at 800 nm, which interacts predominantly with the metallic SRO layers. The electronic excitation relaxes on a subpicosecond timescale through electronic–phonon interaction, thus transferring excess energy into the SRO lattice. This mechanism generates a mechanical stress pattern with a spatial periodicity of $1/d_{SL}$ as only SRO layers are excited. The optically generated stress launches coherent SL phonon motions, *i.e.*, strain waves along the stacking axis of the SL. The latter are connected with changes in interatomic distances in the SL and, thus, give rise to changes in the SL X-ray diffraction pattern. Such changes were followed using diffracting hard X-ray probe pulses (Cu Kα) from the excited sample.

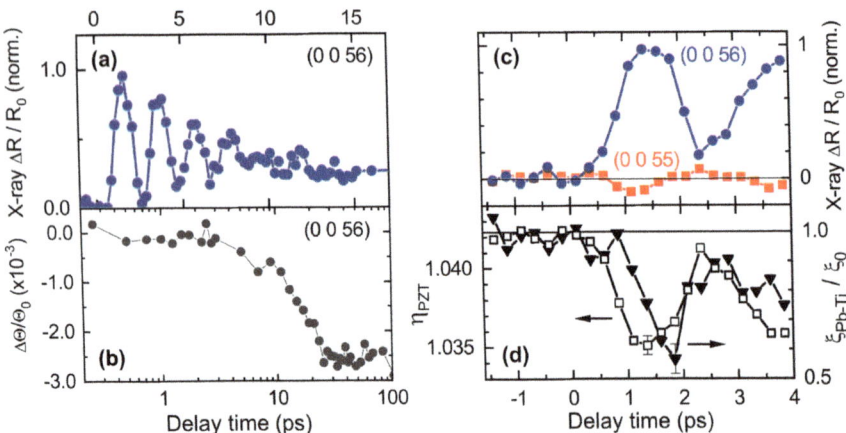

Figure 4.14 (a) Time-resolved X-ray reflectivity change $\Delta R/R_0 = (R - R_0)/R_0$ from a PZT/SRO superlattice on the (0 0 56) superlattice peak after excitation with a femtosecond optical pulse (R, R_0: X-ray reflectivity with and without excitation). The sample displays an oscillatory response with an oscillation period of 2 ps and a damping time on the order of 5 ps. (b) Change in the angular position of the (0 0 56) superlattice peak $\Delta\Theta/\Theta_0 = (\Theta - \Theta_0)/\Theta_0$ with pump–probe delay (Θ, Θ_0: diffraction angle with and without excitation). The angular position remains unchanged for the first 5 ps. The gradual shift at longer delay times is due to an expansion of the superlattice induced by propagating acoustic strain in the sample. (c) Time-resolved X-ray reflectivity change on the (0 0 55) and (0 0 56) superlattice peaks for delay times up to 4 ps. (d) Time evolution of the tetragonality η_{PZT} (open symbols, left ordinate scale) and the soft-mode elongation ξ_{Pb-Ti}/ξ_0 (solid symbols, right ordinate scale) as derived from an analysis of the time-resolved diffraction data.

Time-resolved X-ray data are summarized in Figure 4.14. The change in X-ray reflectivity $\Delta R/R_0$ on the SL peaks (0 0 56) and (0 0 55) is plotted as a function of pump–probe delay in Figure 4.14a and c ($\Delta R = R - R_0$, where R and R_0 are the reflectivity with and without excitation, respectively). There are pronounced oscillations on both SL peaks which display a period of $T = 2$ ps and a damping time on the order of 5 ps. The oscillation phases on the two peaks are similar while the reflectivity changes exhibit an opposite sign. The temporal evolution of the angular position of the SL peak (0 0 56) shows that up to a pump–probe delay of 4 ps, the diffraction angle Θ remains unchanged within the experimental accuracy (Figure 4.14b). At longer delays up to 30 ps, one observes a gradual angular shift to a slightly modified position which is maintained toward long delay times.

The mechanical stress induced by femtosecond optical excitation launches coherent oscillations of layer thickness in the SL. The stress builds up on the subpicosecond relaxation timescale of photoexcited electrons which is substantially shorter than the observed oscillation period of $T = 2$ ps. As a result, a coherent superposition of SL phonon states or strain waves is generated (see Section 4.3.2). As the stress pattern has the same periodicity as the SL and, thus, a wavevector $g_{SL} = 2\pi/d_{SL}$, the generated SL phonon wavepacket consists of states at a wavevector $q = 0$ in the mini-Brillouin zone and represents a longitudinal standing wave in the SL structure, connected with periodic changes in the SL layer thicknesses.

The X-ray diffraction data exhibit strong intensity modulations on SL diffraction peaks whereas their angular position remains unchanged during the first 4 ps. The latter observation demonstrates that the SL period d_{SL} remains unchanged during this period. On the other hand, there is a stress-induced periodic change of layer thickness in the SL. The unchanged angular positions of the SL diffraction peaks demonstrate that the expansion of an excited SRO layer is compensated by a compression of the adjacent PZT layer, thus leaving the SL period d_{SL} and the diffraction angles unchanged. Upon expansion and compression of SRO and PZT layers, their envelope functions (blue and orange lines in Figure 4.13c) shift in opposite directions along the angle coordinate, leading to the observed oscillations of diffracted intensity with different sign on the SL peaks at fixed diffraction angles. In this way, the coherent SL phonon oscillations are mapped onto the X-ray diffraction pattern. The damping of the oscillations on a picosecond time scale (Figure 4.14a) is due to decoherence of the phonon wavepacket.

The angular position of the (0 0 56) SL peak shifts at delay times longer than 4 ps to a smaller diffraction angle (Figure 4.14b). The temporal evolution of this shift on a timescale of 30 ps reflects the propagation of heat in the SL structure which originates from the excess energy supplied by the optical excitation pulse and leads to a thermal expansion of the structure by approximately 0.24%. At the atomic level, this process implies the propagation of acoustic phonons and the 30 ps timescale is in agreement with the propagation time estimated from the velocity of sound and the overall thickness of the SL.

In the present SL structure, the c-axis of the PZT unit cells (Figure 4.13a) is oriented parallel to the stacking axis of the layers. Thus, the change in layer thickness due to the longitudinal coherent phonon elongations is connected with a modification of the tetragonality $\eta_{PZT} = c/a$ of the unit cells, where c and a are the related lattice constants. An elongation along η leads to a change in the anharmonically coupled soft-mode coordinate $\xi_{Pb-Ti} = \xi_{Pb-O} - \xi_{Ti/Zr-O}$. The resulting shift of the centres of positive charge (Pb and Ti cations) and negative charge (O anions) modifies the ferroelectric polarization P of the lattice. In other words, oscillatory motions along η_{Pb-Ti} translate into oscillations of the macroscopic ferroelectric polarization P.

An in-depth analysis of the microscopic lattice elongations along η_{PZT} and ξ_{Pb-Ti} has been presented in ref. 70 where the change in X-ray reflectivity on the (0 0 55) and (0 0 56) SL diffraction peaks has been calculated as a function of η_{PZT} and ξ_{Pb-Ti}. In Figure 4.14d, values of η_{PZT} and ξ_{Pb-Ti}/ξ_0 are plotted as a function of delay time ($\xi_0 = 0.014$ nm: equilibrium value of ξ_{Pb-Ti}) and show that both the tetragonality and the soft-mode elongation display an oscillatory behaviour. In particular, a change in η_{PZT} of less than one percent induces a decrease in ξ_{Pb-Ti}/ξ_0 and, thus, the electric polarization by 50 percent. The maximum change in ζ_{Pb-Ti}/ζ_0 occurs approximately 500 fs after the maximum decrease of η_{PZT}, a behaviour due to a different shape of the anharmonic vibrational potential along the two lattice coordinates and the related propagation times of the phonon wavepacket.

The results discussed here establish a quantitative link between time-dependent microscopic lattice elongations and macroscopic electric polarizations. This new insight lays the ground for optical manipulation of electric properties on ultrafast timescales and may be relevant for future ferroelectric devices. Other dynamic properties of superlattices including magneto-striction[71] have been studied by femtosecond X-ray diffraction as well, underlining the strong potential of ultrafast structure-sensitive methods.

4.5 Phonon-driven Charge Dynamics in Molecular Crystals and Ferroelectrics

The insight into lattice dynamics presented in Section 4.4 is based on time-resolved measurements of diffracted X-ray intensity on few Bragg peaks only. Access to transient electronic charge density requires an analysis of a larger number of Bragg reflections. Here, femtosecond powder diffraction offers the inherent advantage of recording changes on many Debye–Scherrer rings simultaneously and without the need for sample reorientation. This method has paved the way towards ultrafast charge-density analysis, which has been demonstrated for a range of inorganic crystalline materials and molecular crystals. In this section, charge dynamics originating from the excitation of low-frequency phonons, in particular soft modes, is discussed with the help of prototypical results.

The reconstruction of transient charge densities from femtosecond powder diffraction patterns has been described in Section 4.2.3. From simultaneously recorded intensity changes $\Delta I_{hkl}(t)/I_{hkl}^0$ on a multitude of Debye–Scherrer rings, transient charge-density maps of the unit cell $\Delta\rho(\mathbf{r},t) = \rho(\mathbf{r},t) - \rho_0(\mathbf{r})$ are derived using the maximum entropy method (MEM, $\rho(\mathbf{r},t)$: momentary charge density at time t, $\rho_0(\mathbf{r})$: equilibrium charge density). Moreover, the method for reconstructing transient microscopic current densities $\mathbf{j}(\mathbf{r},t)$ from $\Delta\rho(\mathbf{r},t)$ (Section 4.3.4) allows for establishing a direct link between transient microscopic charge densities and macroscopic electric polarization changes. In the following, X-ray powder diffraction results for the prototype material ammonium sulfate [$(NH_4)_2SO_4$, AS] are presented. AS displays a Raman-active soft mode in its paraelectric phase ($T > T_C = 223$ K) and a different infrared-active soft mode in its ferroelectric phase ($T < T_C$). While the A_g soft mode in the paraelectric phase (space group $Pnma$) with a frequency of 50 cm^{-1} (1.5 THz) has been known from temperature-dependent Raman spectra,[72,73] the infrared-active A_1 soft mode in the ferroelectric phase (space group $Pna2_1$) with a frequency of 12 cm^{-1} (0.36 THz) has recently been discovered in femtosecond X-ray powder diffraction experiments performed at a lattice temperature of $T = 200$ K.[74]

AS crystallizes in an orthorhombic lattice structure with four formula units per unit cell (Figure 4.15a). In the paraelectric phase above T_C, the unit cell displays inversion symmetry and mirror planes perpendicular to the crystallographic axes *a*, *b*, and *c*, resulting in a zero net polarization. In the ferroelectric phase below T_C, the inversion centre and the (*ab*) mirror plane perpendicular to the *c*-axis are absent and the tilt of the NH_4 and SO_4 tetrahedra relative to the *c*-axis leads to a net electric dipole moment along *c* (Figure 4.15b and c). The dimensions of the ferroelectric unit cell are $a = 7.8566(3)$ Å, $b = 10.5813(4)$ Å, and $c = 5.9530(2)$ Å. The experimentally observed ferroelectric polarization has a value of $P_{fe}(T = 222 \text{ K}) = 6 \text{ mC m}^{-2}$.[75,76]

In ultrafast diffraction experiments, a 40 μm thick powder sample of AS is electronically excited *via* three-photon absorption of 70 fs pump pulses with a centre wavelength of 400 nm. Promotion to the electronically excited state is connected with the displacive elongation of phonon modes as discussed

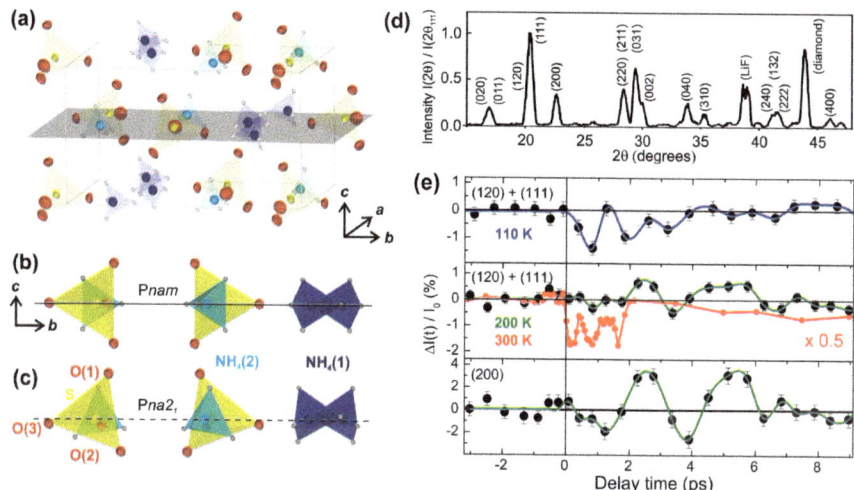

Figure 4.15 (a) Equilibrium crystal structure of ferroelectric ammonium sulfate [$(NH_4)_2SO_4$] viewed along the crystallographic *a*-axis in a ORTEP representation (50% probability ellipsoids). (b and c) Schematic drawing illustrating the orientation of the $(SO_4)^{2-}$ and $(NH_4)^+$ tetrahedra along the *c*-axis in the *para*- (*Pnma*) and ferroelectric (*Pna2$_1$*) phase. (d) X-ray diffraction pattern from the unexcited ferroelectric sample. The normalized diffracted intensity integrated over individual Debye–Scherrer rings is plotted as a function of scattering angle 2Θ, including assignments to lattice plains. (e) Change in diffracted X-ray intensity upon excitation $\Delta I_{hkl}(t)/I_{hkl}^0$ on different Bragg reflections as a function of pump–probe delay in the femtosecond experiments (black symbols). The upper panel shows a transient recorded at a sample temperature of $T = 110$ K while the middle and bottom panels give results for $T = 200$ K. The blue and green lines are a guide to the eye. The red line in the middle panel represents the change in diffracted X-ray intensity measured in the paraelectric phase of AS at $T = 300$ K, scaled by an amplitude factor of 0.5.

in Section 4.3.2 (Figure 4.9). Hard X-ray probe pulses at 0.154 nm (Cu Kα) are diffracted from the excited sample to generate Debye–Scherrer ring patterns at different pump–probe delays.

A powder diffraction pattern from an unexcited ferroelectric AS sample at $T = 200$ K is shown in Figure 4.15d. Integration of the intensity over each diffraction ring gives the diffracted intensities as a function of the scattering angle 2Θ. This result is in agreement with the literature[77] and allows for an assignment of the 15 Bragg peaks to sets of lattice planes. Upon cooling the ferroelectric sample down to $T = 110$ K, there are minor angular shifts of individual peaks connected to a change of the lattice vectors during the phase transition, in particular the crystallographic a axis.[77] A powder diffraction pattern of paraelectric AS measured at $T = 300$ K under similar experimental conditions has been reported in ref. 19.

Upon optical excitation, the angular positions of all reflections are preserved within an experimental angular resolution of 0.1° while the diffracted intensities display pronounced changes $\Delta I_{hkl}(t)$. In Figure 4.15e, the intensity change $\Delta I_{hkl}(t)/I^0_{hkl} = [I_{hkl}(t) - I^0_{hkl}]/I^0_{hkl}$ on selected Bragg peaks is plotted as a function of pump–probe delay ($I_{hkl}(t)$, I^0_{hkl}: intensity diffracted with and without optical excitation). The data recorded at $T = 200$ K exhibit pronounced oscillations with a period of ≈ 3 ps, whereas much faster and rapidly damped oscillations are observed at $T = 110$ K (upper panel of Figure 4.15e) and in the paraelectric phase at $T = 300$ K (red symbols and the line in the middle panel).[19] In the following, we first discuss transient charge density maps for paraelectric AS ($T = 300$ K),[19] followed by an analysis of transient charge density in ferroelectric AS ($T = 200$ K) just below the critical temperature $T_C = 223$ K.[62,74]

4.5.1   Ammonium Sulfate in the Paraelectric Phase

The experimentally observed oscillations at room temperature (red curve in Figure 4.15e) show a frequency of 1.5 THz, identical with that of a Raman-active soft-mode with A_g symmetry of space group *Pnma*.[72] The nuclear motions are connected with a periodic change in the crystal structure and pronounced changes of the electronic charge distribution. The reader is referred to Figure 4.16a for the unit cell of AS in the paraelectric phase at $T = 300$ K. The most-pronounced changes in the transient electron-density map occur in the shaded plane, which is parallel to the c-axis, goes through the centre of the unit cell (*i.e.*, (0.5a, 0.5b, 0.5c)) and is tilted by 60° with respect to the a-axis. Two proton positions of two NH_4^+ ions are included in this plane, and its centre corresponds exactly to the midpoint of the connecting line of oxygen atoms on two opposing SO_4^{2-} units. The stationary electron density $\rho_0(\mathbf{r})$ in this plane (Figure 4.16b) displays exclusively two peaks in electron density, located at the proton positions. The spatial resolution of the diffraction experiment is insufficient for separating the electron density on the hydrogen atoms from that of the nitrogen atoms (*i.e.*, an NH_4^+ ion appears with an almost spherical electron-density distribution).

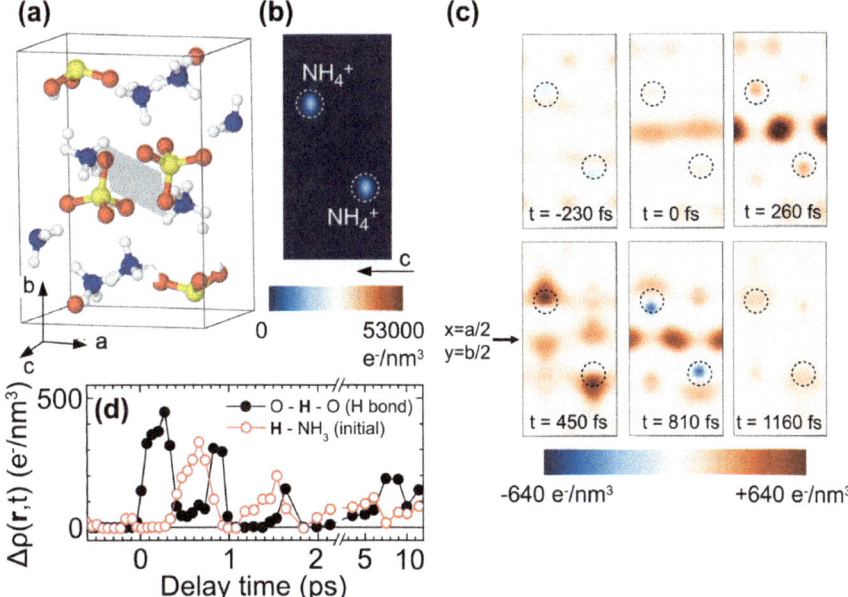

Figure 4.16 (a) Unit cell of ammonium sulfate in the paraelectric phase at room temperature. The shaded plane is parallel to the *c*-axis, goes through $(0.5a, 0.5b, 0.5c)$, is tilted by $60°$ with respect to the *a*-axis and includes the proton positions of two NH_4^+ ions. The centre of this plane corresponds to the midpoint of the connecting line of oxygen atoms of two SO_4^{2-} ions. (b) Stationary electron density $\rho_0(\mathbf{r})$ in this plane. (c) Transient charge density maps $\Delta\rho(\mathbf{r},t) = \rho(\mathbf{r},t) - \rho_0(\mathbf{r})$ for pump–probe delays as indicated. (d) Transient electron densities at the centre of the O–O connecting line (black symbols) and at an initial proton position (open circles).

Transient difference density maps $\Delta\rho(\mathbf{r},t) = \rho(\mathbf{r},t) - \rho_0(\mathbf{r})$ are investigated at six different pump–probe delays (Figure 4.16c). At a delay time of $t = 260$ fs, a pronounced peak in electron density arises exactly at the midpoint of the connecting line between oxygen atoms of two opposing SO_4^{2-} ions. In the electronic ground state, this position is not occupied by an atom of any kind. At $t = 450$ fs, this peak almost vanishes and two weaker electron-density peaks at the initial proton positions of the NH_4^+ ions appear. This behaviour reflects periodic charge motions between such positions (Figure 4.16d). Here, the charge densities at the midpoint position (solid circles) and the initial proton position are plotted as a function of pump–probe delay. The oscillatory charge motions point to a structure change in which a proton from the NH_4^+ ion and an electron from the SO_4^{2-} ion transiently form a hydrogen atom which oscillates back and forth in space between the midpoint of the O–O connection of two opposing SO_4^{2-} ions and the initial proton position in one of the NH_4^+ ions. In the language of chemistry, this concerted motion corresponds to the reversible intra-crystalline chemical reaction

$$(NH_4)^+ + \left[(SO_4)^{2-}\right]^* \rightarrow NH_3 + (HSO_4)^- \leftrightarrow NH_4 + (SO_4)^- \qquad (4.12)$$

with the asterisk * signifying an electronically excited SO_4^{2-} ion, and the double arrow $\leftrightarrow$ signifying the oscillatory motion of the H atom with the soft-mode frequency.[72] Further details of this experiment can be found in ref. 19, in particular (optical pump)–(mid-infrared probe) measurements which confirm the release of a proton from the NH_4^+ ion *via* changes in the vibrational spectrum in the fingerprint range.

4.5.2   Ammonium Sulfate in the Ferroelectric Phase

In the ferroelectric phase below the transition temperature $T_C = 223$ K,[77] AS displays a macroscopic ferroelectric polarization and the unit cell no longer has an inversion centre (Figure 4.17a). Its polar axis is parallel to the crystallographic *c*-axis. The blue arrows in the SO_4^{2-} ion on the right hand side indicate local dipoles between sulphur and oxygen atoms. The stationary electron density $\rho_0(\mathbf{r})$ of ferroelectric AS was derived from stationary X-ray diffraction experiments with single crystals.[74] A sectional view (Figure 4.17b) of the $\rho_0(x, y, z = 0.5)$ electron density (grey plane in Figure 4.17a) in a subvolume around the SO_4^{2-} ion (see blue arrows) shows that the electron density is highest on the sulphur atom and smaller on the two oxygens.

AS in its ferroelectric phase was studied at a temperature of $T = 200$ K, *i.e.*, just below the critical temperature. Experimental results are shown in Figure 4.15; more details can be found in ref. 74. Optical three-photon excitation with femtosecond pulses centred at 400 nm induces coherent vibrational motions of the crystal lattice and a delayed hard X-ray probe pulse scattered from the excited sample generates momentary Debye–Scherrer diffraction patterns. Transient differential electron-density maps $\Delta\rho(\mathbf{r},t)$ were derived from the diffraction data with the help of the MEM discussed in Section 4.2.3.

The experimentally observed coherent lattice motions are dominated by a low-frequency mode with a 2.7 ps period, corresponding to a frequency of 0.36 THz. They are connected with a significant oscillatory charge transfer within the SO_4^{2-} ions, the strongest of which is observed in the $\Delta\rho(x, y, 0.5, t)$ maps for delay times of $t = 2.7$ ps, $t = 3.9$ ps, and $t = 5.1$ ps (Figure 4.17c–e). Electronic charge moves over distances on the order of 1 Å, comparable to a chemical bond length and much larger than the vibrational amplitudes of the low-frequency mode. The latter have been estimated from the positions of the charge centre-of-gravity at the atomic sites, and have values on the order of 10^{-3} Å.[74] The vastly different length scales of nuclear motion and charge relocation are a characteristic feature of soft mode excitations, directly reflecting their hybrid character. Such a behaviour has been observed in a variety of ionic materials and is in line with the theoretical picture developed by Cochran.[58] The decay of charge oscillations on a

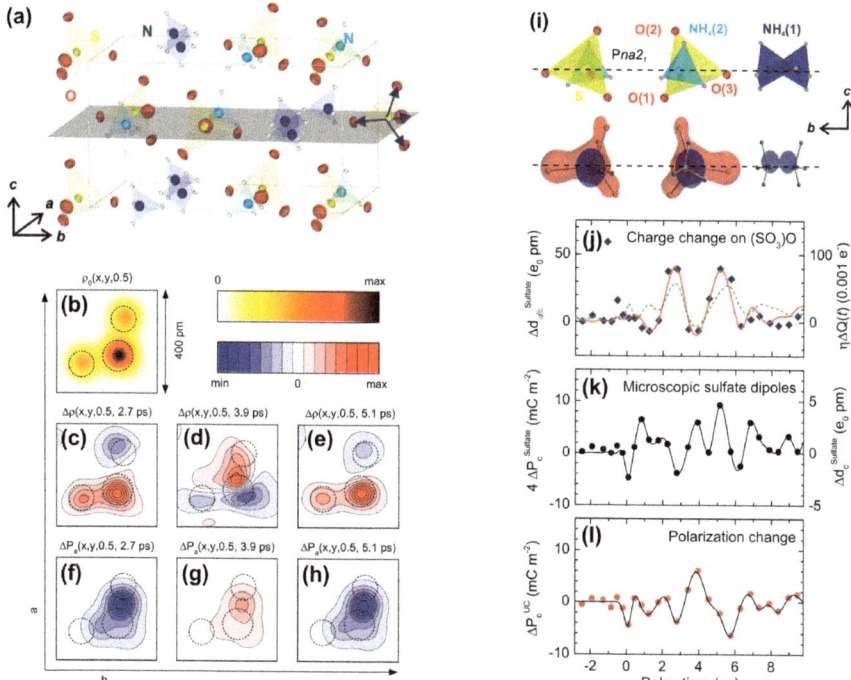

Figure 4.17 (a) Unit cell of ferroelectric ammonium sulfate at $T = 200$ K. (b) Stationary map $\rho_0(x, y, z = 0.5)$ in the grey plane around the sulfate ion with blue arrows [$\rho_0^{max} = 12\,000$ e$^-$ nm^{-3}]. (c–e) $\Delta\rho(x, y, 0.5, t)$ for pump–probe delays as indicated with $\Delta\rho^{max/min} = \pm 75$ e$^-$ nm^{-3}. (f–h) Microscopic polarization changes along the a-axis $\Delta P_a(x, y, 0.5, t)$ with $\Delta P_a^{max/min} = \pm 1$ C m^{-2}. (i) Visualization of the charge density changes along the (ferroelectric) c-axis for $t = 2.8$ ps. (j) Diamonds: transient charge change on the SO$_3$ subgroup within a single sulfate ion. Red: corresponding dipole change $\Delta d_a(t)$ along the a-axis. Green: $\Delta d_b(t)$ along the b-axis. (k) $\Delta d_c(t)$ (right axis) along the c-axis (black). The macroscopic polarization of 4 sulfate ions in a single unit cell (left axis) is very similar to that of the entire unit cell [panel (l)].

timescale of several picoseconds is due to the inhomogeneous distribution of phonon frequencies in the polycrystalline sample.

The series of transient charge density maps derived from the diffraction data allows for a reconstruction of the transient macroscopic electric polarization by the method outlined in Section 4.3.4. In the first step, one derives the microscopic current density $\mathbf{j}(\mathbf{r}, t)$ from the time-dependent charge density with the help of eqn (4.9). The time-integral of $\mathbf{j}(\mathbf{r}, t)$ averaged over the unit cell of ferroelectric AS then gives the transient polarization change $\Delta\mathbf{P}_{el}(\mathbf{r}, t)$ according to eqn (4.8). A detailed inspection of the transient charge-density maps shows that the main contribution to $\Delta\mathbf{P}_{el}(\mathbf{r}, t)$ originates from the anisotropic transfer of electronic charge within the SO$_4^{2-}$ ions (Figure 4.17c–e). The dynamics of local dipole changes $\Delta\mathbf{d}$ in the SO$_4^{2-}$

ions (arrows in Figure 4.17a) can be reconstructed from the microscopic current density $\mathbf{j}(\mathbf{r},t)$ as well. The microscopic current densities provide the microscopic polarization changes along the a axis $\Delta P_a(x,y,0.5,t)=\int_{-\infty}^{t} j_x(x,y,0.5,s)ds$ (see Figure 4.17f–h). The highest microscopic polarization changes occur on the chemical bonds within the SO_4^{2-} ions (*i.e.*, the spatial regions between the dashed circles indicating the atomic positions). Our theoretical treatment confirms this intuitive physical picture and allows at the same time for a quantitative analysis of microscopic polarization changes along various spatial directions, including the change along the a axis $\Delta P_a(x,y,0.5,t)=\int_{-\infty}^{t} j_x(x,y,0.5,s)ds$.

The diamonds in Figure 4.17j represent the transient charge change on the SO_3 subgroup within the SO_4^{2-} ion (right ordinate scale). Integrating the microscopic current density $\mathbf{j}(\mathbf{r},t)$ over time and the subvolume of a single SO_4^{2-} ion allows for calculating transient dipole changes along various axes. The dipole change $\Delta d_a(t)$ along the a axis (red solid line in Figure 4.17j) shows that its transient almost perfectly follows the electronic charge transfer from the upper oxygen atom to the SO_3 subgroup (blue diamonds). The dipole change $\Delta d_b(t)$ along the b axis (green dashed line) is somewhat smaller and shows additional contributions. Although individual SO_4^{2-} ions show large transient dipoles along the a and b directions, they do not create macroscopic polarizations along those directions because, for symmetry reasons (point group $mm2$), the unit cell contains another SO_4^{2-} ion with opposite dipole moments in the a and b directions.

On the one hand, the dipole change $\Delta d_c(t)$ (right ordinate scale in Figure 4.17k) along the c-axis (black solid line in Figure 4.17k) of a single SO_4^{2-} ion is an order of magnitude smaller and shows a temporal behaviour quite different from those along the a and b axes. The reason for the small projection of the SO_4 dipole onto the c axis (Figure 4.17i) is that the transient charge-transfer dipole is essentially oriented along a single S–O bond within the molecular ion and the projection onto the c-axis originates merely from the (weakly) tilted SO_4^{2-} tetrahedrons in the ferroelectric phase of AS.

On the other hand, all transient charge-transfer dipoles of SO_4^{2-} ions within the unit cell of AS point in the same direction and contribute dominantly to the macroscopic polarization $\Delta P_c^{UC}(t)$ of the crystal, which is given by the spatial average of $\Delta P_c(\mathbf{r},t)$ over the unit cell. The left ordinate scale of Figure 4.17k shows the corresponding macroscopic polarization of all four SO_4^{2-} ions belonging to a single unit cell. A comparison with the total macroscopic polarization along the c-axis (Figure 4.17l), which includes contributions from the NH_4^+ units and charge transfer between the (molecular) ions, shows a very similar temporal behaviour and demonstrates a predominant contribution of intramolecular charge transfer dipoles within the SO_4^{2-} ions to the soft-mode polarization in ferroelectric AS. The contribution of nuclear displacements to the total macroscopic polarization according to eqn (4.11) is 30 times smaller than that of the electronic counterpart and can be safely neglected.

4.6 Conclusions

This chapter summarizes the state-of-the-art in femtosecond X-ray diffraction with laser-driven table-top sources. This research area is part of the global effort to map and understand structural dynamics on atomic length and timescales. Advanced large-scale X-ray free-electron lasers offer a hard X-ray flux orders of magnitude higher than table-top sources and have meanwhile found broad application in science. Nevertheless, laser-driven sources enable ultrafast X-ray diffraction experiments in the laboratory frame, allow for the study of subtle ultrafast changes in structures with very high sensitivity, and offer unrestricted measurement periods. Systematic improvements of table-top source technology and implementation of powder diffraction methods have established the new field of femtosecond charge density analysis. Such results have generated new and quantitative insight into the properties of soft modes in polar crystals and ferroelectrics, which are characterized by an interplay of lattice dynamics and spatial relocations of electronic charge on a vastly different length scale. Complementary theoretical work has established a quantitative link between transient macroscopic electric polarizations and charge dynamics at the atomic scale. This connection is highly relevant for designing and characterizing dynamic properties of functional materials, in particular ferroelectrics. The ongoing improvement of both experimental and theoretical methods will allow for new applications in physics and materials research.

References

1. D. von der Linde, K. Sokolowski-Tinten, C. Blome, C. Dietrich, A. Tarasevitch, A. Cavalleri and J. A. Squier, *Z. Phys. Chem.*, 2001, **215**, 1527–1541.
2. A. Rousse, C. Rischel and J. Gauthier, *Rev. Mod. Phys.*, 2001, **73**, 17–31.
3. M. Bargheer, N. Zhavoronkov, M. Woerner and T. Elsaesser, *ChemPhysChem*, 2006, 7, 783–792.
4. M. Chergui and A. H. Zewail, *ChemPhysChem*, 2009, **10**, 28–43.
5. T. Elsaesser and M. Woerner, *J. Chem. Phys.*, 2014, **140**, 020901.
6. C. Bostedt, S. Boutet, D. M. Fritz, Z. Huang, H. J. Lee, H. T. Lemke, A. Robert, W. F. Schlotter, J. J. Turner and G. J. Williams, *Rev. Mod. Phys.*, 2016, **88**, 015007.
7. J. D. Kmetec, C. L. Gordon, J. J. Macklin, B. E. Lemoff, G. S. Brown and S. E. Harris, *Phys. Rev. Lett.*, 1992, **68**, 1527–1530.
8. G. Korn, A. Thoss, H. Stiel, U. Vogt, M. Richardson, T. Elsaesser and M. Faubel, *Opt. Lett.*, 2002, **27**, 866–868.
9. N. Zhavoronkov, Y. Gritsai, M. Bargheer, M. Woerner, T. Elsaesser, F. Zamponi, I. Uschmann and E. Förster, *Opt. Lett.*, 2005, **30**, 1737–1739.
10. C. Reich, P. Gibbon, I. Uschmann and E. Förster, *Phys. Rev. Lett.*, 2006, **84**, 4846–4849.
11. W. Lu, M. Nicoul, U. Shymanovich, A. Tarasevitch, P. Zhou, K. Sokolowski-Tinten, D. Von Der Linde, M. Mašek, P. Gibbon and

U. Teubner, *Phys. Rev. E: Stat., Nonlinear, Soft Matter Phys.*, 2009, **80**, 026404.

12. F. Zamponi, Z. Ansari, C. von Korff Schmising, P. Rothhardt, N. Zhavoronkov, M. Woerner, T. Elsaesser, M. Bargheer, T. Trobitsch-Ryll and M. Haschke, *Appl. Phys. A: Mater. Sci. Process.*, 2009, **96**, 51–58.

13. C. Rischel, A. Rousse, I. Uschmann, P.-A. Albouy, J. P. Geindre, P. Audebert, J. C. Gauthier, E. Foerster, J.-L. Martin and A. Antonetti, *Nature*, 1997, **390**, 490–492.

14. C. W. Siders, A. Cavalleri, K. Sokolowski-Tinten, C. Tóth, T. Guo, M. Kammler, M. H. von Hoegen, K. R. Wilson, D. von der Linde and C. P. J. Barty, *Science*, 1999, **286**, 1340–1342.

15. C. Rose-Petruck, R. Jimenez, T. Guo, A. Cavalleri, C. W. Siders, F. Rksi, J. A. Squier, B. C. Walker, K. R. Wilson and C. P. J. Barty, *Nature*, 1999, **398**, 310–312.

16. K. Sokolowski-Tinten, C. Blome, J. Blums, A. Cavalleri, C. Dietrich, A. Tarasevitch, I. Uschmann, E. Förster, M. Kammler, M. Horn-von Hoegen and D. von der Linde, *Nature*, 2003, **422**, 287–289.

17. M. Bargheer, N. Zhavoronkov, Y. Gritsai, J. C. Woo, D. S. Kim, M. Woerner and T. Elsaesser, *Science*, 2004, **306**, 1771–1773.

18. F. Zamponi, Z. Ansari, M. Woerner and T. Elsaesser, *Opt. Express*, 2010, **18**, 947–961.

19. M. Woerner, F. Zamponi, Z. Ansari, J. Dreyer, B. Freyer, M. Prémont-Schwarz and T. Elsaesser, *J. Chem. Phys.*, 2010, **133**, 064509.

20. F. Zamponi, P. Rothhardt, J. Stingl, M. Woerner and T. Elsaesser, *Proc. Natl. Acad. Sci. U. S. A.*, 2012, **109**, 5207–5212.

21. R. W. Schoenlein, W. P. Leemans, A. H. Chin, P. Volfbeyn, T. E. Glover, P. Balling, M. Zolotorev, K.-J. Kim, S. Chattopadhyay and C. V. Shank, *Science*, 1996, **274**, 236–238.

22. P. Beaud, S. L. Johnson, A. Streun, R. Abela, D. Abramsohn, D. Grolimund, F. Krasniqi, T. Schmidt, V. Schlott and G. Ingold, *Phys. Rev. Lett.*, 2007, **99**, 174801.

23. A. L. Cavalieri, D. M. Fritz, S. H. Lee, P. H. Bucksbaum, D. A. Reis, J. Rudati, D. M. Mills, P. H. Fuoss, G. B. Stephenson, C. C. Kao, D. P. Siddons, D. P. Lowney, A. G. MacPhee, D. Weinstein, R. W. Falcone, R. Pahl, J. Als-Nielsen, C. Blome, S. Düsterer, R. Ischebeck, H. Schlarb, H. Schulte-Schrepping, T. Tschentscher, J. Schneider, O. Hignette, F. Sette, K. Sokolowski-Tinten, H. N. Chapman, R. W. Lee, T. N. Hansen, O. Synnergren, J. Larsson, S. Techert, J. Sheppard, J. S. Wark, M. Bergh, C. Caleman, G. Huldt, D. van der Spoel, N. Timneanu, J. Hajdu, R. A. Akre, E. Bong, P. Emma, P. Krejcik, J. Arthur, S. Brennan, K. J. Gaffney, A. M. Lindenberg, K. Luening and J. B. Hastings, *Phys. Rev. Lett.*, 2005, **94**, 114801.

24. P. Emma, R. Akre, J. Arthur, R. Bionta, C. Bostedt, J. Bozek, A. Brachmann, P. Bucksbaum, R. Coffee, F.-J. Decker, Y. Ding, D. Dowell, S. Edstrom, A. Fisher, J. Frisch, S. Gilevich, J. Hastings, G. Hays, P. Hering, Z. Huang, R. Iverson, H. Loos, M. Messerschmidt,

A. Miahnahri, S. Moeller, H.-D. Nuhn, G. Pile, D. Ratner, J. Rzepiela, D. Schultz, T. Smith, P. Stefan, H. Tompkins, J. Turner, J. Welch, W. White, J. Wu, G. Yocky and J. Galayda, *Nat. Photonics*, 2010, **4**, 641–647.

25. T. Ishikawa, H. Aoyagi, T. Asaka, Y. Asano, N. Azumi, T. Bizen, H. Ego, K. Fukami, T. Fukui, Y. Furukawa, S. Goto, H. Hanaki, T. Hara, T. Hasegawa, T. Hatsui, A. Higashiya, T. Hirono, N. Hosoda, M. Ishii, T. Inagaki, Y. Inubushi, T. Itoga, Y. Joti, M. Kago, T. Kameshima, H. Kimura, Y. Kirihara, A. Kiyomichi, T. Kobayashi, C. Kondo, T. Kudo, H. Maesaka, X. M. Maréchal, T. Masuda, S. Matsubara, T. Matsumoto, T. Matsushita, S. Matsui, M. Nagasono, N. Nariyama, H. Ohashi, T. Ohata, T. Ohshima, S. Ono, Y. Otake, C. Saji, T. Sakurai, T. Sato, K. Sawada, T. Seike, K. Shirasawa, T. Sugimoto, S. Suzuki, S. Takahashi, H. Takebe, K. Takeshita, K. Tamasaku, H. Tanaka, R. Tanaka, T. Tanaka, T. Togashi, K. Togawa, A. Tokuhisa, H. Tomizawa, K. Tono, S. Wu, M. Yabashi, M. Yamaga, A. Yamashita, K. Yanagida, C. Zhang, T. Shintake, H. Kitamura and N. Kumagai, *Nat. Photonics*, 2012, **6**, 540–544.

26. R. Schoenlein, T. Elsaesser, K. Holldack, Z. Huang, H. Kapteyn, M. Murnane and M. Woerner, *Philos. Trans. R. Soc., A*, 2019, **377**, 20180384.

27. H. N. Chapman, P. Fromme, A. Barty, T. A. White, R. A. Kirian, A. Aquila, M. S. Hunter, J. Schulz, D. P. DePonte, U. Weierstall, R. B. Doak, F. R. N. C. Maia, A. V. Martin, I. Schlichting, L. Lomb, N. Coppola, R. L. Shoeman, S. W. Epp, R. Hartmann, D. Rolles, A. Rudenko, L. Foucar, N. Kimmel, G. Weidenspointner, P. Holl, M. Liang, M. Barthelmess, C. Caleman, S. Boutet, M. J. Bogan, J. Krzywinski, C. Bostedt, S. Bajt, L. Gumprecht, B. Rudek, B. Erk, C. Schmidt, A. Hömke, C. Reich, D. Pietschner, L. Strüder, G. Hauser, H. Gorke, J. Ullrich, S. Herrmann, G. Schaller, F. Schopper, H. Soltau, K. Kühnel, M. Messerschmidt, J. D. Bozek, S. P. HauRiege, M. Frank, C. Y. Hampton, R. G. Sierra, D. Starodub, G. J. Williams, J. Hajdu, N. Timneanu, M. M. Seibert, J. Andreasson, A. Rocker, O. Jönsson, M. Svenda, S. Stern, K. Nass, R. Andritschke, C. Schröter, F. Krasniqi, M. Bott, K. E. Schmidt, X. Wang, I. Grotjohann, J. M. Holton, T. R. M. Barends, R. Neutze, S. Marchesini, R. Fromme, S. Schorb, D. Rupp, M. Adolph, T. Gorkhover, I. Andersson, H. Hirsemann, G. Potdevin, H. Graafsma, B. Nilsson and J. C. H. Spence, *Nature*, 2011, **470**, 73–77.

28. C. E. Graves, A. H. Reid, T. Wang, B. Wu, S. de Jong, K. Vahaplar, I. Radu, D. P. Bernstein, M. Messerschmidt, L. Müller, R. Coffee, M. Bionta, S. W. Epp, R. Hartmann, N. Kimmel, G. Hauser, A. Hartmann, P. Holl, H. Gorke, J. H. Mentink, A. Tsukamoto, A. Fognini, J. J. Turner, W. F. Schlotter, D. Rolles, H. Soltau, L. Strüder, Y. Acremann, A. V. Kimel, A. Kirilyuk, T. Rasing, J. Stöhr, A. O. Scherz and H. A. Dürr, *Nat. Mater.*, 2013, **12**, 293–298.

29. I. Schweigert and S. Mukamel, *Phys. Rev. Lett.*, 2007, **99**, 163001.

30. T. E. Glover, D. M. Fritz, M. Cammarata, T. K. Allison, S. Coh, J. M. Feldkamp, H. Lemke, D. Zhu, Y. Feng, R. N. Coffee, M. Fuchs,

S. Ghimire, J. Chen, S. Shwartz, D. A. Reis, S. E. Harris and J. B. Hastings, *Nature*, 2012, **488**, 603–608.

31. S. Priyadarshi, I. González-Vallejo, C. Hauf, K. Reimann, M. Woerner and T. Elsaesser, *Phys. Rev. Lett.*, 2022, **128**, 136402.

32. A. Koç, C. Hauf, M. Woerner, L. von grafenstein, D. Ueberschaer, M. Bock, U. Griebner and T. Elsaesser, *Opt. Lett.*, 2021, **46**, 210–213.

33. J. Weisshaupt, V. Juvé, M. Holtz, M. Woerner and T. Elsaesser, *Struct. Dyn.*, 2015, **2**, 024102.

34. F. Brunel, *Phys. Rev. Lett.*, 2020, **59**, 52–55.

35. Y. Jiang, T. Lee, W. Li, G. Ketwaroo and C. G. Rose-Petruck, *Opt. Lett.*, 2002, **27**, 963–965.

36. K. Huang, M. H. Li, W. C. Yan, X. Guo, D. Z. Li, Y. P. Chen, Y. Ma, J. R. Zhao, Y. F. Li, J. Zhang and L. M. Chen, *Rev. Sci. Instrum.*, 2014, **85**, 113304.

37. J. Weisshaupt, V. Juvé, M. Holtz, S. Ku, M. Woerner, T. Elsaesser, S. Alisauskas, A. Pugzlys and A. Baltuska, *Nat. Photonics*, 2014, **8**, 927–930.

38. L. von Grafenstein, M. Bock, D. Ueberschaer, K. Zawilski, P. Schunemann, U. Griebner and T. Elsaesser, *Opt. Lett.*, 2017, **42**, 3796–3799.

39. S. Hasegawa, R. Takashima, M. Todoriki, S. Kikkawa, K. Soda, K. Takano, Y. Oishi, T. Nayuki, T. Fujii and K. Nemoto, *Rev. Sci. Instrum.*, 2011, **82**, 033301.

40. M. Holtz, C. Hauf, J. Weisshaupt, A. A. H. Salvador, M. Woerner and T. Elsaesser, *Struct. Dyn.*, 2017, **4**, 054304.

41. B. E. Warren, *X-ray Diffraction*, Dover, New York, 1990.

42. J. Als-Nielsen and D. McMorrow, *Elements of modern X-ray physics*, Wiley, 2001.

43. M. Bargheer, N. Zhavoronkov, R. Bruch, H. Legall, H. Stiel, M. Woerner and T. Elsaesser, *Appl. Phys. B: Lasers Opt.*, 2005, **80**, 715–719.

44. J. Stingl, F. Zamponi, B. Freyer, M. Woerner, T. Elsaesser and A. Borgschulte, *Phys. Rev. Lett.*, 2012, **109**, 147402.

45. M. Bocher, *Ann. Math.*, 1906, 7, 81–152.

46. P. Coppens, *X-ray Charge Densities and Chemical Bonding*, Oxford University Press, 1997.

47. N. K. Hansen and P. Coppens, *Acta Crystallogr., Sect. A: Cryst. Phys., Diffr., Theor. Gen. Crystallogr.*, 1978, **34**, 909–921.

48. E. T. Jaynes, *Phys. Rev.*, 1957, **106**, 620–630.

49. E. T. Jaynes, *IEEE Trans. Syst. Sci. Cybern.*, 1968, **4**, 227–241.

50. D. M. Collins, *Nature*, 1982, **298**, 49–51.

51. M. Woerner, M. Holtz, V. Juvé, T. Elsaesser and A. Borgschulte, *Faraday Discuss.*, 2014, **171**, 373–392.

52. S. Smaalen, L. Palatinus and M. Schneider, *Acta Crystallogr., Sect. A: Found. Crystallogr.*, 2003, **59**, 459–469.

53. C. Hauf, A. A. H. Salvador, M. Holtz, M. Woerner and T. Elsaesser, *Struct. Dyn.*, 2019, **9**, 014503.

54. S. L. D. W. T. Pollard, Q. Wang, L. A. Peteanu, C. V. Shank and R. A. Mathies, *J. Phys. Chem.*, 1992, **96**, 6147–6158.

55. M. Cho, M. Du, N. F. Scherer, G. R. Fleming and S. Mukamel, *J. Chem. Phys.*, 1993, **99**, 2410–2428.
56. A. T. N. Kumar, F. Rosca, A. Widom and P. M. Champion, *J. Chem. Phys*, 2001, **114**, 6795–6815.
57. J. H. Hannay, *Eur. J. Phys.*, 1983, **4**, 141.
58. W. Cochran, *Adv. Phys.*, 1960, **9**, 387.
59. R. D. K.-S. W. Zhong and D. Vanderbilt, *Phys. Rev. Lett.*, 1994, **72**, 3618–3621.
60. S. Pal, N. Strkalj, C.-J. Yang, M. C. Weber, M. Trassin, M. Woerner and M. Fiebig, *Phys. Rev. X*, 2021, **11**, 021023.
61. R. Resta, M. Posternak and A. Baldereschi, *Phys. Rev. Lett.*, 1993, **70**, 1010.
62. C. Hauf, M. Woerner and T. Elsaesser, *Phys. Rev. B*, 2018, **98**, 054306.
63. S. A. Stepanov, E. A. Kondrashkina, R. Khler, D. V. Novikov, G. Materlik and S. M. Durbin, *Phys. Rev. B: Condens. Matter Mater. Phys.*, 1998, 57, 4829–4841.
64. C. v. Korff Schmising, M. Bargheer, M. Kiel, N. Zhavoronkov, M. Woerner, T. Elsaesser, I. Vrejoiu, D. Hesse and M. Alexe, *Phys. Rev. B: Condens. Matter Mater. Phys.*, 2006, **73**, 212202.
65. M. Holtz, C. Hauf, A. A. H. Salvador, R. Costard, M. Woerner and T. Elsaesser, *Phys. Rev. B: Condens. Matter Mater. Phys.*, 2016, **94**, 104302.
66. A. M. Glass, D. von der Linde and T. J. Negran, *Appl. Phys. Lett.*, 1974, **25**, 233.
67. R. Hsu, E. N. Malsen, D. du Boulay and N. Ishizawa, *Acta Crystallogr., Sect. B: Struct. Sci.*, 1997, **53**, 420.
68. C. Colvard, T. A. Gant, M. V. Klein, R. Merlin, R. Fischer, H. Morkoc and A. C. Gossard, *Phys. Rev. B: Condens. Matter Mater. Phys.*, 1985, **31**, 2080–2091.
69. A. Krost, G. Bauer and J. Woitok, in *Optical Characterization of Epitaxial Semiconductor Layers*, ed. G. Bauer and W. Richter, Springer, Berlin, 1996, pp. 287–391.
70. C. von Korff Schmising, M. Bargheer, M. Kiel, N. Zhavoronkov, M. Woerner, T. Elsaesser, I. Vrejoiu, D. Hesse and M. Alexe, *Phys. Rev. Lett.*, 2007, **98**, 257601.
71. C. V. Korff Schmising, A. Harpoeth, N. Zhavoronkov, Z. Ansari, C. Aku-Leh, M. Woerner, T. Elsaesser, M. Bargheer, M. Schmidbauer, I. Vrejoiu, D. Hesse and M. Alexe, *Phys. Rev. B: Condens. Matter Mater. Phys.*, 2008, **78**, 060404(R).
72. H. G. Unruh, J. Krüger and E. Sailer, *Ferroelectrics*, 1978, **20**, 3–10.
73. H. G. Unruh, E. Sailer, H. Hussinger and O. Ayere, *Solid State Commun.*, 1978, **25**, 871–874.
74. C. Hauf, A.-A. H. Salvador, M. Holtz, M. Woerner and T. Elsaesser, *Struct. Dyn.*, 2018, **5**, 024501.
75. S. Hoshino, K. Vedam, Y. Okaya and R. Pepinsky, *Phys. Rev.*, 1958, **119**, 405–412.
76. H.-G. Unruh and U. Rüdiger, *J. Phys. Colloq.*, 1972, **33**, C2-77–C2-78.
77. S. Ahmed, A. Shamah, K. Kamel and Y. Badr, *Phys. Stat. Sol. (a)*, 1987, **99**, 131–140.
78. R. E. Cohen, *Nature*, 1992, **358**, 136–138.

CHAPTER 5

Imaging Clusters and Their Dynamics with Single-shot Coherent Diffraction

ALESSANDRO COLOMBO* AND DANIELA RUPP*

ETH Zürich, John-von-Neumann-Weg 9, 8093 Zürich, Switzerland
*Emails: alcolombo@phys.ethz.ch; ruppda@phys.ethz.ch

5.1 Introduction

Diffraction before destruction[1,2] has been a key vision driving the development and construction of extreme ultraviolet (XUV) and X-ray free-electron lasers (FELs; see Chapter 8).[3–9] Here, an interference pattern of a single, free-standing molecule from a single illumination with an intense X-ray pulse is recorded (see Figure 5.1).

The pulse is intense enough to measure a diffraction pattern with sufficient signal-to-noise ratio for retrieving the molecule's structure, and at the same time short enough to outrun the rapid destruction of the molecule due to ionization and charge-driven explosion following X-ray irradiation.

The quest for imaging single free molecules and the concept of diffraction before destruction have attracted widespread attention because of the large number of very important "soft-matter" molecules with crucial roles in life-sciences, such as membrane proteins, for which the standard way of crystallization and subsequent synchrotron-based X-ray diffraction is not feasible.[10] Furthermore, the "snapshot" technique of CDI can be utilized in pump–probe measurements, making photoexcited dynamics in molecules directly visible,[11] such as the first steps of photosynthesis or the molecular mechanisms of vision. Structure determination of bio-samples at FELs has

Theoretical and Computational Chemistry Series No. 25
Structural Dynamics with X-ray and Electron Scattering
Edited by Kasra Amini, Arnaud Rouzée and Marc J. J. Vrakking
© The Royal Society of Chemistry 2024
Published by the Royal Society of Chemistry, www.rsc.org

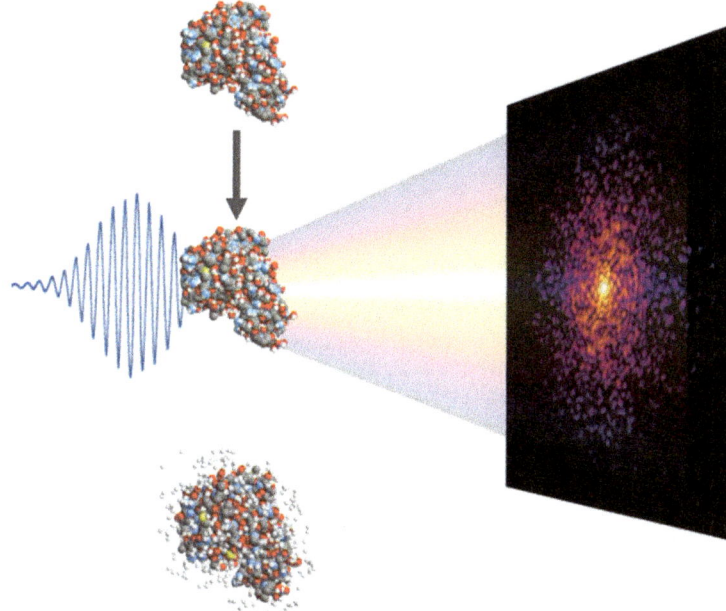

Figure 5.1   Concept of diffraction before destruction. A single protein in free flight is intercepted by an intense coherent X-ray pulse. The diffracted light, captured with a large-area detector, is sufficiently bright for decoding the molecule's structure. The molecule explodes quickly, but the pulse is already over when atomic motion sets in. Reproduced from ref. 1 with permission from Springer Nature, Copyright 2000.

been a story of both success and great challenges.[10] Single viruses, cell organelles and bacteria were successfully imaged[12–14] and hundreds of new molecular structures have been determined *via* serial femtosecond nano-crystallography,[15–17] particularly a large number of previously inaccessible membrane proteins.[†] The first single-shot single-molecule diffraction patterns, however, could be obtained only very recently,[18] and they are still far too weak to provide atomic resolution. The first simulations of single-molecule CDI[1] over two decades ago had already predicted challenging requirements for pulse parameters in terms of intensity and duration, which have come into reach only recently.[19,20] The pioneering theoretical study[1] and subsequent work[21–24] also indicated that "ultrafast radiation damage" would be a main bottleneck of single-molecule CDI, and in accordance, intense X-ray matter interaction has been in the focus of XFEL science from the very beginning.

Atomic clusters have served as prototypical model systems with tunable size and complexity in intense light–matter interaction studies throughout all wavelength regimes. They have also played an important role in developing our current understanding of the intricate processes triggered in any matter at

---

[†] https://blanco.biomol.uci.edu/mpstruc/.

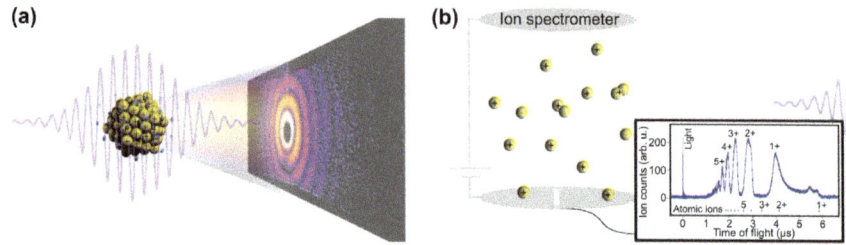

Figure 5.2   Combined single-cluster CDI and ion spectroscopy. (a) The interaction
of a cluster with a short-wavelength pulse produces a diffraction pattern encoding the cluster size and focal intensity. (b) The cluster is also
excited and highly ionized, and it ultimately disintegrates. Measured
ion time-of-flight spectra from single clusters can be sorted by the
diffraction information. An exemplary ion time-of-flight spectrum of a
single xenon cluster shows that high charge states up to 5+ are clearly
separable even though the ions possess high kinetic energies.

the focus of an intense X-ray pulse. In particular, experiments on single
clusters have pushed the field far beyond the mere study of radiation damage.
For the first time, the exact structure of fragile and non-depositable gas-phase
clusters and nanodroplets can be determined. Also, near-background free
spectroscopy from clusters, resolved for cluster size and laser power density,
became possible through single-cluster CDI (see the concept in Figure 5.2).

Arguably, the most exciting aspect is imaging laser-induced dynamics in
single clusters through "X-ray movies", yielding unprecedented insight into
highly excited matter on the nanoscale. In this regard, the current development of intense attosecond pulses from FEL and HHG sources around the
world[19,25–27] opens up an exciting opportunity. The realm of electron
dynamics is reached, a perspective discussed at the end of this chapter.

The core of this chapter provides a step-by-step introduction into the
theory behind the imaging method (see Section 5.2). Here, we aim at
providing a solid starting point for students and scientists interested in CDI
experiments at FELs. We introduce the fundamental building blocks of the
imaging problem, familiarize the reader with available imaging strategies,
and point out common practical pitfalls, often overlooked by scientific
literature. In terms of applications, will emphasize recent results and novel
opportunities for single cluster CDI, while giving concise summaries elsewhere, as actual reviews exist on the CDI method,[28–33] bio-applications of
CDI,[11,12,34–37] CDI of helium nanodroplets,[38,39] and cluster science at
FELs.[40,41]

5.2   A Gentle Introduction to Single-shot Single-particle Coherent Diffraction Imaging

Coherent diffraction imaging (CDI) is an experimental technique that
images isolated samples by collecting their scattering signal. The most

common way the term imaging is understood refers to direct imaging, like for optical or electron microscopes up to telescopes. There, the lenses in the experimental apparatus are responsible for directly providing an image[‡] (and thus spatial properties) of the sample.

However, when imaging is employed with the term diffraction, the perspective changes. CDI is a lens-less imaging technique, whereby no lenses are used in the experiment. This makes CDI the method of choice for those measurements in which lenses cannot be used, like in the case of XUV and (soft) X-ray FELs. The side-effect of the absence of lenses is that the radiation recorded by the detector is no more a direct visualization of the sample's spatial distribution. This renders CDI an indirect imaging technique: the sample's image is retrieved by means of sophisticated analysis methods, which play the role of virtual lenses.

5.2.1 The Inverse Scattering Problem

The theory behind CDI can be conveniently introduced through a simplified view on light–matter interaction, which can be easily described in a semi-classical way. When light interacts with a sample, the oscillating electric field drives the material's electrons. In response, the electrons become emitters of spherical waves that are coherent with the incoming light. This simple view[§] describes a completely elastic scattering, as the photon energy is conserved during the scattering event. The coherent fields scattered by the emitters at different locations in space interfere, giving rise to a diffraction pattern. In a CDI experiment, light is recorded under the so-called far-field condition, which depends on the sample spatial extension and on the employed radiation wavelength. The validity of this condition is not a concern for CDI experiments in the X-ray regime,[46] as explained in ref. 28. We can state that the field $\Psi(\vec{q})$ scattered by a sample seen in far-field is given by:

$$\Psi(\vec{q}) \propto \int \mathrm{d}\vec{r}\, \rho(\vec{r}) e^{i\vec{q}\cdot\vec{r}} \qquad (5.1)$$

Here, $\rho(\vec{r})$ can be addressed as the scattering density, which tells, in the first approximation, the number of scatterers (electrons) present at a given location in space $\vec{r}$. The coordinate $\vec{q}$ is called momentum transfer and, roughly speaking, provides the direction in which the light is emitted. Eqn (5.1) is based on the Born approximation,[28,47] which assumes that light travelling through the sample is not affected by the sample material, apart from the single scattering event (see Section 5.2.3.3). In practice, eqn (5.1)

[‡] In a broad sense. For light scattering, this is usually related to the electronic density distribution. A different example is transmission electron microscopy, where the "image" reveals the atomic potential.
[§] Going into the details of light–matter interaction and scattering is beyond the scope of this chapter. We warmly recommend ref. 42 and 43 for insights into first principles, and ref. 44 and 45 for the link to diffraction theory.

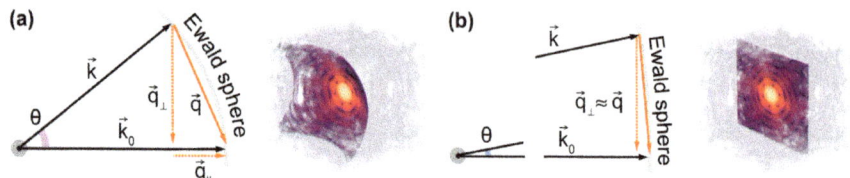

Figure 5.3 Geometry of monochromatic elastic scattering. (a) Relevant quantities in reciprocal space; incoming wavevector $\vec{k}_0$, scattered wavevector $\vec{k}$ for a given scattering angle θ and respective momentum transfer $\vec{q}$ with parallel component $\vec{q}_\parallel$ and orthogonal component $\vec{q}_\perp$. Right: Portion of the Fourier domain accessible *via* monochromatic light scattering, commonly known as the Ewald sphere. (b) Small-angle scattering regime: the parallel component of the momentum transfer $\vec{q}_\parallel$ is negligible. The portion of the Fourier domain experimentally accessible can be approximated by a plane passing through the origin of the reciprocal coordinates.

gives the scattered field as a function of momentum transfer $\vec{q}$ simply as the three-dimensional Fourier transform of the scattering density $\vec{\rho}$. The field $\Psi(\vec{q})$ is not experimentally accessible, and only its intensity $I(\vec{q})$ can be measured, given by

$$I(\vec{q}) = |\Psi(\vec{q})|^2 \propto |\mathcal{F}_{3D}[\rho(\vec{r})](\vec{q})|^2, \tag{5.2}$$

where $\mathcal{F}_{3D}$ denotes the Fourier transform in the three-dimensional spatial domain. Second, only monochromatic light is employed,[¶] which implies that a small fraction of reciprocal coordinates $\vec{q}$ are experimentally accessible, as the recorded data is a two-dimensional surface in the 3D Fourier domain. The relationship between the recorded data and its coordinates in the reciprocal space is sketched in Figure 5.3a. The momentum transfer $\vec{q}$ is defined as the difference between the wavevector of the incident light $\vec{k}_0$ and the wavevector of the scattered light $\vec{k}$ at a given scattering angle[ǁ] θ, *i.e.* $\vec{q} = \vec{k}_0 - \vec{k}$. Thanks to the elastic nature of the scattering process, the photon energy is conserved, implying $|\vec{k}_0| = |\vec{k}|$. As a consequence, the accessible momentum transfer lies on a spherical surface in the Fourier domain, referred to as the Ewald sphere (see right of Figure 5.3a for a 3D render of the Ewald sphere).

It is convenient to separate the momentum transfer $\vec{q}$ into two components. One component is $\vec{q}_\perp$, perpendicular to the beam propagation

---

[¶] The monochromatic assumption is valid in most cases. However, polychromatic[48] or broadband[49] pulses are also employed for some applications, which are not discussed in this section.
[ǁ] Warning: in some contexts, and especially in crystallography, this angle is often called 2θ, where theta is the Bragg angle. Here, we call scattering angle θ the quantity equivalent to twice the Bragg angle.

direction. The second component $\vec{q}_{\parallel}$ is instead parallel to the beam. Their values can be expressed as function of the scattering angle θ:

$$|\vec{q}_{\perp}(\theta)| = |\vec{k}_0|\sin(\theta) \quad |\vec{q}_{\parallel}(\theta)| = |\vec{k}_0| \cdot [1 - \cos(\theta)] \tag{5.3}$$

When the scattering angle reached by the light detector is sufficiently large (see the example shown in Figure 5.3), the curvature of the Ewald sphere allows recording significant information both on the orthogonal plane and on the axis parallel to the beam. This regime is known as wide-angle scattering, where partial 3D information is recorded in a single shot.

The situation drastically changes in the so-called small-angle scattering regime (see the sketch in Figure 5.3b). In such a case, the scattering angle is small enough to neglect the parallel component of momentum transfer $\vec{q}_{\parallel}(\theta)$. In practice, this means that the portion of the Ewald sphere accessible by the detector is small enough to approximate it to a flat slice, orthogonal to the beam propagation direction and passing through the origin of the Fourier domain. It is then possible to take advantage of the Fourier slice theorem, which is a well-known "trick" in the field of computational tomography. Brought into the CDI framework, it allows to implement the following: as long as the recorded data is lying on the plane orthogonal to the beam ($\perp$) in Fourier space, the values on the slice are the 2D Fourier transform of the scattering density ρ projected onto the beam propagation axis ($\parallel$). In mathematical form, this means

$$\mathcal{F}_{3D}[\rho(\vec{r})](\vec{q}_{\perp}, q_{\parallel} = 0) = \mathcal{F}_{2D}\left[\int_{\parallel} \rho(\vec{r})\right](\vec{q}_{\perp}) \tag{5.4}$$

Thus, the actual data recorded by the detector in the small angle case (*i.e.*, the amplitude of the scattered field on the orthogonal plane $I(\vec{q}_{\perp})$) can be expressed as

$$I(\vec{q}_{\perp}) \propto \left|\mathcal{F}_{2D}[\rho_{2D}(\vec{r}_{\perp})](\vec{q}_{\perp})\right|^2, \tag{5.5}$$

where $\rho_{2D}(\vec{r}_{\perp})$ denotes the projection of the scattering density in the parallel direction. Due to the low scattering angle θ, the recorded parallel component of $\vec{q}$ vanishes (see eqn (5.3)). As $\vec{q}_{\parallel}$ carries information in the beam propagation direction (*i.e.*, on the sample depth information), the spatial distribution in the depth direction is, then, lost when only $\vec{q}_{\parallel} = 0$ is accessible.

Eqn (5.2) and (5.5) are a concise solution to the direct scattering problem, where the scattered field Ψ (or its amplitude I) is computed from the knowledge of the scattering density ρ. However, the aim of CDI is to approach the riddle from the opposite direction and retrieve the scattering density ρ from the scattered field amplitude I. This is called the inverse scattering problem, and the way in which it can be approached strongly differs between the wide-angle and small-angle regimes. For wide-angle scattering, partial 3D information about ρ is accessible. However, its main drawback is that the retrieval of the 3D sample structure is highly challenging, as it will be discussed in Section 5.2.3. On the other hand, small-angle diffraction data only give access to a 2D projection of scattering density ρ as suggested by eqn (5.5). Nevertheless, established methods for the retrieval of the density projection ρ exist since many years, and have already proved their consistency and effectiveness. The next section will provide an introduction to these algorithmic methods, giving insights into what can be addressed as the virtual lenses of small-angle CDI: phase retrieval algorithms.

5.2.2   Small-angle Coherent Diffraction Imaging

To approach the topic from the imaging perspective, it is convenient to reformulate eqn (5.1), which presents a solution to the direct scattering problem, in its inverse form for the small angle case, by taking advantage of the Fourier slice theorem

$$\rho(x,y) = \mathcal{F}^{-1}\left[\Psi(q_x, q_y)\right](x,y), \tag{5.6}$$

where $\mathcal{F}^{-1}$ indicates the operation of inverse Fourier transform (IFT).**

As mentioned in the previous section, only information about the modulus of Ψ is experimentally accessible (*i.e.*, $|\Psi| = \sqrt{I}$ where I is the field intensity recorded by the scattering detector). The field phases ϕ are irretrievably lost[††] in the measurement process. It is then convenient to rewrite eqn (5.6) by explicitly separating the modulus and phase components of the scattered field as it follows:

$$\rho(x,y) = \mathcal{F}^{-1}\left[\sqrt{I(q_x, q_y)}e^{i\phi(q_x,q_y)}\right](x,y) \tag{5.7}$$

As long as we trust the fact that any function has a unique Fourier representation (*i.e.*, the Fourier domain is a complete basis), eqn (5.7) makes it clear that, for small-angle CDI, retrieving the phases $\phi(q_x, q_y)$ is equivalent to recovering the scattering density ρ. This is the reason why, in this context, the inverse scattering problem is often called the phase retrieval problem.[29,31,50]

There are several ways to intuitively evaluate the amount of information carried by the field's phase.[‡‡] Here, we try to address this point by studying the result of a random assignment of the phase values in the reciprocal space. In this regard, we temporarily leave the formal mathematical

---

** Here, we introduce some changes in the mathematical notation that will be valid throughout the whole section. From now on, the spatial coordinates will only refer to the 2D orthogonal plane, indicated as x and y for the real space and q_x and q_y for the Fourier domain. Furthermore, the Fourier transform (FT) operation $\mathcal{F}$ acts in the two-dimensional space. Please, also note that we replaced the term $\propto$ with $=$ for the sake of simplicity.

†† Why are the phases lost? To keep it short, detectors measure photon counts, which are proportional to the square field modulus. Are they really irretrievable? Not really, keep on reading.

‡‡ And it turns out that the amount of information provided by the Fourier phases is actually even more than the one from the amplitudes, as nicely investigated in ref. 51.

treatment of the scattering problem and directly dive into its numerical translation.[§§] The first numerical implementation performs a direct scattering simulation, following eqn (5.5). The scattering density $\rho(x,y)$ is encoded into a matrix rho[i, j]. The simulation is numerically achieved by performing a discrete Fourier transform (DFT) on the matrix rho[i, j], and then computing its squared absolute value, as described by the following script:

```
1  # The import of the numpy library is necessary for complex mathematical operations
2  # like the DFT. This line will be omitted in the next code listings
3  import numpy as np
4
5  def get_simulated_pattern(rho):
6      # Calculate the 2D DFT of the input data, and shift the frequency domain
7      # to have the (0,0) component in the centre
8      field = np.fft.fftshift(np.fft.fft2(rho))
9      # Calculate the diffraction pattern as the square modulus of the scattered field.
10     pattern = np.abs(field)**2
11     return pattern
```

This operation produces a two-dimensional matrix, of the same size as the input matrix rho, which provides a simulated diffraction pattern (*i.e.*, a solution to the direct scattering problem). A possible output of this function is shown in Figure 5.4a. We can visualize in real space the effect of assigning random phases to the diffraction pattern *via* the following script:

```
1  def assign_random_phase(pattern):
2      # Create a matrix of random numbers between -pi and pi, of size equal
3      # to the pattern matrix
4      phases = np.random.uniform(-np.pi, np.pi, size = pattern.shape)
5      # Create a field that combines the input pattern with random phases
6      field = np.sqrt(pattern) * np.exp(1j*phases)
7      # Compute the real-space image by inverting the field via an inverse DFT,
8      # after having shifted the frequencies back
9      rho = np.fft.ifft2(np.fft.fftshift(field))
10     return rho
```

Possible outputs of the assign_random_phase function are shown in Figure 5.4b together with the original density placed in the centre. There are two key features that are apparent in Figure 5.4b. First, any information about the overall shape of the pipe and its main features is completely lost when phases are randomly assigned. It is then possible to create an infinite amount of $\rho(x,y)$ having the same diffraction pattern but completely different real-space representations. This already suggests that further information is needed to address the phase problem. The second observation already hints towards which type of information could be employed. In fact, all the scattering densities produced with a random assignment of phases have non-zero values spread around the full matrix rho, while only the correct density is spatially confined to a specific region of the real space.

[§§] The numerical examples will be presented in Python3 language, due to its clarity and popularity. Readers not familiar with Python language should not be afraid, as the examples provided here will be easily understandable without any specific Python knowledge, and all numerical operations will be exhaustively commented. The presented scripts are not to be intended as full standalone scripts, but just snippets for a practical understanding of numerical operations.

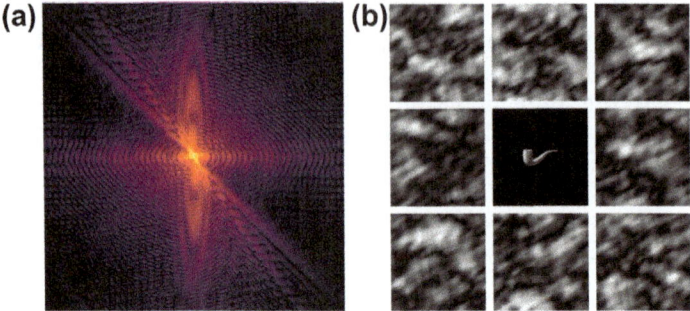

Figure 5.4   The role of the field phases. (a) Simulated diffraction pattern in logarithmic color scale, representing the only experimentally accessible
data. (b) All but the central picture are obtained by assigning random
phases to the diffraction pattern in (a), and then inverting the field to
obtain a spatial distribution by performing an inverse FT. The central
picture in (b) is instead the original image (*i.e.* with the correct phases)
from which the simulation in (a) was calculated.

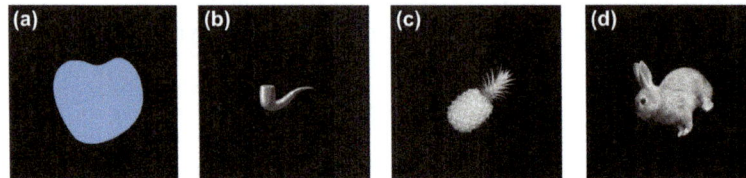

Figure 5.5   The support function. (a) A possible support function is shown. (b–d)
Three different spatial distributions compatible with the support function in (a). It can be demonstrated that only one of them can have the
same diffraction pattern shown in Figure 5.4 (guess which one).

This confined space is described, in this context, as a support function. The
support function is numerically implemented as a binary function, which can
assume only two values, 0 and 1 (see Figure 5.5a). It is worth noting that it is
possible to create an infinite amount of spatial densities that are "contained"
in a support function (see Figure 5.5b–d), similarly to the considerations made
about Figure 5.4. However, it can be mathematically proven[52] that a unique
solution¶¶ to the phase retrieval problem exists if and only if the original
scattering density is spatially confined.[53,54]‖‖ We can then state that

There exists only one spatially confined scattering density with a given Fourier
amplitude, which represents the solution to the phase retrieval problem.

¶¶ The unicity of the solution has to be interpreted in a broad sense, as there are, in practice,
many ambiguous solutions. However, this aspect will be discussed later to avoid interruptions
of the discussion thread.

‖‖ Technically, this spatial confinement is required to satisfy the oversampling condition, such
that the spatial extension of the support function is, in pixels, not greater than half the size of
the full matrix. Jump to Section 5.2.2.5 for further considerations.

Once we "prove" that, under the given conditions, a solution exists,[29,52] the next step is to find it (*i.e.* to actually solve the phase problem). Solving the phase problem means finding the density distribution ρ_{sol} that is spatially confined and whose Fourier amplitude is compatible with the experimental diffraction pattern. For an easier and more intuitive understanding, it is now convenient to change the perspective from which we look at the problem, and describe it in terms of two sets[30,55,56] (see Figure 5.6).

Set I (see the green shaded area in Figure 5.6) is the collection of all spatial distributions that are compatible with the diffraction pattern (*i.e.*, whose squared Fourier amplitude is equal to the scattering data shown in Figure 5.4a). Then, all densities in Figure 5.4b belong to set I. On the other hand, the collection of all the densities which are compatible with a given support function (like the one shown in Figure 5.5a) is referred to as set S (pink shaded area in Figure 5.6). Densities in Figure 5.5b–d all belong to such a set. The solution to the phase problem can be now visualized as the intersection between sets I and S. The expression

$$I \cap S = \{\rho_{sol}\} \tag{5.8}$$

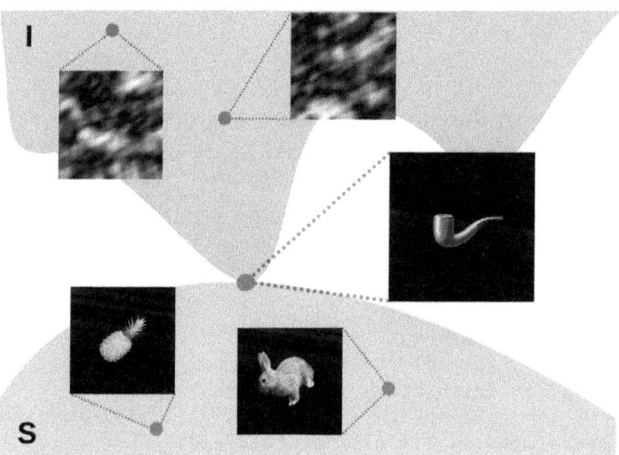

Figure 5.6   Qualitative representation of the phase problem in terms of sets. The 2D coordinates in the picture have to be interpreted as a low-dimensional representation of the values of the density ρ. A point in this space corresponds to a given combination of pixel values (*i.e.*, to a well-defined density matrix). The green set, labelled I, in the upper part of the figure contains all the density distributions that have the same Fourier intensities, as those shown in Figure 5.4b. The pink set S is, conversely, the collection of all spatial distributions that have non-zero values only within the support function, as the ones shown in Figure 5.5. Theoretically, the intersection between the sets contains a single element (here, represented by the pipe) which is the solution to the phase retrieval problem.

is equivalent to stating that the solution ρ_{sol} is unique. Now that the phase problem and its solution have been defined, strategies to find such a solution will be discussed.

5.2.2.1   Phase Retrieval Algorithms

Following eqn (5.8), the solution to the phase problem must satisfy both the intensity constraint and the support constraint. The first step in solving the phase problem is to define an effective way to apply these constraints to a generic density distribution ρ.

For the intensity constraint, we define the operator P_I which acts on a generic spatial density ρ, given by

$$P_I\rho = \mathcal{F}^{-1}\left[\sqrt{I}\,\frac{\tilde{\rho}}{|\tilde{\rho}|}\right], \tag{5.9}$$

where $\tilde{\rho} = \mathcal{F}[\rho]$.*** The operator P_I acts on a density function ρ by calculating its Fourier transform $\tilde{\rho}$, replacing its modulus with the experimental one $\sqrt{I}$, and performing an inverse FT back to the real space. For any density ρ, the result $P_I\rho$ belongs to set I. On the other hand, the operator P_S acts on a generic spatial density ρ as

$$P_S\rho = \rho \cdot S, \tag{5.10}$$

where S is the support function, whose values are $S(x,y) = 1$ for those spatial coordinates where the density is allowed to have non-zero values (for example, see the blue area in Figure 5.5), or $S(x,y) = 0$ elsewhere. Similar to what is observed for P_I, the result of $P_S\rho$ belongs to the set S for any density ρ. The most straightforward numerical implementation of the operators P_I and P_S is given below:

```python
def P_I(rho, modulus):
    # Compute the field of rho
    field = np.fft.fftshift(np.fft.fft2(rho))
    # Replace the current Fourier modulus with the experimental data
    field_new = modulus * field/np.abs(field)
    # Compute the real-space density with the upgraded field.
    rho_new = np.fft.ifft2(np.fft.fftshift(field_new))
    return rho_new

def P_S(rho, support):
    # Compute the real-space density with the upgraded field.
    rho_new = rho * support
    return rho_new
```

*** For the sake of brevity, the dependence on the spatial coordinates x, y for real-space functions and on the reciprocal coordinates q_x, q_y for functions in Fourier domain has been omitted.

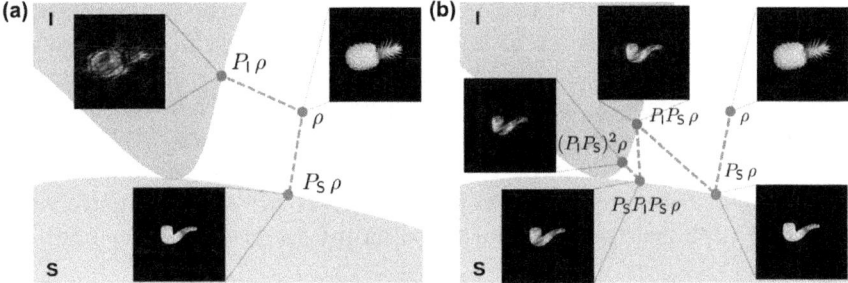

Figure 5.7 (a) Qualitative representation of the action of projectors onto the two sets. The blue dot, which is outside both the sets S and I, stands for a density which is incompatible with both the support and intensity constraints, like the pineapple. The two projector operators provide, as output, the density closest to the pineapple that belongs to the respective sets. (b) Illustration of the alternate action of the two projections on a starting density. The two projectors are applied in an alternate manner to a starting density. The nature of this process, based on projections, is such that the trajectory moves towards a configuration of (local) minimum distance between the two sets.

It is now worth noting that the two operators in eqn (5.9) and (5.10) are actually projectors on the sets I and S, respectively.[†††] A qualitative representation of the action of the two projectors P_I and P_S is given in Figure 5.7a. Their effect is shown on a starting density ρ that does not belong to neither I or S. The outcome of the projector is to produce a density at the closest position of the respective set (*i.e.*, to provide a spatial distribution that satisfies the respective constraint and is as similar as possible to the original density ρ[‡‡‡]). These two projectors represent the fundamental building blocks of phase retrieval algorithms. This can be intuitively deduced by observing the "dynamics" of the alternate application of P_I and P_S onto density ρ, as sketched in Figure 5.7b. This alternate application of the projectors makes the density jump between the two sets. The implications of using the projection operation is that each "jump" is shorter than the previous one (*i.e.*, by using projectors in this way, the values of ρ evolve in a way that minimizes the distance[§§§] between the two sets, I and S).

The first, most trivial, phase retrieval algorithm, which is exactly built in this way, is called the error reduction algorithm[57] due to its property of reducing the distance between the two sets (*i.e.*, the error of the reconstruction) at each iteration. A single iteration of the error reduction (ER) algorithm can be easily expressed, again, in terms of operators as

$$\text{ER} = P_S P_I, \tag{5.11}$$

[†††] This can be easily demonstrated by checking if they are idempotent (which is the definition of a projector). In fact, if they are applied more than once, the result does not change (*i.e.* $P_I^2 \rho = P_I \rho$ and $P_S^2 \rho = P_S \rho$).

[‡‡‡] This statement can be dangerous for the health of mathematicians. Please, handle with care.

[§§§] The concept of minimizing a distance is fundamental and it will be discussed in a later section.

such that a given number n of error reduction iterations can be conveniently written as ER^n. The outcome of a given number of algorithm iterations is addressed as reconstruction. The first algorithm iteration requires a starting density ρ. Such a density, often called starting guess, does not have to satisfy any particular condition (like the pineapple in Figure 5.7b). The iterative nature of the reconstruction process is the reason why this imaging approach is also known as "iterative phase retrieval" (IPR). A straightforward numerical implementation of the ER algorithm is given in the following script:

```
 1  def ER(rho, modulus, support, iterations):
 2      '''
 3      Execute a given number of Error Reduction iterations on the starting guess rho
 4      '''
 5      # Loop over the amount of iterations
 6      for n in range(iterations):
 7          # Apply the intensity projector
 8          rho = P_I(rho, modulus)
 9          # Apply the support projector
10          rho = P_S(rho, support)
11      return rho
```

Although the ER algorithm is elegant and simple to use, it has limited effectiveness. The main culprit is the nature of the intensity constraint, particularly the shape of set I. In fact, the set S has a convenient feature of being a convex set (*i.e.*, any linear combination of densities belonging to the set still belongs to it). This is definitely not the case for set I.[¶¶¶] The different shapes of the two sets can be qualitatively reproduced by using convex and non-convex morphologies (see Figure 5.6). Although not being mathematically rigorous, it helps to incorporate the intuition that such a non-convex shape of set I has the unavoidable effect of creating local minima of the distance between the sets. The ER algorithm, due to its nature, hardly escapes from this trap and always fails to reach the correct solution. This issue was already clear to Fienup, the pioneer of phase retrieval, who in the 1970's already proposed an alternative algorithm: the hybrid input–output (HIO).[58] Again, an iteration of the HIO algorithm can be expressed in terms of projectors as

$$\mathrm{HIO} = P_{\mathrm{S}} P_{\mathrm{I}} + (\mathbf{1} - P_{\mathrm{S}})(\mathbf{1} - \beta P_{\mathrm{I}}), \tag{5.12}$$

where $\mathbf{1}$ is the identity operator. The first part of this expression, $P_{\mathrm{S}} P_{\mathrm{I}}$, is equivalent to the ER algorithm in eqn (5.11). The second part can be interpreted as a "correction". The first important thing to note is that such a correction is only applied to the components of ρ orthogonal to the support set S, thanks to the operation $\mathbf{1} - P_{\mathrm{S}}$.[‖‖‖] In practice, this means that only coordinates of ρ outside the support function are affected. Those values are

---

[¶¶¶] Proving this aspect goes beyond the scope of this chapter, and we warmly suggest to give a read to the enlightening review of phase retrieval algorithms in ref. 30.

[‖‖‖] This operation results in the components of ρ orthogonal to the set S only because P_{S} is an orthogonal projector.

then suppressed by a quantity βP_{I} *via* the operation $\mathbf{1} - \beta P_{\mathrm{I}}$. The HIO operator can be translated into a numerical code in the following way:

```
1  def HIO(rho, modulus, support, iterations, beta):
2    # Loop over the amount of iterations
3    for n in range(iterations):
4      # Apply the intensity projector
5      rho_I = P_I(rho, modulus)
6      # Apply the support projector and the correction to the orthogonal component
7      rho_S = P_S(rho_I, support) + (1-support)*(rho-beta*rho_I)
8      rho = rho_S
9    return rho
```

The hybrid input–output algorithm is still among the most effective and widespread phase retrieval algorithms. Since its conception, several iterative algorithms with improved performances have been developed, based on peculiar combinations of the projector operators. The reader now has the necessary knowledge to enjoy dedicated technical papers about these improved iterative algorithms,[55,56,59,60] starting from a comparison given in ref. 30. Further algorithmic improvements have been achieved by leaving the "pure" application of projectors[61] or even combining the iterative methods with genetic algorithms.[62] When approaching small-angle CDI, the understanding of phase retrieval algorithms is often considered the main obstacle to overcome in setting up a proficient data analysis framework. However, this often turns out to not be the case. Most of the problems concern aspects that are rarely mentioned in the scientific literature and will be discussed in the following sections.

5.2.2.2 Phase Retrieval as an Optimization Problem

The difference between two densities ρ_1 and ρ_2 is represented, under the perspective of sets, as the distance between two "points" in the space of the possible configurations. This space of configurations is the space where all spatial distributions ρ exist, of which I and S are subsets. Such a space is of extremely high dimensionality since for a diffraction pattern of $N \times N$ pixel values, the spatial densities belong to a space of size $\mathbb{R}^{N^2}$.**** The visualization in terms of sets shown in Figure 5.6 is a simplified view of this space of configurations in $\mathbb{R}^2$. The most straightforward way in which a distance can be calculated is through the Euclidean metric. Given the Euclidean norm as $\|\vec{x}\|^2 = \sum_i x_i^2$, the distance between two densities is given by $D[\rho_1, \rho_2] = \|\rho_2 - \rho_1\|$. This way, we can compute how far ρ is from its projection on the two sets I and S (*i.e.*, $D[\rho, P_{\mathrm{I}}\rho]$ for set I and $D[\rho, P_{\mathrm{S}}\rho]$ for set S).

It is possible to "build" functions that evaluate the error of the reconstruction ρ by computing the distance between two alternated projections on the two sets. In particular, for a generic density ρ, we first apply a projection on one of the two sets (*i.e.*, $P_{\mathrm{S}}\rho$ or $P_{\mathrm{I}}\rho$) and then compute the

---

**** In general, scattering densities are complex valued, thus belonging to the space $\mathbb{C}^{N^2}$, but this will be a discussion topic for Section 5.2.2.6.

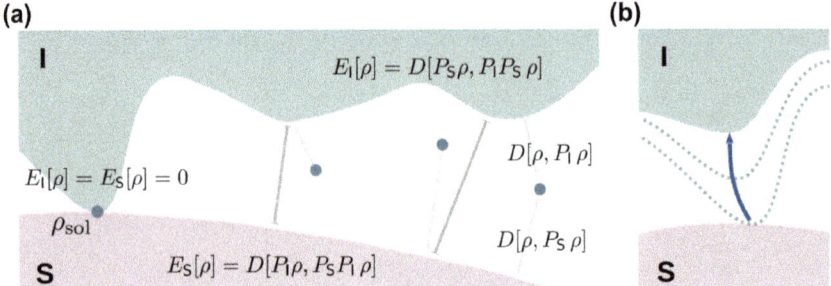

(a)

$E_I[\rho] = D[P_S\rho, P_I P_S\,\rho]$

$E_I[\rho] = E_S[\rho] = 0$

ρ_{sol}

S

$E_S[\rho] = D[P_I\rho, P_S P_I\,\rho]$

$D[\rho, P_I\,\rho]$

$D[\rho, P_S\,\rho]$

I

(b)

I

S

Figure 5.8   (a) The error value of a reconstruction can be visualized in terms of distances (here shown as continuous gray lines) between projections on the two sets I and S. The solution ρ_{sol} is, in theory, the one for which the error is 0, *i.e.*, it belongs to both sets. (b) Due to noise, the two sets do not intersect anymore and the phase problem turns into the problem of finding the density ρ_{sol} with minimal errors.

distance between this projection and the subsequent one on the other set (see Figure 5.8a). These two error values can be numerically computed as

$$E_S[\rho] = D[P_I\rho, P_S P_I\rho] = \sqrt{\sum_{ij}(1-S)|\rho_I|^2},$$

$$E_I[\rho] = D[P_S\rho, P_I P_S\rho] = \sqrt{\sum_{ij}\left(|\tilde{\rho}_S| - \sqrt{I}\right)^2},$$

(5.13)

where $\tilde{\rho}_S$ is the FT of the density projected into set S, ρ_I is the density projected on set I, and S is the support function. In practice, E_S gives the amount of density outside the support function for a density whose diffraction pattern is the experimental one. On the other hand, E_I tells us how far from the experimental data is the Fourier amplitude of a density fully contained in the support function.[††††] The two error functions are, in principle, equivalent, but E_I is often preferred because it is easier to normalize and involves a direct comparison with the diffraction pattern.[‡‡‡‡]

The search for the solution to the phase problem, *i.e.*, the intersection I∩S, is equivalent to the search for the density ρ such that $E[\rho]=0$ (see Figure 5.8a). However, this is not the case when dealing with experimental data that is intrinsically[§§§§] affected by noise. The side-effect of noise is that, in a realistic

[††††] The reader may have noted that E_I is computed in the Fourier space, and not in real space. The Fourier transform operation is unitary, which means that the norm is conserved. Thus, the error can be equivalently computed in real or Fourier space, and the choice of the latter is just for convenience.

[‡‡‡‡] For further discussion on the error function, for example, about its normalization or the inclusion of experimental noise, please refer to ref. 28, 30 and 63.

[§§§§] Do not forget that light detectors count the number of photons that reach each pixel. This means that, at least, Poisson noise affects experimental data. See the dedicated discussion in Section 5.2.2.9.

case, there is no density ρ compatible with both constraints at the same time (*i.e.*, I∩S is an empty set as shown in Figure 5.8b). The phase retrieval problem then becomes the search of the density ρ which minimizes the distance between the two sets, that is,

$$\rho_{\mathrm{sol}} \equiv \{\rho : E[\rho] < E[\sigma] \; \forall \sigma \neq \rho\} \tag{5.14}$$

Eqn (5.14) changes the perspective under which the phase retrieval problem is observed, turning it into a full-fledged optimization problem for the error E. Furthermore, the lowest reachable error, that is, $E[\rho_{\mathrm{sol}}]$, is not known *a priori*. Its value strongly depends on the "structural properties" of the sets I and S, which are in turn connected to the features of the experimental data and the support function, as discussed in the following sections.

5.2.2.3 The Role of the Support Function

The definition of a support function is pivotal for rendering the phase problem solvable. As described in the introduction to this section, in theory the support function does not have to perfectly fit the sample shape, but it just needs to ensure that the oversampling condition is met. For example, this implies that any of the support functions shown in Figure 5.9a is enough to guarantee that the pipe density is the only solution. However, the landscape of the error function strongly depends on how the support function is defined. For example, the different supports A–D in Figure 5.9a are linked to a pictorial representation of their corresponding set S (see Figure 5.9b). As each support function is contained in the larger ones, then the same happens for the sets (*i.e.* $D \subset C \subset B \subset A$).

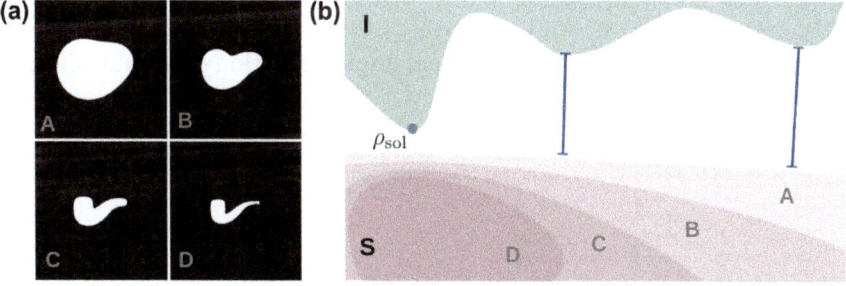

Figure 5.9 The effects of the support shapse on the phase problem. (a) Different support shapes with decreasing size, labeled as A, B, C and D. All of them are compatible with the pipe and are sufficient to ensure the existence of a unique solution. However, the existence of a solution does not tell anything about the possibility to find it. (b) Graphical representation of the change in shape of the set S, depending on the support size. For the looser support function A, the size of set S is bigger, introducing local optima of the error function that vanish for tighter supports.

The two blue lines in Figure 5.9b highlight local minima of the error function when support A is considered. These local minima, far away from the correct solution, are avoided when the support function is made tighter. This effect seems to be of little importance when represented in two dimensions, but it has dramatic consequences when considering the real problem, which lives in a N^2-dimensional space. For this reason, a tight support function must be used for practical applications. The precise knowledge of the support function (*i.e.*, the overall shape of the sample under study) is a strong limitation for the application of CDI, as the sample architecture is, in most cases, unknown. However, a simple and effective solution was provided in the groundbreaking work in ref. 64, which introduced the shrink-wrap algorithm. The concept behind the shrink-wrap is to retrieve the correct support shape along with its respective density. The first step of the algorithm is to perform a smoothing operation on the current density estimate ρ, through a convolution operation with a Gaussian function of width σ. The smoothed density, ρ_σ, is then used as a reference to estimate the new support function:

$$SW[\rho] = \begin{cases} 1 & \text{where } \rho_\sigma > \tau \cdot \max[\rho_\sigma] \\ 0 & \text{elsewhere} \end{cases} \quad (5.15)$$

In practice, a threshold is applied to the smoothed density estimate ρ_σ: the new support function is set to 1 where values of ρ_σ are above this threshold value, and to 0 where they are below. This threshold value is defined as a fraction τ of the maximum value of ρ_σ. The quantity σ is expressed in pixels, and it usually takes values up to a few pixels. Typical values for τ are, instead, around a few % of the maximum value of ρ_σ. This operation is performed once every several iterations of iterative algorithms, and the exact values of σ and τ are dependent on the properties of the sample and the quality of experimental data. A possible numerical implementation of the shrink-wrap algorithm is listed below:

```
 1  # Here we import a gaussian filter for smoothing implemented in the scipy modulus.
 2  from scipy.ndimage import gaussian_filter as smooth
 3
 4  def SW(rho, sigma, tau):
 5    # Apply the gaussian smoothing
 6    rho_smooth = smooth(rho, sigma)
 7    # Initialize the new support with zero values
 8    support = np.zeros(rho.shape)
 9    # Calculate the new support function from the smoothed density
10    support[rho_smooth > tau*np.amax(rho_smooth)]=1
11    return support
```

Thanks to the shrink-wrap algorithm, the initial support can be considerably loose.¶¶¶¶ Different methods can be used to deduce a loose starting

---

¶¶¶¶ The only constraint that it must satisfy is, in principle, the oversampling condition.

support without any particular knowledge of the sample. For a deeper insight into this topic, we warmly recommend ref. 64.

Due to the presence of noise, a practical problem when the support function is retrieved during the reconstruction process is that, for shrinking support sizes, the error value of the solution (*i.e.*, the minimum distance between the two sets) increases (see Figure 5.9). This aspect has three important practical consequences. First, the error value of the reconstruction often increases during the reconstruction process. Second, two reconstructions can be compared based on their error value only if the support size is the same. Third, a wrong reconstruction with a loose support function often has error values lower than the correct reconstruction with a tight support function. Thus, when the support function is also retrieved, the error value as defined in eqn (5.13) is, in general, not a good metric to evaluate the reconstruction quality both in its absolute and relative sense.

5.2.2.4   *Dealing with Missing Data*

A striking feature of experimental data is the lack of diffraction in some areas of the detector. This depends in part due to the structure of the detector itself, organized in separated "tiles", such that stripes of missing data often cover the whole detector surface. However, the always missing portion of experimental data is the one close to the detector centre. For high-intensity light sources in the XUV and X-ray regime, the scattering cross-section of materials is very low, such that only a small fraction of the incoming light undergoes scattering in the sample. For this reason, a large amount of photons (*i.e.*, high radiation intensity) has to be produced to obtain a meaningful diffraction signal in a single laser shot.[||||||] The majority of this radiation simply proceeds past the interaction region, and could irretrievably damage the detector components. For this reason, scattering detectors contain a hole in the middle and cannot record data close to the centre. The spatial coordinates at which scattering data are missing are often encoded in a matrix, known as mask, which has values of 1 where pixels are missing, or 0 otherwise. The action of the intensity projector P_I is then adapted to take the mask M into account

$$P_I \rho = \mathcal{F}^{-1}\left[M \cdot \tilde{\rho} + (M - 1) \cdot \sqrt{I}\frac{\tilde{\rho}}{|\tilde{\rho}|}\right] \tag{5.16}$$

In practical terms, P_I in eqn (5.16) replaces $|\tilde{\rho}|$ with $\sqrt{I}$ only where $M = 0$. This adds an intensity retrieval problem to the phase retrieval problem, as the missing intensities, which cannot be constrained, have to be retrieved along with the phases. The effects of missing diffraction data are, in first approximation, similar to the effects of a looser support (see Figure 5.9).

[||||||] At the end of the story, that is the reason why free electron lasers are necessary.

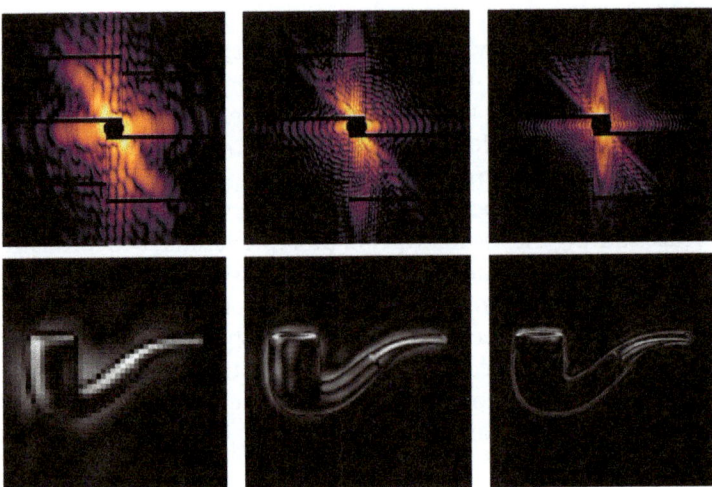

Figure 5.10   Effects of missing diffraction data. Upper row: Diffraction patterns simulated for pipes of increasing size. Missing values in real experimental data are set to zero. Lower row: Pipe image obtained with the masked field intensity in the upper row, giving a real-space qualitative representation of the information available to phase retrieval algorithms, which also have to retrieve the missing scattering amplitude along with the missing phases.

With less available data, the intensity constraint becomes looser and looser, affecting the error landscape and causing the algorithms to stagnate in local minima of the error function.

Furthermore, the Fourier domain has peculiar properties when dealing with isolated samples, and, rephrasing a famous statement from Orwell's animal farm, "all pixels are equal but some are more equal than others". In fact, the central data of the diffraction contain the coordinates of the momentum transfer q close to 0. This corresponds to low-resolution information on the sample.***** This is intuitively shown in simulated diffraction patterns of an increasingly bigger pipe (see the top row in Figure 5.10) that are made more realistic by removing data usually not accessible by the scattering detector. If the field's values in that part are set to 0, and the field is inverted back to the real space (see the bottom row in Figure 5.10), the resulting sample image is roughly a high-pass filtered version of the original one.[††††] In particular, the leftover information for the first, smaller, pipe in Figure 5.10 still contains relevant features of the sample. Increasing the sample size, the overall features of the sample begin to be lost, up to the point where most low-resolution spatial information is completely missing as bigger samples have relatively more information

---

***** For a more careful discussion about resolution, see Section 5.2.2.8.
[††††] In a similar way, a Gaussian smoothing is a low-pass filter (*i.e.*, high q values are suppressed).

closer to the centre of the pattern. It is already possible to guess how phase retrieval algorithms can retrieve the missing intensities (*e.g.*, by carefully looking at **Figure 5.10**). In fact, setting the missing intensities to 0 provides a real-space image of the sample that is no more completely compatible with the support constraint, if the provided support function is close enough to the actual sample shape (*i.e.*, density values different from 0 appear outside the pipe boundaries). This incompatibility, which allows phase retrieval algorithms to restore the missing data thanks to the application of the support projector P_S, reduces with increasing size of the sample (see the bigger pipe in **Figure 5.10**). The larger the sample, the harder the retrieval of missing intensities will be. The missing central data is, for practical applications, the true limiting factor for the maximum size of samples that can be successfully retrieved.

5.2.2.5 The Concept of Oversampling

Oversampling is a key concept in CDI.[54] There are slightly different approaches to interpret the oversampling condition. Here, we proceed in the most intuitive and simple way, while we suggest to refer to ref. 28 and 65 for a more rigorous and correct treatment of the topic. Despite having shown the density functions ρ as real-valued functions (or, better, real valued $N \times N$ matrices), their values are, in general, complex numbers.[‡‡‡‡‡] Not all the values of ρ have to be retrieved, as those outside the support function are constrained to 0, such that only the remaining values inside the support represent the unknowns of the phase problem. Their amount equals to twice the support size in pixels, where the factor of two comes from the complex nature of the density values. If the density values within the support are addressed as the unknowns of the phase problem, the experimental values (*i.e.*, the pixels in the diffraction pattern) are a full-fledged set of equations. The ratio between the amount of pixels in the diffraction data and the the amount of pixels in the support function is known as oversampling ratio σ, given by

$$\sigma = \frac{N^2}{\sum_{i,j} S_{ij}}, \tag{5.17}$$

where the support function is now indicated as a matrix S_{ij} with values 0 or 1, and N is the linear dimension of the diffraction pattern. To render the problem solvable, it is necessary to have more equations than unknowns. This happens only in the case of $\sigma \geq 2$, as the density values to be retrieved belong to $\mathbb{C}$ and two values must be retrieved for each pixel belonging to the support. The oversampling condition described above is necessary, but actually not sufficient to ensure a unique solution. In fact, the set of equations of the problem (*i.e.*, the pixels in the diffraction data) also have to be

---

[‡‡‡‡‡] The reason why the density must be complex will be discussed in the next Section 5.2.2.6.

independent, which means that they must provide additional and not redundant information. For a deeper discussion of this topic, the reader is referred to ref. 28.

5.2.2.6 Complex-valued Density: What CDI Sees

In general, the scattering density ρ is a complex-valued function. Earlier we linked the scattering density ρ with the spatial density of the electrons (which are the fundamental scattering "units" for light). CDI can thus resolve different materials within the sample, thanks to their different electron density. This picture is a very simplified view on light–matter interaction, and does not consider how electrons respond to the incoming radiation. Deriving the behaviour of electrons in their interaction with light is a wide topic and goes well-beyond the scope of this chapter. The reader is recommended to consider ref. 43 and 66 for further details. Here, it is sufficient to state that the response of the electrons to the incoming light is fully encoded in the complex refractive index n. Under some approximations,[43] completely valid§§§§§ in the X-ray regime, we can state that the scattering density $\rho(\vec{r})$ is linked to the sample's spatial distribution of the complex refractive index $n(\vec{r})$ by

$$\rho(\vec{r}) \propto 1 - n(\vec{r})^2 = 1 - [1 - \delta(\vec{r}) + i\beta(\vec{r})]^2 \approx 2[\delta(\vec{r}) - i\beta(\vec{r})], \qquad (5.18)$$

where the complex refractive index is expressed as $n = 1 - \delta + i\beta$. This is a convenient notation when dealing with photon energies in the X-ray regime, where the refractive index of materials is very close to unity.[67] The values of δ and β are then much smaller than 1, allowing the last approximation step in eqn (5.18).¶¶¶¶¶ Here, it is only necessary to note that, in the small-angle regime, only the projection of the scattering density $\rho(x,y)$ is imaged. Its values are related to the optical properties of the sample integrated along the beam propagation direction, as given by

$$\rho(x,y) \propto \int_{\parallel} [\delta(\vec{r}) - i\beta(\vec{r})] = \delta(x,y) - i\beta(x,y) \qquad (5.19)$$

An example of a complex-valued scattering density is given in Figure 5.11. All scattering patterns shown so far were centrosymmetric, *i.e.*, $I(q_x, q_y) = I(-q_x, -q_y)$. This is directly derived from the fact that the Fourier transform of a real-valued function is Hermitian.‖‖‖‖‖‖ This is no more the case for an entirely complex-valued density (*e.g.*, asymmetries of the diffraction data are

---

§§§§§ Almost. See Section 5.2.3.3.
¶¶¶¶¶ A deep understanding of the effects of n in the scattering signal is pivotal for the development for data analysis methods in the wide-angle scattering regime. See Section 5.2.3.3.
‖‖‖‖‖‖ A Hermitian function has the property $f^*(x) = f(-x)$, where * denotes the complex conjugate. The modulus of such functions is then centre-symmetric around 0.

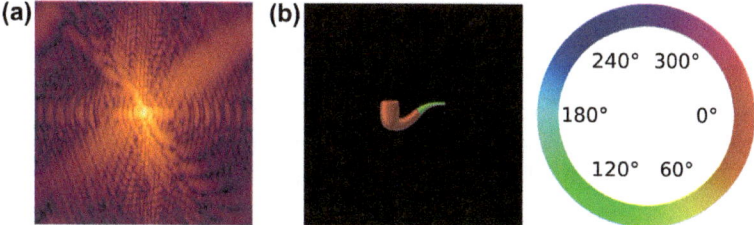

Figure 5.11 A complex-valued density distribution and its simulated diffraction pattern. The real-space phase of the sample is encoded in the hue. We note that an imaginary part different from zero results in a non-centrosymmetric diffraction pattern.

visible in the left panel of Figure 5.11). It is common, and convenient, to restrict the "complexity" of the reconstruction values when performing phase retrieval.[30] In particular, the support projector P_S is often modified to also constrain the positive value of the real part, which means, in numerical terms, that all pixels of ρ that have negative real part are set to zero. In this way, the unknowns of the phase problem (*i.e.*, the density values) are still allowed to be complex, but their value is restricted to half of the complex plane. This constraint is often valid for small variations of the refractive index.[30]

Eqn (5.19) explains why a complex-valued spatial density ρ is required to correctly describe the sample. Still, the story is not over. For example, when an optically homogeneous sample is considered (*i.e.*, made of the same material, or different components with similar optical properties), δ and β maintain constant values within the sample, and zero outside. The result of the integral in eqn (5.19) can be then reformulated in the following way:

$$\rho(x,y) \propto \int_{\parallel} S(\vec{r})[\delta - i\beta] = D(x,y)[\delta - i\beta] = D(x,y)|\delta - i\beta|e^{\Phi} \qquad (5.20)$$

Here, $S(\vec{r})$ can be imagined as a 3D support function which describes the sample's extension, and $D(x,y)$ is a 2D function which describes the sample's depth as a function of projection coordinates. As $\delta - i\beta$ is now a constant complex number, it can be separated into its constant modulus and constant phase Φ components. If we now calculate the diffraction pattern of the density distribution in eqn (5.20), it turns out that the intensity of the scattered field is

$$I(q_x, q_y) \propto \left| \mathcal{F}\left[D(x,y)|\delta - i\beta|e^{\Phi} \right](q_x, q_y) \right|^2 = |\delta - i\beta|^2 \left| \mathcal{F}[D(x,y)](q_x, q_y) \right|^2 \quad (5.21)$$

Thanks to the properties of the Fourier transform, and due to the fact that only the modulus squared of the field is experimentally detectable, any global phase added to the scattering density ρ does not change the scattering pattern. Thus, this information is not accessible *via* phase retrieval, and phase contrast only arises when materials with different ratios between δ and β are inhomogeneously distributed into the sample. The fact that a global real-space phase value does not change the scattering signal is a hint that there are actually infinite solutions to the phase problem, and it only has a unique solution in a broad sense. This is just one among the possible ambiguities, as explained in the next section.

5.2.2.7 Ambiguity of the Solution

As mentioned in the previous section, the existence of a unique solution to the phase problem has to be intended in a broad sense, as there is, in practice, an infinite amount of solutions. This ambiguity arises because there are some properties of the sample which are encoded only in the field's phases, and cannot be thus retrieved. These ambiguities directly derive from the properties of the Fourier transform. A first example was discussed in the previous section using eqn (5.21), where a constant phase factor does not change the diffraction pattern. This is visually presented in Figure 5.12, where the density in Figure 5.12b is obtained by applying a constant phase factor from Figure 5.12a. Furthermore, given a density distribution $\rho(x,y)$, both $\rho(x_0+x, y_0+y)$ and $\rho^*(-x, -y)$ have the same scattered field's amplitudes (where ρ^* stands for the complex conjugate operation). The former means that the diffraction pattern is invariant under a translation of the density to x_0, y_0 (see Figure 5.12c). The latter signifies that the diffraction signal is invariant under coordinate inversion and complex conjugation of the density (see Figure 5.12d). It is easy to show that also any combination of these operations represents a solution (see Figure 5.12e).

In general, this multitude of solutions does not represent a problem, as all of them are easily relatable to each other through simple mathematical operations. When dealing with data analysis based on phase retrieval, it is a common practice to consider the final solution as the average of

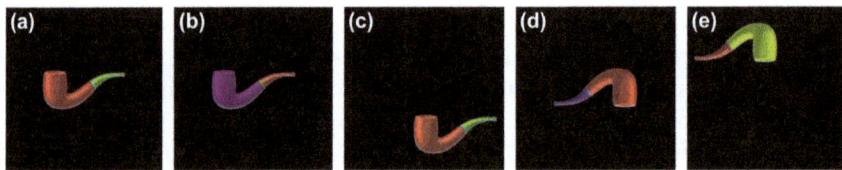

Figure 5.12 Ambiguities of the solution to the phase problem. (a) A complex-valued density function is shown. Its real-space phase is encoded in the hue, following the scheme shown in Figure 5.11. (b) A constant phase value is added to the density. (c) The density is translated. (d) The complex conjugate of the density. Its spatial coordinates are inverted in sign. (e) A density resulting from the combination of the three operations in (b–d). All the densities shown here have the same Fourier amplitudes, and are, thus, all valid solutions to the same phase retrieval problem.

different reconstructions from different starting conditions.[68] As different reconstruction processes will certainly provide different spatial distributions (like the ones in Figure 5.12), it is then necessary to "reshift" these solutions to a common reference before any averaging or comparison. A useful trick is to take advantage of the cross-correlation operation. An intuitive meaning of the cross-correlation $f*g$ between two functions f and g is that its maximum value gives the degree of similarity between the two functions.******

As mentioned before, in the case where two solutions to the same phase retrieval problem ρ_1 and ρ_2 are compared, it is necessary to resolve the ambiguities that exist between the two. If ρ_2 is considered as the reference, the first step is to check which of $\rho_1(x, y)$ and $\rho_1^*(-x, -y)$ is most similar to ρ_2. To do so, two cross-correlations $R(x, y) = \rho_1(x, y) * \rho_2(x, y)$ and $R^*(x, y) = \rho_1^*(-x, -y) * \rho_2(x, y)$ have to be computed. As the maximum value of the cross-correlation provides the similarity between the two densities, it is intuitively clear that if $\max(R) < \max(R^*)$, then it is necessary to "flip" ρ_1 (see Figure 5.12d). Furthermore, the coordinates x, y of the maximum cross-correlation value encode the relative shift between the two functions. For two perfectly overlapping densities, the maximum value is at coordinates $(0,0)$. In general, if the maximum value of the cross-correlation R (or R^*) is placed at coordinates (x_0, y_0), this provides to the amount of spatial shift (in pixels) that must be applied to ρ_1 in order to overlap with ρ_2 (*i.e.*, the translation operation in Figure 5.12c). Finally, the phase of $\max(R)$ (or $\max(R^*)$) yields the phase-shift to add to ρ_1 such that it matches the phase of ρ_2 (*e.g.*, in the case shown in Figure 5.12b). Thanks to this trick, different outcomes of a phase retrieval reconstruction can be spatially compared with each other, and operations like averaging are now meaningful.

An additional, more practical, problem, that affects the phases of the real-space density reconstruction arises from the choice of the coordinates for the centre of the diffraction pattern. In fact, the scattering pattern has to be "prepared" for a phase retrieval algorithm such that, for a pixel resolution $N \times N$, the central peak of the diffraction corresponding to the momentum transfer $(0,0)$ is placed at the coordinate $[N/2, N/2]$ in the scattering matrix (or at $[0,0]$ when frequencies are shifted). When dealing with real data, it is often a difficult task to identify the correct centre, as this data is missing due to the detector hole (see Figure 5.10). The real-space effect for different centre position coordinates of the diffraction (see Figure 5.13), highlights

---

******The cross-correlation can be efficiently computed *via* the convolution theorem. The convolution between two functions f and g, denoted as $f * g$, is given by the inverse FT of their product in Fourier space, *i.e.*,

$$f * g = \mathcal{F}^{-1}[\mathcal{F}[f]\mathcal{F}[g]] \tag{5.22}$$

The cross-correlation between two signals $f(x)$ and $g(x)$ is then the convolution of $f^*(-x) * |g(x)$.

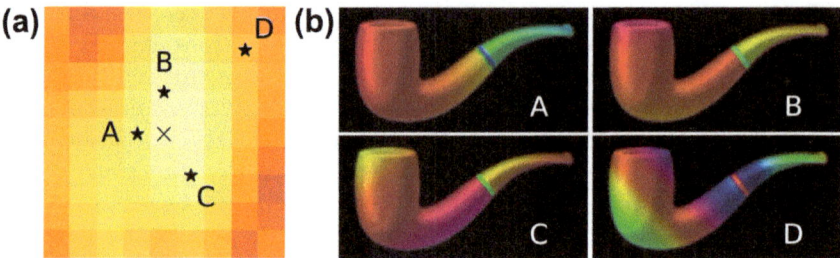

Figure 5.13   Effects of the incorrect centering of diffraction patterns. (a) Close up of the central part of the diffraction pattern. The × marker indicates the correct centre, which produces the "correct" reconstructions in Figure 5.12. Markers A–D indicate different choices of the pattern centre. (b) The effect of the different centre positions on the real-space phases of the pipe.

how it affects the reconstruction result. In particular, such shifts of the diffraction pattern cause a linear drift of the reconstructions phases, which becomes already dramatic even for a centre misplacement of just a few pixels. This effect can be corrected in post-processing, by correctly reshifting the reconstruction FT at the end of the reconstruction process. However, an incorrect diffraction centre position can prevent the convergence of the iterative algorithms, especially when the positivity of the real part of ρ is constrained (see Section 5.2.2.6).

5.2.2.8   The Resolution(s) of CDI

When dealing with the term resolution, it is not uncommon to confuse between the detector resolution (affecting the diffraction data) and the reconstruction resolution (determining the smallest real-space feature that can be resolved, see Figure 5.14).

Recalling the introduction to CDI, it was shown that, in the case of small-angle scattering, only the components of momentum transfer $\vec{q}$ orthogonal to the beam propagation direction $(\vec{q}_\perp)$ are accessible by the detector (see Figure 5.3). This is strictly dependent on how far the scattering detector is from the interaction region,[a] which influences the maximum scattering angle θ_{max} that the detector can record. From simple geometrical considerations, the magnitude of momentum transfer $q = |q_\perp|$ recorded by the detector as a function of the scattering angle is $q(\theta) = k_0 \cdot \sin(\theta) = \frac{2\pi}{\lambda} \cdot \sin(\theta)$, where $k_0 = \frac{2\pi}{\lambda}$ is the magnitude of the wavevector of the incoming radiation at a wavelength λ. The signal recorded by the detector at a given scattering angle

---

[a] The interaction region is where light interacts with the sample, and thus where the scattering events take place.

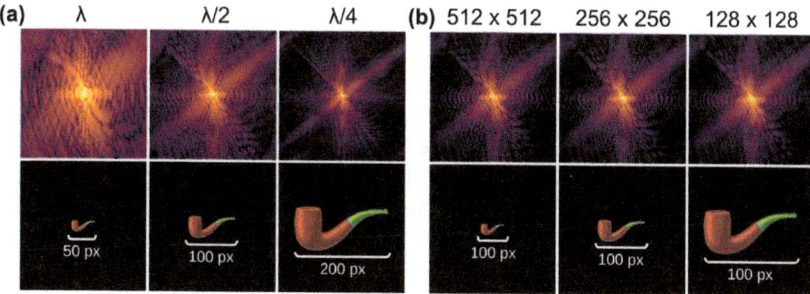

Figure 5.14   The CDI resolutions. (a) Effect of different photon energies on the reconstruction resolution. From left to right, the photon energy is doubled at each step. The real-space reconstruction of the pipe contains an increasing amount of pixels, thus providing information on the sample at higher resolution. The size of the diffraction pattern remains constant at 256×256 pixels. The same effect can be obtained by keeping λ constant and increasing the maximum recorded scattering angle θ_{max}. (b) Effect of different detector resolutions. From left to right, λ and θ_{max} are constant, but the pixel size is doubled. The actual resolution at which the sample is imaged does not change. Sampling at higher frequency in the reciprocal space only adds regions of zeros around the sample. Therefore, the pixel resolution of the detector affects the maximum size of a sample that can be retrieved.

θ, and thus, at a given magnitude of the momentum transfer q, is directly relatable to a real-space length scale D *via* the relationship $D = \dfrac{\pi}{q}$. This relationship (derived from the properties of the Fourier transform and strictly connected to Bragg's law) then gives the smallest "size unit" δ in the real-space, corresponding to the value of D at the maximum recorded scattering angle:

$$\delta = \frac{\pi}{q(\theta_{max})} = \frac{\lambda}{2\,\sin(\theta_{max})} \tag{5.23}$$

This "size-unit" δ is the pixel size of the reconstruction, giving a conversion scale between pixels and metres, also called the half-period resolution. The real-space resolution of CDI is instead often considered to be the full-period resolution, equal to 2δ.[‡‡‡‡‡‡]

Eqn (5.23) nicely highlights the possibility to tune the CDI resolution. The spatial resolution can be enhanced by increasing the maximum recorded scattering angle θ_{max} (by moving the detector closer to the interaction region, but without leaving the small-angle scattering approximation) or by increasing the photon energy (*i.e.*, reducing the wavelength of the incoming radiation). An example of the latter case is given in Figure 5.14a, where the

---

[‡‡‡‡‡‡] The reason for this choice comes from Shannon's sampling theorem.

photon energy is doubled at each step. For the same sample, the respective reconstruction is increased by a factor two each time in terms of pixels, as the physical pixel size is halved.

It is worth noting that the resolution of CDI is not connected to the resolution of the detector. This result, which is quite surprising only at a first glance, is highlighted well by Figure 5.14b. In this example, the diffraction data are acquired under the same experimental conditions (*i.e.*, the detector reaches the same scattering angle), but with different pixel resolution. The detector pixel size does not influence the size (in pixels) of the reconstructed pipe, but only the total size of the matrix in which the pipe is "immersed". Thus, increasing the sampling rate of the scattering signal only increases the zero-value region around the pipe in real-space. Still, the use of scattering detectors with high pixel count carries undeniable advantages. The first is that the number of equations of the phase problem increases, thus augmenting the oversampling ratio σ in eqn (5.17). This allows to image bigger samples at the same spatial resolution.§§§§§§ Thus, as long as the oversampling condition is met (*i.e.*, σ is big enough), it is often convenient to scale down the diffraction data through a binning operation. This does not affect the reconstruction resolution (see Figure 5.14b) and considerably speeds up the reconstruction process.¶¶¶¶¶¶

5.2.2.9 Consequences of Noise

Noise intrinsically affects diffraction data. Two main sources of noise can be distinguished, thermal noise and statistical noise. Thermal noise can be modeled as a uniform, incoherent signal that independently affects each pixel of the diffraction detector. Statistical noise, instead, comes from the intrinsically discrete nature of the radiation. Modern scattering detectors at FEL facilities[70–72] are often capable of single photon sensitivity (*i.e.*, they can distinguish single photon events[73]). This performance is achievable because the strength of the thermal noise, which uniformly affects all the detector pixels, is kept substantially below the signal strength provided by a single photon that impinges on the detector pixel. Here, we only deal with statistical noise, and other sources of signal disturbance are neglected. The single-photon sensitivity of pixel in such detectors renders them "photon

---

§§§§§§ A more rigorous explanation involves the sampling theorem, and a parallelism with the real-space resolution δx. As δx depends on the maximum magnitude of the recorded q, then similarly δ_q (the detector pixels size in Fourier space) affects the maximum size of a retrievable sample. For a more rigorous treatment, see ref. 28.

¶¶¶¶¶¶ For each step of iterative algorithms, a forward and a backward discrete Fourier transform (DFT) must be computed. The computing time of the DFT, when computed *via* the fast Fourier transform,[69] scales as $N^2 \log(N)$, where N is the linear dimension in pixels of the matrix. A simple 2×2 binning operation in the diffraction data reduces the computing time of the DFT by a factor of greater than four.

counters". The statistical behaviour of these photon events for a single detector pixel then follows a Poisson probability distribution.[||||||||] The scattering signal $I_m(q)$ measured using a detector pixel at coordinates $\vec{q}$ can be then modeled as:

$$I_m(q) = \mathcal{P}[I(q)], \tag{5.25}$$

where $I(q)$ is the "exact intensity" (expressed in unit of photons), and $\mathcal{P}[\mu]$ is a Poisson-distributed random number with average μ. Unlike a normal distribution, which requires two parameters (mean μ and standard deviation σ) to produce a random number, the standard deviation of a Poisson number is not a free parameter. It is strictly dependent on the average value, in the form of $\sigma = \sqrt{\mu}$. A common way to express the amount of information contained in a signal in relation to its noise is the signal-to-noise ratio (SNR), defined as the ratio between the average signal and its standard deviation[*******] $\left(i.e., \mathrm{SNR} = \dfrac{\mu}{\sigma}\right)$. Due to the statistical behaviour of photons, the SNR of diffraction data is not constant but depends on the "exact" intensity I:

$$\mathrm{SNR}[I] = \frac{I}{\sqrt{I}} = \sqrt{I} \tag{5.26}$$

The weaker the signal I, the worse the SNR, up to the point where I approaches few counts and the SNR approaches 1 (that is, signal and noise have comparable magnitude). In FEL experiments,[††††††] and especially in the X-ray region where materials are highly transparent to light, it is very common to have large parts of the scattering signal characterized by very low photon statistics, often approaching zero. As the scattering signal for isolated samples strongly decreases with the distance from the central diffraction peak, this low-intensity region is always placed in the outer region of the detector, which collects data corresponding to high q values. High q

---

[||||||||] The Poisson probability distribution is given by

$$P(\mu_p, n) = \frac{\mu_p^n e^{-\mu_p}}{n!} \tag{5.24}$$

For example, the probability of measuring 3 photons for a given exposure time on a pixel whose signal should be, on average, 1.6, is given by $P(1.6,3)$.

[*******] There are actually different ways to define the SNR. This definition is mostly used when dealing with strictly positive signals, like in imaging processing.

[††††††] For applications where the illumination time can be freely decided (like in diffraction experiments at synchrotrons), the SNR value can be indefinitely increased, up to the point when a significant SNR is obtained all over the diffraction data. Such an approach is not viable for FEL diffraction experiments, where the femtoseconds pulse duration is a key feature in both imaging ultrafast dynamics and avoiding signal degradation from the sample damage by the intense ultrashort pulse (the diffract and destroy concept[2]). Furthermore, due to the sample destruction, the same sample usually cannot be imaged with multiple pulses. Thus, the photons available for scattering in FEL experiments are only the ones contained in a single femtosecond pulse.

(a)   **(b)**

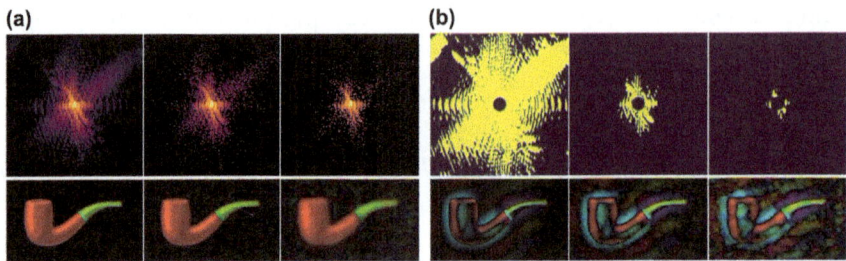

Figure 5.15   The effects of noise. (a) Simulated diffraction patterns with decreasing
signal strength and noise from Poisson statistics. Corresponding
complex densities in the lower row. The effect of losing information
at high scattering angles is to reduce the actual resolution of the
reconstruction, which appears smoother. (b) Realistic data with missing center information. A mask is applied to patterns in (a) and the
regions where the SNR value is above 1 are highlighted in yellow color
(upper row). These can be addressed as areas of meaningful data.
Lower row: Left-over information with similar effects to Figure 5.10.

values correspond to high resolution information, as highlighted by
eqn (5.23). This, in turn, means that noise in diffraction data limits the actual
spatial resolution of CDI (as exemplified by Figure 5.15a). Here, we address
this important topic in a highly simplistic manner, and we refer the reader to
the analysis in ref. 74 for a more complete and correct view on the topic.

The actual CDI resolution is, in this case, not defined by the maximum
scattering angle θ_{max}, but by the maximum scattering angle at which light
can be effectively recorded, θ_{eff}. For example, an increase in θ_{max} would just
decrease the pixel size in real space (see the second and third case shown in
Figure 5.15a) but would not carry an improvement in the effective spatial
resolution (as no additional information would be recorded).

As discussed above, another option to increase resolution is to increase
the photon energy of the radiation. For example, doubling the photon energy
theoretically leads to halving of the pixel size in the real space representation
(see Figure 5.14a). However, the refractive index of materials, and thus their
"scattering strength", dramatically drops with increasing photon energy,
thus lowering the amount of scattered photons and the effective resolution
(see Figure 5.15a), up to the point where increasing the photon energy is
actually counter-effective.[‡‡‡‡‡‡‡]

The effect of low photon statistics is even more dramatic when dealing
with real diffraction data, where the central part of the scattering image is
missing. The same diffraction patterns shown in Figure 5.15a are now rep-
licated in Figure 5.15b, where the central part is now removed to simulate
the detector hole that accommodates the transmitted beam of the laser
pulse. Due to the decrease in the scattering signal with increasing distance

---

[‡‡‡‡‡‡‡]This aspect represents one of the main limiting factors for FEL science towards imaging of
single atoms and molecules, along with the problem of radiation damage.[24]

from the centre, the lower the diffraction brightness, the higher the relative fraction of usable information present in the central portion of the scattering data. When that information is lost due to the detector's hole, the deterioration of the leftover information becomes dramatic. The three cases in Figure 5.15b have, in theory, the same oversampling ratio, as it was defined in eqn (5.17). However, it is now clear that, in the case of noisy data, the amount of pixels containing meaningful information on the sample is much lower than the total number of pixels in the matrix. We can roughly estimate that pixel values are usable only when the signal is distinguishable from the noise (*i.e.*, when the SNR value is greater than a given threshold). These pixels are highlighted in the first row of Figure 5.15b, for a conservative choice of an SNR threshold of one. It is convenient here to define the relative information actually provided by a diffraction pattern as $\eta = \dfrac{\sum \mathrm{SNR}_>}{N^2}$, where $\mathrm{SNR}_>$ is a binary map which has a value of one where the diffraction data has a useful SNR. It is now intuitively clear that the actual oversampling ratio σ required to uniquely identify a solution to the phase problem has to be higher than two and, in particular, it can be estimated at a value of $\dfrac{2}{\eta}$. The value of η cannot be perfectly known *a priori*, because the SNR calculation would require knowing the exact field. Still, it can be roughly estimated by counting the number of pixels in the experimental data that have values higher than one photon count. It is not uncommon in FEL data to have $\eta < 0.2$ (*i.e.*, only 20 percent of the measured data carries significant information). This is the reason why, in most real-case applications, phase retrieval is successful only when the oversampling ratio $\sigma \gg 2$.

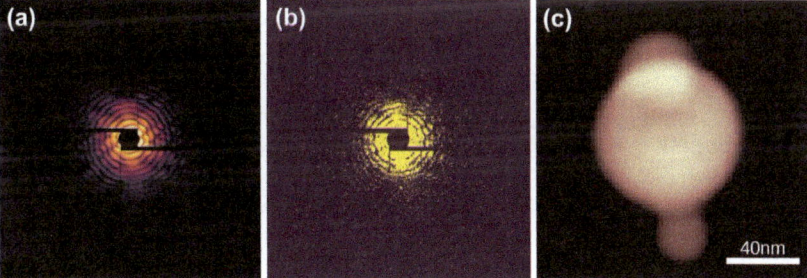

Figure 5.16 Example of real FEL experimental data[76] from the Maloja instrument at SwissFEL.[75] (a) Experimental diffraction pattern. (b) Region of photon counts greater than 1, highlighted in yellow. (c) Absolute value of the reconstruction retrieved from (a). The argon cluster in the reconstruction appears as an agglomerate of 3 spheres of different sizes (see Section 5.3.1). Note the brighter signal in the reconstruction where two spherical sub-clusters overlap, visualizing that small-angle CDI has access to the projection (*i.e.*, integral) of the optical properties along the beam propagation direction. Courtesy of Mario Sauppe (ETH Zürich).

5.2.2.10   A Real-life Example

Here, we apply our observations about noise and missing data on real FEL data and respective phase retrieval results using the example of a representative diffraction pattern (see Figure 5.16) from an experimental campaign at the Maloja endstation of SwissFEL.[75]

One aspect that has not been discussed yet, but is of practical importance, is how to identify relevant and representative patterns out of a large data set. Also rare events can be pivotal for the understanding of the target under study, making them even harder to find. Ideally, this identification and classification of typical and rare patterns is already possible during the experimental campaign. This requires fast methods that can deal with huge data sets. To this end, sophisticated analysis pipelines have been developed in the community.[77,78] Furthermore, machine learning techniques have been recently applied on single-particle CDI data sets to take over the time-consuming manual classification procedures.[79–87]

The pattern discussed here (see Figure 5.16) was obtained with the Jungfrau scattering detector[72,88] from an isolated argon cluster using a photon energy of 1000 eV. The large-area diffraction detector, containing 2048×2048 pixel, was placed at a distance from the interaction region that allows the recording of scattered light up to a scattering angle of 13°.

It is possible to derive *via* eqn (5.23) the equivalent real-space pixel size of the pattern shown in Figure 5.16a. As 1000 eV corresponds to a wavelength $\lambda = 1.24$ nm, the pixel size in real space is $\delta = \dfrac{1.24 \text{ nm}}{2 \sin(13°)} \approx 2.7$ nm, which is the real-space dimension of a single pixel in the reconstruction shown in Figure 5.16c. In reality, the area of meaningful SNR covers only around 5% of the actual matrix surface, which has two main implications. First, a successful reconstruction in these brightness conditions could be attempted when the sample size is small enough to reach an oversampling ratio $\sigma \gtrsim 40$.§§§§§§§ Furthermore, the spatial distribution of this significant data is, as expected, focused around the centre and only reaches one third of the full detector dimension, which corresponds to a scattering angle of roughly 4°. Thus, the effective minimum size of features that can be spatially resolved is around three pixels, and no information below this dimension can be retrieved. This is immediately visible in the reconstruction shown in Figure 5.16, where features that should be in principle sharp, like the cluster edge, are instead smoothed over few pixels.

The discussion made so far is mainly qualitative. A formally correct treatment and evaluation of the actual resolution of CDI would involve not

---

§§§§§§§ An additional problem that may affect experimental data comes from the signal saturation in high-intensity pixels, as detectors are capable of recording signals up to a maximum value. Those areas where the detector's signal is saturated cannot be used for imaging and thus need to be excluded similarly to the missing pixels in the central hole. However, this is a merely technical limitation, and new detectors are less and less affected by this problem, thanks to their high dynamic range.

only a more careful analysis of the SNR, but also an evaluation of the reconstruction process itself.[¶¶¶¶¶¶¶] To take into account the latter aspect, statistical methods that involve many reconstruction processes are employed, like the phase retrieval transfer function.[28,68] This aspect goes beyond the scope of this chapter, and we suggest ref. 28 for further information.

5.2.3 Wide-angle Coherent Diffraction Imaging

As introduced in Section 5.2.1, the wide-angle regime is reached when the component of the recorded momentum transfer parallel to the beam propagation direction $q_\parallel$ becomes non-negligible, and, in reciprocal space, the portion of the Ewald sphere covered by the detector cannot be approximated any more by a flat surface.

Coming from the perspective of the small-angle scattering regime, it means that the Fourier slice theorem is no longer applicable, and not only the 2D projection of the sample refractive index, but its three-dimensional distribution affects the scattering signal. This last sentence is formulated in terms of the direct scattering problem (*i.e.*, how to simulate the scattered field from a density distribution) and tells us that the full knowledge of the 3D sample morphology is required to achieve a wide-angle scattering simulation.

If we approach the topic from the inverse scattering problem point of view, wide-angle CDI allows to capture, in a single two-dimensional diffraction shot, three-dimensional information on the sample. From this view, this technique opens an exciting possibility for the study of ultrafast phenomena by capturing three-dimensional movies of nanomatter. The reason why, so far, no such movies have been experimentally demonstrated is due to intrinsic complications of wide-angle data analysis, which is an active field of research.

5.2.3.1 *Dealing with the Third Dimension*

Similarly to what has been presented in Section 5.2.2.8, we now identify the spatial resolution of wide-angle scattering. It depends on the maximum momentum transfer at which data are recorded, which reaches different magnitudes in the orthogonal direction $q_\perp$ and parallel direction $q_\parallel$. Considering the geometry in Figure 5.3, we find $|\vec{q}_\perp(\theta)| = |\vec{k}_0| \sin(\theta)$ and $|\vec{q}_\parallel(\theta)| = |\vec{k}_0| \cdot [1 - \cos(\theta)]$. The theoretical half-period resolution (see Section 5.2.2.8) on the orthogonal plane $\delta_\perp$ and on the depth axis $\delta_\parallel$ can be then calculated as

$$\delta_\perp = \frac{2\pi}{|\vec{q}_\perp(\theta_{max})|} = \frac{\lambda}{\sin(\theta_{max})}, \quad \delta_\parallel = \frac{2\pi}{|\vec{q}_\parallel(\theta_{max})|} = \frac{\lambda}{1 - \cos(\theta_{max})} \quad (5.27)$$

[¶¶¶¶¶¶¶]The fact that the diffraction data contain information up to a maximum spatial resolution does not imply that phase retrieval algorithms are then really able to achieve it.

Two fundamental considerations must be examined before diving into the data analysis methods. First, significant values for $q_\parallel$ when compared to $q_\perp$ start to appear at $\theta_{max} = 20°$, where $\delta_\parallel \approx 6\delta_\perp$. Due to the $1 - \cos(\theta)$ in the denominator in eqn (5.27), the increase in depth resolution is dramatic for a scattering angle of 30–50°. However, the strong decrease in the scattering signal makes it challenging to measure data at such high scattering angles in practice (see discussion in Section 5.2.2.9).

A second, even more determining aspect is that the information is only partial. In the small-angle scattering regime, knowledge about the 2D amplitude of the field scattered by a 2D sample is almost‖‖‖‖‖‖‖ complete, such that "only" the phases have to be retrieved. When moving to the three-dimensional case, the Fourier representation of the 3D scattering density ρ lies in three dimensions as well, but only a 2D slice of it (the Ewald's sphere, see Figure 5.3) is known.******** Additional strong constraints are, then, necessary to approach 3D imaging from single wide-angle shots. Furthermore, the fact that the intensity constraint is not strong enough has also the practical implication that information cannot be effectively "back-propagated" to the real space. For these reasons, the most effective wide-angle single-shot imaging approach so far attempted is based on fitting forward simulations of constrained sample's models.

5.2.3.2 The Three-dimensional Imaging Problem

Coherent diffraction imaging based on forward fitting is strictly related to the optimization problem encountered in the small-angle inverse scattering problem. In particular, it can be expressed in terms of an optimization problem similarly to eqn (5.14):

$$\vec{p}_{sol} \equiv \{\vec{p} : E[\vec{p}] < E[\vec{s}] \ \forall \vec{s} \neq \vec{p}\} \tag{5.28}$$

Here, the dependence on the matrix ρ has been replaced with a more generic vector $\vec{p}$, which contains the parameters necessary to define the sample

---

‖‖‖‖‖‖‖ Due to the presence of a hole in the detector, as well as noise in the data (see Sections 5.2.2.4 and 5.2.2.9).

******** A direct extension of the small-angle approach to the wide-angle case is, in principle, straightforward from a numerical point of view. The density function ρ and its support function S can be defined as three-dimensional tensors, and 2D phase retrieval algorithms can be directly generalized to the 3D case without substantial modifications. This approach is known in the literature as Ankylography.[89] However, since its disclosure to the scientific community, the method has raised doubts about its consistency and reliability,[90] due intrinsic "dimensional deficiency" that renders the 3D case an ill-posed problem. A straightforward definition of "ill-posed problem" says that such problems either have no solutions, or too many solutions, or the procedure to reach them is unstable (see ref. 91 for a generic introduction to the topic). This deficiency derives from the action of the projector P_I (see Section 5.2.2.1) which, in this case, constrains only a tiny fraction (just on the Ewald sphere) of the total amount of Fourier amplitudes. For this reason, Ankylography turned out to be impractical for real applications, as the loose intensity constraint cannot ensure a unique solution to the phase problem.

properties depending on the model in use. For example, in the case of a simple sphere with uniform refractive index, $\vec{p} = \{r, \delta, \beta\}$ because the architecture of a homogeneous sphere and its response to the incoming light are completely defined by its radius and its refractive index $n = 1 - \delta + i\beta$.

Similarly to the small-angle case, the optimization target E in eqn (5.28) tells us how much the density reconstruction $\rho(\vec{p})$ defined by the set of parameters $\vec{p}$ is compatible with the experimental diffraction pattern I, given by

$$E[\vec{p}] = \|I - S(\vec{p})\|, \tag{5.29}$$

where $S(\vec{p})$ is a function that simulates the diffraction pattern (*i.e.*, that solves the direct scattering problem) for a scattering density defined by the parameters $\vec{p}$.

The minimization problem (eqn (5.28)) of the error function E (eqn (5.29)) makes it clear why this imaging approach is based on fitting of forward simulations. In regard to the fitting aspect, there are different strategies to deal with the minimization of $E[\vec{p}]$ as a function of its parameters $\vec{p} = \{p1, p2, \ldots\}$. This topic is not strictly related to coherent diffraction and will not be treated here. Some applications of peculiar optimization strategies applied to CDI can be found in ref. 62 and 92. Independent of the employed approach, all optimization algorithms require a large amount of evaluations of the optimization target E, which translates into the evaluation of the simulation function S. For this reason, fast and effective simulation methods are fundamental building blocks for wide-angle single-shot three-dimensional coherent diffraction imaging based on fitting of forward simulations, as discussed below.

5.2.3.3   Forward Simulations

Forward simulation methods can be used for solving the direct scattering problem (*i.e.*, for calculating the field scattered by a real-space distribution of a sample's electronic density). In principle, the scattered field can be computed exactly by solving Maxwell's equations for light traveling in the sample. For very simple sample shapes, the field equation can be even solved analytically, like in the case of spheres with a uniform refractive index, or spherical core–shell structures. Their analytical solutions are known as Mie solutions, and come with the great advantage of providing extremely fast and accurate simulations of the scattering signal. However, the main drawback is the limited range of applicability.

Still, Mie simulations are being proficiently used to retrieve information from architecturally simple systems,[95] like uniform nanodroplets of superfluid helium (see Figure 5.17) or core–shell arrangement of neutral and and ionized matter in spherical xenon clusters.[96]

The topological restrictions of Mie simulations can be overcome by numerically solving Maxwell equations, for example, *via* finite difference time domain (FDTD) simulations[97,98] or using Green's function based approaches such as the discrete dipole approximation (DDA).[99,100] However, these simulations

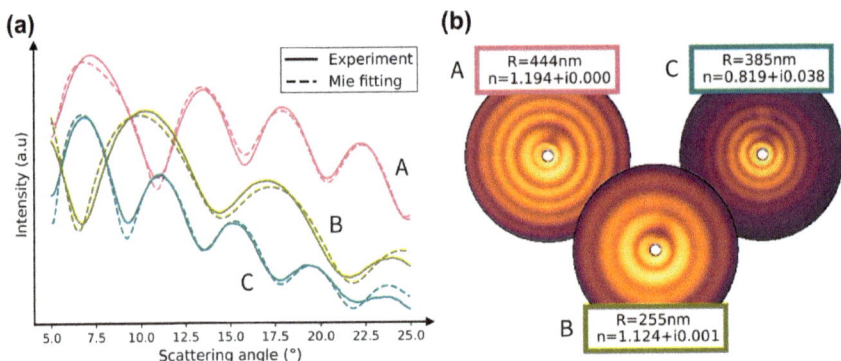

Figure 5.17 Example of wide-angle imaging performed *via* Mie fitting. Helium nanodroplets are imaged[93] at a photon energy of 21 eV at the LDM instrument of FERMI.[94] (a) Experimental radial profiles of three different diffraction patterns, and respective Mie fits. (b) Values of radius R and complex refractive index n resulting from the fitting of the radial profiles are assigned to the respective diffraction pattern.[93] Courtesy of Linos Hecht (ETH Zürich).

methods are unpractical for fitting procedures due to their high computational cost.[⁺⁺⁺⁺⁺⁺⁺] In this context, approximate simulation methods come into play.

Among approximated approaches to the direct scattering problem, the three-dimensional FT of the scattering density ρ (see Section 5.2.1) is by far the most efficient and straightforward, in numerical terms. Still, this method relies on the Born approximation which completely neglects any interaction between light and matter apart from the single scattering events. Thanks to the material's refractive index value being close to one in the X-ray regime, this approximation is sufficient to interpret the output of phase-retrieval results as the projection of the sample's optical properties.[⁺⁺⁺⁺⁺⁺⁺⁺] However, its deviations from the exact field are strong enough, expecially in the XUV energy range and at high scattering angles, to prevent successful imaging attempts.[101]

It has been demonstrated that first-order "corrections" to the Born's approximation restore sufficient similarity between the exact and simulated field to allow forward fitting approaches.[92,101] When a monochromatic plane wave propagates in matter with constant refractive index $n = 1 - \delta + i\beta$ in direction z, its electric field $E(z)$ is described by the plane wave

$$E(z) = E_0 e^{ik_0 nz} = E_0 e^{ik_0(1-\delta+i\beta)z} = E_0 e^{ik_0 z} e^{-\beta z} e^{-ik_0\delta z} = E_0 e^{ik_0 z} C(z, \delta, \beta) \quad (5.30)$$

⁺⁺⁺⁺⁺⁺⁺ The time-to-solution (*i.e.*, for a single simulation) is a pivotal aspect when dealing with fitting routines.

⁺⁺⁺⁺⁺⁺⁺⁺ *I.e.*, for phase retrieval results, the Born approximation means that the retrieved density is an approximation of the refractive index projection. Stronger deviations from this approximation make such an interpretation more incorrect, but do not prevent in most cases the convergence of phase retrieval algorithms to a reconstruction strongly connected with the sample's topological properties.

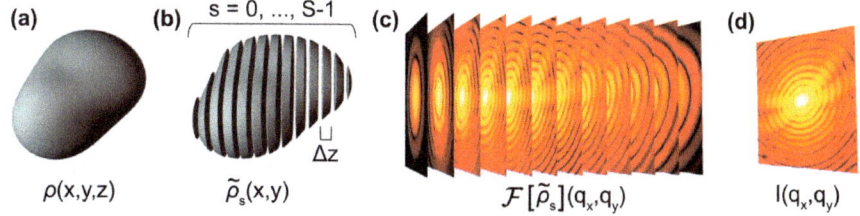

(a) $\rho(x,y,z)$

(b) $s = 0, ..., S{-}1$ Δz $\tilde{\rho}_s(x,y)$

(c) $\mathcal{F}[\tilde{\rho}_s](q_x,q_y)$

(d) $I(q_x,q_y)$

Figure 5.18   MSFT-based simulation method. (a) Rendering of the 3D sample density. (b) The density domain divided into slices. (c) The field scattered from the slices, obtained by combining the slices refractive index with the field shaped by the preceeding ones. (d) The final diffraction pattern is obtained from the sum of all the slices fields. Reproduced from ref. 103, https://doi.org/10.1107/S1600576722008068, under the terms of the CC BY 4.0 license https://creativecommons.org/licenses/by/4.0/.

Here, E_0 is the incoming field amplitude and k_0 is the radiation wavevector. In practical terms, this wave can be split into two components. The first component $E_0 e^{ik_0 z}$ is the equation of a monochromatic field travelling in vacuum (where $\delta = \beta = 0$). The second term $C(z,\delta,\beta) = e^{-\beta z} e^{-ik_0 \delta z}$ is a complex function which encodes the modification of the field's amplitude $e^{-\beta z}$ and phase $e^{-ik_0\delta z}$. The imaginary part β of n is often called the absorption coefficient, and it causes an exponential decrease in the field strength.[§§§§§§§§] Conversely, δ, also known as the refraction coefficient, modifies the "effective wavelength" of the radiation, thus introducing a phase-shift of strength $k_0 \delta z$ with respect to light travelling in vacuum.

It is possible to apply this correction factor C to a simulation made *via* the Born approximation, by subdividing the sample into slices in the real domain (see Figure 5.18). The scattering for each slice is then computed by using an electric field that has been shaped by the previously encountered slices.[103] This approach is based on the multi-slice Fourier transform (MSFT), an efficient method to compute partial 3D Fourier Transforms[104–106] (see also **Figure 5.20** below).

However, the MSFT approach previously presented is again not good enough when moving towards larger values of the refractive index commonly found in the XUV regime and/or close to the material's electronic resonances. This incompatibility comes from two physical effects that are still neglected. On the one hand, eqn (5.30) exactly tells us how light travels within a uniform medium, but does not describe the refraction effects caused by interfaces of different materials (or between the sample and vacuum). In fact, when the refractive index is a function of the spatial coordinates, gradients of $n(\vec{r})$ lead to the deviation of light, such that the approximation of a plane wave is no longer valid within the sample when the

---

[§§§§§§§§] This is the well-known Beer–Lambert law.[102] Johann Heinrich Lambert already noted, back in the 18th century, that light is exponentially absorbed by matter depending on the material thickness.

value of such gradients becomes appreciable. On the other hand, the light scattered by a given sample slice (see Figure 5.18) still interacts with the subsequent slices (*i.e.*, secondary scattering can happen). Simulation methods which, at least partially, include these "second-order" effects are fundamental to deal with wide-angle CDI of samples whose refractive index varies more than ≈5% from vacuum, as highlighted in ref. 103. Improved methods are under development,[107] which would dramatically increase the range of applicability of forward fitting methods for CDI.

5.3   Imaging the Morphology of Clusters

After having laid the theoretical and practical foundations for retrieving structural information from CDI patterns, the remaining three sections aim to give a concise overview of applying CDI to investigate the morphology of clusters (5.3), their interaction with intense XUV and X-ray pulses (5.4), and laser-induced dynamics in nanomatter (5.5). The diffraction-before-destruction principle was first demonstrated on a nanostructure etched into a membrane[108] using a single pulse from the first short-wavelength FEL FLASH in Hamburg.[3,109] Shortly after, the first single-shot single-particle CDI experiments were carried out on aerosols,[110,111] atomic clusters,[112] nano-particles,[113,114] and single viruses.[34]

For many systems, structure determination is typically the main goal of single-particle CDI. The fragility of "soft matter" systems, such as viruses or membrane proteins as a strong motivator for single-particle CDI was already mentioned. Also studies on aerosols, such as soot nanostructures, were guided by the goal of avoiding structural changes during de-position.[110,111,115] In contrast, cluster experiments usually investigate light-induced dynamics, and determining the particle morphology is therefore rather a characterization step. Nevertheless, lightly bound and short-lived rare-gas clusters and nanodroplets cannot be deposited and studied with electron microscopy. Therefore, CDI provided for the first time direct insight into their structure. While metal nanoparticles are much more stable to-wards deposition on a substrate, gas-phase produced metal clusters are also known to change their structure during landing on a surface.[116,117] Results on the morphologies of both cluster types are presented below. We also recommend the review article by Sun and colleagues,[75] which gives a com-pact overview of single-particle CDI experiments, highlighting the Maloja endstation at SwissFEL as an ideally suited instrument for this line of research.¶¶¶¶¶¶¶

¶¶¶¶¶¶¶ In terms of structure determination methods closely related to single-particle CDI, we want to mention experiments at higher photon energies, where X-ray diffraction is sensitive to the crystal lattice of the object under study. Single-shot single-particle Bragg diffraction has been obtained from rare-gas clusters[118] and water/ice micro-droplets.[119,120] Another novel CDI-related technique is in-flight holography[121] using simple-structured strong scatterers such as xenon clusters as reference objects for Fourier transform holography of single viruses.

5.3.1 Understanding the Growth of Rare-gas Clusters

Rare-gas clusters are generated by expanding gas at a certain temperature and pressure from a reservoir through a small nozzle into vacuum.[122,123] In this supersonic jet, the random velocities of the gas atoms in the reservoir are transformed into directed motion, and the relative velocities (and thus the gas temperature) drop dramatically. Accordingly, the trajectory in the phase diagram may cross the vapor pressure curve and condensation sets in.[122] As the atomic density drops along the jet, monomer addition becomes less important, but the clusters can keep growing by coagulation of smaller clusters into larger ones. Theoretical simulations explaining experimental size distributions confirmed this physical picture of monomer addition and coagulation.[124] But without any single-particle imaging methods available for these fragile systems, it remained unclear whether the energy freed in the coagulation events suffices to re-melt the two clusters into a spherical single cluster or whether a non-spherical shape would freeze out (in case of Ne to Xe; He is a different case, treated in the next section). Indeed, CDI of individual xenon clusters revealed that spherical clusters are produced in parallel to twin and triple shapes up to hailstone-like particles[112,125,126] (see representative images from recent high-resolution CDI experiments[75,76] in Figure 5.19a).

5.3.2 Structures of Helium Nanodroplets and Embedded Dopants

Helium nanodroplets can be considered the prototypical model system for light–matter interaction studies as they remain liquid (He refrains from

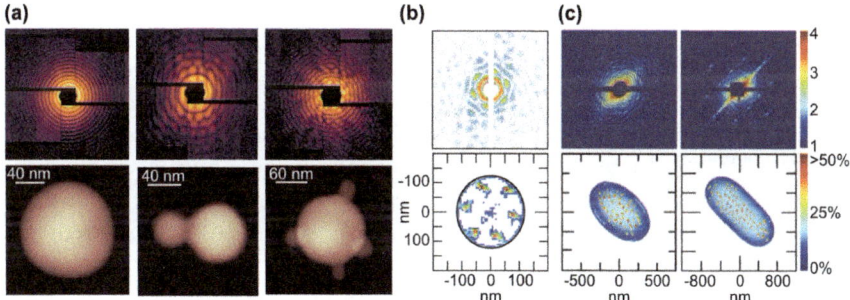

Figure 5.19 Small-angle CDI results on rare-gas clusters and helium droplets. (a) Large argon clusters grow by coagulation. Often, spherical shapes are reached but twin structures or hailstones are also formed. Diffraction data[76] obtained at the Maloja instrument at SwissFEL,[75] courtesy of Mario Sauppe (ETH Zürich). (b) Diffraction pattern and reconstructed density of a spherical ⁴He droplet with a xenon-dressed vortex lattice.[127] Reproduced from ref. 127 with permission from the American Physical Society, Copyright 2016. (c) Patterns and reconstructions of strongly deformed oblate and prolate helium droplets reveal that they also contain vortex lattices.[128] Reproduced from ref. 128 with permission from the American Physical Society, Copyright 2020.

freezing up to high pressures even at 0 K) and can therefore not freeze-out in non-spherical shapes as the other rare-gas clusters do. Also, the electronic structure can be assumed to be simple: the atoms contain only two electrons and are very weakly bound. Nevertheless, helium nanodroplets exhibit a plethora of unexpected and quite fascinating phenomena related to their superfluidity that are a lively topic of current research and have been reviewed recently in a book.[39] Here, we give a brief summary of CDI of helium nanodroplets and hightlight a few examples; for further details, please refer to the CDI chapter[38] in the above-mentioned book and a review paper.[129] The large body of original work,[127,128,130–139] however, testifies that CDI of helium nanodroplets has become a very successful branch of X-ray FEL science on its own.

In the pioneering experiment of the Vilesov group and collaborators,[130] liquid helium was sprayed into vacuum forming nanodroplets. From the diffraction patterns in this and subsequent experiments,[130,132–134] it became obvious that while the majority of droplets were indeed round, some deviated from spherical shape, up to rare, quite strongly deformed shapes. A profound understanding of the shapes of spinning liquid droplets could be developed, both in the presence[135,136] and absence[130,132–134] of friction, tracing them from round to oblate to prolate shapes, rather closely following classical predictions[134] (see examples of oblate and prolate droplets, identified from their features at wide scattering angles in Figure 5.20a). Furthermore, many droplets contained dense lattices of quantum vortices,[127,128,130,131] which could be made visible by doping the helium droplets with xenon atoms (see Figure 5.19b and c).

While quantum vortices in ^{4}He are a fascinating phenomenon, they can also be an obstacle for certain research directions: in the presence of quantum vortices, most atoms or molecules picked up by a droplet get stuck at the vortex sites.[140] This disturbs, for example, the study of how electrons distribute on the surface of a charged droplet[138] and dominates the arrangements of nanostructures forming in the ultracold environment[127,128,130,131] instead of being governed by the mutual interaction between the dopant atoms or molecules.[141] In this regard, a recent collaborative study could clarify regimes for the creation of vortex-free nanodroplets up to a size of 10^8 atoms.[139]

5.3.3   Metal Clusters and Nanoparticles

Early CDI experiments of metal clusters created in a magnetron sputter source[142] using wide-angle CDI (see also Section 5.2.3.3) revealed a number of very symmetric, streaked patterns,[101] which were analysed with forward-fitting based on ideal polyhedron models. The matches indicated that structural motifs such as icosahedra or dodecahedra (which are only energetically favourable at very small sizes) persist up to very large sizes under the rather cold growth conditions applied in the experiment.[101] Only through the development of a more general forward-fitting algorithm,[92,103]

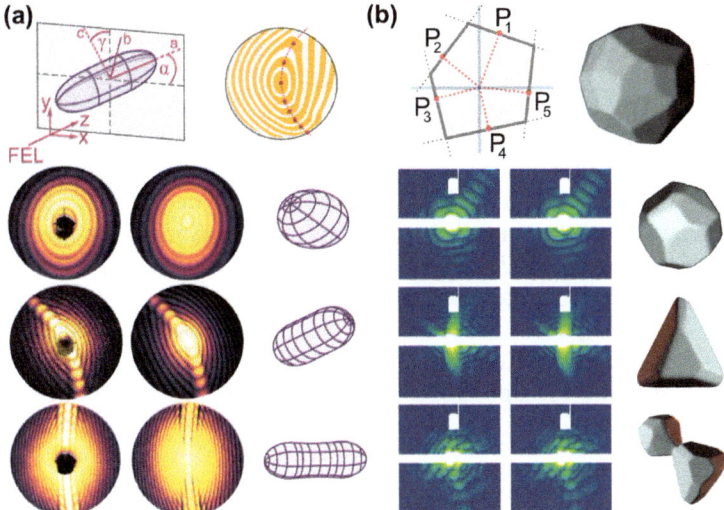

Figure 5.20 Three-dimensional structure determination *via* single-shot wide-angle CDI. (a) Imaging results on spinning superfluid helium nanodroplets. A simple parametric shape model is used and peculiar features produced in a wide-angle diffraction pattern guide the fitting (upper row). Experimental data (left column) and simulated patterns (central column) match well, revealing oblate, elongated architectures of the helium droplets. Reproduced from ref. 134, https://doi.org/10.1103/PhysRevLett.121.255301, under the terms of the CC BY 4.0 license https://creativecommons.org/licenses/by/4.0/. (b) 3D imaging results on silver clusters with the shape parametric model (upper row) constructed by flat facets that define a compact domain in space. Experimental data (left column) and retrieved patterns (center column) and 3D shapes (right column) of single silver clusters and agglomerates. Reproduced from ref. 92 with permission from the author.

it became possible to detect imperfections that result from the rather extreme growth conditions, such as rounded edges, mixed structural motifs, and even agglomerates (see Figure 5.20b).

Besides creating metal clusters in the gas phase, it is also possible to wet-chemically produce metal nanoparticles (and other materials) and spray them into vacuum. This approach has gained importance for time-resolved experiments as both spraying[143] and particle production techniques[144–146] have matured over the last few years. Metal nanoparticles are now available in a broad range of materials (or combinations thereof) and shapes, with narrow distributions of size and shape.[147] While they can be pre-characterized with electron microscopy, the CDI characterizations often show deviations[148] that can be induced by the spray solvent or through the size selection effect of CDI,[149] enhancing the large tail of a broad size distribution.

5.4 Clusters in Intense Short Wavelength Pulses

Our current understanding of the interaction between intense X-ray pulses and nanoscale matter has been largely developed through experimental and theoretical work on clusters. In this section, the experimental and theoretical findings are presented in the framework of a generally accepted physical picture of the dynamics proceeding in three phases. This model was developed in the XUV regime and refined in the X-ray range. We recommend the chapter by Bostedt et al.[40] in ref. 41, where the development is discussed across wavelength regimes from near-edge photon energies to phenomena involving core electrons. Particularly clear insights on X-ray induced dynamics have been gained from the novel method of combined single cluster imaging and ion spectroscopy, which will be also introduced in the following.

5.4.1 The Role of Cluster Experiments at Short Wavelength FELs

Clusters are ideal model systems of solid density but isolated in the gas phase.[150] Compared to bio-molecules, their complexity is drastically reduced. Rare-gas clusters, as the prototypical target for laser-matter interaction experiments, are easily created by expanding gas through a small nozzle into vacuum. Typically, the average cluster size can be chosen via the expansion conditions[122] from single atoms and dimers to a few hundreds, millions, or even 10^{11} atoms per cluster, bridging the regimes of atomic/molecular physics and solid state physics. Also, a wide range of other materials or combinations of materials can be used for creating nanoscale particles that either form in or can be brought into the gas phase. Because of their rather simple handling on the one side and the strong measurement signals obtained on the other, atomic clusters have very regularly been the target of choice in experiments in the start-up and commissioning phases of short-wavelength FELs and their respective endstations.[41,75,94,109,112,151–161]

Already, the first experiments on clusters at the XUV-FEL FLASH[151] started out with a surprise; the ionization states measured were higher than expected.[151,162,163] In the long-wavelength regime, due to inverse bremsstrahlung heating and subsequent impact ionization, the interaction with clusters always produces higher charge states as with atomic gas,[164] but this process scales with the square of the wavelength and thus should have negligible influence in the VUV and XUV regime.[165] Sparked by the first interesting observations, a large number of theoretical works[165–177] and subsequent experimental studies at FELs[95,118,156,157,162,178–198] and HHG sources[163,199,200] were carried out, leading to an ever deeper understanding of the key mechanisms in the short wavelength regime.

5.4.2 Interaction Model in Three Phases

In a widely accepted simple model,[151,165] we separate the intricate dynamics of clusters in intense short-wavelength pulses into three phases (see Figure 5.21). At the beginning of the short-wavelength pulse, the individual atoms of the clusters are photoionized, similar to isolated atoms in a gas, and the cluster environment plays only a minor role. This is in particular true for van der Waals bound rare-gas clusters with an atomic-like electronic structure. Photoionization in the short-wavelength regime, even with very intense pulses, is dominated by sequential single-photon absorption.[201,202] Transient resonances in the ionisation sequence can play an important role.[149,203–206] At higher photon energies, inner-shell holes are created followed by Auger-Meitner decay and thus the creation of additional free electrons.[171]

The second phase starts when electron emission, or 'outer ionization', becomes frustrated and a nanoplasma is formed.[171,180] Still, photons are absorbed and 'inner ionized' electrons are separated from their parent ions but they stay bound to the charge of the cluster compound. Further interaction channels open up, such as electron impact ionization.[173,190] The distinction between inner and outer ionization[207] was already used in the IR regime, where the initial ionization by multiphoton absorption or tunneling leads to electrons without any residual energy that are subsequently accelerated in the slowly changing electromagnetic field.[165,208] Already the first few electrons are thus bound to the cluster compound until they become hot enough to leave the cluster. In the short wavelength regime, the formation of a nanoplasma is instead delayed due to

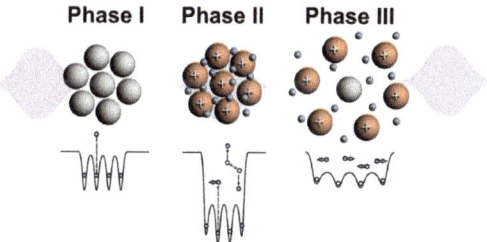

Figure 5.21 Simple picture of laser–cluster interaction in three phases. Phase I: on a few-femtosecond timescale, the atoms are photo-ionized and the cluster environment plays a minor role. Phase II: through charging, the cluster potential becomes deeper and electron emission is frustrated. A nanoplasma forms and evolves during the tens or hundreds of femtosecond pulse. Phase III: the cluster disintegrates within few pico- to nanoseconds. Also, recombination and relaxation processes take place. Potential wells reproduced from ref. 171, https://doi.org/10.1088/1367-2630/13/5/053022, Copyright 2011 Deutsche Physikalische Gesellschaft. Reproduced by permission of IOP Publishing, https://creativecommons.org/licenses/by-nc-sa/3.0/.

significant residual energy of the electrons (difference between photon energy and electron binding energy), which allows them to leave the cluster until it has accumulated a charge creating a potential as deep as the residual energy.[180,209] The turning point of frustrated emission can be estimated *via* the frustration parameter[171] that compares the number of created plasma electrons with the number of electrons that can leave the cluster, the latter being estimated from the simple equation of a probe charge leaving a charged sphere's surface.

In the case of an ongoing irradiation after emission frustration, a cloud of quasi-free electrons is formed which subsequently thermalizes. Its hot tail can boil off,[179,184] resulting in further charging and deepening of the cluster potential. Until ionic motion becomes significant, only a few processes can add or take energy from the electron cloud. In addition to cooling by thermal emission,[184] energy can dissipate through inelastic collisions of electrons with atoms or ions in the cluster (*i.e.*, impact ionization or collisional excitation[190]). An increase in electron temperature results from ongoing photoionization[209] as every new electrons brings in the residual energy, and from recombination of the slowest plasma electrons with cluster ions,[194] which returns the binding energy to the remaining plasma *via* a second involved electron in a three-body collision. The ionic motion can however set in very fast[185] and proceed quite far already during the pulse, in particular at high photon energies, high pulse intensities, and for small cluster sizes; thus the clear separation here is a limitation of the simplified three-phase model.

Finally, during the third phase, when the interaction with the pulse is over, the highly excited cluster nanoplasma will disintegrate. Also, re-combination and relaxation processes are important in this phase.[195,199] While the ionization states reached within clusters and atomic gas during irradiation are similar,[201,210] the final measured ionization states are usually much lower.[151,162] In the dense and fully screened part of the cluster nanoplasma, collisional processes can not only increase the charge states, but also contribute to three-body recombination and thus to a lowering of the charge states.[181,194] However, recombination often occurs into highly excited Rydberg states[200] that can be re-ionized late in the disintegration, *e.g.*, by the constant electric field of the spectrometer, which can thus in-crease the observed ionization states again.[211] Two limit cases, depending on the net charge on the cluster and the temperature of the electron cloud, can be described.[171]

Pure Coulomb explosion occurs when all created electrons have left the cluster and it carries a positive net charge. In this limit case (which is physically possible), the subsequent disintegration is directly driven by the repulsion of the positively charged ions. A characteristic feature of Coulomb explosion is a homogeneous intensity drop (see Figure 5.22a).

In the limit of hydrodynamic expansion (see Figure 5.22b), all created electrons have been captured by the cluster potential and screen the ions. Therefore the dense nanoplasma is net-neutral. However, the electron temperature is non-zero; thus the electron cloud will spill out of the ionic

(a) **(b)**

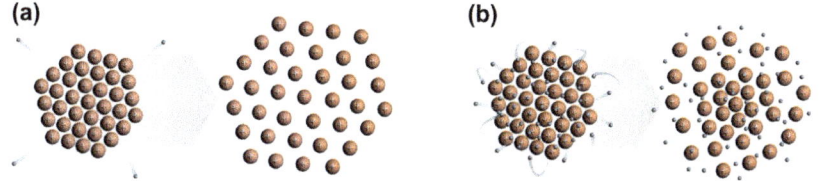

Figure 5.22   Schematics of (a) Coulomb explosion, resulting in self-similar expansion, and (b) hydrodynamic expansion, leading to surface softening.

background, leading to a gradual surface softening[148,192] while inner ions stay screened and still until the ongoing expansion reaches them or they have recombined fully.[193,194] To understand the energy transfer mechanism, also referred to as expansion cooling, a simplified microscopical view is instructive: fast electrons in the nanoplasma can propagate freely through the field-free interior of the cluster, where the ions are screened by other electrons. When they reach the surface, entering vacuum, they leave a positive net-charge behind, pulling them back into the cluster. While they remain outside the cluster, the surface ions are only partially screened by the electron cloud and experience a mutual repulsion. The ions accelerate away from the cluster, flattening the cluster potential. In consequence, the returning electrons will not regain their full previous velocities.

We note that it is of course also Coulombic interaction that drives hydrodynamic expansion. This had been realized early on,[171,212] but the name has led to some confusion.[213] We further note that in reality, a net-neutral nanoplasma can only be an approximation; for a finite electron temperature, some electrons will always be able to leave the cluster compound. A more realistic (and in the case of moderate electron temperatures as created in the XUV regime even pretty accurate) picture is the combined Coulomb explosion of the shell and hydrodynamic expansion of the core.[175,182] Plasma electrons are pulled into the centre of the cluster, fully screening it and leaving an unscreened surface of ions which Coulomb explodes quickly, and the remaining quasi-neutral nanoplasma dissolves by hydrodynamic expansion.

5.4.3   A New Quality with Single Cluster Imaging and Simultaneous Spectroscopy

Conventional cluster experiments on ensembles of clusters suffer from averaging over cluster size distributions and laser intensity profiles, washing out characteristic signatures.[214] A new quality of experiments was achieved by combining single-cluster imaging and simultaneous spectroscopy of ion time-of-flight spectra[112,185,194,198] (see also Figure 5.2). The diffraction pattern allows the determination of particle size and FEL intensity, while ionic residuals of the interaction are measured simultaneously and sorted in post-processing based on the information from the diffraction image. Also, other residuals are possible to be combined with CDI such as electrons, fluorescence, or stimulated emission from clusters.[215]

While ion spectra obtained from ensembles of clusters are dominated by the contribution from the wings of the focal intensity distribution,[185] the brightest single-shot single-cluster spectra from xenon clusters irradiated with intense X-ray pulses yielded only high charge states, giving new insight into suppression of recombination and triggering a discussion of the contribution of re-ionization long after interaction.[199,200] At reduced intensities and using helium nanodroplets as a more weakly interacting target, the full transition from Coulomb explosion to hydrodynamic expansion could be mapped.[198] From the single-cluster spectra of xenon clusters irradiated with an order of magnitude smaller photon energy, the imprint of recombination and its previously unrecognized contribution to plasma heating was found.[194]

5.5   Time-resolved Imaging of Structural Dynamics

The femtosecond snapshots taken with single-shot single-particle CDI make the method also ideally suited for resolving changes in nanoparticles in time and space after irradiation with an optical laser pulse (see Figure 5.23). X-ray induced dynamics can be studied by capturing two consecutive images of the same particle in X-ray-pump X-ray-probe measurements. This, however, requires ways to separate the overlaying diffraction images of pump and probe pulses. Only few attempts have been made so far.[93,118,216] The approach may become more important, as promising pathways have opened up recently with two-color FEL

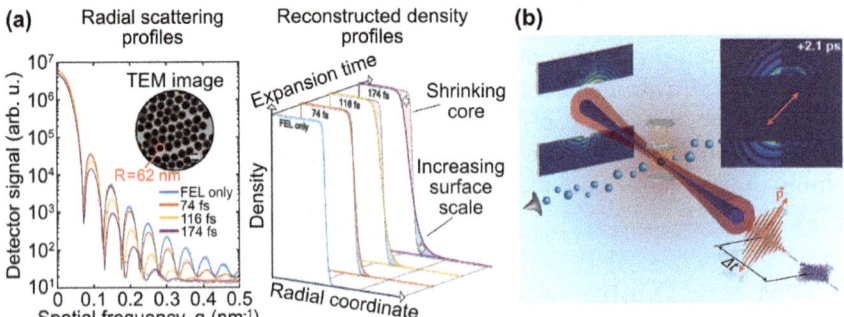

Figure 5.23   (a) Radial profiles of diffraction patterns from NIR excited nanoparticles show a decaying large-angle signal with delay. Thickness of the solid core and softened surface can be reconstructed. Reproduced from ref. 148, https://doi.org/10.1088/1367-2630/ac5e86, under the terms of the CC BY 4.0 license https://creativecommons.org/licenses/by/4.0/. (b) Sketch of the experimental setup and diffraction patterns from NIR excited pure helium droplets. Their clear directionality indicates fast explosion in the polarization direction.[220] Reproduced from ref. 220 with permission from the American Physical Society, Copyright 2022.

pulses with accurately controlled delays[217–219] and CDI detectors that are capable of separating different photon energies.[70–72] Here, we will focus on scenarios using optical pump pulses for excitation of clusters and nanoparticles.

One important aspect for time-resolved imaging of unique structures such as the rare-gas or metal clusters introduced above is the fact that the structure of the original particle is not known and it may be hard to compare events from different delays. A possible strategy is to use repetitive targets, for example, by controlled splitting of liquid jets,[221] or by using wet-chemically produced nanoparticles with a small distribution of shapes and sizes (see the TEM image in Figure 5.23a).

The temporal resolution of such experiments is limited by the pulse durations of optical and short wavelength laser pulses, as well as the temporal jitter between both. Several methods have been developed at FELs to measure the timing jitter for each pair of pulses and to re-sort individual events based on this information.[222,223] This way, the time resolution of laser-pump X-ray-probe experiments at FEL facilities could be lowered to tens of femtoseconds,[224] which is fast enough to "film" the motion of ions or atoms after laser excitation, but not for studying electron dynamics. In this respect, HHG-based sources are still superior. They are comparatively compact, avoiding a large source of jitter, and infrared and shortwave pulses are inherently fixed to sub-femtosecond accuracy with respect to each other due to the generation process. An outlook on the upcoming field of CDI of electron dynamics is given in Section 5.5.4.

5.5.1 Time-resolved Imaging of an Expanding Nanoplasma

The dynamics of clusters,[192,193] nanodroplets[220] and very regular nanoparticles[148] of tens to hundreds of nanometers in diameter, excited with intense near infrared (NIR) pulses, have been investigated with time-resolved CDI. On a timescale of femtoseconds to picoseconds (depending on the laser intensity, the size of the nanostructures, and the respective atomic masses and thus inertia), the surface of the particles softens due to the dominating hydrodynamic expansion. Consequently, the diffraction signal at large angles drops and disappears (to make this plausible, think about the limiting case of a fully smoothed Gaussian distribution; the diffraction pattern consists only of the central maximum). This phenomenon is exemplified in Figure 5.23a.[148] The analysis of the slope and spacing of the diffraction fringes as a function of delay allows to retrieve detailed information on the nanoplasma expansion and direct comparison with modeling.[148,220] For very large, heavy systems, the slow hydrodynamic expansion takes long enough to let the inner part of the cluster nanoplasma recombine fully and disintegrate on a nanosecond time scale as a warm cloud of neutral gas.[193] In the other limit of very light atoms using pure helium nanodroplets, a strong directionality of the nanoplasma evolution along the laser polarization axis is observed[220] (see Figure 5.23b).

5.5.2 Doping Induced Ignition *vs.* Undoped Helium Droplet Explosion

The dynamics induced by an intense NIR laser pulse in a helium nano-droplet can be strongly altered by changing from a pristine droplet to a xenon-doped system.[225–227] The heavier atoms are much more easily ionized by the NIR field, and theoretical modeling indicates that already a few xenon atoms in a large helium droplet suffice to ignite a nanoplasma at much reduced laser intensities. From static CDI experiments of xenon-doped helium droplets (see Figure 5.19b and c), it is known that the distribution of xenon within a helium droplet is governed by the localization of quantum vortices which tend to capture the xenon atoms. Accordingly, after igniting a nano-plasma in helium droplets at more moderate NIR intensities as in ref. 220, the evolution results in distinctly different patterns with particular symmetries[228] (see Figure 5.24).

5.5.3 Resolving Ultrafast Melting and Boiling

Slowly heated matter undergoes the well-known phase transition from solid to liquid to gaseous. For very fast laser-heating, new phenomena such as cavitation of a superheated liquid occur. In a recent experiment,[229,230] Coulomb-driven strong-field effects were avoided by homogeneously heating silver clusters *via* their Mie plasmon resonance[231,232] with a rather weak 400 nm laser pulse[229] (see Figure 5.24b). At delays of hundreds of pico-seconds after irradiation, compact, faceted structures still coexist with large, hollow "nanobubbles" and fragments of exploded bubbles. The results on free plasmon-heated silver nanoclusters mostly agree with previous work on NIR-heated, deposited metal nanoparticles;[233,234] however, the influence of the deposition surface and possible Coulomb-driven effects in the latter

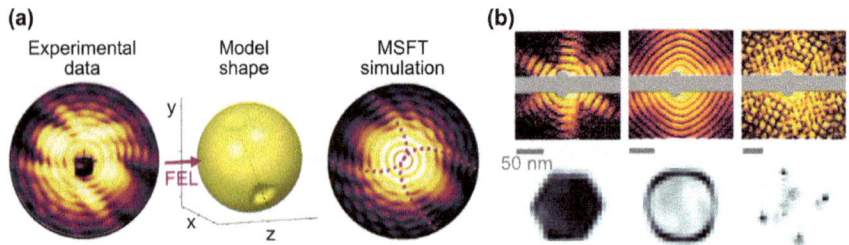

Figure 5.24 (a) Wide-angle diffraction pattern and matching simulation of a doping-ignited helium nanodroplet.[228] Reproduced from ref. 228, https://doi.org/10.1088/1367-2630/aca176, under the terms of the CC BY 4.0 license https://creativecommons.org/licenses/by/4.0/. (b) Hundreds of picoseconds after plasmonic excitation with a rather weak 400 nm laser, silver clusters show faceted shapes, a thin-walled bubble, or fragments.[229] Reproduced from ref. 229 with permission from the authors.

experiments cannot be disentangled from the ultrafast melting process in these experiments, possibly leading to differing observations like off-center bubbles which rapture at much smaller bubble sizes.[234]

5.5.4 Outlook: Towards Attosecond CDI of Electron Dynamics

Until today, most CDI applications rely on the strong "diffraction before destruction" assumption,[2] where the electronic density is considered constant throughout the duration of the X-ray pulse, and the destruction of the sample sets in only afterwards. Electron dynamics occurring during the acquisition of the diffraction patterns are usually neglected or seen as "ultrafast radiation damage". At the same time, it is clear that the process of CDI is intrinsically sensitive to electron dynamics.[235] Elastic light scattering takes place at the electrons bound to the nanoscale object, leading to a change in the scattering response for a strongly altered electron configuration. Experiments on single clusters have shown that ultrafast electronic changes can indeed become visible in an otherwise homogeneous nanoparticle.[95,96] With near-resonant CDI, changes in the diffraction signal due to ionization could be made visible, and even some spatial information on the inhomogeneous distribution of high charge states in the nanoplasma could be extracted.[96] Utilizing the sensitivity of CDI on electronic structures as a tool is however only in its infancy because, until very recently, available FEL pulses had durations of several tens or hundreds of femtoseconds, which were too long to resolve electron dynamics.

Intense XUV pulse trains and even isolated attosecond pulses have been developed at HHG sources around the world within the last decade.[236] In a favourable combination of parameters, single-pulse single-particle HHG-CDI has been demonstrated[133] and first associated time-resolved experiments[237,238] using intense HHG-based XUV sources[163,239] provide a promising outlook of using HHG-based CDI in the study of electron dynamics.

Within the last few years, also the FEL community has seen a leap into the attosecond regime,[19,25–27] driving the emergence of ideas on how to deal with broadband or polychromatic attosecond coherent diffraction images.[48,49,100,240,241] Moreover, theoretical predictions for signatures of ultrafast electron dynamics in the CDI signal[176,235,242–245] show that the extraction of "dynamic scatterer information" will require the development of novel analysis approaches.

The very first work on experimental attoCDI of isolated rare-gas clusters from the group of Tais Gorkhover[149] indicates an enhancement of the scattering signal for the shortest pulse durations, which is assigned to transient resonances. This is very exciting as it may provide better means for single-molecule imaging. X-ray pulses with parameters as projected by simulating successful single-molecule imaging[1] are just now becoming available,[19] and the run for the molecular movie is not over;[18] surprises are to be awaited.

References

1. R. Neutze, *et al.*, *Nature*, 2000, **406**, 752–757.
2. H. N. Chapman, C. Caleman and N. Timneanu, *Philos. Trans. R. Soc., B*, 2014, **369**, 20130313.
3. J. R. Schneider, *J. Phys. B: At., Mol. Opt. Phys.*, 2010, **43**, 194001.
4. P. Emma, R. Akre, J. Arthur, R. Bionta, C. Bostedt, J. Bozek, A. Brachmann, P. Bucksbaum, R. Coffee and F.-J. Decker, *et al.*, *Nat. Photonics*, 2010, **4**, 641–647.
5. E. Allaria, R. Appio, L. Badano, W. Barletta, S. Bassanese, S. Biedron, A. Borga, E. Busetto, D. Castronovo and P. Cinquegrana, *et al.*, *Nat. Photonics*, 2012, **6**, 699–704.
6. T. Ishikawa, H. Aoyagi, T. Asaka, Y. Asano, N. Azumi, T. Bizen, H. Ego, K. Fukami, T. Fukui and Y. Furukawa, *et al.*, *Nat. Photonics*, 2012, **6**, 540–544.
7. W. Decking, S. Abeghyan, P. Abramian, A. Abramsky, A. Aguirre, C. Albrecht, P. Alou, M. Altarelli, P. Altmann and K. Amyan, *et al.*, *Nat. Photonics*, 2020, **14**, 391–397.
8. H.-S. Kang, C.-K. Min, H. Heo, C. Kim, H. Yang, G. Kim, I. Nam, S. Y. Baek, H.-J. Choi and G. Mun, *et al.*, *Nat. Photonics*, 2017, **11**, 708–713.
9. C. J. Milne, T. Schietinger, M. Aiba, A. Alarcon, J. Alex, A. Anghel, V. Arsov, C. Beard, P. Beaud and S. Bettoni, *et al.*, *Appl. Sci.*, 2017, 7, 720.
10. A. Aquila, A. Barty, C. Bostedt, S. Boutet, G. Carini, D. DePonte, P. Drell, S. Doniach, K. Downing and T. Earnest, *et al.*, *Struct. Dyn.*, 2015, **2**, 041701.
11. H. N. Chapman, *Annu. Rev. Biochem.*, 2019, **88**, 35–58.
12. M. M. Seibert, T. Ekeberg, F. R. Maia, M. Svenda, J. Andreasson, O. Jönsson, D. Odić, B. Iwan, A. Rocker and D. Westphal, *et al.*, *Nature*, 2011, **470**, 78–81.
13. M. F. Hantke, D. Hasse, F. R. Maia, T. Ekeberg, K. John, M. Svenda, N. D. Loh, A. V. Martin, N. Timneanu and D. S. Larsson, *et al.*, *Nat. Photonics*, 2014, **8**, 943–949.
14. G. Van Der Schot, M. Svenda, F. R. Maia, M. Hantke, D. P. DePonte, M. M. Seibert, A. Aquila, J. Schulz, R. Kirian and M. Liang, *et al.*, *Nat. Commun.*, 2015, **6**, 5704.
15. H. N. Chapman, P. Fromme, A. Barty, T. A. White, R. A. Kirian, A. Aquila, M. S. Hunter, J. Schulz, D. P. DePonte and U. Weierstall, *et al.*, *Nature*, 2011, **470**, 73–77.
16. P. Fromme, W. S. Graves and J. M. Martin-Garcia, *eLS*, 2020, 1–17.
17. M. O. Wiedorn, D. Oberthür, R. Bean, R. Schubert, N. Werner, B. Abbey, M. Aepfelbacher, L. Adriano, A. Allahgholi and N. Al-Qudami, *et al.*, *Nat. Commun.*, 2018, **9**, 4025.
18. T. Ekeberg, D. Assalauova, J. Bielecki, R. Boll, B. J. Daurer, L. A. Eichacker, L. E. Franken, D. E. Galli, L. Gelisio and L. Gumprecht *et al.*, *bioRxiv*, 2022, preprint, DOI: 10.1101/2022.03.09.483477.

19. J. Duris, S. Li, T. Driver, E. G. Champenois, J. P. MacArthur, A. A. Lutman, Z. Zhang, P. Rosenberger, J. W. Aldrich and R. Coffee, *et al.*, *Nat. Photonics*, 2020, **14**, 30–36.
20. I. Poudyal, M. Schmidt and P. Schwander, *Struct. Dyn.*, 2020, 7, 024102.
21. S. P. Hau-Riege, R. A. London and A. Szoke, *Phys. Rev. E: Stat. Phys., Plasmas, Fluids, Relat. Interdiscip. Top.*, 2004, **69**, 051906.
22. S. P. Hau-Riege, *High-Intensity X-rays-Interaction with Matter: Processes in Plasmas, Clusters, Molecules and Solids*, John Wiley & Sons, 2012.
23. K. Nass, *Acta Crystallogr., Sect. D: Struct. Biol.*, 2019, **75**, 211–218.
24. J. C. Spence, *Struct. Dyn.*, 2017, **4**, 044027.
25. P. K. Maroju, C. Grazioli, M. Di Fraia, M. Moioli, D. Ertel, H. Ahmadi, O. Plekan, P. Finetti, E. Allaria and L. Giannessi, *et al.*, *Appl. Sci.*, 2021, **11**, 9791.
26. M. Beye, M. Gühr, I. Hartl, E. Plönjes, L. Schaper, S. Schreiber, K. Tiedtke and R. Treusch, *Eur. Phys. J. Plus*, 2023, **138**, 193.
27. F. Nolting, C. Bostedt, T. Schietinger and H. Braun, *Eur. Phys. J. Plus*, 2023, **138**, 126.
28. R. A. Kirian and H. N. Chapman, *Synchrotron light sources and free-electron lasers: Accelerator physics, instrumentation and science applications*, 2020, pp. 1337–1397.
29. A. Fannjiang and T. Strohmer, *Acta Numer.*, 2020, **29**, 125–228.
30. S. Marchesini, *Rev. Sci. Instrum.*, 2007, **78**, 011301.
31. Y. Shechtman, Y. C. Eldar, O. Cohen, H. N. Chapman, J. Miao and M. Segev, *IEEE Signal Process. Mag.*, 2015, **32**, 87–109.
32. K. A. Nugent, *Adv. Phys.*, 2010, **59**, 1–99.
33. P. Thibault and V. Elser, *Annu. Rev. Condens. Matter Phys.*, 2010, **1**, 237–255.
34. T. Ekeberg, M. Svenda, C. Abergel, F. R. Maia, V. Seltzer, J.-M. Claverie, M. Hantke, O. Jönsson, C. Nettelblad and G. Van Der Schot, *et al.*, *Phys. Rev. Lett.*, 2015, **114**, 098102.
35. J. Spence, *IUCrJ*, 2017, **4**, 322–339.
36. J. Nelson, X. Huang, J. Steinbrener, D. Shapiro, J. Kirz, S. Marchesini, A. M. Neiman, J. J. Turner and C. Jacobsen, *Proc. Natl. Acad. Sci. U. S. A.*, 2010, **107**, 7235–7239.
37. C. Sanchez-Cano, R. A. Alvarez-Puebla, J. M. Abendroth, T. Beck, R. Blick, Y. Cao, F. Caruso, I. Chakraborty, H. N. Chapman and C. Chen, *et al.*, *ACS nano*, 2021, **15**, 3754–3807.
38. R. M. P. Tanyag, B. Langbehn, T. Möller and D. Rupp, *Molecules in Superfluid Helium Nanodroplets: Spectroscopy, Structure, and Dynamics*, Springer International Publishing Cham, 2022, pp. 281–341.
39. A. Slenczka and J. P. Toennies, *Molecules in Superfluid Helium Nanodroplets: Spectroscopy, Structure, and Dynamics*, Springer Nature, 2022.
40. C. Bostedt, T. Gorkhover, D. Rupp and T. Möller, *Synchrotron light sources and free-electron lasers: Accelerator physics, instrumentation and science applications*, 2020, pp. 1525–1573.

41. E. J. Jaeschke, S. Khan, J. R. Schneider and J. B. Hastings, *Synchrotron light sources and free-electron lasers: accelerator physics, instrumentation and science applications*, Springer, 2016.
42. C. F. Bohren and D. R. Huffman, *Absorption and scattering of light by small particles*, John Wiley & Sons, 2008.
43. D. Paganin *et al.*, *Coherent X-ray optics*, Oxford University Press on Demand, 2006.
44. J. W. Goodman, *Introduction to Fourier optics*, Roberts and Company publishers, 2005.
45. M. Yabashi, K. Tamasaku, K. Sawada, S. Goto and T. Ishikawa, *Synchrotron Light Sources and Free-Electron Lasers: Accelerator Physics, Instrumentation and Science Applications*, 2020, pp. 1123–1159.
46. H. N. Chapman, A. Barty, S. Marchesini, A. Noy, S. P. Hau-Riege, C. Cui, M. R. Howells, R. Rosen, H. He and J. C. Spence, *et al.*, *J. Opt. Soc. Am. A*, 2006, **23**, 1179–1200.
47. M. Born, *Z. Phys.*, 1926, **38**, 803–827.
48. E. Malm, H. Wikmark, B. Pfau, P. Villanueva-Perez, P. Rudawski, J. Peschel, S. Maclot, M. Schneider, S. Eisebitt and A. Mikkelsen, *et al.*, *Opt. Express*, 2020, **28**, 394–404.
49. S. Witte, V. T. Tenner, D. W. Noom and K. S. Eikema, *Light: Sci. Appl.*, 2014, **3**, e163.
50. J. Rosenblatt, *Commun. Math. Phys.*, 1984, **95**, 317–343.
51. R. Millane and W. Hsiao, *Opt. Lett.*, 2009, **34**, 2607–2609.
52. P. Grohs, S. Koppensteiner and M. Rathmair, *SIAM Rev.*, 2020, **62**, 301–350.
53. D. Sayre, *Acta Crystallogr.*, 1952, **5**, 843–843.
54. J. Miao, J. Kirz and D. Sayre, *Acta Crystallogr., Sect. D: Biol. Crystallogr.*, 2000, **56**, 1312–1315.
55. V. Elser, *J. Opt. Soc. Am. A*, 2003, **20**, 40–55.
56. D. R. Luke, *Inverse Probl.*, 2004, **21**, 37.
57. J. R. Fienup, *Appl. Opt.*, 1982, **21**, 2758–2769.
58. J. R. Fienup, *Opt. Lett.*, 1978, **3**, 27–29.
59. H. H. Bauschke, P. L. Combettes and D. R. Luke, *J. Opt. Soc. Am. A*, 2003, **20**, 1025–1034.
60. G. Oszlányi and A. Sütő, *Acta Crystallogr., Sect. A: Found. Crystallogr.*, 2008, **64**, 123–134.
61. C.-C. Chen, J. Miao, C. Wang and T. Lee, *Phys. Rev. B: Condens. Matter Mater. Phys.*, 2007, **76**, 064113.
62. A. Colombo, D. E. Galli, L. De Caro, F. Scattarella and E. Carlino, *Sci. Rep.*, 2017, 7, 1–12.
63. C. Fienup and J. Dainty, *Image Recovery: Theory and Application*, 1987, vol. 231, p. 275.
64. S. Marchesini, H. He, H. N. Chapman, S. P. Hau-Riege, A. Noy, M. R. Howells, U. Weierstall and J. C. Spence, *Phys. Rev. B: Condens. Matter Mater. Phys.*, 2003, **68**, 140101.
65. J. Miao, D. Sayre and H. Chapman, *J. Opt. Soc. Am. A*, 1998, **15**, 1662–1669.

66. C. F. Bohren and D. R. Huffman, *Absorption and Scattering of Light by Small Particles*, Wiley, 1998.
67. B. L. Henke, E. M. Gullikson and J. C. Davis, *At. Data Nucl. Data Tables*, 1993, **54**, 181–342.
68. P. Thibault, V. Elser, C. Jacobsen, D. Shapiro and D. Sayre, *Acta Crystallogr., Sect. A: Found. Crystallogr.*, 2006, **62**, 248–261.
69. J. W. Cooley and J. W. Tukey, *Math. Comput.*, 1965, **19**, 297–301.
70. L. Strüder, S. Epp, D. Rolles, R. Hartmann, P. Holl, G. Lutz, H. Soltau, R. Eckart, C. Reich and K. Heinzinger, *et al.*, *Nucl. Instrum. Methods Phys. Res., Sect. A*, 2010, **614**, 483–496.
71. M. Kuster, K. Ahmed, K.-E. Ballak, C. Danilevski, M. Ekmedžić, B. Fernandes, P. Gessler, R. Hartmann, S. Hauf and P. Holl, *et al.*, *J. Synchrotron Radiat.*, 2021, **28**, 576–587.
72. A. Mozzanica, M. Andrä, R. Barten, A. Bergamaschi, S. Chiriotti, M. Brückner, R. Dinapoli, E. Fröjdh, D. Greiffenberg, F. Leonarski, C. Lopez-Cuenca, D. Mezza, S. Redford, C. Ruder, B. Schmitt, X. Shi, D. Thattil, G. Tinti, S. Vetter and J. Zhang, *Synchrotron Radiat. News*, 2018, **31**, 16–20.
73. C. Brönnimann and P. Trüb, *Synchrotron Light Sources and Free-Electron Lasers: Accelerator Physics, Instrumentation and Science Applications*, 2020, pp. 1191–1223.
74. M. R. Howells, T. Beetz, H. N. Chapman, C. Cui, J. Holton, C. Jacobsen, J. Kirz, E. Lima, S. Marchesini and H. Miao, *et al.*, *J. Electron Spectrosc. Relat. Phenom.*, 2009, **170**, 4–12.
75. Z. Sun, A. Al Haddad, S. Augustin, G. Knopp, J. Knurr, K. Schnorr and C. Bostedt, *Chimia*, 2022, **76**, 529–529.
76. M. Sauppe, A. A. Haddad, A. Colombo, L. Hecht, G. Knopp, K. Kolatzki, B. Langbehn, C. Polat, K. Schnorr, Z. Sun, P. Tümmler, F. Ussling, S. Wächter, A. Weitnauer, J. Zimmermann, M. Zuod, T. Möller, C. Bostedt and D. Rupp, *Imaging the Morphology of Rare Gas Clusters*, https://www.dpgverhandlungen.de/year/2023/conference/samop/part/a/session/27/contribution/39
77. M. F. Hantke, T. Ekeberg and F. R. Maia, *J. Appl. Crystallogr.*, 2016, **49**, 1356–1362.
78. B. J. Daurer, M. F. Hantke, C. Nettelblad and F. R. Maia, *J. Appl. Crystallogr.*, 2016, **49**, 1042–1047.
79. C. H. Yoon, P. Schwander, C. Abergel, I. Andersson, J. Andreasson, A. Aquila, S. Bajt, M. Barthelmess, A. Barty and M. J. Bogan, *et al.*, *Opt. Express*, 2011, **19**, 16542–16549.
80. H. J. Park, N. D. Loh, R. G. Sierra, C. Y. Hampton, D. Starodub, A. V. Martin, A. Barty, A. Aquila, J. Schulz and J. Steinbrener, *et al.*, *Opt. Express*, 2013, **21**, 28729–28742.
81. J. Andreasson, A. V. Martin, M. Liang, N. Timneanu, A. Aquila, F. Wang, B. Iwan, M. Svenda, T. Ekeberg and M. Hantke, *et al.*, *Opt. Express*, 2014, **22**, 2497–2510.

82. T. Ekeberg, S. Engblom and J. Liu, *Int. J. High Perform. Comput. Appl.*, 2015, **29**, 233–243.

83. S. Bobkov, A. Teslyuk, R. Kurta, O. Y. Gorobtsov, O. Yefanov, V. Ilyin, R. Senin and I. Vartanyants, *J. Synchrotron Radiat.*, 2015, **22**, 1345–1352.

84. J. Zimmermann, B. Langbehn, R. Cucini, M. Di Fraia, P. Finetti, A. C. LaForge, T. Nishiyama, Y. Ovcharenko, P. Piseri and O. Plekan, *et al.*, *Phys. Rev. E*, 2019, **99**, 063309.

85. R. Harder, *IUCrJ*, 2021, **8**, 1.

86. L. Wu, S. Yoo, A. F. Suzana, T. A. Assefa, J. Diao, R. J. Harder, W. Cha and I. K. Robinson, *npj Comput. Mater.*, 2021, 7, 175.

87. J. Zimmermann, F. Beguet, D. Guthruf, B. Langbehn and D. Rupp, *npj Comput. Mater.*, 2023, **9**, 24.

88. V. Hinger, A. A. Haddad, R. Barten, A. Bergamaschi, M. Brückner, M. Carulla, S. Chiriotti-Alvarez, R. Dinapoli, S. Ebner and E. Fröjdh, *et al.*, *J. Instrum.*, 2022, **17**, C09027.

89. K. S. Raines, S. Salha, R. L. Sandberg, H. Jiang, J. A. Rodrguez, B. P. Fahimian, H. C. Kapteyn, J. Du and J. Miao, *Nature*, 2010, **463**, 214–217.

90. G. Wang, H. Yu, W. Cong and A. Katsevich, *Nature*, 2011, **480**, E2–E3.

91. S. I. Kabanikhin, *J. Inverse Ill-posed Probl.*, 2008, **17**, 317–357.

92. A. Colombo, S. Dold, P. Kolb, N. Bernhardt, P. Behrens, J. Correa, S. Düsterer, B. Erk, L. Hecht and A. Heilrath, *et al.*, *Sci. Adv.*, 2023, **9**, eade5839.

93. L. Hecht, J. Asmussen, B. Bastian, T. Baumann, L. Ltaief, C. Callegari, A. Colombo, S. De, A. DeFanis, M. di Fraia, S. Dold, T. Fennel, R. Hartmann, K. Kolatzki, S. Krishnan, B. Kruse, A. Lægdsmand, A. Laforge, B. Langbehn, S. Mandal, C. Medina, M. Meyer, T. Möller, R. Moshammer, Y. Ovcharenko, C. Peltz, T. Pfeiffer, P. Piseri, O. Plekan, K. Prince, M. Sauppe, M. Schubert, B. Senftleben, K. Sishodia, F. Stienkemeier, R. Tanyag, P. Tümmler, A. Ulmer, S. Usenko, M. Mudrich and D. Rupp, *Two Color Diffractive Imaging of Helium Nanodroplet Dynamics*, 2022, https://physics.usc.edu/deamn22/Images/DEAMN22 program.pdf

94. V. Lyamayev, Y. Ovcharenko, R. Katzy, M. Devetta, L. Bruder, A. LaForge, M. Mudrich, U. Person, F. Stienkemeier and M. Krikunova, *et al.*, *J. Phys. B: At., Mol. Opt. Phys.*, 2013, **46**, 164007.

95. C. Bostedt, E. Eremina, D. Rupp, M. Adolph, H. Thomas, M. Hoener, A. R. de Castro, J. Tiggesbäumker, K.-H. Meiwes-Broer and T. Laarmann, *et al.*, *Phys. Rev. Lett.*, 2012, **108**, 093401.

96. D. Rupp, L. Flückiger, M. Adolph, A. Colombo, T. Gorkhover, M. Harmand, M. Krikunova, J. P. Müller, T. Oelze and Y. Ovcharenko, *et al.*, *Struct. Dyn.*, 2020, 7, 034303.

97. A. Taflove, *IEEE Trans. Electromagn. Compat.*, 1980, 191–202.

98. C. Varin, C. Peltz, T. Brabec and T. Fennel, *Phys. Rev. Lett.*, 2012, **108**, 175007.

99. E. M. Purcell and C. R. Pennypacker, *Astrophys. J.*, 1973, **186**, 705–714.

100. K. Sander, C. Peltz, C. Varin, S. Scheel, T. Brabec and T. Fennel, *J. Phys. B: At., Mol. Opt. Phys.*, 2015, **48**, 204004.

101. I. Barke, H. Hartmann, D. Rupp, L. Flückiger, M. Sauppe, M. Adolph, S. Schorb, C. Bostedt, R. Treusch and C. Peltz, *et al.*, *Nat. Commun.*, 2015, **6**, 1–7.
102. A. Beer, *Ann. Phys.*, 1852, **162**, 78–88.
103. A. Colombo, J. Zimmermann, B. Langbehn, T. Möller, C. Peltz, K. Sander, B. Kruse, P. Tümmler, I. Barke and D. Rupp, *et al.*, *J. Appl. Crystallogr.*, 2022, **55**, 1232–1246.
104. J. M. Cowley and A. F. Moodie, *Acta Crystallogr.*, 1957, **10**, 609–619.
105. P. G. Self, M. O'keefe, P. Buseck and A. Spargo, *Ultramicroscopy*, 1983, **11**, 35–52.
106. D. Reinhard, B. Hall, D. Ugarte and R. Monot, *Phys. Rev. B: Condens. Matter Mater. Phys.*, 1997, **55**, 7868.
107. P. Tuemmler, B. Kruse, C. Peltz and T. Fennel, *Efficient and accurate simulation of wide-angle single-shot scattering*, 2023, https://www.dpgverhandlungen.de/year/2023/conference/samop/part/a/session/22/contribution/1
108. H. N. Chapman, A. Barty, M. J. Bogan, S. Boutet, M. Frank, S. P. Hau-Riege, S. Marchesini, B. W. Woods, S. Bajt and W. H. Benner, *et al.*, *Nat. Phys.*, 2006, **2**, 839–843.
109. J. Rossbach, J. R. Schneider and W. Wurth, *Phys. Rep.*, 2019, **808**, 1–74.
110. M. J. Bogan, D. Starodub, C. Y. Hampton and R. G. Sierra, *J. Phys. B: At., Mol. Opt. Phys.*, 2010, **43**, 194013.
111. M. J. Bogan, S. Boutet, H. N. Chapman, S. Marchesini, A. Barty, W. H. Benner, U. Rohner, M. Frank, S. P. Hau-Riege and S. Bajt, *et al.*, *Aerosol Sci. Technol.*, 2010, **44**, i–vi.
112. C. Bostedt, M. Adolph, E. Eremina, M. Hoener, D. Rupp, S. Schorb, H. Thomas, A. R. de Castro and T. Möller, *J. Phys. B: At., Mol. Opt. Phys.*, 2010, **43**, 194011.
113. M. J. Bogan, W. H. Benner, S. Boutet, U. Rohner, M. Frank, A. Barty, M. M. Seibert, F. Maia, S. Marchesini and S. Bajt, *et al.*, *Nano Lett.*, 2008, **8**, 310–316.
114. N.-T. D. Loh and V. Elser, *Phys. Rev. E: Stat. Phys., Plasmas, Fluids, Relat. Interdiscip. Top.*, 2009, **80**, 026705.
115. N. Loh, C. Y. Hampton, A. V. Martin, D. Starodub, R. G. Sierra, A. Barty, A. Aquila, J. Schulz, L. Lomb and J. Steinbrener, *et al.*, *Nature*, 2012, **486**, 513–517.
116. K. Koga and K.-I. Sugawara, *Surf. Sci.*, 2003, **529**, 23–35.
117. A. Volk, P. Thaler, M. Koch, E. Fisslthaler, W. Grogger and W. E. Ernst, *J. Chem. Phys.*, 2013, **138**, 214312.
118. K. R. Ferguson, M. Bucher, T. Gorkhover, S. Boutet, H. Fukuzawa, J. E. Koglin, Y. Kumagai, A. Lutman, A. Marinelli and M. Messerschmidt, *et al.*, *Sci. Adv.*, 2016, **2**, e1500837.
119. J. A. Sellberg, C. Huang, T. A. McQueen, N. Loh, H. Laksmono, D. Schlesinger, R. Sierra, D. Nordlund, C. Hampton and D. Starodub, *et al.*, *Nature*, 2014, **510**, 381–384.

120. N. Esmaeildoost, O. Jönsson, T. A. McQueen, M. Ladd-Parada, H. Laksmono, N.-T. D. Loh and J. A. Sellberg, *Crystals*, 2022, **12**, 65.

121. T. Gorkhover, A. Ulmer, K. Ferguson, M. Bucher, F. R. Maia, J. Bielecki, T. Ekeberg, M. F. Hantke, B. J. Daurer and C. Nettelblad, *et al.*, *Nat. Photonics*, 2018, **12**, 150–153.

122. O. F. Hagena, *Rev. Sci. Instrum.*, 1992, **63**, 2374–2379.

123. G. Scoles, *Atomic and molecular beam methods*, Oxford University Press, 1988.

124. J. Soler, N. Garcia, O. Echt, K. Sattler and E. Recknagel, *Phys. Rev. Lett.*, 1982, **49**, 1857.

125. D. Rupp, M. Adolph, T. Gorkhover, S. Schorb, D. Wolter, R. Hartmann, N. Kimmel, C. Reich, T. Feigl and A. R. de Castro, *et al.*, *New J. Phys.*, 2012, **14**, 055016.

126. D. Rupp, M. Adolph, L. Flückiger, T. Gorkhover, J. P. Müller, M. Müller, M. Sauppe, D. Wolter, S. Schorb, R. Treusch, C. Bostedt and T. Möller, *J. Chem. Phys.*, 2014, **141**, 044306.

127. C. F. Jones, C. Bernando, R. M. P. Tanyag, C. Bacellar, K. R. Ferguson, L. F. Gomez, D. Anielski, A. Belkacem, R. Boll and J. Bozek, *et al.*, *Phys. Rev. B*, 2016, **93**, 180510.

128. S. M. O. O'Connell, R. M. P. Tanyag, D. Verma, C. Bernando, W. Pang, C. Bacellar, C. A. Saladrigas, J. Mahl, B. W. Toulson and Y. Kumagai, *et al.*, *Phys. Rev. Lett.*, 2020, **124**, 215301.

129. O. Gessner and A. F. Vilesov, *Annu. Rev. Phys. Chem.*, 2019, **70**, 173–198.

130. L. F. Gomez, K. R. Ferguson, J. P. Cryan, C. Bacellar, R. M. P. Tanyag, C. Jones, S. Schorb, D. Anielski, A. Belkacem and C. Bernando, *et al.*, *Science*, 2014, **345**, 906–909.

131. R. M. P. Tanyag, C. Bernando, C. F. Jones, C. Bacellar, K. R. Ferguson, D. Anielski, R. Boll, S. Carron, J. P. Cryan, L. Englert, S. W. Epp, B. Erk, L. Foucar, L. F. Gomez, R. Hartmann, D. M. Neumark, D. Rolles, B. Rudek, A. Rudenko, K. R. Siefermann, J. Ullrich, F. Weise, C. Bostedt, O. Gessner and A. F. Vilesov, *Struct. Dyn.*, 2015, **2**, 051102.

132. C. Bernando, R. M. P. Tanyag, C. Jones, C. Bacellar, M. Bucher, K. R. Ferguson, D. Rupp, M. P. Ziemkiewicz, L. F. Gomez and A. S. Chatterley, *et al.*, *Phys. Rev. B*, 2017, **95**, 064510.

133. D. Rupp, N. Monserud, B. Langbehn, M. Sauppe, J. Zimmermann, Y. Ovcharenko, T. Möller, F. Frassetto, L. Poletto and A. Trabattoni, *et al.*, *Nat. Commun.*, 2017, **8**, 1–7.

134. B. Langbehn, K. Sander, Y. Ovcharenko, C. Peltz, A. Clark, M. Coreno, R. Cucini, M. Drabbels, P. Finetti and M. Di Fraia, *et al.*, *Phys. Rev. Lett.*, 2018, **121**, 255301.

135. D. Verma, S. M. O. O'Connell, A. J. Feinberg, S. Erukala, R. M. P. Tanyag, C. Bernando, W. Pang, C. A. Saladrigas, B. W. Toulson and M. Borgwardt, *et al.*, *Phys. Rev. B*, 2020, **102**, 014504.

136. A. J. Feinberg, D. Verma, S. M. O'Connell-Lopez, S. Erukala, R. M. P. Tanyag, W. Pang, C. A. Saladrigas, B. W. Toulson, M. Borgwardt and N. Shivaram, *et al.*, *Sci. Adv.*, 2021, 7, eabk2247.

137. R. M. P. Tanyag, C. Bacellar, W. Pang, C. Bernando, L. F. Gomez, C. F. Jones, K. R. Ferguson, J. Kwok, D. Anielski and A. Belkacem, *et al.*, *J. Chem. Phys.*, 2022, **156**, 041102.

138. A. J. Feinberg, F. Laimer, R. M. P. Tanyag, B. Senfftleben, Y. Ovcharenko, S. Dold, M. Gatchell, S. M. O'Connell-Lopez, S. Erukala and C. A. Saladrigas, *et al.*, *Phys. Rev. Res.*, 2022, **4**, L022063.

139. A. Ulmer, A. Heilrath, B. Senfftleben, S. M. O'Connell-Lopez, B. Kruse, L. Seiffert, K. Kolatzki, B. Langbehn, A. Hoffmann and T. M. Baumann *et al.*, *Phys. Rev. Lett.*, 2023, **131**, 076002.

140. F. Coppens, F. Ancilotto, M. Barranco, N. Halberstadt and M. Pi, *Phys. Chem. Chem. Phys.*, 2017, **19**, 24805–24818.

141. R. Tanyag, D. Rupp, A. Ulmer, A. Heilrath, B. Senfftleben, B. Kruse, L. Seiffert, S. O'Connell, K. Kolatzki, B. Langbehn, A. Hoffmann, T. Baumann, R. Boll, A. Chatterley, A. de Fanis, B. Erk, S. Erukala, A. Feinberg, T. Fennel, P. Grychtol, R. Hartmann, S. Hauf, M. Ilchen, M. Izquierdo, B. Krebs, M. Kuster, T. Mazza, K. Meiwes-Broer, J. Montaño, G. Noffz, D. Rivas, D. Schlosser, F. Seel, H. Stapelfeldt, L. Strüder, J. Tiggesbäumker, H. Yousef, M. Zabel, P. Ziolkowski, A. Vilesov, M. Meyer, Y. Ovcharenko and T. Möller, *European XFEL 10th Anniversary Annual Report 2019, Taking snapshots of nanostructures in superfluid helium droplets*, European XFEL GmbH, 2019, p. 20.

142. H. Haberland, M. Karrais and M. Mall, *Z. Phys. D: At., Mol. Clusters*, 1991, **20**, 413–415.

143. J. Bielecki, M. F. Hantke, B. J. Daurer, H. K. Reddy, D. Hasse, D. S. Larsson, L. H. Gunn, M. Svenda, A. Munke and J. A. Sellberg, *et al.*, *Sci. Adv.*, 2019, **5**, eaav8801.

144. O. Pryshchepa, P. Pomastowski and B. Buszewski, *Adv. Colloid Interface Sci.*, 2020, **284**, 102246.

145. Z. Ma, J. Mohapatra, K. Wei, J. P. Liu and S. Sun, *Chem. Rev.*, 2023, **123**(7), 3904–3943.

146. J. E. Ortiz-Castillo, R. C. Gallo-Villanueva, M. J. Madou and V. H. Perez-Gonzalez, *Coord. Chem. Rev.*, 2020, **425**, 213489.

147. K. Ayyer, P. L. Xavier, J. Bielecki, Z. Shen, B. J. Daurer, A. K. Samanta, S. Awel, R. Bean, A. Barty and M. Bergemann, *et al.*, *Optica*, 2021, **8**, 15–23.

148. C. Peltz, J. A. Powell, P. Rupp, A. Summers, T. Gorkhover, M. Gallei, I. Halfpap, E. Antonsson, B. Langer and C. Trallero-Herrero, *et al.*, *New J. Phys.*, 2022, **24**, 043024.

149. S. Kuschel, P. J. Ho, A. A. Haddad, F. Zimmermann, L. Flueckiger, M. R. Ware, J. Duris, J. P. MacArthur, A. Lutman and M.-F. Lin *et al.*, 2022, arXiv, preprint, arXiv:2207.05472.

150. P.-G. Reinhard and E. Suraud, *Introduction to cluster dynamics*, John Wiley & Sons, 2008.

151. H. Wabnitz, L. Bittner, A. De Castro, R. Döhrmann, P. Gürtler, T. Laarmann, W. Laasch, J. Schulz, A. Swiderski and K. Von Haeften, *et al.*, *Nature*, 2002, **420**, 482–485.

152. C. Bostedt, H. N. Chapman, J. T. Costello, J. R. C. López-Urrutia, S. Düsterer, S. W. Epp, J. Feldhaus, A. Föhlisch, M. Meyer and T. Möller, *et al.*, *Nucl. Instrum. Methods Phys. Res., Sect. A*, 2009, **601**, 108–122.

153. B. Erk, J. P. Müller, C. Bomme, R. Boll, G. Brenner, H. N. Chapman, J. Correa, S. Düsterer, S. Dziarzhytski and S. Eisebitt, *et al.*, *J. Synchrotron Radiat.*, 2018, **25**, 1529–1540.

154. C. Bostedt, J. Bozek, P. Bucksbaum, R. Coffee, J. Hastings, Z. Huang, R. Lee, S. Schorb, J. Corlett and P. Denes, *et al.*, *J. Phys. B: At., Mol. Opt. Phys.*, 2013, **46**, 164003.

155. C. Bostedt, S. Boutet, D. M. Fritz, Z. Huang, H. J. Lee, H. T. Lemke, A. Robert, W. F. Schlotter, J. J. Turner and G. J. Williams, *Rev. Mod. Phys.*, 2016, **88**, 015007.

156. Y. Ovcharenko, V. Lyamayev, R. Katzy, M. Devetta, A. LaForge, P. O'Keeffe, O. Plekan, P. Finetti, M. Di Fraia and M. Mudrich, *et al.*, *Phys. Rev. Lett.*, 2014, **112**, 073401.

157. A. LaForge, M. Drabbels, N. B. Brauer, M. Coreno, M. Devetta, M. Di Fraia, P. Finetti, C. Grazioli, R. Katzy and V. Lyamayev, *et al.*, *Sci. Rep.*, 2014, **4**, 3621.

158. M. Yabashi, H. Tanaka, T. Tanaka, H. Tomizawa, T. Togashi, M. Nagasono, T. Ishikawa, J. Harries, Y. Hikosaka and A. Hishikawa, *et al.*, *J. Phys. B: At., Mol. Opt. Phys.*, 2013, **46**, 164001.

159. H. Fukuzawa and K. Ueda, *Adv. Phys.: X*, 2020, **5**, 1785327.

160. E. Seddon, J. Clarke, D. Dunning, C. Masciovecchio, C. Milne, F. Parmigiani, D. Rugg, J. Spence, N. Thompson and K. Ueda, *et al.*, *Rep. Prog. Phys.*, 2017, **80**, 115901.

161. R. Abela, A. Alarcon, J. Alex, C. Arrell, V. Arsov, S. Bettoni, M. Bopp, C. Bostedt, H.-H. Braun and M. Calvi, *et al.*, *J. Synchrotron Radiat.*, 2019, **26**, 1073–1084.

162. H. Wabnitz, A. De Castro, P. Gürtler, T. Laarmann, W. Laasch, J. Schulz and T. Möller, *Phys. Rev. Lett.*, 2005, **94**, 023001.

163. B. Schütte, M. Arbeiter, T. Fennel, M. J. Vrakking and A. Rouzée, *Phys. Rev. Lett.*, 2014, **112**, 073003.

164. T. Ditmire, J. Tisch, E. Springate, M. Mason, N. Hay, R. Smith, J. Marangos and M. Hutchinson, *Nature*, 1997, **386**, 54–56.

165. U. Saalmann, C. Siedschlag and J. Rost, *J. Phys. B: At., Mol. Opt. Phys.*, 2006, **39**, R39.

166. U. Saalmann and J.-M. Rost, *Phys. Rev. Lett.*, 2002, **89**, 143401.

167. R. Santra and C. H. Greene, *Phys. Rev. Lett.*, 2003, **91**, 233401.

168. C. Siedschlag and J.-M. Rost, *Phys. Rev. Lett.*, 2004, **93**, 043402.

169. B. Ziaja, H. Wabnitz, F. Wang, E. Weckert and T. Möller, *Phys. Rev. Lett.*, 2009, **102**, 205002.

170. T. Fennel, K.-H. Meiwes-Broer, J. Tiggesbäumker, P.-G. Reinhard, P. M. Dinh and E. Suraud, *Rev. Mod. Phys.*, 2010, **82**, 1793.

171. M. Arbeiter and T. Fennel, *New J. Phys.*, 2011, **13**, 053022.

172. B. Ziaja, H. Chapman, R. Santra, T. Laarmann, E. Weckert, C. Bostedt and T. Möller, *Phys. Rev. A: At., Mol., Opt. Phys.*, 2011, **84**, 033201.

173. E. Ackad, N. Bigaouette and L. Ramunno, *J. Phys. B: At., Mol. Opt. Phys.*, 2011, **44**, 165102.

174. E. Ackad, N. Bigaouette, S. Mack, K. Popov and L. Ramunno, *New J. Phys.*, 2013, **15**, 053047.

175. C. Peltz, C. Varin, T. Brabec and T. Fennel, *Phys. Rev. Lett.*, 2014, **113**, 133401.

176. B. Kruse, B. Liewehr, C. Peltz and T. Fennel, *J. Phys.: Photonics*, 2020, **2**, 024007.

177. R. Pandit, V. Becker, J. Thurston, K. Barrington, Z. Hartwick, N. Bigaouette, L. Ramunno and E. Ackad, *Phys. Rev. A*, 2023, **107**, 043107.

178. T. Laarmann, A. De Castro, P. Gürtler, W. Laasch, J. Schulz, H. Wabnitz and T. Möller, *Phys. Rev. Lett.*, 2004, **92**, 143401.

179. T. Laarmann, M. Rusek, H. Wabnitz, J. Schulz, A. De Castro, P. Gürtler, W. Laasch and T. Möller, *Phys. Rev. Lett.*, 2005, **95**, 063402.

180. C. Bostedt, H. Thomas, M. Hoener, E. Eremina, T. Fennel, K.-H. Meiwes-Broer, H. Wabnitz, M. Kuhlmann, E. Plönjes and K. Tiedtke, *et al.*, *Phys. Rev. Lett.*, 2008, **100**, 133401.

181. M. Hoener, C. Bostedt, H. Thomas, L. Landt, E. Eremina, H. Wabnitz, T. Laarmann, R. Treusch, A. De Castro and T. Möller, *J. Phys. B: At., Mol. Opt. Phys.*, 2008, **41**, 181001.

182. H. Thomas, C. Bostedt, M. Hoener, E. Eremina, H. Wabnitz, T. Laarmann, E. Plönjes, R. Treusch, A. De Castro and T. Möller, *J. Phys. B: At., Mol. Opt. Phys.*, 2009, **42**, 134018.

183. H. Fukuzawa, X.-J. Liu, G. Prümper, M. Okunishi, K. Shimada, K. Ueda, T. Harada, M. Toyoda, M. Yanagihara and M. Yamamoto, *et al.*, *Phys. Rev. A: At., Mol., Opt. Phys.*, 2009, **79**, 031201.

184. C. Bostedt, H. Thomas, M. Hoener, T. Möller, U. Saalmann, I. Georgescu, C. Gnodtke and J.-M. Rost, *New J. Phys.*, 2010, **12**, 083004.

185. T. Gorkhover, M. Adolph, D. Rupp, S. Schorb, S. W. Epp, B. Erk, L. Foucar, R. Hartmann, N. Kimmel and K.-U. Kühnel, *et al.*, *Phys. Rev. Lett.*, 2012, **108**, 245005.

186. S. Schorb, D. Rupp, M. L. Swiggers, R. N. Coffee, M. Messerschmidt, G. Williams, J. D. Bozek, S.-I. Wada, O. Kornilov and T. Möller, *et al.*, *Phys. Rev. Lett.*, 2012, **108**, 233401.

187. H. Thomas, A. Helal, K. Hoffmann, N. Kandadai, J. Keto, J. Andreasson, B. Iwan, M. Seibert, N. Timneanu and J. Hajdu, *et al.*, *Phys. Rev. Lett.*, 2012, **108**, 133401.

188. H. Fukuzawa, S.-K. Son, K. Motomura, S. Mondal, K. Nagaya, S. Wada, X.-J. Liu, R. Feifel, T. Tachibana and Y. Ito, *et al.*, *Phys. Rev. Lett.*, 2013, **110**, 173005.

189. B. Murphy, T. Osipov, Z. Jurek, L. Fang, S.-K. Son, M. Mucke, J. Eland, V. Zhaunerchyk, R. Feifel and L. Avaldi, *et al.*, *Nat. Commun.*, 2014, **5**, 4281.

190. M. Müller, L. Schroedter, T. Oelze, L. Nösel, A. Przystawik, A. Kickermann, M. Adolph, T. Gorkhover, L. Flückiger and M. Krikunova, *et al.*, *J. Phys. B: At., Mol. Opt. Phys.*, 2015, **48**, 174002.

191. T. Tachibana, Z. Jurek, H. Fukuzawa, K. Motomura, K. Nagaya, S. Wada, P. Johnsson, M. Siano, S. Mondal and Y. Ito, *et al.*, *Sci. Rep.*, 2015, **5**, 10977.

192. T. Gorkhover, S. Schorb, R. Coffee, M. Adolph, L. Foucar, D. Rupp, A. Aquila, J. D. Bozek, S. W. Epp and B. Erk, *et al.*, *Nat. Photonics*, 2016, **10**, 93–97.

193. L. Flückiger, D. Rupp, M. Adolph, T. Gorkhover, M. Krikunova, M. Müller, T. Oelze, Y. Ovcharenko, M. Sauppe and S. Schorb, *et al.*, *New J. Phys.*, 2016, **18**, 043017.

194. D. Rupp, L. Flückiger, M. Adolph, T. Gorkhover, M. Krikunova, J. P. Müller, M. Müller, T. Oelze, Y. Ovcharenko and B. Röben, *et al.*, *Phys. Rev. Lett.*, 2016, **117**, 153401.

195. Y. Kumagai, Z. Jurek, W. Xu, H. Fukuzawa, K. Motomura, D. Iablonskyi, K. Nagaya, S.-I. Wada, S. Mondal and T. Tachibana, *et al.*, *Phys. Rev. Lett.*, 2018, **120**, 223201.

196. Y. Kumagai, Z. Jurek, W. Xu, H. Fukuzawa, K. Motomura, D. Iablonskyi, K. Nagaya, S.-I. Wada, S. Mondal and T. Tachibana, *et al.*, *Phys. Rev. A*, 2020, **101**, 023412.

197. P. J. Ho, B. J. Daurer, M. F. Hantke, J. Bielecki, A. Al Haddad, M. Bucher, G. Doumy, K. R. Ferguson, L. Flückiger and T. Gorkhover, *et al.*, *Nat. Commun.*, 2020, **11**, 167.

198. C. A. Saladrigas, A. J. Feinberg, M. P. Ziemkiewicz, C. Bacellar, M. Bucher, C. Bernando, S. Carron, A. S. Chatterley, F.-J. Decker and K. R. Ferguson, *et al.*, *Eur. Phys. J.: Spec. Top.*, 2021, **230**, 4011–4023.

199. B. Schütte, F. Campi, M. Arbeiter, T. Fennel, M. Vrakking and A. Rouzée, *Phys. Rev. Lett.*, 2014, **112**, 253401.

200. B. Schütte, T. Oelze, M. Krikunova, M. Arbeiter, T. Fennel, M. J. Vrakking and A. Rouzée, *New J. Phys.*, 2015, **17**, 033043.

201. A. Sorokin, S. Bobashev, T. Feigl, K. Tiedtke, H. Wabnitz and M. Richter, *Phys. Rev. Lett.*, 2007, **99**, 213002.

202. N. Gerken, S. Klumpp, A. Sorokin, K. Tiedtke, M. Richter, V. Bürk, K. Mertens, P. Juranić and M. Martins, *Phys. Rev. Lett.*, 2014, **112**, 213002.

203. E. Kanter, B. Kraessig, Y. Li, A. March, P. Ho, N. Rohringer, R. Santra, S. Southworth, L. DiMauro and G. Doumy, *et al.*, *Phys. Rev. Lett.*, 2011, **107**, 233001.

204. B. Rudek, S.-K. Son, L. Foucar, S. W. Epp, B. Erk, R. Hartmann, M. Adolph, R. Andritschke, A. Aquila and N. Berrah, *et al.*, *Nat. Photonics*, 2012, **6**, 858–865.

205. T. Mazza, M. Ilchen, M. Kiselev, E. Gryzlova, T. Baumann, R. Boll, A. De Fanis, P. Grychtol, J. Montaño and V. Music, *et al.*, *Phys. Rev. X*, 2020, **10**, 041056.

206. A. Rörig, S.-K. Son, T. Mazza, P. Schmidt, T. M. Baumann, B. Erk, M. Ilchen, J. Laksman, V. Music and S. Pathak *et al.*, 2023, arXiv, preprint, arXiv:2303.07942.

207. I. Last and J. Jortner, *Phys. Rev. A: At., Mol., Opt. Phys.*, 2000, **62**, 013201.

208. V. P. Krainov and M. B. Smirnov, *Phys. Rep.*, 2002, **370**, 237–331.
209. M. Arbeiter and T. Fennel, *Phys. Rev. A: At., Mol., Opt. Phys.*, 2010, **82**, 013201.
210. L. Schroedter, M. Müller, A. Kickermann, A. Przystawik, S. Toleikis, M. Adolph, L. Flückiger, T. Gorkhover, L. Nösel and M. Krikunova, *et al.*, *Phys. Rev. Lett.*, 2014, **112**, 183401.
211. T. Fennel, L. Ramunno and T. Brabec, *Phys. Rev. Lett.*, 2007, **99**, 233401.
212. P. Mora and R. Pellat, *Phys. Fluids*, 1979, **22**, 2300–2304.
213. I. Last, A. Heidenreich and J. Jortner, *Z. Phys. Chem.*, 2021, **235**, 815–847.
214. M. R. Islam, U. Saalmann and J. M. Rost, *Phys. Rev. A: At., Mol., Opt. Phys.*, 2006, **73**, 041201.
215. A. Benediktovitch, L. Mercadier, O. Peyrusse, A. Przystawik, T. Laarmann, B. Langbehn, C. Bomme, B. Erk, J. Correa and C. Mossé, *et al.*, *Phys. Rev. A*, 2020, **101**, 063412.
216. M. Sauppe, T. Moeller and D. Rupp, *Zeitaufgelöste Dynamik von Clustern in intensiven extrem ultravioletten Doppelpulsen; Time-resolved dynamics of clusters in intense extreme ultraviolet double-pulses*, Desy-door technical report, 2020.
217. A. Lutman, R. Coffee, Y. Ding, Z. Huang, J. Krzywinski, T. Maxwell, M. Messerschvmidt and H.-D. Nuhn, *Phys. Rev. Lett.*, 2013, **110**, 134801.
218. S. Serkez, W. Decking, L. Froehlich, N. Gerasimova, J. Grünert, M. Guetg, M. Huttula, S. Karabekyan, A. Koch and V. Kocharyan, *et al.*, *Appl. Sci.*, 2020, **10**, 2728.
219. E. Prat, P. Dijkstal, E. Ferrari, R. Ganter, P. Juranić, A. Malyzhenkov, S. Reiche, T. Schietinger, G. Wang and A. Al Haddad, *et al.*, *Phys. Rev. Res.*, 2022, **4**, L022025.
220. C. Bacellar, A. S. Chatterley, F. Lackner, C. Pemmaraju, R. M. P. Tanyag, D. Verma, C. Bernando, S. M. O'Connell, M. Bucher and K. R. Ferguson, *et al.*, *Phys. Rev. Lett.*, 2022, **129**, 073201.
221. K. Kolatzki, M. L. Schubert, A. Ulmer, T. Möller, D. Rupp and R. M. P. Tanyag, *Phys. Fluids*, 2022, **34**, 012002.
222. C. Gahl, A. Azima, M. Beye, M. Deppe, K. Döbrich, U. Hasslinger, F. Hennies, A. Melnikov, M. Nagasono and A. Pietzsch, *et al.*, *Nat. Photonics*, 2008, **2**, 165–169.
223. M. B. Danailov, F. Bencivenga, F. Capotondi, F. Casolari, P. Cinquegrana, A. Demidovich, E. Giangrisostomi, M. P. Kiskinova, G. Kurdi and M. Manfredda, *et al.*, *Opt. Express*, 2014, **22**, 12869–12879.
224. E. Savelyev, R. Boll, C. Bomme, N. Schirmel, H. Redlin, B. Erk, S. Düsterer, E. Müller, H. Höppner and S. Toleikis, *et al.*, *New J. Phys.*, 2017, **19**, 043009.
225. T. Liseykina and D. Bauer, *Phys. Rev. Lett.*, 2013, **110**, 145003.
226. A. Mikaberidze, U. Saalmann and J. M. Rost, *Phys. Rev. A: At., Mol., Opt. Phys.*, 2008, **77**, 041201.
227. A. Mikaberidze, U. Saalmann and J. M. Rost, *Phys. Rev. Lett.*, 2009, **102**, 128102.

228. B. Langbehn, Y. Ovcharenko, A. Clark, M. Coreno, R. Cucini, A. Demidovich, M. Drabbels, P. Finetti, M. Di Fraia and L. Giannessi, *et al.*, *New J. Phys.*, 2022, **24**, 113043.

229. S. Dold, T. Reichenbach and A. Colombo, *et al.*, *arXiv*, 2023, preprint, arXiv:2309.00433

230. S. Dold, PhD thesis, Universität Freiburg, 2020.

231. U. Kreibig and L. Genzel, *Surf. Sci.*, 1985, **156**, 678–700.

232. M. A. El-Sayed, *Acc. Chem. Res.*, 2001, **34**, 257–264.

233. J. N. Clark, L. Beitra, G. Xiong, D. M. Fritz, H. T. Lemke, D. Zhu, M. Chollet, G. J. Williams, M. M. Messerschmidt and B. Abbey, *et al.*, *Proc. Natl. Acad. Sci. U. S. A.*, 2015, **112**, 7444–7448.

234. Y. Ihm, D. H. Cho, D. Sung, D. Nam, C. Jung, T. Sato, S. Kim, J. Park, S. Kim and M. Gallagher-Jones, *et al.*, *Nat. Commun.*, 2019, **10**, 2411.

235. D. Popova-Gorelova, *Appl. Sci.*, 2018, **8**, 318.

236. K. Midorikawa, *Nat. Photonics*, 2022, **16**, 267–278.

237. J. C. Zimmermann, *Probing Ultrafast Electron Dynamics in Helium Nanodropletsv with Deep Learning Assisted Diffraction Imaging*, PhD thesis, Technische Universitaet Berlin, Germany, 2021.

238. B. Senfftleben, *Coherent diffractive imaging of electron dynamics in the attosecond domain*, PhD thesis, Technische Universitaet Berlin, Germany, 2023.

239. B. Senfftleben, M. Kretschmar, A. Hoffmann, M. Sauppe, J. Tümmler, I. Will, T. Nagy, M. J. Vrakking, D. Rupp and B. Schütte, *J. Phys.: Photonics*, 2020, **2**, 034001.

240. B. Abbey, L. W. Whitehead, H. M. Quiney, D. J. Vine, G. A. Cadenazzi, C. A. Henderson, K. A. Nugent, E. Balaur, C. T. Putkunz and A. G. Peele, *et al.*, *Nat. Photonics*, 2011, **5**, 420–424.

241. J. Huijts, S. Fernandez, D. Gauthier, M. Kholodtsova, A. Maghraoui, K. Medjoubi, A. Somogyi, W. Boutu and H. Merdji, *Nat. Photonics*, 2020, **14**, 618–622.

242. G. Dixit, J. M. Slowik and R. Santra, *Phys. Rev. Lett.*, 2013, **110**, 137403.

243. D. Popova-Gorelova, D. A. Reis and R. Santra, *Phys. Rev. B*, 2018, **98**, 224302.

244. D. Popova-Gorelova and R. Santra, 2020, arXiv, preprint, arXiv:2012.10334.

245. D. Popova-Gorelova, V. Guskov and R. Santra, 2020, arXiv, preprint, arXiv:2009.07527.

CHAPTER 6

Ultrafast Nanoscale Imaging with High Harmonic Sources

J. ROTHHARDT*[a,b,c] AND L. LOETGERING[a,b]

[a] Helmholtz Institute Jena, Fröbelstieg 3, 07743 Jena, Germany; [b] Institute of Applied Physics, Friedrich-Schiller-University Jena, Albert-Einstein-Straße 15, 07745 Jena, Germany; [c] Fraunhofer Institute for Applied Optics and Precision Engineering, Albert-Einstein-Straße 7, 07745 Jena, Germany
*Email: j.rothhardt@hi-jena.gsi.de

6.1 Introduction

Compact radiation sources based on high harmonic generation (HHG)[1,2] have undergone significant advances during the last decade. Driving lasers with high average powers,[3] cascaded frequency conversion,[4] and repeated recycling of the unconverted part of the driving pulses in enhancement cavities[5] have increased the available average power of these sources to the milliwatt range and beyond[6] at extreme ultraviolet wavelengths. Since the HHG process is phase-locked to the pulsed driving laser field, the generated extreme ultraviolet (XUV) radiation exhibits multiple harmonic orders and femtosecond-to-attosecond pulse durations. Moreover, the inherent synchronisation of the XUV pulses to the driving laser enabled ground-breaking ultrafast spectroscopy experiments in attosecond physics.[7] High harmonic generation additionally results in laser-like beam quality and coherence, which is well-suited for coherent diffraction imaging (CDI).[8]

In Section 6.2, we introduce the basic CDI techniques and demonstrate their ability to access the complex transmission function of a sample with nanoscale resolution. Combining these high-resolution imaging techniques with ultrafast spectroscopy allows us to record nanoscale snapshots of

Theoretical and Computational Chemistry Series No. 25
Structural Dynamics with X-ray and Electron Scattering
Edited by Kasra Amini, Arnaud Rouzée and Marc J. J. Vrakking
© The Royal Society of Chemistry 2024
Published by the Royal Society of Chemistry, www.rsc.org

ultrafast dynamics. Section 6.3 will discuss the underlying principles and present selected experimental demonstrations. The temporal coherence requirement for CDI is in direct conflict with the broad bandwidth required to support few-femtosecond and attosecond pulses. Consequently, ultrafast coherent diffractive imaging on the shortest timescales requires novel imaging approaches. As a result, broadband illumination of the sample must be incorporated into numerical data analysis techniques. Section 6.4 presents various approaches that tackle this challenge, which pave the way for nanoscale imaging of ultrafast dynamics with extreme temporal resolution. A summary and an outlook is given in Section 6.5.

6.2 Methods and State-of-the-art in Coherent Diffractive Table-top XUV Imaging

Classical imaging techniques employ optics to either generate a magnified aerial image of a sample or to produce a small focal spot that is scanned across the sample. These optics can introduce losses and aberrations. For short wavelengths in the XUV and X-ray spectral region in particular, high-quality optics are difficult to fabricate and are, therefore, expensive. Even so, zone plate optics often employed with X-ray radiation still limit the resolution of today's X-ray full-field microscopes to ~ 10 nm.[9] In contrast, coherent diffractive imaging (CDI) techniques avoid using image-forming optics. They have particularly advanced due to the drastic improvement in coherent X-ray flux of synchrotron sources and novel computational approaches to imaging.[10] Since first demonstrations, these coherent methods have experienced remarkable development.[11] Today, the highest spatial resolution X-ray images are obtained with diffractive imaging methods.[12] These techniques have also been implemented with compact HHG sources in recent years. Combining a spatially coherent, femtosecond, short-wavelength light source with CDI techniques provides unique opportunities in ultrafast imaging (*e.g.*, observing ultrafast magnetisation dynamics on the nanoscale). The basic principles of the utilised diffraction-based lensless imaging techniques and selected results from table-top experiments are presented in this section.

6.2.1 Coherent Diffractive Imaging

A typical experimental configuration of CDI (see Figure 6.1) employs a loosely focused beam, in contrast to techniques where the focus is smaller than the object. A diffraction pattern is recorded on a pixelated detector without using image-forming optics downstream of the sample – often termed lensless imaging. Deriving information from the recorded diffraction data necessitates fringe contrast, which requires a high degree of spatial and temporal coherence.

Since HHG sources usually emit many odd-order harmonics, a spectral selection of only a single harmonic line is typically achieved using multilayer

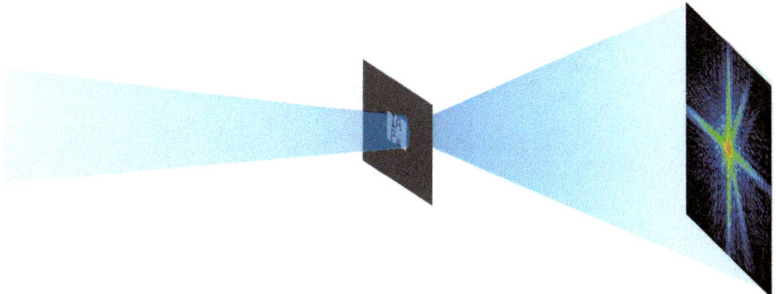

Figure 6.1   Schematic setup of a coherent diffractive imaging (CDI) experiment. An isolated sample is illuminated using a plane or loosely focused coherent wavefront. A detector records diffraction data downstream of the sample. Reproduced from ref. 8, https://doi.org/10.1088/2040-8986/aae2d8, under the terms of the CC BY 3.0 license https://creativecommons.org/licenses/by/3.0/.

mirrors, thin filter foils or grating monochromators between the source and sample, resulting in a nearly monochromatic illumination.

The spatial resolution is given by the Abbe diffraction limit,[13] which is fundamentally restricted to the highest diffraction angle that can be detected at a sufficient signal-to-noise ratio. The detector is usually located in the far-field of the specimen, which connects the measured data and the sample structure by a Fourier transform relation. However, this relation is non-linear, and the recorded intensity lacks phase information, which must be recovered numerically using phase-retrieval algorithms.[14-17] Moreover, additional *a priori* knowledge is required to reconstruct the sample structure unambiguously. This *a priori* knowledge can be available in the form of geometrical or optical properties of the specimen. For instance, bounds on the lateral extent of the sample can be used to restrict the so-called support region, where the specimen is transmissive. This can often be directly estimated from the degree of oversampling or the support of the autocorrelation of the diffraction data.[17] In addition, *a priori* knowledge about material composition, when available, can be used to constrain the phase shift introduced by the specimen.

The first demonstration of CDI with a table-top HHG source was reported in 2007 by Sandberg *et al.*[18] Employing HHG radiation with 29 nm wavelength, a spatial resolution of 214 nm was achieved in this first demonstration of HHG-based CDI (see Figure 6.2a). Table-top CDI setups later demonstrated higher spatial resolution in both transmission and reflection geometries.[19-25] For example, a table-top CDI setup using 18 nm wavelength radiation spatially resolved a binary test sample with a 13 nm spatial resolution (see Figure 6.2b). All results mentioned so far were integrated over $>10^5$ laser pulses, corresponding to detector integration times of minutes to hours. Nonetheless, impressive results have also been achieved with single-shot experiments (see Chapter 5). For example, with a single ~ 20 fs XUV

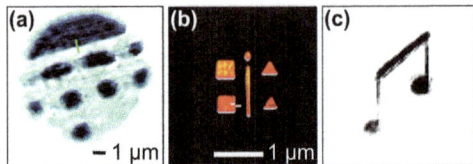

Figure 6.2   Selected results of coherent diffractive imaging experiments with table-top HHG sources. (a) The first table-top CDI image of a masked carbon film has been recorded at 29 nm wavelength. Reproduced from ref. 83 with permission from the American Physical Society, Copyright 2007. (b) A spatial resolution of 13 nm has been reported in a binary test sample using a wavelength of 18 nm. Reproduced from ref. 24 with permission from the Optical Society of America, Copyright 2016. (c) CDI with a single ~20 fs XUV pulse enabled 119 nm resolution on a "harmonic notes"-sample. Reproduced from ref. 21 with permission from the American Physical Society, Copyright 2009.

pulse, a resolution of 119 nm has been achieved (see Figure 6.2c),[21] paving the way for time-resolved imaging of non-repeatable processes.[26]

6.2.2   Fourier-transform Holography

Phase retrieval algorithms, as employed in CDI (see Chapter 5), require significant computational resources and are time-consuming. Their convergence speed and success rate depend on the experimental conditions, such as the coherence and stability of the source, linearity of the detector, and signal-to-noise ratio (SNR) of the measurement.[17,27] Fourier transform holography (FTH) is a robust alternative that overcomes these drawbacks by directly encoding the phase information of the specimen into the diffraction pattern.[28] The experimental setup of FTH (see Figure 6.3) consists of an additional reference wave originating from a reference structure, which interferes with the object wave. A small reference structure is placed close to the sample.

The reference structure can be a pinhole or another aperture type, as further detailed below. In this configuration, a straightforward Fourier transform performs a direct and unambiguous reconstruction of the specimen, which is directly accessible as a cross-correlation of the object wave and the reference wave. The achievable resolution is limited to roughly the size of the reference structure, resulting in a compromise between resolution and reference wave intensity. However, multiple reference structures (see Figure 6.4a),[29] non-redundant arrays[30] or extended ref. 31–33 can be used to improve image contrast and resolution.

Pioneering work in table-top FTH was performed by Sandberg *et al.*, achieving 89 nm resolution[34] in a multiple reference scheme (see Figure 6.4a). Even a higher resolution of 34 nm has been obtained by employing reference holes of only a few wavelengths in diameter

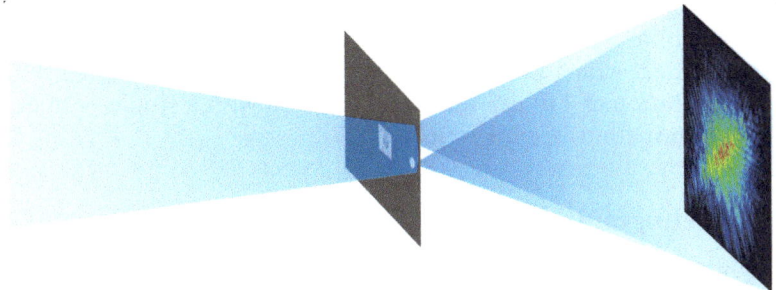

Figure 6.3   Schematic setup of a Fourier-Transform Holography (FTH) experiment. An isolated sample and a reference pinhole are coherently illuminated. The resulting diffraction pattern is recorded behind the sample. Reproduced from ref. 8, https://doi.org/10.1088/2040-8986/aae2d8, under the terms of the CC BY 3.0 license https://creativecommons.org/licenses/by/3.0/.

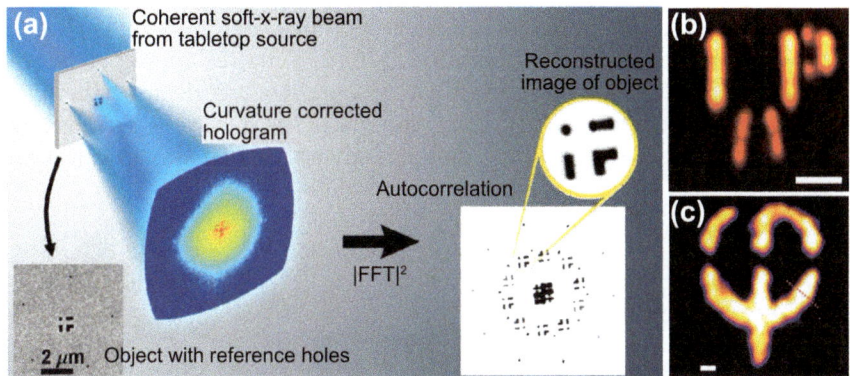

Figure 6.4   (a) Schematic and reconstructed object in a multiple reference FTH experiment. Reproduced from ref. 34 with permission from the Optical Society of America, Copyright 2009. (b) High-resolution XUV image obtained from FTH at 18 nm wavelength. The scale bar is 500 nm. Reproduced from ref. 35, https://doi.org/10.1038/s41598-018-27030-y, under the terms of the CC BY 4.0 license https://creativecommons.org/licenses/by/4.0/. (c) Single-shot XUV image obtained with an extended reference at 32 nm wavelength. The scale bar is 200 nm. Reproduced from ref. 31 with permission from the American Physical Society, Copyright 2010.

(see Figure 6.4b).[35] If the diffraction pattern is measured with sufficient oversampling, iterative phase retrieval algorithms can further refine the image obtained by FTH.[36] In this way, features with a half-period of only 23 nm can be resolved, and waveguiding effects in wavelength-sized features can be observed.[35] Extended reference structures have successfully been used to increase the intensity of the reference wave such that single-shot FTH can be performed with table-top HHG sources (see Chapter 5). This modality, known as holography with extended reference by autocorrelation linear differential operation (HERALDO), relies on the sharp edges of extended references which mimic a differential operation.[37] Using HERALDO, a spatial resolution of 110 nm has been achieved with a single ~20 fs XUV pulse for illumination (see Figure 6.4c).[31]

6.2.3 Ptychography

CDI requires the sample to be limited in size and surrounded by a uniform empty region. A similar problem applies to FTH, where the field of view is even more restricted due to the presence of the reference structure. A method that overcomes these limitations is ptychography. Ptychography can be considered as a scanning version of CDI or, alternatively, as a diffractive extension of scanning transmission microscopy (STM), where a focused beam is scanned across the sample. A pixelated detector measures a sequence of far-field diffraction patterns as the sample is scanned with respect to the illumination, also referred to as the probe. Notably, the scanning is performed such that adjacent positions overlap (see Figure 6.5). By exploring the redundant information gained from overlapping positions, the phase problem can be solved using iterative algorithms where, in addition to the object transmission function, the probe is retrieved in amplitude and phase as well.[38,39] Iterative algorithms enable an on-the-fly

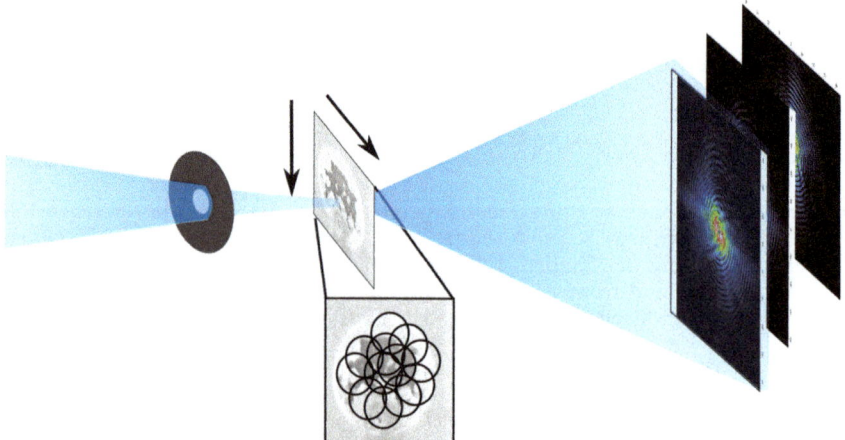

Figure 6.5 Schematic setup of a ptychography experiment in transmission geometry. The illumination, typically referred to as probe field, is usually defined by a pinhole. A series of diffraction patterns are recorded at overlapping-probe positions on the sample and comprise the diffraction dataset, which subsequently is the input data to retrieve the probe field and the complex transmission function of the sample. Reproduced from ref. 8, https://doi.org/10.1088/2040-8986/aae2d8, under the terms of the CC BY 3.0 license https://creativecommons.org/licenses/by/3.0/.

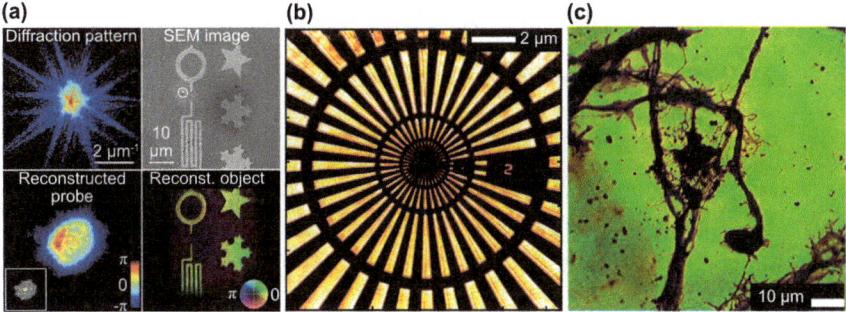

Figure 6.6   Experimental demonstrations of XUV ptychography. (a) XUV ptychography in reflection mode at 29 nm wavelength. Reproduced from
ref. 49 with permission from The Optical Society, Copyright 2014.
(b) XUV ptychography in the transmission mode at 18 nm wavelength.
Adapted from ref. 84, https://doi.org/10.1038/s41598-019-38501-1,
under the terms of the CC BY 4.0 license http://creativecommons.org/
licenses/by/4.0/. (c) XUV imaging of hippocampal mouse neurons at
29 nm wavelength. Adapted from ref. 85, https://doi.org/10.1126/sciadv.
aaz3025, under the terms of the CC BY 4.0 license http://creativecommons.
org/licenses/by/4.0/.

deconvolution of the probe from the sample, resulting in a lateral resolution
smaller than the spot size of the illumination.[40,41] Other algorithmic
extensions allow self-calibration of the experimental setup, including models
that incorporate partial spatial and temporal coherence[42,43] and position
correction.[16,44–46]

While ptychography was initially widely used for X-ray imaging,[47] its use in
combination with high harmonic sources is becoming increasingly popular.[48]
In addition to the transmission geometry (see Figure 6.5), reflection-mode
geometries are feasible and have also been implemented.[49]

Recent results obtained with ptychography using table-top XUV sources
(see Figure 6.6) demonstrated that a reflection-mode ptychography experiment can be performed using radiation with a wavelength of 29 nm
(see Figure 6.6a). A measured diffraction pattern and an electron microscope
image of the sample (see top row of Figure 6.6a), as well as the probe and
object reconstructions from the ptychography scan (see bottom row of
Figure 6.6a), can be obtained. The hue in the object image encodes quantitative phase information, which can be used to extract height information
about the surface topography. Moreover, the transmitted amplitude of a
Siemens-star test sample acquired from a transmission-mode ptychography
scan at 18 nm wavelength (see Figure 6.6b) can resolve features as small
as 45 nm. The application of XUV ptychography to biological samples
(see Figure 6.6c) demonstrated that hippocampal mouse neurons could be
imaged using radiation with a wavelength of 29 nm. The resulting image
combines high resolution with very high contrast, showing the imaging
power of XUV ptychography on a real-world sample.

6.3   Ultrafast Imaging on the Nanoscale

6.3.1   Ultrafast Pump–Probe Imaging – General Considerations

The ultrafast response of a dynamic system can be measured with the pump–probe technique, which is widely used in ultrafast spectroscopy.[50] This technique employs two temporally synchronised light pulses. The "pump" pulse photoexcites the sample. The temporally delayed "probe" pulse – not to be confused with the probe in ptychography, as mentioned above – measures a time-dependent variation of an optical property of the photo-excited sample, such as absorption or reflection. Note that a fast detector is not required, and the temporal resolution is only limited by the duration of the pump and probe pulses. Moreover, both pulses can also have different wavelengths.

At XUV wavelengths, the pump–probe scheme has been used to study ultrafast phenomena on femtosecond (fs) and attosecond timescales, exploiting the temporal synchronisation of high-order harmonics with the driving laser pulse.[51] Pump–probe spectroscopy spatially averages the sample's response across the focal spot and thus does not provide high spatial resolution. However, a combination of pump–probe spectroscopy with XUV imaging techniques could potentially combine sub-(fs) temporal resolution with nanoscale spatial resolution and thus deliver ultrafast snapshots of nanoscale dynamics. This would require multiple diffraction patterns under varying pump–probe delays. In the case of lensless imaging techniques, a series of diffraction patterns are recorded for various pump–probe delays. Additional temporal constraints can help to improve the convergence of phase-retrieval algorithms when recording dynamics.[52]

In the case of a fully reproducible process and a well-stabilised pump–probe delay, each diffraction pattern can be measured by integrating over many probe pulses at each time delay. Even ptychography, which requires additional spatial scanning of the illumination with respect to the object, has been performed in this manner.[53] In this case, a series of diffraction patterns is recorded for each delay. However, if the excitation process is not reproducible, the diffraction pattern must be generated from a single ultrashort XUV pulse, which requires sufficiently high pulse energy (~ 100 nJ) for a single-shot measurement.

6.3.2   Temporal Coherence *Versus* Probe Pulse Duration

Since XUV pulses generated by HHG possess a specific spectral bandwidth, they cannot be considered monochromatic. The product of the temporal duration, τ_P, and the spectral width, $\Delta\omega$, (in frequency space) is referred to as the time-bandwidth product (TBP),

$$\mathrm{TBP} = \tau_P \Delta\omega. \qquad (6.1)$$

The minimum TBP is a property of the Fourier-transform (FT) relation between time and frequency domains, and depends on the spectral shape (*e.g.*, TBP ≥ 0.44 for gaussian-shaped pulses). Due to the additional chirp, the TBP is usually larger than its FT theoretical limit.

Moreover, the spectral bandwidth directly affects the temporal coherence length. Wide spectral bandwidths lead to shorter temporal coherence lengths, resulting in reduced interference contrast at large diffraction angles. The simplest approach to mitigate this problem is to keep all path differences between interfering waves in a coherent imaging set-up smaller than the temporal coherence length, for instance, by limiting the numerical aperture of detection. The resulting minimum spatial resolution Δr can be calculated as

$$\Delta r \geq \frac{a\Delta\lambda}{\lambda}, \qquad (6.2)$$

where a is the field of view on the sample, λ is the wavelength of the radiation source, and $\Delta\lambda/\lambda$ is the relative spectral bandwidth.[54] Note that additional prefactors must be included in eqn (6.2) if the pulse spectrum departs from a Gaussian shape.[55] By substituting eqn (6.1) into eqn (6.2), we find that the product of spatial and temporal resolution is given by

$$\Delta r \tau_P \geq \frac{a\lambda}{2\pi c} \text{TBP}. \qquad (6.3)$$

It is, therefore, essential to keep the field of view sufficiently small to achieve high spatial and temporal resolution. In ptychography, the sample's field-of-view can be extended by scanning the focused beam across the sample, while in CDI and FTH, the sample's field of view is determined by the sample design. Note that lensless imaging beyond this temporal coherence limit is presented in Section 6.4.

6.3.3 Ultrafast Imaging Results

Combining ultrafast pump–probe techniques with nanoscale imaging provides unique opportunities to follow ultrafast dynamics at their natural femtosecond time and nanometre spatial scales. The capabilities of table-top ultrafast nanoscale XUV imaging are illustrated in this sub-section using two examples.

In the first example, the complex thermal and acoustic response of an individual nickel nanoscale antenna was recorded after excitation with a femtosecond laser pulse.[56] This was achieved by combining ptychography with a pump–probe setup. In this setup, one (large) portion of the infrared femtosecond Ti:Sapphire laser pulse was used to generate XUV light *via* HHG, while a second (small) portion was employed as the pump pulse, which induced acoustic and thermal dynamics in two different nickel

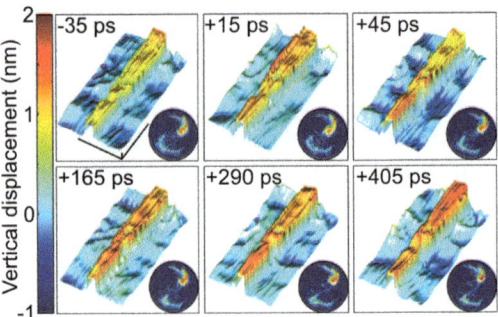

Figure 6.7   Reconstructed snapshots of a nanostructure after excitation with a femtosecond pump pulse. All plots share the scale bars of 5 μm × 5 μm × 0.5 nm. An offset of 1 nm was added to the nanostructure feature for visualisation purposes. Reproduced from ref. 56 with permission from the American Association for the Advancement of Science, Copyright 2018.

nanostructures. At each pump–probe time delay, a full ptychography scan was recorded. This allowed reconstruction of both the amplitude and phase of the sample's reflectivity as a function of pump–probe delay. The phase image is particularly sensitive to changes in the surface topology. A histogram thresholding technique was used to separate the dynamics of the nickel nanostructure from those of the silicon substrate (for further details, see ref. 56). As a result, two-dimensional height maps of the sample's surface could be extracted for different time delays (see Figure 6.7). The temporal evolution is characterised by an expansion at the edges of the nanostructure in combination with a shrinkage within the adjacent substrate regions, reaching a maximum at ∼45 ps. At larger time delays, the surface expansion propagates from the edges toward the centre of the nanostructure. While the reported lateral spatial resolution was on the order of 80 nm, the temporal resolution was estimated to be ∼10 fs, even though the dynamics shown here evolved on a picosecond timescale.

In a second example,[57] the local magnetisation in Co/Pd multilayers was imaged with circularly polarised XUV pulses by exploiting X-ray magnetic circular dichroism (XMCD; see Chapter 7 for details of XMCD). Specifically, the difference between two diffraction patterns recorded with right-hand and left-hand circular polarisation provided the magnetic dichroism signal. A real-space image of the out-of-plane magnetisation can be obtained through holography or phase-retrieval.[58] An ultrafast movie of magnetisation dynamics can be recorded after excitation with a synchronised ultrashort infrared pump pulse. A series of snapshots at different time delays were recorded with ∼10 fs XUV pulses centred at 21 nm wavelength with respect to a 35 fs infrared pump pulse (see Figure 6.8). Imaging of the reversible magnetisation dynamics was conducted stroboscopically at a 1 kHz repetition rate. Here, a rapid demagnetisation (∼200 fs time scale), which results in reduced image contrast, is followed by a slow recovery on a

0 ps   0.2 ps   0.6 ps   1.0 ps   1.5 ps   5.0 ps   60 ps

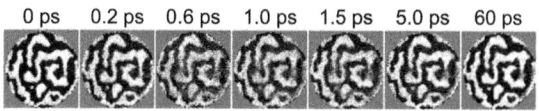

Figure 6.8   A series of snapshot images of a magnetic sample recorded at different time delays with respect to the pump pulse. Reproduced from ref. 57, https://doi.org/10.1038/s41467-021-26594-0, under the terms of the CC BY 4.0 license http://creativecommons.org/licenses/by/4.0/.

few-picosecond timescale. These snapshots were recorded with ~40 nm spatial and ~40 fs temporal resolution.

Both examples presented here demonstrate that ultrafast imaging of structural dynamics in nanomaterials is possible with a combined femtosecond temporal and nanometre spatial resolution. In the future, similar techniques could enable the observation of ultrafast electron and spin dynamics. Advances in laser technology will allow imaging with sub-40 nm HHG sources of sufficiently high photon flux for even higher spatial resolution measurements. Regarding the temporal domain, lensless imaging with attosecond pulses will provide the required temporal resolution to follow the fastest dynamics.[59] However, their large spectral bandwidth calls for novel approaches to lensless imaging, which will be presented in the next section.

6.4   Lensless Imaging Under Broadband Conditions

6.4.1   Attosecond Pulses, Their Spectral Properties and Consequences for Lensless Imaging

As pulse durations approach attosecond timescales, several practical difficulties arise. These span from experimental challenges, such as dispersion control and the design of polychromatic focusing optics, to algorithmic aspects, such as the necessary, appropriate modelling of inverse problems under broadband conditions. The TBP limitation in eqn (6.1) provides a convenient perspective for this discussion. We start by reformulating the energy band in the TBP relation in terms of a wavelength band. To first order, the wavelength band $\Delta\lambda$ of a short pulse of duration Δt can be estimated from eqn (6.1) to be

$$\Delta\lambda \sim \text{TBP} \frac{\bar{\lambda}^2}{c \cdot \Delta t},\tag{6.4}$$

where $\bar{\lambda}$ is the central wavelength, and c is the speed of light. Eqn (6.4) has two practical implications. The first implication concerns the spectral region where attosecond pulses can be practically generated and controlled. Due to the quadratic dependence on the central wavelength, the expected wavelength band rapidly increases towards long central wavelengths, making dispersion compensation a significant challenge. Searching for a central wavelength that results in an estimated spectral width no larger than a fraction F as compared to the central wavelength, we obtain

$$\bar{\lambda} \sim c \cdot \Delta t \cdot \frac{F}{\text{TBP}}. \qquad (6.5)$$

Eqn (6.5) shows that a central wavelength of ~ 14 nm is required to achieve a spectral band occupied by a 200 attosecond pulse that is 10% of its central wavelength. Thus, our first conclusion is that attosecond pulses require operation at XUV or even shorter wavelengths.[60] However, designing efficient focusing optics for polychromatic light in the XUV and soft-X-ray (SXR) spectral range is challenging. While spherical, toroidal and ellipsoidal mirrors allow only moderate focusing of attosecond pulses to micrometre spot sizes,[61] they are not suitable as nanoscale imaging optics due to aberrations and their limited numerical aperture. Likewise, diffractive optical elements (zone plates) cannot be used as imaging optics of broadband light due to the strong chromatic dispersion involved. Thus, the second conclusion is that lensless imaging, as a direct consequence of operation at extremely short wavelengths (≤ 14 nm), is the most feasible route towards imaging at attosecond timescales.

As discussed earlier, lensless imaging at attosecond timescales fundamentally requires broadband operation[62] and algorithmic techniques for polychromatic data inversion. Figure 6.8 illustrates diffraction patterns for varying pulse durations (1 fs, 200 as, and 100 as) from a randomised, nondispersive nanoscale structure, which have been simulated under plane wave illumination with XUV radiation ($\bar{\lambda} = 10$ nm) and a detector numerical aperture of 0.5 (see Figure 6.9). As the pulse duration becomes smaller, from 1 fs to 100 as, the diffraction data experience concentric blurring. This blurring is increasingly pronounced towards high observation angles (edges of the detector), where pulses undergo the largest optical path differences. A crucial step for microscopy at ultrafast timescales, as required by both HHG and free electron lasers, is, therefore, to embed multispectral models and data analysis techniques into diffractive imaging models. Such multispectral models account for reduced contrast in the diffraction data, as described in the next subsection.

(a)   1 fs   **(b)** 200 as   **(c)** 100 as

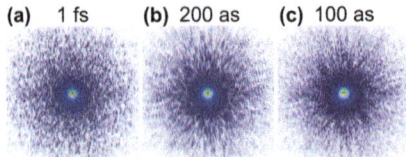

Figure 6.9   Diffraction patterns from a randomised specimen illuminated with ultrafast pulses. (a–c) From panel (a–c), the pulse duration decreases from 1 fs to 100 as, resulting in a continuous, concentric blur and loss of speckle contrast, which is characteristic of broadband diffraction.

6.4.2 Broadband Lensless Imaging Techniques

This section provides an overview of recent advances in broadband diffractive imaging, including single-shot CDI, FTH, and ptychography. We start with a review of broadband CDI. In 2011, Abbey *et al.*[41] demonstrated a polychromatic CDI experiment. In this experiment, an undulator X-ray beam at 1.4 keV with a bandwidth of $\Delta\lambda/\lambda = 2.8\%$ was used to record a polychromatic diffraction pattern of a non-dispersive specimen under a detector numerical aperture of 0.01. Under these conditions, eqn (6.2) predicts a threefold reduction in lateral resolution due to the finite bandwidth of the undulator source as compared to a monochromatic CDI experiment at the same central wavelength. To mitigate this problem, the authors proposed an algorithmic extension to CDI, which models the diffraction intensity as a superposition of several monochromatic but mutually incoherent contributions. This model successfully accounted for the concentric blurring in polychromatic diffraction, allowing for a similar reconstruction quality as observed under monochromatic conditions.

An alternative approach is to spectrally resolve the broadband diffraction data through Fourier transform spectroscopy (FTS). Witte *et al.*[63] generated XUV high-harmonic radiation from two laterally separated and mutually time-delayed femtosecond pulses focused into a gas jet. The time-delay scan provided additional spectral information through FTS. Images of the specimen at different wavelengths could then be numerically retrieved using monochromatic CDI algorithms.

Related strategies were reported by both Huijts *et al.*[64] and Malm *et al.*[65] in 2020. Both groups formulated the process of monochromatising diffraction patterns to bypass the deconvolution problem. This numerical monochromatisation of the broadband diffraction data avoids the need for a time-consuming FTS scan. However, it has the disadvantage of only being applicable to non-dispersive specimens. Moreover, the spectrum of the illumination must be precisely characterised *a priori*.

The second diffractive imaging technique discussed here is broadband FTH. As mentioned in Section 6.2, the diffraction pattern observed in FTH is a coherent superposition between an object wave and a reference wave. The reference pinhole is displaced with respect to the specimen by at least twice the lateral extent of the specimen. This guarantees that the specimen autocorrelation and the specimen-reference cross-correlation can be separated directly from the measured signal. The lateral displacement L of the pinhole with respect to the specimen poses requirements on the temporal coherence length of the specimen. Under the paraxial approximation, the temporal coherence length needs to be smaller than the maximum optical path length difference between the outermost points of the specimen and the reference structure, which leads to the condition

$$\frac{D \cdot L}{2 \cdot z} < \frac{\lambda^2}{\Delta\lambda}, \tag{6.6}$$

where D is the detector size and z is the sample detector distance. If this condition is violated, the contrast of the fringes in the observed far-field diffraction pattern decreases, ultimately compromising the reconstruction quality.[66] Assuming that the temporal coherence length of the illumination is fixed, the most straightforward options to mitigate this problem are to select a smaller specimen or to increase the sample–detector distance. The first solution would limit the field of view and, thereby, the applicability of FTH, while the second option would result in a loss of lateral resolution. Flewett *et al.*[67] proposed placing a grating next to the specimen so that both spatial and temporal coherence properties can be extracted from a broadband FTH experiment. However, this experimental configuration has not experimentally been demonstrated to date. Williams *et al.*[68] proposed an experimental demonstration of broadband FTH where a dedicated specimen was chosen that facilitates the deconvolution of spectral blurring effects. In this implementation, the sample was thin along the direction of the spectral blur that was introduced by the violation of the coherence requirement in eqn (6.6). Moreover, the source exhibited a quasi-discrete spectrum of isolated harmonics. Eschen *et al.*[69] proposed a solution for continuous broadband spectra involving multiple pinholes. Each pinhole introduces spectral blur along the direction of separation to the specimen due to the violation of the coherence requirements. However, thanks to multiple cross-correlations from different pinholes, a high-resolution object can be synthesised by combining only the sharp portions of each reconstruction (see Figure 6.10). This is achieved by directional Fourier-filters applied to the different cross-correlations (see Figure 6.10a). The directional Fourier-filters are chosen to select the Fourier components perpendicular to the direction of blur (see Figure 6.10b). The final object is composed by the summation of these selected Fourier components (see Figure 6.10c) and a final Fourier transform. The final object (see lower panel of Figure 6.10d) features significantly higher resolution compared to the object obtained using the standard FTH approach (see the upper panel of Figure 6.10d). This method thus allows the use of significantly larger spectral bandwidths without sacrificing spatial resolution.

The third diffractive imaging technique discussed here is broadband ptychography. As already introduced in Section 6.2.3, in ptychography, the sample is scanned laterally to illuminate a given specimen region of interest multiple times. This enables ptychography to extract information that cannot be recovered from single-shot techniques, such as the illumination profile.[16,39,70] While broadband conditions were long considered a nuisance for ptychography, in 2013, Batey *et al.* demonstrated that the information in polychromatic diffraction patterns could be used for spectrally-resolved imaging.[43] In this first demonstration, a polychromatic beam was generated from three separate lasers in the visible range. It was shown that objects with strong dispersive properties could be reconstructed through ptychography. Shortly thereafter, Enders *et al.*[71] demonstrated hard X-ray ptychography using a broadband (pink) beam with a peak energy of 16.95 keV and an

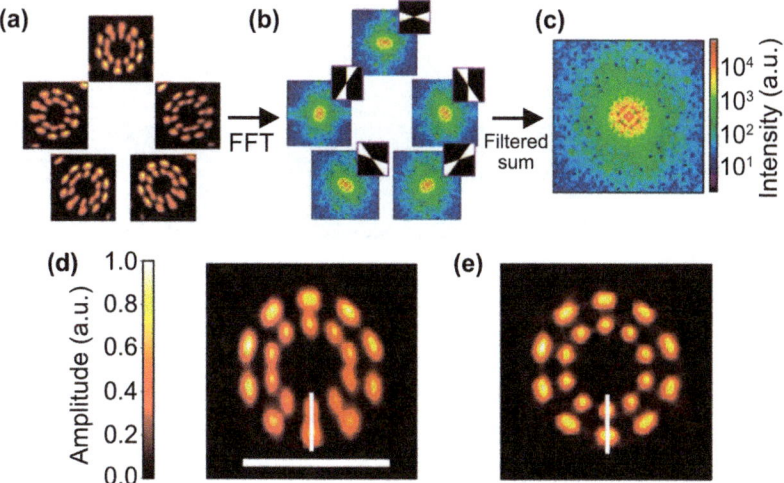

Figure 6.10   Data processing steps in broadband FTH. (a) Cross-correlation terms from separate pinholes. The blur appears orthogonal to the sample-pinhole displacement direction. (b) A weighted average in the Fourier decomposition of the cross-correlation signals allows for a deconvolution of the sample's Fourier spectrum and the effect of blurring due to the broadband nature of the radiation source. (c) Resulting filtered sample spectrum. (d) Cross-correlation term before broadband correction. (e) Recovered sample from the broadband deconvolution approach. Reproduced from ref. 86, https://doi.org/10.1038/s42005-021-00658-5, under the terms of the CC BY 4.0 license http://creativecommons.org/licenses/by/4.0/.

energy bandwidth of 1.5%. The authors employed a different algorithmic approach to account for the loss of contrast in the diffraction pattern due to the broadband nature of the beam. The authors reported that the loss of contrast in broadband ptychography experiments could be addressed by mixed-state ptychography.[42] This technique typically accounts for spatial decoherence by representing the probe by a sum of multiple incoherent spatial modes.[42] As Yao *et al.*[72] observed, mixed state ptychography does not appropriately model the concentric blurring effect in broadband data and can, therefore, only be used to somewhat mitigate the loss of contrast but not to correctly model polychromatic diffraction. Zhang *et al.*[73] reported a multispectral reflection ptychography experiment in which four high-order XUV harmonics were employed for spectrally-resolved imaging. In this experiment, the illumination was focused using an ellipsoidal mirror in grazing incidence to produce a polychromatic focused beam. This approach is flux efficient but does not allow control over the wavefront. In 2021, Loetgering *et al.* performed ptychography with a table-top high harmonic source, where the illumination was generated with a diffractive optical element.[74] The reconstruction of five central wavefronts at different wavelengths (see Figure 6.11) shows an excellent agreement between the

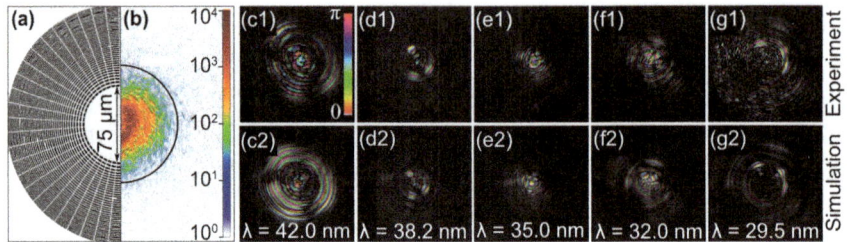

Figure 6.11   Ptychographic wavefront sensing on a multispectral structured beam. (a) SEM of zone plate for XUV beam shaping, (b) 1 out of 252 diffraction patterns in the ptychography scan (the scale bar shows AD counts on a logarithmic scale), (c–g) top row: ptychographic reconstruction. Bottom row: Simulation from the known ZP profile. Only five out of a total of nine reconstructed spectral wavefronts are shown. Reproduced from ref. 74 with permission from The Optical Society, Copyright 2021.

experiment (top) and simulation (bottom), which verifies that multispectral ptychography can adequately characterise polychromatic wavefronts. As a next step, the wavefront information obtained from multispectral ptychography can be used to characterise multispectral samples, surface topographies, and single-short broadband measurements. It is possible that such *a priori* characterisation of a structured wavefront can be used in conjunction with single-shot CDI. Broadband extensions to recent implementation approaches of single-shot CDI[75,76] could enable ultrafast broadband imaging but bypass the complications for dedicated specimen preparation as required in FTH.

6.5   Conclusion and Outlook

Combining table-top high harmonic sources with lensless imaging techniques allows imaging in the XUV spectral region with a spatial resolution approaching 10 nm. Imaging of ultrafast dynamics additionally requires the images to be recorded in a pump–probe-type configuration. In this scenario, the sample is first photoexcited with a pump pulse and subsequently imaged with a temporally delayed probe pulse. Table-top high harmonic sources are suited for such investigations since all generated high-order harmonics are phase-locked and thus well-synchronised to the driving laser pulse. Therefore, the pump pulse (typically the fundamental pulse) and the probe pulse can be selected from the harmonic spectrum. A probe pulse in the XUV spectral region allows the recording of snapshot images at high spatial and temporal resolution. Nanoscale heat transport and ultrafast spin-dynamics have already been observed on (sub-)picosecond timescales. However, the required temporal coherence restricts the applicability of standard lensless imaging methods. Thus, broadband lensless imaging techniques have been proposed and put into practice, extending the applicability of lensless imaging to broadband pulses supporting attosecond durations. In the future, these techniques will have a wide variety of applications in the field of ultrafast dynamics on nanometre spatial and attosecond temporal scales. This opens access to ultrafast electron and spin phenomena, including chemical reactions, charge transport and magnetisation dynamics.

Future advances in this field crucially depend on the availability of shorter wavelength radiation (higher photon energies) from HHG sources. This will allow the resolution of smaller structures and give access to manifold atomic absorption edges in the soft X-ray spectral region. Infrared driving lasers at ~ 2 μm wavelength[77–80] are well-suited for generating soft X-ray radiation by HHG, and scaling these laser systems to high average powers is ongoing.[81] Thus, application-relevant average powers will be available in the soft X-ray spectral region in the near future.

A final ingredient will be the further development of algorithmic techniques. While the first decade of lensless imaging techniques at high harmonic sources was arguably driven by the ever-increasing computing power, the second decade holds promise to incorporate the powerful emerging techniques from the machine learning community. End-to-end optimisation of XUV imaging systems will yield novel illumination optics and efficient detection architectures. Here, the trend goes toward structured illumination[74] and structured detection.[75] The former refers to structuring the illumination wavefront, while the latter implies placing an additional optical modulation element downstream of the specimen, such as a grating or a diffuser. In addition, spectral regularisation techniques and reduced-order models[82] may be used to limit complexity in multispectral lensless imaging.

References

1. M. Ferray, A. L'Huillier and X. Li, *J. Phys. B: At., Mol. Opt. Phys.*, 1988, **21**, L31–L35.
2. A. Mcpherson, G. Gibson, H. Jara, U. Johann, T. S. Luk, I. A. Mcintyre, K. Boyer and C. K. Rhodes, *J. Phys. B: At., Mol. Opt. Phys.*, 1987, **4**, 595–601.
3. S. Hädrich, J. Rothhardt, M. Krebs, S. Demmler, A. Klenke, A. Tünnermann and J. Limpert, *J. Phys. B: At., Mol. Opt. Phys.*, 2016, **49**, 172002.
4. H. Wang, Y. Xu, S. Ulonska, J. S. Robinson, P. Ranitovic and R. A. Kaindl, *Nat. Commun.*, 2015, **6**, 7459.
5. I. Pupeza, C. Zhang, M. Högner and J. Ye, *Nat. Photonics*, 2021, **15**, 175–186.
6. R. Klas, A. Kirsche, M. Gebhardt, J. Buldt, H. Stark, S. Hädrich, J. Rothhardt, J. Limpert, S. Hädrich, J. Rothhardt and J. Limpert, *PhotoniX*, 2021, **2**, 1–8.
7. P. B. Corkum and F. Krausz, *Nat. Phys.*, 2007, **3**, 381–387.
8. J. Rothhardt, G. K. Tadesse, W. Eschen and J. Limpert, *J. Opt.*, 2018, **20**, 113001.

9. B. Rösner, S. Finizio, F. Koch, F. Döring, V. A. Guzenko, M. Langer, E. Kirk, B. Watts, M. Meyer, J. Loroña Ornelas, A. Späth, S. Stanescu, S. Swaraj, R. Belkhou, T. Ishikawa, T. F. Keller, B. Gross, M. Poggio, R. H. Fink, J. Raabe, A. Kleibert and C. David, *Optica*, 2020, **7**, 1602–1608.

10. J. Miao, P. Charalambous, J. Kirz and D. Sayre, *Nature*, 1999, **400**, 342–344.

11. J. Miao, T. Ishikawa, I. K. Robinson and M. M. Murnane, *Science*, 2015, **348**, 249–254.

12. D. A. Shapiro, Y. S. Yu, T. Tyliszczak, J. Cabana, R. Celestre, W. Chao, K. Kaznatcheev, A. L. D. Kilcoyne, F. Maia, S. Marchesini, Y. S. Meng, T. Warwick, L. L. Yang and H. A. Padmore, *Nat. Photonics*, 2014, **8**, 765–769.

13. E. Abbe, *Arch. Mikrosk. Anat.*, 1873, **9**, 413.

14. J. R. Fienup, *Opt. Lett.*, 1978, **3**, 27–29.

15. J. R. Fieneup, *Appl. Opt.*, 1982, **21**, 2758–2769.

16. M. Guizar-Sicairos and J. R. Fienup, *Opt. Express*, 2008, **16**, 7264.

17. J. Miao, D. Sayre and H. N. Chapman, *J. Opt. Soc. Am. A*, 1998, **15**, 1662.

18. R. Sandberg, A. Paul, D. Raymondson, S. Hädrich, D. Gaudiosi, J. Holtsnider, R. Tobey, O. Cohen, M. Murnane, H. Kapteyn, C. Song, J. Miao, Y. Liu and F. Salmassi, *Phys. Rev. Lett.*, 2007, **99**, 098103.

19. M. D. Seaberg, D. E. Adams, E. L. Townsend, D. A. Raymondson, W. F. Schlotter, Y. Liu, C. S. Menoni, L. Rong, C.-C. Chen, J. Miao, H. C. Kapteyn and M. M. Murnane, *Opt. Express*, 2011, **19**, 22470–22479.

20. M. Zürch, J. Rothhardt, S. Hädrich, S. Demmler, M. Krebs, J. Limpert, A. Tünnermann, A. Guggenmos, U. Kleineberg and C. Spielmann, *Sci. Rep.*, 2014, **4**, 7356.

21. A. Ravasio, D. Gauthier, F. R. N. C. Maia, M. Billon, J.-P. P. Caumes, D. Garzella, M. Géléoc, O. Gobert, J.-F. F. Hergott, A.-M. M. Pena, H. Perez, B. Carré, E. Bourhis, J. Gierak, A. Madouri, D. Mailly, B. Schiedt, M. Fajardo, J. Gautier, P. Zeitoun, P. H. Bucksbaum, J. Hajdu and H. Merdji, *Phys. Rev. Lett.*, 2009, **103**, 028104.

22. M. Zürch, C. Kern and C. Spielmann, *Opt. Express*, 2013, **21**, 21131.

23. D. F. Gardner, B. Zhang, M. D. Seaberg, L. S. Martin, D. E. Adams, F. Salmassi, E. Gullikson, H. Kapteyn and M. Murnane, *Opt. Express*, 2012, **20**, 19050–19059.

24. G. K. Tadesse, R. Klas, S. Demmler, S. Hädrich, I. Wahyutama, M. Steinert, C. Spielmann, M. Zürch, T. Pertsch, A. Tünnermann, J. Limpert and J. Rothhardt, *Opt. Lett.*, 2016, **41**, 5170–5173.

25. H. Vu Le, K. Ba Dinh, P. Hannaford and L. Van Dao, *J. Appl. Phys.*, 2014, **116**, 17.

26. D. Rupp, N. Monserud, B. Langbehn, M. Sauppe, J. Zimmermann, Y. Ovcharenko, T. Möller, F. Frassetto, L. Poletto, A. Trabattoni, F. Calegari, M. Nisoli, K. Sander, C. Peltz, M. J. Vrakking, T. Fennel and A. Rouzée, *Nat. Commun.*, 2017, **8**, 1–25.

27. M. Guizar-Sicairos and J. R. Fienup, *J. Opt. Soc. Am. A*, 2012, **29**, 2367.

28. B. Pfau and S. Eisebitt, in *Synchrotron Light Sources and Free-Electron Lasers: Accelerator Physics, Instrumentation and Science Applications*, 2016, pp. 1093–1133.

29. W. F. Schlotter, R. Rick, K. Chen, A. Scherz, J. Stöhr, J. Lüning, S. Eisebitt, C. Günther, W. Eberhardt, O. Hellwig and I. McNulty, *Appl. Phys. Lett.*, 2006, **89**, 1–3.
30. S. Marchesini, S. Boutet, A. E. Sakdinawat, M. J. Bogan, S. Bajt, A. Barty, H. N. Chapman, M. Frank, S. P. Hau-Riege, A. Szöke, C. Cui, D. A. Shapiro, M. R. Howells, J. C. H. Spence, J. W. Shaevitz, J. Y. Lee, J. Hajdu and M. M. Seibert, *Nat. Photonics*, 2008, **2**, 560–563.
31. D. Gauthier, M. Guizar-Sicairos, X. Ge, W. Boutu, B. Carré, J. R. Fienup and H. Merdji, *Phys. Rev. Lett.*, 2010, **105**, 093901.
32. D. Zhu, M. Guizar-Sicairos, B. Wu, A. Scherz, Y. Acremann, T. Tyliszczak, P. Fischer, N. Friedenberger, K. Ollefs, M. Farle, J. R. Fienup and J. Stöhr, *Phys. Rev. Lett.*, 2010, **105**, 043901.
33. V. T. Tenner, K. S. E. Eikema and S. Witte, *Opt. Express*, 2014, **22**, 25397.
34. R. L. Sandberg, D. A. Raymondson, C. La-o-vorakiat, A Paul, K. S. Raines, J. Miao, M. M. Murnane, H. C. Kapteyn and W. F. Schlotter, *Opt. Lett.*, 2009, **34**, 1618.
35. G. K. Tadesse, W. Eschen, R. Klas, V. Hilbert, D. Schelle, A. Nathanael, M. Zilk, M. Steinert, F. Schrempel, T. Pertsch, A. Tünnermann, J. Limpert and J. Rothhardt, *Sci. Rep.*, 2018, **8**, 8677.
36. J. Geilhufe, B. Pfau, C. M. Günther, M. Schneider and S. Eisebitt, *Ultramicroscopy*, 2020, **214**, 113005.
37. M. Guizar-Sicairos and J. R. Fienup, *Opt. Express*, 2007, **15**, 17592–17612.
38. P. Thibault, M. Dierolf, A. Menzel, O. Bunk, C. David and F. Pfeiffer, *Science*, 2008, **321**, 379–382.
39. A. M. Maiden and J. M. Rodenburg, *Ultramicroscopy*, 2009, **109**, 1256–1262.
40. L. W. Whitehead, G. J. Williams, H. M. Quiney, D. J. Vine, R. A. Dilanian, S. Flewett, K. A. Nugent, A. G. Peele, E. Balaur and I. McNulty, *Phys. Rev. Lett.*, 2009, **103**, 243902.
41. B. Abbey, L. W. Whitehead, H. M. Quiney, D. J. Vine, G. A. Cadenazzi, C. A. Henderson, K. A. Nugent, E. Balaur, C. T. Putkunz, A. G. Peele, G. J. Williams and I. Mcnulty, *Nat. Photonics*, 2011, **5**, 420.
42. P. Thibault and A. Menzel, *Nature*, 2013, **494**, 68–71.
43. D. J. Batey, D. Claus and J. M. Rodenburg, *Ultramicroscopy*, 2014, **138**, 13–21.
44. F. Zhang, I. Peterson, J. Vila-Comamala, A. Diaz, F. Berenguer, R. Bean, B. Chen, A. Menzel, I. K. Robinson and J. M. Rodenburg, *Opt. Express*, 2013, **21**, 13592–13606.
45. A. M. Maiden, M. J. Humphry, M. C. Sarahan, B. Kraus and J. M. Rodenburg, *Ultramicroscopy*, 2012, **120**, 64–72.
46. L. Loetgering, M. Du, K. S. E. Eikema and S. Witte, *Opt. Lett.*, 2020, **45**, 2030.
47. F. Pfeiffer, *Nat. Photonics*, 2018, **12**, 9–17.
48. L. Loetgering, S. Witte and J. Rothhardt, *Opt. Express*, 2022, **30**, 4133.
49. M. D. Seaberg, B. Zhang, D. F. Gardner, E. R. Shanblatt, M. M. Murnane, H. C. Kapteyn and D. E. Adams, *Optica*, 2014, **1**, 1–9.

50. M. Maiuri, M. Garavelli and G. Cerullo, *J. Am. Chem. Soc.*, 2020, **142**, 3–15.
51. Y. Mairesse, A. de Bohan, L. J. Frasinski, H. Merdji, L. C. Dinu, P. Monchicourt, P. Breger, M. Kovacev, R. Taïeb, B. Carré, H. G. Muller, P. Agostini and P. Salières, *Science*, 2003, **302**, 1540–1543.
52. Y. H. Lo, L. Zhao, M. Gallagher-Jones, A. Rana, J. Lodico, W. Xiao, B. C. Regan and J. Miao, *Nat. Commun.*, 2018, **9**, 1–10.
53. R. M. Karl, G. F. Mancini, J. L. Knobloch, T. D. Frazer, J. N. Hernandez-Charpak, B. Abad, D. F. Gardner, E. R. Shanblatt, M. Tanksalvala, C. L. Porter, C. S. Bevis, D. E. Adams, H. C. Kapteyn and M. M. Murnane, *Sci. Adv.*, 2018, **4**, 1–8.
54. J. C. H. Spence, U. Weierstall and M. Howells, *Ultramicroscopy*, 2004, **101**, 149–152.
55. B. E. A. Saleh and M. C. Teich, *Fundamentals of Photonics*, Wiley, 2007.
56. R. M. Karl, G. F. Mancini, J. L. Knobloch, T. D. Frazer, J. N. Hernandez-Charpak, B. Abad, D. F. Gardner, E. R. Shanblatt, M. Tanksalvala, C. L. Porter, C. S. Bevis, D. E. Adams, H. C. Kapteyn and M. M. Murnane, *Sci. Adv.*, 2018, **4**, eaau4295.
57. S. Zayko, O. Kfir, M. Heigl, M. Lohmann, M. Sivis, M. Albrecht and C. Ropers, *Nat. Commun.*, 2021, **12**, 1–8.
58. S. Eisebitt, J. Lüning, W. F. Schlotter, M. Lörgen, O. Hellwig, W. Eberhardt, J. Stöhr and A. E. Phenomenology, *Nature*, 2004, **432**, 885–888.
59. F. Siegrist, J. A. Gessner, M. Ossiander, C. Denker, Y. P. Chang, M. C. Schröder, A. Guggenmos, Y. Cui, J. Walowski, U. Martens, J. K. Dewhurst, U. Kleineberg, M. Münzenberg, S. Sharma and M. Schultze, *Nature*, 2019, **571**, 240–244.
60. Z. Chang, *Fundamentals of Attosecond Optics*, CRC Press, Boca Raton, Florida, 2011.
61. H. Pan, C. Späth, A. Guggenmos, S. H. Chew, J. Schmidt, Q. Zhao and U. Kleineberg, *Opt. Express*, 2016, **24**, 16788.
62. K. Sander, C. Peltz, C. Varin, S. Scheel, T. Brabec and T. Fennel, *J. Phys. B: At., Mol. Opt. Phys.*, 2015, **48**, 20.
63. S. Witte, V. T. Tenner, D. W. Noom and K. S. Eikema, *Light Sci. Appl.*, 2014, **3**, e163.
64. J. Huijts, S. Fernandez, D. Gauthier, M. Kholodtsova, A. Maghraoui, K. Medjoubi, A. Somogyi, W. Boutu and H. Merdji, *Nat. Photonics*, 2020, **14**, 618–622.
65. E. Malm, H. Wikmark, B. Pfau, P. Villanueva-Perez, P. Rudawski, J. Peschel, S. Maclot, M. Schneider, S. Eisebitt, A. Mikkelsen, A. L'Huillier and P. Johnsson, *Opt. Express*, 2020, **28**, 394.
66. B. Pfau, C. M. Günther, S. Schaffert, R. Mitzner, B. Siemer, S. Roling, H. Zacharias, O. Kutz, I. Rudolph, R. Treusch and S. Eisebitt, *New J. Phys.*, 2010, **12**, 095006.
67. S. Flewett and S. Eisebitt, *Appl. Opt.*, 2011, **50**, 852–858.
68. G. O. Williams, A. I. Gonzalez, S. Künzel, L. Li, M. Lozano, E. Oliva, B. Iwan, S. Daboussi, W. Boutu, H. Merdji, M. Fajardo and P. Zeitoun, *Opt. Lett.*, 2015, **40**, 3205.

69. W. Eschen, S. Wang, C. Liu, R. Klas, M. Steinert, S. Yulin, H. Meißner, M. Bussmann, T. Pertsch, J. Limpert and J. Rothhardt, *Commun. Phys.*, 2021, **4**, 1–7.

70. P. Thibault, M. Dierolf, O. Bunk, A. Menzel and F. Pfeiffer, *Ultramicroscopy*, 2009, **109**, 338–343.

71. B. Enders, M. Dierolf, P. Cloetens, M. Stockmar, F. Pfeiffer and P. Thibault, *Appl. Phys. Lett.*, 2014, **104**, 171104.

72. Y. Yao, Y. Jiang, J. Klug, Y. Nashed, C. Roehrig, C. Preissner, F. Marin, M. Wojcik, O. Cossairt, Z. Cai, S. Vogt, B. Lai and J. Deng, *J. Synchrotron Radiat.*, 2021, **28**, 309–317.

73. B. Zhang, D. F. Gardner, M. H. Seaberg, E. R. Shanblatt, C. L. Porter, R. Karl, C. A. Mancuso, H. C. Kapteyn, M. M. Murnane and D. E. Adams, *Opt. Express*, 2016, **24**, 18745.

74. L. Loetgering, X. Liu, A. C. C. De Beurs, M. Du, G. Kuijper, K. S. E. Eikema and S. Witte, *Optica*, 2021, **8**, 130.

75. F. Zhang, B. Chen, G. R. Morrison, J. Vila-Comamala, M. Guizar-Sicairos and I. K. Robinson, *Nat. Commun.*, 2016, 7, 1–8.

76. A. L. Levitan, K. Keskinbora, U. T. Sanli, M. Weigand and R. Comin, *Opt. Express*, 2020, **28**, 37103.

77. S. L. Cousin, F. Silva, S. Teichmann, M. Hemmer, B. Buades and J. Biegert, *Opt. Lett.*, 2014, **39**, 5383–5386.

78. M. Gebhardt, T. Heuermann, R. Klas, C. Liu, A. Kirsche, M. Lenski, Z. Wang, C. Gaida, J. E. Antonio-Lopez, A. Schulzgen, R. Amezcua-Correa, J. Rothhardt and J. Limpert, *Light Sci. Appl.*, 2021, **10**, 2021.

79. D. Popmintchev, B. R. Galloway, M. C. Chen, F. Dollar, C. A. Mancuso, A. Hankla, L. Miaja-Avila, G. O'Neil, J. M. Shaw, G. Fan, S. Ališauskas, G. Andriukaitis, T. Balčiunas, O. D. Mücke, A. Pugzlys, A. Baltuška, H. C. Kapteyn, T. Popmintchev and M. M. Murnane, *Phys. Rev. Lett.*, 2018, **120**, 1–6.

80. A. S. Johnson, D. R. Austin, D. A. Wood, C. Brahms, A. Gregory, K. B. Holzner, S. Jarosch, E. W. Larsen, S. Parker, C. S. Strüber, P. Ye, J. W. G. Tisch and J. P. Marangos, *Sci. Adv.*, 2018, **4**, eaar3761.

81. C. Gaida, M. Gebhardt, T. Heuermann, F. Stutzki, C. Jauregui and J. Limpert, *Opt. Lett.*, 2018, **43**, 5853.

82. J. N. Kutz and S. L. Brunton, *Data-Driven Science and Engineering: Machine Learning, Dynamical Systems, and Control*, Cambridge University Press, 2019.

83. R. L. Sandberg, A. Paul, D. A. Raymondson, S. Hädrich, D. M. Gaudiosi, J. Holtsnider, R. Tobey, O. Cohen, M. M. Murnane and H. C. Kapteyn, *Phys. Rev. Lett.*, 2007, **99**, 098103.

84. G. K. Tadesse, W. Eschen, R. Klas, M. Tschernajew, F. Tuitje, M. Steinert, M. Zilk, V. Schuster, M. Zürch, T. Pertsch, C. Spielmann, J. Limpert and J. Rothhardt, *Sci. Rep.*, 2019, **9**, 1–7.

85. P. D. Baksh, M. Ostrčil, M. Miszczak, C. Pooley, R. T. Chapman, A. S. Wyatt, E. Springate, J. E. Chad, K. Deinhardt, J. G. Frey and W. S. Brocklesby, *Sci. Adv.*, 2020, **6**, eaaz3025.

86. W. Eschen, S. Wang, C. Liu, R. Klas, M. Steinert, S. Yulin, H. Meißner, M. Bussmann, T. Pertsch, J. Limpert and J. Rothhardt, *Commun. Phys.*, 2021, **4**, 154.

CHAPTER 7

X-ray Resonant Scattering and Holography with Application to Magnetization Dynamics

B. PFAU* AND S. EISEBITT*

Max Born Institute, Max-Born-Str. 2A, 12489 Berlin, Germany
*Emails: bastian.pfau@mbi-berlin.de; eisebitt@mbi-berlin.de

7.1 Introduction

X-rays provide exceptional possibilities to investigate magnetic phenomena by offering direct access to the magnetic moments of the atoms within a material. X-ray dichroism provides sensitivity to magnetization through the optical properties of the magnetic material and, in particular, the dependence of the material's X-ray absorption on the polarization of the X-ray pulse.[1,2] More generally, the dichroism results from the polarization dependence of the complex-valued elastic scattering cross-section of the atom carrying the magnetic moment. The macroscopic consequence is that X-ray transmission, reflection, and diffraction properties of the magnetic material depend on the relative orientation of the polarization vector of the X-rays to the atomic magnetic moment. Exploiting X-ray dichroism in spectroscopy allows one to distinguish the spin (S) and orbital (L) contributions to the magnetic moment. X-ray dichroism can be observed *via* resonant electronic transitions, predominantly from spin–orbit-split core levels into spin-polarized empty states. Sensitivity to both the ferromagnetic and antiferromagnetic order can be obtained with dichroic effects that are linear or quadratic in the atomic magnetic moment. Together with the ability of X-rays to penetrate through matter, resonant tuning of the X-ray

Theoretical and Computational Chemistry Series No. 25
Structural Dynamics with X-ray and Electron Scattering
Edited by Kasra Amini, Arnaud Rouzée and Marc J. J. Vrakking
© The Royal Society of Chemistry 2024
Published by the Royal Society of Chemistry, www.rsc.org

wavelength into resonance with transitions from suitable core levels, therefore, allows addressing the magnetization of different atomic species selectively and to, for example, selectively probe different layers or atomic sublattices within a magnetic material. This combined atomic and magnetic specificity is the first unique aspect of X-rays compared to other probes of magnetization.

A second distinction arises from the small wavelength of X-rays (nm to Ångström), allowing the detection of nanometre-scale magnetic textures or even atomic-scale magnetic ordering in suitable experiments. Although other magnetic detection techniques also achieve similarly high spatial resolution,[3,4] X-ray methods can uniquely combine high spatial resolution with a high temporal resolution, which are commonly realized with pulsed X-ray sources. High temporal resolution in X-ray experiments is typically provided *via* pump–probe experiments. Typical X-ray pulse durations can reach sub-100 picoseconds (ps) at synchrotron-radiation (SR) sources and sub-100 femtoseconds (fs) at X-ray free-electron lasers (XFELs; see Chapter 8) or high-harmonic generation (HHG) sources.[5] Finally, like other purely photon-based probing techniques, X-rays are insensitive to magnetic fields. Therefore, measurements can be carried out in externally applied magnetic fields of arbitrary strength and direction, adding an essential dimension to the experiments.

Time-resolved X-ray diffraction or imaging experiments provide a spatio-temporal view of magnetization dynamics. In the field of magnetism in the solid state, diffraction methods can be divided into three groups:[6] atomic Bragg diffraction experiments on periodic spin structures determining the magnetic order in a unit cell of the material,[7] diffraction from magnetic (multi)layer systems investigating the structure along the normal of the (two-dimensional, 2D) quasi-infinite layers (reflectometry),[8] and small-angle scattering from magnetic textures in thin (multi)layer systems which generate a lateral structure in the 2D plane. Here, we will only discuss small-angle scattering and related methods, probing lateral, "in-plane" magnetic structures. In this context, the term small angle is typically used to refer to (more or less) diffuse scattering from mesoscale objects (with a characteristic size from a few nm to μm) rather than diffraction from atomic or layer ordering (*i.e.*, few-Å to few-nm). However, the scattering angles are not always small. They can reach more than 10° in the soft-X-ray and the extended ultraviolet (XUV) range, which is where many important atomic resonances providing magnetic contrast are located and where many ultrashort-pulse sources operate.

Traditionally, small-angle X-ray scattering (SAXS) experiments are carried out incoherently, that is, the scattering pattern reflects only ensemble-averaged structural properties of the sample rather than the individual magnetization texture. Today, most modern X-ray sources provide beams with a high degree of coherence with typical coherence lengths of several micrometres, enabling advanced analysis of a scattering pattern that can contain additional fine structure due to interference. In this contribution, we will discuss coherent scattering, photon correlation techniques and, in particular, X-ray holography in their application to the study of magnetization dynamics.

Photon correlation techniques use the interference-based fine structure in the scattering pattern to infer spatiotemporal dynamics without access to the real-space structure of the scattering objects. In contrast, X-ray holography is a full-field imaging technique based on the inversion of a coherent scattering pattern. Holography is complementary to lens-based X-ray full-field microscopy, scanning microscopy and photo-electron microscopy, and it belongs to the established methods for X-ray imaging of nanometre-scale magnetic textures, exploiting magnetic dichroism for image contrast. While these imaging methods successfully access nanosecond and picosecond magnetization dynamics at SR sources, only lensless methods based on coherent scattering have been employed for imaging with femtosecond temporal resolution at ultrashort-pulse sources. Concerning time-resolved nanometre-scale magnetization imaging, we will confine this chapter to time-resolved resonant holography.

In the first part of this chapter, we will provide a methodical introduction to resonant small-angle X-ray scattering and holography to investigate lateral magnetic textures, particularly magnetic domains. In the second part, we will review the application of these methods to investigate magnetization dynamics. We will, therefore, discuss experiments where X-ray magnetic scattering and coherent imaging were exploited in a time-resolved fashion. In a separate section, we will then present the progress achieved so far with X-ray scattering in ultrafast magnetization dynamics, given the significant contribution of X-ray methods to this field.

7.2 Small-angle X-ray Scattering from Magnetic Samples and Related Methods

7.2.1 Resonant X-ray Scattering: Atomic Scattering Factors

Fundamentally, the sensitivity of X-rays to magnetization originates from strong magnetization-dependent terms in the atomic scattering factor for an X-ray photon (with an energy $E = \hbar\omega$) when tuned to suitable resonant electronic transitions. The expansion of the polarization-dependent resonant atomic scattering factor $f(\hbar\omega)$ into charge and magnetic scattering amplitudes is typically written for electric dipole transitions as:[1,9,10]

$$f(\hbar\omega) = \left(\hat{\boldsymbol{e}}_f^* \cdot \hat{\boldsymbol{e}}_i\right) f_c(\hbar\omega) - i\left(\hat{\boldsymbol{e}}_f^* \times \hat{\boldsymbol{e}}_i\right) \cdot \hat{\boldsymbol{m}} f_{m1}(\hbar\omega) + \left(\hat{\boldsymbol{e}}_f^* \cdot \hat{\boldsymbol{m}}\right)\left(\hat{\boldsymbol{e}}_i \cdot \hat{\boldsymbol{m}}\right) f_{m2}(\hbar\omega).$$

$$(7.1)$$

The polarization dependence is given by specific vector products of the complex-valued unit polarization vectors of the incident ($\hat{\boldsymbol{e}}_i$) and scattered ($\hat{\boldsymbol{e}}_f$) X-rays, as well as $\hat{\boldsymbol{m}}$, the unit vector along the magnetic moment of the scattering atom. The polarization vectors are readily expressed using Jones vectors. The asterisk symbol (*) indicates the complex conjugate. The charge $f_c(\hbar\omega)$ and magnetic $(f_{m1}(\hbar\omega), f_{m2}(\hbar\omega))$ atomic scattering factors contain the sums of the transition probabilities of the atomic excitation and decay processes involved in the resonant scattering process. Each contribution to the scattering factor has a specific polarization dependence. The charge scattering factor contains both non-resonant and resonant contributions. In particular, for soft X-rays, the non-resonant atomic scattering factor can be approximated by the number of electrons in the atom, Z. The polarization state upon resonant charge scattering is preserved aside from the well-known reduction of the p-polarized component by a factor $\cos 2\theta$ compared to the s-component. The scattering angle 2θ is enclosed between $\hat{e}_i$ and $\hat{e}_f$ (*i.e.*, between the incident and the scattered beam; see also Figure 7.2 in the next section).

The first magnetic scattering term is linear in $\hat{m}$ and leads to a rotation of the polarization if the incident polarization is linear (or elliptical). For incident circularly polarized X-rays, this term results in a dichroism in X-ray absorption measurements, referred to as the X-ray magnetic circular dichroism (XMCD). On the other hand, the second term, which is quadratic in $\hat{m}$, manifests in a dichroism for linearly polarized X-rays, referred to as X-ray magnetic linear dichroism (XMLD). Here, we will focus on the XMCD part as this contribution typically dominates the magnetic response for ferromagnetic materials, including ferrimagnets with ferromagnetic sublattices or sublayers.

The polarization rotation upon magnetic scattering motivates, in principle, measurements based on polarization analysis, similar to what is predominantly utilized in the optical regime. In particular, this analysis would allow us to clearly distinguish between charge and magnetic scattering. Polarization analysis for soft X-rays, however, is rather inefficient,[11] and, so far, remains unexploited in scattering, let alone imaging experiments. We will here restrict the discussion to intensity-only measurements.

Naturally, the resonant scattering factors show a pronounced and characteristic dependence on the photon energy, as shown by the charge and first magnetic scattering factors for the exemplary case of cobalt across the L_3 and L_2 absorption edges (see Figure 7.1), measured on a Co/Pt multilayer.[13] The scattering factors are composed of a real part and imaginary part which are connected *via* the Kramers–Kronig relation (*e.g.*, $f_c = f_c' + if_c''$ and $f_{ml} = f_{ml}' + if_{ml}''$).

The strong magnetic terms at the L_3 and L_2 edge are characteristic of the magnetic 3d transition metals where the magnetic response arises from resonance with electronic transitions from the spin–orbit-split $2p_{\frac{3}{2},\frac{1}{2}}$ states into the spin-polarized, partly filled 3d states. Due to the clear separation of the transitions from the spin–orbit partners, the XMCD spectrum given by f_{ml}'' can be analyzed by the so-called sum rules to obtain the ratio between the spin and orbital moments.[1,14] Also, transitions from the 3d transition metals' $3p_{\frac{3}{2},\frac{1}{2}}$ levels provide magnetic sensitivity albeit with lower magnetic contrast

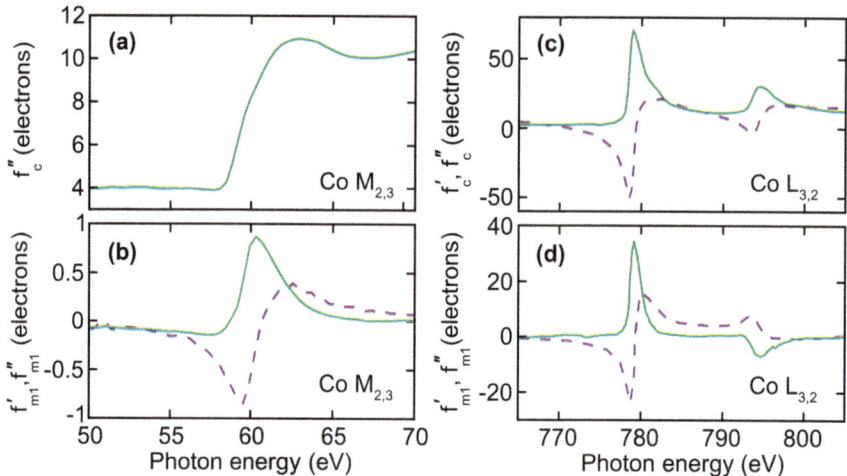

Figure 7.1 Resonant atomic scattering factors of Co at (a and b) the $M_{2,3}$ edge[12] and (c and d) the $L_{3,2}$ edge.[13] Solid lines show the imaginary part, and dashed lines show the real part.

(compared to the charge term) and necessitate a more complicated interpretation of the signal.[12] The smaller spin–orbit coupling, combined with the larger lifetime broadening, leads to spectral overlap of the M_3 and M_2 edges. Furthermore, spectral overlap is also observed for elements adjacent to one another in the periodic table, especially in the real part of the response. Nevertheless, the M edges of 3d transition metals have grown in popularity for time-resolved experiments during the last few years, as short-pulse sources such as XUV-FELs and HHG sources operate in this wavelength regime.

Another important class of magnetic elements is the magnetic rare-earth elements, in which dipole transitions into 4f states offer magnetic contrast. The experimental situation is similar to the transition metals: the M_4 and M_5 edges (3d → 4f) located in the soft-X-ray region provide the most significant contrast.[15] However, also in the case of magnetic rare-earth elements, the N_4 and N_5 edges (4d → 4f) in the XUV are currently gaining attention for applications at laboratory-based short-pulse sources.[16,17] Here, we will not further expand on spectroscopic aspects of magnetic dichroism, which is a broad and vibrant research field. For scattering and imaging applications, it is essential to note that magnetic sensitivity is achieved only when in resonance with particular electronic transitions from a core level, allowing for element-selective probing. For many elements of interest in magnetism, the resonances providing the most substantial magnetic contrast are located in the soft-X-ray range (∼400 eV–2 keV, 3–0.6 nm wavelength). Resonances in the XUV range (∼40–200 eV, 30–6 nm wavelength) offer further experimental opportunities with particular sources.

Finally, a note on the resonant charge scattering factor. Due to its sizable imaginary part (compared to the hard-X-ray range), the penetration depth of soft X-rays and XUV radiation in solids is very short, typically below 100 nm at absorption edges in this spectral range. As a result, transmission experiments are confined to thin films, and the information depth in reflection experiments is also limited to a narrow region close to the surface. In this chapter, we will thus confine ourselves to thin magnetic films and nanostructures as samples.

7.2.2 Magnetization-dependent Scattering

7.2.2.1 Kinematical Scattering

Here, we consider the basic X-ray scattering geometry, considering plane waves (see Figure 7.2). We discuss the frequent case of scattering in transmission with the X-ray beam incident along the normal of a thin film sample. Structural correlations are probed along the direction of the scattering vector $q = k_f - k_i$, where k_i and k_f denote the wavevectors of the incident and scattered waves, respectively. We consider elastic scattering where $k = |k_i| = |k_f| = 2\pi/\lambda$, and λ is the X-ray wavelength. Here, we focus on lateral

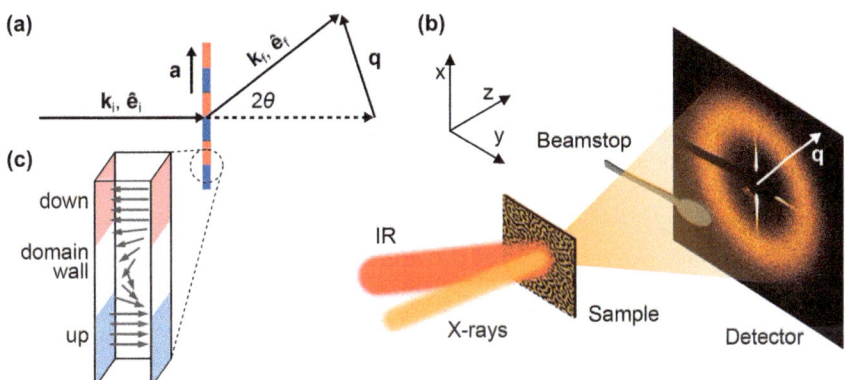

Figure 7.2 Small-angle X-ray scattering geometry. (a) The scattering vector q is given by the difference between the wave vectors of the incident (k_i) and scattered wave field (k_f): $q = k_f - k_i$. In transmission geometry, the scattering is predominantly reflecting in-plane spatial correlations. Textures with in-plane spatial frequency lead to scattering at $q = 2\pi/a$. (b) Typical experimental setup. A sample with labyrinth-like stripe domains is illuminated with an X-ray beam producing a ring-like scattering pattern on the detector. On a detector in the far field, each pixel corresponds to a particular momentum transfer q. The intense, directly transmitted beam is blanked out using a beamstop. Time-resolved pump–probe experiments are realized by adding a synchronized excitation of the sample, for example, an ultrashort IR laser pulse. Reproduced from ref. 18 with permission from Springer Nature, Copyright 2012. (c) Schematic of a typical domain structure of a PMA film. The film consists of domains oppositely magnetized in the out-of-plane direction. The domains are separated by domain walls where the magnetization gradually twists from one direction to the opposite.

magnetic correlations (*i.e.*, scattering vectors which are (approximately) in the plane of the sample), such as the ferromagnetic domains (see Figure 7.2b). The incident and scattered wave fields have the polarization $\hat{e}_i$, and $\hat{e}_f$, respectively. The scattering angle is denoted as 2θ, from which the magnitude of the scattering vector can be calculated as $q = 2k\sin(2\theta/2)$. As we will see below, the scattering at a specific momentum transfer q probes real-space correlations of charge-density variations or variations in the magnetic texture with in-plane wavevector $a = \dfrac{2\pi}{q}\dfrac{q}{q}$ (within the small-scattering-angle approximation).

The wavelength of XUV and soft-X-ray radiation is much larger than intra-atomic distances (*i.e.*, the core–shell radius). Diffraction corresponding to this length scale cannot be detected in the q-range that can be covered, and the q-dependence of the scattering cross-section will be very weak. In this long-wavelength limit, one may integrate the polarization factors in the resonant atomic scattering amplitude (with $|q| = 0$), resulting in the forward scattering amplitude.[1] For circularly polarized light, the resonant scattering amplitude reduces to:

$$f(\hbar\omega) = f_c(\hbar\omega) \mp \hat{e}_k \cdot \hat{m} f_{m1}(\hbar\omega). \tag{7.2}$$

The XMCD polarization dependence is given by $\mp \hat{e}_k \cdot \hat{m}$, where $\hat{e}_k = k_i/|k_i|$ denotes the X-ray propagation direction, and the different signs distinguish right or left circularly polarized X-rays. The contrast is, thus, maximized for a parallel alignment between the X-ray beam and the magnetization. The atomic absorption cross-section is directly related to the imaginary part of the scattering factor by the optical theorem, and is defined as[1]

$$\sigma_{abs} = 2r_e\lambda(f_c'' \mp \hat{e}_k \cdot \hat{m} f_{m1}''), \tag{7.3}$$

where r_e denotes the classical electron radius. The magnetization contribution to the absorption is inverted for opposite helicity of the circular polarization, giving rise to XMCD. In turn, linear polarization can be perceived as the superposition of right and left polarization. As a result, the magnetization contribution in the absorption cancels out precisely. We will see that, in contrast to absorption, this is not the case for the scattering channel, which is why scattering is sometimes employed to obtain XMCD-spectroscopy information using linearly polarized radiation. The atomic scattering cross-section given by $\sigma_{scatt} = \sigma_e|f|^2$ is orders of magnitude smaller than the absorption cross-section due to the small Thomson scattering cross-section of the electron, σ_e, and is typically neglected.[1]

So far, we have only considered atomic resonant absorption and scattering. In solids, significant scattering intensity originates from spatial variations of electron density and magnetic moments on larger length scales. Again, soft X-rays with wavelength $\lambda \gtrsim 1$ nm are insensitive to nearest neighbour atomic distances in solids. However, lateral spin textures or domains in ferromagnetic materials typically form on larger length scales. With the very smallest feature sizes characterized by the material-specific exchange length (typically on the order of >1 nm), it is common that objects of interest cover the range from 10 nm to the micrometre range. Therefore, only small scattering angles frequently have to be considered depending on the target material studied and the wavelength employed.

The scattering from a sample volume is calculated under the Born approximation, which neglects multiple scattering and constitutes the so-called kinematical scattering regime. The scattered intensity at scattering vector q is then provided by the superposition (*i.e.*, the summation) of the scattered waves of all atoms n in the coherence volume in the sample at positions r_n and with unit magnetization $\hat{m}^n$:

$$I_{\pm}(q) \propto \left| \sum_n f^n e^{iqr_n} \right|^2 \tag{7.4}$$

$$\propto \left| \sum_n (f_c^n \mp \hat{e}_k \cdot \hat{m}^n f_{m1}^n) e^{iqr_n} \right|^2. \tag{7.5}$$

If scattering is detected from a larger sample volume than the coherence volume, the coherent sum breaks down into several mutually incoherent parts. The sum in eqn (7.5) can be readily divided into charge and magnetic amplitudes given by

$$I_{\pm}(q) \propto |F_c(q) \pm F_m(q)|^2, \tag{7.6}$$

with $F_c(q)$ and $F_m(q)$ defined as

$$F_c(q) = \sum_n (f_c^{\prime,n} + f_c^{\prime\prime,n}) e^{iqr_n}, \tag{7.7}$$

$$F_m(q) = -\sum_n (f_{ml}^{\prime,n} + f_{ml}^{\prime\prime,n}) \hat{e}_k \cdot \hat{m}^n e^{iqr_n}. \tag{7.8}$$

Expanding eqn (7.6) results in

$$I_{\pm}(q) \propto F_c^* F_c \pm 2\Re[F_c^* F_m] + F_m^* F_m. \tag{7.9}$$

The scattered X-ray intensity comprises a purely charge-based and a purely magnetization-based contribution, as well as an interference term between charge and magnetic scattering. The mixed term vanishes for linear polarization, which can be perceived as a superposition between left and right circularly polarized photons,[10,19] reducing eqn (7.9) to

$$I_{\text{lin}}(q) \propto F_c^* F_c + F_m^* F_m. \tag{7.10}$$

In this case, the scattered intensity is composed of only the incoherent sum of charge and magnetic scattering. Thus, it is important to emphasize that XMCD leads to a magnetic response in the scattering channel for linearly polarized X-rays. Even more, this linear polarization is often preferred in experiments as (i) charge and magnetic scattering functions are separated, and (ii) many X-ray sources do not offer other polarization than linear or only with additional technical challenges.[11] Experimentally, the magnetic scattering contribution can mostly be isolated from charge scattering by following different approaches, such as:

- Subtracting a separately recorded pure charge signal if available, if no charge-magnetic mixing term is present.
- Reducing the charge scattering signal as much as possible by using topographically uniform, high-quality magnetic films and substrates, which do not generate charge scattering into the momentum transfer regions of interest.
- Tailoring the size and shape of topographic sample features (such as lithographic structures or apertures) so that charge and magnetic scattering are well separated in Fourier space. In this situation, the charge scattering on the detector may have to be blocked using suitably designed beamstops or by favourable positioning of the detector, as it may otherwise overwhelm a weaker magnetization signal on a detector with an insufficient dynamic range.[20]

Assuming that only a single chemical element contributes to magnetic scattering $I_{mag}(q)$ under resonant conditions, $I_{mag}(q)$ can further be simplified to

$$I_{mag}(q) \propto F_m^* F_m = \left(f_{ml}'^2 + f_{ml}''^2\right) \Big| \sum_n \hat{e}_k \cdot \hat{m}^n e^{iqr_n} \Big|^2 \qquad (7.11)$$

$$= \left(f_{ml}'^2 + f_{ml}''^2\right) S_m(q). \qquad (7.12)$$

We see that both the real and imaginary parts of the resonant atomic scattering factor cause scattering. The structure factor $S_m(q)$ only contains the Fourier transform of the magnetic structure projected onto the X-ray's propagation direction. This relation establishes a direct connection between the sample's magnetic texture and the scattered intensity. Measuring the time evolution $S_m(q,t)$ in Fourier space, therefore, provides direct information on temporal changes of the sample structure. This relation is the basis of photon correlation techniques, as discussed later in this chapter.

7.2.2.2 *Effective Medium Description*

So far, we have considered atomic positions for retrieving the magnetic structure factor while the probing wavelength and the magnetic structures probed have much larger length scales. Therefore, we will also introduce a complementary, effective medium description of the sample and the scattering process *via* wave propagation and the refractive index. In the X-ray range, the complex refractive index of a material is commonly written as[1,5]

$$n(\hbar\omega, \mathbf{r}, z) = 1 - \delta(\hbar\omega, \mathbf{r}, z) + i\beta(\hbar\omega, \mathbf{r}, z). \tag{7.13}$$

In a thin-film sample, the index of refraction may spatially vary laterally in the film plane ($\mathbf{r} = (x,y)$) and along the depth of the film (z). The optical constants δ and β are directly related to the real and imaginary part of the atomic scattering factor $f(\hbar\omega) = f'(\hbar\omega) + if''(\hbar\omega)$:

$$\delta(\hbar\omega) = \frac{r_e \lambda^2}{2\pi} N_a f'(\hbar\omega), \tag{7.14}$$

$$\beta(\hbar\omega) = \frac{r_e \lambda^2}{2\pi} N_a f''(\hbar\omega). \tag{7.15}$$

Here, N_a denotes the atomic density. For compound materials, the optical constants may be approximated from a weighted sum of the optical constants of the individual elements in the compound – this neglects fine structures at resonances involving transitions to valence states which are modified due to the formation of chemical bonds. Eqn (7.14) and (7.15) can be readily transferred to the resonant case, including magnetic scattering. We can split each optical constant into a charge and a magnetic contribution. Typically, the optical constants are then written as the sum of the charge-based, non-magnetic parts (δ_0 and β_0) and the magnetic corrections ($\Delta\delta$ and $\Delta\beta$). The refractive index in eqn (7.2) then reads for circularly polarized X-rays as

$$\begin{aligned} n(\hbar\omega, \mathbf{r}, z) = 1 - (\delta_0(\hbar\omega, \mathbf{r}, z) &\pm \hat{\mathbf{e}}_k \cdot \hat{\mathbf{m}}(\mathbf{r}, z) \Delta\delta(\hbar\omega, \mathbf{r}, z)) \\ + i(\beta_0(\hbar\omega, \mathbf{r}, z) &\pm \hat{\mathbf{e}}_k \cdot \hat{\mathbf{m}}(\mathbf{r}, z) \Delta\beta(\hbar\omega, \mathbf{r}, z)). \end{aligned} \tag{7.16}$$

We assume that the sample is illuminated with an X-ray beam $\Psi_0(\mathbf{r}, z) = |\Psi_0(\mathbf{r})| e^{i(kz - \omega t)}$ propagating in z direction ($\hat{\mathbf{e}}_k = \hat{\mathbf{e}}_z$) with frequency ω and wavenumber $k = 2\pi/\lambda$. In this framework, $\Psi_0(\mathbf{r}, -d)$ is typically referred to as the illumination function on the sample surface at position $z = -d$, where d denotes the sample thickness. The influence of the sample on the X-ray propagation inside the sample is called the transmission function $T(\mathbf{r})$. Using the projection approximation,[21] the exit wave $\Psi_{\text{exit}}(\mathbf{r})$ behind the sample is a product of $\Psi_0(\mathbf{r}, 0)$ and $T(\mathbf{r})$, given by

$$\Psi_{\text{exit}}(\mathbf{r}) = T(\mathbf{r})\Psi_0(\mathbf{r}, 0) \tag{7.17}$$

with

$$T(\mathbf{r}) = \exp\left[ik \int_{-d}^{0} (n(\mathbf{r}, z) - 1)dz\right], \tag{7.18}$$

where $n(r,z)$ is integrated over sample thickness d. The sample is, thus, represented by a two-dimensional z-projection. The projection can be divided into a charge-based and a magnetization-based part:

$$\int_{-d}^{0} (n(r,z)-1)\mathrm{d}z = \int_{-d}^{0} (\delta_0(r,z)-\mathrm{i}\beta_0(r,z))\mathrm{d}z \mp (\Delta\delta - \mathrm{i}\Delta\beta)\int_{-d}^{0} \hat{e}_k \cdot \hat{m}(r,z)\mathrm{d}z.$$

(7.19)

For the magnetic part, we again employ the assumption that only one atomic species contributes to the contrast under resonant conditions and, thus, the magnetic component of the optical constants is spatially invariant. The integration is then carried out over all atoms of this species across the sample thickness. The result is proportional to the z-component of the total magnetization of the thin film projected along z, which we call $m_z(r)$. This quasi-2D magnetization distribution is the texture probed by magnetic scattering and imaging in the small-angle approximation. We abbreviate the projection integrals with:

$$\mathrm{i}k \int_{-d}^{0} (n(r,z)-1)\mathrm{d}z = \mu_0(r) + \mathrm{i}\phi_0(r) \pm (\Delta\mu + \mathrm{i}\Delta\phi)m_z(r).$$

(7.20)

The exponent in eqn (7.18) is formed by a sum of charge (μ_0 and ϕ_0) and magnetic contributions ($\Delta\mu$ and $\Delta\phi$), and the transmission function can be decomposed into a charge (T_c) and a magnetic part (T_m). The exit wave from eqn (7.17) then reads:

$$\Psi_{\mathrm{exit}}(r) = T_c(r)T_m(r)\Psi_0(r,0).$$

(7.21)

The exit wave at $z=0$ propagates in free space to the detector located in the far-field region. Based on the Fraunhofer approximation, the scattered wavefront $\Psi_{\mathrm{det}}(x_{\mathrm{det}}, y_{\mathrm{det}})$ at the detector is represented by the Fourier transform of $\Psi_{\mathrm{exit}}(r)$. Assuming that the illumination function is a plane wave ($\Psi_0(r,0) = |\Psi_0|$), this is equivalent to a Fourier transform of the transmission function. We here omit the trivial amplitude scaling with the detector distance (z_0) and phase terms which vanish upon detection:

$$\Psi_{\mathrm{det}}(x_{\mathrm{det}}, y_{\mathrm{det}}) = \mathcal{F}[\Psi_{\mathrm{exit}}(r)](q) = |\Psi_0|\mathcal{F}[T(r)](q).$$

(7.22)

Here, the detector coordinates ($x_{\mathrm{det}}, y_{\mathrm{det}}$) are directly linked to the scattering vector as $q = (q_x, q_y) \approx (k/z_0)(x_{\mathrm{det}}, y_{\mathrm{det}})$. The scattering intensity is then given by the magnitude squared of the wave in the far field: $I(q) = |\Psi_{\mathrm{det}}(x_{\mathrm{det}}, y_{\mathrm{det}})|^2$. For thin films with weak X-ray interaction, the exponential function in eqn (7.18) can be Taylor expanded to linear order. As a result, the transmission function is expressed as a sum of charge and magnetic contribution (see eqn (7.19)) and will lead to equivalent findings for the scattered intensity as in eqn (7.9) and (7.10). While this approximation typically holds for homogeneous thin films in the absence of thickness variations as a function of **r**, it will become invalid once the topographic structure with associated absorption variation is superimposed on the magnetic structure. For example, using absorption masks in the sample geometry to effectively shape the illumination function (see Section 7.2.4) enables a coherent scattering pattern to be converted into a hologram.

7.2.3 X-ray Small-angle Scattering from Magnetic Domains

7.2.3.1 *Magnetic Multilayers with Perpendicular Anisotropy*

In the normal-incidence transmission geometry considered here, magnetic scattering based on XMCD contrast preferentially probes the out-of-plane component (m_z) of a thin film's magnetization. Contributions from in-plane components of the magnetization are negligible for sufficiently small scattering angles. Consequently, scattering-based methods have predominantly focused on thin films with perpendicular magnetic anisotropy (PMA), where the energetically preferred magnetic axis (easy axis) is perpendicular to the film. Many ferromagnetic PMA films intrinsically form magnetic domains of alternating polarization ("up" and "down"), divided by domain walls (see Figure 7.2c). The domains reduce the large stray-field energy originating from the film's out-of-plane magnetization, competing with locally increased energy due to magneto-crystalline anisotropy and exchange coupling.[22] Inside the domain walls, the magnetic moments continuously rotate from up to down orientation, leading to canted adjacent moments against the ferromagnetic exchange coupling. It is important to note that, in many cases, these domains form (i) in the absence of any topography and (ii) independently from the material's grain structure.[13] These materials thus exhibit textures of a purely magnetic nature and, therefore, provide a rich playground to probe dynamics that are also purely magnetic, be it local or collective.

The second key aspect of these materials is the flexibility in the design of their properties. PMA films are often produced as multilayer films, typically consisting of layers of a ferromagnet (*e.g.*, Fe, Co, and Ni) in alternation with a heavy metal (*e.g.*, Pt, Pd, Ta, and Au). At the interface between both materials, spin–orbit coupling leads to a magnetic anisotropy perpendicular to the interface.[23] If the magnetic layers are sufficiently thin, the PMA may dominate over the in-plane shape anisotropy, and the magnetic moments preferentially align in the out-of-plane direction. The freedom of choosing the multilayers' composition, structure, and deposition parameters offers a vast parameter space to tailor the properties of the films, such as coercivity, saturation magnetization, and domain size, to the application case. Due to these opportunities, multilayers with PMA attract fundamental research and are also considered for application in magnetic data-storage technologies.[24]

Even more possibilities arise from further variations in the multilayer stacking. The insertion of particular spacer layers provides control of the coupling between the magnetic layers (*e.g.*, the sign of the coupling), enabling artificial structures of ferromagnetically and antiferromagnetically aligned layers to be produced.[22] In recent years, multilayers with magnetic layers asymmetrically stacked between two different materials (*e.g.*, a heavy metal and a dielectric material) have gained much attention. In these multilayers, antisymmetric exchange at the layer interfaces (so-called Dzyaloshinskii–Moriya interaction, DMI)[25] leads to the formation of chiral and sometimes topologically non-trivial spin textures.[26,27] Spin-polarized currents and laser pulses can manipulate the topological magnetization textures, of which skyrmions are currently the most prominent examples, providing new routes for potential technological applications.[28–30]

Against this backdrop, it is unsurprising that almost all research examples presented in the second part of this chapter use magnetic domains in PMA multilayers as objects of investigation. The opposite magnetic orientation in adjacent domains provides a strong magnetic scattering contrast, and the typical dimensions of the domains are in the probing range of soft X-rays and XUV radiation. We will first introduce some fundamental properties of the scattering signal from static magnetic domain patterns as a basis for the understanding of the measurement of their dynamics.

7.2.3.2   *Static Scattering from Stripe Domains*

The domain pattern expressed by ferromagnetic PMA films is very characteristic for a specific system, comprising patchy patterns, stripes and bubbles. The domain width of stripe domain patterns is a direct consequence of the balance between the competing magnetic interactions.[22] The domain pattern may either be isotropic in the sample plane, exhibiting a labyrinth-like arrangement of the stripes, or – at the other extreme – be aligned into parallel stripes of almost uniform orientation (see Figure 7.3).

In both cases, the alternating variation of the magnetic scattering factor (due to the alternating sign of $\hat{e}_k \cdot \hat{m}(r)$) behaves like a diffraction grating and leads to pronounced intensity maxima in suitably resonant X-ray scattering. For labyrinth domains, the scattering maximum is azimuthally isotropic, leading to a ring-like scattering pattern (see Figure 7.2). Typically, only the first diffraction order stands out, and higher orders are weak or completely suppressed (see Figure 7.3).[13,31] In contrast, higher diffraction orders are typically detectable for aligned domain patterns. The even orders cancel when up and down domains have the same width, as in a fully demagnetized state with equal areas covered by up and down domains. However, even diffraction orders emerge outside this demagnetized state when the film has a non-zero net magnetization.[31] The diffraction of aligned domains is oriented perpendicular to the stripes.

When introducing magnetic scattering, we have, so far, assumed coherent X-ray radiation. In this case, we find a direct relationship between the magnetic real-space structure of a sample and its scattering pattern (see eqn (7.11)). If the sample is spatially homogeneous on a larger scale (*i.e.*, if it expresses the same mesoscale structural correlations everywhere), scattering

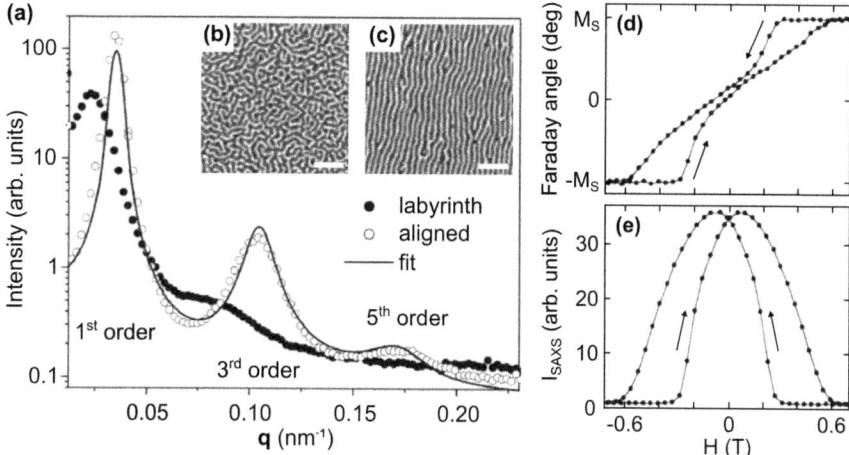

Figure 7.3 Scattering signal from magnetic labyrinth and stripe domains. (a) SAXS intensity as a function of scattering vector q for a [Co/Pt] multilayer sample exhibiting labyrinth (b) and aligned (c) stripe domains. The insets (b and c) show magnetic force microscopy images of the real-space domain structure. Scalebars, 1 μm. Reproduced from ref. 31 with permission from Elsevier, Copyright 2003. (d) Typical magnetic hysteresis for a similar [Co/Pt] sample as in (a–c), recorded with Faraday magnetometry. The characteristic shearing originates from the formation and growth of domains. (e) Corresponding scattering intensity I_{SAXS} during a field sweep in an applied field H. The scattering vector is close to the position of the intensity maximum (1st order). Reproduced from ref. 13 with permission from the American Physical Society, Copyright 2001.

with non-coherent X-rays also provides valuable spatial information. In this mode, one can consider the whole illuminated sample area to be partitioned into many coherent areas whose scattering patterns incoherently sum up on the detector. As a result, only ensemble-average information on the sample properties is available.

Two parameters extracted from the scattering pattern commonly attract the main interest – the scattering intensity, typically measured over the first correlation peak (see Figure 7.3), and the position of the intensity maximum of the first correlation peak. Supposing the magnetization of the domains (outside of the walls) is $+M_z$ or $-M_z$, the scattering intensity is proportional to M_z^2. In many time-resolved experiments, the scattering intensity directly measures the system's average local magnetization magnitude within a domain. In addition, the intensity scales linearly with the number of scatterers in the illuminated volume (*i.e.*, with the number of domains). The behaviour becomes particularly obvious when sweeping through the magnetic hysteresis of a domain-forming multilayer (see Figure 7.3d and e). While the magnetic scattering completely disappears in magnetic saturation, the scattering intensity reaches its maximum in the coercive field when the number of domains is maximized, but the net magnetization vanishes.

For a given spatial arrangement of magnetic domains, the scattering signal can, thus, be used as a measure of the square of the magnetization magnitude within the domains. We would like to stress that in such an experiment, the XMCD contrast mechanism (*i.e.*, the term with linear dependence on m in eqn (7.1)) can be exploited with linearly polarized X-rays. One way to rationalize this is the following line of thought: decomposing linearly polarized X-rays into their two circularly polarized components, one realizes that these two components see an object with inverted magnetic contrast at every point of the sample (see eqn (7.2)). Following Babinet's principle, both components generate the same scattering pattern, with both polarizations adding up incoherently. The difficulty of generating circularly polarized X-rays at elevated photon energies at many sources explains why a class of resonant SAXS experiments does not exploit the momentum transfer information. Instead, such experiments typically focus on the evolution of magnetization with time, as described in Section 7.4.3.

In the SAXS pattern, the position of the correlation peak's maximum (q_{max}) is commonly related to the average real-space periodicity of the domains (d_{dom}) as $d_{dom} = 2\pi/q_{max}$. In the demagnetized state, up and down domains have the same average widths given by $d_{dom}/2$. However, the typically sizable variation of domain widths in a sample significantly influences the shape of the correlation peak. In particular, it leads to an obvious broadening and the suppression of higher-order correlation peaks (particularly for labyrinth-like domains), and it also influences the position of the maximum intensity.[32]

Modelling the small-angle scattering of stripe domains that goes beyond these two parameters is hardly used. Hellwig *et al.* developed an analytical model for aligned stripe domains that can reproduce the experimental data very well (see Figure 7.3a).[31] The model is based on the diffraction of a plain grating and takes into account both the existence of domain walls of finite size and a Gaussian distribution of the domain widths. For labyrinth-like stripe domains, a phenomenological model based on gamma-distributed domain widths has shown convincing agreement with the experiment.[32] In particular, the model brings forth a certain discrepancy between the simplified estimate of the width of the domains through the position of the correlation peak and the actual average width of the underlying gamma distribution. However, both models rely on a one-dimensional representation of the domain pattern. The actual two-dimensional shape of the domain pattern can hardly be discerned from the incoherent diffraction pattern. For instance, it was recently shown that a short-range ordered lattice of round, bubble-like domains leads to a ring-like diffraction pattern very similar to that of labyrinth domains.[30] To determine the domain configuration with certainty, real-space imaging methods are needed.

7.2.4 X-ray Holography

7.2.4.1 *Geometry of Fourier-transform Holography*

The limitations of incoherent small-angle scattering (*i.e.*, the inability to distinguish different individual sample structures possessing the same ensemble average properties) can be overcome by imaging methods that take advantage of the direct relationship between the sample's real-space structure and its Fraunhofer diffraction pattern, mathematically provided by the Fourier transform. Holography belongs to these full-field imaging methods, which rely on the inversion of a sample's diffraction pattern to form its real-space image. While these methods do not require focusing optics for image formation (lensless imaging; see Chapters 5 and 6), coherent radiation is needed to retrieve the real-space structure from the diffraction pattern.

The fundamental problem when trying to invert a coherent scattering pattern to the real-space image of the structure generating the pattern is to find a solution to the so-called imaging phase problem. This problem arises because only the intensity of an X-ray wave field can be detected since the electric field oscillates too fast to be detected on a 2D detector.[33] As a result, the phase of the wave field is lost upon detection, making a direct inversion of the sample's Fraunhofer diffraction pattern to real space impossible.

Holography is a way to solve this phase problem. The basic idea of holography is to coherently mix the scattering signal from the sample with a known reference wave forming an interference pattern, called the hologram. This hologram encodes the relative phase information between sample scattering and the reference wave into detectable intensity variations. Holography is a two-step process: first, the hologram is recorded, and afterwards, it can be optically or digitally reconstructed to a real-space image of the sample. Modern holographic microscopy methods use digital reconstruction to avoid aberrations from the optical reconstruction system and allow for fully digital image processing and analysis.

Significant challenges for holography with X-rays arise from the optical properties of matter: compared to the visible spectral range, it is much more challenging to split, focus and steer X-ray beams. Given the small wavelength, interference fringes have a small spacing and are thus hard to detect due to the finite spatial resolution of 2D detectors in digital holography. Furthermore, even for narrow-bandwidth radiation, the absolute longitudinal coherence length is small, limiting tolerable path length differences (and hence also maximum scattering vectors) for which interference can be observed. For soft X-rays, absorption is relatively strong, and imaging in the transmission mode is restricted to thin samples (typically on the order of 100 nm). These limitations promote X-ray holography geometries with an in-line arrangement of all elements (optics, sample, and detector) along the optical axis, keeping deflection angles and geometrical path-length differences small. In the X-ray regime, two different holography geometries are established: (i) in-line holography with a small source point upstream of the largely transmissive sample providing both the sample illumination and reference wave, and (ii) Fourier-transform holography (FTH) where a point source located in the sample plane is the origin of a spherical reference wave. In-line holography is applied to biological samples, while FTH focuses on condensed-matter samples, including magnetic materials.[34] Therefore, we will here restrict our attention to the FTH geometry.

In contrast to the first experiment demonstrating sub-100 nm resolution imaging with X-rays,[36] absorption masks are now commonly used to realize the FTH geometry. The key idea of mask-based FTH with soft X-rays is to use an absorption mask with (at least) two apertures in the sample plane, splitting the incident radiation into two beams: (i) the beam illuminating the sample and (ii) the reference beam (see Figure 7.4). The reference aperture produces a spherical wave that is slightly off-axis from the scattered object wave. Notably, both apertures have to be close enough to guarantee mutually

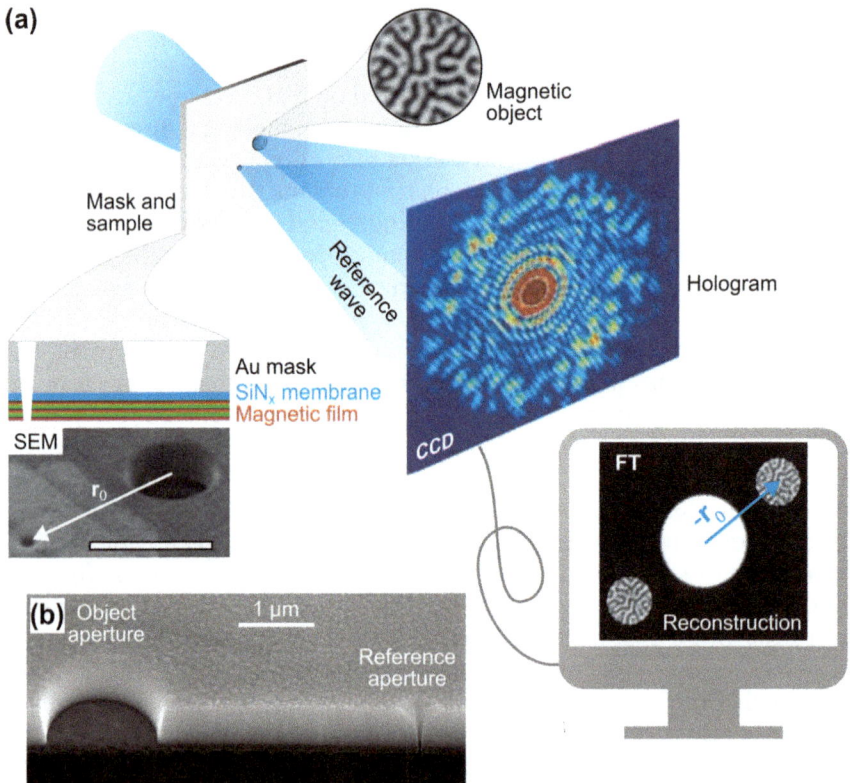

Figure 7.4 (a) Setup of mask-based Fourier-transform holography with soft X-rays. The sample is masked with two apertures splitting the beam into two parts, the sample scattering and the reference wave, which interfere at the detector. A real-space image of the sample is numerically retrieved by taking the Fourier transform of this hologram. Reproduced from ref. 35 with permission from Springer Nature, Copyright 2004. (b) Scanning electron microscopy (SEM) image of a cross-section of a combined sample and mask structure. The gold mask on top (light grey) contains the aperture for the field of view (left), which stops at the SiN_x support membrane, and the reference aperture (right), which is produced through the whole sample. For this particular sample, the thin magnetic film (dark grey) on the opposite side of the membrane is additionally patterned into an array of square islands.

coherent illumination. A transversal coherence length of several micro-metres is reached at most undulator beamlines of recent SR facilities and XFELs for soft X-rays. Furthermore, the small separation also keeps the angle between both beams very small, which leads to interference fringes wide enough to be detected by a pixelated detector.

Practically, the soft-X-ray FTH geometry has been very conveniently real-ized by combining the wavefront-dividing absorption mask and the actual sample to be imaged into a single unit supported by an X-ray transparent membrane, such as a SiN$_x$ membrane.[34,35] On one side of the membrane, the two apertures are produced into a metal film that is opaque to soft X-rays elsewhere (see Figure 7.4). This absorber film is typically made of a heavy element such as Au with a thickness on the order of 1 μm. The nano-patterning is most commonly realized with a focused ion beam (FIB), which allows the sample to be deposited on the opposite side of the substrate. In research on magnetism, the samples investigated include continuous magnetic thin films,[35,37–40] nanopatterned films,[41] and dispersed[42] or single objects.[43] These samples can be produced by placing an absorption mask on top of a sample (see Figure 7.4b). In this approach, the microscopic field of view is fixed to the pre-defined aperture in the metal film. Concepts of a separated absorption mask which is moveable with respect to the sample have also been demonstrated to overcome the restriction of a fixed field of view.[44] However, the monolithic approach primarily benefits from its in-herent stability and dispenses with the need for highly stable and precise positioning stages to manipulate either the mask or the sample.

7.2.4.2 Image Formation

To avoid confusion when discussing image formation, we will use the fol-lowing nomenclature when talking about samples that have been outfitted with an FTH mask (see Figure 7.4b): the entire structure is the "sample", and the area to be imaged is the "object", defined *via* the "object aperture" of the mask. The "reference aperture" is also part of the sample but separate from the object aperture.

To mathematically understand the image formation in FTH, we again describe the object with eqn (7.21), where we decomposed the total trans-mission of the object into a charge (T_c) and magnetic (T_m) part when reson-antly illuminated with circularly polarized X-rays. The charge term particularly includes the diffraction from the object aperture in the absorber mask, defining the field of view. The charge term, thus, equals zero outside the aperture. Note that the reference aperture will be added later. In addition, all charge contributions from the object itself enter T_c. The magnetic part is only provided by the object and is considered to be small, allowing for a Taylor expansion of eqn (7.18) up to linear order using the abbreviations defined in eqn (7.20):

$$T_m(r) = 1 \pm \Delta T_m(r) \tag{7.23}$$

with

$$\Delta T_{\mathrm{m}}(r) = (\Delta\mu + i\Delta\phi)m_z(r). \tag{7.24}$$

Transmission-mode holography is again sensitive to the out-of-plane component of magnetization, which makes the method particularly suitable for PMA magnetic films. For normal incidence and in the limit of scattering angles approaching zero, it is exclusively sensitive to the object's perpendicular component of magnetization. Nevertheless, also in-plane moments were already imaged in a tilted configuration.[45,46]

In addition to the spatially structured transmission of the object, we have to add the source of the reference wave to the transmission function. In order to produce a spherical reference wave, this source is approximated by a delta function with a (complex) transmission t_0 to account for the finite size of this source. The reference source is laterally offset from the object and located at a position r_0 with respect to the object. Commonly, the reference aperture is realized as a "through hole" (*i.e.*, it is not covered by the membrane and magnetic material. See Figure 7.4b). The total transmission of the object and reference is then given by $T(r) = T_c(r)(1 \pm \Delta T_{\mathrm{m}}(r)) + t_0\delta(r - r_0)$. The hologram (*i.e.*, the coherent scattering from the masked sample on the detector) is again readily obtained as the magnitude squared of the Fourier transform of the transmission function (again, we omit all trivial intensity scaling terms):

$$I_{\mathrm{H}}(q) = |\mathcal{F}[T_c(r)(1 \pm \Delta T_{\mathrm{m}}(r)) + t_0\delta(r - r_0)]|^2 \tag{7.25}$$

$$\begin{aligned} &= |\mathcal{F}[T_c(r)(1 \pm \Delta T_{\mathrm{m}}(r))]|^2 \\ &\quad + \mathcal{F}^*[t_0\delta(r - r_0)]\mathcal{F}[T_c(r)(1 \pm \Delta T_{\mathrm{m}}(r))] \\ &\quad + \mathcal{F}[t_0\delta(r - r_0)]\mathcal{F}^*[T_c(r)(1 \pm \Delta T_{\mathrm{m}}(r))] + |\mathcal{F}[t_0\delta(r - r_0)]|^2. \end{aligned} \tag{7.26}$$

The real-space image of the object is numerically reconstructed simply by taking the inverse Fourier transform of the hologram, $\mathcal{F}^{-1}[I_{\mathrm{H}}(q)]$. The inverse Fourier transform of a diffraction pattern is typically referred to as its Patterson map. Each term in eqn (7.26) forms a product that turns into a convolution after the inverse Fourier transform. The first and the last term yield the auto-correlations of the object and the reference, respectively, and appear in the origin of the Patterson map. However, the second and third terms transform into cross-correlations between the object and reference of the mathematical form:

$$\begin{aligned} p_\pm(r) &= t_0^*\delta(-r - r_0) * T_c(r)(1 \pm \Delta T_{\mathrm{m}}(r)) \\ &= t_0^* T_c(-r - r_0)(1 \pm \Delta T_{\mathrm{m}}(-r - r_0)) \end{aligned} \tag{7.27}$$

where the * operator indicates the convolution. In this way, the object's transmission function is reproduced in the Patterson map around the position $-r_0$, forming a direct image of the object. The scalar factor t_0^* only leads to a trivial scaling of the amplitude and, possibly, to a constant phase offset. In other words, the object's position, magnitude, and phase are imaged relative to the reference. The third term in eqn (7.26) produces a redundant but complex conjugated image of the object at r_0 and is called the twin image.

As the images are formed by a convolution between the reference aperture and the object, the spatial resolution of the image is directly related to the spatial extent of the reference aperture, which is a limiting factor. Therefore, this aperture has to be produced as a tiny nanometre-sized hole to reach the desired resolution. Today's FIB technology allows routine manufacturing of apertures with sub-50 nm diameter into metal films of ≥ 1 μm thickness.[47] Unfortunately, smaller references also lead to smaller intensity of the reference beam (reflected by t_0) and, thus, to a smaller signal-to-noise ratio of the image. This dilemma is overcome by more sophisticated schemes for producing the reference wave, such as multiple openings,[48] potentially arranged as uniformly redundant arrays[49,50] or other shapes such as slits[51,52] and Fresnel zone plates.[53] Furthermore, a resolution limitation due to the reference aperture size can be mitigated based on an image reconstruction considering the imaging transfer function rather than a single inverse Fourier transform.[54] Alternatively, the resolution can be increased by combining FTH with coherent diffraction imaging[55] (see Chapter 6), where the numerical phase retrieval based on iterative algorithms exploits (ideally trivial) *a priori* information to invert the coherent diffraction pattern.[50,56–59] In any event, a separate limiting factor for the spatial resolution is the maximum momentum transfer up to which the coherent diffraction pattern constituting the hologram is recorded.

The magnetic contrast in the image is provided by $\Delta T_m(r)$ in eqn (7.27), where the different signs again correspond to the helicity of the circular polarization. The interference between the reference wave and this magnetic part is only present for circularly polarized X-rays and will vanish with linear polarization in the perpendicular incidence and small-angle scattering geometry considered here. This is because the circular polarization is retained both for the object wave after interaction with the magnetic material as well as for the reference wave diffracted at an open reference aperture not containing magnetic material (see eqn (7.1)). Therefore, interference of object and reference waves can occur.[19] The magnetic contrast of the object $\Delta T_m(r)$ can (partly) be isolated from a typically dominant charge contrast $T_c(r)$ by recording holograms with opposite helicity of the circularly polarized X-rays. The difference between both reconstructions removes the charge offset from the image:

$$\frac{1}{2}\left(p_+(r) - p_-(r)\right) = t_0^* T_c(-r - r_0)\Delta T_m(-r - r_0). \tag{7.28}$$

However, T_c remains as a multiplicative factor in the image. First, T_c defines the field of view. Second, T_c naturally reduces the magnitude of the image reconstruction and constantly adds a phase offset with respect to the X-ray transmission through homogeneous sample layers of the object as substrate or magnetic layers. If T_c is inhomogeneous across the field of view, then the image of the magnetic texture will be influenced by the charge contribution. However, in practice, topographic variations are typically small and not a limiting factor for applying the method in nanomagnetism.

7.2.4.3 Application of FTH

After the development of masked-based Fourier-transform holography (FTH) by Eisebitt *et al.*[35] in 2004, the method quickly evolved into a reliable imaging tool, particularly for thin magnetic film samples. The FTH concept has two main advantages distinguishing the method from other established soft-X-ray imaging methods. First, mask-based FTH is inherently insensitive against spatial sample drifts due to its monolithic design of the FTH mask and object. Even for a drift of the whole sample by several tens of micrometres during the hologram exposure, FTH can still provide nanometre-scale resolution. Second, FTH provides a lot of free space around the sample due to the lensless imaging principle – no other image-forming optics, such as Fresnel zone plates, are required in the vicinity of the sample. Together, these two properties make it easy to create custom sample environments, such as high magnetic fields (>1 T)[37,43,60] and temperature control,[61,62] realizing excitations, for example, with magnetic field,[63] current[29,64] or laser pulses[30,65] as well as performing temporally[57,63,66] or spectrally[59] multi-dimensional measurements with inherent overlap between consecutive images. Today, masked-based FTH is routinely and productively used at synchrotron-radiation sources, where highly coherent soft-X-ray radiation with circular polarization is available at many undulator beamlines. The spectrum of applications ranges from investigating magnetic domains in PMA films,[37–40,61] over magnetic nanopatterns[41,42,60] and topological textures as magnetic skyrmions[29,30,63,65] to even meteoroid material.[43] Also, research outside magnetism was addressed, such as imaging biological specimens[67–69] and phase separation during the insulator–metal phase transition in a correlated material.[59,62] Applications of FTH at ultra-short pulse sources in the sub-picosecond regime and, in particular, time-resolved measurements will be discussed in the next section.

7.3   Applications to Magnetization Dynamics

Magnetization dynamics span many orders of magnitude in timescales, even when only considering the mesoscopic length scale (*i.e.*, micrometres to nanometres). For example, the stability of data written on a magnetic hard drive is considered on a timescale of years, while laser-excited magnetization dynamics happens on a femtosecond timescale.

Magnetization dynamics below the nanosecond regime, often considered "slow", is typically described in terms of fluctuations and noise. Typical examples include superparamagnetism, magnetic domain switching, Barkhausen noise during magnetization reversal, or fluctuations of domain-wall positions. Magnetic systems can often be considered quasi-static on this timescale as the magnetization always appears in equilibrium, even when resting in a metastable state (*e.g.*, due to an external field). The elementary dynamic processes happen on a much faster timescale than the dynamics observable on the mesoscale. Still, these thermally activated processes have enormous technological relevance as they determine, *e.g.*, the stability of magnetically stored data and the error rate of magnetic writing. Coherent X-ray methods can potentially study such "slow" mesoscale fluctuation dynamics and have already contributed to this field. We will review experiments in the first part of this section.

On shorter timescales and correspondingly shorter length scales, magnetization dynamics is described using a micromagnetic approach incorporating short-range exchange interaction, anisotropies, and long-range magnetostatic effects. Micromagnetics is a continuum theory based on the fundamental constraint that $|M(r,t)| = M_s$ (*i.e.*, that the magnitude of the local magnetization vector always remains equal to the constant saturation magnetization M_s), leaving two degrees of freedom – the rotation angles. Thereby, the local magnetization $M(r,t)$ is treated as a quasi-continuous macrospin representing many underlying atomic spins. As a result, micromagnetics, represented by the phenomenological Landau–Lifshitz–Gilbert (LLG) equation, describes the collective motion of magnetic moments under an applied magnetic field and its own magnetic anisotropy field. There is a wealth of these low-energy collective phenomena successfully described by micromagnetics, and many methods exist which allow the detection of such dynamics. Here, we can only refer to textbooks and reviews for a comprehensive introduction.[70–73] Synchrotron-radiation-based pump–probe methods are well established for imaging such collective dynamics down to the sub-nanosecond timescale with high spatial resolution. In the second part of this section, we will present work based on resonant X-ray holography as well as efforts to study fluctuating states with coherent scattering.

On ultrafast timescales of picoseconds and femtoseconds, the focus is often placed on dynamics after an impulsive excitation, transiently modifying the size and direction of magnetic moments and, in effect, possibly permanently changing the lateral spin texture. Here, the macrospin concept has to be replaced by an atomistic description of the spin dynamics. The actual processes occurring are a subject of research. They depend both on the nature of the excitation and the system under investigation, and include, for example, spin-flips mediated by scattering events in the presence of nonequilibrium electronic distributions and spin-dependent transport and redistribution of electrons on the timescale of the presence of electric and magnetic fields. The latter explicitly includes high-frequency field transients which can be present in the vicinity of a laser pulse.

It is often useful to divide the system into the three subsystems of "electrons", "spins", and "lattices", acting as reservoirs for energy and angular momentum. The coupling within and between the subsystems, allowing for the exchange of energy and angular momentum, is then used to explain typical relaxation times of dynamics observed. Such (spin-dependent) transport phenomena have to be considered, particularly in the presence of heterogeneity of the specimen on length scales matching inelastic mean free path lengths for transport, for example, of hot electrons. With relevant length scales ranging from interatomic distances into the nanometre regime, X-ray scattering methods provide the sensitivity and spatial resolution to particularly address these phenomena. Examples of processes on the few-nanometre length scale are discussed in Section 7.4.

7.3.1 Quasi-static and Real-time Experiments

7.3.1.1 Quasi-static and Real-time Imaging

Studying the dynamics of a magnetic system requires suitable time-resolved detection. The most straightforward approach is to observe the system's evolution in real-time by taking snapshots at a certain temporal rate. To represent in-plane dynamics in a quasi-2D thin film system, a 2D detector is ideal for mapping the real-space or reciprocal-space evolution of the system unless the processes under study are *a priori* known to possess high symmetry within the sample plane. Given the currently commercially available 2D detector technology, this real-time detection is restricted to relatively slow processes, typically on a scale of hundreds of milliseconds to seconds when exploiting soft-X-ray scattering or imaging to probe the magnetic system. For hard X-rays, image sampling rates exceeding 1 MHz have been demonstrated for projection imaging of a non-magnetic system.[74] We note that for the detection of soft X-rays with less energy per photon available for detection as compared to hard X-rays, CCD detectors with high quantum efficiency but slow readout are commonly used. A push towards faster soft X-ray detectors based on CMOS technology or integrating individual readout electronics within every sensor pixel is ongoing, but only a few systems are currently available.[75–78]

For processes driven by an external parameter such as an applied field or temperature, experiments can be carried out quasi-statically by incrementally driving the parameter and imaging the system between the steps. This mode of operation was extensively used to image magnetic reversal and hysteresis in nanomagnetic materials with FTH.[37,39,41–43,60,61,79]

Projection imaging through detecting a magnetic Bragg peak in specular reflection geometry was recently used to image antiferromagnetic domains. Here, phase contrast across a domain wall was exploited to image antiphase domains in a collinear antiferromagnet.[80] With a spatial resolution of a few micrometres and sub-second temporal resolution, thermal fluctuations of antiferromagnetic domain walls could be followed in real space with this approach.[81]

7.3.1.2 Correlation Methods Based on Coherent X-ray Scattering

Direct imaging of magnetization is, so far, unsuitable for studying typical magnetic fluctuations on the nanometre scale in thermal equilibrium, despite the possibility of steering fluctuation rates by changing the temperature of the sample. X-ray imaging at sub-50 nm resolution (*via* any method) requires signal averaging typically over many seconds. However, detecting these fluctuations provides a direct way to probe the underlying microscopic magnetic interactions. Incoherent X-ray scattering allows for faster data acquisition but offers access to only average sample properties integrated over the illuminated sample area. In particular, equilibrium dynamics remain hidden under incoherent X-ray illumination as ensemble average properties stay constant in thermal equilibrium. In contrast, coherent scattering is sensitive to spatial and temporal fluctuations of magnetization (and, of course, likewise, the electron density) even if ensemble-averaged properties do not change as a consequence of these fluctuations. This sensitivity is due to the direct relation between the real-space structure and its coherent far-field diffraction pattern, mathematically provided by the Fourier transform (eqn (7.22)). In particular, scattering from a disordered sample leads to distinct and sharp variations of

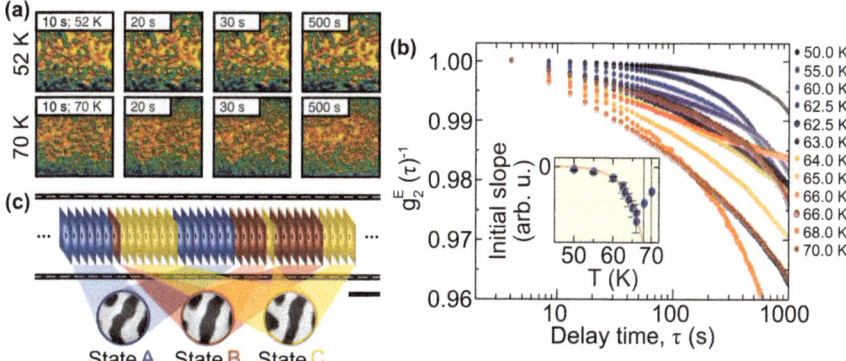

Figure 7.5 Correlation-based X-ray scattering and imaging techniques. (a and b) XPCS measurement of antiferromagnetic domain fluctuations in a holmium layer. (a) Speckle patterns and their temporal evolution at two different temperatures. The domain dynamics lead to continuous local intensity changes in the speckle pattern. (b) Corresponding correlation functions show a faster decay of correlations with increasing temperature (in particular *via* the initial slope). Reproduced from ref. 87 with permission from the American Physical Society, Copyright 2011. (c) The principal idea of the time-resolved coherent correlation imaging method. Consecutively recorded coherent scattering patterns (frames) are temporally assigned to specific states *via* correlation and clustering techniques. Selectively averaging over frames of the same state results in a high-resolution reconstruction of a real-space image of the state—here, fluctuating ferromagnetic domains in a multilayer. Reproduced from ref. 89, https://doi.org/10.1038/s41586-022-05537-9, under the terms of the CC BY 4.0 license http://creativecommons.org/licenses/by/4.0/.

the intensity, known as speckles (see Figure 7.5a). If the arrangement of scattering centres (*e.g.*, the magnetic texture) changes with time, the corresponding speckle pattern will also change. Measuring intensity fluctuations of the speckles using a correlation analysis allows the detection of the underlying dynamics in the sample, a method referred to as X-ray photon correlation spectroscopy (XPCS). Temporal correlations are typically quantified with the help of normalized intensity (*i.e.*, second order) autocorrelation function $g^{(2)}(\boldsymbol{q},t)$ that compares pairs of measurements with the time lag t between both measurements:[33,82–84]

$$g^{(2)}(\boldsymbol{q},t) = \frac{\langle I(\boldsymbol{q},t')\rangle\langle I(\boldsymbol{q},t'+t)\rangle_T}{\langle I(\boldsymbol{q},t')\rangle_T^2}. \tag{7.29}$$

The brackets denote averaging over all times t' during the whole measurement time T. Examples of such temporal autocorrelation functions (see Figure 7.5b) show that the intensity autocorrelation provides direct access to the time-dependent analogue of the structure factor $S(\boldsymbol{q})$ through the intermediate scattering function $f(\boldsymbol{q},t) = S(\boldsymbol{q},t)/S(\boldsymbol{q},0)$ as given by

$$g^{(2)}(\boldsymbol{q},t) = 1 + \beta_{\mathrm{C}}|f(\boldsymbol{q},t)|^2, \tag{7.30}$$

where β_{C} denotes the speckle contrast, a measure of the degree of coherence. The intermediate scattering function contains information about collective dynamics, and its decay in time is typically modelled in an exponential form:

$$f(\boldsymbol{q},t) = \exp\left[-\left(\frac{t}{\tau(\boldsymbol{q})}\right)^{\alpha(\boldsymbol{q})}\right]. \tag{7.31}$$

Here, $\tau(\boldsymbol{q})$ is typically referred to as the correlation time or relaxation time, and $\alpha(\boldsymbol{q})$ is an exponent that stretches or compresses the exponential decay. For example, $\alpha = 1$ in the case of classical diffusion dynamics (Brownian motion), $\alpha > 1$ in super-diffusive regimes, and $\alpha < 1$ for many examples of glassy dynamics. Replacing the time average in eqn (7.29) with averaging over equivalent $\boldsymbol{q}$-regions in the scattering patterns allows calculation of the two-time correlation of the general form $\langle I(\boldsymbol{q},t_1)I(\boldsymbol{q},t_2)\rangle$ and, therefore, even to detect non-equilibrium dynamics.[83] We refer to review articles for a more detailed introduction to these photon correlation methods.[82–84]

The first study employing these correlation methods in magnetism research was a quasi-static experiment searching for memory effects in reversing a ferromagnetic film forming domains.[85] In contrast to imaging experiments, this technique makes it possible to gain sensitivity over a significantly larger field of view of the sample and, thus, average the correlation function over a larger number of speckles. The first time-resolved XPCS study on a magnetic material was published by Shpyrko *et al.* in 2007, providing the detection of antiferromagnetic domain fluctuations in chromium at low temperatures.[86] These domains form spin-density waves with different orientations along the crystallographic axes. As the spin-density waves are directly coupled to a charge-density wave, they can be detected through a charge contrast without resonant magnetic scattering.

In 2011, the first XPCS experiment exploiting direct resonant magnetic contrast was published.[87] This experiment probed the dynamics of anti-ferromagnetic domains in a holmium layer. It has to be noted that the experiment was carried out in reflection geometry using speckles appearing in a superstructure peak emerging from the helical order of the holmium film (see Figure 7.5). As can be seen from the speckle patterns, the speed of the domain-wall dynamics changes with temperature. At 52 K, the pattern is almost static, while it continuously changes at 70 K. This is also reflected by the steeper decay of the correlation function at higher temperatures (see Figure 7.5b). Recently, fluctuations of domain walls were also investigated for an artificial antiferromagnet based on a spin-ice system.[88]

The photon correlation technique introduced so far operates in Fourier space, allowing us to analyse detector frames with significantly lower photon counts than imaging data and, therefore, operate at higher repetition rates. Recently, a method called coherent correlation imaging (CCI) was developed that combines this major advantage of correlation methods with coherent imaging. The method provides direct insight into real-space dynamics in an untriggered "movie mode" at a temporal resolution below the exposure time required to acquire a single image. The technique relies on a classification of 2D scattering patterns recorded in a time series to identify and "time stamp" states of an evolving system (see Figure 7.5c for an illustration of this process). Once classified, sufficient statistics can be accumulated to form a real-space image without compromising the temporal resolution of the individual time stamps. The method was demonstrated by investigating the stochastic reorganization of domain walls in a ferromagnetic material in thermal equilibrium dictated by the landscape of pinning potentials magnetic stray fields.[89]

7.3.2 Nanosecond Collective Dynamics

Pump–probe measurements are employed to investigate magnetization dynamics on timescales faster than those possible with real-time detection. Here, a pump stimulus triggers the dynamics, and a probe pulse measures the system's response after a certain pump–probe time delay. To record a complete time series, the process must be identically repeated multiple times while varying the pump–probe delay. Most experiments build up statistics for each time delay by integrating over many acquisition shots to achieve the required signal-to-noise ratio. The repeatability, excitation frequency, and minimum time resolution vary significantly depending on the X-ray source employed. Synchrotron-radiation (SR) sources operate with repetition rates on the order of 100 MHz, and the pump–probe cycle is repeated millions or billions of times. The pulse duration of the SR X-ray bunches limits the time resolution to typically 30 ps to 100 ps. The magnetization dynamics is typically triggered electronically by radio-frequency currents or current pulses through micrometre-scale striplines, generating magnetic fields or spin torques. In contrast, experiments using XFELs or HHG sources commonly reach a temporal resolution of $\leq$100 fs in all-optical experiments, where a femtosecond laser pulse triggers the ultrafast dynamics. We extensively discuss these experiments in Section 7.4.

Nanosecond collective magnetization dynamics is intensely investigated by classical X-ray imaging methods at SR sources, such as scanning transmission X-ray microscopy (STXM),[71] full-field transmission microscopy,[90] and photoemission electron microscopy (PEEM).[91] In particular, a number of STXM endstations have specialized in time-resolved magnetization imaging based on an asynchronous excitation scheme allowing for an efficient recording of many time frames independently of the filling mode of the storage ring. The magnetism research addressed with these X-ray microscopes comprises magnetic domain-wall dynamics, the dynamics of topological textures, in particular, vortices and skyrmions, spin-torque oscillators, and spin-wave dynamics. A review of this body of work is beyond the scope of this chapter. Even three-dimensional imaging of a vortex gyration was demonstrated recently.[92] On the other hand, laser excitation replacing electric current pulses, enabling much shorter impulsive excitation, was first realized by PEEM-based time-resolved imaging, opening X-ray imaging at SR sources to the field of photo-excited dynamics.[93,94]

7.3.2.1   Time-resolved X-ray Holography Experiments

So far, scattering-based methods have played only a minor role in this research field, most likely because dedicated permanent SR endstations providing fully developed setups for time-resolved measurements are not available. Therefore, it is unsurprising that only two time-resolved X-ray holography studies were published by expert groups for the method. In the first pioneering work, the gyration of a magnetic skyrmion was imaged after it had been displaced from its equilibrium position inside a magnetic nanodisk by a magnetic field pulse (see Figure 7.6a–c).[63] The study made experimental progress in two ways: (i) it is the first experiment where repeatable nanoscale collective magnetization dynamics were observed in a magnetic film with perpendicular anisotropy. These films typically show a complex magnetic anisotropy landscape due to the sensitivity of the anisotropy to structural properties such as local crystal orientation, interface roughness, stress, and layer thicknesses.[95] As a result, most time-resolved magnetic X-ray imaging work was and is carried out with in-plane magnetized films. (ii) The study demonstrated the tracking of the skyrmion with 3 nm precision (see error bars in Figure 7.6c), which is again possible due to the inherent insensitivity to the drift of mask-based X-ray FTH. Based on this accuracy, the authors could model the trajectory of the skyrmion's gyrotropic motion, taking inertia into account. The corresponding inertial mass was found to be associated with a breathing mode of the skyrmion.

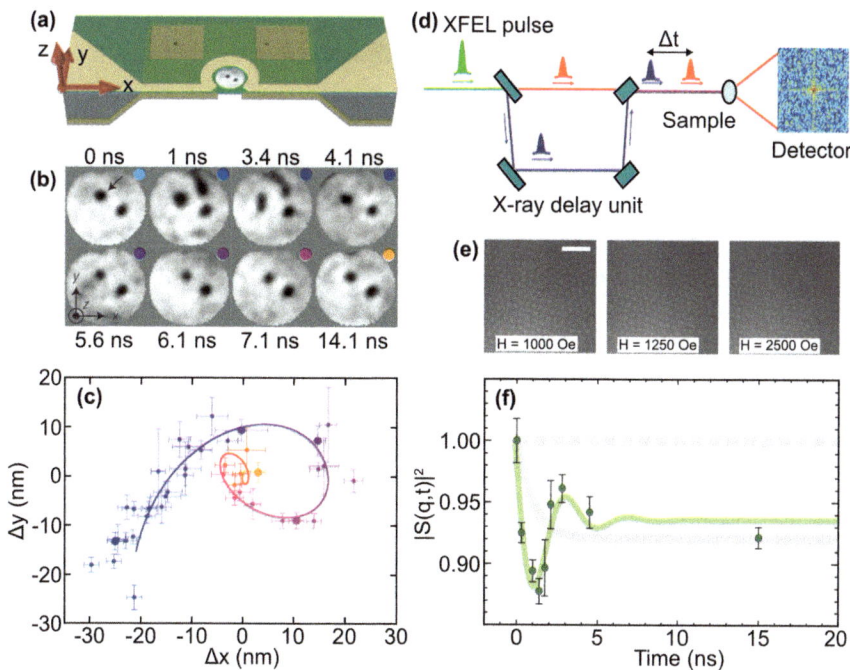

Figure 7.6 Nanosecond collective magnetization dynamics investigated with X-ray holography and XPCS. (a–c) Picosecond time-resolved imaging of the gyration dynamics of a skyrmion inside a magnetic disk of 550 nm diameter. (a) Sample layout showing a microcoil stripline surrounding the disk to create a transient magnetic field from an electrical current pulse, triggering the dynamics. (b) Holographically retrieved images of the magnetization dynamics. The skyrmion of interest and the time steps are indicated. (c) The trajectory of the skyrmion gyration tracked with 3 nm precision (data points). A fit (line) to the data is applied to better illustrate the skyrmion gyration, assuming a finite topological mass of the skyrmion. Reproduced from ref. 63 with permission from Springer Nature, Copyright 2015. (d–f) XFEL-based XPCS measurement of spontaneous fluctuations of a skyrmion lattice. (d) The split-pulse concept for ultrafast XPCS measurements. Reproduced from ref. 97 with permission from the Optical Society, Copyright 2009. (e) Skyrmion lattice observed in GdFe using transmission X-ray microscopy at an applied field as indicated. Scale bar, 1 μm. Reproduced from ref. 98 with permission from the American Physical Society, Copyright 2017. (f) Correlation function (here given as the square of the intermediate scattering function $S(\boldsymbol{q},t)$) of the spontaneous fluctuations of a skyrmion lattice in GdFe at 210 mT due to thermal motion. Reproduced from ref. 99, https://doi.org/10.1103/PhysRevResearch.3.033249, under the terms of the CC BY 4.0 license https://creativecommons.org/licenses/by/4.0/.

A second time-resolved X-ray holography experiment in the nanosecond regime studied the dynamics of a vortex core that forms in a micrometre-sized magnetic square.[96] The experiment was exceptional in that the source for the holographic reference wave was produced as a thin slit. This so-called HERALDO (holography with extended reference autocorrelation by linear differential operator) geometry allows tilting of the sample with respect to the incoming X-ray beam. In this way, images with sensitivity to in-plane magnetic moments can be recorded in addition to pure out-of-plane magnetic contrast.

With the advent of diffraction-limited storage rings, which deliver almost fully transversally coherent soft X-rays, it can be expected that the relevance and usage of coherent-scattering-based imaging methods will also increase for time-resolved measurements. Supported by state-of-the-art numerical image reconstruction algorithms, coherent-scattering measurements promise a significant improvement in the spatial resolution breaking the 10 nm limit under routine operation. Regarding time-resolved investigations, it must be noted, however, that the increased transversal coherence comes at the expense of the electron bunch length and pulse duration, which is typically significantly above 100 ps in diffraction-limited storage rings, limiting the achievable temporal resolution.

7.3.2.2   Ultrafast Speckle Correlations

Pump–probe experiments rely on repetitive dynamics and are, thus, unsuitable for studying stochastic magnetic fluctuations. While correlation-based X-ray scattering methods are ideally suited to detect fluctuations, soft-X-ray area detectors have not reached nanosecond temporal resolution, yet. However, Gutt et al. have developed an approach that circumvents this dilemma, at least at XFEL sources.[97] The idea is based on splitting the probing X-ray pulse into two pulses and introducing a variable time delay between both pulses (see Figure 7.6d). The fluctuating sample is probed with both pulses, and both coherent scattering patterns are recorded within the same acquisition cycle. The information on the local (i.e., beyond ensemble-averaged properties) sample structure is again encoded in its speckle pattern formed on the detector. If this structure changes within the delay time between both pulses, then the associated speckle pattern also changes. The speckle patterns from two subsequent pulses are incoherently added to the detector if the overlapping patterns are recorded in a single exposure. A temporal change of the system is encoded as a function of momentum transfer through the reduction in speckle contrast. The temporal correlation function characterizing the system's dynamics is then retrieved from the delay dependence of the speckle contrast.

This approach was already employed to detect spontaneous fluctuations in a skyrmion lattice emerging in a Fe/Gd multilayer.[99,100] In these materials, a densely packed hexagonal lattice of skyrmions (see Figure 7.6e) emerges in a pocket of the magnetic-field-temperature phase diagram of lateral spin textures.[98] The nanosecond correlation measurements reveal an inherent, collective mode of underdamped oscillation within this lattice (see the correlation function in Figure 7.6f). Moreover, a slowing down of the fluctuations in the skyrmion lattice is observed when pushing the system through a field-driven, continuous phase transition, ultimately reaching a uniformly magnetized state.

The instrumental demands of this correlation method are very high as, so far, it can only be applied at sources delivering sufficient intensity to determine correlations from two superimposed single shots. Currently, this is only feasible at XFEL sources. Moreover, pulse pairs with variable delay must be provided by the accelerator or optically with a beamsplitter device. However, the technique provides unique opportunities through fluctuation measurements on the nanometre length scale and nanosecond to femtosecond timescale to directly study dynamics ranging from low-energy fluctuations in equilibrium to highly excited non-equilibrium states.

7.4   Ultrafast Magnetization Dynamics

The discovery by Beaurepaire *et al.* in 1996 that laser pulses can efficiently manipulate magnetic order on a femtosecond timescale[101] opened a new research field in modern magnetism.[102,103] The experimental findings in this field range from femtosecond demagnetization over laser-induced switching of the magnetization to the optical manipulation of extended spin textures. Understanding the fundamental mechanisms of relevance on different lengths and timescales is an ongoing endeavour. At the same time, the ultrafast optical manipulation of magnetization opens perspectives for future applications in spintronics, data storage, and processing. Since the availability of sources for ultrashort X-ray pulses, X-ray methods have significantly contributed to the understanding of the underlying interactions between spins, charges, and lattice, and the momentum flow between them.[104] In the soft X-ray regime, the capability to investigate such interactions is often facilitated by the sensitivity to elemental spin and orbital moments.[1,105,106] In this respect, the strength of resonant scattering and imaging methods is to add a spatially resolved view of the ultrafast processes, such as highlighting the effects of (spin-dependent) transport phenomena or the formation of nanoscale to mesoscale textures. This section will review the scientific achievements in ultrafast magnetism using X-ray scattering and coherent imaging methods. We decided to divide the experiments into two classes. The first class comprises studies on ultrafast demagnetization where X-ray scattering is predominantly employed as a sensitive spectroscopic probe of (element specific) magnetization, while spatial information in real or reciprocal space is not exploited. In contrast, the second class of experiments takes advantage of spatial information in studying ultrafast spin dynamics.

Regarding the latter case, we confine the following discussion to dynamics appearing in the lateral dimension of thin film systems. We do note, however, that resonant X-ray scattering is a powerful approach to temporally trace the magnetic order that is also perpendicular to the sample in thin multilayer films.[107,108] In this field, laser-based laboratory experiments have just started complementing experiments at large-scale X-ray facilities in investigating sub-nanosecond dynamics.[16,109]

7.4.1 Pioneering Experiments at XFELs

Since 2005, free-electron lasers have delivered radiation in the XUV to X-ray range with unique properties.[110–112] The sources provide ultrashort X-ray pulses of $\sim$100 fs down to 1 fs pulse duration with microjoule to millijoule pulse energies, leading to extreme peak pulse intensities (see Chapter 8). The radiation has a high degree of transverse coherence, and its wavelength is freely tunable. Polarization control at XUV and soft X-ray undulators is becoming increasingly available.

The continuous repetition rate at XFELs with normal conducting acceleration cavities is typically 10–120 Hz. At superconducting accelerators, MHz bursts are currently available, with machines offering MHz repetition rates in a "continuous-wave" mode going into user operation. Today's XFELs provide sufficient photon flux (10^{12} photons per pulse and more) to achieve appreciable signal-to-noise ratios in single-shot X-ray diffraction or imaging measurements. However, such an intense probe pulse severely alters or destroys the sample. The magnetic properties can be permanently modified even without the X-ray pulse reaching the ablation threshold.[113–115] In order to record a series of different time delays in such a single-shot mode, one has to perform repetitive measurements with an attenuated or defocused beam, or a set of identical samples is needed, each following the same evolution after the pump, for example, by laser excitation.[116] Naturally, such identical sample conditions are difficult to fulfil for samples with statistical defects or heterogeneity.

Holography and coherent scattering methods particularly benefit from the high coherent flux provided by XFEL sources. For example, the first single-shot FTH images were recorded when the first XUV-FEL, FLASH[110] in Hamburg, Germany, became available for user operation, albeit without magnetic contrast.[50,117,118] First demonstrations of resonant magnetic scattering from a static sample with FEL radiation in the XUV[113] and soft-X-ray regime[119] were subsequently followed by first time-resolved pump–probe magnetic scattering studies (see Section 7.4.4.).[18,120] The first single-shot hologram, imaging the magnetization of ferromagnetic domains with the corresponding real-space images (see Figure 7.7a), was recorded at LCLS, Stanford, USA,[114] using thin-foil magnetic polarizers based on dichroic absorption to provide circularly polarized X-rays.[121] While this experiment on a static domain pattern was performed with soft X-rays tuned to the Co L_3 absorption edge, the first femtosecond time-resolved FTH experiment was then conducted at the seeded XUV-FEL FERMI[112] in Trieste, Italy, in resonance with absorption at the Co $M_{2,3}$ edge.[66] In this experiment, circular polarization was provided using an APPLE-II-type undulator, and a time series of four images were recorded (see Figure 7.7b and Section 7.4.4 for more details).

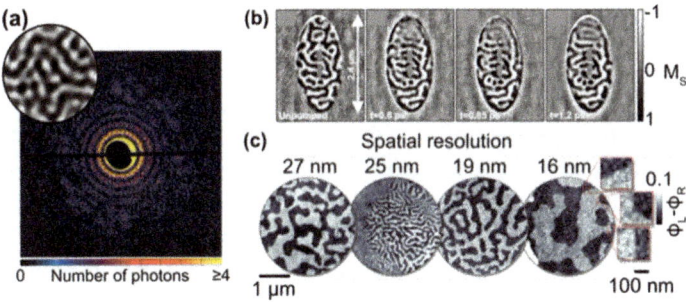

Figure 7.7 Pioneering experiments of imaging magnetic texture using ultrashort X-ray pulses. All experiments have imaged magnetic domains in Co-based multilayers. (a) First femtosecond single-shot hologram recorded at an XFEL. The inset displays the reconstruction of an image. Reproduced from ref. 114 with permission from the American Physical Society, Copyright 2012. (b) First femtosecond time-resolved resonant X-ray imaging with nanometre-scale resolution again based on X-ray holography. The sequence visualizes local demagnetization induced by an optical excitation structured by near-field diffraction. Reproduced from ref. 66 with permission from the American Physical Society, Copyright 2014. (c) X-ray holography with sub-wavelength resolution obtained using an HHG source. The reconstruction was significantly improved by numerical phase retrieval, finally achieving a single-pixel (diffraction-limited) resolution as demonstrated in the insets. Reproduced from ref. 57, https://doi.org/10.1038/s41467-021-26594-0, under the terms of the CC BY 4.0 license http://creativecommons.org/licenses/by/4.0/.

FERMI was also the first FEL facility to deliver pulses with two different photon energies originating as harmonics from the same seed wavelength.[122] This mode of operation enabled two atomic species across the sample to be resonantly studied simultaneously. FTH inherently disperses different wavelengths in image reconstruction.[117] If the wavelength separation between two discrete harmonics is sufficiently large, then it is possible to record an FTH image with two photon energies simultaneously. Willems *et al.* and Weder *et al.* demonstrated this possibility by imaging magnetic domains in a Co/Pt multilayer simultaneously at the Co $M_{2,3}$ edge and the Pt N_7 edge through static measurements.[123,124]

7.4.2 Pioneering Experiments with HHG Radiation

In recent years, laboratory-based, laser-driven HHG instruments have been developed as an alternative source to XFELs for coherent XUV radiation[125] (see Chapter 6) and soft X-rays[126,127] with femtosecond or even sub-femtosecond pulse duration.[128] The intensity of a single HHG pulse is orders of magnitude lower than that of XFEL sources, rendering single-shot HHG imaging of magnetic textures difficult, even though proof-of-principle single-shot results of non-magnetic high-contrast objects have been published based on extended references[129] or coherent diffractive imaging.[130,131] Still, the average HHG photon flux (10^{10}–10^6 photons per eV per s at the source strongly depending on the wavelength) is sufficient to conduct time-resolved pump–probe experiments based on magnetic scattering or imaging. The repetition rate in such experiments is typically on the order of a few kHz, and many thousands of pulses are required to form a scattering pattern with sufficient signal-to-noise. Time-resolved magnetic scattering was first demonstrated by Vodungbo *et al.* by monochromatizing broadband HHG radiation in the XUV to the Co $M_{2,3}$ resonance (21 nm wavelength) using multilayer mirrors.[132] Only recently, pump–probe magnetic scattering at a much smaller wavelength of 8 nm corresponding to the Tb N-edge was observed.[17]

All these experiments, however, directly exploit the linear polarization of the XUV radiation produced by the HHG source when driven by linearly polarized laser pulses. In order to provide circularly polarized radiation needed for XMCD-contrast-based imaging (rather than scattering), two approaches were applied: first, a multiple-reflection-based phase shifter for the XUV radiation,[133,134] and second, the direct generation of circularly polarized harmonics by bi-chromatic counter-rotating circularly polarized driver laser pulses, as theoretically predicted in 1995[135] and implemented through various approaches. Implementation *via* the so-called MAZEL-TOV method[136] was finally used by Kfir *et al.* to record the first FTH image of a ferromagnetic domain pattern in a Co/Pd multilayer at an HHG source.[56] In their setup, a toroidal grating monochromatized the HHG radiation to the Co XUV resonance. The same group also presented the first time-resolved FTH imaging results using an HHG source, tracing laser-induced changes in a ferromagnetic domain pattern with 40 fs temporal resolution.[57] Combining FTH with numerical phase retrieval methods, the group achieved a sub-wavelength resolution of 16 nm in static images (see Figure 7.7c).

7.4.3   Magnetic Scattering as a Spectroscopic Probe

Spectroscopic X-ray studies on magnetic materials are predominantly carried out in an XMCD[14,106,134] or MOKE geometry,[137] that is, the intensity directly transmitted through or reflected from the magnetized material is detected. Typically, this intensity signal of the specular transmitted or reflected beam with zero momentum transfer (apart from the reflection) is then recorded as a function of photon energy, applied magnetic field, and time delay after a pump in the case of time-resolved experiments in order to extract the magnetization as a function of time. In particular, with broadband HHG sources, the possibility to efficiently combine time resolution with spectroscopic information makes these geometries very attractive.[16,138,139]

For thin film samples with perpendicular magnetic anisotropy (PMA) forming magnetic domains with a well-defined in-plane correlation length, detecting the magnetically scattered radiation at a suitable momentum transfer provides an alternative, ideally background-free detection channel.

As discussed earlier, the magnetic-domain pattern can be perceived as a diffraction grating whose resonant magnetic scattering contrast is directly related to the magnetization of the "grating bars" (see Figure 7.3). In particular, for uniform films, it is straightforward to produce samples where all charge variations (*e.g.*, from grains, substrate roughness, or membrane frames) are either small enough or geometrically well-defined such that the associated charge scattering is negligible or well-separated in reciprocal space from the magnetic information in the scattering pattern. The intensity of this scattering signal can then be used to trace the magnitude of m_z^2 within the domains in time-resolved experiments. Often, the momentum-transfer information is not used, and hence we refer to these experiments as "spectroscopic" in nature. This approach has several practical advantages: (i) the magnetization can be probed with linearly polarized X-rays; (ii) an external magnetic field does not have to be applied as the sample returns to a domain state in remanence, even after demagnetization; and (iii) at X-ray sources with large shot-to-shot intensity fluctuations such as unseeded XFELs, these fluctuations can be corrected for through a suitably tailored static charge scattering signal that can be detected simultaneously with the magnetic scattering.[115,140,141]

In a prototypical study, magnetic scattering from domains in a Co/Pd multilayer was used to address whether ultrafast laser-triggered demagnetization can proceed without the laser pulse being absorbed in the material that is demagnetized.[142] Following previous work,[143] the magnetic film was covered with an optically opaque aluminium (Al) layer. Incident femtosecond IR pulses absorbed in the Al layer still led to an ultrafast demagnetization of the adjacent magnetic film, albeit with a delayed and slower response compared to a control experiment with direct IR excitation of the magnetic film. These observations can only be explained by considering an energy transport mechanism mediated through hot electrons first excited in the Al layer which then travel into the magnetic film. Excited electron distribution in the magnetic film is, thus, sufficient to trigger ultrafast demagnetization.

In experiments with intense femtosecond X-ray pulses, one can expect to operate beyond the non-perturbative regime due to interaction with the X-ray (probe) pulse alone. In the first experiments on resonant magnetic scattering and imaging with intense XFEL pulses without any other stimulus, a reduction of the X-ray scattering signal with increasing X-ray peak fluence on the sample was observed both in the XUV[144] and the soft-X-ray range.[114] X-ray pulses with fluences of the order 1 J cm^{-2} almost entirely suppress the magnetic signal.[144,145] While solid samples are structurally damaged at such X-ray fluences in the soft X-ray regime, diffract-before-destroy concepts with ultrashort X-ray pulses, as initially discussed in the context of structural analysis of biomolecules (see Chapter 1),[146,147] were also discussed for the investigation of nanostructures with magnetic order. Understanding the breakdown of the magnetic scattering signal is thus of fundamental interest and relevance for the preparation and interpretation of resonant scattering experiments at FELs. Three main lines of interpretation have been suggested so far: first, an X-ray-induced ultrafast demagnetization of the material during the interaction with the X-ray pulse,[114,115,148] second, loss of XMCD contrast due to transient changes in the electronic structure,[144,149] and, third, the onset of stimulated emission of X-rays reducing the signal from the $q \neq 0$ scattering channel.[145,147,150,151] In particular, for the latter coherent process, the lifetime of the core hole state compared to the X-ray pulse length is a critical parameter to dominate over the spontaneous emission. Along this line, experiments have strongly indicated that non-coherent effects (demagnetization and electronic damage) prevail for scattering in resonance with core excitations at the $M_{2,3}$ edges of the 3d transition metals (*e.g.*, the lifetime of Co 3p core holes is 0.26 fs) at least for the FEL pulse duration (70 fs, FWHM) used so far.[115] In contrast, coherent amplification of the forward scattered radiation at the expense of the scattered X-rays was proposed to be relevant for the scattering of shorter pulses (down to 2.5 fs single-spike pulses) in resonance with the longer-lived core excitations at the $L_{3,2}$ edges (*e.g.*, the lifetime of Co 2p core holes is 1.4 fs),[144,150,151] in addition to electronic damage.[148] Regardless of the relative importance of the mechanisms debated, it has become evident that the resonant magnetic scattering signal deviates from being strictly proportional to m_z^2 at fluences of the ultrashort X-ray pulses above 10 mJ cm^{-2}.

7.4.4 Ultrafast Lateral Spin Dynamics and Spin Transport

As outlined in the previous section, magnetic systems with PMA forming mesoscopic labyrinth-like or aligned stripe domains in remanence have been the subject of intense research in ultrafast magnetism. This research was initially inspired by studies on layered magnetic heterostructures which demonstrated that spin currents, travelling along the sample normal between different adjacent layers in the heterostructure, ignited by the ultrafast excitation of a ferromagnet play a vital role during ultrafast demagnetization. On the one hand, spin currents are induced by ultrafast demagnetization;[152] on the other hand, they can also be responsible for the demagnetization[142,143,153] or, at least, speed it up.[154] In layered heterostructures, magnetic heterogeneity is naturally coupled to chemical and structural heterogeneity, as the different layers consist of different materials. Spin currents, thus, have to transverse an interface. In contrast, lateral transport through domain walls within the plane of a (possibly also multilayered) material offers the possibility to study spin transport between differently magnetized domains in chemically homogeneous media. Resonant small-angle X-ray scattering can provide sensitivity to a possible spatial rearrangement of spin texture within the plane of a thin-film-based sample.

The first time-resolved experiments on Co-based multilayers already found indications for spin-transport also in the lateral dimension.[18,132] In particular, an ultrafast structural modification of the spin texture was detected proceeding on the same timescale as the ultrafast demagnetization through a shift of the magnetic correlation peak in the SAXS from the magnetic domains.[18] The proposed explanation of this modification is related to different mean free-path lengths and lifetimes of hot minority and majority electrons in a ferromagnet,[153,155] resulting in a transient spin accumulation at the domain walls with an associated domain-wall broadening. The latter is detected through a shift of the correlation peak in the SAXS with XMCD contrast. While this picture was also qualitatively supported by the first time-resolved direct imaging experiments,[66] recent investigations with a higher temporal and spatial resolution (40 nm) could not detect such an effect on similar samples[57] at an incident IR pump fluence about a factor of six lower than that in ref. 18 and 66. In scattering experiments claiming an even higher spatial sensitivity, the presence of lateral spin diffusion over distances larger than 3 nm was excluded.[20] Here, a spatial sensitivity of about $\lambda/7$ using $\lambda = 21$ nm radiation was achieved by monitoring the transient changes in diffraction intensity at an (initially) symmetry-forbidden diffraction order from transient magnetic gratings of different periodicity.

Progressing research on spin dynamics in a landscape of ferromagnetic domains suggests that the transient shift of the correlation peak in their scattering pattern depends on the actual sample used and the IR excitation fluence. Results range from no shift in the correlation peak on a sub-ps timescale[17,18] to shifts with picosecond delay after the excitation.[141,156] Correspondingly, alternative explanations for the observed shift were developed, in particular, ultrafast domain dilation resulting from hot electron–magnon scattering[156] and ultrafast modification of the uniaxial anisotropy also leading to a transient domain-wall broadening.[141] Recent experiments have even indicated that the presence of the shift depends on the actual alignment of the domains in a labyrinth-like or aligned pattern.[157] To date, this observation challenges all theories proposed so far, and further research in this direction is needed to learn more about the ultrafast response of lateral spin textures to optical excitation.

Lateral spin currents have also been identified to play a role during all-optical switching (AOS) of magnetization in a ferrimagnetic alloy. In AOS, the magnetization is reversed by an ultrashort laser pulse, without the assistance of an external magnetic field. While the initial observations of helicity-dependent AOS were limited to specific rare-earth–transition-metal (RE–TM) alloys,[158] in 2014, it was discovered that all-optical control of magnetic order is also feasible in a broader class of materials, including RE–TM multilayers, heterostructures, synthetic ferrimagnets,[159] and ferromagnetic multilayers (Co/Pt) or granular films (FePt).[160] Moreover, all-optical control of magnetic order is also possible with linearly polarized light.[161] In the case of RE–TM materials, XMCD measurements revealed that a transient ferromagnetic state mediates the switching,[105] which occurs due to different demagnetization time constants of the two ferrimagnetic sublattices.

In the work of Graves *et al.*, however, time-resolved resonant scattering from a GdFeCo film suggests an additional, non-local channel for demagnetization (see Figure 7.8). Based on resonant X-ray charge scattering

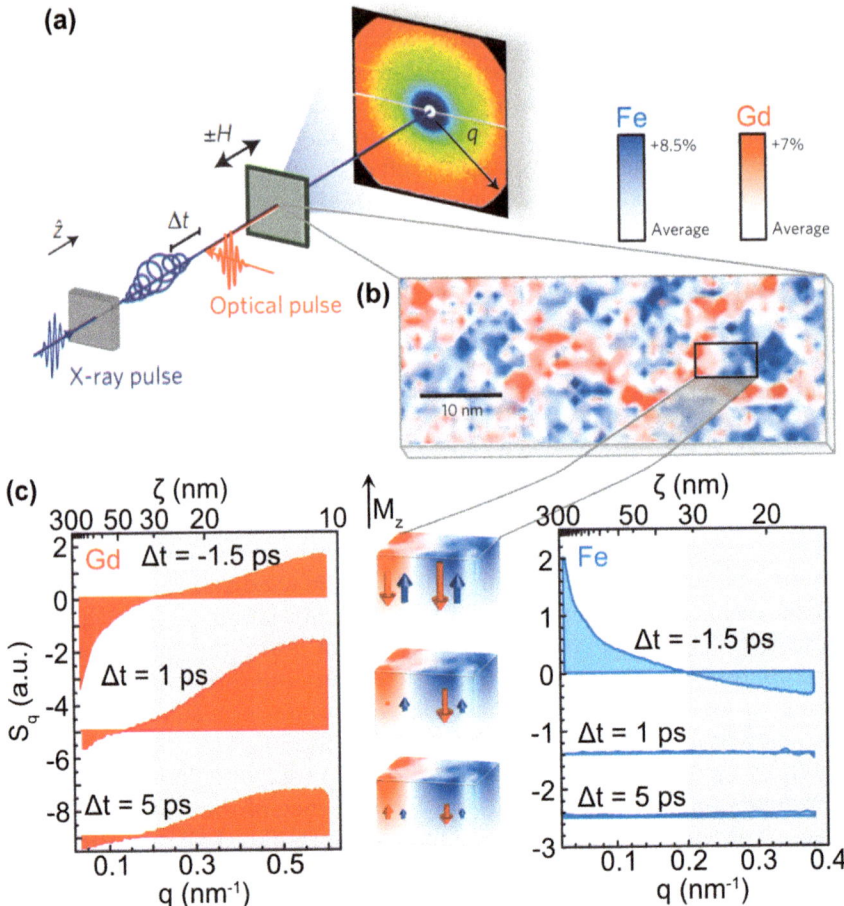

Figure 7.8 Time-resolved SAXS experiment to follow all-optical magnetic switching in ferrimagnetic GdFeCo. (a) Schematic of the experimental setup of the pump–probe measurements at an XFEL. (b) Nanometre-scale variation of the composition in the magnetic film. The image was recorded in a separate measurement by scanning transmission electron microscopy with element-sensitive energy-dispersive X-ray spectroscopy mapping the nanoscale composition: Gd-rich (red) and Fe-rich (blue) areas. (c) Resonant SAXS with sensitivity to the magnetic moments of Gd atoms (left, red) and Fe atoms (right, blue). Delay times Δt are indicated. Both sublattices demagnetize on average (low-q information). However, the magnetization dynamics in the two sublattices are remarkably different at high momentum transfer q, corresponding to small spatial distances ξ (grey background shading). This difference is explained in the schematics in the centre showing local variation in the spin distribution within the Gd- and Fe-rich regions. Spins flow from Fe-rich to Gd-rich areas, leading to a strong Gd magnetization contrast between Gd- and Fe-rich areas and, thus, strong scattering at high q in the Gd channel. The spin flow promotes magnetic switching in the Gd-rich areas. Adapted from ref. 120 with permission from Springer Nature, Copyright 2013.

(see Figure 7.8a) and electron microscopy (scanning transmission electron microscopy, STEM, combined with energy-dispersive X-ray spectroscopy, EDX), the authors identified lateral heterogeneity of the alloy's composition, leading to nanometre-scale Gd-rich and Fe-rich regions (see Figure 7.8b). In turn, transient modifications of the magnetic X-ray scattering indicated spin transport from Fe-rich regions to initially oppositely magnetized Gd-rich regions within the first picosecond after the optical excitation (see Figure 7.8c). This spin flow reverses the magnetization in the Gd-rich areas, significantly promoting the switching. This study impressively demonstrates how the combination of spectral, femtosecond temporal, and nanometre spatial information *via* pump–probe resonant magnetic scattering sheds light on the role of nanometre-scale heterogeneity in ultrafast magnetic phenomena.

Finally, it has recently been demonstrated that it is possible to achieve sensitivity to the chirality of spin textures in the sample in dichroic, resonant X-ray scattering experiments with a fixed angle of incidence and 2D detection. The chirality of the system, such as the sense of rotation of magnetic moments in a domain wall, becomes accessible through observation of the dichroism in the scattering pattern for scattering with opposite circular polarization in suitable geometries.[162] A first pump–probe experiment at the FERMI FEL investigated, in particular, the remagnetization dynamics of a magnetic system where right-handed Nèel-type domain walls were energetically preferred due to the presence of Dzyaloshinskii–Moriya interaction and found the dynamics of the recovery of magnetic order to be chirality-dependent.[163,164]

7.4.5 Remagnetization Dynamics After Optical Excitation

The experiments on ultrafast magnetization dynamics discussed so far focused on the magnetic system's response at early times (femtoseconds to few picoseconds) after the photo-excitation, including demagnetization, magnetization switching, and hot-electron (spin) currents. Recently, a growing interest emerged in the subsequent remagnetization phase on the picosecond to nanosecond timescale. Here, pattern formation on the nanometre scale extending into the mesoscale triggered by optical pump pulses has been studied. This is a complex field, as on these longer spatial and temporal scales, effects beyond a unit cell of the material or beyond the volume defined by the inelastic mean free path of hot electrons are involved. Long-range interactions such as magnetic dipole interaction and longer-wavelength collective excitations, as well as macroscopic properties such as heat conductivity, play an increasing role in a particular sample geometry. The goal of many investigations is to understand and control the formation of particular patterns of magnetic domains. On the one hand, photo-induced changes may be restricted to local modifications of the magnetic texture while retaining ensemble-averaged properties such as domain width and orientation. On the other hand, the excitation may even lead to global changes in preferred domain orientation, periodicity or topology, and a change in the total magnetization of a given sample area in a switching event.

In the case when a shallow energy barrier separates different states of a system, one has to be aware that an X-ray probe pulse of high fluence, such as that used for single-shot resonant scattering and imaging experiments at XFELs, may induce modifications. We would like to discriminate between the changes in the magnetization pattern from structural modifications and the magnetic properties of thin (multi)layer-based magnetic materials from a change in the spatial magnetization pattern in an otherwise unaltered material. In the former case, damage occurs at higher (peak) power deposited in the material and has, for example, been witnessed through modifications associated with changes in the in-plane correlation length of magnetic domains due to a permanent change in the magnetic anisotropy in the area irradiated with single XUV pulses.[113]

Purely magnetic and, hence, non-permanent changes, induced by single pulses of soft X-rays, could be seen in the first nanoscale images of magnetic structures recorded with a single femtosecond pulse.[114] In later experiments, operating in a regime where the soft X-ray pulse is not changing but solely probing the sample, local reordering of a ferromagnetic domain pattern on the sub-μm scale due to an IR pump pulse has been witnessed, again exploiting FTH for image formation.[66,123]

The first experiment explicitly studying spatial aspects of remagnetization after IR-triggered ultrafast demagnetization extending into the nanosecond regime was carried out by Bergeard *et al.* at LCLS in 2015.[165] In this experiment, the photo-excitation led to an almost complete magnetization loss of domains prepared in well-aligned stripes. The magnetization recovers within 5 ns; however, the domain pattern has transformed into a labyrinth state, which has a slightly lower magnetostatic energy than a pattern of parallel stripe domains, with a small remaining orientation anisotropy, as witnessed from the resonant SAXS signal. The small remaining alignment is at first oriented according to the initial domain direction and, after about ~7 ns, orients along a weak external magnetic field while the sample is potentially still at high temperature with a reduced magnetic anisotropy. This first SAXS-based ultrafast remagnetization experiment already disclosed the typical nature of this kind of magnetization dynamics as being at the boundary between atomistic and continuous descriptions of magnetization. While the process starts in a highly excited non-equilibrium state, the domain reorientation on the nanosecond timescale was described with Landau–Lifschitz–Gilbert dynamics.

The nucleation, interaction, and growth of ferrimagnetic domains in GdFeCo were studied by Iacocca *et al.*, triggered by photo-excitation starting from a magnetically saturated state in an applied out-of-plane magnetic field.[166] Note that this material, in principle, allows for single-shot AOS to occur. Transient short-range magnetic order emerges on the picosecond timescale, as reflected by a distinct magnetic SAXS signal broadly distributed in reciprocal space. Assisted by atomistic and micromagnetic modelling, the authors relate the appearance of this transient magnetic order to a rapid localization and coalescence of magnons occurring during the system's rapid passage from an almost paramagnetic state after demagnetization back to a magnetically ordered state. In their work, the authors speculate about the possibility of forming topological magnetic defects observed previously[167] through a mechanism similar to the one underlying the magnon coalescence.

This process of nucleating a phase containing topologically non-trivial magnetic skyrmions from a photo-induced high-temperature state was investigated at European XFEL by Büttner *et al.* in 2021 (see Figure 7.9).[30]

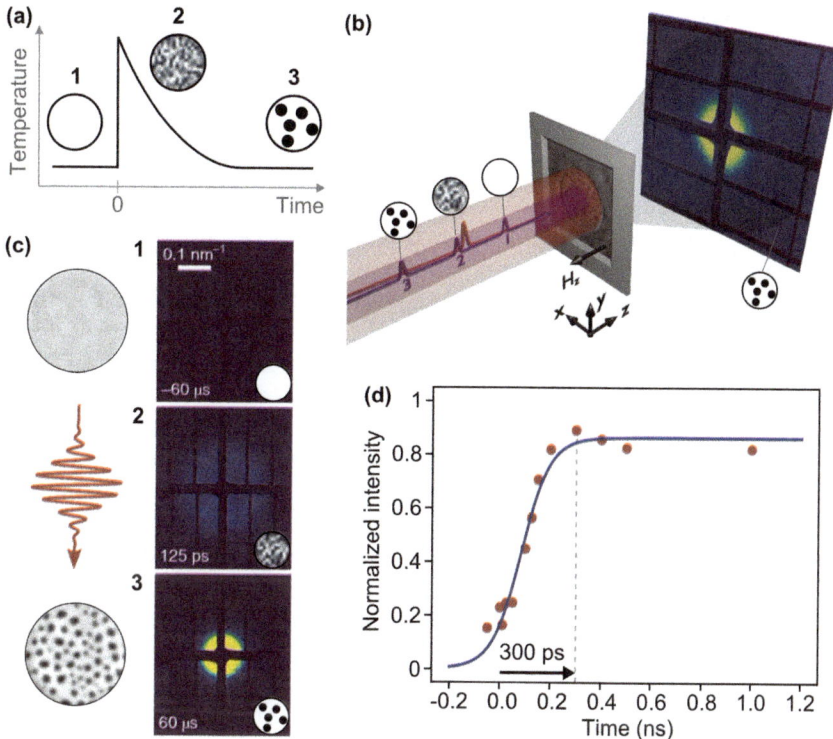

Figure 7.9 Time-resolved resonant SAXS experiment following the nucleation of a topological skyrmion phase. (a) Schematics of the topological phase transition. The sample is heated with a laser pulse from the saturated state (1) into a fluctuation state (2) where topological switching occurs. After cooling down, the stable skyrmion state (3) is established. (b) Schematics of the XFEL experiment. Each state of the sample (1–3) is probed by an XFEL pulse (blue). The transient state is recorded with a time delay to the laser pulse (red) exciting the sample. (c) Typical diffraction patterns for the three states (right) and corresponding real-space images recorded statically with soft-X-ray holography in a separate experiment. The field of view is 1.5 μm. (d) Evolution of the scattering intensity after laser excitation. The plateau reached after 300 ps indicates the formation of the skyrmion phase. Reproduced from ref. 30 with permission from Springer Nature, Copyright 2021.

A static X-ray holography experiment has revealed that a single laser pulse can transform a homogeneously magnetized ferromagnetic film into a dense pattern of skyrmions stabilized by antisymmetric exchange and dipolar interaction (see Figure 7.9c).[30,65,166] In the time-resolved XFEL experiment, the corresponding SAXS patterns were recorded (see Figure 7.9b): the pattern from the initial state, which is empty due to the lack of skyrmions (1), the pattern from the final state, which shows high scattering intensity from the skyrmions created (3), and, most importantly, the transient pattern capturing the creation process which was recorded with a delay up 1 ns after the laser excitation (2) (see Figure 7.9c). The transient signal is characterized by scattering over a broad q-range with increasing intensity after the IR laser pulse (see Figure 7.9d). The scattering signal was interpreted as a signature for a high-temperature fluctuation phase, promoting topological switching and, thereby, the nucleation of the skyrmion phase (see Figure 7.9a). This phase is established 300 ps after photoexcitation. The topological magnetic texture subsequently coarsens until the skyrmions' equilibrium size and density are reached.

These experiments highlight the essential role X-ray scattering methods have begun to play in researching pattern formation connected to ultrafast demagnetization, switching, and remagnetization processes. So far, these techniques are unique in providing nanometre-scale spatial sensitivity to magnetic texture formation on ultrafast timescales. The ability to probe functionalized heterostructures and multilayers of tens of nanometre thickness is particularly important in this context. Combined with elemental sensitivity, this allows for dissecting the dynamics on the nanometre scale with respect to the individual constituents. While, so far, mainly incoherent scattering has been used to follow the evolution of the structure factor of the transient textures, interference-based methods such as speckle correlation analysis and coherent X-ray imaging will allow for the distinction of individual magnetic configurations and hence enable the identification of processes invisible within an ensemble average.

7.5   Outlook and Conclusion

Resonant magnetic X-ray scattering and X-ray holography allow access to the dynamics of magnetization structures in either reciprocal or real space, respectively. When recorded with suitably short pulses of soft X-rays, they can provide access to ultrafast dynamics down to the femtosecond scale in pump–probe experiments. Exclusively focusing on time-resolved work, experiments performed at synchrotron-radiation sources, free-electron X-ray lasers and laser-driven high harmonic generation sources were considered in this chapter. We have discussed examples investigating fundamental processes influencing magnetization dynamics in thin-film systems, such as hot-electron (spin) currents, magnon coalescence or the nucleation and topology-specific dynamics of magnetic skyrmions. The discussion was restricted to experiments probing lateral dynamics in thin film systems, but we note that especially at synchrotron-radiation sources, perpendicular changes in magnetization and magnetic order have also been investigated in time-resolved experiments. Recently, the first experiment of this kind could be performed at a laser-driven plasma source,[109] paving the way for studies of transient magnetic order down to the 10 picosecond timescale outside large-scale facilities and also for photon energies above 1 keV.

With the information on high spatial frequencies in a sample residing at large momentum transfer and with the associated intensity in scattering patterns and FTH holograms being low, the key limiting factor for achieving a combined high spatiotemporal resolution in the sub-50 nm–sub-50 fs regime is typically the photon flux in a suitable bandwidth available for resonant scattering. There have been tremendous advances in photon flux and stability from HHG sources for this kind of work, pushing to shorter wavelengths and making the first scattering and imaging experiments on magnetic structures in this spatiotemporal regime possible. This has recently enabled the first scattering studies on lanthanide-based magnetic materials at around 150 eV photon energy, corresponding to approximately 8 nm wavelength. On the other hand, this being the state of the art also illustrates the formidable task of reaching the L-absorption edges of 3d metal atoms above 700 eV with a tuneable femtosecond source enabling resonant scattering. The facts that L-edge near-edge absorption and circular dichroism in these important constituents of magnetic materials (i) provide detailed spectroscopic information with strong magnetic contrast and (ii) that the inelastic mean free path of hot electrons in metallic magnetic materials is only on the order of a few nanometres underline the importance of such a source development to study the laser-induced magnetization phenomena discussed here.

At XFELs, these and higher photon energies are easily reached with a high number of photons per sub-100 fs pulse. Here, challenges are associated with reaching the highest temporal resolution and sensitivity in pump–probe experiments gave the typically pronounced fluctuations of FEL beam parameters. This is especially difficult when experiments are supposed to be non-perturbative regarding the probing X-ray beam, considering radiation damage to samples consisting of thin solid films. In this respect, the advent of FELs operating at kHz or even MHz repetition rates with evenly spaced pulses and potentially complete polarization control will provide new opportunities to advance into the few-nanometre–few-femtosecond regime in the study of ultrafast magnetization phenomena on the nanoscale.

References

1. J. Stöhr and H. C. Siegmann, *Magnetism*, Springer-Verlag, Berlin Heidelberg, New York, 2006.
2. G. van der Laan and A. I. Figueroa, *Coord. Chem. Rev.*, 2014, **277–278**, 95–129.
3. E. Marchiori, *et al.*, *Nat. Rev. Phys.*, 2021, 1–12.

4. D.-T. Ngo and L. T. Kuhn, *Adv. Nat. Sci.: Nanosci. Nanotechnol.*, 2016, **7**, 045001.

5. D. Attwood and A. Sakdinawat, *X-Rays and Extreme Ultraviolet Radiation: Principles and Applications*, Cambridge University Press, Cambridge, 2nd edn, 2017.

6. J. Fink, *et al.*, *Rep. Prog. Phys.*, 2013, **76**, 056502.

7. L. Paolasini, *École thématique de la Société Française de la Neutronique*, 2014, **13**, 03002.

8. S. Macke and E. Goering, *J. Phys.: Condens. Matter.*, 2014, **26**, 363201.

9. J. P. Hannon, *et al.*, *Phys. Rev. Lett.*, 1988, **61**, 1245–1248.

10. J. B. Kortright, *J. Electron Spectrosc. Relat. Phenom.*, 2013, **189**, 178–186.

11. C. von Korff Schmising, *et al.*, *Rev. Sci. Instrum.*, 2017, **88**, 053903.

12. F. Willems, *et al.*, *Phys. Rev. Lett.*, 2019, **122**, 217202.

13. J. B. Kortright, *et al.*, *Phys. Rev. B: Condens. Matter Mater. Phys.*, 2001, **64**, 092401.

14. C. T. Chen, *et al.*, *Phys. Rev. Lett.*, 1995, **75**, 152–155.

15. J. Prieto, *et al.*, *Appl. Phys. A: Mater. Sci. Process.*, 2005, **80**, 1021–1027.

16. M. Hennecke, *et al.*, *Phys. Rev. Res.*, 2022, **4**, L022062.

17. G. Fan, *et al.*, *Optica*, 2022, **9**, 399–407.

18. B. Pfau, *et al.*, *Nat. Commun.*, 2012, **3**, 1100.

19. S. Eisebitt, *et al.*, *Phys. Rev. B: Condens. Matter Mater. Phys.*, 2003, **68**, 104419.

20. D. Weder, *et al.*, *Struct. Dyn.*, 2020, **7**, 054501.

21. K. A. Nugent, *Adv. Phys.*, 2010, **59**, 1–99.

22. O. Hellwig, *et al.*, *J. Magn. Magn. Mater.*, 2007, **319**, 13–55.

23. M. T. Johnson, *et al.*, *Rep. Prog. Phys.*, 1996, **59**, 1409–1458.

24. B. Tudu and A. Tiwari, *Vacuum*, 2017, **146**, 329–341.

25. A. Fert, V. Cros and J. Sampaio, *Nat. Nanotechnol.*, 2013, **8**, 152–156.

26. S. Woo, *et al.*, *Nat. Mater.*, 2016, **15**, 501–506.

27. W. Jiang, *et al.*, *Phys. Rep.*, 2017, **704**, 1–49.

28. K. Litzius, *et al.*, *Nat. Phys.*, 2017, **13**, 170–175.

29. F. Büttner, *et al.*, *Nat. Nanotechnol.*, 2017, **12**, 1040–1044.

30. F. Büttner, *et al.*, *Nat. Mater.*, 2021, **20**, 30–37.

31. O. Hellwig, *et al.*, *Phys. B: Condens. Matter.*, 2003, **336**, 136–144.

32. K. Bagschik, *et al.*, *Phys. Rev. B*, 2016, **94**, 134413.

33. B. Lengeler, *Naturwissenschaften*, 2001, **88**, 249–260.

34. B. Pfau and S. Eisebitt, in *Synchrotron Light Sources and Free-Electron Lasers*, ed. E. Jaeschke, S. Khan, J. R. Schneider and J. B. Hastings, Springer International Publishing, Cham, 2015, pp. 1–36.

35. S. Eisebitt, *et al.*, *Nature*, 2004, **432**, 885–888.

36. I. McNulty, *et al.*, *Science*, 1992, **256**, 1009–1012.

37. O. Hellwig, *et al.*, *J. Appl. Phys.*, 2006, **99**, 08H307.

38. D. Stickler, *et al.*, *Phys. Rev. B: Condens. Matter Mater. Phys.*, 2011, **84**, 104412.

39. C. Tieg, *et al.*, *Appl. Phys. Lett.*, 2010, **96**, 072503.

40. S. Streit-Nierobisch, *et al.*, *J. Appl. Phys.*, 2009, **106**, 083909.

41. B. Pfau, *et al.*, *Appl. Phys. Lett.*, 2011, **99**, 062502.

42. C. M. Günther, *et al.*, *Phys. Rev. B: Condens. Matter Mater. Phys.*, 2010, **81**, 064411.
43. R. Blukis, *et al.*, *Geochem. Geophys. Geosyst.*, 2020, **21**, e2020GC009044.
44. D. Stickler, *et al.*, *Appl. Phys. Lett.*, 2010, **96**, 042501.
45. C. Tieg, *et al.*, *Opt. Express*, 2010, **18**, 27251–27256.
46. T. A. Duckworth, *et al.*, *New J. Phys.*, 2013, **15**, 023045.
47. E. Malm, *et al.*, *Opt. Express*, 2022, **30**, 38424–38438.
48. W. F. Schlotter, *et al.*, *Appl. Phys. Lett.*, 2006, **89**, 163112.
49. C. M. Günther, *et al.*, *J. Opt.*, 2017, **19**, 064002.
50. S. Marchesini, *et al.*, *Nat. Photonics*, 2008, **2**, 560–563.
51. T. A. Duckworth, *et al.*, *Opt. Express*, 2011, **19**, 16223–16228.
52. D. Zhu, *et al.*, *Phys. Rev. Lett.*, 2010, **105**, 043901.
53. J. Geilhufe, *et al.*, *Nat. Commun.*, 2014, **5**, 3008.
54. J. Geilhufe, *et al.*, *Ultramicroscopy*, 2020, **214**, 113005.
55. H. N. Chapman and K. A. Nugent, *Nat. Photonics*, 2010, **4**, 833839.
56. O. Kfir, *et al.*, *Sci. Adv.*, 2017, **3**, eaao4641.
57. S. Zayko, *et al.*, *Nat. Commun.*, 2021, **12**, 6337.
58. R. L. Sandberg, *et al.*, *Opt. Lett.*, 2009, **34**, 1618–1620.
59. A. S. Johnson, *et al.*, *Sci. Adv.*, 2021, **7**, eabf1386.
60. B. Pfau, *et al.*, *Appl. Phys. Lett.*, 2014, **105**, 132407.
61. T. Hauet, *et al.*, *Phys. Rev. B: Condens. Matter Mater. Phys.*, 2008, **77**, 184421.
62. L. Vidas, *et al.*, *Nano Lett.*, 2018, **18**, 3449–3453.
63. F. Büttner, *et al.*, *Nat. Phys.*, 2015, **11**, 225–228.
64. L. Caretta, *et al.*, *Nat. Nanotechnol.*, 2018, **13**, 1154–1160.
65. K. Gerlinger, *et al.*, *Appl. Phys. Lett.*, 2021, **118**, 192403.
66. C. von Korff Schmising, *et al.*, *Phys. Rev. Lett.*, 2014, **112**, 217203.
67. E. Guehrs, *et al.*, *Opt. Express*, 2009, **17**, 6710–6720.
68. E. Guehrs, *et al.*, *New J. Phys.*, 2012, **14**, 013022.
69. T. Gorkhover, *et al.*, *Nat. Photonics*, 2018, **12**, 150–153.
70. A. Barman, *et al.*, *J. Appl. Phys.*, 2020, **128**, 170901.
71. H. Stoll, *et al.*, *Front. Phys.*, 2015, **3**, 26.
72. A. D. Kent, H. Ohldag, H. A. Dürr and J. Z. Sun, Magnetization Dynamics, in *Handbook of Magnetism and Magnetic Materials*, ed. M. Coey and S. Parkin, Springer International Publishing, Cham, 2020, pp. 1–33.
73. R. E. Camley, *Magnetism of Surfaces, Interface, and Nanoscale Materials*, Elsevier, Amsterdam, 2016.
74. P. Vagovič, *et al.*, *Optica*, 2019, **6**, 1106.
75. S. Cartier, *et al.*, *J. Instrum.*, 2014, **9**, C05027.
76. A. Marras, *et al.*, *J. Synchrotron Radiat.*, 2021, **28**, 131–145.
77. M. Porro, *et al.*, *IEEE Trans. Nucl. Sci.*, 2021, **68**, 1334–1350.
78. C. Léveillé, *et al.*, *J. Synchrotron Radiat.*, 2022, **29**, 103–110.
79. B. Pfau, *et al.*, *J. Appl. Phys.*, 2017, **122**, 043907.
80. M. G. Kim, *et al.*, *Nat. Commun.*, 2018, **9**, 5013.
81. M. G. Kim, *et al.*, *Sci. Adv.*, 2022, **8**, eabj9493.

82. M. Sutton, *C. R. Phys.*, 2008, **9**, 657–667.
83. A. Madsen, A. Fluerasu and B. Ruta, in *Synchrotron Light Sources and Free-Electron Lasers*, ed. E. Jaeschke, S. Khan, J. R. Schneider and J. B. Hastings, Springer International Publishing, Cham, 2018, pp. 1–30.
84. F. Perakis and C. Gutt, *Phys. Chem. Chem. Phys.*, 2020, **22**, 19443–19453.
85. M. S. Pierce, *et al.*, *Phys. Rev. Lett.*, 2003, **90**, 175502.
86. O. G. Shpyrko, *et al.*, *Nature*, 2007, **447**, 68–71.
87. S. Konings, *et al.*, *Phys. Rev. Lett.*, 2011, **106**, 077402.
88. X. M. Chen, *et al.*, *Phys. Rev. Lett.*, 2019, **123**, 197202.
89. C. Klose, *et al.*, *Nature*, 2023, **614**, 256–261.
90. P. Fischer, *et al.*, *Mat. Today*, 2006, **9**, 26–33.
91. G. Schönhense *et al.*, in *Advances in Imaging and Electron Physics*, ed. P. Hawkes, Elsevier, 2006, vol. 142, pp. 159–323.
92. S. Finizio, *et al.*, *Nano Lett.*, 2022, **22**, 1971–1977.
93. L. Gierster, *et al.*, *Ultramicroscopy*, 2015, **159**, 508–512.
94. L. Le Guyader, *et al.*, *Nat. Commun.*, 2015, **6**, 5839.
95. F. Büttner, *et al.*, *Phys. Rev. B: Condens. Matter Mater. Phys.*, 2013, **87**, 134422.
96. N. Bukin, *et al.*, *Sci. Rep.*, 2016, **6**, 36307.
97. C. Gutt, *et al.*, *Opt. Express*, 2009, **17**, 55–61.
98. S. A. Montoya, *et al.*, *Phys. Rev. B*, 2017, **95**, 224405.
99. M. H. Seaberg, *et al.*, *Phys. Rev. Res.*, 2021, **3**, 033249.
100. V. Esposito, *et al.*, *Appl. Phys. Lett.*, 2020, **116**, 181901.
101. E. Beaurepaire, *et al.*, *Phys. Rev. Lett.*, 1996, **76**, 4250–4253.
102. J. Walowski and M. Münzenberg, *J. Appl. Phys.*, 2016, **120**, 140901.
103. A. Kirilyuk, A. V. Kimel and T. Rasing, *Rev. Mod. Phys.*, 2010, **82**, 2731–2784.
104. C. Dornes, *et al.*, *Nature*, 2019, **565**, 209–212.
105. I. Radu, *et al.*, *Nature*, 2011, **472**, 205–208.
106. M. Hennecke, *et al.*, *Phys. Rev. Lett.*, 2019, **122**, 157202.
107. N. Thielemann-Kühn, *et al.*, *Phys. Rev. Lett.*, 2017, **119**, 197202.
108. V. Chardonnet, *et al.*, *Struct. Dyn.*, 2021, **8**, 034305.
109. D. Schick, *et al.*, *Optica*, 2021, **8**, 1237–1242.
110. E. A. Seddon, *et al.*, *Rep. Prog. Phys.*, 2017, **80**, 115901.
111. W. Ackermann, *et al.*, *Nat. Photonics*, 2007, **1**, 336–342.
112. E. Allaria, *et al.*, *Nat. Photonics*, 2012, **6**, 699–704.
113. C. Gutt, *et al.*, *Phys. Rev. B: Condens. Matter Mater. Phys.*, 2010, **81**, 100401.
114. T. Wang, *et al.*, *Phys. Rev. Lett.*, 2012, **108**, 267403.
115. M. Schneider, *et al.*, *Phys. Rev. Lett.*, 2020, **125**, 127201.
116. A. Barty, *et al.*, *Nat. Photonics*, 2008, **2**, 415–419.
117. B. Pfau, *et al.*, *New J. Phys.*, 2010, **12**, 095006.
118. C. M. Günther, *et al.*, *Nat. Photonics*, 2011, **5**, 99–102.
119. C. Gutt, *et al.*, *Phys. Rev. B: Condens. Matter Mater. Phys.*, 2009, **79**, 212406.

120. C. E. Graves, *et al.*, *Nat. Mater.*, 2013, **12**, 293–298.
121. B. Pfau, *et al.*, *Opt. Express*, 2010, **18**, 13608–13615.
122. G. De Ninno, *et al.*, *Phys. Rev. Lett.*, 2013, **110**, 064801.
123. D. Weder, *et al.*, *IEEE Trans. Magn.*, 2017, **53**, 6500905.
124. F. Willems, *et al.*, *Struct. Dyn.*, 2017, **4**, 014301.
125. R. A. Bartels, *et al.*, *Science*, 2002, **297**, 376–378.
126. T. Popmintchev, *et al.*, *Science*, 2012, **336**, 1287–1291.
127. M. van Mörbeck-Bock *et al.*, *High Power Lasers and Applications*, 2021, pp. 11–16.
128. M. Chini, K. Zhao and Z. Chang, *Nat. Photonics*, 2014, **8**, 178–186.
129. D. Gauthier, *et al.*, *Phys. Rev. Lett.*, 2010, **105**, 093901.
130. A. Ravasio, *et al.*, *Phys. Rev. Lett.*, 2009, **103**, 028104.
131. E. Malm, *et al.*, *Opt. Express*, 2020, **28**, 394.
132. B. Vodungbo, *et al.*, *Nat. Commun.*, 2012, **3**, 999.
133. B. Vodungbo, *et al.*, *Opt. Express*, 2011, **19**, 4346–4356.
134. F. Willems, *et al.*, *Phys. Rev. B: Condens. Matter Mater. Phys.*, 2015, **92**, 220405.
135. S. Long, W. Becker and J. K. McIver, *Phys. Rev. A*, 1995, **52**, 2262–2278.
136. O. Kfir, *et al.*, *Nat. Photonics*, 2015, **9**, 99–105.
137. C. La-O-Vorakiat, *et al.*, *Phys. Rev. Lett.*, 2009, **103**, 257402.
138. S. Mathias, *et al.*, *Proc. Natl. Acad. Sci. U. S. A.*, 2012, **109**, 4792–4797.
139. F. Willems, *et al.*, *Nat. Commun.*, 2020, **11**, 871.
140. M. Schneider, *et al.*, *Nat. Commun.*, 2018, **9**, 214.
141. M. Hennes, *et al.*, *Phys. Rev. B*, 2020, **102**, 174437.
142. B. Vodungbo, *et al.*, *Sci. Rep.*, 2016, **6**, 18970.
143. A. Eschenlohr, *et al.*, *Nat. Mater.*, 2013, **12**, 332–336.
144. L. Müller, *et al.*, *Phys. Rev. Lett.*, 2013, **110**, 234801.
145. B. Wu, *et al.*, *Phys. Rev. Lett.*, 2016, **117**, 027401.
146. R. Neutze, *et al.*, *Nature*, 2000, **406**, 752–757.
147. H. N. Chapman, *et al.*, *Nat. Phys.*, 2006, **2**, 839–843.
148. D. J. Higley, *et al.*, *Nat. Commun.*, 2019, **10**, 5289.
149. K. J. Kapcia, *et al.*, *npj Comput. Mater.*, 2022, **8**, 1–9.
150. J. Stöhr and A. Scherz, *Phys. Rev. Lett.*, 2015, **115**, 107402.
151. Z. Chen, *et al.*, *Phys. Rev. Lett.*, 2018, **121**, 137403.
152. A. Melnikov, *et al.*, *Phys. Rev. Lett.*, 2011, **107**, 076601.
153. M. Battiato, K. Carva and P. M. Oppeneer, *Phys. Rev. Lett.*, 2010, **105**, 027203.
154. G. Malinowski, *et al.*, *Nat. Phys.*, 2008, **4**, 855–858.
155. M. Aeschlimann, *et al.*, *Phys. Rev. Lett.*, 1997, **79**, 5158–5161.
156. D. Zusin, *et al.*, *Phys. Rev. B*, 2022, **106**, 144422.
157. N. Z. Hagström, *et al.*, *Phys. Rev. B*, 2022, **106**, 224424.
158. C. D. Stanciu, *et al.*, *Phys. Rev. Lett.*, 2007, **99**, 047601.
159. S. Mangin, *et al.*, *Nat. Mater.*, 2014, **13**, 286–292.
160. C.-H. Lambert, *et al.*, *Science*, 2014, **345**, 1337–1340.
161. T. A. Ostler, *et al.*, *Nat. Commun.*, 2012, **3**, 666.

162. J.-Y. Chauleau, *et al.*, *Phys. Rev. Lett.*, 2018, **120**, 037202.

163. N. Kerber, *et al.*, *Nat. Commun.*, 2020, **11**, 6304.

164. C. Léveillé, *et al.*, *Nat. Commun.*, 2022, **13**, 1412.

165. N. Bergeard, *et al.*, *Phys. Rev. B: Condens. Matter Mater. Phys.*, 2015, **91**, 054416.

166. E. Iacocca, *et al.*, *Nat. Commun.*, 2019, **10**, 1756.

167. S.-G. Je and P. Vallobra, *et al.*, *Nano Lett.*, 2018, **18**, 7362–7371.

CHAPTER 8

Free Electron Lasers for X-ray Scattering and Diffraction

M. DUNNE,* R. W. SCHOENLEIN, J. P. CRYAN AND
T. J. A. WOLF

SLAC National Accelerator Laboratory, Linac Coherent Light Source,
2575 Sand Hill Rd, Menlo Park, CA 94025, USA
*Email: mdunne@slac.stanford.edu

8.1   Introduction

The recent emergence of X-ray free-electron lasers (XFELs) represents a revolution in X-ray science that can potentially transform the field for the 21st century. XFELs can generate X-ray beams with a peak brightness of nine-to-ten orders of magnitude higher and an average brightness of three orders of magnitude higher compared to a synchrotron source. Pulse durations can be as short as 0.2 femtoseconds (fs) to 200 fs, with a time-bandwidth product close to the Fourier transform limit, over a spectral range from 0.2 to >30 keV, and with almost complete spatial and temporal coherence. Advanced operating modes can deliver pairs of pulses or multiple-pulse sequences with a temporal separation ranging from sub-fs to microseconds (μs) and with variability of the relative wavelength, polarization, and spectral bandwidth in addition to the precise time spacing between pulses.

The combination of high average brightness (a good proxy for signal-to-noise ratio, SNR) with high peak brightness provides a means to determine the structures of complex molecules at very high resolution from inverse techniques such as X-ray diffraction and coherent scattering, and from real-space imaging techniques such as phase contrast imaging. Each XFEL

Theoretical and Computational Chemistry Series No. 25
Structural Dynamics with X-ray and Electron Scattering
Edited by Kasra Amini, Arnaud Rouzée and Marc J. J. Vrakking
© The Royal Society of Chemistry 2024
Published by the Royal Society of Chemistry, www.rsc.org

pulse can extract a large amount of information from a single interaction with the sample. This is particularly powerful for systems that undergo irreversible transitions or phenomena sensitive to initial conditions, such as material fracture or the propagation of shock waves in complex media. Similarly, the information from individual pulses provides an avenue to investigate materials that are difficult to crystallize, that are susceptible to radiation damage from prolonged irradiation, or that require measurement under ambient (*e.g.*, physiologically relevant) conditions such as membrane proteins using the new method of serial femtosecond crystallography (SFX).

Of primary significance is the ultrafast, pulsed nature of the X-rays, which invites their use as a probe of sample dynamics on timescales appropriate for observing electronic motion, molecular dynamics, and the emergence of complex spin, orbital, and lattice couplings that give rise to macroscopic quantum mechanical properties of materials. Unlike most other ultrafast probes, X-rays provide chemical site specificity for unambiguously identifying the structural evolution of the systems under investigation. Similarly, in quantum materials science, X-ray scattering provides direct information on lattice displacements and heterogeneous anomalies (*e.g.*, defects). X-ray scattering also delivers momentum resolution that is inaccessible using long-wavelength probes. This ability to map the dynamical function of complex systems on timescales ranging across 12 orders of magnitude (fs to ms), and under conditions that stretch from near-equilibrium to exotic excited states, has opened up a fundamentally new era for the investigation of the world around us.

This chapter will review some of the XFEL methods developed over the first decade of operation. It will explore the emerging era of high repetition rate beam delivery that is set to transform this field further (moving from ~ 100 Hz operation to kHz and even MHz repetition rates). The chapter is organized as follows. Section 8.2 briefly introduces XFEL physics and characteristic features of XFEL facilities worldwide. Sections 8.3–8.5 outline the science impact of X-ray scattering and diffraction methods using XFEL sources in the areas of (a) fundamental processes in atoms and molecules and gas-phase chemistry (Section 8.3), (b) condensed-phase chemistry and catalysis (Section 8.4), and (c) structural dynamics in complex materials (Section 8.5).

8.2   XFEL Facilities and Performance

The promising approach of using free relativistic electrons to generate coherent radiation was recognized by Motz *et al.* in the 1950s.[1,2] This early work provided a basis for the concept of a FEL, introduced in 1971 by Madey[3] and demonstrated by his group in 1977.[4] These important early FEL developments provided a foundation for the key conceptual advances in the 1980s by Kondratenko and Saldin *et al.*,[5,6] Bonifacio and Pellegrini *et al.*[7–9] and others, that ultimately led to the development of X-ray FELs (XFELs).

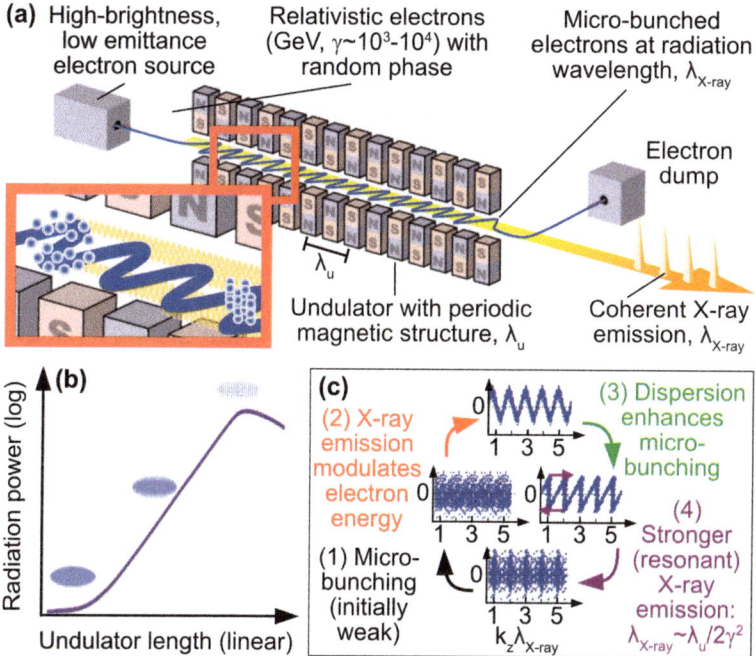

Figure 8.1   Illustration of the self-amplified spontaneous emission (SASE) process for XFELs. (a) Propagation of a relativistic electron bunch through an undulator. Reproduced from ref. 10, https://en.wikipedia.org/wiki/Free-electron_laser#/media/File:FEL_principle.png, under the terms of the CC BY SA 3.0 license https://creativecommons.org/licenses/by-sa/3.0/. (b) Exponential growth (and saturation) of X-ray power along the length of the undulator due to micro-bunching of the electron beam in the presence of the radiated X-ray field (SASE process). (c) Illustration of the collective instability leading to micro-bunching of the electron beam and coherent emission of X-rays.

The first self-consistent predictive XFEL theory[7] identified the conditions necessary for the emergence of a collective instability in the coupled system consisting of an undulator (periodic magnetic array), a relativistic electron beam, and the emission field (see Figure 8.1a).[10] The "collective instability" is a powerful self-organization process in which the electron beam (with an initial random time-distribution of electrons within a bunch), in the presence of the emission field, self-organizes into micro-bunches at the scale of the emission wavelength, at which point the emission becomes coherent (see Figure 8.1b and c). The entire process is called self-amplified spontaneous emission (SASE). Importantly, these early theoretical projections for SASE were valid even to the Ångström scale, and established the conditions necessary to realize an XFEL in the hard X-ray regime.

The first XFEL to generate hard X-rays, the Linac Coherent Light Source (LCLS), began operation in 2009 at the SLAC National Laboratory in California. The facility was created using an electron accelerator with a pulse repetition rate of 120 Hz, a beam energy of 10–17 GeV, bunch charges of 0.02 to 0.3 nC, peak currents of up to 5 kA, and transverse slice emittance in the 0.2 to 0.7 μm range. The SASE process is established by passing the accelerated electrons through a precision-tuned, 130 m-long periodic magnetic undulator. An illustration of the layout of the LCLS-II-HE system (see Figure 8.2a) consists of superconducting (SC) and copper (Cu) linacs with electron bunch compressors (BC1 and BC2), undulator beamlines (see Figure 8.2b), and an X-ray instrument suite. The efficiency of the SASE process maximizes the pulse energy and tuning range of XFELs, and creates spatially coherent X-ray beams. However, the inherently stochastic nature of SASE gives rise to multiple longitudinal modes (within the XFEL bandwidth, ∼ few eV), and thus limits the temporal coherence. Advanced operating modes for controlling temporal coherence include self-seeding (see Figure 8.2c and d) where an initial SASE X-ray pulse is monochromatized part-way along the undulator chain using a grating and monochromator optics for soft X-rays (see Figure 8.2c) or using a diamond Bragg crystal for hard X-rays (see Figure 8.2d), with a chicane bypass for the electron beam. The monochromatized X-rays seed the growth of a specific spectral mode in the subsequent amplification stage. This monochromatization can reduce the bandwidth from the SASE level of $\Delta\omega/\omega \sim 0.2\%$ (*e.g.*, 20 eV at 10 keV) to $\sim 10^{-4}$ (~ 1.5 eV at 10 keV), increasing the spectral brightness by an order of magnitude or more. This self-seeding process can be implemented in a soft X-ray self-seeding layout[11] (see Figure 8.2c) and hard X-ray self-seeding layout[12] (see Figure 8.2d).

Today, five hard X-ray FELs operate worldwide (LCLS, SACLA, PAL-XFEL, SwissFEL, and European XFEL), and others are under construction (LCLS-II/HE, SHINE). The SACLA XFEL in Japan began user operation in 2012 following first lasing in 2011, and is based on a compact pulsed-RF C-band Cu-linac (normal conducting) with a maximum electron energy of 8.5 GeV and a maximum repetition rate of 60 Hz.[13,14] The third hard X-ray FEL in the world, PAL-XFEL in Korea, began user operations in 2017 following first lasing in 2016. Similar to LCLS and SACLA, PAL-XFEL is based on a pulsed-RF Cu-linac (10 GeV) with a maximum repetition rate of 60 Hz.[15,16] The SwissFEL XFEL in Switzerland began pilot experiments in 2018 following first lasing in the X-ray range in 2017.[17] Similar to the first three hard X-ray FELs described above, SwissFEL is based on a pulsed-RF Cu-linac (5.8 GeV) operating at a repetition rate of 100 Hz, to generate X-rays of 1 Å wavelength using the lowest electron beam energy necessary for operation at this wavelength. The European XFEL (EuXFEL) facility in Germany began operation in 2017, and is the first hard X-ray XFEL to be based on super-conducting accelerator technology (17.5 GeV linac energy).[18–20] In addition to the high peak X-ray brightness, the EuXFEL is designed to achieve high average brightness by operating in a burst-mode at 10 Hz, in which each burst contains a maximum of 2700 pulses (at an intra-burst repetition rate of 4.5 MHz) for an effective maximum average repetition rate of 27 kHz.

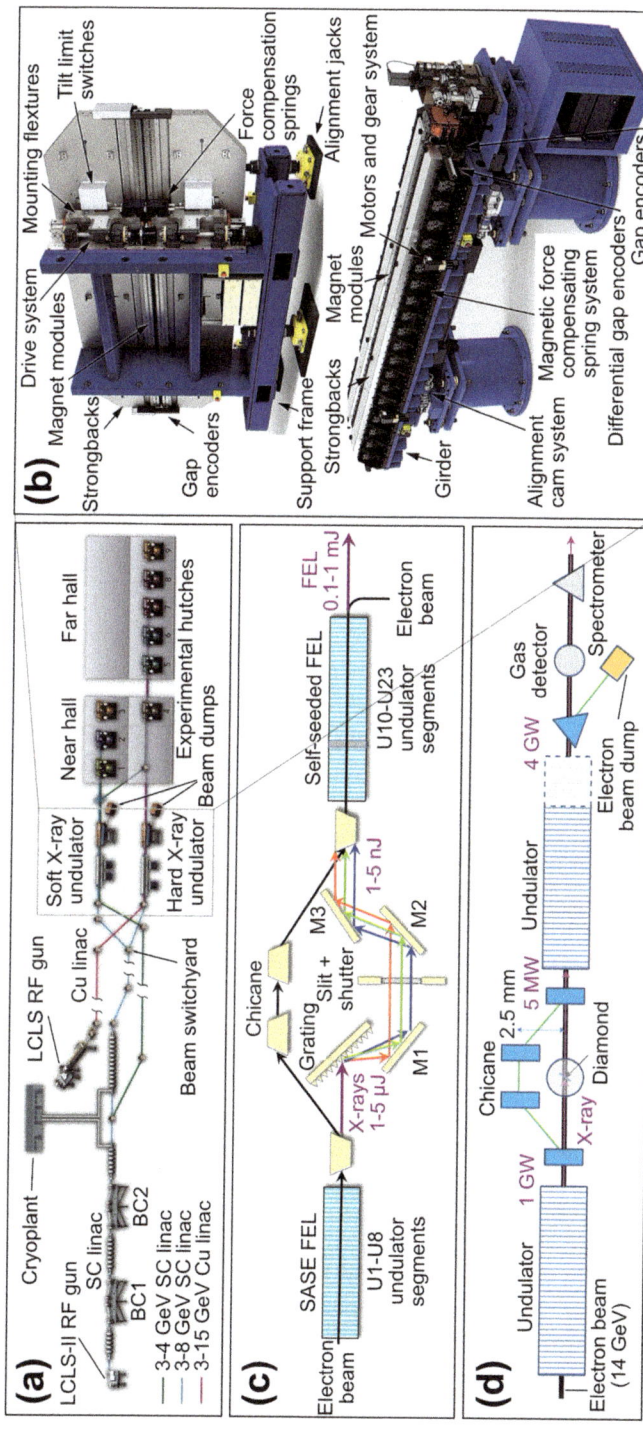

Figure 8.2 (a) Layout of an XFEL facility, in this case the LCLS-II-HE system, showing the combination of superconducting (SC) and copper (Cu) linacs with electron bunch compressors (BC1 and BC2), undulator beamline(s), and an X-ray instrument suite. (b) Undulator units for the soft X-ray (top) and hard X-ray (bottom) beamlines at LCLS. (c) Schematic view of soft X-ray self-seeding,[11] showing the selection of a narrow photon bandwidth using a grating-mirror system (M1, M2, and M3) that introduces a seed onto the electron beam that is delayed *via* a magnetic chicane. Reproduced from ref. 11 with permission from the American Physical Society, Copyright 2015. (d) Schematic of hard X-ray self-seeding,[12] in which the selection of the narrow-band photon spectrum is achieved using a Bragg crystal such as diamond (111). Reproduced from ref. 12 with permission from Springer Nature, Copyright 2012.

Table 8.1   Typical photon beam parameters for XFELs.

Parameter	Value
Pulse duration	0.2–200 fs
Pulse energy	0.1–10 mJ
Fractional spectral bandwidth	• 0.2–2% (SASE) • 10^{-4} (seeded, or *via* a X-ray monochromator)
Repetition rate	• 10–120 Hz (Cu linac based systems such as LCLS-1, SACLA, PAL and SwissFEL) • Pulsed 4.5 MHz in 10 Hz bunches, 27 kHz average rate (European XFEL) • Continuous 1 MHz operation (LCLS-II and SHINE)
Diameter	50–100 μm (natural beam width), typically focussed to between 10 μm and <100 nm
Photons per pulse	10^{11}–10^{14}
Average brightness	10^{20}–10^{24} photons per s per $mrad^2$ per mm^2 per 0.1% bw
Peak brightness	10^{32}–10^{34} photons per s per $mrad^2$ per mm^2 per 0.1% bw

For a recent overview of XFEL facilities and parameters, see ref. 21 and 22 (see Table 8.1 for a summary of photon beam parameters).

While the first generation of XFELs has delivered unprecedented peak brightness, the average brightness is modest, similar to that of a synchrotron source. Furthermore, it has been found that many experiments require attenuation of the peak intensity to avoid perturbation of the sample by the X-ray probe. In these cases, signal accumulation times often become prohibitive at 10 to 120 Hz, thus rendering many experiments impractical. A new generation of XFELs will overcome this limitation by exploiting superconducting accelerator technology to provide ultrafast X-ray pulses at high repetition rates ($\sim$ MHz). These facilities will either operate in a burst mode (*e.g.*, up to 2700 pulses at 4.5 MHz in 10 Hz bunches at the European XFEL, as mentioned earlier), or with continuous streams and programmable time structures (LCLS-II and SHINE). Representative values of average and peak beam brightness (see Figure 8.3) highlight the fundamentally new performance regime created by XFEL sources providing both high peak brightness ($>10^{31}$ ph per s per mm^2 per $mrad^2$ per 0.1% BW) and high average brightness ($>10^{24}$ ph per s per mm^2 per $mrad^2$ per 0.1% BW).

More advanced modes of operation include eSASE (enhanced SASE), which can deliver a high current spike capable of generating a very short X-ray pulse (typically 200–400 attoseconds) with very high coherent bandwidth (on the order of 5 eV at 1 keV), opening up the study of a wide range of electronic wavepacket dynamics in excited state atomic systems. Generating attosecond pulses at LCLS (see Figure 8.4a for the layout) was recently demonstrated with measurements of pulse durations reconstructed from angular streaking (see Figure 8.4b and c).[23] Here, the electron beam

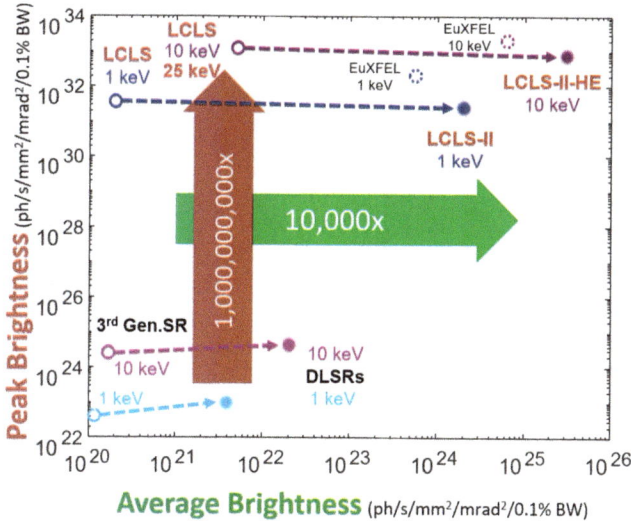

Figure 8.3   Characteristic performance of XFEL and synchrotron facilities showing the peak and average X-ray brightness for two representative photon energies.

travels through a long-period (35 cm) wiggler, developing a single-cycle energy modulation. A magnetic chicane turns the energy modulation into a density spike and is sent to the LCLS undulator to generate sub-femtosecond X-ray pulses. Angular laser-streaking of X-ray photoelectrons using a velocity map imaging (VMI) spectrometer (see the inset of Figure 8.4a, which shows a circularly-polarized infrared pulse overlapped with an attosecond X-ray pulse in a target gas) is used to reconstruct the pulse profile in the time domain.[24] In this angular streaking approach, X-ray photoelectrons (*e.g.*, from a neon core shell) are dressed with the electric field of a circularly polarized IR laser pulse. This encodes characteristic angle and frequency-dependent modulations of the resulting photoelectron spectrum (see the top-right inset of Figure 8.4a), with sub-optical-cycle precision.

In general, tunable isolated soft X-ray attosecond pulses can be generated with high intensities possessing a pulse energy more than 10^6-fold larger than any other source of isolated attosecond pulses in the soft X-ray spectral region. Moreover, such pulses can also have a peak power exceeding 100 GW and a coherent bandwidth of several eV in the 500–1000 eV spectral range.[23]

Of prime importance in XFEL experiments is to record the X-ray characteristics on a pulse-by-pulse basis, given the inherent variability in the most common (SASE) mode of operation. Recent studies have employed pulse-by-pulse spectral measurements with ghost imaging methods for transient X-ray absorption spectroscopy.[25,26] In the time domain, one key example is the ability to measure the pulse duration to a resolution of <1 fs RMS in a non-invasive manner using a transverse electron beam deflector to measure the

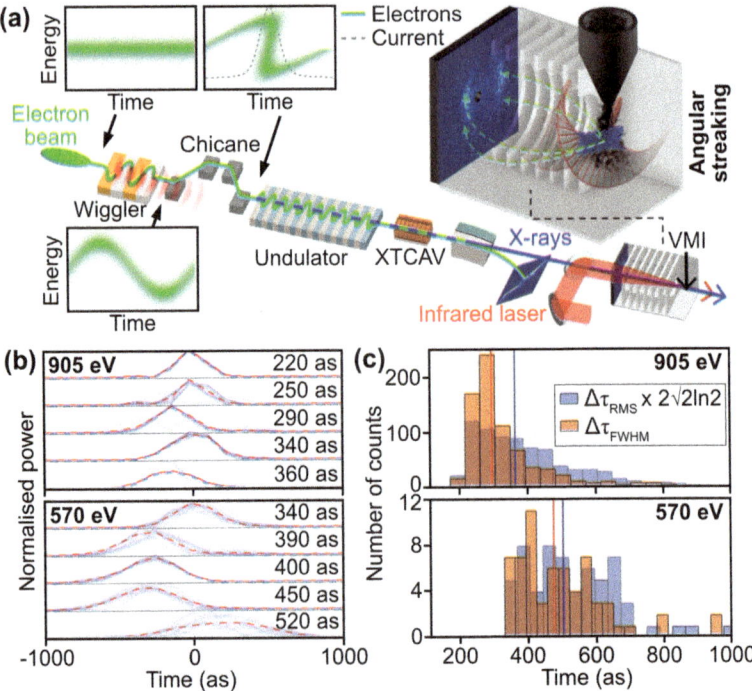

Figure 8.4 (a) Schematic representation of attosecond X-ray generation from LCLS. See main text for details. (b) Angular streaking measurements using a velocity map imaging (VMI) spectrometer (inset) in which a circularly-polarized infrared pulse is overlapped with the attosecond X-ray pulse in a target gas of neon (for the 905 eV pulses) and CO_2 for the 570 eV pulses. The shaded blue lines are the solutions from the reconstruction algorithm initiated by random seeds, and the red lines are the most probable solution. The labelled number is the averaged $\Delta\tau_{FWHM}$.[23] (c) Reconstructed temporal profile of pulse from angular streaking data shown in panel (b). Reproduced from ref. 23 with permission from Springer Nature, Copyright 2019.

energy loss induced by the lasing process. This so-called X-band Transverse CAVity (XTCAV)[27,28] provides a precise record of the electron time-energy distribution (see Figure 8.5) which is a direct fingerprint of time profile of the generated X-ray pulse. This is combined with precise monitoring of the X-ray beam's transverse spatial profile and spectral properties, along with timing systems that monitor the synchronization of the XFEL pulse to a pump pulse (*e.g.*, external optical laser pulse) that is often used to initiate an ultrafast process. Synchronization at the 10 to 100 fs level is typical.[29]

In the following sub-sections, brief examples are given of XFEL's applicability in advancing important areas of science, with characteristic experimental methods for each area, while the subsequent sections provide more extensive details, recent examples, and ideas of important future science opportunities.

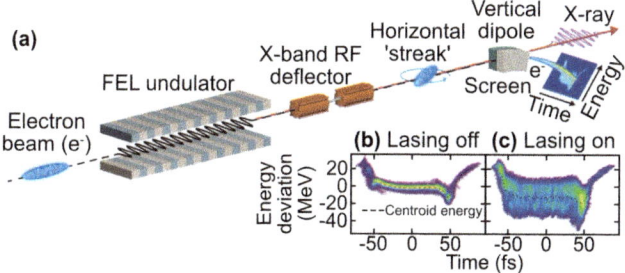

Figure 8.5   (a) Schematic of the X-band Transverse CAVity (XTCAV) temporal pulse diagnostic. (b and c) Characteristic XTCAV signals, showing the electron beam energy in the absence and presence of XFEL lasing, allowing the temporal structure of the X-ray pulse to be extracted.[27,28] Reproduced from ref. 28 with permission from Springer Nature, Copyright 2014.

8.2.1   Fundamental Dynamics of Energy and Charge

Charge migration, redistribution and localization, even in simple molecules, are not well understood at the quantum level. These processes are central to complex processes such as photosynthesis, catalysis, and bond formation/dissolution that govern all chemical reactions. Indirect evidence points to the importance of quantum coherences and coupled evolution of electronic and nuclear wavefunctions in many molecular systems. However, we have not been able to directly observe these processes to date, and they are beyond the description of conventional chemistry models. High-repetition-rate soft X-rays from XFEL facilities enable new dynamic molecular reaction microscope techniques that directly map charge distributions and reaction dynamics in the molecular frame. Complementing this, new nonlinear X-ray spectroscopy methods offer the potential to map quantum coherences in an element-specific way for the first time.

Experimental approaches:
- Dynamic molecular reaction microscopy (for example, see ref. 30 and 31).
- Time-resolved photoemission spectroscopy (for example, see ref. 30 and 31).
- Time-resolved hard X-ray scattering (for example, see ref. 32–35).
- New nonlinear X-ray spectroscopy methods (for example, see ref. 36).

8.2.2   Catalysis and Photo-catalysis

Understanding catalysis and photo-catalysis is essential for directed design of new systems for chemical transformation and solar energy conversion that are efficient, chemically selective, robust, and based on earth-abundant elements. XFELs can reveal the critical (and often rare) transient events in these multi-step processes, from light harvesting and charge separation to charge migration and subsequent accumulation at catalytically active sites. Time-resolved, high-sensitivity, element-specific spectroscopy can provide a direct view of charge dynamics and chemical processes at interfaces, making it possible to pinpoint where charge carriers are lost (within a molecular complex or device) — a crucial bottleneck for efficient solar energy conversion. Such approaches can capture rare chemical events in operating catalytic systems across multiple time and length scales. The joint use of hard and soft X-ray pulses opens the possibility to follow chemical dynamics (*via* spectroscopy) concurrent with structural dynamics (substrate scattering) during heterogeneous catalysis.

Experimental approaches:
- Time-resolved resonant inelastic X-ray scattering and absorption spectroscopy (for example, see ref. 22 and 37).
- Time-resolved X-ray photoelectron spectroscopy.
- Simultaneous soft X-ray spectroscopy and hard X-ray scattering (for example, see ref. 22 and 32–35).
- X-ray photon correlation spectroscopy (for example, see ref. 38 and 39).
- New nonlinear X-ray spectroscopy methods (for example, see ref. 36).

8.2.3   Emergent Phenomena in Quantum Materials

There is an urgent need to understand and ultimately control the exotic properties of new materials, ranging from superconductivity to ferroelectricity and magnetism. These properties emerge from the correlated interactions of the constituent matter components of charge, spin, and phonons. They are not well-described by conventional band models that underpin present semiconductor technologies. Fully coherent X-rays from an XFEL enable new high-resolution spectroscopy approaches to map the collective excitations that define these new materials in unprecedented detail. Ultrashort X-ray pulses and optical fields facilitate new coherent light–matter approaches for manipulating charge, spin, and phonon modes to advance our fundamental understanding and point the way to new approaches for materials control.

Experimental approaches:
- Time-resolved and high-resolution resonant inelastic X-ray scattering (for example, see ref. 37 and 40).
- Time-resolved X-ray dichroism and coherent scattering/imaging.
- Time- and spin-resolved hard X-ray photoemission.
- X-ray photon correlation spectroscopy (for example, see ref. 38 and 39).

8.2.4   Nanoscale Materials Dynamics, Heterogeneity and Fluctuations

The properties of functional materials are often defined by interfaces, heterogeneity, imperfections, and fluctuations in charge and/or atomic structures. Models of ideal materials often break down when describing the properties that arise from these complex, non-equilibrium conditions. Ultrashort X-ray pulses from an XFEL can provide element-specific snapshots of materials dynamics to characterize transient non-equilibrium and metastable phases. Programmable trains of soft X-ray pulses at high repetition rates can be used to characterize spontaneous fluctuations and heterogeneities at the nanoscale across many decades, while coherent hard X-ray scattering provides unprecedented spatial resolution of the material structure, its evolution, and relationship to functionality under operating conditions.

Experimental approaches:
- X-ray photon correlation spectroscopy (for example, see ref. 38 and 39).
- Time-resolved X-ray scattering.

8.2.5 Matter in Extreme Environments

Unpredictable material phases and properties also emerge under extreme temperature, pressure, and applied fields. Understanding the behaviour of matter under these extreme conditions is essential to improve the function of materials in extreme environments, such as those required for fusion energy, and to advance our understanding of planets and stars in the universe. Penetrating hard X-ray pulses allow the unique characterization of unknown structural phases. The spatial and temporal resolution of an XFEL enable direct comparison with theoretical models relevant for inertial-confinement fusion and planetary science.

Experimental approaches:
- Time-resolved X-ray scattering.
- Time-resolved X-ray Thomson scattering/X-ray spectroscopy (for example, see ref. 41 and 42).
- Simultaneous measurement using soft X-rays (spectroscopy) and hard X-rays (scattering).

8.2.6 Revealing Biological Function

Dynamic changes profoundly influence the biological function in protein conformations and interactions with molecules and other complexes — processes that span many decades. Such dynamics are central to the function of biological enzymes, cellular ion channels comprising membrane proteins, and macromolecular machines responsible for transcription, translation and splicing, to name just a few examples. X-ray crystallography at modern synchrotrons has transformed the field of structural biology by routinely resolving simple macromolecules at the atomic scale. XFEL facilities have already demonstrated a major advance in this area by resolving the structures of macromolecules that were previously inaccessible, using the new approaches of serial nano-crystallography and diffract-before-destroy with high-peak-power X-ray pulses. The high repetition rate of emerging XFELs (such as the European XFEL, LCLS-II-HE and SHINE) heralds another major advance by revealing biological function through the ability to follow the dynamics of macromolecules and interacting complexes in real time and in native environments. Advanced solution scattering and coherent imaging techniques can be used to characterize, at the sub-nanometre scale, the conformational dynamics of heterogeneous ensembles of macromolecules. Spontaneous fluctuations of isolated complexes and conformational changes may be initiated by specific molecules, environmental changes, or other stimuli. The ability to generate two-colour hard X-ray pulses enables new phasing schemes for nano-crystallography. It can resolve atomic-scale structural dynamics of biochemical processes that are often the first step leading to larger-scale protein motions.

Experimental approaches:
- Time-resolved X-ray scattering and spectroscopy (for example, see ref. 34, 35, 43 and 44)
- Time-resolved resonant inelastic X-ray scattering.

8.3 Fundamental Ultrafast Processes in Atoms and Molecules, and Gas-phase Chemical Dynamics

Understanding the fundamental interactions of strong X-ray fields with matter, mapping the atomic-scale structural dynamics of molecules, visualizing chemical reactions, transformations, as well as making and breaking of chemical bonds were identified early on as some of the primary science drivers for the development of XFELs, based largely on the promise of X-ray probing of molecular structural dynamics on the fundamental time and length scales of atomic motion.

Overarching themes and overcoming grand challenges in the area of gas phase chemistry and atomic and molecular physics include:

(1) *Understanding and controlling fundamental charge dynamics in atoms and simple molecules, with particular goals of linking (X-ray) experimental observables with theory to advance fundamental understanding and predictive theoretical treatments of*
- correlation effects and quantum coherence in the very early steps of chemistry,
- non-adiabatic transitions (*e.g.*, conical intersections) and quantum coupling between molecular structural dynamics and electronic transitions, and their role in mediating energy flow in molecular systems, and
- probing and controlling electron motion in strong-fields.

The timescale for this motion is set by the energetic separations between electronic states, which evolves on the femtosecond-to-attosecond scale. This implies that electronic motion occurs on sub-optical-cycle timescales and requires higher frequency probes to observe. Moreover, due to the atomic-site specificity offered by ultrafast X-ray pulses, electronic excitations can be precisely tracked using spectroscopic and structural (scattering) probes. The unique capabilities of an XFEL, including tunable attosecond pulses (and pulse sequences) provide a qualitative advance in our ability to track electronic motion on fundamental time and spatial scales. In addition, the high intensity offered by an XFEL opens the door to nonlinear X-ray spectroscopy methods, providing even deeper insight into the role of electronic coherence, electron–electron interactions, and electron–nuclear coupling in photo-driven chemistry. Highly differential (multi-dimensional) observables, which can only be obtained with high repetition rate XFEL sources, will enable better comparison with theoretical predictions.

(2) *Understanding and controlling excited-state photochemistry in isolated quantum systems (gas-phase molecules) of increasing complexity, targeting systems with broad chemical and biochemical relevance.*

Ultrafast organic photochemistry provides nearly limitless opportunities for applications in synthetic chemistry and energy storage. Light-induced chemistry can be controlled to a high level in space and time. Additionally, the required energy for a photochemical transformation can be delivered directly to a target molecule, which allows it to take place under mild thermal conditions. Ultrafast photochemical reactions are comparably selective since many unintended reaction channels cannot compete on an ultrafast timescale – creating promising new opportunities for efficient chemical storage of light energy. However, it is difficult to predict photochemical reactivity based solely on the reactant structure (in all but the simplest cases). In ground state organic chemistry, the established structure–function relationship of functional groups (such as amino or carbonyl groups) underpins the great success of synthetic organic chemistry. The formulation of similar photochemical functional groups is still in its infancy.

Establishing structure–reactivity relationships for ultrafast photochemistry is complicated because the reaction mechanisms involve coupled electron–nuclear dynamics through conical intersections (CIs), presenting significant challenges for both experimental methods and quantum chemical theories. Desired experimental observables are selectively sensitive to correlated electronic or nuclear changes during nonadiabatic dynamics. Here, ultrafast X-rays (see Chapter 9) and relativistic electrons (see Chapter 12) offer distinct advantages. XFEL studies have demonstrated the ability to map transient changes in valence electron density, with spatial resolution, *via* element- and site-specific soft X-ray spectroscopy, complemented by the structural sensitivity of hard X-ray and relativistic electron diffraction methods. Example goals include

- capturing small (low quantum yield) photochemical channels that represent important reaction pathways, which are obscured by conventional ensemble-based methods,
- understanding the influence of reactant conformations on photochemical dynamics and reaction pathways, and
- capturing key intermediates along excite-state reaction trajectories in photochemistry.

(3) *Tracking excited-state dynamics in complexes and nanoscale systems beyond isolated gas-phase molecules, including*
  - understanding the influence of the local environment on ultrafast photochemistry and
  - the study of the boundary between the molecular regime and that of macroscopic, condensed matter physics, which defines nanoscale physics. Research into the properties of nanoscale systems has the potential to lead to breakthroughs in fundamental science and novel applications and technologies.

Most photochemical processes with relevance to chemical applications happen in the solution phase. Compared to investigations of photochemical dynamics in isolated molecules, studies of such reactions are complicated by the need to differentiate between solute dynamics, solute–solvent interactions, and solvent dynamics. An important approach to investigate the contributions from solute–solvent interactions to a photochemical reaction is the comparison of the dynamics of the isolated molecule in the gas phase and the solvated molecule in solution phase. Soft X-ray XAS provides tremendous opportunities for such studies, since it is one of the few methods that can be used in the solution and gas phase. Moreover, the high repetition rate provided by emerging XFEL facilities will increase the sensitivity of gas phase experiments to the single molecule level. This sensitivity increase will enable measurements that are currently unfeasible in the gas phase. Many photochemically active species in the solution phase are charged, preventing the creation of dense molecular beams in the gas phase. High repetition rate XFEL studies will enable soft X-ray XAS in dilute ionic targets. Furthermore, transient structural information (*e.g.*, from charged transition metal complexes) can be obtained from Coulomb explosion imaging (CEI) experiments, which can be directly compared to information from solution phase X-ray diffraction studies.

(4) *Advancing chemical-physics studies of model gas-phase molecules toward comprehensive studies of gas-phase chemical reactions associated with combustion, reactive chemical flows, aerosol chemistry and related areas.*

High-temperature chemical reactions play a critical role in a wide range of applications, such as (i) the oxidation of renewable and carbon-free fuels for power generation and propulsion, (ii) the environmental and health impact of soot and particulate matter that is formed as a by-product of combustion and wildland fires, (iii) the high-temperature flame-synthesis of nanomaterials, such as carbon black and TiO_2, and (iv) heterogeneous reactions in high-pressure environments, involving CO_2-adsorption, solvent/solute systems, and astrophysical reactions. Common to these chemical reactions are precursor radical chemistry, short-lived transition states, and rate-controlling collision processes. Although progress has been made in examining these processes at the macroscopic level, the limited spatial and temporal resolution of commonly employed analysis tools and optical measurement techniques significantly restrict our ability to investigate the underlying molecular processes at the fundamental structural-dynamic level. As such, the capabilities of an XFEL for enabling precision measurements of chemical processes and their structural-dynamic response at atomic spatial scales will have a substantial impact. The combination of femtosecond temporal resolution, high repetition rate, high coherence, and unprecedented high average X-ray photon flux, offers unique opportunities for generating new fundamental insight into the equilibrium and non-equilibrium reaction pathway of exothermic reaction processes in the gas-phase and high-pressure systems.

8.3.1 Mapping Atomic Structural Dynamics of Excited-state Chemical Reactions

Time-resolved X-ray scattering is a powerful approach for mapping the structural dynamics of excited-state chemical reactions at the atomic scale. The pattern of X-rays scattered from the sample is a Fourier transform projection of the molecular electron density distribution. Ultrafast X-ray pulses capture a scattering snapshot (or time-lapsed snapshots in a pump/probe configuration) that reflects the transient molecular structure.[45] One important example is the application of ultrafast X-ray scattering (UXS) to map the photoinduced ring-opening of electrocyclic reactions, a textbook problem in organic chemistry and the basis of photobiological reactions in vitamin D synthesis. Although the relaxation from photoexcited electronic states during ring-opening has been investigated previously, the underlying changes in atomic morphology were not resolved. Pioneering XFEL experiments demonstrated the potential of time-resolved hard X-ray scattering to reveal transient molecular structures during ultrafast chemical reactions (see Figure 8.6).[46] In a laser-pump and X-ray scattering probe measurement layout (see Figure 8.6a), the characteristic scattering intensities from photoexcited 1,3-cyclohexadiene as a function of momentum transfer, q, were measured for a range of pump–probe time delays (see Figure 8.6b). A sub-200 fs transient increase in scattering intensity was observed in the

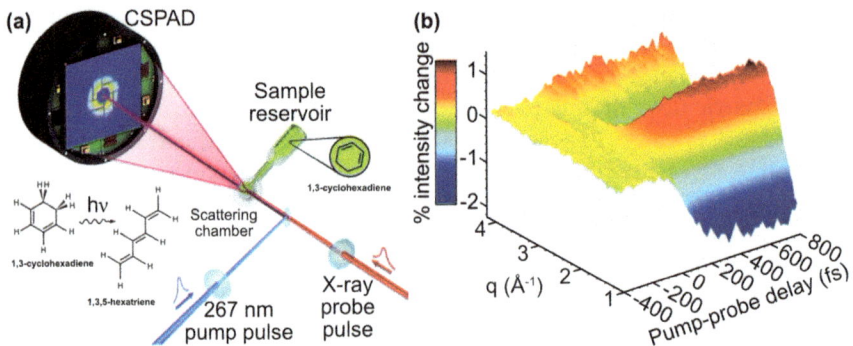

Figure 8.6   (a) Experimental setup at the LCLS X-ray pump–probe (XPP) instrument[32,33] for measuring molecular photoexcited dynamics initiated by an optical 267 nm laser pulse and subsequently probed by hard X-ray scattering. (b) The differential transient X-ray diffraction pattern of ring-opening in 1,3-cyclohexadiene to 1,3,5-hexatriene as a function of pump–probe delay time and scattering angle, q, following excitation at 267 nm.[46] Reproduced from ref. 46 with permission from the American Physical Society, Copyright 2015.

2.1–2.5 Å^{-1} region, and a delayed transient decrease in scattering intensity was observed in the 2.9–3.1 Å^{-1} region. The evolution of scattering signals reflects changing distances in non-neighbouring carbon atoms, and is compared directly with trajectory calculations of the ring-opening reaction.[46] This UXS field has recently been complemented by measurements using ultrafast electron diffraction facilities, including the MeV-UED instrument at LCLS.[47–50] A recent example is the first measurements of the conformation-dependent dynamics of the photochemical electrocyclic ring opening of α-phellandrene.[51]

Although experiments have successfully mapped trajectories of molecular photoexcited states in relatively simple model complexes, the technique is still in its infancy. One significant limitation is the range and fidelity of X-ray scattering data obtained from dilute gas-phase molecules, often comprising elements with low atomic mass. The spatial resolution is strongly dependent on the X-ray photon energy. For example, the differential scattering data (see Figure 8.6) were obtained with a photon energy of 9.5 keV which, for geometric reasons, correspond to X-ray scattering recorded only to $q \sim 4\,\text{Å}^{-1}$, greatly limiting the structural detail that can be resolved. A recent upgrade of the LCLS undulators has significantly increased the available photon energies allowing for a spatial resolution of ~1 Å or better ($q \sim 6\,\text{Å}^{-1}$, 15 keV). A further increase in spatial resolution will provide much deeper insight. Still, the scattering signals diminish at higher photon energies due to decreasing elastic scattering cross-sections, increasing inelastic scattering cross-sections, and limited hard X-ray flux at 120 Hz. Moreover, for more complex, larger molecules with multiple reaction pathways, the target densities often must be more dilute to create and maintain stable samples.

8.3.2 Revealing Correlated Changes in Valence Charge Density and Molecular Structures

The close coupling of electronic and nuclear degrees of freedom underpins much of excited-state photochemistry, giving rise to nonadiabatic dynamics that often mediate excited-state reaction pathways. Such processes often cannot be described within the Born–Oppenheimer approximation. Understanding the underlying mechanisms requires the investigation of both electronic and nuclear degrees of freedom, on equal footing, and with clearly separable experimental observables. Ultrafast X-ray scattering methods, coupled with advanced theory, offer a promising approach for significant advances on this long-standing challenge.

As an example of recent significant advances in this area, intramolecular charge transfer and the associated changes in molecular structure in *N,N'*-dimethylpiperazine have been tracked simultaneously for the first time using femtosecond gas-phase X-ray scattering (see Figure 8.7).[52] The molecules are optically excited to the 3p state at 200 nm. Following rapid relaxation to the 3s state, distinct charge-localized (3sL) and charge-delocalized (3sD) species related by charge transfer are observed. The experiment determines the molecular structure of the two species, with the redistribution of electron density accounted for by a scattering correction factor. The initially dominant charge-localized state has a weakened carbon–carbon bond and reorients one methyl group compared with the ground state. Subsequent charge transfer to the charge-delocalized state further elongates the carbon–carbon bond, creating an extended 1.634 Å bond and reorients the second methyl group. At the same time, the bond lengths between the nitrogen and the ring-carbon atoms contract from an average of 1.505 Å to 1.465 Å. The experiment determines the overall charge transfer time constant for approaching the equilibrium between charge-localized and charge-delocalized species to be 3.0 ps.

In related studies, recent ultrafast X-ray scattering experiments have captured the initial atomic-scale redistribution of valence electron density in

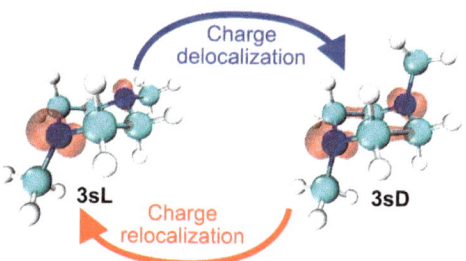

Figure 8.7 Charge transfer between charge-localized (3sL) and charge-delocalized (3sD) species of the gaseous molecule *N,N'*-dimethylpiperazine (DMP) in the 3s electronic state. Charge is localized predominantly on the planar nitrogen in 3sL, while equally distributed across both nitrogen atoms in 3sD.[52] Figure adapted from ref. 52 with permission from National Academy of Sciences USA, Copyright 2021.

a photo-excited heterocyclic molecule (1,3-cyclohexadiene, see Chapter 9) for the first time.[53] These results illustrate the power of ultrafast X-ray scattering for mapping the earliest steps in photochemistry with unprecedented spatial and temporal resolution.

High-repetition-rate XFELs will provide a qualitative advance in mapping correlated changes in valence charge density and molecular structures. The ~ 1000-fold improvement in data acquisition times will enable dramatic improvements in signal-to-noise ratios, even in regions where scattering data are sparse. Operating at 12 to 20 keV photon energies will nearly double the range of observable momentum transfer to 6–10 Å^{-1}. This will transform our ability to determine the transient atomic structures, and valence charge densities, of complex molecules at extremely high accuracy for comparison with theory. Importantly, ultrafast X-ray scattering can access the frontiers of time resolution, for example, in the range of 10 fs where the stereo-specificity of an electrocyclic reaction can be determined.

8.3.3 Creating and Probing Coherent Wavepackets of Valence Electrons

Recent advances in generating high-intensity sub-femtosecond pulses in the X-ray regime have opened new approaches for understanding charge dynamics on fundamental time and length scales. Scattering of high peak power attosecond X-ray pulses (~ 100 GW with several eV of coherent bandwidth) from simple molecules can create an electronic excitation (coherent wavepacket) localized near a single atom. This exploits the element specificity of core excitations, and the nonlinear process of impulsive stimulated X-ray Raman scattering (ISXRS).[54–56] The theoretical challenge of describing a coherent superposition of electronic states is complex. Still, it has great potential to provide fundamental new insight (linked with experimental observables) into charge correlation and quantum coherences. Semi-classical models describing ultrafast molecular dynamics, such as isomerization following photoexcitation (see, for example, ref. 57), under assumptions of a single electronic state (or non-adiabatic coupling between closely lying states) and neglecting coherence between electronic states. There is also a vast literature of theoretical models of coherent charge dynamics (electronic wavepackets) in molecular systems (see ref. 58 and references therein), but typically based on the assumption of a fixed nuclear geometry. Only recently has there been any attempt to consider coupled motion.[59,60] Current state-of-the-art calculations consider the coupling of several nuclear degrees of freedom for a finite number of electronic excited states, in very simple systems.[61,62] Experimental observations are essential to validate these new theoretical approaches, and extend these efforts towards more complex systems.

In one example, intense attosecond pulses from LCLS have been used to create a localized electronic wavepacket in nitric oxide (NO) molecules for

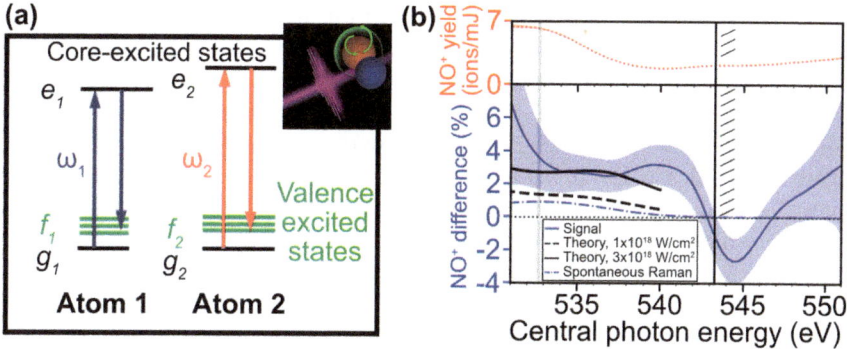

Figure 8.8 (a) Illustration of impulsive stimulated X-ray Raman scattering (ISXRS) in which localized valence excitations are created and probed *via* resonant processes at specific atoms. The inset shows the ISXRS in NO molecule using an attosecond X-ray pulse (magenta) which excites a valence electron wavepacket (green) at the oxygen site (red). (b) Central photon energy dependence of the total yield of NO^+ ions (red) and NO^+ difference signal (solid blue, 1σ error shaded). The total yield of NO^+ ions demonstrates the energetic position of the oxygen 1s → 2π* resonant feature (vertical grey shaded). The difference signal is directly proportional to the number of valence excited state neutral molecules. A vertical dashed bar indicates the oxygen K-edge (at ∼543 eV). Simulated excited-state neutral yields for peak intensities of 10^{18} W cm^{-2} (dashed black) and 3×10^{18} W cm^{-2} (solid black), and expected spontaneous Raman rate (dot dashed blue).[63] Reproduced from ref. 63 with permission from American Physical Society, Copyright 2020.

the first time (see Figure 8.8).[63] The nonlinear impulsive stimulated X-ray Raman scattering (ISXRS) process (see Figure 8.8a) involves a single impulsive interaction to excite a coherent superposition of electronic states. Observing the subsequent electronic motions is crucial to understand the role of electronic coherence in chemical processes. Electronic population transfer *via* ISXRS using broad bandwidth (5.5 eV FWHM) attosecond X-ray pulses has been demonstrated. The impulsive excitation is resonantly enhanced by the oxygen 1s → 2π* resonance of NO as shown in Figure 8.8b.[63] The photon energy dependence of the total yield of NO^+ ions (shown in red) demonstrates the energetic position of the oxygen 1s → 2π* resonant feature (vertical grey shaded line). The NO^+ difference signal (shown in blue, 1σ error shaded) is directly proportional to the number of valence excited state neutral molecules and shows good agreement with theory predictions for ISXRS in the 533–540 eV range (black solid and dashed lines).[63]

Beyond the initial excitation, soft X-ray spectroscopy methods are a powerful means to probe valence electron dynamics. The core binding energy and core-to-valence absorption spectrum provide a sensitive measure of localized electron density around various atoms in a molecule or molecular complex. This atomic site-specificity will significantly aid comparisons between time-domain measurements and theoretical models due to the distinct spectral fingerprints of dynamics at different atomic sites.

In addition to spectroscopic observables of ultrafast charge motion, further development of high intensity attosecond hard X-ray pulses will enable diffractive imaging of attosecond charge motion. Naively, one might assume that the X-ray diffraction signal from a time-dependent charge density is simply given by the instantaneous charge density when the X-ray pulse arrives. This picture is flawed, and the measured scattering pattern depends critically on the coherences and inelastic scattering components[64,65] which presently cannot be resolved outside the bandwidth of the probe pulse. The measured diffraction signal is a mixture of elastic contributions from the ground and excited states, inelastic Stokes and anti-Stokes processes, and a mixed elastic/inelastic contribution that encodes the quantum coherence.[65] Disentangling these terms is non-trivial and will take considerable experimental and computational effort. Thus, developing new methods for imaging charge motion will be critically tied to developing novel momentum- and energy-resolved scattering techniques.

8.3.4   Challenges and Opportunities in Combustion Science

The benefits of modern combustion technologies (*e.g.*, reliable electricity, rapid transportation, *etc.*) are accompanied by negative consequences such as combustion particulates, photochemical smog, and anthropogenic climate change. It is increasingly important to utilize combustion with greater efficiency and fewer harmful impacts. For example, in the transportation sector, new energy sources such as biomass-derived fuels offer an opportunity to optimize the fuel stream for new highly efficient engines, and to develop novel fuels that will help reduce greenhouse gas emissions. Accordingly, there is an increasing need for predictive models of engine combustion that are accurate from the scale of molecules and electrons through the macroscopic scale of engine cylinders.

The unique capabilities of XFELs could be instrumental in creating the science base for predictive combustion models. Because combustion relies on a complex interrelationship of chemistry and turbulent fluid mechanics, it exhibits inhomogeneities and correlations across a wide range of length and time scales. One aspect of this challenge related to soot formation (see Figure 8.9) is the byproduct of gas-phase combustion and pyrolysis of hydrocarbons in the reaction zone of a flame. These reactions involve multiple pathways and intermediate states of the reactants. Still, they can often be classified by their time and length scales – spanning from atoms and molecules to aggregates on >100 nm scale (see Figure 8.9).[66] Understanding and thereby controlling the inception/nucleation and growth processes of soot particles is important both for direct impact to human health, and for affecting the Earth radiation balance. It is unclear if nucleation and condensation processes lead to soot particles held together by dispersion forces, or if another mechanism is at play. The leading theory is that resonant stabilized cyclic radicals drive chain reactions that initiate and propagate the chemistry that leads to solid carbon formation under high-temperature, hydrocarbon-rich conditions.[67] The bonding forces that drive the agglomeration of these incipient particles into larger mature soot is also poorly understood.

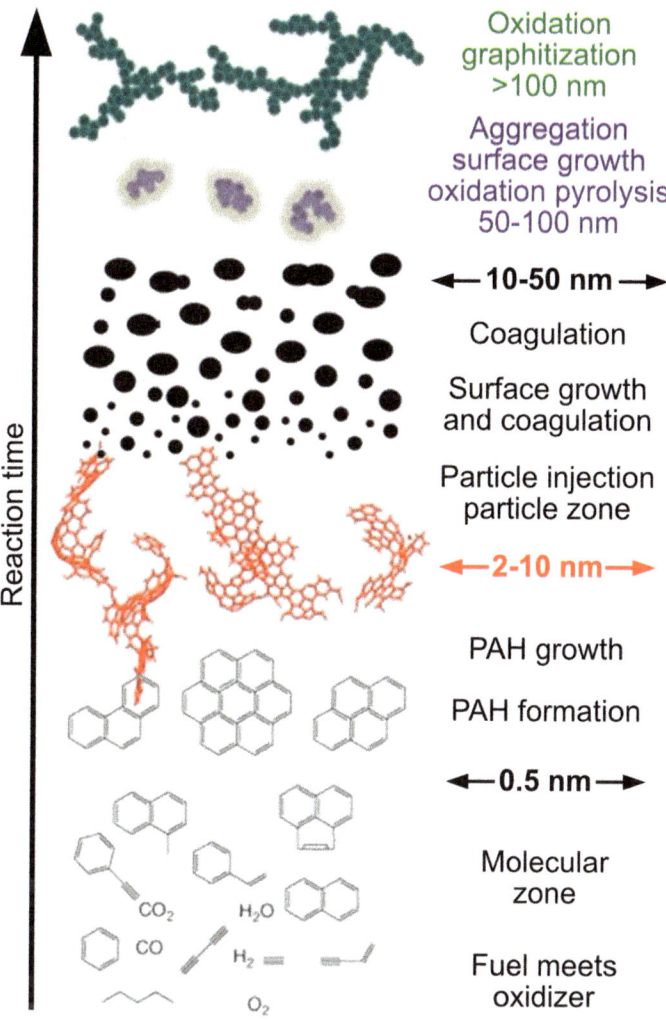

Oxidation
graphitization
>100 nm

Aggregation
surface growth
oxidation pyrolysis
50-100 nm

← **10-50 nm** →

Coagulation

Surface growth
and coagulation

Particle injection
particle zone

← **2-10 nm** →

PAH growth

PAH formation

← **0.5 nm** →

Molecular
zone

Fuel meets
oxidizer

Reaction time

CO_2 H_2O
CO H_2
O_2

Figure 8.9 Schematic of soot formation and evolution in a flame, starting from the production of hydrocarbon radicals during initial fuel oxidation.[66,68,69] These radicals lead to higher molecular-weight growth and polycyclic aromatic hydrocarbons (PAHs)[68] that grow to form incipient nanoparticles in the size range of 2–10 nm (red).[70,71] Coagulations, hydrogen loss and surface growth continue until these agglomerated particles become covalently bound aggregates of mostly carbonaceous particles with a fine structure resembling polycrystalline graphite (green).[69] Subsequent oxidation of these graphitic particles proceeds at "active" sites[72] on the graphitic surface to form CO and CO_2.[73] Molecular structures and particles are not drawn to scale. Adapted from ref. 66 with permission from Springer Nature, Copyright 1994. Figure was modified with molecular species predicted to initiate soot formation (Violi, personal communication) and extended to include aggregation, graphitization, and oxidation of mature graphitic soot.

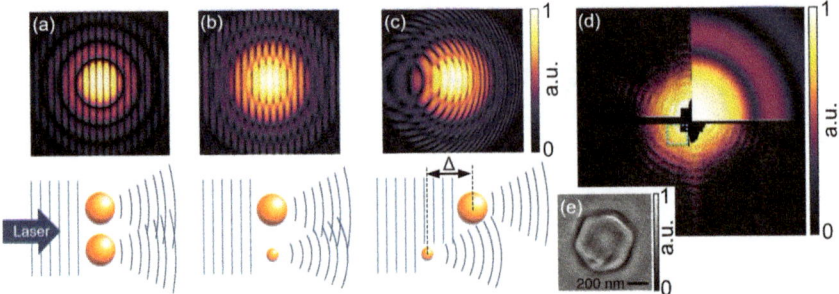

Figure 8.10   In-flight holography. (a) The sizes of the spheres are encoded in the circular diffraction pattern while the lateral distance between them is encoded in the vertical lines (analogous to Young's double-slit). (b) Modification of the diffraction pattern as the size of one of the spheres decreases. (c) When the longitudinal distance Δ changes, the diffraction pattern exhibits curved lines. In this case, the small sphere can be regarded as a reference. (d) Hologram of a Mimi virus. The envelope of the reference is shown in the upper right quadrant. (e) A 2D projection of the reconstruction.[75] Reproduced from ref. 75 with permission from Springer Nature, Copyright 2018.

Early experiments at the LCLS demonstrated the potential for XFELs to reveal the shape and structure of individual aerosol particles in their native environment, using the method of coherent X-ray diffractive imaging.[68–74] The recent development of in-flight X-ray holography represents a promising advance for *in situ* imaging of soot formation across relevant length scales (see Figure 8.10). This coherent imaging method similarly opens novel avenues to study cloud formation and catalytic processes on both the nanoscale and at the single-particle level. The first X-ray 3D holograms of free non-periodic nanoparticles were recorded at the LCLS,[75] and image interpretation required only two steps of Fourier inversion (without any predefined assumptions), which is orders of magnitude fewer than usually required. In this approach, the scattering objects' relative positions are encoded in the vertical interference fringes, while the relative sizes are encoded in the circular diffraction pattern (see Figure 8.10a–c). Randomly placed nanoclusters are used as reference X-ray scatterers to encode relative phase information into the diffraction patterns of a virus (see Figure 8.10d). The resulting hologram contains an unambiguous three-dimensional map of the virus (see Figure 8.10e). This approach unlocks the benefits of holography for ultrafast X-ray imaging of nanoscale, non-periodic systems. It paves the way for direct observation of complex electron dynamics down to the attosecond timescale.

8.4   Condensed Phase Chemistry and Catalysis

Charge separation, migration, redistribution, and localization are central to efficient and robust energy storage and retrieval from chemical bonds, as well as more general photo-driven chemical transformations through photosynthesis and catalysis. It has long been recognized that these processes in molecular systems are mediated by the close coupling between electronic and nuclear degrees of freedom (vibronic coupling). However, this coupling is not well understood at the quantum level, even in simple molecules. Although ultrafast optical and IR methods provide important insight into the timescales of these nonequilibrium dynamics, we have not been able to directly observe these processes to date on their fundamental timescales together with the required chemical specificity. Furthermore, conventional theoretical models based on step-wise equilibrium (*e.g.*, intramolecular vibrational energy redistribution, IVR, followed by inter-system crossing, ISC) or separation of timescales (*e.g.*, Born–Oppenheimer approximation) are inadequate in describing electronic excited-state reactions that are inherently non-equilibrium in nature. Advanced time-resolved X-ray scattering and spectroscopy methods enabled by XFELs (closely coupled with new theoretical approaches) will drive a qualitative advance in our understanding of excited-state chemical transformations, leading to design principles for controlling these processes through directed synthesis.

One major scientific objective underpinning approaches to artificial photosynthesis is understanding how the rapid evolution of excited-state charge-density and spin-density distributions and their interaction with the solvent environment determine non-radiative relaxation and chemical reaction pathways. Charge-transfer dynamics in transition-metal complexes and assemblies for light harvesting and photocatalysis are of particular interest. Here, the challenge is establishing design principles for achieving excited-state lifetimes and catalytic performance exploiting earth-abundant 3d metals presently exhibited in complexes based on precious 4d and 5d metals.

Three general areas of condensed phase chemistry where XFEL facilities have had a significant impact to date are:

- Mapping the evolution of chemical bonds: charge distributions, oxidation states and frontier orbitals.
- Atomic scale structural studies of excited-state chemical dynamics: revealing relaxation pathways, and coupling of atomic structures, electronic structures, and solvent dynamics.
- Mapping reaction pathways and transition states in surface chemistry.

Results to date point to significant future science opportunities in increasingly complex chemical systems, such as photosensitizers, photocatalysts, and heterogeneous catalysis. This increased impact is enabled by emerging advances in the repetition rate and average power that support new capabilities such as

- Advanced time-resolved spectroscopy including valence-to-core emission spectroscopy (VtC), high energy resolution fluorescence detection XANES (HERFD-XANES), and 2D RIXS maps of dilute solutes and low-concentration reactive species on surfaces.

- X-ray scattering over an extended q-range (finer spatial resolution) for mapping subtle structural motions and more direct structural refinement from scattering data.
- Multimodal studies combining X-ray scattering and spectroscopy.
- Studies of coupled electronic and atomic structural dynamics of complex molecular assemblies at low concentrations, including relevant solvent structural dynamics.

In this section, we focus on advanced X-ray scattering and multimodal X-ray methods (scattering and spectroscopy), as well as the significant science impact that the application of these methods have had on condensed-phase chemistry and catalysis enabled by XFELs.

8.4.1 Atomic Scale Structural Dynamics of Molecules, Chemical Reactions, and Solvation

Following the structural dynamics of molecules in real time and on the atomic scale provides powerful new insight into chemical reactions. This has become possible for the first time with ultrafast hard X-ray pulses from XFELs. Furthermore, much of chemistry occurs in a solvent (often aqueous) environment, and the structural dynamics of the solvation shell often play an integral role in determining reaction pathways. Ultrafast X-ray diffuse scattering studies at LCLS, combined with quantum mechanics/molecular mechanics (QM/MM) simulations, reveal the structural dynamics of excited photocatalysts and further distinguish the contributions from the solute

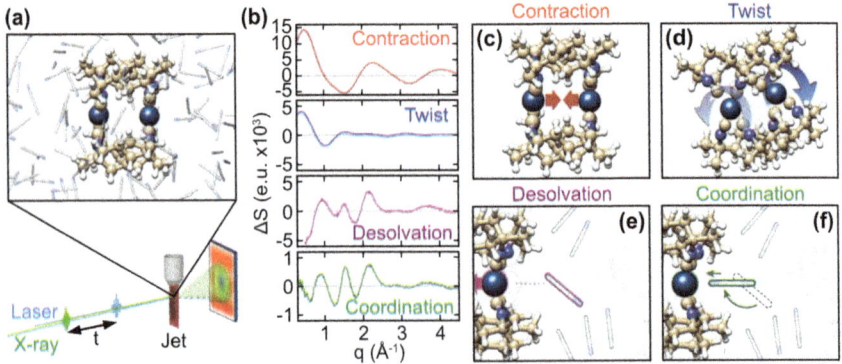

Figure 8.11 (a) Experimental setup for solution-phase X-ray scattering. The inset shows a snapshot of $[Ir_2(dimen)_4]^{2+}$ in acetonitrile solution from QM/MM simulations. (b) Four main transient components that describe the q-dependent time-resolved scattering data. (c–f) Four distinct structural dynamics of the solute and solvent that correspond with the signals in (b).[76] Reproduced from ref. 76, https://doi.org/10.1038/ncomms13678, under the terms of the CC BY 4.0 license https://creativecommons.org/licenses/by/4.0/.

molecule and the solvent. One important illustrative example is an LCLS study of the model photocatalytic molecular system $[Ir_2(dimen)_4]$ (see Figure 8.11a) photoexcited using an optical laser and probed by the X-ray FEL.[76] Analysis of the solute–solvent pair distribution function (see Figure 8.11b) revealed details of the solvent structural dynamics (see Figure 8.11e and f) around the catalytically active Ir site. In particular, large translational motions (see Figure 8.11c), strongly coupled to rotations (see Figure 8.11d), lead to much slower excited-state dynamics than those associated with the standard picture of solvation in polar solvents.[76]

In a more recent study of the molecular photocatalyst $[Pt_2(POP)_4]$, tailored non-resonant optical excitation is used to create a coherent non-equilibrium ground state population (vibrationally cold hole), allowing the ground-state potential surface (along the Pt–Pt coordinate) to be mapped on the sub-Ångström scale for the first time.[77] This represents an important step toward understanding and ultimately controlling photocatalysts.

8.4.2 Multi-modal Studies of Excited-state Chemical Dynamics: Revealing Relaxation Pathways, and Coupling of Atomic Structures, Electronic Structures, and Solvent

Multi-modal ultrafast X-ray studies have been pioneered at LCLS (see Figure 8.12), demonstrating the power of combining hard X-ray scattering and spectroscopy to map coupled electronic and atomic structural dynamics in molecular complexes and assemblies. Figure 8.12 shows a characteristic multi-modal experimental setup, with optical pump and X-ray probe beams interacting in a flowing sample. Forward time-resolved X-ray scattering

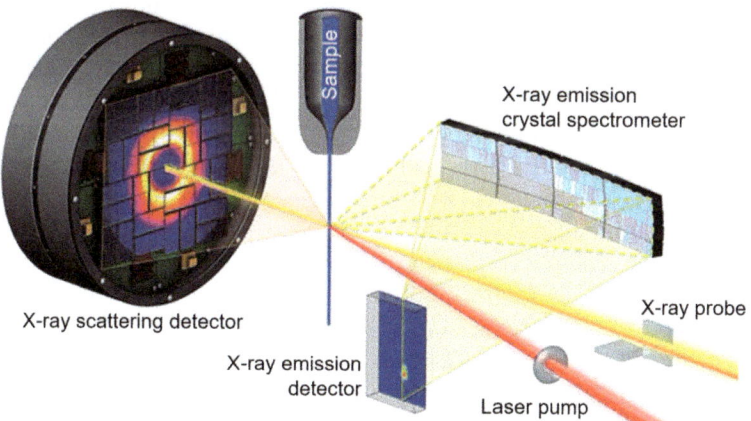

Figure 8.12 Schematic of the experimental setup enabling both X-ray diffuse scattering or diffraction (using a large area detector), and X-ray emission, XES (using a von Hamos XES spectrometer).[78] Reproduced from ref. 78 with permission from the Royal Society of Chemistry.

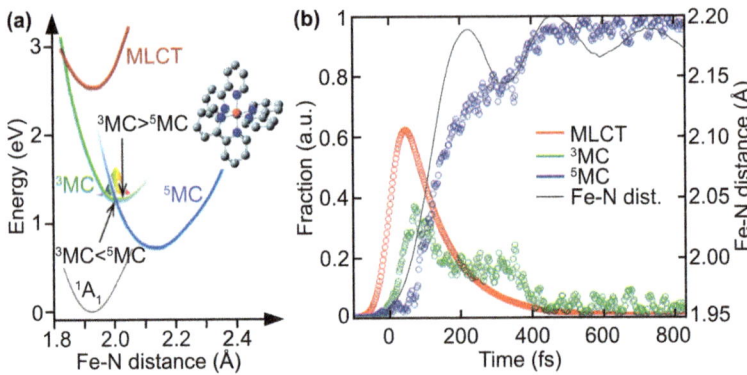

Figure 8.13   Hard X-ray scattering and XES studies of the $[Fe(bpy)_3]^{2+}$ light harvesting complex. (a) Potential energy surfaces indicating the initial metal-to-ligand charge-transfer (MLCT) excited state, and subsequent metal-centered (MC) states. (b) Measured electronic and atomic structural dynamics compared to a simulation from the same potential energy surfaces.[78] Reproduced from ref. 78 with permission from the Royal Society of Chemistry.

signals are captured in an area detector. Time-resolved X-ray emission spectra are captured in a spectrometer composed of bent crystal collection optics and a second area detector.[78]

Multi-modal ultrafast X-ray studies using a model iron-based molecular light harvester, $[Fe(bpy)_3]$, have directly quantified the initial coupling of electronic and structural configurations with atomic resolution and spin-state specificity.[78] Time-resolved X-ray scattering follows the evolution of the Fe–N bond distance (see Figure 8.13), while XES reveals the transition from the initial metal-to-ligand charge-transfer (MLCT) state, and the interconversion between metal-centered ^3MC and ^5MC states. This study revealed, for the first time, the coherent structural motion (modulation of the Fe–N bond distance) and its direct coupling to coherent electronic transitions observed between ^3MC and ^5MC states (see, for example, at ~ 300 fs in Figure 8.13b) at the intersection between the respective potential energy surfaces at an Fe–N distance of ~ 2.15 Å.

LCLS multi-modal studies of an iron-based photo-sensitizer further illustrate the scientific potential of this approach.[79] Combined time-resolved XES and solution X-ray scattering measurements of the $[Fe(bmip)_2]^{2+}$ complex reveal the coherent oscillations of an Fe–ligand stretching vibration on a ^3MC excited-state potential surface. These nonequilibrium dynamics mediate the branching pathways between a longer-lived charge-transfer state (desirable for photovoltaics and photo-redox catalysis) and an undesirable relaxation pathway (through the ^3MC state) in which the captured photo-excitation energy is dissipated as heat.

In addition to the effects of coupled intra-molecular structural dynamics, charge transfer and relaxation processes in light harvesting complexes and

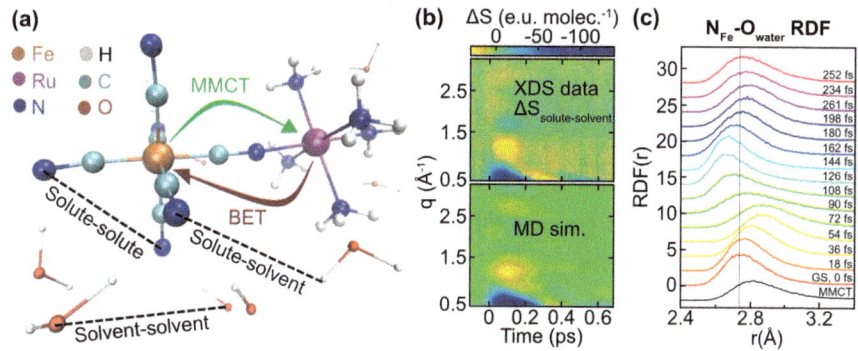

Figure 8.14 (a) Mixed-valence aqueous $Fe^{II}Ru^{III}$ molecule undergoing photoexcitation with an 800 nm pulse that induces metal-to-metal charge transfer (MMCT) from the Fe to the Ru center, followed by ultrafast back-electron transfer (BET). (b) X-ray scattering data and molecular dynamics (MD) simulations. The interatomic distances probed by the elastically scattered X-rays are classified into solute–solute (intramolecular), solute–solvent, and solvent–solvent atom pair distances. (c) MD simulations of N_{Fe}–O_{water} radial distribution function (RDF) as a function of time following the initial MMCT transition. The black vertical line marks the position of the first-solvation-shell peak in the ground state. The RDF of the equilibrated MMCT state is shown for reference (black RDF).[80] Reproduced from ref. 80 with permission from Springer Nature, Copyright 2021.

photocatalysis are substantially influenced by solvent reorganization. Multimodal ultrafast X-ray scattering and spectroscopy studies enabled by XFELs provide important new insight into this long-standing science challenge.

Recent LCLS studies in a model mixed-valence complex for photocatalysis illustrate this area's challenges and important future science opportunities. The solvated cyanide-bridged mixed-valence bi-metallic charge-transfer complex, Fe^{II}–Ru^{III}, serves as a model system for understanding the complex couplings between electronic and atomic degrees of freedom in the solute and the surrounding solvent mediates the electron transfer (and back electron transfer) processes (see Figure 8.14). Combined time-resolved X-ray scattering (see Figure 8.14b) and spectroscopy studies of solvated Fe^{II}–Ru^{III} at LCLS demonstrated that the dominant back electron transfer process is coupled to coherent translational motions of the first solvation shell.[80] Comparison with non-equilibrium MD simulations (see Figure 8.14b and c) demonstrate that the observed coherent translational motions arise from hydrogen bonding changes between the solute and nearby water molecules, in response to the initial photoexcitation (see oscillatory changes in the peak position with increasing delay time in Figure 8.14c). This result provides an atomistic view of coherent solvent reorganization (on the tenths of Å length scale and <100 fs time scale), mediating ultrafast intramolecular electron transfer.

An important future science opportunity is to extend this class of research to increasingly complex chemical systems that are of broad interest as photosensitizers and photocatalysts. This requires much greater measurement sensitivity and resolution (larger q-range available from hard X-rays) to directly reveal subtle structural changes from measurements of dilute solutions, with less reliance on model calculations for interpretation.

8.4.3 Dynamic Studies of Biochemistry Enabled by Advanced XFEL Methods

In nature, biological catalysts or enzymes enable organisms to perform thermodynamically highly-demanding chemical transformations under ambient conditions, close to the thermodynamic limit, often involving only very subtle structural distortions coupled with changes in electronic structures (*e.g.*, oxidation or spin state). Sensitive time-resolved X-ray spectroscopy and scattering at XFELs are powerful tools for understanding enzyme function and related biochemical processes in operating environments. In addition to accessing fundamental dynamics, ultrafast X-ray pulses can probe molecular and electronic structures before the onset of X-ray induced redox and related damage mechanisms that are a significant limitation for synchrotron studies at room temperature.

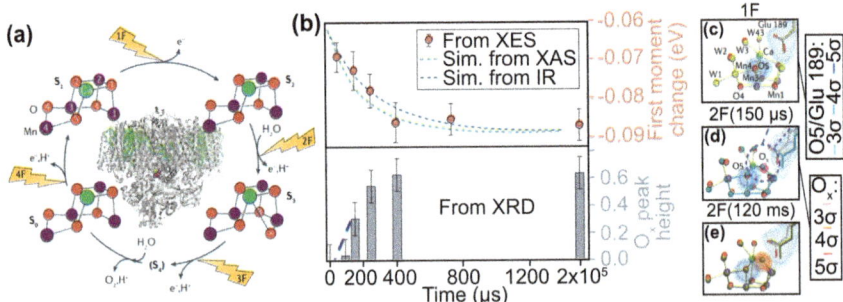

Figure 8.15   Combined XES and X-ray diffraction (XRD) studies of photosystem II.[22] (a) Illustration of the structure of the protein and the four-step catalytic cycle (Kok cycle), triggered by 4-photon absorption events (flashes 1F–4F). The structure of the catalytic Mn_4Ca cluster (O-red, Mn-purple, and Ca-green), obtained from XRD, is shown for each of the stable states S_0, S_1, S_2 and S_3. (b) Results from time-resolved XES and XRD measurements, compared with kinetic simulations based on infrared (IR) and X-ray absorption (XAS) measurements. XRD and XES data show concomitant Mn oxidation and insertion of a new oxygen (OX) in the Mn cluster on the 250 µs timescale during the $S_2 \rightarrow S_3$ transition. (c–e) The S_1 state structure (1F) is shown (C light grey, O-red, Mn-purple, and Ca-green) together with the S_2 state structure at different time points: 2F (150 µs) and 2F (200 ms), which is the S_3 state structure. The electron density is contoured at 3, 4 and 5σ around the O5 and O_X atoms of the Mn cluster and Glu189, a critical mobile amino acid side chain. Reproduced from ref. 22 with permission from Springer Nature, Copyright 2021.

The impact of this approach is illustrated by studies on the oxygen-evolving cluster (Mn_4Ca) responsible for water splitting in photosystem II (PS II) through the four-photon, four-electron process comprising the Kok cycle (see Figure 8.15a). LCLS time-resolved multi-modal XES and X-ray diffraction (XRD) measurements provided critical information on the valence state of the Mn atoms (see Figure 8.15b), correlated with changes in atomic distance in the catalytic cluster through the oxygen evolution reaction cycle.[81–84] The structural changes of the Mn_4Ca cluster occur between the S_1 and S_3 states (see Figure 8.15c–e). In studies of PS II, XES was an essential control to track the intactness of the proteins during structural crystallography measurements.

8.4.4 Structural Dynamics in Nanocatalysts *Via* Advanced XFEL Scattering Methods

Heterogeneous catalysis is essential in chemical processes for the chemical industry. It is often the limiting factor in many approaches for renewable energy (*e.g.*, fuel cells, sustainable chemical syntheses, solar fuels, *etc.*). Although a thermodynamic picture of catalytic activity for many processes is well developed, the details of the chemical dynamics and related factors that determine selectivity remain poorly understood. Optimizing selectivity requires knowledge of competing mechanisms that span a range of timescales and occur at different active sites under operating catalytic conditions.

In functioning heterogeneous catalysts, the materials are neither static nor homogeneous. The evolution of atomic and electronic structures, linked with inhomogeneities, strain, and morphology changes through catalytic cycling ultimately determine functionality. Knowing the time evolution of the atomic and electronic structure of substrates and reactants, particularly near elusive transition states, is critical for developing a predictive understanding for design of new catalysts.

Today, we are unable to develop a complete picture of this structural evolution on the femtosecond to picosecond timescale relevant to atomic motion or the nanosecond to millisecond timescale characteristic of diffusion and materials evolution. LCLS has enabled the study of simple surface reactions, and reported the first observation of a surface transition state.[85] These groundbreaking studies on ideal crystals, with reactants prepared at high concentrations in a vacuum, demonstrate the potential for XFELs to fully understand chemical reactions on surfaces. However, these studies do not address "working" catalysts typically of low dimensionality (*e.g.*, nanoclusters of metals on an oxidic support), where coupled electronic and atomic structure fluctuations become increasingly important.

High repetition rate XFELs will enable completely new approaches for simultaneously following the atomic and electronic structure of heterogeneous catalysts. Of particular interest are multimodal methods that combine time-resolved X-ray spectroscopy (*e.g.*, EXAFS, XES, RIXS, or PES) to probe the

330

Chapter 8

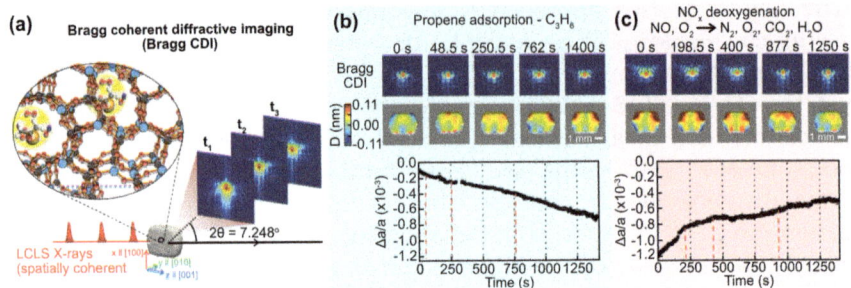

Figure 8.16 (a) Experimental schematic of time-resolved Bragg coherent diffractive imaging (Bragg CDI) of zeolite catalysts in an operating environment. The inset shows the zeolite structure with inhomogeneous distribution of active sites. (b and c) Bragg CDI images and projected displacement field maps with corresponding the strain-rate coefficients during (b) propene adsorption (in the light blue panel) and (c) NO_x deoxygenation process (in the light magenta panel), respectively. During the propene adsorption, the scattering patterns in (b) show that the fringes surrounding the central Bragg peak expand to higher wave vectors from 250.5 s to 1400 s. In contrast, during the NO_x deoxygenation process (c) the scattering patterns change significantly, then abruptly return to the initial state at 877 s. Panel (c) reproduced from ref. 88, https://doi.org/10.1038/s41467-020-19728-3, under the terms of the CC BY 4.0 license http://creativecommons.org/licenses/by/4.0/.

electronic structure and chemical environment, with time-resolved coherent X-ray diffraction to map changes in the atomic structure. Demonstration experiments in the soft X-ray range at FLASH,[86] LCLS,[73] and EuXFEL[87] have highlighted the promise of this approach for characterizing heterogeneous ensembles of nanoparticles at the atomic scale using hard X-rays.

An important recent example highlights the new insight provided by transient coherent X-ray scattering methods applied to zeolite catalysts in operating environments (see Figure 8.16).[88] Zeolites are three-dimensional aluminosilicates having unique properties arising from the size and connectivity of their sub-nanometre pores, the Si/Al ratio of the anionic framework, and the charge-balancing cations (see inset Figure 8.16a). The inhomogeneous distribution of the cations affects their catalytic performance because it influences the intra-crystalline diffusion rates of the reactants and products. However, conventional analytical tools have not yet observed the structural deformation regarding inhomogeneous active regions during the catalysis.

Time-resolved Bragg coherent diffraction imaging (CDI; see Chapters 5 and 6) using X-rays was used to investigate the internal deformations in Cu-exchanged ZSM-5 zeolite crystals during the deoxygenation of nitrogen oxides with propene (see Figure 8.16a).[88] Figure 8.16b and c show Bragg CDI images and projected displacement field maps with corresponding strain-rate coefficients during propene adsorption and NO_x deoxygenation processes. During propene adsorption, the scattering patterns show an expansion of fringe patterns to higher wave vectors. In contrast, during the NO_x deoxygenation process, the scattering patterns change significantly, and then abruptly return to the initial state at 877 s.[88]

The results show that the interactions between the reactants and the active sites lead to an unusual strain distribution, evolving throughout the adsorption (see Figure 8.16b) and deoxygenation processes (see Figure 8.16c). This is confirmed by density functional theory simulations, originating from the inhomogeneous distributions of Cu ion active sites. These observations provide new insight into the role of structural inhomogeneity in zeolites during catalysis and will inform the future design of zeolites for targeted applications.

8.5 Understanding and Controlling Structural Dynamics in Complex Materials

The discovery, understanding, and control of novel phases of complex matter are at the core of modern material and condensed matter physics. Femtosecond pump–probe methods investigate the impulse response of materials to tailored excitations (disentangling coupled degrees of freedom). They are a powerful approach to create pathways towards new phases of matter. Time resolved X-ray diffraction is a well-established method to reveal atomic structural dynamics in complex matter on fundamental timescales. The additional insight from multi-modal methods (*e.g.*, X-ray diffraction and spectroscopy) significantly enhances the science impact in key applications.

Furthermore, complex matter is often characterized by mesoscale heterogeneities (*i.e.*, material domains exhibiting different properties) that underpin important macroscopic behaviour. Such mesoscale texture is often associated with intrinsic dynamics on relevant length scales. These coupled structural dynamics substantially influence the properties and functionality of complex matter.[89,90] Examples include structural phase transitions such as in phase-change materials, polarization switching in multiferroics, nanoscale energy transport, and fluctuating electronic orders in correlated materials. Here, heterogeneity can take the form of nanoscale phase separation (domains), crystal defects, or even engineered nanostructures. The characteristic length scales span from sub-nm to >1 μm, and the corresponding dynamics range from <100 fs for ballistic phenomena to >1 ns for diffusive processes.

Studies of heterogeneous materials to date have been largely limited to either spatially-resolved measurements at equilibrium (*i.e.*, averaging over spontaneous fluctuations), or time-resolved measurements averaging across many domains and/or nanostructures. However, new X-ray capabilities and methods allow significant advances in our understanding of complex matter by simultaneously characterizing mesoscale heterogeneity and associated structural dynamics on fundamental time and length scales. Advanced time-resolved diffuse X-ray scattering and coherent scattering/imaging techniques provide access to the relevant length scales, and XFELs in particular provide access to the ultrafast timescales.

Finally, spontaneous fluctuations and heterogeneity are pervasive in complex matter, and are often central to their functional properties. In contrast to the pump–probe approaches for creating and characterizing nonequilibrium material phases, direct observation of spontaneous dynamics and rare events near equilibrium is essential for a deeper understanding in many systems. The important timescales for such dynamics can span many decades, and are directly coupled with structural changes that span from the atomic scale to mesoscale.

Coherent, ultrafast hard X-ray pulses from XFELs provide unprecedented opportunities, particularly exploiting X-ray photon correlation spectroscopy (XPCS) and related X-ray speckle visibility spectroscopy (XSVS) methods to observe the stochastic dynamics of atomic and molecular motion over the relevant range of timescales, giving access to higher-order spatial and temporal correlation functions. However, there are significant experimental challenges in measuring stochastic dynamics since, unlike pump–probe measurements, temporally uncorrelated fluctuations are the relevant observable. This places stringent limitations on the intensity of X-ray pulses that can be used without perturbing the essential correlations of interest.

In this section, we focus on advanced X-ray scattering methods, including time-resolved Bragg diffraction, diffuse X-ray scattering, coherent X-ray scattering, and multimodal X-ray methods (scattering and spectroscopy), and the significant science advances in understanding and controlling complex matter enabled by XFELs – in three key areas:

- Understanding and controlling structural dynamics in complex materials driven out of equilibrium by tailored excitations.
- Understanding the role of low-energy collective modes and correlation phenomena in determining the important properties of quantum materials.
- Understanding stochastic dynamics and structural fluctuations often determining real-world materials' functional properties.

8.5.1 Structural Dynamics in Complex Matter Driven Out of Equilibrium

Manipulating the structure and function of materials remains one of the ultimate challenges of modern condensed matter physics. While ultrafast pulses can create novel phases not accessible in equilibrium, the dynamics are often simplified by assuming that they evolve on a potential energy surface described by a few degrees of freedom with a well-defined spatial periodicity. Experimentally, ultrafast X-ray (and electron) diffraction studies are sensitive to structural dynamics on the atomic length-scale, however, they measure averages over many unit cells in ordered materials, and thus

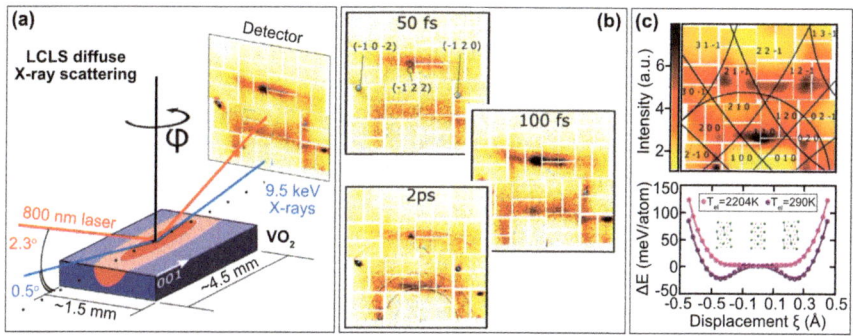

Figure 8.17 (a) Experimental schematic of time-resolved diffuse X-ray scattering of the photo-induced insulator–metal transition (IMT) in VO_2. (b) Time-resolved snapshots of the diffuse X-ray scattering intensity taken at 50 fs, 100 fs, and 2 ps following photo-excitation. (c) Hard X-ray scattering captures large regions of momentum and shows an increase in the diffuse intensity simultaneously with a decrease of the super-structure Bragg peaks.[91] Panels (a–c) are reproduced from ref. 91 with permission from the American Association for the Advancement of Science, Copyright 2018.

are less sensitive to random displacements. Femtosecond X-ray diffuse scattering overcomes this limit to reveal the dynamics of the structural transitions on all length-scales.

One important example of the insight provided by time-resolved X-ray diffuse scattering are studies of the structural dynamics of the photo-induced insulator-to-metal transition in the prototypical correlated oxide VO_2 (see Figure 8.17a).[91] Despite numerous time-resolved X-ray (and electron) diffraction studies,[92–94] time-resolved X-ray diffuse scattering snapshots (see Figure 8.17b) surprisingly revealed significant new insight into structural transition – that it proceeds by uncorrelated disordering of the vanadium ions from their initial dimerized distribution, rather than the previously proposed synchronized motion along an optical phonon mode. Comparison of LCLS scattering results with *ab initio* molecular dynamics simulations (see Figure 8.17c) show that a highly-anharmonic, flat potential energy surface allows the quasi-rutile structure in the photoexcited state to develop on femtosecond timescales, by disrupting the vanadium pairs and populating a continuum of modes that enables the system to reach ergodicity within 150 fs. More generally, these results question the common interpretation of many ultrafast measurements in terms of motion along a well-defined reaction coordinate, or order parameter, on a potential energy surface and provide a new level of understanding regarding how complex systems can reach ergodicity on an ultrafast timescale.

A second important example of new scientific insight into non-equilibrium structural dynamics in complex matter is recent time-resolved X-ray scattering studies mapping the structural evolution of phase change materials (PCM), such as $Ag_4In_3Sb_{67}Te_{26}$,[95] widely used for computer memory applications.

Logic states are stored in two stable material phases: a low resistance crystalline phase, and a high resistance amorphous (glassy) phase. The "glassy" phase must be resistant to crystallization at ambient temperatures, while at the same time, the crystalline phase must form rapidly upon heating (as this limits the memory speed). A significant open question is what are the atomic structural dynamics (temperature-dependent atomic mobility) that control the nanosecond crystallization kinetics, and how can material design principles be developed to optimize these properties?

The high peak brightness of ultrafast XFEL pulses combined with time-resolved wide-angle scattering provides an atomic scale view of the melt-quenching and crystallization process for the first time (see Figure 8.18a).[95] Characteristic X-ray scattering patterns for the amorphous (glassy) phase were measured (see Figure 8.18b), with dominant momentum transfer (scattering rings) indicated by q_1 and q_2, and for the crystalline phase. The time evolution of the scattering patterns and dominant momentum-transfer components (see Figure 8.18c) were studied in detail. A careful comparison of X-ray scattering results with *ab initio* molecular dynamics simulations revealed that the key to PCM functionality is a temperature-dependent liquid–liquid phase transition – with a "strong" liquid characterized by rigid bonds and low atomic mobility (inhibiting crystallization), and a "fragile" liquid characterized by flexible bonds and high atomic mobility (facilitating rapid crystallization).[95] The fragile liquid is further characterized by continuous six-fold coordination of the nearest atomic neighbours, while the strong liquid results from a Peierls distortion that breaks the nearest-neighbour

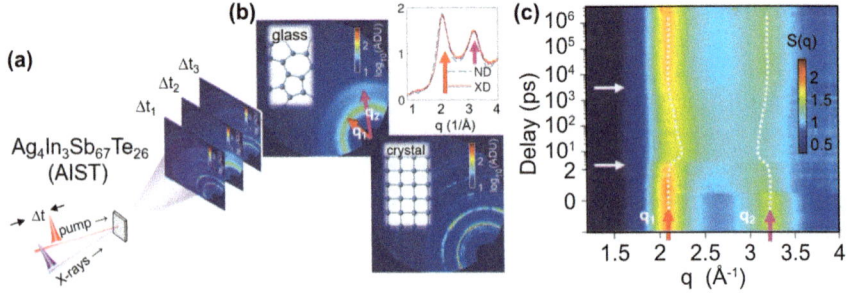

Figure 8.18   (a) Experimental schematic of time-resolved X-ray scattering in the phase-change material (PCM) $Ag_3In_3Sb_{67}Te_{26}$, AIST. An ultrafast optical pulse initiates a transient temperature change, followed by a delayed X-ray scattering probe of the atomic structure. (b) Characteristic X-ray scattering patterns for the amorphous (glassy) phase, with dominant momentum transfer (scattering rings) indicated by q_1 and q_2, and for the crystalline phase. (c) Time evolution of the dominant scattering peaks. The dependence of these dynamics as a function of fluence (not shown), and comparison with *ab initio* molecular dynamics modeling, reveals the presence of a temperature-mediated strong-liquid to fragile-liquid phase transition.[95] Reproduced from ref. 95 with permission from the American Association for the Advancement of Science, Copyright 2019.

symmetry. This deeper understanding of the relationship between atomic structure and kinetics establishes important guidance for tailored PCM design, namely, materials with low kinetic activation energy for the liquid–liquid phase transition (compared to the melting temperature) to provide the largest thermal operating window between logic phases.

8.5.2 Low-energy Collective Modes and Correlation Phenomena in Complex Matter

Understanding the coupling between charge and lattice degrees of freedom is a central challenge for correlated materials, particularly unconventional superconductors. In FeSe and related pnictide superconductors, electron–phonon coupling is thought to enhance electron–electron correlations, but direct experimental evidence has been lacking. Multi-modal time-resolved studies of FeSe by Gerber *et al.*[96] using X-ray scattering combined with time-resolved ARPES (see Figure 8.19a) directly characterized the bulk electronic structure and electron–phonon coupling for the first time. The oscillatory coherent vibrational responses (see Figure 8.19b and c) from both the band structure and lattice displacements (as triggered by impulsive optical excitation) are combined in a high-precision "coherent lock-in" measurement of the electron–phonon coupling strength in the THz regime. This technique provides sufficient time, space, energy and orbital resolution to enable the first model- and assumption-free measurement of the electron–phonon coupling strength. Significantly, the electron–phonon coupling in FeSe was shown to be far larger than predicted, providing new direct guidance to theory. Since electron–phonon coupling impacts superconductivity exponentially, this enhancement suggests a

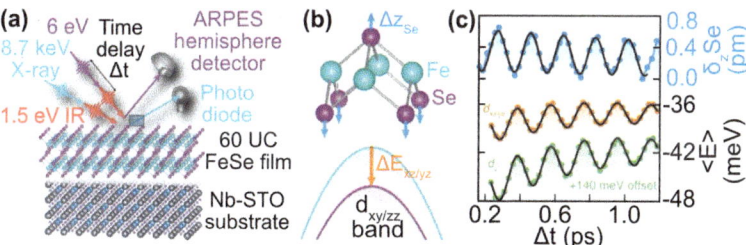

Figure 8.19   (a) Time-resolved ARPES and X-ray diffraction measurements reveal electron–phonon coupling in the correlated material FeSe following perturbative excitation of coherent A_{1g} phonon modes with an 800 nm femtosecond optical pulse (*i.e.*, a photon energy of 1.55 eV). (b) Schematic of the A_{1g} phonon mode, which periodically modulates the electronic band energies ($d_{xz/yz}$ band). (c) Oscillations of the selenium displacement (blue) measured by time-resolved X-ray diffraction and the momentum-averaged energy shift (orange, green), measured by time-resolved ARPES.[96] Panels (a–c) are reproduced from ref. 96 with permission from the American Association for the Advancement of Science, Copyright 2017.

pathway towards novel superconducting states, where electron–phonon and electron–electron interactions act in concert. In the prototypical super-conducting cuprate YBCO, time-resolved X-ray scattering studies by Man-kowsky *et al.*[97] determined the subtle atomic structural distortions (driven by nonlinear IR-driven ionic Raman scattering) as the basis for a transient enhancement of superconductivity.

8.5.3 Stochastic Dynamics, Structural Fluctuations, and Rare Events in Complex Matter

Recent experiments highlight the potential for advanced coherent X-ray scattering methods with ultrafast XFEL pulses to map the nanoscale stochastic fluctuations of matter that mediate transport, polarization, and mechanical properties in various materials.[98] As first proposed more than a decade ago by Gutt *et al.*,[99] a pair of femtosecond X-ray pulses with adjust-able time separation (from femtoseconds to nanoseconds) probes the nanoscale order of a sample *via* coherent scattering (see Figure 8.20). The measured observable is the integrated X-ray speckle pattern from both pulses of the incident pair. The speckle contrast (normalized change in the intensity distribution in the summed speckle pattern), as a function of the time delay between the two pulses, is equivalent to the intensity correlation g_2 measured in more conventional sequential XPCS experiments (see Figure 8.20b).[100,101] This approach, X-ray speckle visibility spectroscopy

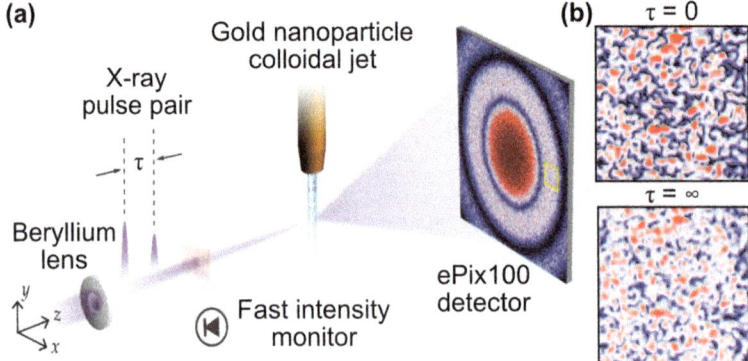

Figure 8.20   (a) Experimental schematic of two-pulse X-ray speckle visibility spec-troscopy (XSVS) experiments. A pair of X-ray pulse with variable time separation are focused on a colloidal jet of gold nanoparticles. A fast upstream detector monitors the relative intensities of the two pulses, and an X-ray area detector measures the sum of the scattering from both incident pulses. Data are collected as a function of the pulse time delay (τ), and analyzed to extract the speckle contrast. (b) Example simulated speckle patterns from the region of interest (yellow square) illustrating the expected change in contrast for $\tau = 0$ and $\tau = \infty$.[98] Reproduced from ref. 98 with permission from the American Physical Society, Copyright 2021.

(XSVS), when combined with femtosecond XFEL pulse pairs, accesses fluctuation timescales (fs–ns) that are far beyond the capabilities of fast X-ray detectors. Demonstration XSVS studies at LCLS map nanosecond colloidal dynamics in a free-flowing liquid jet. The anisotropy of the speckle visibility is directly related to the flow rate and the nonuniform flow field within the jet.[98] These results represent an important step in realizing the full potential of XFELs for mapping stochastic fluctuations in complex materials.

An essential enabling technology for XSVS studies is the development of stable X-ray optical systems for creating pulse pairs with variable delay (<100 fs to ps), while preserving the mutual coherence and alignment of the pulses during delay cycling (and over long measurement times). Several split and delay systems have been developed and demonstrated in the past few years at LCLS.[102–105] One recent example of an amplitude split and delay system based on diamond gratings (for pulse splitting and combining) and six-channel cut Bragg crystals to create a very stable time delay (see Figure 8.21a).[102] The optical system creates pairs of femtosecond X-ray pulses with mutual coherence that is significantly higher than previously

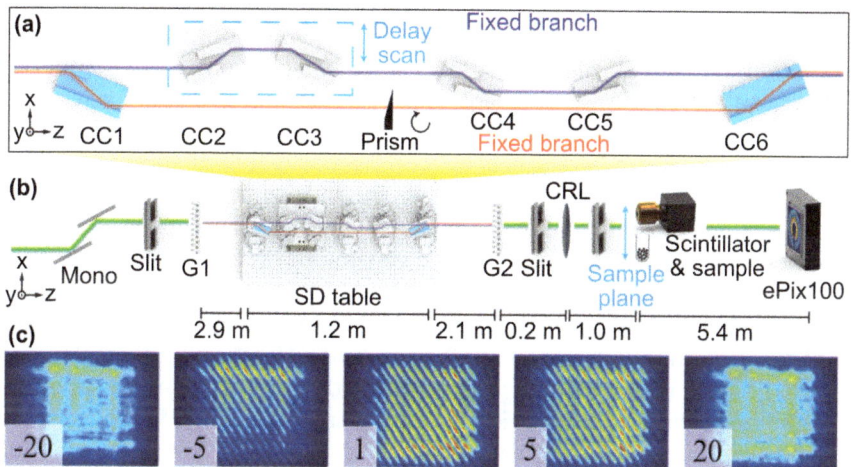

Figure 8.21 (a) Optical schematic of an XFEL split and delay consisting of six channel-cut crystals (CC). Two pairs of channel-cut crystals are translated simultaneously to adjust the pulse delay. A prism (glassy carbon) in the fixed branch creates a small crossing angle between the two beams. (b) Experimental layout for characterizing the coherence and stability of the pulse pair created by the X-ray split and delay. The X-ray pulse is first split by the diamond grating (G1), and then transits the split and delay optics, before being re-combined in a second diamond grating (G2). (c) Measured interference fringes as a function of time delay (shown on bottom-left in fs).[102] Reproduced from ref. 102, https://doi.org/10.1103/PhysRevResearch.3.043050, under the terms of the CC BY 4.0 license https://creativecommons.org/licenses/by/4.0/.

demonstrated systems. Characterization of this mutual coherence (see the experimental layout in Figure 8.21b) provided highly stable interference fringe measurements as a function of the time delay between the two branches of the split and delay (see Figure 8.21c). Furthermore, this approach ensures high relative alignment stability between the two branches preserved over continuous cycling of the time delay. Importantly, this design is relatively insensitive to pointing drift in the incident X-ray beam.

Coherent scattering studies of stochastic dynamics using XFELs is further enabled by significant advances in detectors (for example, see ref. 106 and references therein) and analysis algorithms (for example, see ref. 107 and references therein). The availability of high-repetition rate XFELs will drive another qualitative advance in this field. At a fundamental level, the ability to discern structural changes at short time intervals scales inversely with the square of the X-ray source brightness ($\Delta t \sim B^{-2}$). Thus, the ~ 1000-fold increase in average brightness from high-repetition rate XFELs will translate to dramatic improvements in measurement sensitivity and accessible time-scales. At a practical level, accessing these timescales depends on multiple-pulse measurement schemes with moderate peak brightness to ensure that the X-ray pulses do not perturb the stochastic dynamics of interest. Thus, the high average coherent power of the next generation of XFELs, with pro-grammable pulses at a high repetition rate, will enable studies of spontaneous fluctuations at the atomic scale from μs (or longer) down to fundamental femtosecond timescales, thus opening up important new areas of science.

8.6   Conclusion

The emergence of X-ray free-electron lasers promises to transform the field of X-ray science for the 21st century. The generation of high brightness, nearly fully coherent, ultrafast X-ray pulses now enables sensitive probing of struc-tural dynamics in matter on the fundamental time scales of atomic motion. Developing tunable sub-fs coherent X-ray pulses further enables element-specific probing of the electronic structure and charge dynamics of the fun-damental timescales of electronic motion. The first decade of XFEL operation has already opened many new fields of science, and progress in XFEL capabilities (including methods and instrumentation) and the corresponding science impact is expanding at a remarkable pace. In this chapter, we have highlighted key science applications in the areas of (a) fundamental processes in atoms and molecules and gas-phase chemistry, (b) condensed-phase chemistry and catalysis, and (c) structural dynamics in complex materials. While this chapter focuses on science impact using advanced X-ray scattering and diffraction methods enabled by XFELs, this is not intended to be comprehensive. Notably, many other areas of science are exploiting com-plementary capabilities of XFELs (*e.g.*, for time-resolved spectroscopy, X-ray wave-mixing and nonlinear interactions, creation of novel states of mater, *etc.*). The science opportunities opened by XFELs are widely recognized, as evidenced by the proliferation of XFEL facilities worldwide.

Acknowledgements

This work was supported by the US Department of Energy, Office of Science, Basic Energy Sciences under contract no. DEAC02-76SF00515. We gratefully acknowledge insightful discussions with SLAC colleagues A. Aquila, M. Kling, A. Summers, T. Driver, and Stanford colleague M. Ihme.

References

1. H. Motz, *J. Appl. Phys.*, 1951, **22**, 527–535.
2. H. Motz, W. Thon and R. N. Whitehurst, *J. Appl. Phys.*, 1953, **24**, 826–833.
3. J. M. J. Madey, *J. Appl. Phys.*, 1971, **42**, 1906–1913.
4. D. A. G. Deacon, L. R. Elias, J. M. J. Madey and G. J. Ramian, *et al.*, *Phys. Rev. Lett.*, 1977, **38**, 892–894.
5. A. M. Kondratenko and E. L. Saldin, *Part. Accel.*, 1980, **10**, 207–216.
6. E. Saldin, E. Schneidmiller and M. V. Yurkov, *The physics of free electron lasers*, Springer Science & Business Media, 1999.
7. R. Bonifacio, C. Pellegrini and L. M. Narducci, *Opt. Commun.*, 1984, **50**, 373–378.
8. C. Pellegrini, *Eur. Phys. J. H*, 2012, **37**, 659–708.
9. C. Pellegrini, A. Marinelli and S. Reiche, *Rev. Mod. Phys.*, 2016, **88**, 015006.
10. https://en.wikipedia.org/wiki/Free-electron_laser.
11. D. Ratner, R. Abela, J. Amann and C. Behrens, *et al.*, *Phys. Rev. Lett.*, 2015, **114**, 054801.
12. J. Amann, W. Berg, V. Blank and F.-J. Decker, *et al.*, *Nat. Photonics*, 2012, **6**, 693–698.
13. T. Ishikawa, H. Aoyagi, T. Asaka and Y. Asano, *et al.*, *Nat. Photonics*, 2012, **6**, 540–544.
14. M. Yabashi, H. Tanaka and T. Ishikawa, *J. Synchrotron Radiat.*, 2015, **22**, 477–484.
15. I. S. Ko, H.-S. Kang, H. Heo and C. Kim, *et al.*, *Appl. Sci.*, 2017, **7**, 479.
16. H.-S. Kang, C.-K. Min, H. Heo and C. Kim, *et al.*, *Nat. Photonics*, 2017, **11**, 708–713.
17. C. Milne, T. Schietinger, M. Aiba and A. Alarcon, *et al.*, *Appl. Sci.*, 2017, **7**, 720.
18. M. Altarelli, *The European X-ray Free-electron Laser Technical Design Report*, DESY Hamburg, 2006.
19. T. Tschentscher, Layout of the X-Ray Systems at the European XFEL, XFEL.EU TR-2011-001, 2011, DOI: 10.3204/XFEL.EU/TR-2011-001, https://bib-pubdb1.desy.de/record/90829.
20. W. Decking and T. Limberg, *European XFEL Post-TDR Description*, Internal Report, 2013.
21. R. Schoenlein, T. Elsaesser, K. Holldack and Z. Huang, *et al.*, *Philos. Trans. R. Soc., A*, 2019, **377**, 20180384.

22. U. Bergmann, J. Kern, R. W. Schoenlein and P. Wernet, *et al.*, *Nat. Rev. Phys.*, 2021, **3**, 264–282.
23. J. Duris, S. Li, T. Driver and E. G. Champenois, *et al.*, *Nat. Photonics*, 2020, **14**, 30–36.
24. N. Hartmann, G. Hartmann, R. Heider and M. S. Wagner, *et al.*, *Nat. Photonics*, 2018, **12**, 215–222.
25. S. Li, T. Driver, O. Alexander and B. Cooper, *et al.*, *Faraday Discuss.*, 2021, **228**, 488–501.
26. T. Driver, S. Li, E. G. Champenois and J. Duris, *et al.*, *Phys. Chem. Chem. Phys.*, 2020, **22**, 2704–2712.
27. Y. Ding, C. Behrens, P. Emma and J. Frisch, *et al.*, *Phys. Rev. Spec. Top.-Accel. Beams*, 2011, **14**, 120701.
28. C. Behrens, F. J. Decker, Y. Ding and V. A. Dolgashev, *et al.*, *Nat. Commun.*, 2014, **5**, 3762.
29. N. H. Lindner, G. Refael and V. Galitski, *Nat. Phys.*, 2011, 7, 490–495.
30. P. Walter, T. Osipov, M.-F. Lin and J. Cryan, *et al.*, *J. Synchrotron Radiat.*, 2022, **29**, 957–968.
31. LCLS TMO Instrument, https://lcls.slac.stanford.edu/instruments/neh-1-1.
32. M. Chollet, R. Alonso-Mori, M. Cammarata and D. Damiani, *et al.*, *J. Synchrotron Radiat.*, 2015, **22**, 503–507.
33. LCLS XPP Instrument, https://lcls.slac.stanford.edu/instruments/xpp.
34. M. Liang, G. J. Williams, M. Messerschmidt and M. M. Seibert, *et al.*, *J. Synchrotron Radiat.*, 2015, **22**, 514–519.
35. LCLS CXI Instrument, https://lcls.slac.stanford.edu/instruments/cxi.
36. LCLS NEH 1.2 or Tender X-ray Instrument (TXI), https://lcls.slac.stanford.edu/instruments/neh-1-2.
37. LCLS ChemRIXS and qRIXS Instruments, https://lcls.slac.stanford.edu/instruments/neh-2-2.
38. R. Alonso-Mori, C. Caronna, M. Chollet and R. Curtis, *et al.*, *J. Synchrotron Radiat.*, 2015, **22**, 508–513.
39. LCLS XCS Instrument, https://lcls.slac.stanford.edu/instruments/xcs.
40. LCLS DXS Instrument, https://lcls.slac.stanford.edu/instruments/dxs.
41. B. Nagler, B. Arnold, G. Bouchard and R. F. Boyce, *et al.*, *J. Synchrotron Radiat.*, 2015, **22**, 520–525.
42. LCLS MEC Instrument, https://lcls.slac.stanford.edu/instruments/mec.
43. R. G. Sierra, A. Batyuk, Z. Sun and A. Aquila, *et al.*, *J. Synchrotron Radiat.*, 2019, **26**, 346–357.
44. LCLS MFX Instrument, https://lcls.slac.stanford.edu/instruments/mfx.
45. B. Stankus, H. Yong, J. Ruddock and L. Ma, *et al.*, *J. Phys. B: At., Mol. Opt. Phys.*, 2020, **53**, 234004.
46. M. P. Minitti, J. M. Budarz, A. Kirrander and J. S. Robinson, *et al.*, *Phys. Rev. Lett.*, 2015, **114**, 255501.
47. SLAC Mev UED, https://lcls.slac.stanford.edu/instruments/mev-ued.
48. J. Yang, X. Zhu, T. J. A. Wolf and Z. Li, *et al.*, *Science*, 2018, **361**, 64–67.
49. T. J. A. Wolf, D. M. Sanchez, J. Yang and R. M. Parrish, *et al.*, *Nat. Chem.*, 2019, **11**, 504–509.

50. J. Yang, X. Zhu, J. P. F. Nunes and J. K. Yu, *et al.*, *Science*, 2020, **368**, 885–889.
51. E. G. Champenois, D. M. Sanchez, J. Yang and J. P. F. Nunes, *et al.*, *Science*, 2021, **374**, 178–182.
52. H. Yong, X. Xu, J. M. Ruddock and B. Stankus, *et al.*, *Proc. Natl. Acad. Sci. U. S. A.*, 2021, **118**, e2021714118.
53. H. Yong, N. Zotev, J. M. Ruddock and B. Stankus, *et al.*, *Nat. Commun.*, 2020, **11**, 2157.
54. J. D. Biggs, Y. Zhang, D. Healion and S. Mukamel, *Proc. Natl. Acad. Sci. U. S. A.*, 2013, **110**, 15597–15601.
55. S. Mukamel, D. Healion, Y. Zhang and J. D. Biggs, *Annu. Rev. Phys. Chem.*, 2013, **64**, 101–127.
56. N. Rohringer, *Philos. Trans. R. Soc., A*, 2019, **377**, 20170471.
57. B. G. Levine and T. J. Martínez, *Annu. Rev. Phys. Chem.*, 2007, **58**, 613–634.
58. I. K. Alexander and S. C. Lorenz, *J. Phys. B: At., Mol. Opt. Phys.*, 2014, **47**, 124002.
59. D. Mendive-Tapia, M. Vacher, M. J. Bearpark and M. A. Robb, *J. Chem. Phys.*, 2013, **139**, 044110.
60. S. Lünnemann, A. I. Kuleff and L. S. Cederbaum, *Chem. Phys. Lett.*, 2008, **450**, 232–235.
61. V. Despré, N. V. Golubev and A. I. Kuleff, *Phys. Rev. Lett.*, 2018, **121**, 203002.
62. A. Marciniak, V. Despré, V. Loriot and G. Karras, *et al.*, *Nat. Commun.*, 2019, **10**, 337.
63. J. T. O'Neal, E. G. Champenois, S. Oberli and R. Obaid, *et al.*, *Phys. Rev. Lett.*, 2020, **125**, 073203.
64. G. Dixit, O. Vendrell and R. Santra, *Proc. Natl. Acad. Sci. U. S. A.*, 2012, **109**, 11636–11640.
65. J. R. Rouxel, D. Keefer and S. Mukamel, *Struct. Dyn.*, 2021, **8**, 014101.
66. H. Bockhorn, *Soot Formation in Combustion: Mechanisms and Models*, Springer-Verlag, Berlin, 1994.
67. K. O. Johansson, M. P. Head-Gordon, P. E. Schrader and K. R. Wilson, *et al.*, *Science*, 2018, **361**, 997–1000.
68. H. Richter and J. B. Howard, *Prog. Energy Combust. Sci.*, 2000, **26**, 565–608.
69. K.-H. Homann, *Angew. Chem. Int. Ed.*, 1998, 2434–2451.
70. H. Wang, *Proc. Combust. Inst.*, 2011, **33**, 41–67.
71. A. Violi, *Combust. Flame*, 2004, **139**, 279–287.
72. J. R. Hahn, H. Kang, S. M. Lee and Y. H. Lee, *J. Phys. Chem. B*, 1999, **103**, 9944–9951.
73. B. R. Stanmore, J. F. Brilhac and P. Gilot, *Carbon*, 2001, **39**, 2247–2268.
74. N. D. Loh, C. Y. Hampton, A. V. Martin and D. Starodub, *et al.*, *Nature*, 2012, **486**, 513–517.
75. T. Gorkhover, A. Ulmer, K. Ferguson and M. Bucher, *et al.*, *Nat. Photonics*, 2018, **12**, 150–153.
76. T. B. van Driel, K. S. Kjær, R. W. Hartsock and A. O. Dohn, *et al.*, *Nat. Commun.*, 2016, **7**, 13678.

77. K. Haldrup, G. Levi, E. Biasin and P. Vester, *et al.*, *Phys. Rev. Lett.*, 2019, **122**, 063001.

78. K. S. Kjær, T. B. Van Driel, T. C. B. Harlang and K. Kunnus, *et al.*, *Chem. Sci.*, 2019, **10**, 5749–5760.

79. K. Kunnus, M. Vacher, T. C. B. Harlang and K. S. Kjær, *et al.*, *Nat. Commun.*, 2020, **11**, 634.

80. E. Biasin, Z. W. Fox, A. Andersen and K. Ledbetter, *et al.*, *Nat. Chem.*, 2021, **13**, 343–349.

81. M. Ibrahim, T. Fransson, R. Chatterjee and M. H. Cheah, *et al.*, *Proc. Natl. Acad. Sci. U. S. A.*, 2020, **117**, 12624–12635.

82. J. Kern, R. Chatterjee, I. D. Young and F. D. Fuller, *et al.*, *Nature*, 2018, **563**, 421–425.

83. J. Kern, R. Tran, R. Alonso-Mori and S. Koroidov, *et al.*, *Nat. Commun.*, 2014, **5**, 4371.

84. J. Kern, R. Alonso-Mori, R. Tran and J. Hattne, *et al.*, *Science*, 2013, **340**, 491–495.

85. H. Öström, H. Öberg, H. Xin and J. LaRue, *et al.*, *Science*, 2015, **347**, 978–982.

86. I. Barke, H. Hartmann, D. Rupp and L. Flückiger, *et al.*, *Nat. Commun.*, 2015, **6**, 6187.

87. K. Ayyer, P. L. Xavier, J. Bielecki and Z. Shen, *et al.*, *Optica*, 2021, **8**, 15–23.

88. J. Kang, J. Carnis, D. Kim and M. Chung, *et al.*, *Nat. Commun.*, 2020, **11**, 5901.

89. P. W. Anderson, *Science*, 1972, **177**, 393–396.

90. E. Dagotto, *Science*, 2005, **309**, 257–262.

91. S. Wall, S. Yang, L. Vidas and M. Chollet, *et al.*, *Science*, 2018, **362**, 572–576.

92. A. Cavalleri, C. Tóth, C. W. Siders and J. A. Squier, *et al.*, *Phys. Rev. Lett.*, 2001, **87**, 237401.

93. P. Baum, D.-S. Yang and A. H. Zewail, *Science*, 2007, **318**, 788–792.

94. V. R. Morrison, R. P. Chatelain, K. L. Tiwari and A. Hendaoui, *et al.*, *Science*, 2014, **346**, 445–448.

95. P. Zalden, F. Quirin, M. Schumacher and J. Siegel, *et al.*, *Science*, 2019, **364**, 1062–1067.

96. S. Gerber, S.-L. Yang, D. Zhu and H. Soifer, *et al.*, *Science*, 2017, **357**, 71–75.

97. R. Mankowsky, A. Subedi, M. Forst and S. O. Mariager, *et al.*, *Nature*, 2014, **516**, 71–73.

98. Y. Sun, G. Carini, M. Chollet and F.-J. Decker, *et al.*, *Phys. Rev. Lett.*, 2021, **127**, 058001.

99. C. Gutt, L.-M. Stadler, A. Duri and T. Autenrieth, *et al.*, *Opt. Express*, 2009, **17**, 55.

100. F. Lehmkuhler, J. Valerio, D. Sheyfer and W. Roseker, *et al.*, *IUCrJ*, 2018, **5**, 801–807.

101. J. Carnis, W. Cha, J. Wingert and J. Kang, *et al.*, *Sci. Rep.*, 2014, **4**, 6017.

102. H. Li, Y. Sun, J. Vila-Comamala and T. Sato, *et al.*, *Phys. Rev. Res.*, 2021, **3**, 043050.
103. Y. Sun, M. Dunne, P. Fuoss and A. Robert, *et al.*, *Phys. Rev. Res.*, 2020, **2**, 023099.
104. H. Li, Y. Sun, M. Sutton and P. Fuoss, *et al.*, *Opt. Lett.*, 2020, **45**, 2086–2089.
105. Y. Sun, N. Wang, S. Song and P. Sun, *et al.*, *Opt. Lett.*, 2019, **44**, 2582–2585.
106. A. Bergamaschi, A. Mozzanica and B. Schmitt, *Nat. Rev. Phys.*, 2020, **2**, 335–336.
107. Y. Sun, V. Esposito, P. A. Hart and C. Hansson, *et al.*, *Appl. Sci.*, 2021, **11**, 10041.

CHAPTER 9

Time-resolved X-ray Scattering of Excited State Structure and Dynamics

H. YONG,[a] A. KIRRANDER[b] AND P. M. WEBER*[c]

[a] Department of Chemistry and Biochemistry, University of Califormia San Diego, La Jolla, California 92093, USA; [b] Physical and Theoretical Chemistry Laboratory, Department of Chemistry, University of Oxford, South Parks Road, Oxford OX1 3QZ, UK; [c] Department of Chemistry, Brown University, Providence, Rhode Island 02912, USA
*Email: peter_weber@brown.edu

9.1 Introduction

Directly observing atomic motions in molecules during chemical dynamics has long constituted one of the grand challenges in chemistry.[1] The emergence of X-ray free-electron lasers (XFELs) with their ultrashort pulse durations and extreme brightness[2,3] and the near-parallel development of ultrafast electron sources (see Chapters 11–16), have made ultrafast scattering measurements of molecular dynamics in free gas-phase molecules possible. Although chemical dynamics on ground electronic state potential energy surfaces is conceptually well understood and easier to model computationally, the fact remains that excited states play important roles in chemical processes, especially in the field of photochemistry. Molecules are often more reactive in their excited states than in the ground state because activation energies can be lower and because electronic relaxation processes can insert a great deal of internal energy into vibrational modes.[4] Additionally, excitation to excited states can induce, after rapid electronic relaxation, fast

Theoretical and Computational Chemistry Series No. 25
Structural Dynamics with X-ray and Electron Scattering
Edited by Kasra Amini, Arnaud Rouzée and Marc J. J. Vrakking
© The Royal Society of Chemistry 2024
Published by the Royal Society of Chemistry, www.rsc.org

kinetic or dynamic processes on the ground electronic surface. Yet, it is challenging to directly observe the rapid molecular dynamics in excited states in real time, far away from the equilibrium, because mapping the atomic motions induced by electronic excitations requires atomic scale spatial and femtosecond temporal resolution.

X-ray scattering is sensitive to the electronic charge density whereas electron diffraction measures the total (electronic and nuclear) charge density (see Chapter 3). However, the core electrons track the nuclei closely, and both X-ray and electron scattering can thus be employed toward the real-space imaging of transient molecular structures. Even though X-ray scattering has a much smaller cross-section and, at the photon energies currently available, covers a smaller range of scattering angles compared to electron diffraction, the high flux and high energy X-ray pulses of XFELs make up for these shortcomings. Importantly, ultrafast X-ray scattering has achieved a better temporal resolution (~ 30 fs) than MeV-UED as the latter is limited by the intrinsic space–charge interactions between electrons within an electron pulse. This makes time-resolved X-ray scattering a unique tool for measuring ultrafast processes such as the rapid redistribution of electron density upon photoexcitation. The relative merits and strengths of the two methods have been discussed extensively,[5–8] and so the present chapter focuses only on ultrafast X-ray scattering. Although important studies have investigated condensed matter and solutions,[9–17] this review focuses on samples in the gas phase. In the absence of a solvent, investigations of free molecules reveal the pure and unperturbed chemical dynamics of molecular species. High quality scattering patterns can be obtained and quantitatively compared to theory.

Static X-ray scattering is traditionally used as an essential tool for determining molecular structures in the ground electronic state of molecules at equilibrium.[18] Since X-ray scattering arises from the interaction of molecular electrons with the electromagnetic field of the incoming X-rays, it is a sensitive probe of the electron density distribution in a molecule. However, as mentioned above, because the core electrons in atoms are usually centered tightly around the nuclei, the geometry (nuclear structure) of molecules can be determined with X-ray scattering. It is important to also mention that while time-resolved spectroscopy methods are fundamentally limited by the time–energy uncertainty relation, X-ray scattering does not have that limitation because it measures time and space, which are not complementary \variables. Thus, time-resolved X-ray scattering can, in principle, measure chemical reactions with both high spatial and temporal resolution without fundamental restrictions. With existing technology at XFELs, ultrafast gas-phase X-ray scattering experiments employing a photon energy of 9.5 keV have achieved a spatial resolution of 0.01 Å and a temporal resolution of 30 fs.[24] All these features make ultrafast X-ray scattering a promising technique for unveiling nuclear and electron dynamics during complex chemical reactions.

The first time-resolved gas-phase X-ray scattering experiment with femtosecond time resolution was demonstrated in 2015 when a "molecular movie" of the ring-opening reaction of 1,3-cyclohexadiene (CHD) was successfully recorded.[19] Since then, this ultrafast gas-phase X-ray scattering setup has been developed,[20,21] resulting in many new and previously unattainable insights about excited state molecular systems. For instance, ultrafast nuclear motions in molecules during chemical reactions have been studied extensively,[22–25] enabling the determination of polyatomic molecular structures in electronically excited states.[24,26] Specific signatures of excited states have also been measured, including the identification of the initially populated electronic state[27] and the direct measurement of the redistribution of the molecular electron density immediately after photoexcitation.[28] Further examples of ultrafast gas-phase X-ray scattering measurements include important chemical processes and related properties such as chemical kinetics,[29–31] multiphoton processes,[32,33] anharmonicities and correlations.[34]

9.2 Experimental Implementation and Data Processing

9.2.1 Experimental Implementation

The experimental scheme for ultrafast gas-phase X-ray scattering experiments follows the pump–probe methodology. An optical pump laser excites molecules in the gas phase and a subsequent X-ray pulse probes the molecules, and a series of scattering images is then measured on an array detector (see Figure 9.1). The time delay between the optical pump and X-ray probe pulses is controlled by an electronic delay stage, and the timing jitter is monitored shot-by-shot with a spectrally encoded cross correlator ('time tool'), achieving sub-10 fs root mean square (RMS) resolution.[35]

The gas pressure in the sample cell must be carefully adjusted to obtain the optimal signal-to-noise ratio (SNR) for the pump–probe scattering experiment. The total scattering signal scales linearly with the gas pressure, so that for gas-phase structure determination of ground state molecules, higher pressure is always desirable if the formation of clusters is avoided. In pump–probe experiments with optical excitation, however, the laser beam can be attenuated by the sample as it traverses the interaction region. If this attenuation happens, the scattering signal from downstream molecules will not contain as much pump–probe signal while still contributing to the total scattering signal and therefore to the noise of the measurement. It is not advisable to compensate for this attenuation by increasing the optical pump pulse energy because that would more likely lead to undesired multi-photon processes. For typical absorption cross-sections, the optimum gas pressure is in the range of a few Torr, corresponding to one trillion molecules in the scattering interaction region (with a 30 μm FWHM beam focus and 2.4 mm path length).[21] With about one trillion photons in each X-ray pulse produced at LCLS, only a small fraction ($\approx 10^{-7}$) of X-ray photons is scattered by the target sample (see Figure 9.2).

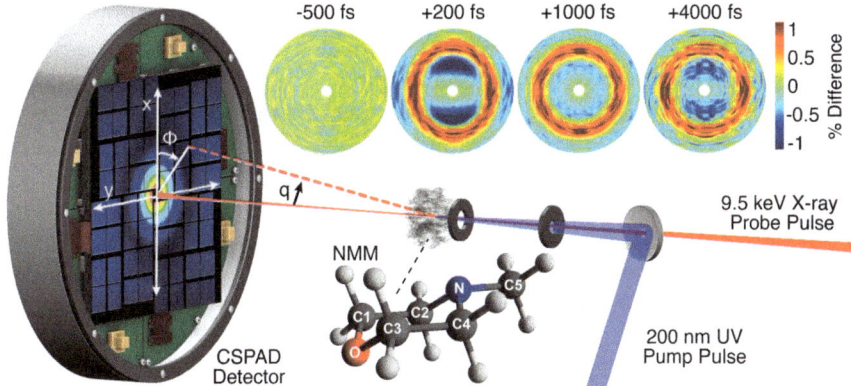

Figure 9.1 Schematic of a typical experimental setup for time-resolved gas-phase X-ray scattering. The target gas-phase molecule (pictured here as *N*-methylmorpholine, NMM) is excited using a 200 nm UV pump pulse and probed *via* the scattering of 9.5 keV X-ray probe pulses. A series of snapshots were measured on a CSPAD detector at various time delays between the pump and the probe. The percent difference scattering patterns as a function of momentum transfer magnitude *q* and azimuthal angle ϕ for several time delays are shown along the top of the figure. Reproduced with permission from ref. 24.

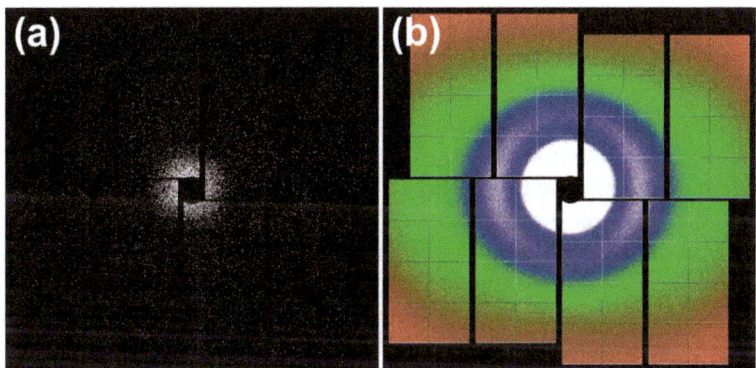

Figure 9.2 Typical scattering pattern of gas-phase quadricyclane molecules with 15.2 keV X-rays of flux 10^{14} photons per second at LCLS (without pump laser). (a) Scattering pattern from a single shot measurement. (b) Integrated pattern measured over hundreds of shots.

Static X-ray diffraction images of quadricyclane molecules with 15.2 keV X-rays of flux 10^{14} photons per second at LCLS show that very few scattered photons are detected from the single-shot scattering pattern (see Figure 9.2a), while integration of hundreds of X-ray shots reveals an image rich in X-ray scattering structures (see Figure 9.2b). As a result, considerable care must be taken to avoid scattering of the primary X-ray beam by any window or air.

The first successful ultrafast gas-phase X-ray scattering experiment studied the ring-opening of CHD,[19] and was performed at the X-ray pump–probe (XPP) instrument[36] of the linac coherent light source (LCLS). The experiment used a windowless cell for the primary X-ray beam and careful elimination of the background X-rays.[20] More ideally, the imaging detector is enclosed in the vacuum, as is now implemented at the coherent X-ray imaging (CXI) instrument[37] at the LCLS. Other improvements in the experimental implementation include a re-designed short-pathlength windowless scattering cell, careful optimization of the sample density, and normalization of the shot-by-shot X-ray intensity,[21] allowing experiments with the CXI instrument to achieve an exceptionally high SNR and sensitivity to very small ($\sim 0.1\%$) changes in the scattering signal. These developments and advances in the experimental methodology are described in ref. 21.

9.2.2 Data Processing

The measured time-resolved pump–probe scattering signals are conveniently expressed as percent differences,[19,38]

$$\%\Delta I(q, \phi, t) = 100 \times \frac{I_{on}(q, \phi, t) - I_{off}(q, \phi)}{I_{off}(q, \phi)}, \tag{9.1}$$

where q is the magnitude of the momentum transfer vector, ϕ is the azimuthal angle on the detector, $I_{on}(q,\phi,t)$ represents the pump-laser-on scattering signal at a pump–probe delay time t, and $I_{off}(q,\phi)$ represents the pump-laser-off reference scattering signal. The percent difference expression not only accentuates small changes in the scattering pattern over time but also eliminates many experimental factors that multiplicatively affect both laser-on and laser-off scattering signals. These factors are discussed in Section 2.4 of ref. 21, which include details on pixel noise, the attenuation of the scattered signal by the beryllium exit window, scattering intensity corrections due to the linear polarization of the X-rays at LCLS (X-ray polarization factor), and the detector planarity (geometric correction factor).

It is important to note that there are several other experimental artifacts that do not cancel out when using the percent difference signal. This includes the shot-by-shot X-ray intensity fluctuations of the self-amplified spontaneous emission (SASE) pulses at non-seeded XFELs. To take this factor into account, the transmitted X-ray intensity after the gas sample is monitored using a photodiode mounted downstream of the detector. The X-ray scattering signals are then corrected using the measured photodiode value shot-by-shot before averaging. In addition, any experimental uncertainties introduced by the pump laser such as laser intensity fluctuations and changes in the laser/X-ray spatial overlap remain and must be handled carefully. Finally, even though background scattering from apparatus components is eliminated in the numerator of eqn (9.1), it still remains in the denominator and therefore needs to be minimized. It has been shown that the current experimental design assisted by careful data processing can yield exceptional high-quality data with experimental background signals at least three orders smaller than the desired scattering signal.[21]

To measure the scattering patterns as a function of q and ϕ accurately, it is important to calibrate the detector geometry with regard to factors such as the position of the detector centre and the sample-to-detector distance. For this purpose, a least-squares optimization between the two-dimensional theoretical reference image generated from a calculated ground-state molecular geometry and the experimentally measured laser-off scattering pattern, $I_{off}(q,\phi)$, is performed. To directly compare the measured absolute scattering signals with the theoretical ones, the aforementioned factors including the X-ray polarization and the geometric correction factors, which are eliminated in the percent difference signal (eqn (9.1)), must now be included. The calibration procedure optimizes the parameters that define the centre of the detector, the azimuthal angle of the detector relative to the X-ray polarization direction, and the distance of the detector relative to the interaction region.

9.3 Observing Excited-state Molecular Systems in Real Time

Excited-state molecular systems usually display a complex interplay between the electron density distributions and the nuclei.[39,40] In the Born–Oppenheimer approximation, the electrons create a potential energy landscape that guides the trajectory of the nuclei, while the electron density itself evolves as the nuclei move along the potential energy surface during the chemical reaction. The Born–Oppenheimer approximation breaks down in the presence of non-adiabatic coupling, particularly in the vicinity of conical intersections where two or more potential energy surfaces of equal symmetry intersect, strongly coupling the electronic and nuclear motions. The determination of nuclear motions and the rearrangement of electron density distributions is thus essential for understanding the excited-state chemical reactions and processes. With the capability to observe both nuclear and electron dynamics in molecules, time-resolved X-ray scattering offers unique views of coupled electron–nuclear dynamics that are complementary to spectroscopic methods. In this section, we describe various important new insights and observables for excited-state molecular systems that have been unveiled by state-of-the-art ultrafast gas-phase X-ray scattering experiments.

9.3.1 Transition Dipoles and Multi-photon Processes with Anisotropic X-ray Scattering

The time-energy uncertainty relationship implies that molecular absorption that induces ultrafast excited state dynamics is inherently broad. Consequently, traditional spectroscopic investigations often remain ambiguous even about the identity of the specific excited state that is initially populated. A direct measurement of the optical transition dipole moment can therefore provide essential guidance for the accurate assignment of the initially excited electronic state and successful interpretation of the observed light-induced dynamics.

It is known from both theoretical predictions[41,42] and experimental measurements[43,44] that gas-phase X-ray scattering from aligned molecules can have angle-dependent scattering patterns due to the intrinsic anisotropy of the molecular geometry. However, for an ensemble of free molecules with random orientations, the X-ray scattering patterns are isotropic (other than the polarization factor) and can be conveniently analyzed using rotationally-averaged scattering signals. Due to various limitations of molecular alignment techniques and complexities associated with intro-ducing a third alignment laser,[43,44] almost all pump–probe gas-phase X-ray scattering experiments so far measure randomly orientated molecules.

Nevertheless, the pump–probe X-ray scattering patterns from isotropic ensembles of free molecules can be anisotropic.[23,27] The angle-dependent signal (see insets of Figure 9.1) usually appears at very early delay times (≈ 200 fs) before quickly disappearing thereafter (≈ 1 ps), even though it sometimes reappears at much longer delay times (≈ 4 ps). This effect is caused by the interaction of the linearly polarized optical light with the gas-phase ensemble and subsequent rotational motions of the molecule in the laboratory frame.[45,46] Because the optical pump laser is linearly polarized, those molecules whose transition dipole moment vector (TDMV) aligns with the polarization of the laser pump pulse in the laboratory frame will be preferentially excited. This creates a molecular ensemble with an anisotropic population of excited-state molecules as well as an anisotropic population of unexcited ground-state molecules, leading to angle-dependent scattering signals. Subsequent rotational dephasing and rephasing of the molecules after photoexcitation are responsible for the disappearance and recurrence of the anisotropic scattering signal. In general, such anisotropic scattering signals can be conveniently decomposed with $2n$ even order Legendre poly-nomials for an n-photon absorption process. The zeroth order term, called the isotropic term, contains all the information in the molecular frame that can be simplified as the isotropic rotationally-averaged signal, while additional information about the transition dipole moment[27] and multi-photon pro-cesses[33] can be extracted from higher-order, anisotropic terms.

The orientation of the optical TDMV can be determined using the aniso-tropic component of ultrafast pump–probe X-ray scattering signals, allowing for the identification of the initially excited electronic state prepared by optical excitation.[27] The concept is illustrated with *N*-methyl morpholine (NMM) molecules (see Figure 9.3). Single-photon excitation with a pulsed 200 nm laser can, at least in principle, excite any one or a mixture of the three energetically close-lying 3p Rydberg states.

The resulting pump–probe percent difference simulated scattering pat-terns (see Figure 9.3) are markedly different for the three TDMV orientations

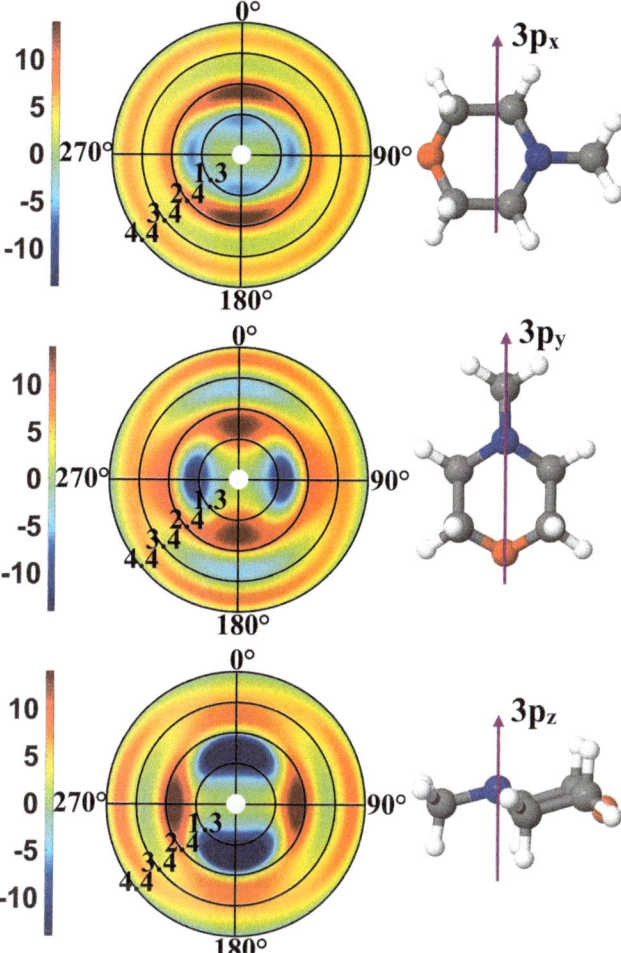

Figure 9.3 Simulated percent difference scattering patterns for NMM molecules excited to the $3p_x$, $3p_y$, and $3p_z$ electronic Rydberg states. The orientation of the TDMV relative to the molecule is shown in the right column. A $\cos^2 \theta$ distribution with respect to the laser polarization axis is assumed for the excited state, and orientations due to rotation about the laser polarization axis are averaged out. In the right column, the orientation of the TDMV in the molecular frame is indicated using a purple arrow, as calculated from MRCI(2,5)/6-311+G(d). Reproduced from ref. 27 with permission from American Chemical Society, Copyright 2018.

that correspond to the three 3p states. This is because the intrinsic orientation of the TDMV in the molecular frame determines the orientations of the excited-state population of molecules in the laboratory frame, which is then reflected in the angular-dependence of the scattering signal. By comparing the simulated two-dimensional (2D) patterns (see Figure 9.3) with the

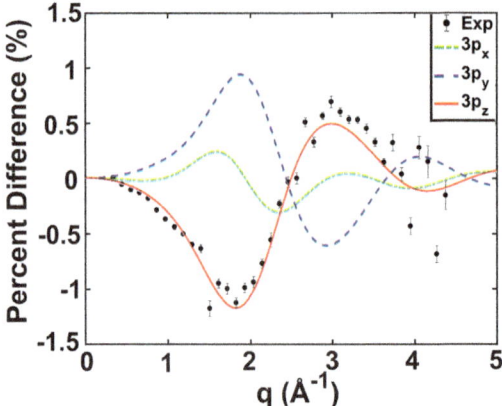

Figure 9.4   Anisotropic signal derived from experimental results at a pump–probe delay time of $t = 150$ fs. The three theoretical signals are derived from calculated data (see Figure 9.3). Reproduced from ref. 27 with permission from American Chemical Society, Copyright 2018.

experimental results (see Figure 9.1), it is straightforward to see that only the simulated pattern for $3p_z$ excitation shows the same symmetry as the experiment. To quantitatively compare the measured and calculated data, the second-order Legendre polynomial term, which represents the anisotropic scattering components, is extracted from the original 2D scattering pattern. The result unambiguously determines that the optical excitation is predominantly to the $3p_z$ state (see Figure 9.4) around 150 fs after initial photoexcitation of the NMM molecule, with almost no admixture of the other two states.

For multi-photon excitations, the anisotropy of ultrafast X-ray scattering signals has been used by Natan *et al.* to resolve and disentangle different multiphoton processes in a single experiment.[33] They show the decomposed experimental anisotropy terms up to order $n = 10$ in the Legendre polynomial for gas-phase iodine molecules (I_2) excited by 520 nm laser pulses, with noticeable signals up to the $n = 8$ term. This suggests that multi-photon processes up to 4-photon absorption take place for I_2 on interaction with a high intensity 520 nm optical pulse ($\sim 5 \times 10^{11}$ W cm^{-2}). One should note that an anisotropic scattering signal from an nth order term also has projections to all $k < n$ order Legendre polynomials. However, by analyzing all higher order terms, various reaction channels that are excited by different multiphoton transitions can be filtered out. The anisotropy information can thus be used as a powerful toolbox to differentiate and trace multiple excitation pathways that occur simultaneously.

9.3.2   Excited-state Electron Densities in Real Space

The first step in all photochemical and photophysical processes, photoexcitation, has conventionally been studied using spectroscopic tools which measure the transitions between different states.[47–49] It has been proposed theoretically that pump–probe scattering experiments have the potential to image the dynamic changes in the electron density upon photoexcitation.[41,50–53] The first direct experimental observation of the initial redistribution of electron density in a molecule upon photoexcitation (1,3-cyclohexadiene, CHD, optically excited to an electronic 3p Rydberg state using a 200 nm pump pulse) was recently achieved using ultrafast gas-phase X-ray scattering (see Figure 9.5).[28] The difference radial distribution function (see Figure 9.5a), ΔRDF(r), is obtained from the experimental difference signal through a sine transform. It shows, in real space, the rapid redistribution of the electron density at $\sim$25 fs after photoexcitation. The depletion of the electron density at small distances (0–3 Å) and the broad increase at larger distances (4–9 Å) reflects the diffuse character of the 3p Rydberg state and is in good agreement with the theoretically calculated electron density differences (see the inset of Figure 9.5a). A good agreement also exists between the experimental and theoretical percent difference scattering signals (see Figure 9.5b), providing further evidence that the changes in electron density due to transitions between electronic states are clearly observed.

The experiment introduced here is aided by the fact that CHD is a relatively small organic molecule without heavy elements, that the 3p state of CHD has a relatively long lifetime ($\sim$200 fs),[30,54] and that the change in the electronic structure of the 3p Rydberg state is large while the changes in

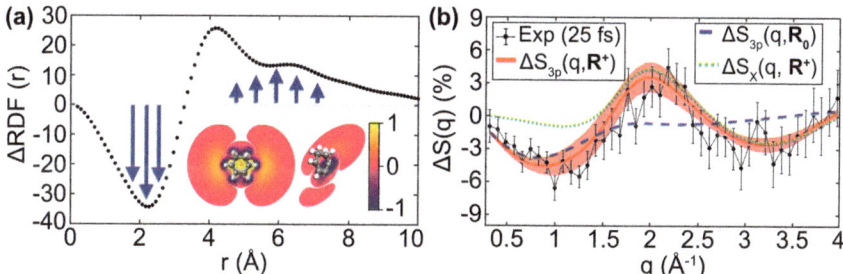

Figure 9.5 Experimental and theoretical signals of CHD. (a) The real-space difference radial distribution function, ΔRDF(r), obtained from the experimental data at 25 fs pump–probe delay time. The blue arrows point to the depletion and increase in electron density at short and long electron distances, respectively. The inset shows the corresponding contour slices of the electron density difference from electronic structure calculations. (b) The isotropic component of the experimental signal at 25 fs delay time is shown in black with 1σ error bars. The corresponding theoretical $\Delta S_{3p}(q,\mathbf{R}^+)$ signal for the electronic 3p state is shown in red with the shaded region accounting for the sampling of geometries in the excited state. For comparison, theoretical signals for the ground electronic state (X) at the 3p geometry, $\Delta S_X(q,\mathbf{R}^+)$, and for the excited 3p state at equilibrium geometry, $\Delta S_{3p}(q,\mathbf{R}_0)$, are included. Reproduced from ref. 28, https://doi.org/10.1038/s41467-020-15680-4, under the terms of the CC BY 4.0 license, http://creativecommons.org/licenses/by/4.0/.

molecular geometry are small. With the ongoing improvements in XFEL sources[55] and development of scattering theories for data analysis,[56–58] it can be expected that in the near future, ultrafast X-ray scattering will be able to measure electron density distributions in valence excited states and to image time-evolving electron dynamics.

9.3.3 Determination of Excited-state Molecular Geometry

The determination of excited-state molecular geometries from the direct inversion of scattering patterns is challenging because of (i) the fundamental phase problem[59] in all X-ray scattering techniques including X-ray crystallography, (ii) the limited range of the scattering momentum transfer vectors observed in the experiment, and (iii) the geometry may be undefined for instance due to strong dispersion of the wavepacket, depending on circumstances (see, *e.g.*, ref. 60). For gas-phase scattering experiments, the problem is further compounded by the rotational averaging due to randomly oriented molecules in the sample and the intrinsic structural complexity of polyatomic molecules. Following the procedures of conventional gas-phase diffraction experiments,[61] the experimental scattering patterns are compared to calculated patterns and the molecular structure is determined when satisfactory agreement is reached. We note that only the intensity information of scattered X-rays is used in the analysis of current ultrafast X-ray scattering experiments. It has been shown that the phase information in diffraction patterns of a virus can be measured through techniques like X-ray Fourier holography imaging[62] (see Section 6.2.2 in Chapter 6). It might be possible to extend such techniques to the molecular scale.

9.3.3.1 Least-squares Refinement

Traditionally, a least-squares refinement of structural parameters is used to determine the molecular structures from scattering data.[61] Based on a hypothetical molecular structure, a matrix of interatomic distances, R_{ij}, is constructed for molecules in their ground electronic states. Then, the independent atom model (IAM, see Chapter 3) is invoked to calculate scattering patterns from the interatomic distance matrix. Least-squares refinement of selected adjustable parameters, such as a set of bond lengths and angles, is performed to retrieve the best-fitting structural parameters by minimizing the difference between calculated and experimental scattering signals. This method works well also for transient species created by optical excitation to a dissociative state, provided that the fragments are in their ground electronic state and have only a few degrees of freedom, or when only specific coordinates are of interest.[29,63] This concept has been adopted to determine the structure of transient dimethylamine radicals (DMA) measured by ultrafast gas-phase X-ray scattering[29] (see Figure 9.6). Keeping the C–H distances in DMA fixed, the best fit between the measured

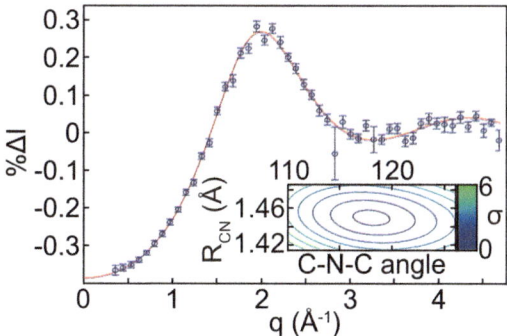

Figure 9.6   Percent difference patterns of the DMA transient obtained from the isotropic signal of the experimental data at 1 ns (blue circles) and computed using an IAM model from the best-fitting structure (red line). Inset: Residuals from the fit for different values of the bond length and bond angle. Contour lines are at 0.2, 1, 2, 3, 4, 5, and 6σ. Reproduced from ref. 29 with permission from John Wiley & Sons, Copyright 2019 Wiley-VCH Verlag GmbH & Co. KGaA, Weinheim.

and calculated percentage difference patterns yields the precise DMA structure with a C–N–C bond angle of $118 \pm 4°$ and a C–N bond length of 1.45 ± 0.02 Å.

9.3.3.2   Structure Pool Analysis

For more complex, polyatomic molecules, the least-squares refinement approach becomes difficult and almost impossible to implement computationally. The choice of the independent adjustable parameters becomes problematic for nonlinear molecules with more than four atoms, because an N-atom nonlinear molecule has $\dfrac{N(N-1)}{2}$ interatomic distances while only $3N-6$ geometrical parameters are needed. As the number of atom–atom interatomic distances scales with N^2, and the number of geometrical parameters scales linearly with N, there are correlations between structural parameters chosen for refinement that can lead to multiple solutions as well as possibly unphysical structures. The correlation problem becomes more pronounced for molecular systems that are far from their equilibrium,[34] such as molecules in excited electronic states. To circumvent this complexity, a novel structure determination method has been developed[24,26] that is capable of determining precise excited-state molecular structures of polyatomic molecules with more than four non-hydrogenic atoms[24,31] (see Figure 9.7). The method compares the measured scattering patterns against the simulated patterns corresponding to a large pool of molecular structures to determine the full set of structural parameters. It consists of three important steps:[26] creating a structure pool, calculating scattering patterns and determining molecular structures.

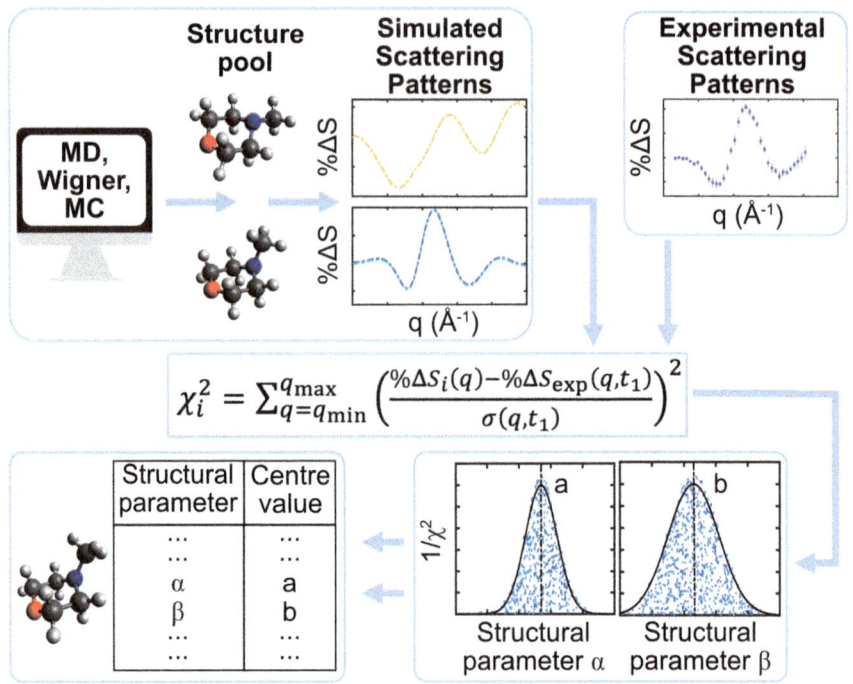

Figure 9.7 Concept of the method for determining molecular structures in excited electronic states from experimental scattering patterns. Reproduced from ref. 26 with permission from the Royal Society of Chemistry.

9.3.3.2.1 Creating a Structure Pool. The first step is to create a pool of trial structures that are in the vicinity of the target structure in the large structural parameter space. It has been found that one million trial structures are enough to reach convergence for molecular systems like NMM (15 degrees of freedom excluding hydrogen atoms).[24] A larger number of structures would be necessary for molecular systems with more degrees of freedom. Three sampling methods for creating structure pools have been introduced,[26] including molecular dynamics sampling (MD pool), Wigner sampling (Wigner pool) and Monte Carlo based sampling (MC pool). The MD pool is created by calculating molecular dynamics trajectories that propagate on potential surfaces that resemble the measured chemical dynamics. The structures in the pool are extracted from MD trajectories without reference to their time sequence. As the ultimate goal of the structure pool is to provide many structure that are in the vicinity of the correct target structure, one could sample many physically viable structures by displacing their geometries to provide the opportunity to find unexpected structures. This can be carried out either by sampling geometries from a Wigner distribution[64] (Wigner pool), or by using a Monte Carlo based approach to randomly create chemically viable structures (MC pool). Further details of the three sampling methods can be found in Section 2.2.1 of ref. 26. In principle, the method for creating the structure pool is not limited to the three methods introduced here, as the concept mentioned earlier (see Figure 9.7) itself is largely independent of the method employed as long as the created pool is sufficiently expansive in the vicinity of the sought structure and dense enough to yield good matches to the experimental patterns. The choice of the sampling method is thus partially a matter of convenience and should depend on the experiment at hand. For example, the MD pool's sampling over a large section of the available geometries is particularly useful in molecular systems that involve a coherent dynamic motion across a significant part of the potential energy surface.[24] However, as the MD simulation itself is relatively computationally expensive, the Wigner and MC pools are better choices when the target system is an equilibrium excited-state structure that is near the minimum of a potential surface.[26,31]

9.3.3.2.2 Calculating Scattering Patterns.

With the structure pool at hand, a simulated scattering pattern needs to be calculated for every structure in the pool. The theoretical percent difference scattering signal of an excited-state molecular system can be written as[28]

$$\%\Delta S_{exc}(q, \mathbf{R}') = 100 \times \frac{I_{exc}^{vib}(q, \mathbf{R}') - I_X(q, \mathbf{R}_0)}{I_X(q, \mathbf{R}_0)}, \quad (9.2)$$

where $I_{exc}^{vib}(q, \mathbf{R}')$ is the excited-state scattering intensity including vibrational excitation and $I_X(q, \mathbf{R}_0)$ is the scattering intensity of the ground-state molecule, with $\mathbf{R}'$ and $\mathbf{R}_0$ being the equilibrium nuclear geometries of the excited-state and the ground-state molecule, respectively. The theoretical percent difference signal can be directly related to the experimental percent difference signal (eqn (9.1)) using $\%\Delta I = \gamma \%\Delta S_{exc}$, where γ is the fraction of molecules that are optically excited, a scalar quantity that can be determined from the experimental analysis. It has been shown previously[26,28] that by inserting two null contributions, $0 = I_X^{vib}(q, \mathbf{R}') - I_X^{vib}(q, \mathbf{R}')$ and $0 = I_X(q, \mathbf{R}') - I_X(q, \mathbf{R}')$, eqn (9.2) can be rewritten as

$$\%\Delta S_{exc}(q, \mathbf{R}') = 100 \times \left(\frac{I_{exc}^{vib}(q, \mathbf{R}') - I_X^{vib}(q, \mathbf{R}')}{I_X(q, \mathbf{R}_0)} + \frac{I_X^{vib}(q, \mathbf{R}') - I_X(q, \mathbf{R}')}{I_X(q, \mathbf{R}_0)} + \frac{I_X(q, \mathbf{R}') - I_X(q, \mathbf{R}_0)}{I_X(q, \mathbf{R}_0)} \right)$$

$$= \Delta S^{elec}(q, \mathbf{R}') + \Delta S_{vib}^{nucl}(q, \mathbf{R}') + \Delta S_0^{nucl}(q, \mathbf{R}'), \quad (9.3)$$

where $\Delta S^{elec}(q, \mathbf{R}')$ represents the electronic contribution which is the difference between the excited and ground electronic state scattering signal in the molecular geometry $\mathbf{R}'$, assuming that the electronic excitation does not affect the molecular vibrations. The term $\Delta S_{vib}^{nucl}(q, \mathbf{R}')$ describes the contribution from the change in molecular vibrations upon laser excitation at a given structure $\mathbf{R}'$. The third term $\Delta S_0^{nucl}(q, \mathbf{R}')$ accounts for a change in the scattering signal that stems solely from the change in molecular geometry from $\mathbf{R}_0$ to $\mathbf{R}'$ while remaining at the ground electronic state.

In principle, the abovementioned structure determination approach (see Figure 9.7) can be implemented with any choice of method to calculate the scattering patterns and to construct the theoretical percent difference scattering signals. For example, an *ab initio* calculation can be adopted to directly calculate the X-ray scattering terms $I_{exc}^{vib}(q, \boldsymbol{R}')$ and $I_X(q, \boldsymbol{R}_0)$ that are required for eqn (9.2).[41,57,65] Unfortunately, the *ab initio* calculations are computationally expensive and almost impossible to be implemented for a large structure pool with more than one million structures. With the on-going developments of computationally efficient methods for predicting the scattering patterns,[53,66,67] this current computational bottleneck could be addressed in the near future.

Eqn (9.3) can be adopted instead to circumvent the current complexity. Previous studies have shown that the $\Delta S^{elec}(q, \boldsymbol{R}')$ term for electronic excitation to a Rydberg state is observable and nearly independent of the molecular geometry for reasonably small variations in the structure.[28] This suggests that the $\Delta S^{elec}(q, \boldsymbol{R}')$ term can be treated as a constant correction term[24] that can be simulated accurately with high-level *ab initio* X-ray scattering calculations, while the time-dependent scattering signal is mainly attributed to the change in the nuclear geometry, $\Delta S_0^{nucl}(q, \boldsymbol{R}')$. The IAM can adequately describe the $\Delta S_0^{nucl}(q, \boldsymbol{R}')$ term despite being a rather crude approximation, in part due to the strongly bound core electrons. While neglecting the specific effects of any distortion in the valence electron density distribution, the IAM nicely captures the scattering difference caused by the evolving molecular geometry and avoids potential systematic errors that might be introduced by inaccuracies of *ab initio* electronic structure methods. Additionally, the IAM offers computational simplicity and efficiency so that it can easily be applied to a large pool of structures. The influence of vibrational state distributions, $\Delta S_{vib}^{nucl}(q, \boldsymbol{R}')$, has been found to be negligible within the current experimental range of scattering vectors and a detection limit of $\sim 0.05\%$, even when the molecules are assumed to have a comparatively high internal vibrational energy.[26] Nevertheless, eqn (9.3) offers a means to include the effects of changes in the vibrational distribution once this becomes observable with further improvements in the experimental technique.

9.3.3.2.3 Determining Molecular Structures.
The last step is to determine molecular structures from experimental patterns. For each calculated theoretical percent difference scattering pattern from the structure pool, the χ^2 deviation from the experimental pattern is calculated as[26]

$$\chi_i^2 = \sum_{q=q_{min}}^{q_{max}} \left(\frac{\%\Delta S_i(q) - \%\Delta S_{exp}(q, t_1)}{\sigma(q, t_1)} \right)^2, \tag{9.4}$$

where $\%\Delta S_i(q)$ is the computed percent difference pattern for structure i in the pool, $\%\Delta S_{exp}(q,t_1)$ is the experimental percent difference scattering pattern $\%\Delta I(q,t)$ divided by the excitation fraction γ at delay time t_1, and $\sigma(q,t_1)$

represents the experimental uncertainty of $\%\Delta S_{exp}(q,t_1)$, calculated as the statistical counting noise. Each structure in the pool is associated with a specific χ_i^2 value calculated using eqn (9.4), representing how well the structure's scattering pattern agrees with the experimentally measured pattern. To determine the best structure, the inverse of the χ_i^2 values (*i.e.*, χ_i^{-2}) is plotted against molecular structure parameters such as the interatomic distances, bond angles or torsional angles, for all structures. By looking at the complete distributions instead of only picking the lowest χ_i^2 structures, artifacts associated with the sparse sampling can be largely overcome. Given the randomness inherent in the generation of the structure pools, it is not surprising that for any value of a structure parameter, there are many structures that give poor fits between the simulated and experimental scattering patterns and therefore high values of χ_i^2. Those poor fits in the simulated and measured scattering data lead to small values of χ_i^{-2} and fall beneath the envelope of the χ_i^{-2} overall distribution as a function of the structural parameters α and β (see blue dots below the black envelope of the example distributions in Figure 9.7). Retaining only the best-fitting structure for each value of the structure parameter (*i.e.*, the structure with the largest χ_i^{-2} value), the envelope of the distribution assumes a normal or skewed normal distribution.[24] The peak of the envelope is then taken as the determined value for each structural parameter, and a complete set of structural parameters can be constructed to determine the molecular structure. One should note that the functional form for describing the envelope of the χ_i^{-2} distribution and the bin widths of structural parameters should be carefully chosen, as discussed in detail in ref. 26.

Conceptually, the plot of χ_i^{-2} is a surface in a multi-dimensional space as it depends on the $3N - 6$ dimensions for a non-linear molecule with N atoms. The best-fitting structure should be the extremal point on this surface. Because it is impractical to visualize the entire multi-dimensional space, it is reduced to a series of one-dimensional fits that can be viewed as projections of the multi-dimensional surface onto a specific structural parameter. One should note that a list of determined structural parameters does not necessarily correspond to a physically realizable structure because of correlations between structural parameters.[68] For the method described in this section, the correlations between structural parameters have been shown to be largely preserved since each χ_i^{-2} value is calculated from a 3D geometrical structure.[26] Besides overcoming the problem of correlations among different structural parameters of the molecule, the structure pool analysis has two further advantages compared to the traditional least-squares refinement approach. First, it prevents the analysis from converging to a structure that is physically or chemically impossible. Second, the analysis does not require assumptions regarding the molecular symmetry, making it applicable to relatively complicated polyatomic molecular systems. For example, the method has been utilized to record an experimental "molecular movie" uncovering the coherent vibrational motions in excited-state NMM.[24] The full time-dependent structural parameters of NMM from 0 to 4 ps were determined (see Figure 9.8).

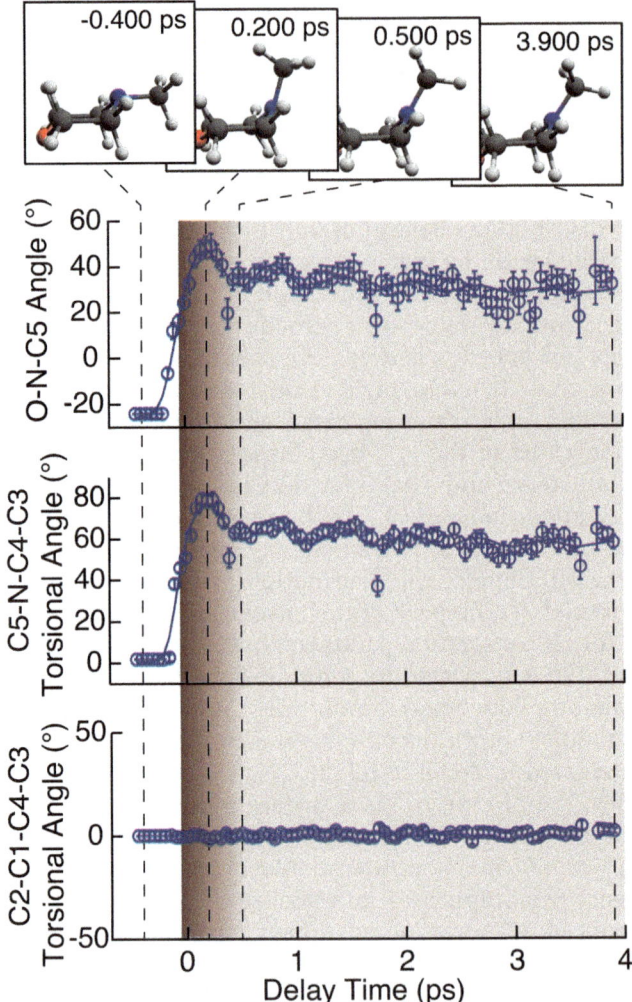

Figure 9.8 Time-dependence of selected structural parameters of NMM following 200 nm excitation. The O–N–C5 angle, the C5–N–C4–C3 torsional angle and the C2–C1–C4–C3 torsional angle (refer to Figure 9.1 for atom lables) are extracted from the structural determination and are shown along with their respective 1σ error bars. The dynamic fits to the respective vibrational motions of the O–N–C5 angle and C5–N–C4–C3 torsional angle are also shown as solid lines. The approximate lifetime of the initially excited $3p_z$ Rydberg state is shown as a dark red shaded region, which corresponds to the 3s state when the color is lighter. Representative molecular structures for selected time points are also shown. Reproduced with permission from ref. 24.

This study also determined the equilibrium structure of NMM in the excited 3s electronic state after the damping of the coherent vibrations,[24,26] with a precision in the parameters on the order of 0.01 Å.

9.3.3.3   Weighted Molecular Dynamics Trajectories

The structure pool analysis method can be applied to time-dependent dynamics problems, although in implementations to date, each time point is analyzed individually for the molecular structure. Dynamic molecular structures evolve continuously, so that structures at neighboring time points must be related, provided that the time points are finely spaced enough. The 2015 study of the ring-opening dynamics of CHD explicitly took this into account.[19] About 100 quantum molecular dynamics trajectories (see Figure 9.9) were calculated using the multiconfigurational Ehrenfest method[69] with potential energies and nonadiabatic couplings obtained on-the-fly at the SA3-CAS(6,4)/cc-pVDZ level of theory. All trajectories were compared with the time-dependent experimental data, resulting in a weight for each trajectory that was determined from a multi-start nonlinear least-square optimization routine with a finite-difference gradient.[19] The overall distribution of weighted trajectories represents the shape of the wavepacket and determines the associated structures for each individual species at any given time delay.

9.3.4   Ultrafast Chemical Reaction Dynamics

The first gas-phase X-ray scattering study of chemical reaction dynamics with femtosecond time resolution explored the ring-opening dynamics of CHD with 267 nm excitation.[19] Optical excitation at 267 nm (see green arrows in Figure 9.10) prepares the molecule in the 1B valence state. The wavepacket slides down the 1B potential energy surface and transitions through two conical intersections (CI) to the ground state of the ring-open 1,3,5-hexatriene (HT) product.[70] Interestingly, a 200 nm excitation of CHD to the 3p state results in a different reaction pathway, leading to a kinetic ring-opening reaction on the ground electronic surface that will be discussed in the next section.[30]

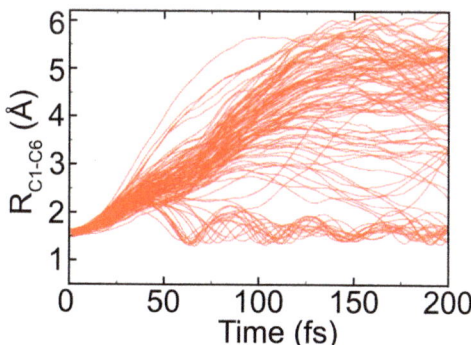

Figure 9.9   Calculated 100 trajectories for CHD ring-opening characterized by the terminal carbon C1–C6 distance as a function of time. Reproduced from ref. 19 with permission from the American Physical Society, Copyright 2015.

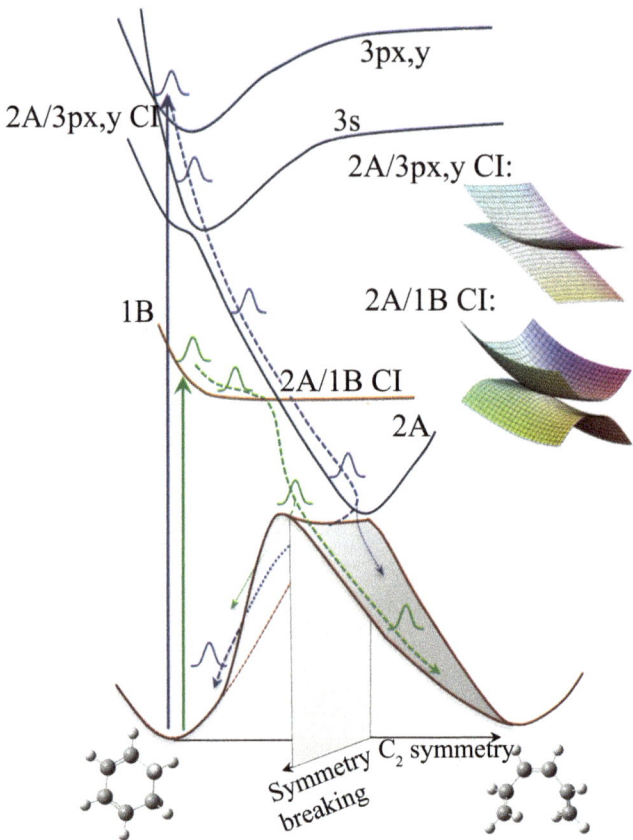

Figure 9.10   The potential energy surfaces of CHD allow for reversion of the reactant molecule to the ground state. The path through the 2A/1B conical intersections (CIs) can either deflect the wavepacket away from the symmetry plane (1B excitation, leading preferentially to the HT product) or focus it onto the symmetry plane ($3p_{xy}$ excitation, leading preferentially to the hot CHD). Reproduced from ref. 30, with permission from the Authors, Copyright 2019.

The dynamic ring-opening reaction of CHD, a prototypical example of an electrocyclic reaction, has been studied extensively through spectroscopic experiments.[71–73] Even though the time scales were well known, the time-evolving molecular structures during the reaction remained experimentally undetermined. By using ultrafast X-ray scattering, the molecular motions associated with the ring-opening reaction of CHD became accessible by the measurement of X-ray scattering patterns at several selected time points (see Figure 9.11). To model the dynamic motions of the molecule, the 100 calculated trajectories previously mentioned (see Figure 9.9) were weighted and compared to the experimental signals using the method described in Section 9.3.3.3. The calculated scattering signals (colored lines) agree well with the experimental data (black lines) in the first 250 fs of the

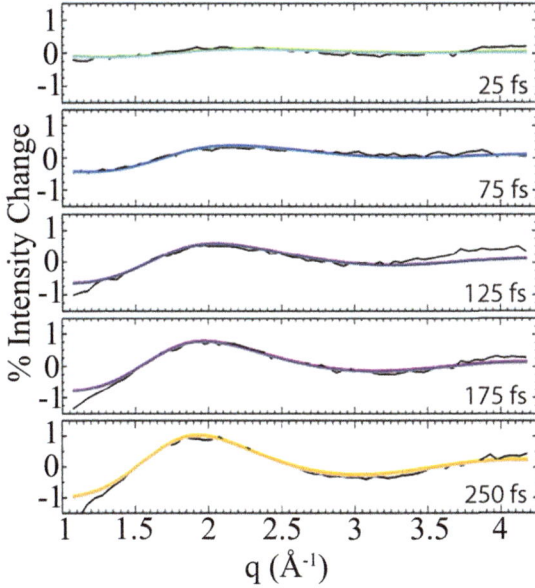

Figure 9.11 Experimental (black lines) and computational (colored lines) scattering signals for the first 250 fs of the ring-opening reaction of CHD at 267 nm excitation. Reproduced from ref. 19 with permission from the American Physical Society, Copyright 2015.

reaction (see Figure 9.11). From the analysis, the time-dependent molecular structures of both ring-opening and ring-closed channels are extracted, with the branching ratio determined as $3:2$ (*i.e.*, 60% HT yield).[19] A later theoretical study[74] using extended multistate complete active space second-order perturbation (XMS-CASPT2) surface hopping found the quantum yield for HT formation to be $47 \pm 8\%$, which is in reasonable agreement with the ultrafast X-ray scattering study.

For simple molecular systems, specifically diatomic molecules such as I_2, frequency-resolved X-ray scattering signals obtained from the temporal Fourier transform of the time-resolved X-ray scattering patterns can be used to characterize bound and dissociative motions of the molecule.[25] For example, the ultrafast X-ray scattering signal of dissociating I_2 following 520 nm excitation was studied in both the time domain and frequency domain (see Figure 9.12). In the positive angular frequency region, there are two dissociation channels (white lines) which have positive slopes. The final velocities for the two dissociations were determined to be 16.4 ± 0.2 Å ps^{-1} and 19.9 ± 0.2 Å ps^{-1}, respectively. There is also a bound state motion, which appears as a horizontal bright line in the positive angular frequency region at $\omega = 11.6 \pm 1.1$ THz in Figure 9.12b. Frequency-resolved X-ray scattering provides a useful interpretation of the ultrafast X-ray scattering signal for diatomic systems. Implementations for more complex polyatomic molecules, and in molecular systems with multiple competing dissociation pathways, remain to be developed.[75]

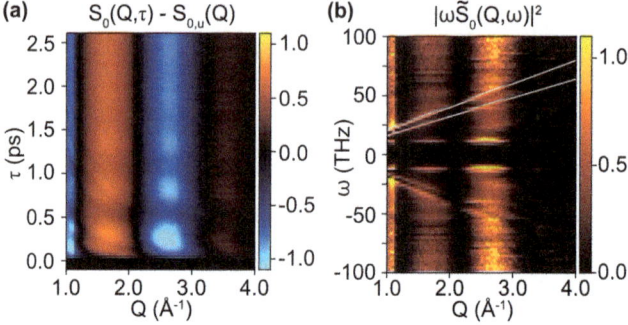

Figure 9.12   (a) Isotropic component of the experimental time-resolved X-ray scattering following 520 nm excitation. Here, $S_{0,u}(Q)$ is the isotropic scattering signal of unexcited molecules before time zero. (b) Power spectrum of (a). Reproduced from ref. 25 with permission from the American Physical Society, Copyright 2019.

9.3.5   Ultrafast Chemical Kinetics

Dynamic motions of molecules as discussed above entail the movements of wavepackets across potential energy surfaces that are concerted across all molecules of an ensemble. Kinetic reactions occur when the wavepackets have dephased and the reaction dynamics are better described using statistical models. In kinetic reactions, only the reactants, any transients, and the reaction products are observed, while the passing through transition states remains obscured. Nevertheless, a great deal of information can be obtained by studying chemical kinetics on ultrafast time scales. Time-dependent X-ray scattering patterns measured from femtosecond to picosecond or further to nanosecond time scales have been applied to study the kinetics of photoinduced chemical reactions.[29–31] Because chemical kinetics can usually be modeled with rate equations, the time-dependent scattering signals can be viewed as orthogonal contributions with the time dependence following the kinetic process and q dependence arising from the patterns of the individual transient species. The time-dependent X-ray scattering signal can then be modeled as[30,31]

$$\%\Delta I_{iso}(q,t) = \gamma \left(\sum_{\alpha} \%\Delta S_{\alpha}(q) F_{\alpha}(t) \right) \times g(t), \qquad (9.5)$$

where $\%\Delta I_{iso}(q,t)$ is the isotropic component of $\%\Delta I(q,\phi,t)$, $\%\Delta S_{\alpha}(q)$ represents the isotropic percent difference scattering pattern of transient structure α which can be treated as adjustable parameters, while $F_{\alpha}(t)$ is the corresponding time-dependent population as determined from the kinetic scheme. The scalar γ is the excitation probability and $g(t)$ is a Gaussian function that characterizes the temporal instrument response. By fitting the experimental data using eqn (9.5), the time constants of the kinetic scheme and the scattering patterns, $\%\Delta S_{\alpha}(q)$, of individual transient species can be determined simultaneously from a global fit. This concept has been used to study the photoinduced ground-state ring-opening kinetics of CHD[30] and the intramolecular excited-state charge transfer of *N,N'*-dimethylpiperazine.[31]

While excitation of the CHD molecule at 266 nm results in dynamic motions, excitation at 200 nm, which leads to the 3p Rydberg state, results in an electronically excited state with a short, but measurable lifetime (see Figure 9.13).[30] In addition to the initially excited state and the subsequent electronic decay on the femtosecond time scale, the experimental data show a gradually evolving signal from picosecond regime up to 1 ns. This suggests that a kinetic reaction at the hot ground state of the system is involved following the electronic decay of the initially excited 3p state.

By using eqn (9.5) to fit the experimental data, a full reaction scheme of the kinetics can be unveiled (see Figure 9.10). The analysis determined that the initially excited 3p state decays in 208 ± 11 fs to the electronic ground state.

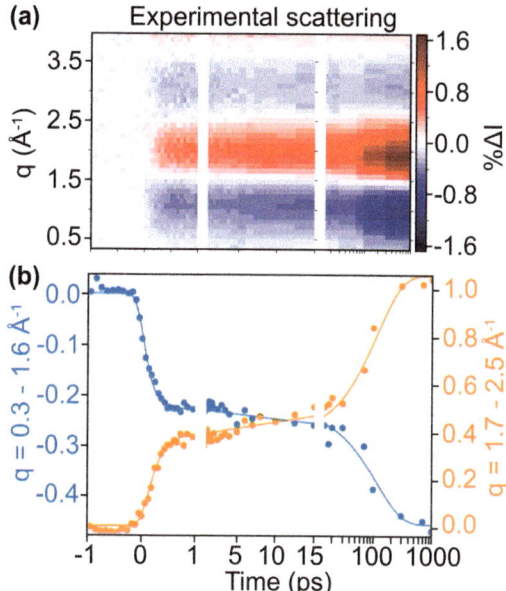

Figure 9.13 (a) The isotropic component of the time-dependent experimental percent difference scattering signal of CHD as a function of time and the absolute value of the scattering momentum transfer. (b) Averages over two q ranges (dots) and the kinetic fit (lines). Blue dots are averaged over q range of 0.3–1.6 Å^{-1} while orange dots are averaged over a q range of 1.7–2.5 Å^{-1}. The panels are divided into three time segments: one from −1 to 1 ps to show the ultrafast temporal response to the pump laser pulse; the time range from 1 ps to 15 ps showing the initial ground-state population; and the time range from 15 ps to 1 ns (on a log scale) giving the increase in HT population as the molecules equilibrate on the ground state potential energy surface. Reproduced from ref. 30, with permission from the Authors, Copyright 2019.

During this process, $76 \pm 3\%$ of the molecules were found to decay back to the vibrationally hot, ring-closed ground-state CHD, while the remainder undergoes a rapid ring-opening reaction to form hot HT. A thermal ring-opening reaction on the ground electronic state surface then occurs and an equilibrium between the hot CHD and hot HT is reached with the ring-opening and ring-closing time constants determined to be 174 ± 13 ps and 355 ± 45 ps, respectively. The analysis also yields the scattering patterns of the separate hot CHD and HT products. To accurately model the scattering patterns of the hot products, a novel method based on molecular dynamics trajectories was developed.[34] This is necessary for vibrationally hot molecules because large amplitude vibrational motions are usually anharmonic, and correlated distances and distance shifts must be included when modeling their scattering patterns (see Figure 9.14).

In *N,N'*-dimethylpiperazine (DMP), previous photoelectron investigations have shown that excitation to the 3p state will lead to rapid relaxation to the charge-localized (3sL) and charge-delocalized (3sD) conformers in the 3s state on the femtosecond time scale. The charge transfer then proceeds as the molecules evolve on the 3s potential energy surface (see Figure 9.15). An equilibrium between 3sL and 3sD conformers is reached with an overall time constant of 2.65 ps.[76] This kinetic scheme was recently confirmed through an ultrafast X-ray scattering experiment.[31]

Using the previously determined reaction kinetics and eqn (9.5), the scattering patterns of the 3sL and 3sD products were extracted (see Figure 9.16). By using the structure determination method based on structure pools as described in Section 9.3.3.2, complete molecular structures of the charge-localized and the charge-delocalized species in the 3s state were determined. It was found that charge transfer weakens the carbon–carbon bond to an unusual 1.634 Å bond length while the bond lengths between the

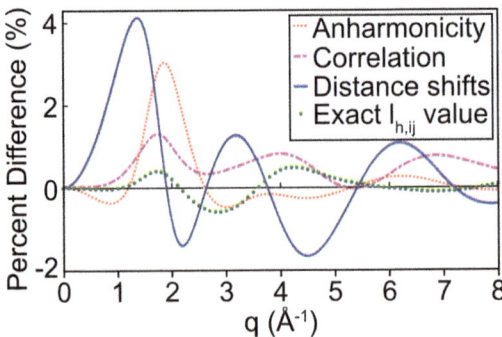

Figure 9.14   Contributions to the percent difference scattering signals of the CHD, after 200 nm excitation and subsequent relaxation into thermal (~ 2870 K) vibrations. Shown are the contributions arising from the distance shifts, anharmonicity, correlations between atom–atom pair distances and exact vibrational amplitude. Reproduced from ref. 34 with permission from AIP Publishing, Copyright 2019.

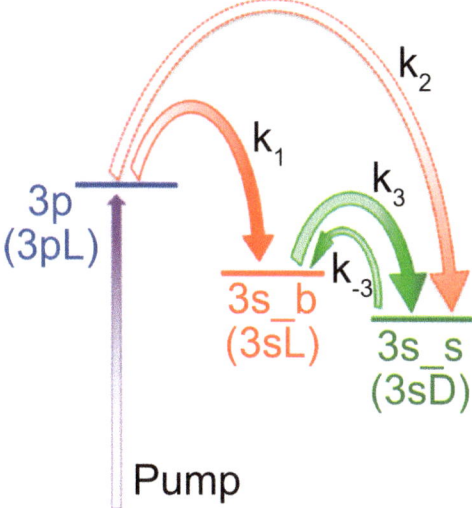

Figure 9.15 Reaction pathway for Rydberg-excited *N,N'*-dimethylpiperazine (DMP). Reproduced from ref. 76 with permission from American Chemical Society, Copyright 2013.

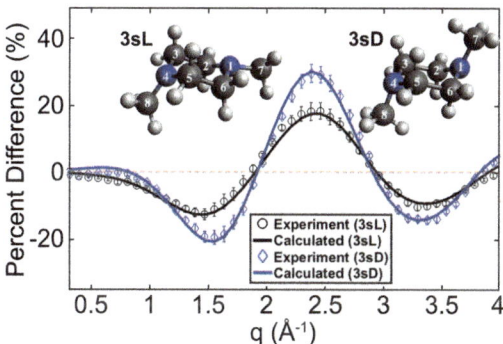

Figure 9.16 Experimental and calculated percent difference isotropic scattering patterns and molecular structures of *N,N'*-dimethylpiperazine (DMP) in the charge-localized 3sL and the charge-delocalized 3sD conformers. The experimental results (circles and diamonds) are extracted from the kinetics fit with 3σ uncertainties and divided by the excitation fraction γ, determined from the fit. Calculated scattering patterns (solid lines) are for the experimentally determined optimal structural parameters with electronic contributions included. (inset) Representative geometries of 3sL and 3sD. Reproduced from ref. 31 with permission from the National Academy of Sciences.

nitrogen and the ring-carbon atoms contract from an average of 1.505 to 1.465 Å.[31] This demonstrates that the ultrafast X-ray scattering can resolve the changes in the molecular structure that arise from charge transfer and can provide valuable benchmarks for the evaluation of computational electronic structure methods.[77,78]

9.4 Summary

The advent of XFEL facilities has had a transformative impact on the study of excited-state molecular structures and chemical reaction dynamics. One of the experimental techniques enabled by XFELs is ultrafast time-resolved gas-phase X-ray scattering. Accurate measurements of excited-state electron densities and molecular geometries provide new insights into chemical bonding and molecular dynamics, which offer essential benchmarks for the further development of theory and computational methods in dynamics and electronic structure theory. Ultrafast time-resolved, gas-phase X-ray scattering can resolve both nuclear and electron motions during chemical reactions. Further advances, including higher energy photons and high repetition rate XFELs such as LCLS-II,[55] the ongoing development of ultrafast X-ray scattering theory[56,58,79–81] and advanced structural inversion methods,[82–84] promise future ultrafast X-ray scattering experiments that directly map the coupled electron and nuclear motions, providing a detailed and comprehensive view of chemical reactions in real time.[85,86]

Acknowledgements

The research of PMW and his group benefited from funding by the US Department of Energy, Office of Science, Basic Energy Sciences, award # DE-SC0017995 and DE-SC0020276, and from the National Science Foundation, award # CHE-2309434. In addition, AK acknowledges funding from the Engineering and Physical Sciences Research Council (EP/V049240/2, EP/V006819/2, EP/X026698/1 and EP/X026973/1), the Leverhulme Trust (RPG-2020-208), and a natural sciences fellowship at the Swedish Collegium for Advanced Study with financial support from the Erling-Persson Family Foundation and the Knut and Alice Wallenberg Foundation.

References

1. A. A. Ischenko, P. M. Weber and R. J. D. Miller, *Chem. Rev.*, 2017, **117**, 11066.
2. P. Emma, R. Akre, J. Arthur, R. Bionta, C. Bostedt, J. Bozek, A. Brachmann, P. Bucksbaum, R. Coffee, F.-J. Decker, Y. Ding, D. Dowell, S. Edstrom, A. Fisher, J. Frisch, S. Gilevich, J. Hastings, G. Hays, P. Hering, Z. Huang, R. Iverson, H. Loos, M. Messerschmidt, A. Miahnahri, S. Moeller, H.-D. Nuhn, G. Pile, D. Ratner, J. Rzepiela, D. Schultz, T. Smith, P. Stefan, H. Tompkins, J. Turner, J. Welch, W. White, J. Wu, G. Yocky and J. Galayda, *Nat. Photon.*, 2010, **4**, 641.
3. S. Liu, W. Decking, V. Kocharyan, E. Saldin, S. Serkez, R. Shayduk, H. Sinn and G. Geloni, *Phys. Rev. Accel. Beams*, 2019, **22**, 060704.

4. L. Serrano-Andres and J. J. Serrano-Perez, in *Handbook of computational chemistry*, ed. J. Leszczynski, Springer, Dordrecht, 2012, ch. 6, pp. 1–88.
5. J. C. H. Spence, *Struct. Dyn.*, 2017, **4**, 044027.
6. M. Stefanou, K. Saita, D. V. Shalashilin and A. Kirrander, *Chem. Phys. Lett.*, 2017, **683**, 300.
7. L. Ma, H. Yong, J. D. Geiser, A. Moreno Carrascosa, N. Goff and P. M. Weber, *Struct. Dyn.*, 2020, 7, 034102.
8. J. R. Rouxel, D. Keefer and S. Mukamel, *Struct. Dyn.*, 2021, **8**, 014101.
9. C. Rose-Petruck, R. Jimenez, T. Guo, A. Cavalleri, C. W. Siders, F. Rksi, J. A. Squier, B. C. Walker, K. R. Wilson and C. P. J. Barty, *Nature*, 1999, **398**, 310.
10. R. Neutze, R. Wouts, S. Techert, J. Davidsson, M. Kocsis, A. Kirrander, F. Schotte and M. Wulff, *Phys. Rev. Lett.*, 2001, **87**, 195508.
11. J. Davidsson, J. Poulsen, M. Cammarata, P. Georigiou, R. Wouts, G. Katona, F. Jacobson, A. Plech, M. Wulff, G. Nyman and R. Neutze, *Phys. Rev. Lett.*, 2005, **94**, 245503.
12. H. Ihee, M. Lorenc, T. K. Kim, Q. Y. Kong, M. Cammarata, J. H. Lee, S. Bratos and M. Wulff, *Science*, 2005, **309**, 1223.
13. M. Bargheer, N. Zhavoronkov, M. Woerner and T. Elsaesser, *Chem. Phys. Chem.*, 2006, 7, 783.
14. J. H. Lee, M. Wulf, S. Bratos, J. Petersen, L. Guerin, J.-C. Leicknam, M. Cammarata, Q. Kong, J. Kim, K. B. Møller and H. Ihee, *J. Am. Chem. Soc.*, 2013, **135**, 3255.
15. D. Arnlund, L. C. Johansson, C. Wickstrand, A. Barty, G. J Williams, E. Malmerberg, J. Davidsson, D. Milathianaki, D. P. DePonte, R. L Shoeman, D. Wang, D. James, G. Katona, S. Westenhoff, T. A White, A. Aquila, S. Bari, P. Berntsen, M. Bogan, T. B. van Driel, R. B. Doak, K. S. Kjær, M. Frank, R. Fromme, I. Grotjohann, R. Henning, M. S. Hunter, R. A. Kirian, I. Kosheleva, C. Kupitz, M. Liang, A. V. Martin, M. M. Nielsen, M. Messerschmidt, M. M. Seibert, J. Sjöhamn, F. Stellato, U. Weierstall, N. A. Zatsepin, J. C. H. Spence, P. Fromme, I. Schlichting, S. Boutet, G. Groenhof, H. N. Chapman and R. Neutze, *Nat. Methods*, 2014, **11**, 923.
16. M. Levantino, G. Schirò, H. T. Lemke, G. Cottone, J. M. Glownia, D. Zhu, M. Chollet, H. Ihee, A. Cupane and M. Cammarata, *Nat. Commun.*, 2015, **6**, 6772.
17. J. G. Kim, S. Nozawa, H. Kim, E. H. Choi, T. Sato, T. W. Kim, K. H. Kim, H. Ki, J. Kim, M. Choi, Y. Lee, J. Heo, K. Y. Oang, K. Ichiyanagi, R. Fukaya, J. H. Lee, J. Park, I. Eom, S. H. Chun, S. Kim, M. Kim, T. Katayama, T. Togashi, S. Owada, M. Yabashi, S. J. Lee, S. Lee, C. W. Ahn, D.-S. Ahn, J. Moon, S. Choi, J. Kim, T. Joo, J. Kim, S. Adachi and H. Ihee, *Nature*, 2020, **582**, 520.
18. B. E. Warren, *X-Ray Diffraction*, Addison-Wesley, New York, 1968.
19. M. P. Minitti, J. M. Budarz, A. Kirrander, J. S. Robinson, D. Ratner, T. J. Lane, D. Zhu, J. M. Glownia, M. Kozina, H. T. Lemke, M. Sikorski, Y. Feng, S. Nelson, K. Saita, B. Stankus, T. Northey, J. B. Hastings and P. M. Weber, *Phys. Rev. Lett.*, 2015, **114**, 255501.

20. J. M. Budarz, M. P. Minitti, D. V. Cofer-Shabica, B. Stankus, A. Kirrander, J. B. Hastings and P. M. Weber, *J. Phys. B: At., Mol. Opt. Phys.*, 2016, **49**, 034001.

21. B. Stankus, H. Yong, J. M. Ruddock, L. Ma, A. Moreno Carrascosa, N. Goff, S. Boutet, X. Xu, N. Zotev, A. Kirrander, M. Minitti and P. M. Weber, *J. Phys. B: At., Mol. Opt. Phys.*, 2020, **53**, 234004.

22. B. Stankus, J. M. Budarz, A. Kirrander, D. Rogers, J. Robinson, T. J. Lane, D. Ratner, J. Hastings, M. P. Minitti and P. M. Weber, *Faraday Discuss.*, 2016, **194**, 525.

23. J. M. Glownia, A. Natan, J. P. Cryan, R. Hartsock, M. Kozina, M. P. Minitti, S. Nelson, J. Robinson, T. Sato, T. van Driel, G. Welch, C. Weninger, D. Zhu and P. H. Bucksbaum, *Phys. Rev. Lett.*, 2016, **117**, 153003.

24. B. Stankus, H. Yong, N. Zotev, J. M. Ruddock, D. Bellshaw, T. J. Lane, M. Liang, S. Boutet, S. Carbajo, J. S. Robinson, W. Du, N. Goff, Y. Chang, J. E. Koglin, M. P. Minitti, A. Kirrander and P. M. Weber, *Nat. Chem.*, 2019, **11**, 716.

25. M. R. Ware, J. M. Glownia, N. Al-Sayyad, J. T. O'Neal and P. H. Bucksbaum, *Phys. Rev. A*, 2019, **100**, 033413.

26. H. Yong, A. Moreno Carrascosa, L. Ma, B. Stankus, M. P. Minitti, A. Kirrander and P. M. Weber, *Faraday Discuss.*, 2021, **228**, 104.

27. H. Yong, N. Zotev, B. Stankus, J. M. Ruddock, D. Bellshaw, S. Boutet, T. J. Lane, M. Liang, S. Carbajo, J. S. Robinson, W. Du, N. Goff, Y. Chang, J. E. Koglin, M. D. J. Waters, T. I. Sølling, M. P. Minitti, A. Kirrander and P. M. Weber, *J. Phys. Chem. Lett.*, 2018, **9**, 6556.

28. H. Yong, N. Zotev, J. M. Ruddock, B. Stankus, M. Simmermacher, A. Moreno Carrascosa, W. Du, N. Goff, Y. Chang, D. Bellshaw, M. Liang, S. Carbajo, J. E. Koglin, J. S. Robinson, S. Boutet, M. P. Minitti, A. Kirrander and P. M. Weber, *Nat. Commun.*, 2020, **11**, 2157.

29. J. M. Ruddock, N. Zotev, B. Stankus, H. Yong, D. Bellshaw, S. Boutet, T. J. Lane, M. Liang, S. Carbajo, W. Du, A. Kirrander, M. P. Minitti and P. M. Weber, *Angew. Chem., Int. Ed.*, 2019, **58**, 6371.

30. J. M. Ruddock, H. Yong, B. Stankus, W. Du, N. Goff, Y. Chang, A. Odate, A. Moreno Carrascosa, D. Bellshaw, N. Zotev, M. Liang, S. Carbajo, J. E. Koglin, J. S. Robinson, S. Boutet, A. Kirrander, M. P. Minitti and P. M. Weber, *Sci. Adv.*, 2019, **5**, eaax6625.

31. H. Yong, X. Xu, J. M. Ruddock, B. Stankus, A. Moreno Carrascosa, N. Zotev, D. Bellshaw, W. Du, N. Goff, Y. Chang, S. Boutet, S. Carbajo, J. E. Koglin, M. Liang, J. S. Robinson, A. Kirrander, M. P. Minitti and P. M. Weber, *Proc. Natl. Acad. Sci. U. S. A.*, 2021, **118**, e2021714118.

32. P. H. Bucksbaum, M. R. Ware, A. Natan, J. P. Cryan and J. M. Glownia, *Phys. Rev. X*, 2020, **10**, 011065.

33. A. Natan, A. Schori, G. Owolabi, J. P. Cryan, J. M. Glownia and P. H. Bucksbaum, *Faraday Discuss.*, 2021, **228**, 123.

34. H. Yong, J. M. Ruddock, B. Stankus, L. Ma, W. Du, N. Goff, Y. Chang, N. Zotev, D. Bellshaw, S. Boutet, S. Carbajo, J. E. Koglin, M. Liang, J. S.

Robinson, A. Kirrander, M. P. Minitti and P. M. Weber, *J. Chem. Phys.*, 2019, **151**, 084301.

35. M. R. Bionta, N. Hartmann, M. Weaver, D. French, D. J. Nicholson, J. P. Cryan, J. M. Glownia, K. Baker, C. Bostedt, M. Chollet, Y. Ding, D. M. Fritz, A. R. Fry, D. J. Kane, J. Krzywinski, H. T. Lemke, M. Messerschmidt, S. Schorb, D. Zhu, W. E. White and R. N. Coffee, *Rev. Sci. Instrum.*, 2014, **85**, 083116.

36. M. Chollet, R. Alonso-Mori, M. Cammarata, D. Damiani, J. Defever, J. T. Delor, Y. Feng, J. M. Glownia, J. B. Langton, S. Nelson, K. Ramsey, A. Robert, M. Sikorski, S. Song, D. Stefanescu, V. Srinivasan, D. Zhu, H. T. Lemke and D. M. Fritz, *J. Synchrotron Radiat.*, 2015, **22**, 503.

37. M. Liang, G. J. Williams, M. Messerschmidt, M. M. Seibert, P. A. Montanez, M. Hayes, D. Milathianaki, A. Aquila, M. S. Hunter, J. E. Koglin, D. W. Schafer, S. Guillet, A. Busse, R. Bergan, W. Olson, K. Fox, N. Stewart, R. Curtis, A. A. Miahnahri and S. Boutet, *J. Synchrotron Radiat.*, 2015, **22**, 514.

38. R. C. Dudek and P. M. Weber, *J. Phys. Chem. A*, 2001, **105**, 4167.

39. D. R. Yarkony, *Rev. Mod. Phys.*, 1996, **68**, 985.

40. L. J. Butler, *Annu. Rev. Phys. Chem.*, 1998, **49**, 125.

41. T. Northey, N. Zotev and A. Kirrander, *J. Chem. Theory Comput.*, 2014, **10**, 4911.

42. A. Moreno Carrascosa, T. Northey and A. Kirrander, *Phys. Chem. Chem. Phys.*, 2017, **19**, 7853.

43. J. Küpper, S. Stern, L. Holmegaard, F. Filsinger, A. Rouzée, A. Rudenko, P. Johnsson, A. V. Martin, M. Adolph, A. Aquila, S. Bajt, A. Barty, C. Bostedt, J. Bozek, C. Caleman, R. Coffee, N. Coppola, T. Delmas, S. Epp, B. Erk, L. Foucar, T. Gorkhover, L. Gumprecht, A. Hartmann, R. Hartmann, G. Hauser, P. Holl, A. Hömke, N. Kimmel, F. Krasniqi, K.-U. Kühnel, J. Maurer, M. Messerschmidt, R. Moshammer, C. Reich, B. Rudek, R. Santra, I. Schlichting, C. Schmidt, S. Schorb, J. Schulz, H. Soltau, J. C. H. Spence, D. Starodub, L. Strüder, J. Thøgersen, M. J. J. Vrakking, G. Weidenspointner, T. A. White, C. Wunderer, G. Meijer, J. Ullrich, H. Stapelfeldt, D. Rolles and H. N. Chapman, *Phys. Rev. Lett.*, 2014, **112**, 083002.

44. T. Kierspel, A. Morgan, J. Wiese, T. Mullins, A. Aquila, A. Barty, R. Bean, R. Boll, S. Boutet, P. Bucksbaum, H. N. Chapman, L. Christensen, A. Fry, M. Hunter, J. E. Koglin, M. Liang, V. Mariani, A. Natan, J. Robinson, D. Rolles, A. Rudenko, K. Schnorr, H. Stapelfeldt, S. Stern, J. Thøgersen, C. H. Yoon, F. Wang and J. Küpper, *J. Chem. Phys.*, 2020, **152**, 084307.

45. P. M. Felker and A. H. Zewail, *J. Chem. Phys.*, 1987, **86**, 2460.

46. P. M. Felker, *J. Phys. Chem.*, 1992, **96**, 7844.

47. O. Geßner, A. M. D. Lee, J. P. Shaffer, H. Reisler, S. V. Levchenko, A. I. Krylov, J. G. Underwood, H. Shi, A. L. L. East, D. M. Wardlaw, E. T. H. Chrysostom, C. C. Hayden and A. Stolow, *Science*, 2006, **311**, 219.

48. W. Li, A. A. Jaroń-Becker, C. W. Hogle, V. Sharma, X. Zhou, A. Becker, H. C. Kapteyn and M. M. Murnane, *Proc. Natl. Acad. Sci. U. S. A.*, 2010, **107**, 20219.

49. A. R. Attar, A. Bhattacherjee, C. D. Pemmaraju, K. Schnorr, K. D. Closser, D. Prendergast and S. R. Leone, *Science*, 2017, **356**, 54.

50. M. Ben-Nun, T. J. Martínez, P. M. Weber and K. R. Wilson, *Chem. Phys. Lett.*, 1996, **262**, 405.

51. A. Debnarova, S. Techert and S. Schmatz, *J. Chem. Phys.*, 2010, **133**, 124309.

52. A. Kirrander, *J. Chem. Phys.*, 2012, **137**, 154310.

53. R. M. Parrish and T. J. Martínez, *J. Chem. Theory Comput.*, 2019, **15**, 1523.

54. C. C. Bühler, M. P. Minitti, S. Deb, J. Bao and P. M. Weber, *J. At. Mol. Phys.*, 2011, **2011**, 637593.

55. A. Halavanau, F. J. Decker, C. Emma, J. Sheppard and C. Pellegrini, *J. Synchrotron Radiat.*, 2019, **26**, 635.

56. K. Bennett, M. Kowalewski, J. R. Rouxel and S. Mukamel, *Proc. Natl. Acad. Sci. U. S. A.*, 2018, **115**, 6538.

57. A. Moreno Carrascosa, H. Yong, D. L. Crittenden, P. M. Weber and A. Kirrander, *J. Chem. Theory Comput.*, 2019, **15**, 2836.

58. M. Simmermacher, A. Moreno Carrascosa, N. E. Henriksen, K. B. Møller and A. Kirrander, *J. Chem. Phys.*, 2019, **151**, 174302.

59. G. Taylor, *Acta Crystallogr.*, 2003, **D59**, 1881.

60. V. Makhija, A. E. Boguslavskiy, R. Forbes, K. Veyrinas, I. Wilkinson, R. Lausten, M. S. Schuurman, E. R. Grant and A. Stolow, *Faraday Discuss.*, 2021, **228**, 191.

61. I. Hargittai, in *Stereochemical applications of gas-phase electron diffraction*, ed. I. Hargittai and M. Hargittai, Wiley-VCH, Weinheim, 1988, ch. 1, pp. 2–51.

62. T. Gorkhover, A. Ulmer, K. Ferguson, M. Bucher, F. R. N. C. Maia, J. Bielecki, T. Ekeberg, M. F. Hantke, B. J. Daurer, C. Nettelblad, J. Andreasson, A. Barty, P. Bruza, S. Carron, D. Hasse, J. Krzywinski, D. S. D. Larsson, A. Morgan, K. Mühlig, M. Müller, K. Okamoto, A. Pietrini, D. Rupp, M. Sauppe, G. van der Schot, M. Seibert, J. A. Sellberg, M. Svenda, M. Swiggers, N. Timneanu, D. Westphal, G. Williams, A. Zani, H. N. Chapman, G. Faigel, T. Möller, J. Hajdu and C. Bostedt, *Nat. Photon.*, 2018, **12**, 150.

63. J. Yang, M. Guehr, X. Shen, R. Li, T. Vecchione, R. Coffee, J. Corbett, A. Fry, N. Hartmann, C. Hast, K. Hegazy, K. Jobe, I. Makasyuk, J. Robinson, M. S. Robinson, S. Vetter, S. Weathersby, C. Yoneda, X. Wang and M. Centurion, *Phys. Rev. Lett.*, 2016, **117**, 153002.

64. J. P. Dahl and M. Springborg, *J. Chem. Phys.*, 1988, **88**, 4535.

65. T. Northey, A. Moreno Carrascosa, S. Schafer and A. Kirrander, *J. Chem. Phys.*, 2016, **145**, 154304.

66. T. Northey and A. Kirrander, *J. Phys. Chem. A*, 2019, **123**, 3395.

67. N. Zotev, A. Moreno Carrascosa, M. Simmermacher and A. Kirrander, *J. Chem. Theory Comput.*, 2020, **16**, 2594.

68. T. F. Havel, in *Encyclopedia of computational chemistry*, ed. P. R. Schleyer, N. L. Allinger, T. Clark, J. Gasteiger, P. A. Kollman, H. F. Schaefer and P. R. Schreiner, John Wiley and Sons, Ltd., 2002.

69. K. Saita and D. V. Shalashilin, *J. Chem. Phys.*, 2012, **137**, 22A506.
70. S. Deb and P. M. Weber, *Annu. Rev. Phys. Chem.*, 2011, **62**, 19.
71. M. Garavelli, C. S. Page, P. Celani, M. Olivucci, W. E. Schmid, S. A. Trushin and W. Fuss, *J. Phys. Chem. A*, 2001, **105**, 4458.
72. N. Kuthirummal, F. M. Rudakov, C. L. Evans and P. M. Weber, *J. Chem. Phys.*, 2006, **125**, 133307.
73. C. C. Pemberton, Y. Zhang, K. Saita, A. Kirrander and P. M. Weber, *J. Phys. Chem. A*, 2015, **119**, 8832.
74. I. Polyak, L. Hutton, R. Crespo-Otero, M. Barbatti and P. J. Knowles, *J. Chem. Theory Comput.*, 2019, **15**, 3929.
75. I. Gabalski, M. R. Ware and P. H. Bucksbaum, *J. Phys. B: At., Mol. Opt. Phys.*, 2020, **53**, 244002.
76. S. Deb, X. Cheng and P. M. Weber, *J. Phys. Chem. Lett.*, 2013, **4**, 2780.
77. X. Cheng, Y. Zhang, E. Jónsson, H. Jónsson and P. M. Weber, *Nat. Commun.*, 2016, 7, 11013.
78. M. Gałynska, V. Ásgeirsson, H, Jónsson and R. Bjornsson, Localized and delocalized states of a diamine cation: A critical test of wave function methods, 2020, arXiv, Preprint, https://arxiv.org/pdf/2007.06125 (accessed 15 July 2020).
79. G. Dixit, O. Vendrell and R. Santra, *Proc. Natl. Acad. Sci. U. S. A.*, 2012, **109**, 11636.
80. A. Moreno Carrascosa, M. Yang, H. Yong, L. Ma, A. Kirrander, P. M. Weber and K. Lopata, *Faraday Discuss.*, 2021, **228**, 60.
81. H. Yong, D. Keefer and S. Mukamel, *J. Am. Chem. Soc.*, 2022, **144**, 7796.
82. M. Asenov, N. Zotev, S. Ramamoorthy and A. Kirrander, Inversion of ultrafast X-ray scattering with dynamics constraints, in *Machine Learning and the Physical Sciences*, Vancouver, Canada, 2020, p. 7, https://ml4physicalsciences.github.io/2020/
83. M. Zhang, S. Zhang, Y. Xiong, H. Zhang, A. A. Ischenko, O. Vendrell, X. Dong, X. Mu, M. Centurion, H. Xu, R. J. D. Miller and Z. Li, *Nat. Commun.*, 2021, **12**, 5441.
84. A. Hosseinizadeh, N. Breckwoldt, R. Fung, R. Sepehr, M. Schmidt, P. Schwander, R. Santra and A. Ourmazd, *Nature*, 2021, **599**, 697.
85. M. Simmermacher, N. E. Henriksen, K. B. Möller, A. Moreno Carrascosa and A. Kirrander, *Phys. Rev. Lett.*, 2019, **122**, 073003.
86. D. Keefer, F. Aleotti, J. R. Rouxel, F. Segatta, B. Gu, A. Nenov, M. Garavelli and S. Mukamel, *Proc. Natl. Acad. Sci. U. S. A.*, 2021, **118**, e2022037118.

Photoelectron Diffraction

T. JAHNKE*[a] AND D. ROLLES*[b]

[a] European XFEL, Holzkoppel 4, 22869 Schenefeld, Germany; [b] J.R. Macdonald Laboratory, Department of Physics, Kansas State University, Manhattan, KS 66506, USA
*Emails: till.jahnke@xfel.eu; rolles@phys.ksu.edu

10.1 Introduction

Accessing and visualizing the microscopic domain consisting of atoms and molecules in a successively increasing level of detail is a long-lasting, yet ongoing research endeavour. Several techniques have been developed to gather information on the three-dimensional (3D) structure of matter as demonstrated, for example, with different types of electron microscopes. Other approaches rely on the scattering of X-rays or electrons at the sample, and on reconstructing the microscopic position-space information from the momentum-space information contained in the diffraction patterns. In the following, we describe a particular type of electron diffraction, known as X-ray photoelectron diffraction (XPD), where the photoelectric effect is employed to generate photoelectrons inside the sample of interest, which are then diffracted by the sample and measured by a detector. Historically, this approach was first used to investigate small (di-atomic and tri-atomic) molecules absorbed on a surface (see ref. 1–3 for some early reviews) before being extended to studies of molecules in the gas phase, which are the focus of the following discussion.

10.2 Photoelectron Scattering

Photoelectron diffraction is conceptually closely related to other types of electron diffraction, such as ultrafast electron diffraction (UED; see Chapters 11,

Theoretical and Computational Chemistry Series No. 25
Structural Dynamics with X-ray and Electron Scattering
Edited by Kasra Amini, Arnaud Rouzée and Marc J. J. Vrakking
Published by the Royal Society of Chemistry, www.rsc.org

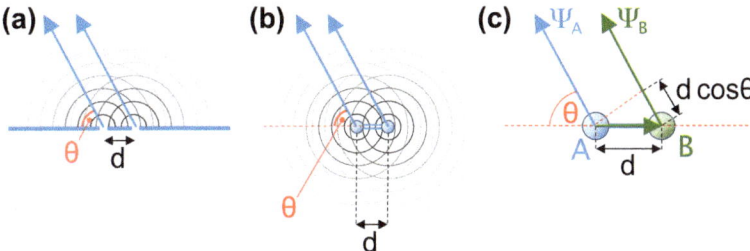

Figure 10.1 (a and b) Analogy between Young's double slit (a) and a homonuclear molecule emitting a photoelectron wave (b). (c) The related concept of photoelectron diffraction, explained on the example of a di-atomic molecule AB with bond length d. After absorption of an X-ray photon, an inner-shell electron is emitted as a photoelectron wave Ψ_A that is localized on atom A. Scattering of this photoelectron wave on atom B creates a scattered wave Ψ_B. In the far field, the interference of the direct wave and the scattered wave with a path length difference $(d + d \cos \theta)$ produces a photoelectron angular distribution $I(\theta) = |\Psi_A + \Psi_B|^2$ with characteristic oscillations that encode the bond length d.

12 and 14) or laser-induced electron diffraction (LIED; see Chapter 13). However, photoelectron diffraction is typically treated in a rather different theoretical framework that describes the molecular-frame photoelectron angular distribution (MFPAD) as a superposition of an infinite number of partial waves, as discussed in more detail in Section 10.3. Nonetheless, it is helpful to consider the molecular photoelectron emission process in a schematic picture that highlights the scattering and diffraction character (see Figure 10.1), and allows for a more intuitive understanding of the expected scattering patterns.[4–8]

The (core–shell) photoionization of a di-atomic molecule (see Figure 10.1b) can be considered a scenario very similar to that of Young's double slit (see Figure 10.1a). In both cases, two waves[†] emerge from two source points resulting in the well-known interference pattern in the far-field. The photoelectron diffraction concept (see Figure 10.1c) follows the same idea: assuming that an inner-shell photoelectron is emitted from a well-localized position inside the molecule (in this case, the position of atom A), this simple picture suggests that the MFPAD can be interpreted as a diffraction pattern resulting from the interference of the directly emitted wave Ψ_A and a scattered wave Ψ_B, originating from the neighbouring atom B (for a poly-atomic molecule, a scattered wave would originate from each atom in the molecule except the emitter atom). The contribution from single scattering only (on atom B; see green arrows in Figure 10.1c) can also be easily extended if higher scattering orders are included. For simplicity, the following discussion and simulations assume that multiple scattering can be neglected, which is the case for sufficiently high photoelectron kinetic energies.

[†] In case of molecular photoionization, the photoelectron is considered an electron wave in this picture.

Interestingly, the interference of a direct ("reference") wave, Ψ_A, and a scattered wave, Ψ_B, in the far field can also be interpreted as a hologram, which should enable a holographic reconstruction of the molecular structure from the photoelectron diffraction pattern.[3,9] However, despite the conceptual appeal of this approach, in particular for time-resolved experiments, an experimental demonstration of photoelectron holography for gas-phase molecules has not been achieved to date.

Photoelectron diffraction (see Figure 10.1c) resembles the double slit interference but with a crucial distinction that an additional path length d is introduced as the wave originating from atom A travels to atom B before it is scattered. This has an interesting consequence that the path length difference perpendicular to the molecular axis (*i.e.*, at an emission angle $\theta = 90°$) is d, while it is $2d$ in the direction of the emitter atom ($\theta = 0°$), and zero in the direction of the scattering atom ($\theta = 180°$). In other words, the MFPAD of a di-atomic molecule (in the single scattering limit) will always have an interference maximum along the molecular axis but pointing away from the emitter atom (this effect is sometimes also referred to as "forward focusing"), while the positions of the other maxima and minima depend on the relationship between the photoelectron's de-Broglie wavelength (*i.e.*, its kinetic energy) and the bond length d.

In order to visualize these conclusions by simulating an actual MFPAD, we need to first make some assumptions regarding the symmetry of the emitted and scattered photoelectron waves, which we can expand in terms of the spherical harmonics Y_m^l. For simplicity's sake, we shall consider the scattered wave to always be a spherical wave, Y_0^0, centred around atom B, but we shall perform our simulations for emission of either a spherical wave or a p-wave, Y_0^1, from atom A. The former yields the most intuitive diffraction patterns, while the latter is a more realistic choice for photoemission from a core orbital such as the carbon or oxygen 1s level.

Simulated MFPADs of a photoelectron with a kinetic energy of 250 eV emitted from the carbon atom in a CO molecule for both of the two scenarios described above are considered (see Figure 10.2a) together with a two-dimensional representation of the simulated photoelectron intensity as a function of the emission angle θ and the photoelectron kinetic energy (see Figure 10.2b).

Both figures show the characteristic "forward-focusing" maximum in the direction of the oxygen atom ($\theta = 180°$), while the number and angular position of the other interference maxima and minima depend on the electron kinetic energy, as expected. The shape of the MFPAD is significantly modified and some of the clear and intuitive diffraction effects are obscured when modeling the scattering for emission of a p-wave, but the resulting MFPADs are qualitatively more similar to those obtained experimentally, which are shown in Section 10.3. A more quantitatively correct simulation of the MFPADs requires a more realistic description of the molecular potential in order to correctly evaluate the (angle-dependent) scattering phase shifts, which is often carried out using multiple scattering methods.[10,11] To zeroth

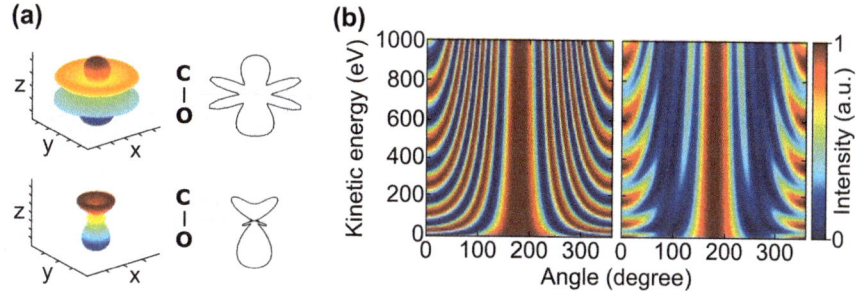

Figure 10.2   (a) Simulated MFPAD of a photoelectron with a kinetic energy of 250 eV emitted from the carbon atom in a CO molecule with the molecular axis oriented along the z axis and the carbon atom pointing in the positive z direction. The top row shows the simulation assuming emission of a spherical (Y_0^0) wave, the bottom row assuming emission of a p (Y_0^1) wave polarized along the z-direction. The panels on the left show the 3D photoelectron intensity distributions and the panels on the right show polar plots of the photoelectron intensity in the yz plane as a function of the emission angle θ. (b) Two-dimensional representation of the simulated photoelectron intensity in the MFPAD of a CO molecule as a function of the emission angle θ and the photoelectron kinetic energy. The left plot is simulated for emission of a spherical (Y_0^0) wave, the right plot for a p (Y_0^1) wave polarized along the molecular axis. Reproduced from ref. 8 with permission from the author.

order, this effect can be included by adding an additional phase shift to the scattered wave Ψ_B. It should be noted that while this phase shift affects the exact positions of the maxima and minima of the diffraction pattern, it does not result in additional lobes in the angular distribution. Some quantitative insights can therefore still be extracted from the simplified modelling provided here, such as the fact that the intensity oscillations at $\theta = 0°$ appear when integer multiples of the photoelectron's de Broglie wavelength are equal to twice the bond distance. This oscillating behaviour is confirmed experimentally, for example, for the case of inner-shell photoionization of carbon monoxide (CO; see Figure 10.3). A more general application of this concept, known as the extended X-ray absorption fine structure (EXAFS), is widely used as a tool for determining the local structure in condensed-phase samples.[12] Note also that while the intensity in the "forward" direction (towards the oxygen atom) is mostly flat as expected, a strong peak is seen at low kinetic energy at the position of the so-called "shape resonance", which can be attributed to multiple scattering in the molecular potential.[13] Since the contribution from multiple scattering quickly decays with increasing kinetic energy, this low-energy feature is the most prominent multiple scattering signature. Similar to EXAFS, its condensed-phase equivalent, known as near edge X-ray absorption fine structure (NEXAFS), is also a common tool for X-ray-based structure determination.[14]

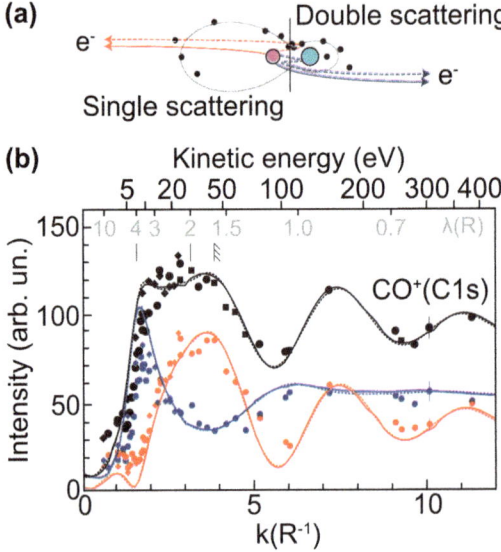

Figure 10.3   (a) Sketch of the photoelectron diffraction process in a di-atomic molecule. The black circles show the experimentally determined MFPAD of a CO molecule after carbon 1s ionization. (b) Photoelectron intensity in the MFPAD of a CO molecule after carbon 1s ionization as a function of the photoelectron kinetic energy (top axis), the photo-electron momentum k (bottom axis), and the de Broglie wavelength λ in units of the bond length R (second top axis, grey). The red and blue data points are the intensity in the carbon and oxygen direction, respectively, and the black data points are the sum of both. The solid lines are results of a partially-relaxed-core Hartree Fock calculation. Reproduced from ref. 53 with permission from Springer Nature, Copyright 2008.

10.3   Molecular-frame Photoelectron Angular Distributions

The previous section described the concept of the emitted photoelectron being considered as a wave that is scattered by the molecular potential. While this concept provides a very intuitive picture of photoelectron diffraction imaging and its possible opportunities for structure determination, studies in the gas phase have emerged historically from a different point of view that will be presented in this section.

The photoelectric effect releases an electron into the continuum. Most prominently, as employed in spectroscopic studies, the kinetic energy of the electron can be used to gather information on the emitter of the photo-electron. In addition, however, the photoelectron has an angular emission distribution, which is a measure for the angular momentum of the emitted electron. In the atomic case, the angular emission distribution, $\frac{\partial \sigma}{\partial \Omega}$, where

σ is the partial cross section and Ω is the solid angle element, is given by a rather simple expression, and the angular features (within the dipole approximation) are solely determined by a single parameter β (the "anisotropy parameter") acting as a coefficient in the second-order Legendre polynomial:[15]

$$\frac{\partial \sigma}{\partial \Omega} = \frac{\sigma}{4\pi}\left(1 + \frac{\beta}{2}\left(3 \cdot \cos^2 \theta' - 1\right)\right) \tag{10.1}$$

Here, angle θ' refers to the emission direction of the photoelectron with respect to the polarization vector of a linearly polarized photon employed for ionization (as opposed to angle θ, which was introduced before and which referred to the emission angle with respect to the molecular axis). Eqn (10.1) is valid for the photoionization of atoms and randomly oriented (non-chiral) molecules, and its simplicity is rooted in the principles of parity and angular momentum conservation. The situation becomes more complex when spatially oriented or aligned molecules are considered. In these cases, the electron angular emission distribution may consist of more structured features, which cannot be described by a single parameter β anymore. These features correspond to the diffraction pattern described in the previous section. In a mathematical description, they are related to contributions of higher angular momenta, and the electron angular distribution can be expanded in terms of spherical harmonics[11] given by

$$\frac{d\sigma}{d\Omega} = \sum_{l=0}^{l_{\max}} A_{lm} Y_{lm}(\theta, \phi), \tag{10.2}$$

where A_{lm} represents the coefficients of the expansion, and $l_{\max}$ is the highest-order non-negligible contribution. The higher-order angular momenta are possible since angular momentum conservation is achieved by a corresponding (counter-)rotation of the whole molecule.[16] Eqn (10.2) is equally valid for randomly oriented molecules if the emission distribution is considered in the molecular frame of reference. Accordingly, these distributions are typically referred to as "molecular-frame photoelectron angular distributions" (MFPADs) in the domain of gas-phase studies. The field has been reviewed extensively in the past, see, for example, ref. 17–20.

A more general view on the information possibly encoded in MFPADs can be obtained by considering the ionization amplitude, which, in the dipole approximation and to first order perturbation theory, is given by

$$D = \langle \Phi_{\mathrm{n}}(\vec{r}) | \hat{\mu}(\vec{r}) | \chi_{\mathrm{k}}(\vec{r}) \rangle, \tag{10.3}$$

where Φ_{n} is the initial state, $\hat{\mu}$ is the dipole operator, and χ_{k} is the photoelectron's final state in the continuum. Each of these three contributions to the ionization amplitude has a distinct impact on the measured MFPADs and the applicability of photoelectron diffraction imaging on single molecules in the gas phase.[21] For photoelectron diffraction imaging, it would be favourable if only the final-state contributions from scattering by the molecular potential prevailed. The angular contributions due to the initial state and the dipole operator can be minimized, for example, by preferably ionizing electrons from the K-shell and inspecting the MFPADs after averaging over all orientations of the molecule with respect to the polarization vector of the ionizing light. In addition, in the high energy limit (or "Born limit"), the final state in the continuum can be approximated by plane waves. In this limit, the scattering amplitude approaches zero, and the MFPAD directly images the square of the electron's initial state Φ_n.[21]

It shall be briefly noted here that not only information on molecular properties, but the full quantum mechanical information on the molecular photoionization process is imprinted in the MFPAD. The so-called "complete experiments" have been proposed (and performed) in which a minimal set of parameters (*e.g.*, the amplitudes and phases of the complex prefactors in eqn (10.2)) is extracted from the measured data.[22–25] In the spirit of "complete experiments", a particularly elegant parametrization of eqn (10.2) can be achieved by introducing a set of four "F-functions" (F_{00}, F_{20}, F_{21} and F_{22}[26,27]) that allow extracting the complete information on the dipole matrix element from measured MFPADs. In particular, the F-function method uses experimental data for a wide range of molecular orientation with respect to the light's polarization vector as an input, which is of great advantage as achieving sufficient statistics is often a problem in corresponding experiments. For example, molecular-frame photoionization time delays were recently measured using this scheme.[28,29] The contribution F_{00} corresponds to the polarization-averaged MFPAD (*i.e.*, the molecular-frame angular distribution that occurs after integrating over all spatial orientations of the molecule). Returning from this detour on MFPADs back to photoelectron diffraction imaging, it should be noted that several works have demonstrated that the complete information on the molecular geometry (and thus, the information sought after when employing photoelectron diffraction imaging) is contained in this contribution to the photoionization amplitude.[30–32]

An early example of measured MFPADs[33] (see Figure 10.4) demonstrates how richly structured these distributions can be and how sensitive the structure is to the photoelectron energy. However, the photoelectron diffraction scheme is not applicable at such low electron energies, and the complex interference pattern observed (see Figure 10.4) can be attributed to a molecular shape resonance. For comparison, the results of the modelling described in Section 10.2 for an electron with a kinetic energy of 15 eV (see Figure 10.5a) show much less structure than the experimentally observed MFPADs since the single-scattering model does not describe the shape resonance. At much higher kinetic energies, the (single) scattering model yields angular distributions that are closer to the measured ones. The corresponding MFPAD recorded at 633 eV electron energy (and employing circularly polarized photons; see Figure 10.5b) shows that the forward focusing peak

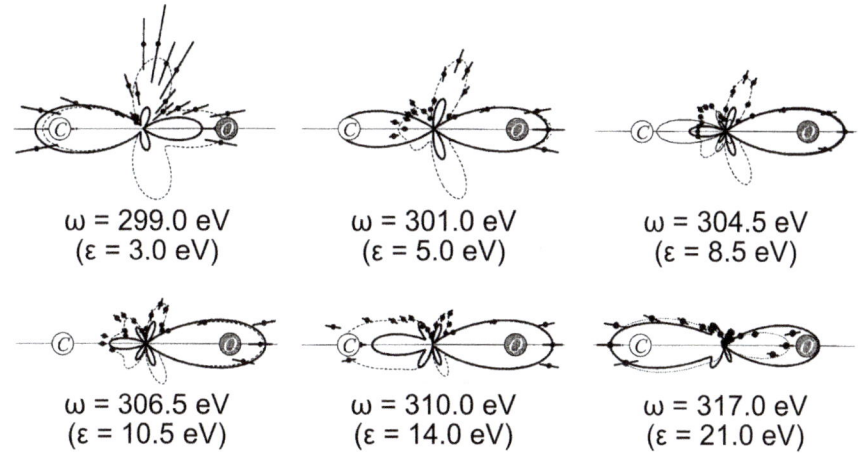

ω = 299.0 eV
(ε = 3.0 eV)

ω = 301.0 eV
(ε = 5.0 eV)

ω = 304.5 eV
(ε = 8.5 eV)

ω = 306.5 eV
(ε = 10.5 eV)

ω = 310.0 eV
(ε = 14.0 eV)

ω = 317.0 eV
(ε = 21.0 eV)

Figure 10.4   MFPADs of K-shell photoelectrons emitted from the carbon atom of a CO molecule. The panels depict the measured distribution (dots with error bars), a fit to the data employing a sum of Legendre polynomials (dashed line), and the results of two different theoretical modelling schemes (full lines). Reproduced from ref. 33 with permission from IOP Publishing Ltd, Copyright 2000.

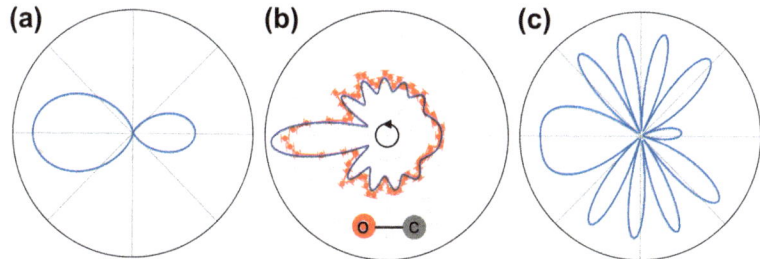

Figure 10.5   (a) MFPAD of an electron of 15 eV kinetic energy as obtained from the scattering approach described in Section 10.2 using a p-wave. Courtesy of Sven Grundmann (personal communication). (b) MFPAD of a C-K-shell photoelectron with an energy of 633 eV. Dots: experiment, line frozen core Hartree–Fock modelling. Adapted from ref. 35, with permission from the Royal Society of Chemistry. (c) MFPAD of a 633 eV electron according to the scattering approach (using an s-wave). Courtesy of Sven Grundmann (personal communication).

towards the oxygen atom is clearly visible and accompanied by several diffraction fringes. The tilt of the angular distribution is caused by the helicity of the ionizing light, as beautifully described in ref. 34. The outcome of the scattering model[‡] (see Figure 10.5c) is closer to the measured distribution, but

---

[‡] We have included the tilt of the angular distribution (which occurs due to the circular polarization of the ionizing light) in accordance with the findings provided in ref. 35.

it yields the wrong position of the interference fringes because it does not include the scattering phase shifts, as discussed in Section 10.2.

If instead of di-atomic molecules, systems of higher complexity are examined, the structural information contained in MFPADs is no longer restricted to one dimension. For example, electrons that are emitted from the carbon K-shell of methane directly image the 3D structure of the molecule in a certain regime of kinetic energies.[36] An extensive theoretical study reported in ref. 37 highlights explicitly the relation between the 3D geometry and polarization-averaged MFPADs. An example for a 3D system (see Figure 10.6a–c) shows the molecular-frame photoelectron angular distribution of an oxygen K-shell electron emitted from methyloxirane (C_3H_6O, see sketch in Figure 10.6a). Here, the molecule is fixed in space (*i.e.*, it is oriented with respect to the propagation direction of the circularly polarized light; see sketch in Figure 10.6a). The calculated and measured 3D

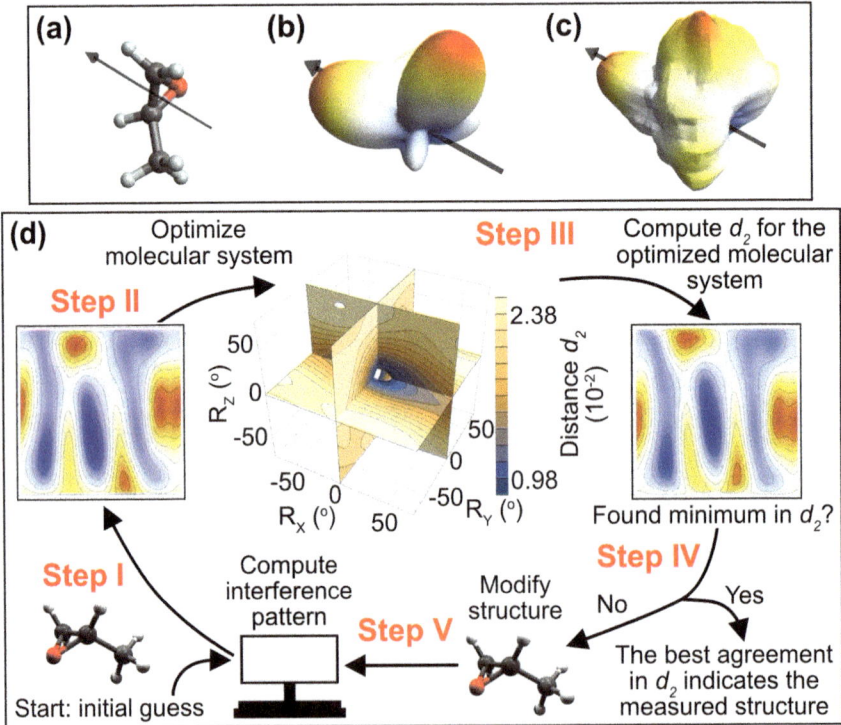

Figure 10.6   (a–c) MFPAD of an oxygen K-shell electron emitted from methyloxirane (a), simulation (b) and measured distribution (c). A sketch of the molecule and its orientation with respect to the propagation direction of the circularly polarized X-ray beam is shown on the left. (d) Sketch of the iterative approach to extract structural information from measured MFPADs. Adapted from ref. 82, with permission from the American Physical Society, Copyright 2021 and ref. 39 with permission from the Royal Society of Chemistry.

MFPAD distributions (see Figure 10.6b and c, respectively) demonstrate that it is also possible to measure MFPADs of much larger molecules than the di-atomics discussed so far.

The sensitivity of MFPADs to even small details of the molecular geometry was demonstrated in experiments already more than a decade ago. For example, in ref. 38, an asymmetry of the MFPAD of oxygen 1s photoelectrons emitted from CO_2 molecules is reported and attributed to the asymmetric-stretch vibrational motion of the molecule. Here, the kinetic energy of the photoelectron is only 11.5 eV (see Figure 10.6) – very far away from the regime where the intuitive single-scattering picture introduced in Section 10.2 is expected to hold. However, the angular distribution is highly structured due to multiple scattering processes, suggesting that it contains a large amount of information on the molecular potential and, thus, the molecular geometry. An iterative approach targeting the extraction of exactly that information is presented in ref. 39, demonstrating that even small details of the structure of the (chiral) methyloxirane molecule are accessible *via* measured MFPADs. Brief details of this approach are given in the following (see Figure 10.6d for a sketch). The iterative loop starts with an initial guess of the molecular structure (step I), which is employed to model the MFPAD belonging to that molecular geometry using full *ab initio* calculations (step II). By computing a distance-parameter d_2 (step III), the modelled MFPAD is compared to the measured molecular-frame angular distribution (step IV). Then the molecular structure of the molecule is slightly altered in the simulation, and the MFPAD of this altered structure is recomputed (step V). By evaluating and minimizing the distance-parameter, the geometrical structure which yields the MFPAD with closest correspondence to the measured one is obtained. The study demonstrates that this approach can help determine the bond length between distinct atoms of the molecule with an accuracy of better than 5%. Obviously, other methods for structure determination are already much more accurate nowadays, but this proof of principle work suggests that photoelectron diffraction imaging may become indeed a valuable tool for time-resolved studies in the future as this technique examines single molecules and allows taking ultrafast "snapshots" of the current molecular geometry. Section 10.6 will elaborate in more detail on this topic.

The current state-of-the-art suggests that the evaluation of MFPADs – or photoelectron diffraction imaging data in general – is indeed a promising approach for structure determination of single molecules in the gas phase. However, this technique is still in its early stages of development, and only first proof-of-principle studies have been performed so far. While the investigation of somewhat trivial examples of di-atomic molecules has became routine during the last decade, the technique is now being applied to molecules consisting of ten to twenty atoms. Depending on the amount of approximations that are introduced, the corresponding *ab initio* simulations of the diffraction patterns can now handle molecules of similar size given today's computational power. Up to now, such simulations are vital for extracting the structural information from the measured patterns.

10.4 Achieving Spatial Orientation of Molecules in the Gas Phase

From the point of view of experiments, the question arises how to measure MFPADs or the photoelectron diffraction patterns given that molecules in the gas phase are typically randomly oriented in space. A natural approach to this problem would be an active spatial orientation of the molecule prior to ionization. This can be achieved by employing strong laser fields. Two possible routes have emerged here, which are known as adiabatic alignment and impulsive alignment. In the former case, a (long) strong laser pulse is used to align the molecules adiabatically due to their polarizability. Impulsive alignment relies on (short) strong laser pulses, which induce a defined rotation of the molecule; alignment of the molecule occurs periodically after distinct time intervals even over long times after the initial laser pulses. A comprehensive review on these approaches can be found in ref. 40. Alternatively, active alignment of molecules can also be achieved by making use of DC electric fields generated by multipole structures that act on a molecular beam.[41]

A second, conceptually different approach to the problem consists of avoiding an active orientation/alignment of the molecules. Instead, the molecular orientation at the instant of the photoionization event is determined *a posteriori* from the information gathered in a coincidence experiment. This approach is feasible if the molecule dissociates during or quickly after the ionization process. In many cases (most obvious for di-atomic molecules), the emission direction of the fragments corresponds to the direction of the initial molecular bond. This requires that the fragmentation occurs rapidly as compared to the possible rotational motion of the molecule – a requirement typically referred to as the "axial recoil approximation".[42] This approach is typically employed in cases where one or several charges are generated *via* Auger decay after the initial (inner-shell) photoionization. When investigating larger molecules, the number of charged fragments produced after single-photon ionization is usually not high enough to infer directly the spatial orientation of the molecule. However, it was demonstrated recently that a breakup of a methyloxirane molecule into three fragments (and their coincident detection) is sufficient to define a molecular frame of reference in the experiment and to determine the orientation of the molecule within this coordinate frame by comparison to theory.[43] Alternatively, X-ray multiphoton ionization by intense XFEL pulses can be used to highly ionize polyatomic molecules such that their orientation can be determined.[44,45]

10.5 Electron–Ion Coincidence Experiments

The first measurements of MFPADs of gas-phase molecules were performed using traditional spectrometers for angle-resolved photoelectron and ion spectroscopy using small entrance apertures to define the emission angle of photoelectrons and fragment ions.[46,47] These were time consuming experiments since they required many coincidence measurements to be performed sequentially at different detection angles, which were then "stitched together" to obtain an MFPAD. A modified variant of the approach used in ref. 47 implementing a toroidal electron spectrometer[48] was employed in the early 2000s for the detection of molecular-frame Auger electron angular distributions.[49]

With the emergence of 3D momentum resolving spectrometers, it quickly became evident that MFPAD measurements could be carried out much more efficiently if the ionic fragments[50–53] or both the fragment ions and photoelectrons were collected in the full solid angle in a momentum resolving fashion (see Figure 10.7). Especially, the latter made it possible to extract 3D MFPADs for all possible orientations of the molecular axis with respect to the polarization vector from a single coincidence measurement.[54,55] This experimental scheme can be found in the literature by many different names

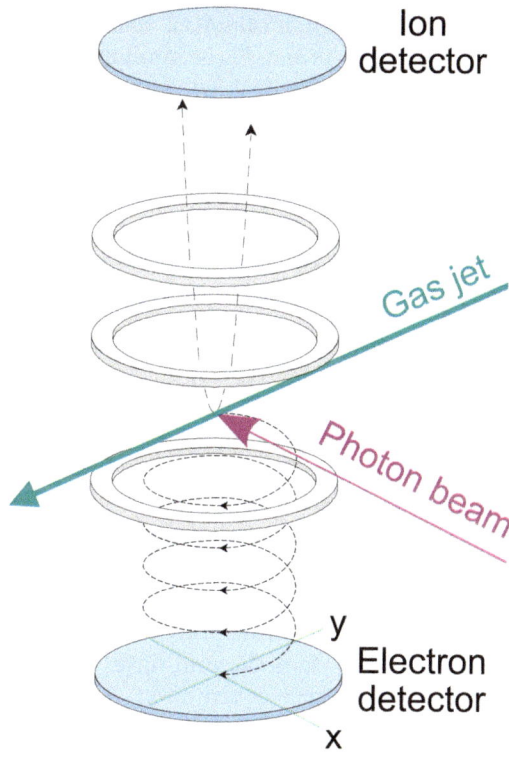

Figure 10.7 Schematic of a COLTRIMS setup for coincident ion and electron momentum imaging. Two time- and position-sensitive detectors image the momenta of ions and electrons emitted from the interaction region defined by the intersection of a gas jet and a photon beam. Electric and magnetic fields guide the charged particles to the two detectors. Reproduced from ref. 83, with permission from American Chemical Society, copyright 2020.

such as "COld Target Recoil Ion Momentum Spectroscopy" (COLTRIMS),[56,57] "Reaction Microscope" (REMI),[58] "Vector Correlation",[59,60] "dissociation dynamics detector system",[61,62] and several others. At the heart of this approach is a supersonic (molecular) beam, which is intersected at a right angle by the ionizing photon beam. Ions and electrons generated in photoionization are guided by an electric field to two time- and position-sensitive large-area MCP detectors. An additional homogeneous magnetic field is often used to confine the electrons in a cyclotron motion within the spectrometer volume (see Figure 10.7).

By measuring the positions of impact and the flight times of the particles, their initial vector momenta are reconstructed. The particles are measured in coincidence, and the experimental conditions are chosen such that they can be attributed to photoionization events of single molecules. From the measured momenta, all derived quantities such as kinetic energies or emission angles are retrieved.

Another popular approach to measure single particle momenta, which has also been used for the investigation of MFPADs,[8,63] is "velocity map imaging" (VMI).[64,65] While the working principle is similar to that of the aforementioned approaches, the VMI scheme does not yield full 3D information on the particles' momenta, but instead only a 2D projection. If the momentum distribution is rotationally symmetric, the missing information can be reconstructed employing an inversion algorithm.[66,67] In other cases, the use of tomographic approaches is an option.[68]

10.6   Time-resolved Photoelectron Diffraction

While MFPAD measurements are a popular approach to study molecular photoionization, especially using synchrotron radiation, rather little attention was paid in the early days to their interpretation in terms of photoelectron diffraction – mostly because more precise and less experimentally challenging methods exist for determining the equilibrium structure of gas-phase molecules. However, this situation changed with the emergence of intense, short-pulse X-ray light sources, most notably X-ray free-electron lasers (XFELs), that have opened up the possibility to use X-ray photoelectron diffraction to study transient molecular geometries *via* pump–probe experiments.[9,69] It should be noted that time-resolved pump–probe studies of MFPADs have also been conducted using optical femtosecond laser sources,[70–72] however, without directly linking the observed temporal evolution of the angular distributions to the evolving geometric structure in a quantitative manner.

Given the low repetition rate (≈ 100 Hz) of the first XFELs, active laser alignment combined with VMI detection of the electron angular distributions was used in the pioneering XPD experiments at XFELs[73–78] since *a posteriori* alignment *via* electron–ion coincidence measurements was deemed unfeasible. An early example of photoelectron diffraction on adiabatically laser-aligned 1-ethynyl-4-fluorobenzene molecules (see Figure 10.8) shows that the MFPADs – displayed as photoelectron angular distribution differences

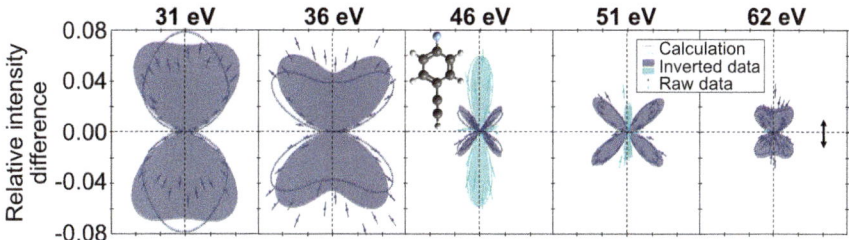

Figure 10.8 Fluorine 1s photoelectron angular distribution differences obtained by subtracting the photoelectron angular distributions of adiabatically laser-aligned and unaligned 1-ethynyl-4-fluorobenzene (C_8H_5F, see inset) for different photoelectron kinetic energies. Positive and negative values of the difference are plotted in cyan and blue, respectively. The distributions obtained from both the "raw" and the inverted Velocity Map Imaging data are shown. The polarization axes of alignment laser and XFEL were parallel and are indicated by the arrow. The dotted lines are results of a density functional theory calculation. Reproduced from ref. 74 with permission from the American Physical Society, Copyright 2013.

between aligned and unaligned molecules shown for five different kinetic energies of the fluorine 1s photoelectron – possess a clear dependency on the photon energy. This is well reproduced by density functional theory calculations (see dotted lines) that have been performed for the same degree of molecular angular alignment as achieved in the experiment.

However, it quickly became apparent that even the highest degree of molecular alignment that could practically be reached with these methods still led to considerable angular averaging that washed out much of the interference structure, and that the strong laser field present for adiabatic alignment has unwanted effects both on the photoelectrons and on the molecular dynamics of interest.[6,75] The development of the next generation of XFELs with tens to hundreds of kHz repetition rates has therefore sparked new hope that time-resolved photoelectron diffraction using photoelectron–ion coincidence techniques will soon be a viable addition to the portfolio of ultrafast structure determination methods for gas-phase molecules. Indeed, first proof-of-principle experiments at the European XFEL (see Figure 10.9) have demonstrated that time-resolved XPD can be used to image the Coulomb explosion of O_2 molecules.[79] Rather than relying on a timed pump–probe scheme, this experiment used the sequential absorption of two photons at random times during the XFEL pulse to first trigger the Coulomb explosion of the molecule, and to then release the electron that is probing its fragmentation. The information on the internuclear distance (or the time delay between the two absorption events) is extracted from the recorded coincidence data by inspecting the kinetic energy release of the fragmentation. Such an approach has also been used in the past to perform femtosecond-resolved studies at synchrotron facilities,[80] which provide X-ray pulses of typically ~ 100 ps duration, highlighting the wealth of information

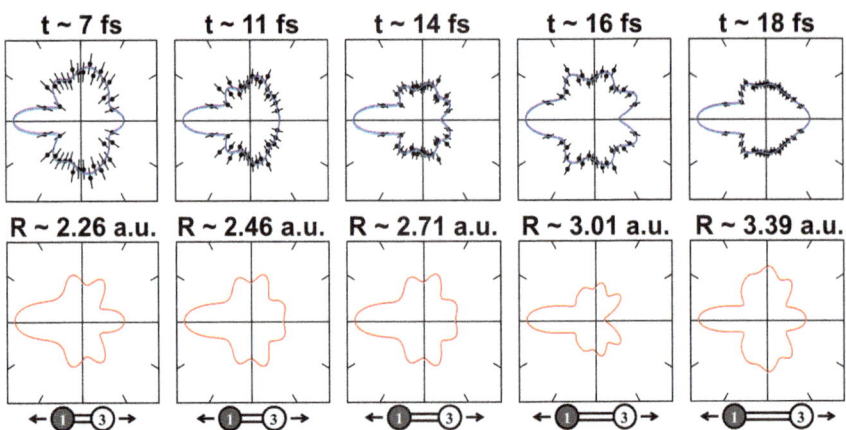

Figure 10.9 Photoelectron diffraction imaging of the breakup of an oxygen molecule. Top row: Measured polarization-averaged MFPAD for different internuclear distances between the two oxygen atoms. The blue line is a fit using a sum of Legendre polynomials up to $l = 5$. Bottom row: Corresponding angular distributions obtained from *ab initio* calculations. Reproduced from ref. 79, https://doi.org/10.1103/PhysRevX.10.021052, under the terms of the CC BY 4.0 license https://creativecommons.org/licenses/by/4.0/.

that can be extracted from coincidence measurements. The MFPADs measured at the European XFEL (see top row of Figure 10.9) and the corresponding theoretical modelling results (see bottom row) show an increase in O–O bond length with increasing time as the O_2 molecule fragments in a Coulomb explosion. This is reflected in the MFPADs, as detailed in Section 10.2. As the bond length d increases, further maxima are expected to appear on the right (*i.e.*, in direction of the emitter atom (see label "3" in the insets of Figure 10.9), and the existing maxima/nodes move more towards the scatterer (labelled "1"). The progression (see Figure 10.9) starts with a maximum at the emitter atom, which gradually turns into a minimum, and finally changes to a further maximum in the rightmost panel.

The findings by Kastirke *et al.* suggest a promising future for time-resolved photoelectron diffraction imaging, in particular in connection with the aforementioned high-repetition-rate X-ray free-electron lasers. Indeed, first pump–probe photoelectron diffraction imaging measurements were performed just recently using X-rays from the European XFEL, but the data analysis is still ongoing as this text is being written. With the technological advances that are already planned for the next two years, measuring full three-dimensional diffraction images of ever more complex molecules should be routinely possible. For example, roaming hydrogen atoms, as suggested by Ota *et al.*,[81] will be caught in the act not only in computer modeling, but also in fully time-resolved photoelectron diffraction imaging experiments. Similarly, predictions by Boll *et al.*[74] show that the geometries of transition states in photochemical reactions lead to MFPADs that are drastically

Equilibrium	Doubled F-C bond	Transition state 1	Transition state 2

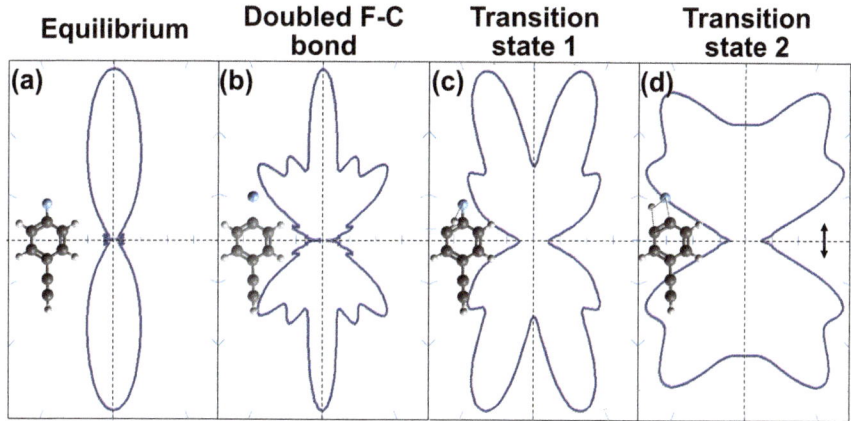

Figure 10.10   (a and b) Calculated fluorine (1s) photoelectron angular distributions for the same molecule as shown in Figure 10.8 at a photoelectron kinetic energy of 45 eV obtained from DFT calculations for (a) the equilibrium geometry, (b) for a molecule with twice the equilibrium bond length between fluorine and the benzene ring. (c and d) Similar to panels (a and b) but for two proposed transition states (c and d), whose geometries are shown in the insets (light blue: fluorine, dark grey: carbon, and light grey: hydrogen). The molecules are one-dimensionally aligned with the F–C axis parallel to the X-ray polarization as indicated by the arrow in panel (d). Reproduced from ref. 74 with permission from the American Physical Society, Copyright 2013.

different from those of the equilibrium geometry, especially if the structural changes occur in the direct vicinity of the emitter atom (see Figure 10.10). In principle, as high-repetition-rate X-ray sources become more widely available, such time-resolved photoelectron diffraction experiments should be feasible not just for gas-phase molecules but also for condensed-phase samples and molecules on surfaces, which would reconnect the photoelectron diffraction technique back to its original starting point.

Acknowledgements

We are grateful to our mentors Uwe Becker, Horst Schmidt-Böcking, Reinhard Dörner and Charles Fadley and their group members in the early 2000s for the many insightful discussions about MFPADs and photoelectron diffraction, as well as to our own group members and collaborators in the years since who have helped develop and refine our understanding of these phenomena. We thank Denis Anielski, who performed the model calculations shown in Section 10.2 and Sven Grundmann, who performed the modelling shown in Figure 10.5. TJ would like to thank Nikolai Cherepkov for initial education on molecular photoelectron emission.

References

1. C. S. Fadley, *Prog. Surf. Sci.*, 1984, **16**, 275–388.
2. D. P. Woodruff and A. M. Bradshaw, *Rep. Prog. Phys.*, 1994, **57**, 1029.
3. C. S. Fadley, *Prog. Surf. Sci.*, 1997, **54**, 341.
4. D. Rolles, PhD thesis, Technische Universität Berlin, 2005.
5. M. Kazama, J.-I. Adachi, H. Shinotsuka, M. Yamazaki, Y. Ohori, A. Yagishita and T. Fujikawa, *Chem. Phys.*, 2010, **373**, 261–266.
6. R. Boll, PhD thesis, Universität Heidelberg, 2014.
7. D. Rolles, R. Boll, S. R. Tamrakar, D. Anielski and C. Bomme, *Proc. SPIE*, 2014, **9198**, 91980O.
8. D. Anielski, PhD thesis, Universität Hamburg, 2020.
9. F. Krasniqi, B. Najjari, L. Strüder, D. Rolles, A. Voitkiv and J. Ullrich, *Phys. Rev. A: At., Mol., Opt. Phys.*, 2010, **81**, 033411.
10. D. Dill and J. L. Dehmer, *J. Chem. Phys.*, 1974, **61**, 692–699.
11. D. Dill, *J. Chem. Phys.*, 1976, **65**, 1130–1133.
12. J. J. Rehr and R. C. Albers, *Rev. Mod. Phys.*, 2000, **72**, 621.
13. M. N. Piancastelli, D. W. Lindle, T. A. Ferrett and D. A. Shirley, *J. Chem. Phys.*, 1987, **86**, 2765.
14. J. Stöhr, *NEXAFS Spectroscopy*, Springer, 1992.
15. V. Schmidt, *Rep. Prog. Phys.*, 1992, **55**, 1483.
16. H. C. Choi, R. M. Rao, A. G. Mihill, S. Kakar, E. D. Poliakoff, K. Wang and V. McKoy, *Phys. Rev. Lett.*, 1994, **72**, 44.
17. K. L. Reid, *Ann. Rev. Phys. Chem.*, 2003, **54**, 397–424.
18. K. L. Reid, *Mol. Phys.*, 2012, **110**, 131–147.
19. A. Stolow and J. G. Underwood, *Adv. Chem. Phys.*, 2008, **139**, 497–584.
20. A. Yagishita, *J. Electron Spectrosc. Relat. Phenom.*, 2015, **200**, 247–256.
21. M. Waitz, R. Bello, D. Metz, J. Lower, F. Trinter, C. Schober, M. Keiling, U. Lenz, M. Pitzer, K. Mertens, M. Martins, J. Viefhaus, S. Klumpp, T. Weber, L. P. H. Schmidt, J. Williams, M. Schöffler, V. Serov, A. Kheifets, L. Argenti, A. Palacios, F. Martín, T. Jahnke and R. Dörner, *Nat. Commun.*, 2017, **8**, 2266.
22. U. Heinzmann, *J. Phys. B: At., Mol. Opt. Phys.*, 1980, **13**, 4367.
23. N. A. Cherepkov, *Adv. At. Mol. Phys.*, 1983, **19**, 395–447.
24. H.-J. Beyer, J. B. West, K. J. Ross, K. Ueda, N. M. Kabachnik, H. Hamdy and H. Kleinpoppen, *J. Electron Spectrosc. Relat. Phenom.*, 1996, **79**, 339–342.
25. N. A. Cherepkov, G. Rasee, J. Adachi, Y. Hikosaka, K. Ito, S. Motoki, M. Sano, K. Soejima and A. Yagishita, *J. Phys. B: At., Mol. Opt. Phys.*, 2000, **33**, 4213–4236.
26. R. R. Lucchese, A. Lafosse, J. C. Brenot, P. M. Guyon, J. C. Houver, M. Lebech, G. Raseev and D. Dowek, *Phys. Rev. A: At., Mol., Opt. Phys.*, 2002, **65**, 020702.
27. A. Lafosse, J. C. Brenot, P. M. Guyon, J. C. Houver, A. V. Golovin, M. Lebech, D. Dowek, P. Lin and R. R. Lucchese, *J. Chem. Phys.*, 2002, **117**, 8368.

28. F. Holzmeier, J. Joseph, J. C. Houver, M. Lebech, D. Dowek and R. R. Lucchese, *Nat. Commun.*, 2021, **12**, 7343.

29. J. Rist, K. Klyssek, N. M. Novikovskiy, M. Kircher, I. Vela-Perez, D. Trabert, S. Grundmann, D. Tsitsonis, J. Siebert, A. Geyer, N. Melzer, C. Schwarz, N. Anders, L. Kaiser, K. Fehre, A. Hartung, S. Eckart, L. P. H. Schmidt, M. S. Schöffler, V. T. Davis, J. B. Williams, F. Trinter, R. Dörner, P. V. Demekhin and T. Jahnke, *Nat. Commun.*, 2021, **12**, 6657.

30. H. Fukuzawa, R. R. Lucchese, X.-J. Liu, K. Sakai, H. Iwayama, K. Nagaya, K. Kreidi, M. S. Schöffler, J. R. Harries, Y. Tamenori, Y. Morishita, I. H. Suzuki, N. Saito and K. Ueda, *J. Chem. Phys.*, 2019, **150**, 174306.

31. F. Ota, K. Yamazaki, D. Sébilleau, K. Ueda and K. Hatada, *J. Phys. B: At., Mol. Opt. Phys.*, 2021, **54**, 024003.

32. F. Ota, K. Hatada, D. Sébilleau, K. Ueda and K. Yamazaki, *J. Phys. B: At., Mol. Opt. Phys.*, 2021, **54**, 084001.

33. S. Motoki, J. Adachi, Y. Hikosaka, K. Ito, M. Sano, K. Soejima, A. Yagishita, G. Raseev and N. A. Cherepkov, *J. Phys. B: At., Mol. Opt. Phys.*, 2000, **33**, 4193–4212.

34. H. Daimon, S. Imada and S. Suga, *Surf. Sci.*, 2001, **471**, 143–150.

35. I. Vela-Peréz, F. Ota, A. Mhamdi, Y. Tamura, J. Rist, N. Melzer, S. Uerken, G. Nalin, N. Anders, D. You, M. Kircher, C. Janke, M. Waitz, F. Trinter, R. Guillemin, M. N. Piancastelli, M. Simon, V. T. Davis, J. B. Williams, R. Dörner, K. Hatada, K. Yamazaki, K. Fehre, P. V. Demekhin, K. Ueda, M. S. Schöffler and T. Jahnke, *Phys. Chem. Chem. Phys.*, 2023, **25**, 13784–13791.

36. J. B. Williams, C. S. Trevisan, M. S. Schöffler, T. Jahnke, I. Bocharova, H. Kim, B. Ulrich, R. Wallauer, F. Sturm and T. N. Rescigno, *Phys. Rev. Lett.*, 2012, **108**, 233002.

37. E. Plesiat, P. Decleva and F. Martín, *Phys. Rev. A: At., Mol., Opt. Phys.*, 2013, **88**, 063409.

38. N. Saito, K. Ueda, A. D. Fanis, K. Kubozuka, M. Machida, R. D. I. Koyano, A. Czasch, L. Schmidt, A. Cassimi, B. Z. K. Wang and V. McKoy, *J. Phys. B: At., Mol. Opt. Phys.*, 2005, **38**, L277–L284.

39. K. Fehre, N. M. Novikovskiy, S. Grundmann, G. Kastirke, S. Eckart, F. Trinter, J. Rist, A. Hartung, D. Trabert, C. Janke, M. Pitzer, S. Zeller, F. Wiegandt, M. Weller, M. Kircher, G. Nalin, M. Hofmann, L. P. H. Schmidt, A. Knie, A. Hans, L. B. Ltaief, A. Ehresmann, R. Berger, H. Fukuzawa, K. Ueda, H. Schmidt-Böcking, J. B. Williams, T. Jahnke, R. Dörner, P. V. Demekhin and M. S. Schöffler, *Phys. Chem. Chem. Phys.*, 2022, **24**, 26458–26465.

40. H. Stapelfeldt and T. Seideman, *Rev. Mod. Phys.*, 2003, **75**, 543.

41. T. P. Rakitzis, A. J. V. D. Brom and M. H. M. Janssen, *Science*, 2004, **303**, 1852–1854.

42. R. N. Zare, *Mol. Photochem.*, 1972, **4**, 1–37.

43. K. Fehre, N. M. Novikovskiy, S. Grundmann, G. Kastirke, S. Eckart, F. Trinter, J. Rist, A. Hartung, D. Trabert, C. Janke, G. Nalin, M. Pitzer, S. Zeller, F. Wiegandt, M. Weller, M. Kircher, M. Hofmann, L. P. H. Schmidt, A. Knie, A. Hans, L. B. Ltaief, A. Ehresmann, R. Berger,

H. Fukuzawa, K. Ueda, H. Schmidt-Böcking, J. B. Williams, T. Jahnke, R. Dörner, M. S. Schöffler and P. V. Demekhin, *Phys. Rev. Lett.*, 2021, **127**, 103201.

44. X. Li, A. Rudenko, M. S. Schöffler, N. Anders, T. M. Baumann, S. Eckart, B. Erk, A. De Fanis, K. Fehre, R. Dörner, L. Foucar, S. Grundmann, P. Grychtol, A. Hartung, M. Hofmann, M. Ilchen, C. Janke, G. Kastirke, M. Kircher, K. Kubicek, M. Kunitski, T. Mazza, S. Meister, N. Melzer, J. Montano, V. Music, G. Nalin, Y. Ovcharenko, C. Passow, A. Pier, N. Rennhack, J. Rist, D. E. Rivas, I. Schlichting, L. P. H. Schmidt, P. Schmidt, J. Siebert, N. Strenger, D. Trabert, F. Trinter, I. Vela-Perez, R. Wagner, P. Walter, M. Weller, P. Ziolkowski, A. Czasch, D. Rolles, M. Meyer, T. Jahnke and R. Boll, *Phys. Rev. Res.*, 2022, **4**, 013029.

45. R. Boll, J. Schäfer, B. Richard, K. Fehre, G. Kastirke, M. Abdullah, N. Anders, T. Baumann, A. Czasch, S. Eckart, B. Erk, A. D. Fanis, L. Foucar, S. Grundmann, P. Grychtol, A. Hartung, M. Hofmann, M. Ilchen, L. Inhester, C. Janke, Z. Jurek, M. Kircher, K. Kubicek, M. Kunitski, X. Li, T. Mazza, S. Meister, N. Melzer, J. Montano, V. Music, G. Nalin, Y. Ovcharenko, C. Passow, A. Pier, N. Rennhack, J. Rist, D. Rivas, D. Rolles, I. Schlichting, L. Schmidt, P. Schmidt, M. Schöffler, J. Siebert, N. Strenger, D. Trabert, F. Trinter, I. Vela-Perez, R. Wagner, P. Walter, M. Weller, P. Ziolkowski, R. Dörner, S.-K. Son, A. Rudenko, M. Meyer, R. Santra and T. Jahnke, *Nat. Phys.*, 2022, **18**, 423–428.

46. A. V. Golovin, N. A. Cherepkov and V. V. Kuznetsov, *Z. Phys. D: At., Mol. Clusters*, 1992, **24**, 371–375.

47. E. Shigemasa, J. Adachi, M. Oura and A. Yagishita, *Phys. Rev. Lett.*, 1995, **74**, 359–362.

48. R. Guillemin, E. Shigemasa, K. L. Guen, D. Ceolin, C. Miron, N. Leclercq, P. Morin and M. Simon, *Rev. Sci. Instrum.*, 2000, **71**, 4387–4394.

49. R. Guillemin, E. Shigemasa, K. L. Guen, D. Ceolin, C. Miron, N. Leclercq, P. Morin and M. Simon, *Phys. Rev. Lett.*, 2001, **87**, 203001.

50. F. Heiser, O. Gessner, J. Viefhaus, K. Wieliczek, R. Hentges and U. Becker, *Phys. Rev. Lett.*, 1997, **79**, 2435.

51. U. Becker, O. Gessner and A. Rüdel, *J. Electron Spectrosc. Relat. Phenom.*, 2000, **108**, 189–201.

52. D. Rolles, M. Braune, S. Cvejanovic, O. Geßner, R. Hentges, S. Korica, B. Langer, T. Lischke, G. Prümper, A. Reinköster, J. Viefhaus, B. Zimmermann, V. McKoy and U. Becker, *Nature*, 2005, **437**, 711–715.

53. B. Zimmermann, D. Rolles, B. Langer, R. Hentges, M. Braune, S. Cvejanovic, O. Geßner, F. Heiser, S. Korica, T. Lischke, A. Reinköster, J. Viefhaus, R. Dörner, V. McKoy and U. Becker, *Nat. Phys.*, 2008, **4**, 649–655.

54. A. Landers, T. Weber, I. Ali, A. Cassimi, M. Hattass, O. Jagutzki, A. Nauert, T. Osipov, A. Staudte, M. Prior, H. Schmidt-Böcking, C. Cocke and R. Dörner, *Phys. Rev. Lett.*, 2001, **87**, 013002.

55. T. Jahnke, T. Weber, A. Landers, A. Knapp, S. Schössler, J. Nickles, S. Kammer, O. Jagutzki, L. Schmidt and A. Czasch, *et al.*, *Phys. Rev. Lett.*, 2002, **88**, 073002.

56. R. Dörner, V. Mergel, O. Jagutzki, L. Spielberger, J. Ullrich, R. Moshammer and H. Schmidt-Böcking, *Phys. Rep.*, 2000, **330**, 95–192.
57. T. Jahnke, T. Weber, T. Osipov, A. L. Landers, O. Jagutzki, L. P. H. Schmidt, C. L. Cock, M. H. Prior, H. Schmidt-Böcking and R. Dörner, *J. Electron Spectrosc. Relat. Phenom.*, 2004, **141**, 229.
58. J. Ullrich, R. Moshammer, A. Dorn, R. Dörner, L. P. H. Schmidt and H. Schmidt-Böcking, *Rep. Prog. Phys.*, 2003, **66**, 1463.
59. A. Lafosse, M. Lebech, J. C. Brenot, P. M. Guyon, O. Jagutzki, L. Spielberger, M. Vervloet, J. C. Houver and D. Dowek, *Phys. Rev. Lett.*, 2000, **84**, 5987.
60. M. Lebech, J. C. Houver and D. Dowek, *Rev. Sci. Instrum.*, 2002, **73**, 1866–1874.
61. M. Lavolle, *Rev. Sci. Instrum.*, 1999, **70**, 2968–2974.
62. M. Gisselbrecht, A. Huetz, M. Lavolle, T. J. Reddish and D. P. Seccombe, *Rev. Sci. Instrum.*, 2005, **76**, 013105.
63. M. Takahashi, J. P. Cave and J. H. D. Eland, *Rev. Sci. Instrum.*, 2000, **71**, 1337–1344.
64. H. Helm, N. Bjerre, M. J. Dyer, D. L. Huestis and M. Saeed, *Phys. Rev. Lett.*, 1993, **70**, 3221.
65. A. T. J. B. Eppink and D. H. Parker, *Rev. Sci. Instrum.*, 1997, **68**, 3477.
66. C. Bordas, F. Paulig, H. Helm and D. Huestis, *Rev. Sci. Instrum.*, 1996, **67**, 2257–2268.
67. M. J. J. Vrakking, *Rev. Sci. Instrum.*, 2001, **72**, 4084–4089.
68. M. Wollenhaupt, M. Krug, J. Köhler, T. Bayer, C. Sarpe-Tudoran and T. Baumert, *Appl. Phys. B: Lasers Opt.*, 2009, **95**, 647–651.
69. M. Kazama, T. Fujikawa, N. Kishimoto, T. Mizuno, J.-I. Adachi and A. Yagishita, *Phys. Rev. A: At., Mol., Opt. Phys.*, 2013, **87**, 063417.
70. J. A. Davies, R. E. Continetti, D. W. Chandler and C. C. Hayden, *Phys. Rev. Lett.*, 2000, **84**, 5983.
71. O. Gessner, A. M. D. Lee, J. P. Shaffer, H. Reisler, S. V. Levchenko, A. I. Krylov, J. G. Underwood, H. Shi, A. L. L. East, D. M. Wardlaw, E. Chrysostom, C. C. Hayden and A. Stolow, *Science*, 2006, **311**, 219–222.
72. C. Z. Bisgaard, O. J. Clarkin, G. Wu, A. M. D. Lee, O. Gessner, C. C. Hayden and A. Stolow, *Science*, 2009, **323**, 1464–1468.
73. J. Cryan, J. Glownia, J. Andreasson, A. Belkacem, N. Berrah, C. Blaga, C. Bostedt, J. Bozek, C. Buth, L. DiMauro, L. Fang, O. Gessner, M. Guehr, J. Hajdu, M. Hertlein, M. Hoener, O. Kornilov, J. Marangos, A. March, B. McFarland, H. Merdji, V. Petrović, C. Raman, D. Ray, D. Reis, F. Tarantelli, M. Trigo, J. White, W. White, L. Young, P. Bucksbaum and R. Coffee, *Phys. Rev. Lett.*, 2010, **105**, 083004.
74. R. Boll, D. Anielski, C. Bostedt, J. Bozek, L. Christensen, R. Coffee, S. De, P. Decleva, S. Epp, B. Erk, L. Foucar, F. Krasniqi, J. Küpper, A. Rouzée, B. Rudek, A. Rudenko, S. Schorb, H. Stapelfeldt, M. Stener, S. Stern, S. Techert, S. Trippel, M. Vrakking, J. Ullrich and D. Rolles, *Phys. Rev. A: At., Mol., Opt. Phys.*, 2013, **88**, 061402(R).
75. R. Boll, A. Rouzée, M. Adolph, D. Anielski, A. Aquila, S. Bari, C. Bomme, C. Bostedt, J. D. Bozek, H. N. Chapman, L. Christensen, R. Coffee, N. Coppola, S. De, P. Decleva, S. W. Epp, B. Erk, F. Filsinger, L. Foucar,

T. Gorkhover, L. Gumprecht, A. Hömke, L. Holmegaard, P. Johnsson, J. S. Kienitz, T. Kierspel, F. Krasniqi, K.-U. Kühnel, J. Maurer, M. Messerschmidt, R. Moshammer, N. L. M. Müller, B. Rudek, E. Savelyev, I. Schlichting, C. Schmidt, F. Scholz, S. Schorb, J. Schulz, J. Seltmann, M. Stener, S. Stern, S. Techert, J. Thøgersen, S. Trippel, J. Viefhaus, M. Vrakking, H. Stapelfeldt, J. Küpper, J. Ullrich, A. Rudenko and D. Rolles, *Faraday Discuss.*, 2014, **171**, 57–80.

76. D. Rolles, R. Boll, M. Adolph, A. Aquila, C. Bostedt, J. D. Bozek, H. N. Chapman, R. Coffee, N. Coppola, P. Decleva, T. Delmas, S. W. Epp, B. Erk, F. Filsinger, L. Foucar, L. Gumprecht, A. Hömke, T. Gorkhover, L. Holmegaard, P. Johnsson, C. Kaiser, F. Krasniqi, K.-U. Kühnel, J. Maurer, M. Messerschmidt, R. Moshammer, W. Quevedo, I. Rajkovic, A. Rouzée, B. Rudek, I. Schlichting, C. Schmidt, S. Schorb, C. D. Schröter, J. Schulz, H. Stapelfeldt, M. Stener, S. Stern, S. Techert, J. Thøgersen, M. J. J. Vrakking, A. Rudenko, J. Küpper and J. Ullrich, *J. Phys. B: At., Mol. Opt. Phys.*, 2014, **47**, 124035.

77. K. Nakajima, T. Teramoto, H. Akagi, T. Fujikawa, T. Majima, S. Minemoto, K. Ogawa, H. Sakai, T. Togashi, K. Tono, S. Tsuru, K. Wada, M. Yabashi and A. Yagishita, *Sci. Rep.*, 2015, **5**, 14065.

78. S. Minemoto, H. Shimada, K. Komatsu, W. Komatsubara, T. Majima, S. Miyake, T. Mizuno, S. Owada, H. Sakai, T. Togashi, M. Yabashi, P. Decleva, M. Stener, S. Tsuru and A. Yagishita, *J. Phys. Commun.*, 2018, **2**, 115015.

79. G. Kastirke, M. S. Schöffler, M. Weller, J. Rist, R. Boll, N. Anders, T. M. Baumann, S. Eckart, B. Erk, A. D. Fanis, K. Fehre, A. Gatton, S. Grundmann, P. Grychtol, A. Hartung, M. Hofmann, M. Ilchen, C. Janke, M. Kircher, M. Kunitski, X. Li, T. Mazza, N. Melzer, J. Montano, V. Music, G. Nalin, Y. Ovcharenko, A. Pier, N. Rennhack, D. E. Rivas, R. Dörner, D. Rolles, A. Rudenko, P. Schmidt, J. Siebert, N. Strenger, D. Trabert, I. Vela-Perez, R. Wagner, T. Weber, J. B. Williams, P. Ziolkowski, L. P. H. Schmidt, A. Czasch, F. Trinter, M. Meyer, K. Ueda, P. V. Demekhin and T. Jahnke, *Phys. Rev. X*, 2020, **10**, 021052.

80. H. Sann, T. Havermeier, C. Müller, H.-K. Kim, F. Trinter, M. Waitz, J. Voigtsberger, F. Sturm, T. Bauer, R. Wallauer, D. Schneider, M. Weller, C. Goihl, J. Tross, K. Cole, J. Wu, M. S. Schöffler, H. Schmidt-Böcking, T. Jahnke, M. Simon and R. Dörner, *Phys. Rev. Lett.*, 2016, **117**, 243002.

81. F. Ota, S. Abe, K. Hatada, K. Ueda, S. Díaz-Tender and F. Martín, *Phys. Chem. Chem. Phys.*, 2021, **23**, 20174–20182.

82. K. Fehre, N. M. Novikovskiy, S. Grundmann, G. Kastirke, S. Eckart, F. Trinter, J. Rist, A. Hartung, D. Trabert, C. Janke, G. Nalin, M. Pitzer, S. Zeller, F. Wiegandt, M. Weller, M. Kircher, M. Hofmann, L. P. H. Schmidt, A. Knie, A. Hans, L. B. Ltaief, A. Ehresmann, R. Berger, H. Fukuzawa, K. Ueda, H. Schmidt-Böcking, J. B. Williams, T. Jahnke, R. Dörner, M. S. Schöffler and P. V. Demekhin, *Phys. Rev. Lett.*, 2021, **127**, 103201.

83. T. Jahnke, U. Hergenhahn, B. Winter, R. Dörner, U. Frühling, P. V. Demekhin, K. Gokhberg, L. S. Cederbaum, A. Ehresmann, A. Knie and A. Dreuw, *Chem. Rev.*, 2020, **120**, 11295–11369.

CHAPTER 11

The Many Facets of Ultrafast Electron Diffraction and Microscopy: Development and Applications

C.-Y. RUAN

Department of Physics and Astronomy, Michigan State University, East Lansing, MI 48824, USA
Email: ruanc@msu.edu

11.1 Introduction

In the last two decades, the rapid development of time-resolved scattering methodologies utilizing femtosecond X-ray pulses through either table-top (see Chapter 4) or free electron laser (FEL) systems (see Chapter 8), as well as the versatile electron-based femtosecond scattering and microscopy setups (see Chapters 12–15), has led to a new area of investigating structural dynamics. These ultrafast structural probes enabled studies across all phases of matter, ranging from gaseous molecules and liquids to crystalline solids, surfaces, nanomaterials, as well as monolayer films and interfaces.

While significant synergies between ultrafast electron diffraction (UED; see Chapters 10 and 12–14) and ultrafast X-ray diffraction (UXRD; see Chapters 4–6 and 9) have been discussed in some depth, here, we will highlight the complementary aspects in the applications designed to hone the strengths of the electron-based technologies to create dynamical contrast. We will provide a motivation for electron-based technologies by

Theoretical and Computational Chemistry Series No. 25
Structural Dynamics with X-ray and Electron Scattering
Edited by Kasra Amini, Arnaud Rouzée and Marc J. J. Vrakking
© The Royal Society of Chemistry 2024
Published by the Royal Society of Chemistry, www.rsc.org

discussing and comparing two types of structural probes: electron diffraction and microscopy. In particular, we will discuss the emerging opportunities that take advantage of the manipulability of the electron beam with electron optics, providing a variety of probes for gas-phase molecules (Section 11.2), surfaces and nanostructures (Section 11.3), quantum materials (Section 11.4), and *in situ* investigation of complex structures and interfaces under a microscope setting (Section 11.5). We will also discuss how these technologies fare with different electron beam energies, and how space-charge effects are overcome with the new development of precisely synchronized RF optics for phase-space manipulation of electron beams, enabling multi-modality observations based on the optimization of merit-based brightness (Section 11.6).

11.1.1   Comparison of X-ray and Electron Scattering

The Fourier transform of the probe-independent structure factors, $S_{\rho\rho}(s, t)$, extracted from the time-resolved X-ray or electron diffractograms produces the pair-correlation function $G(r, t)$. This function encodes the 3D atomic structures of the systems which can be retrieved through a careful comparison and analysis of calculated to measured scattering data. Here, s is the pixelated scattering momentum transfer wavevector given by $s = 4\pi/\lambda \sin(\theta/2)$, where θ is the scattering angle and λ is the electron de Broglie wavelength given by $\lambda = 1.22643/(V_0 + 0.97845 \times 10^{-6}V_0^2)$ nm, where V_0 is the electron beam accelerating voltage.[1] The pair correlation function is given by $G(r, t) = \langle \rho(\mathbf{r}', t)\rho(\mathbf{r}'', t)\rangle$, where the bracket denotes ensemble averaging from the probed volume V and over the repetitive probe cycles, and $\mathbf{r} = \mathbf{r}' - \mathbf{r}''$ gives the distance between the pairs averaged. The common form of the scattering cross-section recorded within the detector pixelated solid angle Ω from the X-ray and electron scattering approaches is given by

$$\frac{d\sigma_i(\mathbf{s}, t)}{d\Omega} = \frac{1}{(2\pi)^3}|f_i|^2 S_{\rho\rho}(\mathbf{s}, t), \tag{11.1}$$

where f_i is the respective (electron or X-ray) atomic scattering factor.

One key aspect that sets the electron and X-ray scattering techniques apart is the different (Thompson and Rutherford, respectively) scattering mechanisms. It is instructive to know that the electron and X-ray atomic scattering factors (f_e and f_x, respectively) are interconvertible through the Mott formula given by

$$f_e(s) = \left(Z - \tilde{f}_x(s)\right)\frac{8\pi^2 m^2}{h^2 s^2}, \tag{11.2}$$

where Z is the atomic number and $f_x(s) = \tilde{f}_x(s)\dfrac{e^2}{m_e c^2}$ (hence, only f_x is given in the International Table of Crystallography[2]). Therefore, the two technologies are equivalent but differ only in their practical implementation. We can judge this from two different perspectives. First, from the Mott formula (see eqn (11.2)), one can see that in electron scattering, the signal strength is significantly enhanced not just by the presence of the nuclei that scatter strongly from the incoming waves but also due to the nature of the Rutherford scattering that give a higher scattering cross-section than that of X-rays (Thompson scattering). As a matter of fact, the enhancement factor, given by $\left(\dfrac{f_e(s)}{f_x(s)}\right)^2$, is on the order of 10^6 at $s = 10$ Å^{-1}, a nominal cut-off range of scattering that one uses to extract useful structure information by following the two approaches. This high atomic scattering strength for the electrons has a direct consequence on its scope of applications. We note especially that, depending on the effectively sampled region of the probe, the pair correlation function $G(r, t) = \langle \rho(\mathbf{r}', t)\rho(\mathbf{r}'', t)\rangle$ to be solved by inverting the diffractograms naturally carries the information that depends on the system size – not just the averaged atomic distances but also the fluctuations in the excited structures that vary in space and time induced by the pump pulses. The large penetration depth of X-rays frequently means that UXRD can resolve longer range phenomena and correlations than UED. Although the scattering cross-section of electrons is much greater than that for X-rays, the scattering fall off for X-rays is actually not so abrupt as for electrons. Given good single crystals for either experiment, it would be possible to collect relatively higher intensity data at larger s range in X-ray diffraction. As the cut-off angle that yields the maximum recorded momentum range (s_{max}) is one important factor to determine small bond variation and to extract the thermal factor from scattering, the X-ray approach could extract a finer structure than that of electrons when it is possible to gather data at large momentum transfer such as those available at high beam energy synchrotron-based X-ray sources.[3]

On the other hand, the short penetration depth of electrons combined with their high scattering strength allows UED to examine thinner samples, ranging from gas molecules (see also Chapters 10, 12 and 13) to nanomaterials and thin films. Moreover, UED is also relatively more sensitive to scattering of hydrogen atoms than in X-ray diffraction experiments. This can be easily seen in the Mott formula, where scattering from the atomic nucleus is described using a delta-function weighted by its positive charge Z, and hence its Fourier transform, after correcting for the $1/s^2$ factor, is a constant value over all reciprocal space. Meanwhile, the electron cloud surrounding the nucleus acts as a screening field and can be approximated by Gaussian functions in cross-sections,[4] given by $\tilde{f}_x(s)$. Since the charge cloud contribution decays more rapidly in s for lighter elements, the ratio of scattering factors of hydrogen to carbon is not greatly different at small angles for electron scattering, making it more suitable for studying organic or carbon-rich materials.[4] In addition, inelastic processes (*i.e.*, the scattering of the incoming electron against charge cloud exciting the sample species; see Chapter 3) are better distinguished at small angles away from elastic scattering at large angles, the latter possessing atomic structure information (coherent elastic scattering) in electron scattering (see Section 11.5).

Moreover, recent MeV UED system developments (see Chapter 12) currently lead to higher beam brightness and better avoidance of the space-charge-induced effects; however, table-top-scale UED systems operating at 20–200 keV are still indispensable, versatile tools for multi-modal and multi-perspective investigations offering imaging, diffraction, and spectroscopy capabilities all in the same setup. In addition, the recent understanding of the nature and reversal of space-charge-effects has pushed the resolution and the sensitivity of keV-scale UED instrumentation very close to those achieved by MeV systems (see Section 11.6). With these distinctions in mind, we will give a survey of the multi-faceted applications of UED, focusing in the 30–200 keV energy range, together with a brief discussion on recent developments of the core technologies that are poised to further improve the performance of keV-scale UED.

11.2   Dawn of Ultrafast Molecular Movies: Gas-phase UED Development

One major goal behind developing UED techniques is driven by the advances of photochemistry and photobiology[5] – understanding the transition state structure and ensuing dynamical evolution heralds the key to controlling chemical reactions. Before the development of UED came to fruition, knowledge about the transient states relied heavily on the spectroscopy experiments performed with probe wavelengths ranging from the ultraviolet to the infrared and far-infrared range. While these measurements are performed on the femtosecond (fs) timescale, providing snapshots of frozen localized structures in space (wave packets) and allowing observation of their evolution in time, the large wavefield covers all atoms indiscriminately and the atomic dynamics are inferred from the dynamics through optical signals.

11.2.1   Historical Overview of UED

For complex molecular structures, the positions of all atoms at a given time can only be obtained if the probe pulse is able to capture interferences from all atoms.[6] In fact, it is because of the large electron scattering cross-section and the higher sensitivity to detect light atoms in organic molecules that prompted Mark and Wierl in 1930[7] to use electrons (instead of X-rays[8]) to study gas-phase molecular structures. It is due to these two crucial technological advantages that the gas phase UED methodology currently has tremendous growth potential to realize the dream of many chemists: capturing robust snapshots of molecular movies at the fundamental limits (10 fs–0.01 Å) for exploring the excited free energy surface, building on not just the direct structural resolution but also the sensitivity to discriminating different reaction channels from excited landscape under the pump control, and as a result providing an effective feedback mechanism for devising strategies for laser-selective chemistry[9] or even quantum control of chemical reactions.

The bulk of early UED development has been pioneered in Zewail's group at Caltech starting in the early 1990s,[10] where the technological advances centred on merging the diffraction approach with newly developed ultrafast laser technology and high-speed streak cameras (also see pioneering work on picosecond–nanosecond timescales by Mourou, Williamson, Li and Elsayed-Ali;[11,12] Shafer, Ewbank, and Ischenko[13]), with the aim of extracting as intense and as short pulses possible through photoemission from the fs laser pulses.[14] This development spanned over three generations[6,10,15,16]— the first robust UED system was completed in the year 2000.[16] However, even with the latest technologies, the conducting gas-phase UED experiment was very time consuming, typically requiring several hours to collect a single timeframe with only a few thousand electrons per pulse in the probe beam to obtain a resolution just shy of 1 picosecond (ps) while operating at a 1 kHz pump–probe repetition rate (f_{pp}). Nonetheless, reliable data were measured with a spatiotemporal resolution of 0.005 Ångström and ~ 1 ps from many topical systems involving radiationless transitions,[17,18] structures in non-concerted organic reactions,[16,19,20] structures in non-concerted organometallic reactions,[19,21] structures of carbene intermediates,[15,22] nonequilibrium states and structures,[23] and conformational dynamics on complex energy landscapes.[24] In many ways, these early efforts serve to push the field into new territories and highlight the essential questions that can be unveiled with a new technical capability for structure dynamics. Further improvement in the time resolution *via* MeV UED development (see Chapter 12) to the sub-ps regime has not only successfully revisited many topics earlier but also provided essential new insight into nonadiabatic dynamics involving the initial population of excited state potential energy surfaces with improved temporal resolution. Despite these new advances, many core issues that became quickly apparent at the early stage (*e.g.*, the velocity mismatch problem,[25] space-charge effects,[6] and diffraction difference/product-only approach[6,25] for structure refinements, including an appropriate treatment of nonequilibrium and vibrationally hot structures for the transient dark states[23]) have become even more relevant today considering that these issues can now be solved directly with the new, improved techniques.

11.2.2 Retrieving Excited State Structures *Via* the Pump–Probe Technique

Changes in the molecular structure can, in general, be retrieved from the recorded electron diffraction patterns. However, the total scattering intensity, I, is a sum of contributions from individual atoms (incoherent atomic scattering, I_A) superimposed with interference terms from all atom–atom pairs (coherent molecular interferences, I_M) in the framework of the independent-atom model (*i.e.*, the independence of the electronic potentials of each atom in the molecule is assumed). Since there is no long-range order in gases to enhance coherent interferences, the incoherent atomic scattering from gases is orders of magnitude higher. For the purpose of structural determination, only the coherent interference part (I_M) is of interest because it contains information on internuclear separations over N atoms in the molecule given by

$$I_M(s) = C \sum_{i}^{N} \sum_{j \neq i}^{N} |f_i||f_j| \exp\left(-\frac{1}{2}\langle l_{ij}^2 \rangle s^2\right) \cos\left(\eta_i - \eta_j\right) \frac{\sin(sr_{ij})}{sr_{ij}}, \qquad (11.3)$$

where $\langle l_{ij}^2 \rangle$ is the mean squared amplitude of vibration between the distance pair r_{ij} between atoms i and j, thereby probing structural changes undergoing different relaxation processes. Note here, with scattering from isotropic samples, the Fourier counterpart of the pair correlation function $G(r, t)$ is the scaled molecular scattering function given by $sM(s) = sI_M(s)/\langle |f_a||f_b| \rangle$. Although the molecular scattering function contains all of the structural information about the molecule, a more intuitive interpretation of experimental results is achieved by taking the Fourier (sine) transform of $sM(s)$ and examining the radial distribution function, $G(r, t)$, given by

$$G(r, t) = \int_0^{s_{max}} sM(s)\sin(sr)\exp\left(-ks^2\right)\mathrm{d}s, \qquad (11.4)$$

where k is the damping constant. The exponential damping term filters out the artificial high frequency oscillations in $G(r)$ caused by the cutoff at s_{max}. The radial distribution curve reflects the relative density of internuclear distances in the molecule. In the last generation of gas phase UED [UED-3] systems, the available experimental scattering intensity typically ranges from $s_{min} = 1.5$ and up to $s_{max} \sim 15$ Å^{-1}. For the range from 0 to s_{min}, the theoretical scattering intensity, $sM_{theo}(s)$, is appended to avoid distortions of the radial distribution baseline. It should be noted that all data analyses and structural refinements shall be performed on $sM(s)$ and not on $G(r)$ because of inaccuracies that could potentially be introduced into $G(r)$ through improper choice of k.

Notably, since the fraction of molecules undergoing a change is small (typically 10% or less), the recorded diffraction patterns contain large contributions from non-reacting parent molecules. Therefore, to accentuate the diffraction signals arising from structural changes occurring over the course of reaction, one typically employs the diffraction–difference method[6] where a reference image is used to obtain the diffraction–difference signal, $I(t; t_{ref}; s)$, from the relation:

$$\Delta I(t; t_{ref}; s) = I(t; s) - I(t_{ref}; s). \qquad (11.5)$$

Finally, we extract the molecular diffraction signal resulting only from the transient species *via* the transient-only or the transient-isolated $sM(s)$.

In this case, the reactant diffraction signal ($sM_r(s)$, obtained at a negative delay time) is scaled by the fractional change in $sM_r(t; t_{ref}; s)$, and added to the diffraction difference signals obtained at positive delay times, thereby cancelling out the parent reactant contribution:

$$\Delta sM(t; t_{ref}; s) + \Delta p_r(t; t_{ref}) \cdot sM_r(t; t_{ref}; s) = \sum_{\alpha \neq r} \Delta p_\alpha(t; t_{ref}) \cdot sM_\alpha(t; s). \quad (11.6)$$

From these equations, information on population dynamics and branching ratios of the product species from each reaction channel can be retrieved, allowing to significantly constrain the parameter spaces for determining the transient structures. One naturally hopes to extract such information independently, for example, *via* optical spectroscopy techniques preferably conducted *in situ* with UED measurements. As a historical note, a mass spectroscopy system to identify the reaction products has been designed into the UED-3 system to aid in this effort. Nonetheless simultaneous UED and spectroscopy measurements are currently unavailable, and so the current protocols based on the diffraction–difference approach provides population and structural information at the same time.

Historically, gas phase electron diffraction studies heavily relied on the *ab initio* structure to assist in structure determination.[26] The ground-state DFT structure is typically found quite reliably with modern day computing; however, such structures are intrinsically vibration-free and to accurately reconstruct the diffraction pattern, which is relied upon to determine the unknown reaction intermediates and products, one needs to consider an appropriate encoding of temperature effects and vibrationally-induced nongeometrically consistent motifs (*e.g.*, shrinkage effects[26]). Since highly-excited vibrational states lead to decoherence effects in the ensemble-averaged structure factor, it is treated with an exponential damping factor. This translates to a Lorentz-broadening in $G(r)$ peaks after Fourier transform. In the ground state, this is justified with Boltzmann population statistics.[26] However, the vibrations associated with an excited transient species are initially non-Boltzmann-like, rooted in the initial nonequilibrium vibronic excitations. There has been a strong interest in resolving the intramolecular vibrational-energy redistribution (IVR) involving quantum beating, jamming, and nonexponential decay with UED[27] (*e.g.*, see the review article by Ischenko and Aseyev[28]). Only with a dataset of sufficient s-range—to distinguish a distance variation of 0.1 Å, an s_{max} of up to 15 Å^{-1} would need to be extracted—could such issues be fully disentangled.

In recent efforts, *ab initio* molecular dynamics (AIMD) calculations[29] have played the historical role of the *ab initio* structure in guiding the structure refinements for transient species obtained in UED. As part of the structural refinement protocol, AIMD calculations provide credible initial motifs as well as branching ratios for modelling the vibrationally-hot transient structures through ensemble-averaging the AIMD trajectories. AIMD-assisted modelling has played an essential role in understanding the otherwise difficult-to-resolve transient species and nonequilibrium structures elaborated using the new MeV-UED apparatus[30] (see Chapter 12).

11.3   UED for Surfaces and Interfaces

The lower energy electron probe (15–30 keV), especially when directed at a grazing incidence angle, is exclusively sensitive to the surfaces and surface-supported nanostructures, thus giving specific information on nano-sciences that the traditional transmission (transmission electron microscopy, TEM) and reflection-based (X-ray) diffraction techniques cannot offer. Its ultrafast variant opens a new field of investigating dynamical processes involving nanomaterials, surfaces, and interfaces on the 1–10 nm scale – a regime of concern for many practical applications that few other techniques can provide. One well-known issue is the emergence of noncrystallographic structural types[31,32] with a tendency to form closed shells exhibiting magic numbers in size distribution.[33] The atomic structures of these nanocrystals also highly depend on the supporting substrate and surface terminations,[34,35] exemplified by contraction[36] and twin boundary formation[37] from the relaxation of surface strains. With these new structural forms, novel properties often emerge that are connected with specific size and morphology. The capabilities for fabricating material hierarchy with specific properties *via* nanosynthesis or self-assembly open interesting prospects for the use of nanoparticles as building blocks for new devices with high specificity, speed, and density, which can be deployed in diverse fields, such as electronics,[38–40] photonics,[41] magnetism,[42] catalysis,[43] and sensing.[40]

For example, the reflection high-energy electron diffraction (RHEED) patterns from in-plane and out-of-plane coherence allow the identification of atomically thin surface structures with sub-monolayer sensitivity. The exclusive monolayer sensitivity is evidenced in the arc shaped patterns referred to as Laue zones as given in the left panel of Figure 11.1a.

In this case, the patterns reflect exclusively the geometry of the intersection between the Ewald sphere and the continuous reciprocal lattice rods (relrods) with no nodes due to the lack of ordering. By increasing the angle of incidence θ_i, the deeper penetration gives rise to increased accessibility to the buried layers, with the intensity peaking upon meeting the specular reflection at the scattering angle $\theta_t = 2\theta_i$ (see the left panel of Figure 11.1). One version of surface-study-based UED[11] inheriting such properties can be used to study the reconstructed surface which is typically sensitive to the free energy derived from the interaction between the adsorbate–substrate and adsorbate–adsorbate interactions under the UHV environment.[44,45] Another direction of research focuses on the nanostructures or the adsorbates themselves instead of atomically flat surfaces. In this case, the supported nanostructures do not create a continuous film, but rather form puddles or crystallites.[46] At a grazing incidence, typically below 1° or less, the incident beam can effectively penetrate such nanoscale structures, generating transmission-type diffraction patterns[47] (see the right panel in Figure 11.1). Similar transmission features can also be obtained from a rough surface

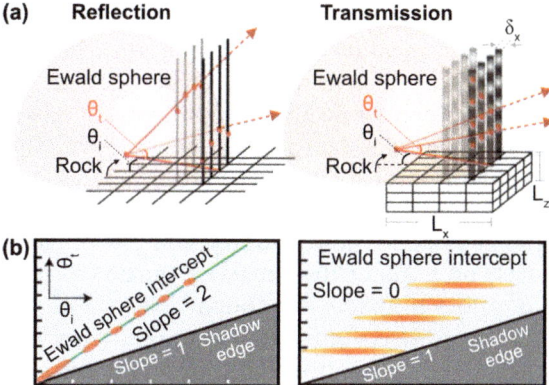

Figure 11.1 Surface electron diffraction pattern inder different conditions. (a) Ewald sphere construction in grazing incidence geometry. By tilting (rocking) the angle of incidence between the electron beam and the surface, the Ewald sphere intercepts the reciprocal lattice rods (relrods) at different heights. The coherent diffraction condition is satisfied when the intercept is at the reciprocal lattice node. The left panel shows diffraction from a 2D surface. Because there is almost no structure within each relrod, the diffraction exhibits ring patterns revealing the in-plane translational order but no out-of-plane order. The right panel shows diffraction from nano-materials with reciprocal nodes broadened by the finite size L_Z of the sample. Sampled by the elongated Ewald sphere intercepts, the diffraction pattern exposes the multiple periodic modulations along the relrods, indicative of the lattice periodicity. (b) Corresponding rocking curve maps. Reproduced from ref. 195 with permission from Springer Nature, Copyright 2014.

with nano-sized step edges[48,49] or uneven surfaces frequently encountered from exfoliated surfaces with van der Waals 2D materials. As it is likely that both the surface-type and the nanostructure-type diffraction could occur, to distinguish various scattering features from complex surfaces, one could employ the rocking map constructed in Figure 11.1b for the respective cases. Here, the ratio $a \equiv \theta_t/\theta_i$ reflected in the traces above the shadow edge is exclusive for different scattering types; for transmissive diffraction, the ratio is near zero instead of 2 (reserved for surface scattering). A particular intriguing scenario is simply using the atomically flat surface as a support over which different types of nanoparticles can be dispersed as a means for studying structural dynamics on the nanometre scales.[50] In this case, the scattering from the top surface of the substrate, especially when treated with a mono-buffer molecular/amorphous layer, often gives only the diffuse background.

11.3.1 The Development of Ultrafast Electron Nano-crystallography (UEnC)

Ultrafast electron nanocrystallography (UEnC) aims to address the rich dynamical phenomena (including structural and charge relaxation) embodied by nanoscale materials, interfaces, and surfaces. It builds on the earlier discussed structure identification schemes and the different dynamical contrasts on ultrafast timescales to discern active phenomena on surfaces often involving all the interconnected components. It also involves the development of new schemes for data acquisition and reduction along with a robust 3D refinement protocol for solving nanostructures.[51] Here, we first introduce the probing concept of UEnC tailored for dynamics of nanostructures based on nanometre-scale "powder-diffraction", where the nanoparticles are dispersed sparsely on a solid substrate (see Figure 11.2a–c).

Although it is possible to disperse nanoparticles on ultrathin amorphous films in a TEM grid, it is found that the instability of the film can introduce artifacts in the diffraction data, or even cause damage upon intense laser illumination of the thin film where much lower thermal relaxation is expected. To avoid such vulnerability and to generalize the investigation to include interfacial dynamics, we choose to anchor the nanoparticles on a firm solid surface and use a grazing incidence electron probe to interrogate particles. In addition, we employ a buffer ad-layer composed of a self-assembled monolayer (SAM) of aminosilane molecules to elevate the nanoparticles above the solid surface, which serves three

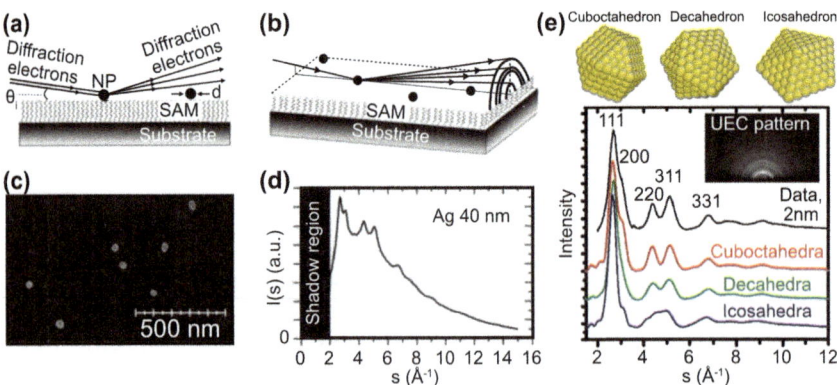

Figure 11.2   Concept of ultrafast electron nanocrystallography to probe surface-supported nanostructures. (a) Scanning electron microscopy image of Ag nanoparticles ($d = 40$ nm) dispersed on a Si(111) surface. (b and c) At a grazing incident angle $\theta_i \approx 1$–$5°$, the electron beam is used to diffract from the nanoparticles supported on a surface with a scattering solid angle up to 2π. To minimize the interference from the substrate scattering, a "soft" buffer layer is employed to elevate the nanoparticle from the substrate while suppressing the background scattering. (d) Total diffraction curve from such a diffraction geometry. (e) Experimentally obtained structure factor $S(s)$ for 2 nm gold nanoparticles, compared to simulated curves from the known non-crystallographic motifs prevalent in nanostructures. The resemblance of experimental curve to a cuboctahedra structure is apparent. Adapted from ref. 51 with permission from Oxford University Press, Copyright 2009.

important purposes:[47] (i) it suppresses the scattering from the substrate that would otherwise produce a strong background signal, thus overwhelming the diffraction signal from the nanostructures, (ii) allows transmission diffraction signals to be collected unobstructedly over a 2π solid angle, and (iii) controls the rates of heat and charge dissipation of the nanoparticles. To ensure that the outgoing diffracted beams have a relatively low probability of being intercepted by another nearby nanoparticle, an optimal particle density of $\sin^2\theta_i/d^2$ is desirable, where d is the diameter of the nanoparticle. Figure 11.2c shows the SEM image of a typical sample arrangement for UEnC satisfying all the above criteria using 40 nm Ag nanoparticles sparsely dispersed ($\sim$10 μm^2 per particle) on a Si-111 substrate on top of the SAM layer.[52] The diffraction signal retrieved from this sample over $\sim$1 minute integration time (see Figure 11.2d) possesses sufficient signal-to-noise ratios (SNRs) for quantitative studies, despite the nanoparticles occupying less than 1% of the area. This capability demonstrates the high sensitivity to nanostructure responses in contrast to X-ray and neutron diffraction approaches in which much higher volume and density are typically required.

11.3.1.1   Refinement of Nanostructure Dynamics Based on Imminent Causality

The resulting pattern from the nanostructures can be analysed as if they were obtained from a thin film at normal incidence. One prototypical example is the studies of size-dependent nanoparticle phase transformations.[53] With this approach, nanostructures as small as 2 nm have been successfully investigated by UEnC[47] (see Figure 11.2e). Here, the power pattern deduced by radially integrating the 2-cone above the shadow-edge provides sufficient quality for it to be discerned against the reciprocal fingerprints from close-packed clusters with distinctly different symmetries represented by cuboctahedra, decahedra, and icosahedra. The cubic structure (see Figure 11.3a) as revealed from visually inspecting the locations of the intensity peaks can be further confirmed by calculating its $G(r)$ Fourier pair (see Figure 11.3b).

In this representation, all the well-distinguishable pair-distance peaks above 2.8 Å given by $G(r)$ are in close agreement with the Au–Au distance table based on a face-centre-cubic (FCC) motif (see Figure 11.3a), which constitutes the internal lattice repetition of a cuboctahedra. Nonetheless, in the nanocrystalline context, the correlation runs out over the size of the particles. Beyond 15 Å, the pair correlation could only emerge from the coherent interferences between the atoms located near the surface. Therefore, this finite-scale sensitivity, as given by $G(r, t)$, here offers a unique tool to probe the surface structure dynamics that is expected to vary significantly from those of the bulk. In addition, from the small angle patterns near the shadow edge, the UEnC approach also probes features representative of ordered structures of the SAM (see Figure 11.3c and d) as the oriented patterns instead of rings.

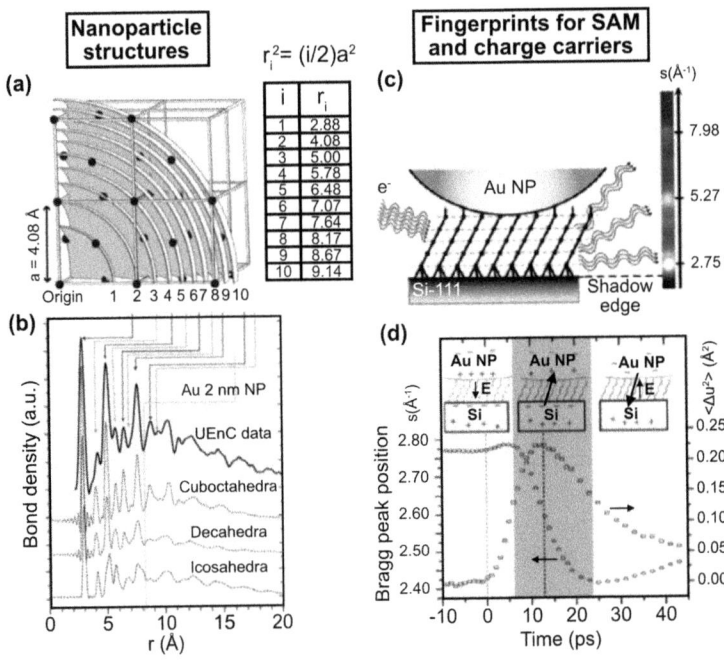

Figure 11.3   UEnC data retrieval based on diffraction from nanocrystals or self-assembled molecular (SAM) layers. (a) FCC coordination shells corresponding to interatomic distances r_i, calculated based on the bond order i and the Au lattice constant, $a = 4.08$ Å. (b) Experimental modified radial distribution functions of static Au nanoparticles (NPs) along with theoretical predictions for cuboctahedra, decahedra, and icosahedra. (c) Schematic representation of scattering from the SAM and corresponding transmission-type diffraction patterns captured on the detector. (d) Dynamical signatures from the SAM-associated Bragg peak movements as well as the mean-squared vibration amplitudes deduced from the intensity of the SAM-peaks based on the Debye–Waller analyses, suggesting a charging and discharging of the nanoparticle. Adapted from ref. 51 with permission from Oxford University Press, Copyright 2009.

Using the different diffraction patterns, the structural changes after the pulsed laser heating of the nanoparticles were determined (see Figure 11.4a and b). We explore the complementary structural insight in the analyses deduced by the distributions of the two radially-averaged (sine) Fourier-transform pairs, $sM(s, t)$ and $G(r, t)$, similar to the treatment performed for gas-phase UED experiments.

First, we examine the radially-averaged $sM(s, t)$ distributions given in the full frame of the ground state and difference frames for the excited structures based on the diffraction difference method. The melting occurs in three stages (see Figure 11.4a): (i) between 0–20 ps, the intensity of the Bragg peaks is reduced but does not disappear entirely, which presumably is caused by the loss of medium-range order for 2 nm gold nanocrystals – this

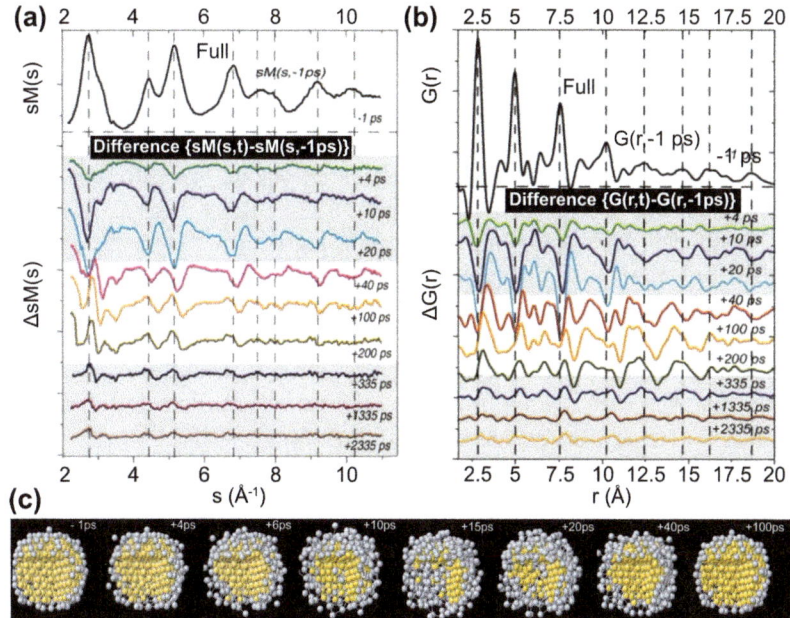

Figure 11.4 Modelling of nonperiodic structural changes from UEnC. (a) $\Delta sM(s, t)$ difference curves, showing loss of long-range order, evident from a drop in Bragg peak intensity (0–20 ps), followed by the coexistence of disordered domains with partial long-range order (20–200 ps). From 335 ps onward, the existence of an enhanced transient order at an elevated temperature is observed. (b) Complementary $\Delta G(r, t)$ difference curves, showing an initial partial adjustment in the local bond-orders (1–10 ps), followed by most significant lattice disordering during 10–40 ps with signatures of surface melting in the full suppression of the correlation peak above 15 Å. (c) Snap-shot images of the nanoparticles deduced based on a structure refinement using the progressive Reverse Monte Carlo approach. Adapted from ref. 51 with permission from Oxford University Press, Copyright 2009.

can be explained by either the thermal disorder (Debye–Waller effects) or by a decrease in persistent length caused by partial melting;[54,55] (ii) between 20–200 ps, the behaviour of the Bragg peaks may be related with the coexistence of the disordered region and the partially ordered domains; (iii) at 335 ps and onwards, the negative peaks largely disappear, while some of the weak positive peaks corresponding to the ground-state cuboctahedra spectrum become stronger. The restructuring behaviour and the movements in the peak positions indicate an appearance of the transient state at elevated temperatures.

In contrast to gas-phase UED, the significant increase in the number of atoms in the nanostructures and the aperiodic nature of the dynamics involving phase coexistence prohibit a simple 3D structural model to be employed for refinement. A solution to this nanostructure refinement problem is based on a progressive reverse Monte Carlo (PRMC) algorithm,[51] which uses simulated annealing for structural refinement with a perturbative iteration scheme.[56] The reverse Monte Carlo (RMC) algorithm, known to be a global structural solver but prone to yielding non-unique solutions, typically relies on physical constraints to guide the search.[56] We have implemented a scheme to restrain the RMC search for solving the UEnC nanostructural problem, based on the difference curves with a rolling reference point. We start with a supercell model of a 2.5 nm cuboctahedral nanocrystal with a cubic cell size of 50 Å. The PRMC algorithm iteratively fits both the simulated $G(r)$ and $sM(s)$ with experimental counterparts based on adjusting the supercell structure. This is carried out by choosing the time interval between neighbouring delays in the UEnC experiments to match the rate of change such that the structural difference between adjacent delays is small. This maintains the structural correlation between the rolling reference structure and the imminent structure, thereby allowing changes to be tracked progressively and reliably within a restrictive search space, without need for a global search. Using this approach, the 3D representation of the transient atomic structures of gold nanocrystals is constructed as given in Figure 11.4. These 3D images show the features of surface "premelting"[54,55] rather than inhomogenerous disordering. With increasing temperature, the surface layer becomes thicker, causing the compression of the crystal core. At the peak of the surface melting in the interval of 15–20 ps, no more than 25% of the atoms are structured in the ordered form at the core region. Using the core size as the reaction front, it is possible to trace the speed of the melting process and to observe the re-solidification process.

11.3.1.2   Structural Signatures Associated with Interfacial Charge-transfer at Excited Interfaces

Here, we take advantage of the fact that the linker molecules taking part in the anchoring of nanoparticles tend to orient themselves in a manner that is sufficiently ordered, such that very clear diffraction spots from the linker molecules are present in the patterns. By gating on diffraction signals emerging from the SAM region, one can determine the molecular structure and its transient response to the imposing interfacial charge carrier responses. This is illustrated in Figure 11.3c, where sharp transmission-type scattering patterns encoding the in-layer structure of the aminosilane molecules are shown. As these interfacial signatures are obtained from the same diffraction images that give the structures of the nanoparticles, they guide the understanding of correlated dynamics between the interfacial charging, thermal transport, and structure dynamics.

We examine instead 20 nm Au nanoparticles where, with the insulating SAM, the larger particle size presents an effective capacitive coupling with the semiconducting substrate, presenting a case for an elementary nano-electronics building block. Upon examining the positions of the Bragg reflections from the SAM, we observe a clear refractive-type response characterized by the nonreciprocity in the Fourier-transform-pair, namely, the shift identified in the lower order reflection is actually more rather than less, violating the reciprocity[57,58] given by the Fourier transform between the structure factor and correlation function governing any structure-related changes.[49] The origin of this intriguing phenomenon, which bewildered the early works of UED,[51,59,60] was traced to an electric field between the particle and substrate that emerged from an interfacial charge transfer assisted by laser pulses, causing the scattered beam to deflect from the surface. From the perspective of electron optics,[61] the scattered waves emerged from the substituent crystalline planes with their optical path and phase altered while maintaining their phase difference in the direction of a Bragg-diffracted beam. With the goal of restoring the reciprocity by imposing a photovoltage $V_s(t)$ across the SAM one can estimate the field strength and the capacitive effect established at the interface. A generic formula based on the parallel capacitor model[62] is shown to be effective in treating the interfacial charge transfer dynamics,[52] for different scattering types as characterized by the rocking curve slope $a = \delta\theta_t/\delta\theta_i$, given by

$$
\begin{cases}
a = 0: & \chi = \dfrac{\left(\theta_t \Delta_B - \theta_i \theta_t + \Delta_B^2/2\right)^2 - \theta_i^2 \left(\theta_t + \Delta_B\right)^2}{\left(\theta_i + \theta_t\right)^2}, \\
a = 1: & \chi = -\Delta_B(\Delta_B + 2\theta_t)
\end{cases}
\tag{11.7}
$$

where $\chi = V_s/V_0$ with eV_0 being the incident beam energy, θ_i is the probe incidence angle and Δ_B is the angle of deflection associated with a Bragg-scattered beam located at θ_t (see also Figure 11.1). For $a = 2$, implying $\theta_i = \theta_0$ in RHEED geometry, and the measured Δ_B in experiments is not affected by V_s. We note a more elaborated approach considering geometry beyond simple capacitor scheme is given in ref. 63. Here, the shift of θ_i due to charging of the Si surface is corrected based on the corresponding movement of the shadow edge. In this case, the initial downward deflection in the beams diffracted from the SAM is clear evidence of electrons advancing from the Si interface states into the nanoparticle through the molecular wires. Based on the Coulomb refraction shift for 20 nm Au nanoparticles, with $dV_s/ds \approx 20$ V Å^{-1}, we estimate a charging of $250\ e$ at the peak of charging.

We also track another signature from the SAM based on the transient anisotropic mean-squared amplitude of vibration, $\langle \Delta u_{hkl}^2 \rangle$, given by Debye–Waller Factor (DWF) analysis with

$$
\langle \Delta u_{hkl}^2 \rangle = -\ln\left(I_{hkl}(t)/I_{hkl}(t_{ref} < 0)\right)/s^2.
\tag{11.8}
$$

We find that the transiently enhanced C–C vibrations normal to the interface are intimately coupled to the interfacial carrier dynamics probed through the refractive effect, with similar rise and decay time.

11.3.2   Ultrafast Photovoltages and Charge Transport Dynamics at the Interface

Broadly speaking, characterizing the photoinduced charge carrier dynamics in nanostructures and interfaces is relevant for understanding fundamental interfacial charge-transfer processes in the emerging technologies of nano-electronics,[38–40] solar photovoltaics,[64–66] and quantum-dot-based photonics.[67–70] These processes are nonetheless difficult to characterize on ultrafast time-scales. While several powerful ultrafast optical techniques have been applied to investigate electron dynamics at the interfaces,[71–74] they are usually not well suited to directly derive the through-space photoconductivity at the interface. Through the sensitivity of the diffracted electron beams to the electric fields on site, the ultrafast electron scattering approach effectively provides valuable insights into various phenomena. In this section, we highlight the contact-free ultrafast photovoltammetry (UPV) approach derived from UEnC measurements,[52,62,63] for probing the local photovoltage dynamics and their connections with the structural relaxation and changes in the local electronic properties at interfaces and in nanostructures.

To illustrate the methodology, we carried out controlled experiments probing the transient photovoltages for the Si/SiO_2 substrate and the linker molecules that anchored the nanoparticles (NPs) to the substrate discussed earlier.[51] The charging dynamics will now be examined using the effective circuit model to describe the interconnected nanoparticle/SAM/semiconductor structure shown in Figure 11.5a.

Again, the interface here is illuminated with 800 nm near-infrared laser pulses. The total transient shift of the scattering angle includes two serial voltage drops: one across the Si/SiO_2 surface (V_B), and the other across the SAM (V_M), as depicted by the effective RC-circuit in Figure 11.5a. To isolate the voltage across the linker molecule, the contribution from V_B must be subtracted out. The $V_B(t)$ associated with a bare Si/SiO_2 surface is probed with three diffraction beams corresponding to θ_0 of 3.50°, 4.02°, and 4.63°, which are utilized to extract the electromotive force (emf) from the Si surface, following eqn (11.7). The result, given by the refraction shift Δ_B, shows an exponential downturn followed by a 200 ps recovery (see black symbols in Figure 11.5b). This result is plotted against the refraction shift due to photovoltage $V_S(t)$ measured through the SAM in the Si/SAM/Au–NP interconnect (see red symbols in Figure 11.5b). By taking the difference between V_B and V_S, we obtain the photovoltage across the molecular linkers V_M (see blue symbols in Figure 11.5b). Remarkably, the molecular transport dynamics deduced here is nearly identical to the thermal transport dynamics deduced based on the Debye–Waller analysis reported in Figure 11.3d, indicating that the heightened vibrations found during the ultrafast charging and discharging of the particles are mediated by current across the linker molecules.

To investigate this further, the transient V_M was investigated over a range of laser fluence (see Figure 11.5c).[75] Here, $V_M = V_B - V_S$ (sign change) is taken to highlight the initial hot carrier current launched from nanoparticles.

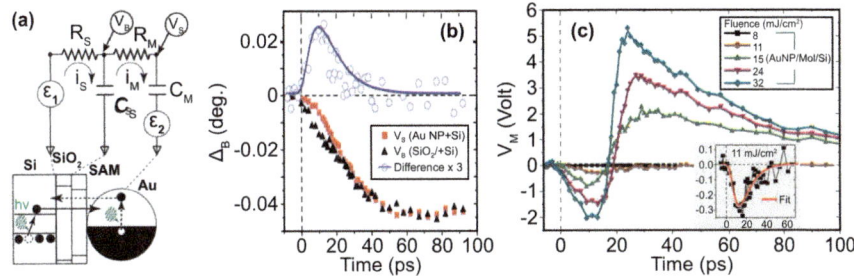

Figure 11.5 Hot carrier driven interfacial charge-transfer at the Si/SAM/Au–NP interface. (a) An effective circuit model depicting the transient surface voltage $V_s(t)$ measurement via the refraction shift of the diffracted beams through SAM. The R_S, C_S, R_M, and C_M are the effective resistance and capacitance of the substrate (S) and the SAM (M), respectively. (b) The overall refraction shift determined by SAM diffracted beam (labelled V_S), the background (labelled V_B) obtained from SiO_2/Si interface, and the molecular charge transport contribution, obtained by subtracting V_B from V_S. (c) The photovoltage V_M determined for different fluences, showing the electron currents being shuffled between the semiconductor substrates and the gold nanoparticles across the molecular interface. Inset, the V_M obtained at $F = 11$ mJ cm^{-2} is fit to an RC model with nearly equal time constants of 8.1 ps for charging/discharging. Adapted from ref. 195 with permission from Springer Nature, Copyright 2014.

The initial negative $V_M(t)$ across the SAM obtained at 11 mJ cm^{-2} again indicates net positive charging of the gold nanoparticles, driven mainly by the hot carriers. Based on the charging and discharging time of 8.1 ps (see the inset of Figure 11.5c), we obtain a resistance $R_M = 2.74$ MΩ using $C = 2.92 \times 10^{18}$ Farad. It is rather interesting to compare this molecular resistance with the steady-state value of 12.5 MΩ obtained by applying a bias voltage across the molecular interface[76] using 10 nm (thus, a SAM area four times smaller) Au nanoparticles. The results show that the molecular resistivity obtained from two different methods is nearly identical. At higher fluences, this trend is heightened up to $F = 32$ mJ cm^{-2} when a saturation is reached and the reversal time exhibits a minimum value of 3 ps. From examining this trend, a critical value of approximately 0.8 V nm^{-1} can be determined as the threshold for conductance switching.

We note that the small incidence angle here gives rise to a wave front mismatch between the pump and probe beams, leading to a reduced temporal resolution. Such an issue has been addressed by synthetically depressing the specimen into horizontal stripes[77] and by employing laser wavefront tilting to rematch the arrival time between the pump and probe for the stretched region.[48] One should also expect that employing electron optics to isolate the scattering paths in the context of ultrafast electron microscopy[78] (see discussion in Section 11.6) will lead to vastly improved temporal resolution and site specificity in UPV measurements.

11.4 Ultrafast Electron Crystallography (UEC)

Crystallography has been the gold standard for structural determination. The protocols for solving complex structures for bulk crystals using X-ray and neutron scattering to better than 0.01 Å resolution are well-established. Yet, femtosecond crystallography remains a niche area due to the unavailability of femtosecond hard X-ray sources to regular users. Only a few bright femtosecond hard X-ray beamlines currently exist in the free-electron laser facilities (see Chapter 8). On the other hand, it is relatively straightforward to extend UED technologies into an ultrafast electron crystallography (UEC) system. As the standard crystallography approach relies on sharp Bragg peaks to index the crystalline orders, it has been more reserved to the study of large crystals with 'infinite' long-range periodicity well-suited to the X-ray or neutron approaches. Electron crystallography has been traditionally conducted using a high-voltage electron microscope.[4] The high energy beam (80–200 keV) can effectively penetrate sub-micrometre sized small crystals or thin films, equally producing sharp Bragg peaks. To this end, the high-energy transmission-based ultrafast electron crystallography (UEC) serves as a synergistic counterpart to X-ray or neutron crystallography. Meanwhile, the high-energy beam also produces a large Ewald sphere, giving access to a great number of Bragg peaks in the patterns created from even a singular specimen tilt angle, eliminating the need to rock the specimen to obtain the full pattern from the Brillouin zone (BZ). This avoids the drudgery of continuously rotating the solid crystals in order to sample a sufficient number of symmetry-based crystalline structure form factors F_{hkl} in order to solve complex structures. Furthermore, unlike gas-phase UED and UEnC experiments, the high-energy transmission-based UEC system typically does not suffer greatly from the pump–probe wave front mismatch issues. Hence, it becomes more straightforward to obtain high temporal resolution once the space-charge-affiliated issues are overcome – a subject we shall discuss in Section 11.6. Here, we will primarily focus on the science cases enabled by high momentum-resolution fs electron crystallography.

11.4.1 Scattering from Periodic Potentials: Basic Theory

Scattering from single crystals of sufficient quality, namely, in a situation where the crystal size is much greater than the coherent length of the probe X_c, which in turn should be much greater than the unit cell of the crystals, electron crystallography provides similar information as that given by X-ray crystallography. The crystallographic analysis is based on sampling the complete set of Bragg reflections directed along the key symmetry planes, annotated by the reciprocal lattice vector $G_{hkl} = h\mathbf{b}_1^* + k\mathbf{b}_2^* + l\mathbf{b}_3^*$, where h, k, and l are the Miller indices. Due to the long-range ordering, such Bragg peaks are delta-function-like and easily identifiable. From intensity of such

Bragg reflection arrays, one seeks to identify the crystalline form factor, given by the interferences from the atoms within the unit cell:

$$F_{hkl} = \sum_{\text{unit cell}} f_i e^{i\mathbf{G}_{hkl}\cdot\mathbf{r}} = \sum_{\text{unit cell}} f_i e^{2\pi i[hx_i + ky_i + lz_i]}. \qquad (11.9)$$

In eqn (11.9), f_i is the electron atomic scattering factor and (x_i, y_i, z_i) are the fractional coordinates of atoms from the unit cell. Given that $f_i(s)$ is the Fourier counterpart of the atomic potential, in this format, one could easily see that the inverse transformation of eqn (11.9) gives the potential distribution of the entire unit cell (as mentioned in Chapter 4):

$$\rho(\mathbf{r}) = V^{-1} \int ds F_{hkl}(s) e^{-i\mathbf{s}\cdot\mathbf{r}_i}. \qquad (11.10)$$

One could adopt the theoretically deduced atomic scattering factors to normalize the form factor: $\tilde{F}_{hkl} = F_{hkl} / \left\langle \sum_i f_i \right\rangle$. As this divides out the equivalences of the charge density distribution given by the form factor, the inverse Fourier transform then gives the delta-function-like atomic positions $\rho_R(\mathbf{r})$ in the unit cell.

With electron crystallography, there is a well-established protocol for determining the structures of molecular crystals.[4] The electron crystallography for solving the structure in organic (and biological) systems is augmented by the favourable ratio between the electron scattering factor of hydrogen and other light elements (C, O, N, *etc.*) when compared to the counterparts using X-ray or neutrons, making electron-based approach favourable for studying small molecular crystals that can be hard to crystallize in larger than micrometre sizes. The *ab initio* structure determination as outlined in eqn (11.10) can, however, only be retrieved from the imaging plane when multiple scattering effects are absent in a TEM (see Section 11.5). Without the correct imaging optics, the diffraction patterns obtained from the crystallography setup only record the probability intensity, given by $I_{hkl} = F_{hkl}F_{hkl}^*$. Since only the magnitude $|F_{hkl}|$ can be retrieved directly, the omission of the relevant phase factor ϕ $(F_{hkl}(\mathbf{s}) = |F_{hkl}(\mathbf{s})|e^{i\phi_{hkl}})$ prevents the *ab initio* structure determination directly from Fourier-inversion of the diffraction pattern.

To circumvent this phase problem, the crystallography approach typically relies on the refinement protocol to solve the molecular structures based on a plausible "structural model", which nonetheless could be well-informed by the pair distribution function (or Patterson map) that one obtains by conducting inverse Fourier transform of I_{hkl}. In UEC studies, the primary interest may lie in the dynamics for systems where the atomic structures have been solved (*i.e.*, the phase factor is known). In this case, the phase problem for the excited state can be reduced by studying the difference pattern $\Delta I_{hkl}(\mathbf{s}) = F'_{hkl} F'^*_{hkl} - F_{hkl}F^*_{hkl}$ to reduce the complexity of modelling.

To see how this is accomplished, one first considers the dynamical contrast from a known structure with the displacement vector $\Delta\mathbf{r}_{ij}$. We write the new form factor for the excited state following the displacements as $F'_{hkl} = \sum f_i e^{i\mathbf{G}_{hkl}\cdot(\mathbf{r}_i + \Delta\mathbf{r}_i)}$, and arrive at

$$\Delta I_{hkl} = 2\left[\sum_{i\neq j}\tilde{f}_i f_j\left(\cos\left(\mathbf{G}_{hkl}\cdot\left(\mathbf{r}_{ij} + \Delta\mathbf{r}_{ij}\right)\right) - \cos\left(\mathbf{G}_{hkl}\cdot\mathbf{r}_{ij}\right)\right)\right], \qquad (11.11)$$

with $\mathbf{r}_{ij} = \mathbf{r}_i - \mathbf{r}_j$. Since $\Delta\mathbf{r}_{ij} \ll \mathbf{r}_{ij}$, and for $s\cdot\Delta\mathbf{r}_{ij} \ll 1$, the equation simplifies to

$$\Delta I_{hkl} = -2\left[\sum_{i\neq j}f_i f_j \sin\left(\mathbf{G}_{hkl}\cdot\mathbf{r}_{ij}\right)\left(\mathbf{G}_{hkl}\cdot\Delta\mathbf{r}_{ij}\right)\right]. \qquad (11.12)$$

Given that $\sin(\mathbf{G}_{hkl}\cdot\mathbf{r}_{ij})$ is already known, the structure determination is reduced to refine a set of projected distances; for a system with a small number of atoms in a unit cell, this is quite a manageable problem. Notice here that the information content regarding $\Delta\mathbf{r}_{ij}$ is obtained over the long range along the reciprocal lattices, so one expects to obtain a better precision in the determination of small displacements when higher order Bragg peaks can be obtained. From a theoretical perspective aiming to understand the driving force, the underlying geometry of the free-energy landscape is encoded in the differential of the mean trajectory: $\Delta\mathbf{r}_{ij}(t)$ over time, which is similar to solving the dynamical molecular structures in gas phase UED experiments.[6,14] To track this global trajectory from diffraction patterns, one utilizes eqn (11.11) to derive the normalized change in the structure factor: $m_{hkl}(t) = \langle\Delta I_{hkl}(t)/I_{hkl}(t<0)\rangle$, with the bracket denoting the stochastic averaging along the trajectory, to compare with the experimental results. The damping factor pertaining to the stochastic deviations in the interatomic distances described in the gas phase diffraction is manifested here as the Debye–Waller factor (DWF) as also discussed in Section 11.3. A significant body of work in the transient molecular structure determination utilizing the femtosecond electron crystallography pioneered by Miller's group may constitute such an attempt for making "molecular movies" in the condensed matter systems. We refer the interested readers to relevant review articles.[79–81]

The goal of this UEC section, however, is geared towards an emerging area driven by recent significant efforts in the material and physics communities to understand the complex phenomena of phase transitions in quantum materials. In this context, unlike in typical crystallography, the interested entities are those beyond the crystalline unit cells; for example, the spontaneous formation of charge or spin-density waves, ferroelectric domains, or magnetic orders occurs upon exerting external influences across a critical point.[82–84] As these topics dwell on phenomena over a much larger length scale than the unit cells, the perfect ordering hypothesis, which is the foundation for conventional crystallography, breaks down. Nonetheless, as these large-scale phenomena derive their properties from the symmetry-breaking process on the microscale, the dynamics of phase transitions necessarily involve the emergence of correlated behaviour, which underpins the critical phenomena in the quantum many-body systems (see Chapter 2). The ultrafast coherent crystallography approach can simultaneously probe the relevant multiscale dynamics and is expected to play an essential role in understanding the emergent phenomena in different quantum materials.

11.4.2   Scattering from Structures with Long-range Broken-symmetry Orders

The quantum materials, sometimes also referred to as strongly correlated electron materials, are featured by their complex phase diagrams where multiple electronic phases often occur adjacently. While such complexity is rooted in the active interactions between multiple microscopic degrees of freedom—lattice, charge, spin, and orbital—competing ground states are of macroscopic nature characterized by broken symmetry[83,84] (see Chapter 2) and their evolutions over the external control parameters constitute the core problem in discussing phase transitions. Here, we describe how the UEC approach can capture the broken-symmetry "field" projected in the lattice parameters, notably in terms of breaking the translational or rotational symmetry of the underlying atomic lattices. In this section, we introduce a framework to treat ultrafast scattering from the nonequilibrium states of quantum materials. In this case, the broken-symmetry order expressed in the lattice field is given by a distinct structure factor $S_\eta(s, t)$ associated with the new broken-symmetry state with a characteristic ordering wav-vector $\mathbf{Q}_\eta$ and amplitude η_0 and phase ϕ_η described using a complex order parameter $\eta = \eta_0 e^{i(\mathbf{Q}_\eta \cdot \mathbf{r} + \phi_\eta)}$.

At the fundamental level, studying how a nonequilibrium many-body system spontaneously organizes into a broken-symmetry phase is a problem of broad interest from condensed matter[85–87] to high-energy physics.[88–90] The UEC pump–probe platform offers the opportunities for studying non-equilibrium physics of phase transition. More specifically, we ask how a quantum material containing long-range broken-symmetry states may effectively switch states under an ultrafast "quench"[91,92] enforced by a laser pulse in a reaction pathway distinctively different from a thermal state.[93,94] This scenario is not that different from femtochemistry[5] where the ultrafast electronic excitation sets the new bonding energy landscape before the molecular nuclear dynamics can follow. Given the separation of timescales, the impulsive unveiling of the new energy landscape sets forth the ensuing conformational dynamics wherein the dynamics of electrons follow those of the nuclei adiabatically. However, in the many-body systems, the system cannot order instantaneously. Such nonequilibrium many-body dynamics shall be captured directly in the evolution of the corresponding structure factors resolved in momentum, phase and amplitude.

11.4.2.1   Order Parameter Dynamics

Because the order parameter is defined at a different vector $\mathbf{Q}_\eta$, our main concern focuses on the new structure factor $S_\eta(\mathbf{k}; t)$ introduced in the reciprocal subspace by the broken-symmetry state with the momentum wavevector $\mathbf{k} = \mathbf{q} - \mathbf{Q}_\eta$ assigned to the new ordering field. As such, the dynamics of the many-body states are separable in the cryptographic sense from those of the crystalline lattice captured by the structure factor $S_G(\mathbf{q}; t)$ at $\mathbf{G}_{hkl}$. There has been a rich literature discussing the order-parameter dynamics in the context of the correlation function, modelled using the Landau–Ginzburg equation.[95] Taking the example of the charge-density wave (CDW) order, the appropriate long-wavelength properties of the CDW state are governed by the Landau-type effective Hamiltonian building on the CDW-order parameter field η:

$$H = \int dr \left(f_0(\eta) + \alpha_{ij} \frac{\partial \eta}{\partial x_i} \frac{\partial \eta}{\partial x_j} \right). \tag{11.13}$$

Here, $f_0(\eta)$ represents the free-energy density in the homogeneous state limit. The gradient term with rigidity tensor α_{ij} describes the energy cost in establishing an inhomogeneous domain where interfacial energy is given by $\alpha_{ij} \frac{\partial \eta}{\partial x_i} \frac{\partial \eta}{\partial x_j}$. Of particular interest is the η^4 – Mexican-hat potential for the Ising-like spontaneously broken-symmetry (SSB) field appropriate for describing the incommensurate CDW state free energy, given by[96–99]

$$f_0 = \frac{1}{2} A_2 |\eta|^2 + \frac{1}{4} A_4 |\eta|^4, \tag{11.14}$$

where $A_2 = a(x - x_c)$ with the distance to the critical point $(x - x_c)$ serving as the external control parameter to shape the free energy across x_c. This field model is connected with the crystallographic probe by means of the correlation function of the field parameter given by the structure factor. In the dynamical context, the ordering field $\eta(r, t)$ is mapped into the symmetry-breaking lattice field $u(r, t)$ with the relationship $u_\eta = \frac{\kappa}{2}(\eta + \text{c.c.})$ where $\kappa \sim 0.1–0.2$ Å is the typical coefficient.[100] Then, the two-point correlation function of the field is given by the structure factor $S_{\eta\eta}(\mathbf{k}) = \kappa^2 \langle \eta_\mathbf{k} \eta_\mathbf{k}^* \rangle$. This connection is commonly discussed in the thermal transition case for CDW.[101] Near the critical point T_c, the correlation in the field is strengthened when approached from both sides of the transition (see Figure 11.6).[100]

Above T_c, the field potential minimum is zero, i.e., $\langle \eta \rangle = 0$. Nonetheless, the enhanced correlation is described by the two-point correlation function $\langle \eta_\mathbf{k} \eta_\mathbf{k}^* \rangle = \frac{k_B T}{V} \left(\frac{1}{f_{\eta\eta} + \alpha_{ij} k^2} \right)$, which is a Lorentzian with fluctuating field mode $\eta_\mathbf{k}$ assuming a Boltzmann distribution in the system volume V. Here, the field mode line-width is given by $\Delta k_L = \left(\frac{a(T - T_c)}{\alpha_{ij}} \right)^{1/2}$. One further observes that as T_c is approached, the line width gets narrower, which is an indicator that the fluctuations at long wavelength become more and more

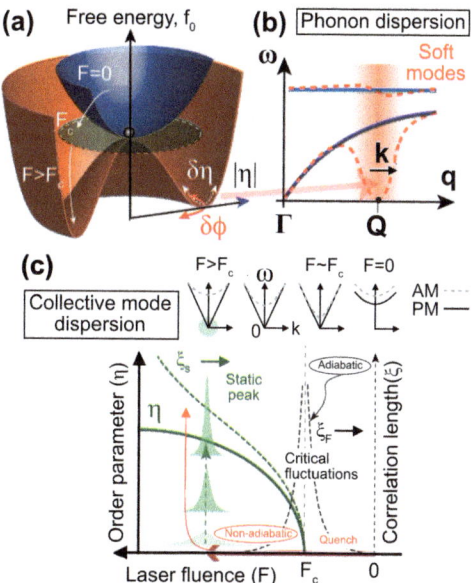

Figure 11.6 Landau–Ginzburg free-energy surface for spontaneous symmetry breaking and the order parameter dynamics. (a) The 2D free-energy landscape
that defines the order parameter field at different temperature. The
arrow directions show a temperature quench. The field fluctuations are
depicted in the change of the amplitude ($\delta\eta$) and the phase ($\delta\varphi$). (b) The
corresponding lattice phonon dispersion curves that couple to the
landscape modification. The lattice softening, driven by the temperature quench, occurs at momentum wavevector **Q** of the long-range state.
(c) The phase ordering kinetics orchestrated by the fluctuation fields.
The near-equilibrium scenario is depicted in black curves. The non-
equilibrium one, depicted in green, is driven by a deep quench, where
$T \ll T_c$. The dispersion curves for the amplitude mode (AM) and phase
mode (PM) of the CDW state are depicted at the top. Reproduced
from ref. 100, https://doi.org/10.5802/crphys.86, under the terms of the
CC BY 4.0 license https://creativecommons.org/licenses/by/4.0/.

correlated, a precursor for the phase transition. Below T_c, the correlation
in the fluctuation wave also exists but in a slightly different form [$\Delta k_L =
(2a(T_c - T)/\alpha_{ij})^{1/2}$]. However, since the field can now support a static phase, the
scattering factor will have two components: the static order parameter with a
temperature-dependent amplitude $\eta_0(T) = \sqrt{\dfrac{a}{A_4}}(T_c - T)^{1/2}$ and the diffuse
scattering spreads over the field mode spectrum characterized by finite momentum **k** (see Figure 11.6c). The corresponding partial differential equation for
the dynamics of the CDW order parameter will follow the time-dependent
Landau–Ginzburg (TD–LG) equation (also referred to as the model A dynamics in
the Hohenberg–Halperin classification or Langevin equation),[102–104] given by

$$\frac{\partial \eta(\mathbf{r}, t)}{\partial t} = -\frac{1}{\gamma}\frac{\partial f_0}{\partial \eta} - \alpha_{ij}\frac{\partial \eta}{\partial x_i}\frac{\partial \eta}{\partial x_j} + \varsigma(\mathbf{r}, t), \tag{11.15}$$

where γ is the phenomenological time constant and $\varsigma(r, t)$ is the Gaussian white noise. The model is most effective in describing the relaxational quench dynamics of the order parameter.[85] However, the equation of motion can include the system reactive responses in the coherent regime, including the spontaneously driven amplitude modes in CDW.[105] For such a treatment, a phenomenological mass term (β) is added with the new dynamical equation given by[105]

$$\beta \frac{\partial^2 \eta}{\partial^2 t} + \gamma \frac{\partial \eta}{\partial t} = -\frac{\partial f_0}{\partial \eta} - \alpha_{ij} \frac{\partial \eta}{\partial x_i} \frac{\partial \eta}{\partial x_j} + \varsigma(\mathbf{r}, t), \qquad (11.16)$$

where the characteristic frequency $\omega_0 = \left(\frac{k}{\beta}\right)^{1/2}$ is defined at the local minimum with potential $\tilde{f}_0 = \frac{k}{2}(\eta - \eta_0)^2$. Over the timescale $t \gg \gamma$ where the damping of the coherent dynamics kicks in, the system evolution becomes relaxational and the evolution is veered into the Langevin dynamics.[102–104]

11.4.2.2 *A Prototypical Example of a Nonequilibrium Phase Transition: The Rare-earth Tritelluride System*

A prototypical example of the nonequilibrium spontaneous symmetry breaking investigated through the UEC approach is the rare-earth tritelluride (RTe_3) compound which has a unit cell structure of alternating layers of square Te lattice planes and coupled RTe slabs (see Figure 11.7a).

The near perfect $C4$ symmetry of the Te plane exposes this system to in-plane charge-density wave (CDW) formations along the crystallographic c and a-axes.[106–108] However, due to the weak coupling between the two non-equivalent square Te nets,[109] the weak preference of symmetry-breaking typically drives the system into a c-CDW state (*i.e.*, a stripe phase along the c-axis, upon lowering the temperature below the critical temperature, T_c). This generic single-phase-dominated feature is commonly observed across the equilibrium phase diagram for the RTe_3 family (see Figure 11.7b). This is why the recent discoveries of a checkerboard phase upon light excitation in the light members of RTe_3 ($LaTe_3$[110] and $CeTe_3$[111]) that do not normally host the a-CDW phase has sparked interest in understanding the nonequilibrium symmetry breaking process that makes this transition possible.

Here, we show the simple nonequilibrium field perturbation induced from ultrafast laser quench described in the TD–LG framework can easily capture the surprising a-CDW state formation and the essential dynamical features as observed by UEC. In TD–LG modelling, first we consider two independent SSB fields, η_c and η_a, that share the common T_c based on the C_4 symmetry. However, the two ordering fields vying for the spectroscopic weight become unstable near T_c. Indeed, this competitive scenario has been evidenced by inelastic X-ray scattering probing the pre-transitional field

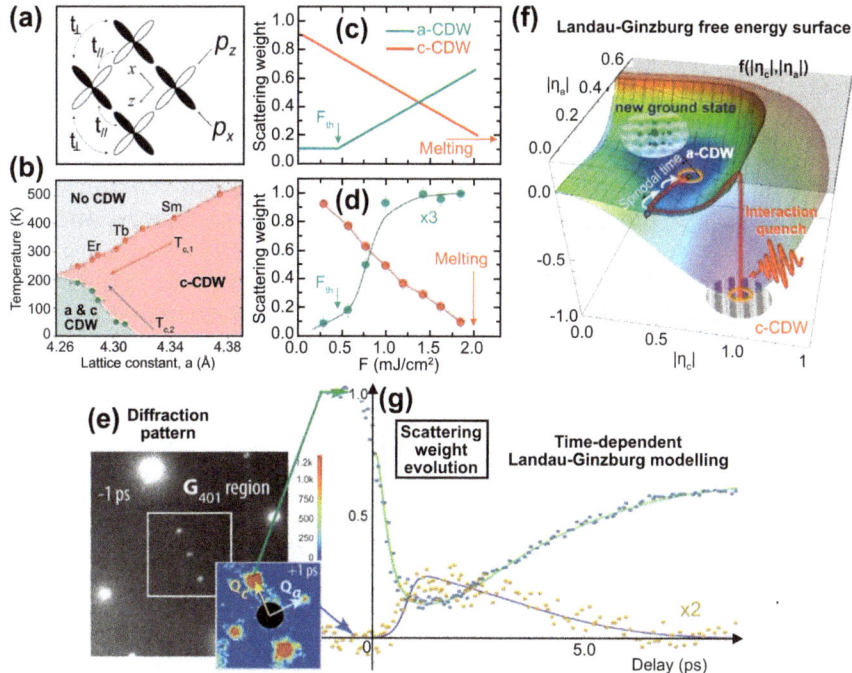

Figure 11.7 TD–LG modelling of the order-parameter dynamics involving two CDW orders in CeTe$_3$. (a) The frontier orbitals made up by p$_x$ and p$_z$ with inter-orbital coupling $t_\parallel$ and $t_\perp$ responsible for the Fermi surface. (b) The phase diagram of ReTe$_3$. (c and d) The theoretically predicted and measured transition curves. (e) Diffraction patterns before (−1 ps; black/white) and after (+1 ps; coloured) applying laser pulses. (f) The unfolding of the energy landscape upon laser quench. (g) The data and TD–LG trajectory calculations for the dual-state order-parameter evolutions.[111] Reproduced from ref. 100, https://doi.org/10.5802/crphys.86, under the terms of the CC BY 4.0 license https://creativecommons.org/licenses/by/4.0 and from ref. 111, https://doi.org/10.1038/s41467-020-20834-5, under the terms of the CC BY 4.0 license https://creativecommons.org/licenses/by/4.0.

fluctuations.[112,113] With the two fields repulsively coupled, one can write a single free-energy equation for the coupled fields[109,114] as

$$f_0(|\eta_c|, |\eta_a|) = \sum_{l=c,a} \left(\frac{1}{2} a_l \left(T^{(l)} - T_c \right) |\eta_l|^2 + \frac{1}{4} A_l |\eta_l|^4 \right) + \frac{1}{2} \tilde{A} |\eta_c|^2 |\eta_a|^2, \quad (11.17)$$

with coupling $\tilde{A} |\eta_c|^2 |\eta_a|^2$ ($\tilde{A} > 0$). We note that this scenario appears to be rather common in quantum materials hosting multiple broken-symmetry ground states.[115–118] So, the approaches described here should be applicable to these systems as well. Before considering the effect of ultrafast quenching, one could already notice that in a competitively coupled scenario, once the system adopts a single dominant ground state, the effective critical point for the competing phase is renormalized by an amount $-\dfrac{\tilde{A}}{a_a}|\eta_c|^2$, following eqn (11.17). This leads to a renormalizable free-energy landscape with a new critical point $T_c^* = T_c - \dfrac{\tilde{A}}{a_a}|\eta_c|^2$ for the competing a-CDW phase. In our experiment with CeTe$_3$, $T_c \sim 500$ K and accordingly we can reproduce the equilibrium phase diagram with a selection of Landau parameters and initial state $\eta_c = 1$; in doing so, the T_c^* projected for a-CDW is at a negative temperature consistent with Figure 11.7b, thus precluding it from the phase diagram. We now consider the scenario under an ultrafast laser quench. With the same parametrization of repulsive coupling, one intriguingly finds that if η_c is suppressed to 75% under the nonequilibrium condition, the T_c^* for a-CDW field can be raised to above room temperature, hence introducing a-CDW (see Figure 11.7c). Remarkably, this laser-quench mediated phase diagram for the two competitive phases is reproduced experimentally (see Figure 11.7d).

We can now solve the dynamical equation of motion for the two CDW states based on eqn (11.16). In Figure 11.7f, we plot the free energies before and after applying laser quench with a fluence $F = 1.85$ mJ cm^{-2}. Indeed, the introduced energy landscape following quench now contains a local minimum that supports both a- and c-CDW orders. The solutions from the dynamical equation are given for the trajectories of $\eta_c(t)$ and $\eta_a(t)$ in Figure 11.7g where we compare the data (symbols) and the theoretical predictions (lines). We find that an agreement between the two can be achieved by assigning an appropriate damping time τ_e of ~ 4.5 ps (see Figure 11.7g). The fluctuation effect from the quench is absorbed at the effective temperatures of the two systems. Without knowing how the system might thermalize microscopically, we solve the equation of motion with a temperature profile evolution $T^{(l)}(t)$ for the two states based on the effective relaxation time τ_e (for thermalization). Further justification on the different field effective temperature evolution is backed by the measurements of the fluctuation wave amplitudes; see ref. 111 and the discussion below. One noticeable characteristic reproduced by solving the TD–LG is the two-step dynamical sequence with rapid downhill quench of the c-CDW phase followed by a delayed build-up of the a-CDW phase. The dynamics here are given primarily for the integrated intensity of the structure factors, which are mapped into the Landau–Ginzburg equation $m_l(t) = \eta_l^2(t)$. The experimental measurements of the structure factor profile evolution give further details on the field integrity, measured in terms of the correlation length ξ_c and ξ_a (see Figure 11.8).

The success of this phenomenological modelling provides a clear demonstration for the landscape-driven dynamics on the non-equilibrium timescale, and hence the controllability, for CDW systems. The reduction of the central peak intensity as well as the correlation length ξ_c here directly reflects a rapid destruction of the predominant static order parameter into

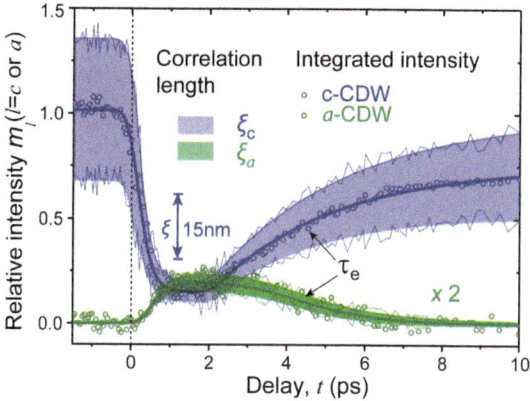

Figure 11.8   A comparison between the evolution of the static c-CDW and the emerging a-CDW order parameters based on integrated intensity. The correlation length (ξ) of each state is also presented in the shaded envelop along with the order parameter evolution. Reproduced from ref. 111, https://doi.org/10.1038/s41467-020-20834-5, under the terms of the CC BY 4.0 license https://creativecommons.org/licenses/by/4.0.

fluctuating field modes (retrieved from the structure factor at $\mathbf{Q}_c$ – a signature for a nearly instantaneous creation of a new free energy basin). It is interesting to note that the field-mode fluctuations are also excited along the a-CDW field, and such field modes eventually condense on the new free-energy basin around (η_a, η_c) minimum (see Figure 11.7f) through condensation growth, evidenced from the increasing $m_a(t)$ and $\xi_a(t)$ in Figure 11.8.

11.4.3   The Microscopic Picture

While the coarse-grained phenomenological model could capture the system evolution, more microscopic details can be extracted by analyzing the $\mathbf{q}$-resolved dynamics encoded in the different structure factors. Below, we illustrate how such fine details could inform the fundamental physics behind the development of the specific order-parameter field.

11.4.3.1   *Basic Principle*

In order to further deduce from the crystallographic data the relevant microscopic dynamics, such as the lattice vibrations derived from the random stochastic phonons or those associated with CDW field-mode fluctuations sensitive to the free-energy landscape, we first need to learn how to identify these respective features in the structure factors. To examine these effects, we reformulate the lattice system form factor with the CDW element, written as

$$F(\mathbf{s}) = \sum f_L \delta(\mathbf{r} - \mathbf{L} - u_L(t)) e^{i\mathbf{s}\cdot\mathbf{r}} d\mathbf{r}, \tag{11.18}$$

where $f_L = \sum_i \rho_i e^{-i s \cdot \varrho_i}$, and ϱ_i is the unmodified mean relative position of the atom in the unit cell representing the form factor of the undistorted lattice. Specifically, here $\mathbf{u}_L = \mathbf{u}_q + \mathbf{u}_\eta$ considers two different types of lattice field: the lattice phonons ($\mathbf{u}_q$) or collective field modes of the broken-symmetry phase ($\mathbf{u}_\eta$) with momentum wavevector $\{\mathbf{q}\}$ and $\{\mathbf{k}\}$ given by their relative location to the respective Bragg scattering centered at G_{hkl} and $G_{hkl} + Q$ (see Figure 11.7e), *i.e.*, $\mathbf{q} = s - G_{hkl}$ and $\mathbf{k} = s - G_{hkl} - Q$. To encode the lattice vibration effects into the structure factor, we first express the displacement field coupling to the phonon branches by $\mathbf{u}_q(\mathbf{r}, t) = \sum_q u_{q,0} \hat{\mathbf{e}}_q \sin\left(\mathbf{q}\cdot\mathbf{r} - \omega_q t + \phi_q\right)$.

Here, the phonon mode spectrum is governed by the dispersion curves $\omega_q(\mathbf{q})$. The spectral integration across the dispersion curve assuming Boltzmann distribution leads to a damping factor e^{-M_q}, with $M_q = \sum_q \frac{1}{4}\left(s \cdot \mathbf{u}_{q,0}\right)^2$, in the form factor.

With this correction factor and the renormalization term due to the CDW distortive field, the structure factor for the main lattice Bragg peak is given by

$$S_{G_{hkl}}(\mathbf{s}) = FF^* = \delta(\mathbf{s} - G_{hkl})e^{-2M_q}\left|f_L J_0\left(\mathbf{s} \cdot \mathbf{u}_\eta\right)\right|^2, \tag{11.19}$$

where e^{-2M_q} is the Debye-Waller factor (DWF). Secondly, for the consideration of fluctuating field modes, we write $\mathbf{u}_\eta(r, t) = \mathbf{u}_{0,\eta}\hat{\mathbf{e}}_\eta(1 + \delta\hat{A}(r, t))\sin[Q\cdot\mathbf{r} + \delta\phi(r, t)]$ to include the dynamical components $\delta\hat{A}(r, t)$ and $\delta\phi(r, t)$, representing the amplitude and phase modes in the field equation. Similar to the phonons, these dynamical components, referred to as fluctuation waves (to distinguish them from the phonons), can be expanded over the momentum spectra: $\delta\hat{A}(\mathbf{r}, t) = \sum_{k'} \hat{A}_{0,k'} \sin(\mathbf{k}' \cdot \mathbf{r} - \omega_{k'}t)$ and $\delta\phi(\mathbf{r}, t) = \sum_k \phi_{0,k} \sin(\mathbf{k} \cdot \mathbf{r} - \omega_k t)$. One can show that the spectral integration over the respective dispersion curves (see Figure 11.6c) produces a damping effect on the superlattice intensity very similarly to the role of DWF on the lattice Bragg intensities. Specifically, to see their impacts on the super-lattice structure factor, we give the form factor that includes the amplitude field modes with momentum wavevector $\mathbf{k}$ (explicit to the second order contribution):[100]

$$F(\mathbf{s}; \mathbf{k}) = e^{-M_q} \sum_{L,n} e^{-i(\mathbf{s} + n\mathbf{Q})\cdot L} J_n\left(\mathbf{s} \cdot \mathbf{u}_\eta\right)\left\{1 + \frac{|n|}{2i}\eta_k\left[e^{i(\mathbf{k}\cdot L - \omega_k t)} + e^{-i(\mathbf{k}\cdot L - \omega_k t)}\right]\right.$$
$$\left. - \frac{|n|(|n| - 1)}{8}\eta_k{}^2\left[2 + e^{i(2\mathbf{k}\cdot L - 2\omega_k t)} + e^{-i(2\mathbf{k}\cdot L - 2\omega_k t)}\right] + \cdots\right\} \tag{11.20}$$

One can see that the respective damping factor contribution from the specific field mode at momentum $\mathbf{k}$ is derived from wave summing carried out from the Jacobi–Anger generating function expansion over the Bessel function J_n, which transfers the scattering weight from the central static peak at $F(\mathbf{s}; \mathbf{k}=0)$ to the spectrum at $\mathbf{k}$ in the diffuse scattering factor $F'(\mathbf{s}; \mathbf{k})$. The effect leads to peak broadening in the structure factor $S_Q(\mathbf{k})$. A similar argument can be made for the phase modes.[100] The impact from these modification factors cannot be treated as the regular thermal phonon excitations which typically constitute a weak diffusive background. Due to the correlated nature of the field fluctuations, especially when excited near a critical point, field modes typically render a Lorentzian profile comparable to the central static peak and from which the critical correlation length ξ_F can be deduced from the linewidth (see Figure 11.6). The derivation here shows that these field mode contributions are an integral part of the total structure factor, *i.e.*, the integrated intensity calculated for the total scattering at $\mathbf{Q}$ (including both the static component and diffuse field modes). Thereby, exciting these field modes does not impact the strength of the lattice Bragg peak identified at $\mathbf{G}_{hkl}$. By carrying out the momentum integration for the respective structure factors, one can evaluate the impacts of the fluctuations on the experimentally observed results. For the main lattice, one obtains

$$m_{\mathbf{G}_{hkl}} = \int S_{\mathbf{G}_{hkl}}(\mathbf{s})d\mathbf{s} = |f_{\mathrm{L}}|^2 e^{-2M_q} \prod_l |J_0(\mathbf{G}_{hkl} \cdot \mathbf{u}_{0,\,\eta_l})|^2, \qquad (11.21)$$

with l indexing the presence of multiple broken-symmetry orders within the Brillouin zone (BZ). Similarly, for the superlattices, the integration gives:

$$m_{\mathbf{Q}_l} = \int S_{\mathbf{Q}_l}(\mathbf{s})d\mathbf{s}f = |f_{\mathrm{L}}|^2 e^{-2M_q} |J_1(\mathbf{G}_{hkl} \cdot \mathbf{u}_{0,\,\eta_l})|^2. \qquad (11.22)$$

Based on eqn (11.21) and (11.22), two observations can be made relating the correlated and uncorrelated effects:

(i) The two expressions carry the same DWF e^{-2M_q}. This simply means that $S_{\mathbf{G}_{hkl}}$ and $S_{\mathbf{Q}_l}$ deduced from the same BZ pick up the same incoherent effects from lattice vibrations.

(ii) The two intensities carry complementary Bessel functions J_0 and J_1 with the argument given by the projected order-parameter field along the respective reciprocal momentum. This indicates that a complementary exchange of scattering weight occurs between $S_{\mathbf{G}_{hkl}}$ and $S_{\mathbf{Q}_l}$ whenever the CDW ordering field is adjusted.

We can designate the DWF as exclusively from exciting independent vibrational modes and relate the mean-square (MS) value of incoherent lattice vibrations to M_q given by

$$u_{hkl}^2 = 2M_q / |\mathbf{G}_{hkl}|^2. \qquad (11.23)$$

We note in this designation, the DWF becomes a true indicator for the temperature only when the equipartition principle applies. However, the equation equally applies to the nonequilibrium regime with a caveat that $\langle \Delta u_{hkl}{}^2 \rangle$ represents a stochastic amplitude averaged from all active modes, which does not require satisfying the Boltzmann statistics. In this case, the DWF might be used to gauge the kinetic energy flowing into the subsystem manifold as an effective local temperature.

Of concern is how one can decouple this DWF from the symmetry-associated contribution pertaining to the phase transition. To this end, with appropriate consideration of multi-Q contributions, one can independently obtain the respective order parameter dynamics *via* evaluating $h(t) = \dfrac{m_{Q_l}(t)}{m_G(t)}$ and $g(t) = \dfrac{m_{Q_a}(t)}{m_{Q_c}(t)}$, where the contribution from the DWF is eliminated. Specifically,

$$h(t) = \frac{\left| J_1\left(\mathbf{G}_{hkl} \cdot \hat{\mathbf{e}}_{\eta_l} u_{0,\eta_l}(t)\right) \right|^2}{\prod_l \left| J_0\left(\mathbf{G}_{hkl} \cdot \hat{\mathbf{e}}_{\eta_l} u_{0,\eta_l}(t)\right) \right|^2} \tag{11.24}$$

and

$$g(t) = \left| \frac{J_1\left(\mathbf{G}_{hkl} \cdot \hat{\mathbf{e}}_{\eta_a} u_{0,\eta_a}\right)}{J_1\left(\mathbf{G}_{hkl} \cdot \hat{\mathbf{e}}_{\eta_c} u_{0,\eta_c}\right)} \right|^2 , \tag{11.25}$$

Given the polarization of the CDW state $\hat{\mathbf{e}}_{\eta_l}$, the order parameter $u_{0,\eta_l}(t)$ can be retrieved and used to deduce the DWF at $\mathbf{G}_{hkl}$.

11.4.3.2   *Experimental Observations*

In the following, we describe how one may successfully extract the useful microscopic details from the evolution of structure factors. We use the crystalline lattice and order-parameter field momentum designators in terms of $\{\mathbf{q}\}$ and $\{\mathbf{k}\}$ to differentiate the two types of lattice vibrations. In Figure 11.9a, we pictorially depict these processes in the spontaneous symmetry-breaking phase transition for establishing the new a-CDW upon laser quench. We take the viewpoint from the a-CDW field for observing the event sequences, where upon first receiving the laser quench, the general underpinning lattice is excited with fluctuations given by $\langle \Delta u_{\mathrm{L}}{}^2 \rangle = \langle \Delta u_{\mathrm{q}}{}^2 \rangle + \langle \Delta u_{\mathrm{k}}{}^2 \rangle$ if the excitations in the two subsystems are considered as independent. In this case, the fluctuation effects (manifested in DWF) grow exponentially as the lattice heats up – on the timescale of electron-phonon coupling. Meanwhile, the field fluctuations under a new free-energy landscape (see Figure 11.8) now allows the new order parameter in the a-CDW field to build up, coalescing at the new field minimum. According to eqn (11.21), the

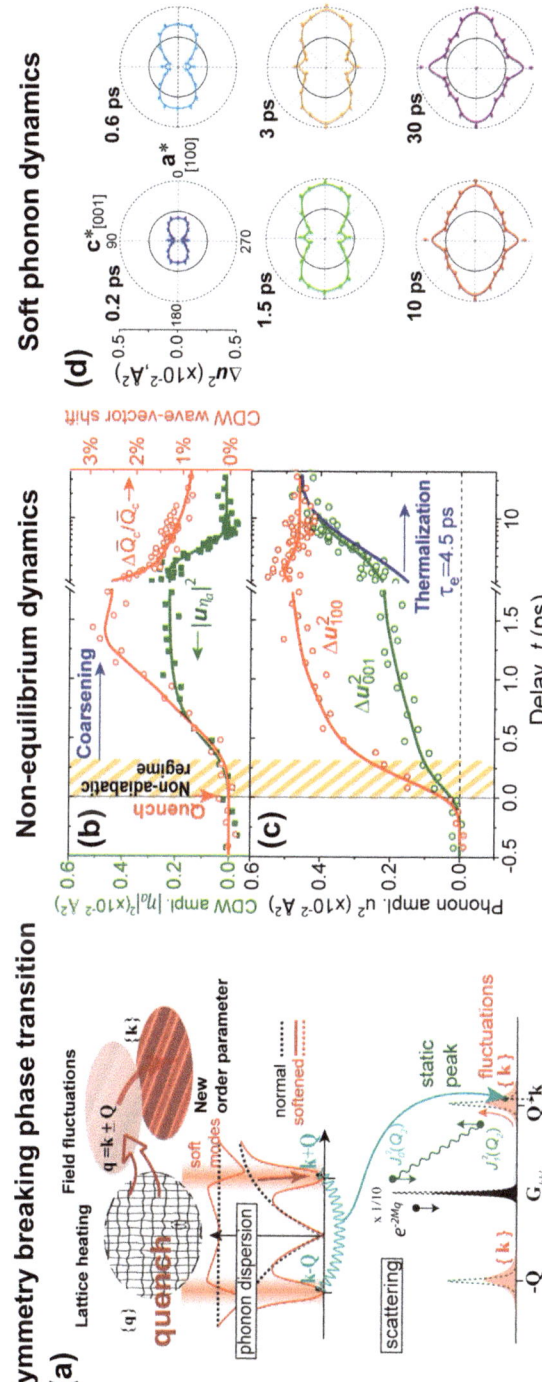

Figure 11.9 Scattering weight redistribution from exciting phonons and fluctuation waves during quantum material phase transitions. (a) (top) The anticipated change in the phonon dispersion curves associated with a symmetry-breaking phase transition to form the a-CDW state. The phonon modes near the wavevector ($\mathbf{Q}$) of the emerging a-CDW is softened (shaded area) upon laser quench. Scattering weight transfer between a lattice Bragg peak (S_G) and CDW satellite (S_Q) at different stages of symmetry-breaking phase transformation. The process starts with the excitation of phonons upon laser quench; the transfer of scattering weight from the lattice peak (S_G) to the CDW (S_Q) peak due to pairing of phonons; and finally the nonequilibrium phase ordering. (b) The order parameter field evolution examined via $|u_{0,\eta_a}|^2$ and $\Delta\bar{\mathbf{Q}}_c/\bar{\mathbf{Q}}_c$ respectively for a- and c-CDWs. Here, $\bar{\mathbf{Q}}_c$ is the mean wavevector including c- and c^+ contributions. (c) The vibrational MS phonon amplitude changes projected along [001] and [100] were obtained from the Debye–Waller analysis. (d) UEC characterization of soft mode excitations upon laser-induced broken-symmetry phase transitions in $CeTe_3$. The mean-squared (MS) phonon amplitude changes obtained using crystalline Bragg peaks at $|G_{hkl}| \sim 6\ r.l.u.$ shown in polar coordinates, indicating strong anisotropy. Reproduced from ref. 111, https://doi.org/10.1038/s41467-020-20834-5, under the terms of the CC BY 4.0 license https://creativecommons.org/licenses/by/4.0.

process decreases $m_{G_{hkl}}$ in the direction of the field mode polarization and promotes m_{Q_l} at the respective $\mathbf{Q}_l$ from the coherent factors J_0 and J_1. Additionally, the incoherently excited lattice phonons supress both intensities in a way not specific to the orientation, given by $\langle \Delta u_q^2 \rangle$.

11.4.3.2.1 Inter-system Vibrational Energy Redistribution and Nonthermal Evolution of Order.

Now we apply these scattering-weight-transfer rules developed based on the adiabatic process to understand the kinetic energy flows into different subsystem manifolds and the nonequilibrium formation of the CDW order as observed through UEC experiments on CeTe$_3$. In Figure 11.9b and c, the respective buildups of the incoherent soft modes connected with the broken-symmetry phase transition are extracted along with the new order parameter (designated as the static amplitude $|u_{\eta_a}|^2$) and they are plotted as a function of time to understand the nonadiabatic and athermal nature of the phase transition. Specifically, an evaluation of the kinetic energy flowing into the two CDW subsystems following the quench is given by observing the lattice soft phonon amplitude dynamics along two perpendicular directions in Figure 11.9c. Here, one can clearly identify the prompt excitations of the lattice modes in the two CDW fields and their continuing growth well after the formation of the new CDW order. Taking the MS amplitude identified here to be representative of the effective field temperature $T^{(l)}$, and in this case $T^{(c)}$ and $T^{(a)}$ evolve separately, we may conclude the formation of the a-CDW proceeds under a highly nonequilibrium condition where $T^{(c)} \gg T^{(a)}$. Correspondingly, the timescale where the two amplitudes become equal might be ascribed as the thermalization time between the two order parameter fields.[111]

Intriguingly, here a well distinguished delay in the formation of the new order against creating the corresponding field-mode fluctuations is observed. This behaviour apparently violates the adiabatic scattering weight transfer rules given above, *i.e.*, the suppression of S_Q that is field mode-related should spontaneously translate to a growth of S_G. This breakdown of the adiabatic scattering rules observed here reflects that there is a hidden energy transduction step that is not taken into account in the scattering weight transfer mechanism.

We characterize this process as an inter-system vibrational coupling, mediated by a nonlinear two-phonon process that combines the counterpropagating acoustic phonons into a standing-wave field mode. The very notion was first suggested by Overhauser[119,120] in the context of charge-density waves in chromium. The process is depicted in Figure 11.9a (green curves) for illustrating this process in the symmetry-breaking phase transition. Here, the interconversion between the phonon modes and the fluctuation wave channels is described as a "coherent superposition" of two phonon modes having wavevector $\mathbf{k} + \mathbf{Q}$ and $\mathbf{k} - \mathbf{Q}$ to create an intermediate field mode at momentum $\mathbf{k}$. As the intermediate field modes have yet to develop the phase coherence necessary to be integrated into the new

CDW system, the scattering from such modes is only picked up in the lattice structure factor, but not in the CDW structure factor, as observed in the early timescale.

Following this, one might contemplate utilizing the nonadiabatic soft phonons to probe the driving force behind the formation of the CDW in RTe_3. In Figure 11.9d, we analyse the field phonon fluctuations following the quench along different azimuthal angles by utilizing the higher order Bragg reflections,[111] where a larger contrast ratio is available for such perturbative dynamics; see Section 11.4.1. From the soft-phonon MS amplitude contour, we first recognize that the most excited phonon fields adopt a two-fold lobe directed initially in the direction of a^*, witnessing the perturbative responses from suppressing the existing c-CDW. Similarly, accompanying the formation of a-CDW, the two-fold lobe along the c^* axis also emerges with a smaller initial lobe amplitude consistent with the magnitude of the new order. The results here clearly show the precursive soft phonons connected to the field modes to be transverse modes, *i.e.*, the phonon polarization vector is perpendicular to the respective Q_l. We note, such field mode designation is consistent with the recent studies from a joint X-ray inelastic scattering and DFT.[112] Interestingly, a recent Raman study of the phonon spectra around the critical point of $GdTe_3$ and $LaTe_3$ also shows a similar CDW soft-mode azimuthal distribution (they designated that to the Higgs or amplitude modes),[121] which is clearly distinguished from that of the lobes associated with the phonon mode that instead shows the p_x–p_z orbital four-fold symmetry (see Figure 11.7a) and does not undergo significant change during phase transition. Fundamentally, the signals isolated in Figure 11.9d by UEC are soft phonons coupled to such Higgs modes, driven by the two-phonon process; this coupling may be mediated by the quantum process of nesting at a singular nesting wavevector (Q and $-Q$) hotspot. A caveat that emerges from the cross-polarization Raman studies is that the CDW composition in RTe_3 might involve the wave-vectors at Q_c and $c^* - Q_c$, which were previously identified based on two alternative Fermi surface nesting scenarios,[109] leading to two quantum interference pathways instead of just one. Similar investigation employing UEC instead of Raman might shed important light on the origin question and the physical manifestation of the complex CDW fields.

11.5 Ultrafast Electron Microscopy

Transmission electron microscopy (TEM) systems have been one of the most utilized characterization tools for research in material sciences and nano-technologies. They stand out as genuine multi-modal platforms with the embodiment of extensive electron-optical components, which, unlike those employed for X-rays or neutrons, are entirely field-based. The direct field control of optical systems completely forgoes the need to physically adjust distances to maintain focusing and tuning under a variety of application circumstances. This flexibility allows one to create a highly adaptable system for optimizing the TEM for diffraction and direct imaging modalities that could overcome the phase problem. In addition, the ability to select inelastically scattered electrons for element-sensitive microanalysis enables analytical works that directly relate the structure and properties together. It is this highly integrated feature that makes TEM a uniquely powerful instrument in addressing many complex issues that singularly specialized toolsets might not fully grasp. Historically, TEM systems have been consistently refreshed with the latest developments in source, specimen manipulation, and detector technologies to address rising scientific and technological challenges.[122] Incorporating high-temporal resolution into a continuous wave (CW) TEM represents a significant next-step forward,[122] establishing an ultrafast electron microscope (UEM) which will be a breakthrough in the ultrafast science toolset development, uniting structure, property, and dynamics characterizations.[123]

Among the more specialized ultrafast electron-based setups discussed in earlier sections and other chapters in this book, UEC and UEM setups stand closest in how the probe interacts with the specimen, and share many core technologies in the source/illuminating systems and detector technologies to boost the sensitivity (see Section 11.6 for more details). Yet, the richer information obtained through multimodalities enabled by UEM relies on isolating information contents often by slicing off only a narrow part of the scattered particle phase space. These advanced features thrive on balancing between resolution and the SNR achievable. It is thus easy to appreciate that obtaining the fullest utilities from a UEM system, an increased current efficiency in the illuminating beam is the most central element.[123] Nonetheless, the pervasive space-charge issue associated with any increase in the number of particles in the probe beam forces the UEM development to bifurcate into two branches. In the first branch, which is pioneered by Zewail *et al.*,[124,125] one triggers the photoemission from a tip with a low intensity femtosecond laser pulse to generate a wave packet of a photoelectron. Since it operates on the one-electron-at-a-time paradigm also held in conventional TEM, it elegantly maintains much of the characteristics that traditional TEMs offer, such as the emitter brightness and the electron optics, and makes up the required currency by operating at a very high repetition rate (f_{pp} ranges from 1–100 MHz)[125] (see Section 11.6). The stroboscopic approach is poised to advance the fields of nanophotonics,[126] plasmonics, and micro-analysis,[107] where a high f_{pp} is well-suited to these studies by taking advantage of the combined high temporal (≤ 1 ps) and spatial (≤ 1 nm) resolutions. In the second branch, with the aim of addressing irreversible changes occurring in materials processes, single-shot operation is necessary.[122] To meet this goal, the dynamical TEM (DTEM)[127] development employs an extremely high bunch charge (up to 10^9 electrons) while side-stepping the space-charge effects through using a much longer pulse (10–100 ns) to extract the pulses, effectively maintaining the single-particle-stream picture at a relatively high beam energy (*e.g.*, 200 keV). The DTEM system or its mode-of-operation have been adopted for *in situ* studies of irreversible reaction fronts[128] and shock-induced material deformation.[129]

For practical consideration, there are lots of opportunities left in the middle of the two opposite ends of time-resolved electron microscope technologies, which may be accessed by overcoming the space-charge effects face-on with a new design of microscope optics taking advantage of the relevant technologies for phase-space manipulation long employed in the design of electron linear accelerators.[130] Specifically, utilising high-bunch-charge electron sources, the new ultrafast microscope platform employs a radio frequency (RF) cavity as a longitudinal condenser lens to tune the space-charge-dominated phase-space structure of a relatively long bunch to refocus the electron bunches in the energy or time domains. Conceptually, this is equivalent to the transverse refocussing performed using the conventional magnetic lenses for achieving tight cross-over or higher coherence (smaller angular spread).[131] The synergistically combined optical adjustments will allow the phase space structures of the incidence bunches to be realigned for the best performance of diffraction, imaging, and spectroscopy without sacrificing the beam dose. We note separate technological advances also adopted the RF cavity in a TEM for slicing the beam to significantly improve the temporal resolution and enabling new types of spectroscopy, see Chapter 15 (by Borrelli *et al.*). The high bunch charge approach gives the new UEM system the flexibility to operate in a broad range of repetition rates while maintaining a very high temporal resolution.

To see how ultrafast diffraction techniques mesh with imaging and spectroscopy under high bunch charge limits, we first discuss the core elements of the imaging principles building on the finite phase space area of the high-density bunch created in the new UEM system.

11.5.1 Multi-purpose UEM System

The major potential advantage in using electron scattering data, compared to X-ray or neutron diffraction, is that electrons can be focused using an electromagnetic lens; thus, both images and diffraction patterns can be utilized for structural analysis under the same platform. Importantly, the different modalities are controlled by the post-specimen optics and aperture placement (see Figure 11.10) with additional sets of (intermediate and projection) lenses. Each lens has its own image and back focal planes. The lens combination offers great flexibility in creating either image or diffraction signals from a large field of view or data for the microanalysis of small regions, through employing appropriate apertures on intermediate imaging/diffraction planes. Below, we illustrate the basic electron-optical concepts that play a key role in achieving multimodality and how they are applied in UEM.

11.5.1.1 Electron Optics: Fourier Pairs

One key concept uniting the different resolution concepts in the optical arrangements is the Fourier transform pairs[132] linking the two entities of

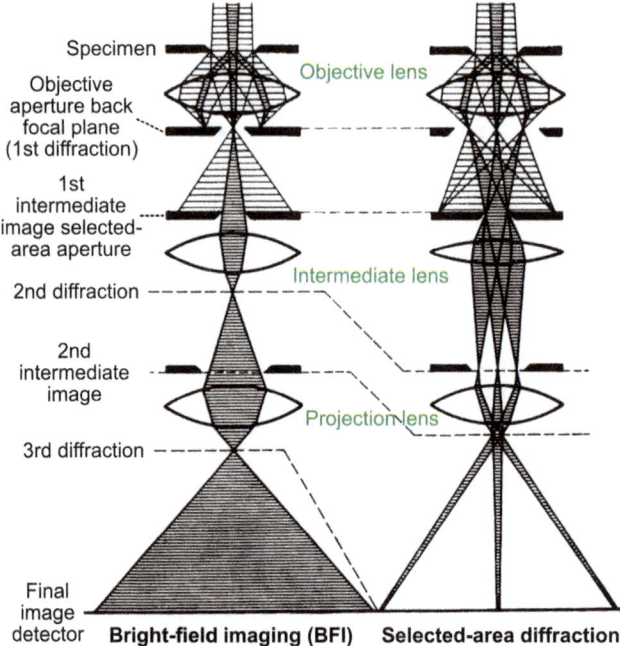

Specimen

Objective lens

Objective aperture back focal plane (1st diffraction)

1st intermediate image selected-area aperture

Intermediate lens

2nd diffraction

2nd intermediate image

Projection lens

3rd diffraction

Final image detector **Bright-field imaging (BFI)   Selected-area diffraction**

Figure 11.10   Schematic drawing and ray diagrams for a TEM in (left) the bright-field mode and (right) selected-area electron diffraction (SAED) mode. Reproduced from ref. 134 with permission from Springer Nature, Copyright 2008.

diffraction and imaging rooted in Abbey's theory for image formation.[133] Here, we discuss their implementation considering a generic imaging arrangement of a TEM, schematically drawn in Figure 11.10, under a fully coherent illuminating field.[134] In this case, the exit wave emerges from the specimen carrying the information on the potential field, $\phi(X,Y)$ or $\phi(\mathbf{R}_\parallel)$ with $\mathbf{R}_\parallel$ representing the coordinates in the X,Y-plane of the specimen, and is brought to a focus on the back focal plane (BFP) of the objective lens. At the BFP, which may be regarded as a plane at infinity on which parallel lines intersect, a diffraction pattern is created there with little cross-talk between the scattered rays from different regions. Meanwhile, with a follow-on lens system (projection lens) to bring the rays back to the plane conjugate to the object plane – that is, the image plane – the interference process is inverted, and the object exit wave function is re-created.[135]

11.5.1.2   Phase Contrast Imaging Principle

We may consider this more quantitatively under the Fourier optics principle. As the electron is scattered from the Coulomb potential of the object, the main contrast mechanism from a unitary probe electron wave traversing through a relatively thin object is a phase change that is proportional to the projected potential field $\phi(\mathbf{R}_{\|})$ for the exit wave. Here, $\phi(\mathbf{R}_{\|}) = \int \phi(\mathbf{r})\mathrm{d}z$. This expression, known as the "weak phase object approximation" in TEM,[132] is derived from the Born series by taking the limit of small angle given by

$$\Psi(\mathbf{R}_{\|}) = \mathrm{e}^{-i\sigma\phi(\mathbf{R}_{\|})} \cong 1 - i\sigma\phi(\mathbf{R}_{\|}) \tag{11.26}$$

with σ expressing the interaction of the electron beam with the sample (*i.e.*, $\sigma = 2\pi m_e \lambda/h^2$). It is easy to see the Fourier pair relation by considering scattering from a singular atom. As here ϕ represents the atomic Coulomb potential, we have $\mathrm{FT}\{\sigma\phi_i\} = f_i$, the atomic scattering factor as given in Section 11.1. In the scattering from a periodic potential, the Fourier transform of the exit wave directly gives the crystallographic form factor F_{hk} at the BFP:

$$\mathrm{FT}\{\Psi(\mathbf{R}_{\|})\} = \delta - i\sigma F_{hk}. \tag{11.27}$$

Hence, in this ideal phase contrast scenario, the obtained image gives the integrated charge density along the beam axis (z) sampling the continuous $\phi(\mathbf{R}_{\|})$; whereas in crystallography, these will be observed in the transform of $\phi(\mathbf{R}_{\|})$ as diffracted intensities: $I(h,k) = |F(h,k)|^2$ aligned with different symmetry axes.[4]

11.5.1.3   Multi-modalities from Dynamical Contrasts

Injecting ultrafast time resolution into the typical TEM modalities will not only allow ultrafast imaging and diffraction patterns to be obtained for studying material processes, but also the delicate control of timing between the pump and probe pulses can create a new contrast mechanism at various stages of the material's responses, which can enable imaging and spectroscopy in ways not available in the CW case. Here, we will discuss the uses of dynamical contrast to enhance the imaging capabilities as well understanding of the material processes not available from UEC investigation. These multimodal operations could be performed without changing the basic optical pump and the electron illumination geometry, so the information retrieved from different dynamical modalities is entirely correlated, allowing researchers to piece together information not available from a special-purpose ultrafast probe.

While the different optical settings define the modalities and field of view, the basic information content available for dynamical contrast is largely controlled by the objective lens aperture coupled to the illumination system. For example, a smaller aperture can be used to isolate just the central beam, so that all the strongly diffracting beams will not enter the imaging system to provide contrast in a bright field image (BFI), showing details such as crystal bends. Similar isolation of a single diffraction spot to produce a darkfield image can show, for example, the regions where the specific order parameters exist which contribute to the signal in the specific crystalline orientations represented by the recorded bend contour imaging contrast.

Generally, either elastically or inelastically scattered electrons may be deployed in these imaging modalities. Especially, spectral imaging that depends on inelastic scattering provides insights into the electronic structure and chemical identity of solids in addition to providing atomic-structure information.[136] However, spectral imaging can carry a significant penalty on the SNR when only a small sector of the inelastic spectrum is focused upon in forming an image.[137] That is why until now, spectral imaging, like in the case of selected-area or micro-diffraction, remains a challenging topic in UEM applications.[138] Generally, an improvement in the SNR by increasing the local flux of electrons with convergent beams is the simplest solution[1,139] but shall be balanced with a somewhat reduced spatial and sometimes even temporal resolutions, with the ceiling set by the beam source brightness; see discussion in Section 11.5.4. However, the benefits of these advanced analytical and micro-imaging technologies access areas that cannot be matched by many other techniques.[140]

11.5.1.4   Resolutions Presented by Lens Aberrations

The examples given in Section 11.5.1.2 exemplify the simplest lens pair to interconvert the Fourier transform pair from diffraction to imaging and back under "ideal illumination". While the Fourier transform pair carries the same amount of information when passing information across the TEM column, one needs to worry about the lens aberration effects with the waves traveling in directions at higher angles to the axis being passed on to the detector. For example, spherical aberration perturbs the wave front by a fourth-order term, so the slope of the wave front that gives the deviation in the direction of the ray is by a third-order term in the divergence angle. Applying an aperture to limit the range will not only reduce the aberration but also lead to the loss of information and the reduction of the fidelity due to a reduction in beam intensity.

Eqn (11.26) and (11.27) deal with only the coherent wave function, ignoring two very important effects that lead to blurring in image formation, namely, chromatic aberration, and beam divergence. Beam divergence is the result of using a converged incident beam to increase the illumination brightness. Chromatic aberrations are caused by instabilities in the objective-lens current supply and by the spread of energy in the incident beam, caused by the high voltage power supply instabilities and space-charge effects. Under these circumstances, the image formation is not a perfect reconstruction of the object. The deviation is represented by the phase contrast transfer of the lens typically written in real space as a spread function $t(\mathbf{R}_\parallel)$, which convolutes with the exit wave to cause a distorted wave front: $\Psi(\mathbf{R}_\parallel) \cong [1 - i\sigma\phi(\mathbf{R}_\parallel)] \otimes t(\mathbf{R}_\parallel)$, where $\otimes$ denotes a convolution. The effect might be easier to consider in Fourier space, where the spread function is described by[135]

$$t(U) = A(U)e^{i\chi(U)}. \tag{11.28}$$

Here, U depicts beam spreading following the literature convention, given by $U \equiv s/2\pi$, while $A(U)$ is the band pass function associated with the objective aperture. The second exponential term encodes the effect from the lens where the most important term is the spherical aberration and distortion due to defocusing (Δf), and the phase factor is given by $\chi(U) = \pi\lambda\Delta f U^2 + \frac{\pi}{2}C_s\lambda^3 U^4$. One can see that the spread function $t(U)$ limits the precision of the information to be passed on by the lens system.[135]

Next, we will consider the imprecision in the encoding of the object function $\phi(\mathbf{R}_\|)$. We note that within the weak phase object approximation, the measured intensity contrast is given by $I(\mathbf{R}_\|) = |\Psi(\mathbf{R}_\|) \otimes t(\mathbf{R}_\|)|^2 = 1 + 2\sigma\phi(\mathbf{R}_\|) \otimes s(\mathbf{R}_\|)$, where the smearing effect is given only by the imaginary part of the spread function, $s(\mathbf{R}_\|)$. In general, in dealing with a thicker sample and inclusion of the absorptive effect by the object, the higher order terms in the Born series shall be considered, which means that both real $[c(\mathbf{R}_\|)]$ and imaginary $[s(\mathbf{R}_\|)]$ components of the spread function $t(\mathbf{R}_\|)$ will contribute. The image intensity is then given by[135]

$$I(\mathbf{R}_\|) = 1 + 2\sigma\phi(\mathbf{R}_\|) \otimes s(\mathbf{R}_\|) - \sigma^2\phi^2(\mathbf{R}_\|) \otimes c(\mathbf{R}_\|)$$
$$+ [\sigma\phi(\mathbf{R}_\|) \otimes s(\mathbf{R}_\|)]^2 + [\sigma\phi(\mathbf{R}_\|) \otimes c(\mathbf{R}_\|)]^2 \tag{11.29}$$

The relative importance of the various terms in this expression will vary with the defocus and the objective-aperture size. For low resolution imaging given by the use of a small objective aperture (only the region with small U will be passed on to the detector), one instead obtains[135]

$$I(\mathbf{R}_\|) = 1 - \sigma^2\phi^2(\mathbf{R}_\|) \otimes c(\mathbf{R}_\|) + [\sigma\phi(\mathbf{R}_\|) \otimes c(\mathbf{R}_\|)]^2. \tag{11.30}$$

In this case, the sign of $\sigma\phi(\mathbf{R}_\|)$ is lost, and so the positive and negative deviations from the average projected potential can give similar contrast. Eqn (11.29) and (11.30) are found to be more relevant in solving the imaging/diffraction contrasts for UEM applications under the practical circumstances to boost the SNR.

11.5.2   Beam Delivery for Imaging and Diffraction: Practical Considerations

In practical situations, the illumination wave in a TEM cannot be fully coherent and this will also couple to the imperfection of the lens, leading to defocussing effects. Furthermore, a more sophisticated TEM system relies on not just the high beam coherence but also high current efficiency to perform imaging and diffraction with tuneable field-of-view and for various analytical purposes (see Section 11.5.1.3). Below we will consider these factors especially under the scenarios of high-intensity beam originated from finite-sized photo-electron emitters to understand the design principle and the resolutions that the new high-throughput UEM system hopes to achieve.

11.5.2.1 Coherence in Beam Delivery

Since this review will focus on high throughput UED/UEM systems using a finite-sized electron pulse containing a large number of electrons, the topic dealing with partial coherent illumination is important. We first introduce the concept of coherence and discuss the strategy in the beam delivery system to increase the coherent flux with specific emitters.

11.5.2.1.1 Coherence. Based on the Zernike–Van Crittea theorem,[141] one can estimate the source coherence width as

$$X_c = \frac{F\alpha_c}{\pi} = \lambda/2\pi\theta_s, \qquad (11.31)$$

with the coherence angle $\alpha_c = \lambda/\theta_s$. If the coherence width X_c of the source is larger than the size of the convergence-limiting aperture (C2), given by $D = 2F\alpha_0$ with α_0 the half angle extended by the convergence-limiting aperture, the electron probe might be considered as fully coherent.[1] This is frequently the case for a field emission gun (FEG) source; however, there is a limit to how many electrons that can be photo-extracted from a point source under a single laser pulse. For high-current-efficiency operation, here one faces a compromise in reduction of the transverse coherence to achieve a higher throughput. This scenario, typically resulted in a choice of a source area of $\sim$100 μm, where $\alpha_c \ll \alpha_0$ and the illumination wave at C2 might be treated as partially coherent. In fact, a photoemission source in this regime is quite similar to using a LaB$_6$ thermionic source; hence, the performance there serves as a good benchmark for considering high throughput operation.

We now turn our attention to scenarios employing partial coherent illumination from high-intensity photo-emitters and the impact that a finite source phase space (emittance) has on the imaging and diffraction resolution by calculating the transfer function. Figure 11.11 shows the schematic ray diagram of the electron-optical systems, showing the simplest optical arrangement for the delivery system of a TEM (see left panel). The size of the microdiffraction probe depends on the demagnification of the probe forming lenses, on the electron source size, and focus settings which includes the aperture size and spherical aberrations – a subject that is crucial for conducting the coherent UEC experiments (see Section 11.4) with a microscope.

Realising the full scope of imaging capabilities with a CW TEM system requires the further development of an electron source with a higher beam coherence that is achievable with field emitters of increasingly reduced size (see middle panel in Figure 11.11). In such a setup, the beam illumination

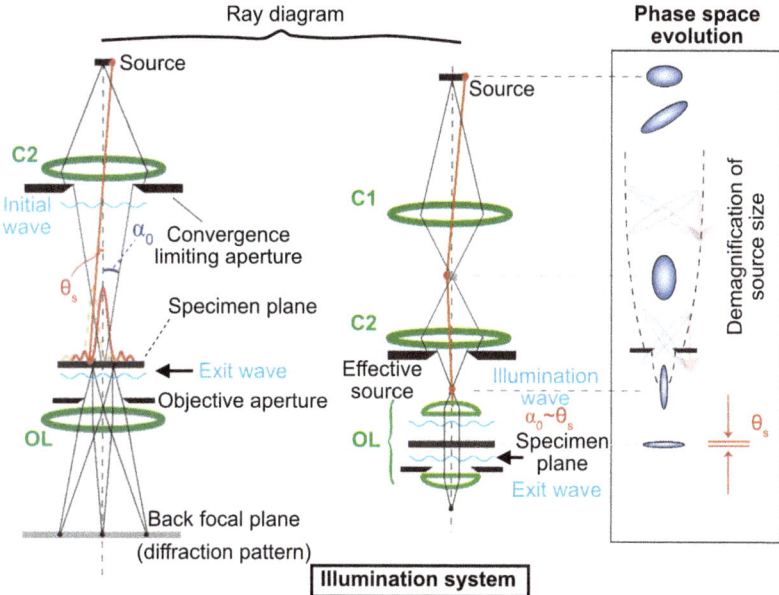

Figure 11.11   Illumination system for TEM. (left) A basic system to illustrate the effect of finite source angle θ_s and α_0 on the incidence beam convergence profile.[1] (right) Improved coherence in the illuminating wave front can be achieved using a condenser to reduce the source size and place the cross-over at the focal distance of the objective prefield. The panel inset shows how this strategy is employed in the case of the pulsed electron source with finite-size emission.

utilizes the combination of a condenser lens $(C1 + C2)$ and an objective lens (OL) pre-field to create a condensed and parallel illumination for coherent diffraction and imaging. The scenario depicted here is applicable for a hairpin or LaB_6 thermionic sources[134] and is therefore applicable for pulsed electron generation from a finite-sized emission disk (~ 100 μm) employed in high-throughput UEM experiments. In the latter case, the electron bunch property can be described in a finite phase(trace) space along the $p_x(\alpha) - x(r)$ coordinate (see right panel in Figure 11.11). Here, the source angle θ_s can be defined by the stochastic velocity spread at the emitters, which could be manipulated through the demagnification process using the $C1 + C2$ lens pair to achieve a better coherent beam delivery.

11.5.2.1.2   Strategy for Phase-space Manipulation.   Generally, given the finite-sized phase space area (or emittance), as one squeezes the probe size (see the inset of Figure 11.11b), θ_s will increase (*i.e.*, the probe becomes less coherent). By placing the cross-over at the focal distance (typically a few mm) of the objective pre-field to demagnify the incident beam, the beam delivery system strikes a balance of maximizing the electron dose with a moderate compromise in coherence to operate TEM imaging and diffraction modalities. This conventional phase-space manipulation strategy along the transverse dimension could also apply along the beam direction (the longitudinal direction) to operate high-bunch-charge UEC and UEM experiments. Here, similar effects from aberrations and defocussing also occur (see Section 11.6), leading to imprecision in the illumination. Such instabilities couple to the chromatic aberration, leading to degradation of temporal coherence. The inherent electron pulses' stochastic phase space structures are ultimately responsible for incoherent illumination.[142] Like the case described in the horizontal (x,y) plane, an appropriate demagnification of the pulse longitudinal footprint will reduce the incoherence effect.

Figure 11.12 shows examples of this phase-space manipulation employing an RF cavity synchronised to the pulsed photoemission. The cavity acts as a longitudinal condenser lens that can either compress the phase-space projection along z or p_z axes, resulting in an ultrashort or monochromatized illumination wave. In the sub-sections below, we shall look into the impacts of different finite-size phase space extensions on the UEM performance and their optimization strategy in the design of a high-throughput UEM system and their applications.

11.5.2.2   Brightness and Incoherence Envelope

The discussion of imperfect or partial coherence can be complicated for photoelectrons. However, for our main purpose of considering coherent illumination, the effects might be understood from the stochastically filled phase spaces of the electron bunch created at the photocathode in the transverse and longitudinal directions. In our cases, one can assume that the illumination of the specimen is the same as it would be if the condenser aperture were an incoherent source illuminating the specimen. The observed intensity is the sum of the intensities produced by all points of the source considered separately. Even if the beam were focused to give a wave as close to a plane wave as possible, there would be some convergence

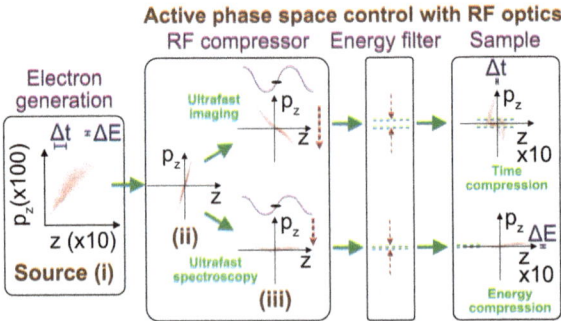

Figure 11.12   Active phase-space manipulation by a longitudinal condenser lens exemplified by an synchronize RF-cavity.

because of the finite source size. This leads to spatial incoherence in the illumination. Furthermore, when the probe particles arrive with a different energy at the sample, the observed intensity is the sum of the intensities for all different wavelengths considered separately. The effect causes temporal incoherence in the illumination. To consider these effects, we first give the conventional treatment for the effects arising from spatial and temporal incoherence together with an expansion of the transfer function to include their respective spread functions $E_S(k)$, and $E_T(k)$ in momentum space:

$$t(k) = E_S(k)E_T(k)A(k)e^{i\chi(k)}. \tag{11.32}$$

The specific spread functions were given by Frank.[142] The spatial incoherence envelope is given by

$$E_S(k) = \exp\left\{-\left[\pi\sigma_s\left(\Delta f k + C_S\lambda^2 k^3\right)\right]^2\right\}, \tag{11.33}$$

where σ_s is the sigma of the Gaussian modelling of the convergent illumination cone at the specimen, which is related to the convergence half angle $\alpha_0 = \sqrt{2\ln(2)}\sigma_s$. The temporal incoherence envelope is given by

$$E_T(k) = \exp\left\{-0.5\left[\pi\sigma_T\lambda k^2\right]^2\right\}, \tag{11.34}$$

with σ_T depicting the beam energy stochastic spread coupled to the chromatic aberration

$$\sigma_T = C_c\left[\left(\frac{\sigma_E}{E_0}\right)^2 + \left(2\frac{\sigma_I}{I_0}\right)^2\right]^{1/2}, \tag{11.35}$$

where σ_E and σ_I are the root-mean-square (RMS) deviations of the beam Gaussian energy and the objective current spreads, respectively, and C_c is the chromatic-aberration constant. We note, unlike the scenario discussed in Section 11.5.1.4, the limitation for the degradation in resolution arises from the fundamental limit set by the stochastic phase space of the source and not the optics of the TEM. Yet, an improperly handled enlargement in the phase-space projection will couple to the aberration coefficients, leading to a compounded impact. Eqn (11.32)–(11.34) can provide guidance for the choice of the source and the placement of the optics to best implement the phase-space manipulation employed in UEM to improve its spatial and temporal resolutions.

Unlike the CW system, used to describe the photoemitters employed in a UEM, the source brightness is best characterized by the phase-space density of the electron bunch. To better understand the deliverable imaging resolution in practical applications, we can restict ourselves to the more relevant 4D (instead of 6D) density as given by

$$\rho_{4D} = \frac{N_e}{\tilde{\varepsilon}_x} \tilde{\varepsilon}_y, \tag{11.36}$$

where N_e is the emitted particle number, and $\tilde{\varepsilon}_{x,y}$ denotes the transverse normalized emittance defined by $\tilde{\varepsilon}_x = \frac{1}{m_0 c} \sqrt{\sigma_x^2 \sigma_{p_x}^2} = \frac{1}{m_0 c} \sqrt{\langle x^2 \rangle \langle p_x^2 \rangle - \langle x p_x \rangle^2}$. Since $\tilde{\varepsilon}_{x,y}$ is invariant with respect to a change in the beam energy, ρ_{4D} is a conserved quantity once the beam is fully extracted from the cathode region. Building on this, we can make a direct connection to the instrument's performance that is well-benchmarked in CW TEM systems. In the CW regime, the beam brightness is one key figure of merit often defined as the current per unit area and solid angle, given by

$$B_{CW} = \frac{I_e}{(\pi d \alpha)^2}, \tag{11.37}$$

with I_e representing the current emission, d representing the beam diameter, and α representing the convergence half angle. Table 11.1 gives the typical B_{CW} and the energy spread ΔE_{CW} in several common types of TEM electron sources.

Alternatively, for a pump–probe-based UEM system, the beam brightness is described by the phase-space density multiplied by the repetition rate f_{pp}, which can be linked to the CW brightness defined for a TEM by

$$B_{CW} = \rho_{4D} f_{pp} \left(\frac{\gamma \beta}{\pi 2 \ln(2)} \right)^2. \tag{11.38}$$

Using eqn (11.36)–(11.38), which are considered in the framework of the electron bunch phase space, we can show that this CW brightness is intimately connected to the other important figures of merit for imaging and diffraction. Under the bunch illumination, the electron dose is given by $D_e = N_e / \{2 \ln(2)\} \pi \sigma_x \sigma_y$, where $\sigma_{x,y}$ represents the transverse Gaussian beam width at the specimen. Using eqn (11.31) describing the coherence length, we give the normalized transverse phase-space area as

$$\tilde{\varepsilon}_x \tilde{\varepsilon}_y = \frac{N_e}{16 \pi^3} \left(\frac{\beta \gamma \lambda}{\ln(2)} \right)^2 \frac{1}{D_e X_c^2}. \tag{11.39}$$

Note that the condenser lens system in a TEM reduces the effective source size such that X_c becomes larger down the column at the expense of reducing the dose D_e. Typically, the illumination exceeds the region of interest in order to image local features; selected-area diffraction provides a measure of "local emittance" (or slice emittance) $\tilde{\varepsilon}_x^{loc}$, which is smaller than the beam emittance. This means that the local measurement based on diffraction from features smaller than the beam waist could not directly report the total

Table 11.1 Beam parameters for pulsed and CW sources, derived based on ref. 143, 144 and 134, 140 respectively; values in brackets give the equivalent value for the parameter if the beam were operated in the alternative format, CW or pulsed.

Symbol		Photoemission Pierce	Photoemission FEG	CW LaB$_6$	CW FEG
N_e	Particle number	10^6	1	1	1
eV_0 (keV)	Beam energy	100	100	100	100
v (m s^{-1})	Beam velocity	1.64×10^8	1.64×10^8	1.64×10^8	1.64×10^8
γ	Relativistic factor	1.20	1.20	1.20	1.20
β	Relativistic factor	0.548	0.548	0.548	0.548
λ_e (Å)	Electron de Broglie wavelength	0.037	0.037	0.037	0.037
ρ_{4D} (mm^2 mrad^2)	4D phase-space density	1.01×10^{11}	1.80×10^{10}	2.81×10^4	1.80×10^{10}
f_{PP} (kHz)	Pump–probe repetition rate	1	1	(9.82×10^{10})	(3.07×10^7)
J (A m^{-2})	Beam current density	2.04×10^{-2}	3.26×10^1	2.00×10^5	1.00×10^9
I (mA)	Total beam current	1.60×10^{-10}	1.60×10^{-16}	1.57×10^1	4.91×10^{-3}
d (µm)	Source diameter	1.00×10^2	2.50×10^{-3}	1.00×10^1	2.50×10^{-3}
ε_x (µm)	Transverse emittance	3.14×10^{-3}	7.46×10^{-6}	5.97×10^{-3}	7.46×10^{-6}
α_s (rad)	Source angle	1.33×10^{-3}	1.26×10^{-1}	2.52×10^{-2}	1.26×10^{-1}
B_{cw} (A cm^{-2} sr^{-1})	CW beam brightness	(3.67×10^1)	(6.52×10^0)	1.00×10^6	2.00×10^8

emittance of the bunch. However, from the performance perspective, it is the "brightness", which is based on ρ_{4D}, that matters. In fact, assuming an incompressible electron fluid in phase space (Liouville theorem) for the illumination, we arrive at a form for universal measurement conducive to the local measurement:

$$B_{\mathrm{CW}} = \frac{4\pi}{\lambda^2} D_{\mathrm{e}} X_{\mathrm{c}}^2 f_{\mathrm{pp}}. \qquad (11.40)$$

11.5.2.3 Practical Consideration: Current Efficiency and Repetition Rate

For considering the performance of a TEM, the beam brightness is not the only key figure of merit. For analytical work, the current efficiency is as important if not more critical, for some of the more stringent works, such as element-sensitive probing and low-contrast dynamics. The brightness is the common denominator for conceiving resolutions under various modalities only if the current efficiency can support the specific modality for reaching a sufficient SNR over a reasonable time frame of observation. For example, high B_{CW} available from a field-emission gun (FEG) makes it an ideal tool for work requiring a small probe or very high spatial resolution but is typically limited to a small field of view. Meanwhile, the thermionic emission LaB_6 gun can provide a very large current, particularly useful for surveying the material properties over a large area for functional analysis.[140] For conceiving the type of UEM source for target applications and the f_{pp} suited for the applications, it is thus useful to draw on the estimates of the beam brightness and current efficiency that can be delivered from the high-bunch-charge UEM photo-emitters, using the FEG and LaB_6 gun as benchmark systems. Such results are generated for a moderate f_{pp} of 1 kHz (see Table 11.1) using the relevant source emittance parameters derived from the experiments[143,144] and literature.[134,140] It is interesting to note that in general the photoemitter becomes brighter than a LaB_6 emitter with an f_{pp} larger than 30 MHz. This is an important benchmark as the image resolution from a LaB_6-based source can typically reach 1 nm without aberration correlation, which is a critical length scale for studying nanoscale dynamics.

We are now ready to consider multi-modality observations on the ultrafast timescale based on these source characteristics. We take into account the degraded performance caused by the lens aberration and defocusing, which are convolved with the incoherence envelopes that one can derive based on the beam phase-space parameters as given in Section 11.5.2.2. For illustration, we consider a nominal region for UEM operation where one focuses on running on a high temporal resolution (~ 100 fs) and a moderate f_{pp} of 1 kHz, achievable with a Pierce cathode biased to 80 keV and slicing the emission down to a bunch-charge (N_{e}) of 10^5 delivered to the specimen to maintain the core brightness. Under this condition, considering the decoherence effects and lens aberrations contributing to the transfer function

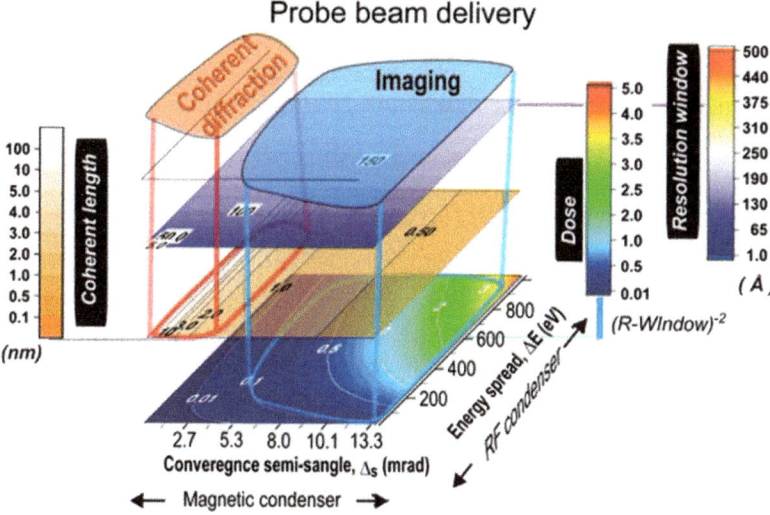

Figure 11.13 Single-bunch performance maps based on the effective electron dose (D_e'), real-space resolution (R), and the probe coherence length (X_c) calculated based on the electron bunch phase space parameters for source $N_e = 10^5$ with nominal transverse and longitudinal emittance of 10 mm mrad in both cases.[143,144] The maps are simulated for the following settings: $eV_0 = 80$ keV, $C_S = C_C = 2$ mm. Two variables: the convergence semi-angle α and bunch energy spread (FWHM) are given by tuning the condenser C2 and the RF longitudinal lens in the illumination system.

based on the bunch emittances, only specific operation windows for multimodality measurements are possible (see Figure 11.13).

In particular, we depict the key performance figures for real-space resolution (R) and the coherence length (X_c) under a high bunch charge, with the relevant SNR set by the effective electron dose (D_e') over the resolution window set by the bunch brightness and f_{pp}. For practical ultrafast imaging experiments, a resolution in the range of 10–150 nm (see the blue circled area in Figure 11.13) is chosen considering the constraint of achieving an effective electron dose of at least 100 e, giving a SNR of 10 for resolving changes at the 10% level. For diffraction experiments, the main concern however is the illuminating beam coherence length X_c, which is set primarily using the condenser lens (C2). For selected-area diffraction measurements, a site specificity of 50–150 nm extended by the image mode resolution R is chosen (see the red circle area in Figure 11.13). Allowing a larger illuminating area will permit a greater integrated dose ranging from 0.01–0.1 e μm^{-2}, and therefore, a long exposure time and/or high f_{pp} is generally expected. For coherent diffraction with X_c larger than 10 nm at $f_{pp} \leq 1$ kHz, in general, requires an illumination area of ≥ 1 μm × 1 μm. In general, the performance will be improved upon increasing the beam energy for coherent diffraction and imaging, whereas a lower beam energy may be preferred for electron spectroscopy and for providing conditions for joint multimodality studies. We note based on the relevant parameters listed in Table 11.1, increasing the TEM energy from 200 keV to 2 MeV – a regime where the accelerator-based UED system operates – the B_{CW} increases by ~ 20 times mainly due to the relativistic boost (see eqn (11.40)).

11.5.3 Inelastic Scattering and Electron-energy-loss Spectroscopy

Another particularly useful feature for ultrafast studies is the inelastic electron scattering contrast, which plays a crucial role in understanding the electronic structures and collective excitations of sample materials. In fact, inelastic scattering is not a negligible effect when comparing the integrated electron scattering cross-section to the elastic electron scattering contribution. The total inelastic electron scattering cross-section exceeds the elastic electron scattering cross-section of elements lighter than copper, and for scattering mainly concentrated to small scattering angles.[136] Applying the principles of conservation of energy and momentum to inelastic electron scattering, one can determine the best angular range to collect signals for either core-level and the low-loss regimes that provide information on the chemical composition and the valence-band electronic excitations. A characteristic angle $\theta_E = E_l/E_0$, where E_l and E_0 $(= \gamma m_0 v^2)$ are the energy loss and relativistic beam energy, respectively, provides the guidance for the setup for probing the dynamical loss function from inelastically scattered electrons. The literature on the inelastic scattering of electrons is vast. We confine ourselves here to the most probable inelastic processes for keV-scale electrons traversing thin specimens that have a bearing on the topics related to material sciences and quantum materials, such as plasmonics and insulator-to-metal transitions, primarily in the low loss regions.

11.5.3.1 *Loss Function from Inelastically Scattered Electrons*

To see the correlation between the loss function and the scattering angle for designing the ultrafast experiments, one applies the simple expression of the Bethe theory[145,146] given by

$$\frac{d^2\sigma}{d\Omega dE_l} = \frac{e^4}{E_i E_l} \frac{df(\theta, E_l)}{dE_l} \frac{1}{\theta^2 + \theta_E^2}, \qquad (11.41)$$

where the differential loss function is defined as

$$\frac{df(\theta, E_l)}{dE_l} = \frac{2mE_l}{\hbar^2 s^2} \left| \left\langle f \left| \sum \exp(i\mathbf{s} \cdot \mathbf{r}_i) \right| i \right\rangle \right|^2, \qquad (11.42)$$

where i and f denote the initial and final states accessible by the fast electron, respectively. In the case of low loss, where $E_l \ll E_0$, by using the first

Born approximation, the transition matrix (see bra-ket term in eqn (11.42)) is given by a dipole transition and sums over all possible transitions accessible by fast electrons. In this limit, one thus can link the dipole field transition to the Drude–Lorentz model where the dielectric function summing over all the oscillators j is given by

$$\varepsilon(\omega) = 1 + \frac{e^2}{m\varepsilon_0} \sum_j \frac{n_j}{\omega_j^2 - \omega^2 - i\omega/\tau}, \tag{11.43}$$

where n_j is the number of electrons able to oscillate at the eigenfrequency ω_j, and τ describes the relaxation time due to friction in the electron gas.[147,148] The expression thus contains the contributions from free oscillators (such as the Drude free electrons) and the bound oscillators. This has a significant ramification in ultrafast measurements. As in the optical transition (where the transient reflection or transmittance measurement carries information on the dielectric responses at the bandwidth accessible by the optical probe), the transmittance measurement *via* the fast electron through the materials also carries similar information over a larger bandwidth. Such information is crucial for measuring the dynamical dielectric response of the materials system over an ultrafast timescale, such as insulator-to-metal transitions and surface plasmon resonance (SPR)-associated effects. In this regime, one expects a sufficiently high SNR as the integrated oscillator strength changes significantly during the formation of a new long-range-ordered phase with gap formation. During the insulator-to-metal transition, a significant part of the dielectric function is from the increase in the carrier density (n_e) in the band-to-band transition, hence giving rise to an increase in the plasmon peak that could extend over a large energy window. In the case of scattering from nanostructures, the surface plasmon resonance would also be a prominent feature but spanning across a narrower energy window in the low loss regime. The near-field SPR feature has been recently utilized in the form of photoinduced near-field electron microscopy (PINEM; see Chapter 16)[149,150] that has been applied in the prototype studies of nano-plasmons,[151] oxide-state mapping,[152] and electronic phase transitions,[153] which was only through high f_{pp} measurements using the single-electron UEM system.

In summary, practical ultrafast electron microscopy is a matter of achieving the optimum compromise between several conflicting require-ments: the illumination should be made as coherent as possible by de-creasing the illumination aperture sizes; the number of electrons at the source should be kept sufficiently away from the virtual cathode limit for improving temporal/spectroscopic resolutions and yet the signal-to-noise ratio in the image should be kept as high as possible by maximizing the number of electrons recorded per resolution element emission current; the exposure time should be minimized to prevent image smearing owing to specimen drift; and the instabilities in the beam deliver and optical systems.

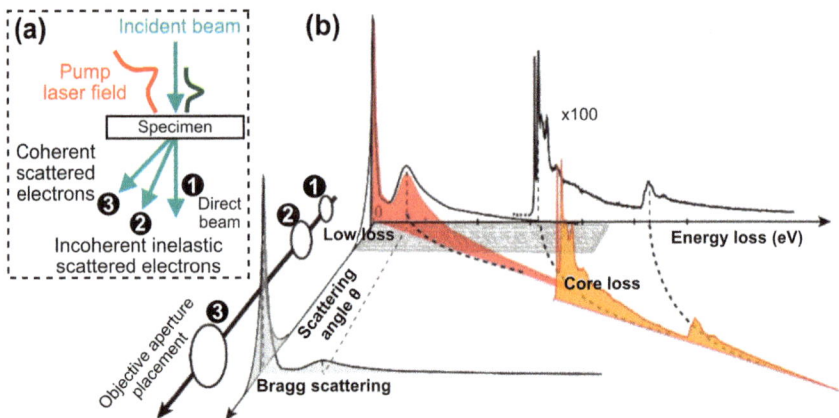

Figure 11.14   Selection of different scattering channels for multimodal characterization in a TEM. (a) Pictorial depiction of the three main scattering
channels utilized for conducting the imaging and scattering measurements. (b) Schematic differential scattering cross-section as a function
of energy loss and scattering angle θ.[147] Angular dependence for
regions of interest for inelastic losses (including low losses and core
losses), and the elastic scattering as part of Bragg peak are outlined.

11.5.4   Applications of a High-throughput UEM System

The UEM instrument configuration with high-throughput femtosecond electron pulses offers great flexibility in creating different dynamical contrasts
about the excited material systems upon photoexcitation. Figure 11.14 indicates the angular dependence in the elastic and inelastic processes of interest
(see also Chapter 3). The elastic scattering includes large-angle Rutherford
scattering from the nucleus contributing to Bragg scattering, whereas the inelastically scattered channels with a low-loss (for electron kinetic energies of a
few tens of electron volts) are primarily located within 0.2 mrad in a 100 keV
electron microscope.[136] The distinct angular ranges from these different
scattering channels offer the possibility to isolate the information content of
interest with the objective aperture for conducting ultrafast imaging, diffraction, and spectroscopy under the same illumination settings, forming
multi-message stereo viewpoints on the same events. This aspect is amplified
by achieving the ultrafast temporal resolution. Below, we give some examples
where diffraction and imaging difference approaches based on the reference
signal retrieved at negative pump–probe delay time could result in new dynamical contrasts that expand the utilities for ultrafast studies.

11.5.4.1   *Coherent Imaging and Diffraction in UEM:*
Complementarity

The capability to zoom-in on the relevant dynamical contrast *via* angular selection using the objective aperture (see Figure 11.10) also reduces the range

of useful information that can be passed on for further analyses. This can be easily seen from eqn (11.28), where the (objective) aperture function $A(U)$ sets a boundary on the collected information from $U_{\min}$ to $U_{\max}$. In the imaging modality, further considering the effects from the incoherence envelopes, the detector gives a modified Fourier transform pattern of the exit wave:

$$I(\mathbf{R}_\parallel, t) = \left[FT\left\{ A(U)E_S(U)E_T(U)e^{i\chi(U)}\Psi(U,t) \right\}^2 \right].$$ (11.44)

Meanwhile, the diffraction pattern is given by

$$D(U,t) = c[A(U)E_S(U)E_T(U)]^2 |\Psi(U,t)|^2.$$ (11.45)

Structural details are primarily captured by the coherent structure factor which provides the resolvable distance that is roughly given by the Nyquist frequency of its Fourier spectrum (*i.e.*, $\Delta R \sim [(U_{\max} - U_{\min})/2]^{-1}$ with the U-range set physically by the aperture). Additionally, the spatial and temporal incoherence carried by the envelope function will smear some details contained in the structure factor within the available information range. Fortunately, the diffraction does not explicitly suffer from the wave front distortion caused by the spherical aberration and hence will benefit from a large aperture size to reduce the Nyquist frequency. Here, the incoherence envelope limits the resolution for diffraction only in its field of view (set by spatial coherence length X_c) in the context of the pair correlation length. Therefore, the best way to retrieve the structure factor is still from coherent diffraction. On the other hand, the imaging modality strives to reach an improved real-space resolution by reducing the aperture size to avoid aberrations and may benefit from the needed boost in the dose from a tighter focusing geometry in the pulse delivery; however, this increases the effects from the incoherence envelopes. The practical UEM imaging modality must strike a balance between the two opposing factors (as also depicted in Figure 11.13) with a larger source size to offer a higher current efficiency in new UEM systems. This suggests that it is intrinsically more difficult to obtain atomic resolution when a large number of electrons are employed. At a moderate resolution, the relevant features (defects, dislocations, step-edge, and morphology) gain their contrast not from the structure factor directly but rather from the coarse-grained contrast mechanism associated with individual modality. Nonetheless, imaging offers sufficient real-space resolution in the feature distribution, which is not available directly from a wide-field-view diffraction pattern. This complementarity is highly useful for UEM applications.

In Figures 11.15 and 11.16, we show the typical resolutions achievable with the current high-throughput UEC and UEM systems that share similar beam delivery characteristics. We employ both crystallography and direct imaging modalities to probe the specimen of TaS_2 to show the complementarity. In both cases, the pulses are compressed to ~ 100 fs, which sets a stringent limit on the obtainable spatial resolution in imaging modalities,

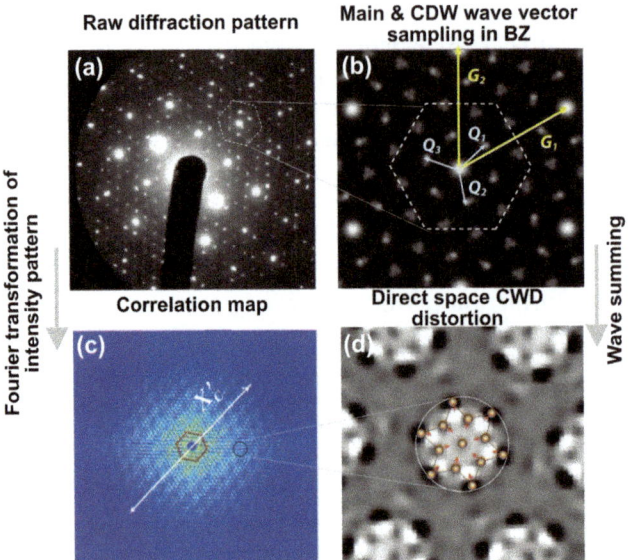

Figure 11.15   Diffraction pattern of TaS$_2$ and reconstruction of order parameter function. (a) Diffraction pattern obtained at 100 keV. (b) Features identified within the first Brillouin zone region give the irreducible representation of the key wavevector components (1st and higher orders) and their respective intensities, which can be used to perform wave summing to preproduce the lattice superstructures of the density waves. Here, the 1st-order CDW satellites are marked by the respective momentum wavevector Q_m, whereas the main lattice peaks are marked with the reciprocal unit cell wavevectors G_1 and G_2. (c) The real-space pair-correlation map $G(r',r)$, showing the long-range textures of the density waves modulating the main lattice intensities. $X_c' \approx 20$ nm gives the limit of resolvable pair distance. (d) The 13-atom supercell of the density wave in David-Star shape, presented in the lattice distortion map. Reproduced from ref. 100, https://doi.org/10.5802/crphys.86, under the terms of the CC BY 4.0 license https://creativecommons.org/licenses/by/4.0.

while not having as much impact in the UEC experiments ($X_c > 20$ nm under temporal compression;[111] also see Figure 11.7e). For coherent diffraction measurements, one sets the condenser to place the beam cross-over at the focal distance of the objective pre-field to create a near parallel illumination of the specimen (see Figure 11.11). With the coherent diffraction setup, we investigate the nonequilibrium CDW phase transition (see also Section 11.4.2.2 with UEC methodology). In the TaS$_2$ case, the density-wave system is commensurate with the building block consisting of the 13-atom David-Star (DS) cluster. The three-fold symmetric structure is clearly identifiable in the patterns. Critical to the crystallography approach is the adequate sampling of the Fourier fingerprints within the crystalline reciprocal BZ in order to solve a complex structure. Here, in the room temperature phase, the so-called near-commensurate CDW (NC-CDW),[154–159]

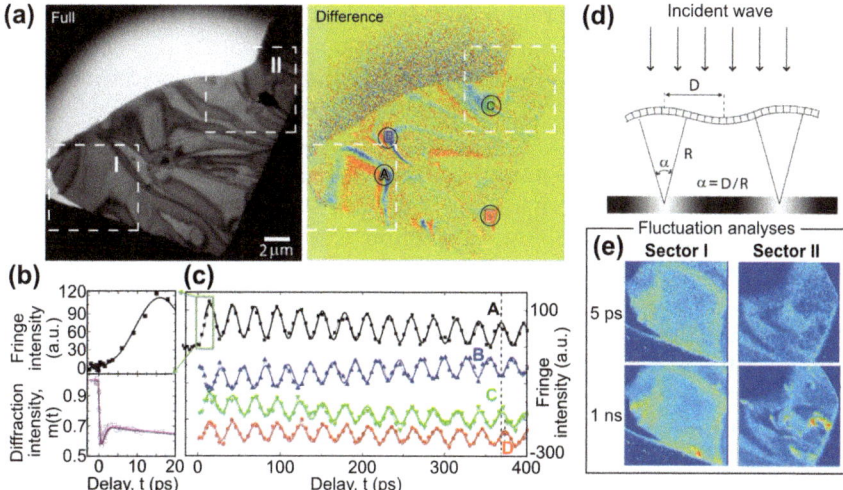

Figure 11.16 Ultrafast electron microscopy of bend contour dynamics in exfoliated 1T-TaS$_2$ thin films. (a) The bright-field image (BFI) of the TaS$_2$ thin film taken under the UEM in full and difference images. The full image is taken at $t = -20$ ps; while the difference image is taken by subtracting the negative-time full image with one obtained at $t = 15.5$ ps. (b) Bend contour formation from two-beam interferences.[53] (c) The BFI intensity oscillation versus the CDW satellite intensity evolution in the first 20 ps upon applying a 1.5 mJ cm^{-2} near-infrared laser pulse. (d) Contrast intensity oscillations at selected regions of the image (see panel (a)). Region A, B and C have the same time period (T) of 27.20 ± 0.05 ps. A and B are out of phase. Area D has a slightly different time period of 27.60 ± 0.08 ps, which is 1.5% higher than the other areas. After 14 cycles of oscillation, we can see a clear offset from area D. The longitudinal speed of sound along the c-axis is $v \approx 3$ nm ps^{-1}.[54,55] The sample thickness can be estimated from $d = vT/2 \approx 40$ nm. The 1.5% difference in the time periods gives ~ 0.59 nm difference in thickness, which is almost exactly 1 van der Waals layer of TaS$_2$. (e) Fluctuation analysis conducted on the micrographs obtained in the pump-probe sequences reflects surface wave dynamics which are not entirely synchronized with the pump pulses at long times. Reproduced from ref. 100, https://doi.org/10.5802/crphys.86, under the terms of the CC BY 4.0 license https://creativecommons.org/licenses/by/4.0.

the DS clusters organize into a hexagonal domain texture[160,161] with triply split fundamental and higher-order multi-$\mathbf{Q}_m$ components representing its Fourier spectrum within the BZ identified at each reciprocal lattice $\mathbf{G}_{hk}$ (see Figure 11.15b). With the well-known structure of the unexcited ground state from imaging,[160,162] where relevant phase information is readily retrieved, one can obtain the excited CDW structures from coherent wave-summing[154,155,163] – an operation equivalent to the FT term within the square bracket of eqn (11.44) with the complex CDW wave given by

$$\Psi(U, t) = \sum_m A_m e^{i(\mathbf{Q}_m \cdot \mathbf{R} + \phi_m)} \sum_{hk} A_{hk} e^{i(\mathbf{G}_{hk} \cdot \mathbf{R} + \phi_{hk})}, \qquad (11.46)$$

with m denoting the different CDW branches and A_m representing the complex wave amplitude derived from the structure factor $(I_m = |A_m|^2)$. To recreate the domain state, a total of 15 wave constituents located within the BZ are coherently coupled[161] (see Figure 11.15d for the distorted DS cluster state depicted in the difference map). The clear, atomically resolved picture obtained from the crystallography approach as shown here is a mere demonstration of accurately deriving the satellite $F_{hk,m}$ from the specific Bragg reflections. In this picture, the crystalline correlation is frequently considered to be infinite (*i.e.*, no inhomogeneity in the wave reconstruction is assumed).

We contrast this to the result from a direct Fourier transform from the intensity pattern, which gives the correlation map $G(\mathbf{r}'')$ with $\mathbf{r}'' = \mathbf{r} - \mathbf{r}'$ (see Figure 11.15c). Clearly, the incoherence envelope from the illuminating beam sets a limit for probing the long-range system as evidenced in the visibility running off over $|\mathbf{r}''| > X_c' \sim 20$ nm. This X_c' is given by the width from the relevant structure factor $(S_{\text{CDW}}$, in this case), rather than the actual beam spatial coherence length X_c as other factors, such as the aperture function, energy instabilities and the detector point spread, also contribute to the width of structure factor. In contrast to Figure 11.15d, the atomic unit cell is barely resolvable, which is a direct result of the Nyquist frequency $(4\pi/s_{\text{max}})$ set by Figure 11.15a. It is instructive to point out that, if the effect from lens aberrations (see eqn (11.44)) can be kept small, the resolution given here is representative of what is achievable with ultrafast imaging under the same coherent illumination condition. This, while clearly not the same as those achievable through the crystallographic approach, is still capable of resolving basic DS building blocks, reflecting the fundamental limits of imaging experiments in the depth of contrast and SNR highly susceptible to the optical arrangements.

11.5.4.2   Dynamical Bend Contours and Thickness Fringes

In a moderate resolution range (10–100 nm), the most visible features from imaging thin objects are frequently those derived from the fringe contrast, obtained *via* bright-field or dark-field imaging – in the latter a small aperture is used to limit the diffraction to only one peak to the imaging system. As the diffracted beam is only excited along certain crystal inclination, the selected diffracted beam from the nearby bended plane (or change in thickness) will contribute to the intensity modulations, leading to a two-beam interference pattern reflecting the thin object's morphology. Some defects such as edge dislocations can be detected through moiré magnification or thickness fringes. Monitoring the fringe evolution and dynamics thus gives essential information to probe the property evolutions based on the system inhomogeneity, morphology, and the emerging acoustic waves that can trace their roots to the ultrafast timescales.

Figure 11.16 shows the dynamical contrast in imaging modalities that one can deliver through similar temporally compressed high-intensity bunches obtained using the generation II UEM system at MSU for the charge-density waves materials (see Figure 11.15).

In the BFI modality, given in panel a, one obtains the microscopic images of a 40 nm, exfoliated TaS_2 specimen placed on a TEM grid. The micrographs, as given in the full frame [left, $I(\mathbf{R}_{\parallel}, t_{ref})$ with $t_{ref} = -20$ ps] and the difference frame [right, $I(\mathbf{R}_{\parallel}, t) - I(\mathbf{R}_{\parallel}, t_{ref})$; $t = 15.5$ ps], show the morphology of the system and its dynamical contrast upon shining fs laser. In these BFI images, the baseline of the intensity level is given by the mass-thickness contrast[56]—e.g., the small debris left on top of region II is respectively darker than the general area. However, even in regions with the same thickness, the two-beam diffraction effect can create fringes from the sample curvature.[56] This interference-mediated contrast is sensitive to the inclination of the sample relative to the beam (panel b). In the regions of bright or dark fringes, selected by a local tilt angle, the diffraction contrast is enhanced drastically.[56] This is utilized to measure the generation of acoustic waves. Panels c and d underscore the formation of acoustic standing waves that couple to the bend contours.[57-64] In this case, clocking on the frequency associated with regional contour oscillation provides a measurement of sample thickness with a high precision.[58,63] This is shown in comparing the results from different regions, marked in A–D in Figure 11.16a. Taking the oscillation out to 400 ps, the oscillation from region D is obviously out of sync with the rest. From the phase difference, one can determine a delay of 6 ps being developed and gives a local film thickness of 0.59 nm (one Ta layer) higher than the other areas. Of particular interest is to compare the out-of-plane crystal oscillation with the photoinduced CDW order parameter dynamics, which one obtains by switching to the diffraction mode, as shown in Figure 11.16b, to probe the embedded density-wave ordering. Both experiments are conducted under the same sample pump conditions (the ~ 1 mJ cm^{-2}, 50 fs near-infrared pulse illumination in nearly normal incidence at the repetition rate of 1 kHz). It is interesting to see the complex dynamics unfold at different timescales. Moreover, analysing the statistics of fluctuation in the imaging model offer insight into the local fluctuations triggered on the surface acoustic waves building on the large data sets of the repeated experiments.

11.6 Core Enabling Technologies

In the last decade, constructive dialogs[122,123,164] between the ultrafast electron diffraction, traditional electron microscopy, and the accelerator physics communities have proven to be extremely fruitful in creating a blossoming development in both UED and UEM instrumentation. Among the most visible progress is the incorporation of the accelerator technologies, especially those used in the RF linear accelerators, into the UED or UEM systems that has presented new pathways in dealing with space-charge effects. The advantages include a greatly improved temporal resolution and/or boosted bunch charges that yield unprecedented SNR, exemplified in the construction of the second generation dynamic TEM (DTEM) at LLNL,[127] and the incorporation of the RF cavity compression in keV UED beam-lines,[143,165–168] and the MeV-UED systems with relativistic RF photoinjectors at SLAC,[169] BNL,[170] DESY,[171] and Shanghai Jiaotong University,[172] among many others. Specifically, the dual-purpose deployment of the RF cavity as a longitudinal condenser and a monochromator in an electron microscope was proposed and pursued at MSU (see earlier sections).

It is instructive to point out that, while considering the strategy for bunch compression and the employment of high bunch charges, the best approach should be judged based on the instrument parameters, such as the energy scale, system stability, and the initial structure of the electron bunch. The accelerator-based pulse-compression technologies mostly have been developed for high-voltage electron accelerators, through the deployment of either bunching cavities[173,174] or properly placed static magnetic fields in a chicane[175,176] and achromat compressor.[177] Many of the concepts may be directly translated into the UED or UEM systems at lower beam energies. For example, the resonance cavity compression approaches have been deployed for the long pulse delivered at higher bunch charges at the low frequencies (≤1 GHz), or slightly shorter bunch at multi-GHz compact RF cavity systems, to take advantages of the lower power needed to operate them. Alternatively, optically based THz-driven keV/MeV bunch compressors have also been demonstrated,[178,179] which due to their shorter wavelength might be better utilized for circumstances of much shorter pulse input (typically ≤ few ps), but potentially poised to reach attosecond performance due to their jitter-free settings.

In this section, we shall return to the related core technologies that drive the new generations of the UEC/UEM instrumentation. This includes the high precision calibration of RF optics for beam dynamics control along with the rapid diagnosis toolsets developed to characterize the electron bunch phase-space parameters *in situ*, which in turns allow one to design and operate new sources with higher bunch charges.

11.6.1 Phase Space Manipulation

We will focus on phase-space manipulation to reach fs pulse compression in the limit of high bunch charge in the UEC/UEM beam delivery systems discussed in Sections 11.4 and 11.5. Generally, the precision in phase-space manipulation critically depends on the knowledge of the incident pulse structure, particularly, the pulse length and velocity chirp. Due to the strong space-charge forces experienced by the particles within the bunch, the pulse phase-space structure must be treated as a dynamical system that is constantly evolving along the beamline. Therefore, in running sophisticated UEC/UEM instrumentation, the need to adjust the electron optical elements upstream renders a necessity to compensate the changes in accordance with the (RF) optical systems downstream, which is one of the most challenging tasks in implementing the multi-modal experiments.

11.6.1.1 Dynamical System Characterization

Here, we describe the methodology employed at the MSU RF-UEC/UEM facilities for coping with the constant restructuring of the bunch phase space, aiming to provide *in situ* feedbacks for predictive control of the optical system to not only achieve but also maintain the optimization. This rapid *in situ* diagnosis for a wide range of tuning required for running the UEM applications builds on the *in situ* capability of performing coherent diffraction experiments. In such diagnosis runs, we insert high-quality single crystals into the specimen holder where the long-range periodic structure of the lattice serves as an "atomic grating". As in the optical grating case, the coherent diffraction from periodic long-ranged lattice renders delta-like Bragg peaks (Section 11.4). In this case, using electrons with a de Broglie wavelength as short as 0.037 Å, the momentum structure of the incident waves can be characterized in greater detail. The cross-dimensional characteristics of crystallography provides the mapping between the longitudinal properties in the energy and time domains into the transverse plane detector responses.[143]

To cope with the varying demands of UEM applications, we coin the bunch characterization protocols in a beam-energy-invariant fashion based on the phase-space structures expressed in the rest frame of bunch: (z, v_z). Furthermore, the de facto time-zero for the RF field is placed to match with the origin $(0, 0)$ of this coordinate system, that is, the centre-of-mass of the electron bunch precisely passes the centre of the RF gap when the running RF phase ϕ becomes 0. Under such a circumstance, the relative position of the particle (z) is mapped into the running phase $\phi = f_0 t$, where f_0 is the resonance frequency, based on $z = v_z t$ in the picture presented in Figure 11.17.

Within this framework, we can describe the velocity chirp in the rest frame $(v_z$ and $z)$:

$$a = \frac{dv_z}{dz}. \tag{11.47}$$

In the role of a longitudinal condenser lens, the field gradient from the synchronized RF cavity imparts a velocity chirp that compensates for space-charge dispersion effects (experienced during cathode-cavity propagation) to achieve a temporal focus at the specimen as the electron pulse propagates from the cavity. To understand how our diagnosis protocol is utilized to determine both the pulse and RF system characteristics in the beam delivery, we specify three key stages in our characterization (see Figure 11.17): (i) the pre-cavity stage, where the space-charge-driven bunch has a positive chirp $(a_0 > 0)$ entering the RF gap; (ii) the momentum kick exerted in the gap

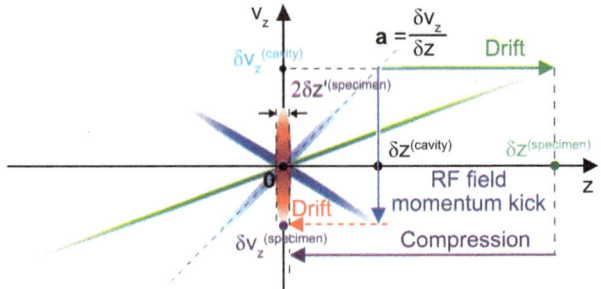

Figure 11.17 Longitudinal phase space manipulation for squeezing the bunch duration depicted in the rest frame of the bunch.

period; specifically given here is the scenario for time compression where the chirp of the bunch is shifted to a negative value; (iii) the drift period, where the negatively chirped bunch reaches longitudinal cross-over upon passing the specimen.

In the rest frame of the bunch, the action of the phase-space manipulation is given by the normalized field gradient (proportional to the RF amplitude ε_0), defined here as

$$\eta = -\frac{1}{v_0}\left(\frac{\delta v_z}{\delta\phi}\right), \tag{11.48}$$

where v_0 is the bunch velocity. For a finite-sized bunch, the velocity carried by each individual particle within the bunch is mapped into the phase of the particle. Provided a linear gradient field, as given by eqn (11.48), can be delivered by the RF cavity for the entire period of the bunch, then the linear chirp possessed by the bunch shall be maintained during the course of pulse compression. Specifically, the MSU UEM beamline is operated at 1.013 GHz (13th harmonics of the frequency comb generated at the fs laser oscillator, 77.923 MHz). The long wavelength cavity (~30 cm) provides a long period of linear gradient, which is particularly well-suited for compressing high bunch charge pulses (*i.e.*, containing up to 10^7 electrons) into the femtosecond regime or ≤1 eV.[131,143] We will show that the atomic grating effectively determines the key pulse characteristics (chirp, pulse spread and relative timing), and the RF system parameters (field strength, and phase).

In Figure 11.18, we show the results obtained in the injector/beam deliver sector of the first generation UEM system at MSU. Here, from the *in situ* coherent diffraction pattern, we seek to achieve precise characterization of the bunch from running the two main RF controller tuning parameters: the electric field amplitude (ε_0) and RF phase (ϕ). First, the effect from tuning the RF phase is studied with the correlated changes in the diffraction pattern (see Figure 11.18a). In this case with RF timing zero

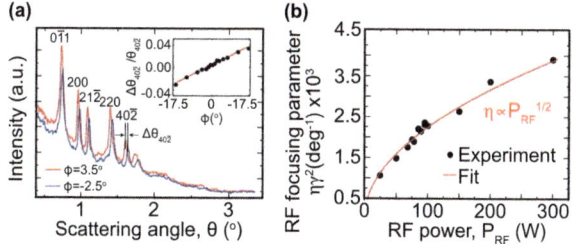

Figure 11.18 Characterization of the RF focusing control parameter η. (a) The angular shift of the diffraction curves recorded on the camera induced by an adjustment of the RF phase (ϕ) taken at an RF power (P_{RF}) of 75 W. The specific angular shift ($\Delta\theta_{40\bar{2}}$) associated with the ($40\bar{2}$) reflection is utilized to determine the focusing parameter η. As shown in the inset, the normalized angular shift $\Delta\theta_{40\bar{2}}/\theta_{40\bar{2}}$ is proportional to the RF phase ϕ. (b) The measured slope ($\eta\gamma^2$) versus the applied P_{RF} (solid symbols). Reproduced from ref. 143, https://doi.org/10.1063/1.4999456, under the terms of the CC BY 4.0 license https://creativecommons.org/licenses/by/4.0.

set to the de facto zero-crossing, the patterns obtained at a positive ϕ shrink to a small angular span, indicative of the bunch gaining energy. We could understand this based on the action exerted by η, given in eqn (11.48). Here, the linear RF gradient field (expressed in η) proportionally adjusts the particle velocity according to particle z. The action shifts the bunch velocity chirp to negative, hence compressing the bunch downstream irrespective of the phase. However, a distinction is made in the absolute velocity. With a net particle phase $\langle\phi\rangle > 0$ (or $\langle z\rangle < 0$ where $z = -v_z t$ for each particle in the rest frame), more particles in the back of the bunch gain energy than those in the front that lose energy, increasing the bunch velocity as a whole.

We now discuss how the relatively simple atomic grating characterization as given in Figure 11.18a can already render the conditions for reaching RF focusing conditions in the time and energy domains. Specifically, the slope in Figure 11.18a is utilized to provide the field strength:

$$\eta = \frac{1}{\theta_{hkl}} \left(\frac{\partial\theta_{hkl}}{\partial\phi}\right). \tag{11.49}$$

Under the coherent diffraction circumstance, the main source for the energy spread comes from the velocity chirp, which is observed *via* the smearing of the recorded Bragg peaks on the pixelated detector. Connecting the determined field strength with the condition minimizing pulse energy spread, denoted by η_E, one can reach the characterization of one extrema scenario (*i.e.*, the energy compression), easily carried out without requiring running the time-resolved measurement.[143] Based on eqn (11.47)–(11.49), one can see that the conditions for η_E are directly linked to the pre-cavity chirp, and so is the field strength η_t, needed to obtain the time compression, given by:

$$
\begin{cases}
\eta_t = \dfrac{a_0 + v_0/L_{\text{lens-sample}}}{2\pi f_0} \\[3mm]
\eta_E = \dfrac{a_0}{2\pi f_0}
\end{cases}
\tag{11.50}
$$

The two extrema field strengths are separated by a constant $v_0/(2\pi f_0 L_{\text{lens-sample}})$ – which only weakly depends on the defocusing under the scenario of a tight cross-over or the presence of the repulsive space-charge field against the formation of a tight cross-over. Irrespective of the effects at play, eqn (11.50) may offer a reliable prediction for the control from relatively straightforward characterization given by Figure 11.18a.

In practice, one needs to calibrate the RF system to properly place η_t and η_E by adjusting the input RF power for the optical control. Based on the mapping of the laboratory-frame energy (velocity) gain according to eqn (11.49) by applying a phase offset, one arrives at[143]

$$
\eta = \frac{e\varepsilon_0}{\gamma^3 \pi f_0 m v_0} \sin\left(\frac{\pi f_0 d}{v_0}\right)\cos\phi,
\tag{11.51}
$$

where d is the gap size. It is easy to see that by operating under the zero-crossing condition, eqn (11.49) leads to a direct characterization of the RF field ε_0. It is important to point out that to convert the slope obtained in Figure 11.18a to the beam energy shift informed by the diffraction pattern, one needs to consider the length contraction, that is $\gamma^2 \Delta V_0 = \Delta v_z$, where γ is the relativistic Lorentz factor. This correction $\left(i.e., -\dfrac{\gamma^2 \Delta v_0}{v_0} = \dfrac{\Delta\theta_{hkl}}{\theta_{hkl}}\right)$ is important for the appropriate calibration of the RF power delivery, where the power in the gap is directly given by $P_{RF}^0 = \varepsilon_0^2/2$, whereas the actual required power at the input port (P_{RF}) will necessarily depend on the loss function at the cavity, the coupling coefficients, and detuning effects, described operationally by $\eta\gamma^2 = k\sqrt{P_{RF}}$, where k is the RF lens constant. In Figure 11.18b, we present the results demonstrating the linearity between the measured field and the square root of the applied RF power at the input port, from which the lens constant k can be extracted and used to properly set the required η_t and η_E for the compression.

11.6.1.2   Characterization of the Bunch Brightness

Based on the detailed results from operating the system in the two extrema cases, we can determine the bunch brightness from the atomic grating approach. The phase-space density (ρ_{4D}) is primarily given by the coherent length (X_c) and the electron dose (D_e) obtained directly *via* the coherent diffraction pattern itself; see eqn (11.38)–(11.40) which are true statements for the beam down the column after different stages of slicing. Here, the energy spread from the detector point spread function (PSF) needs to be deconvoluted from the coherence length measurement, given by the overall width $w^2 = w_{PSF}^2 + w_E^2$ (in the limit where high quality crystal with crystalline correlation length $\gg X_c$). The energy spread component can be isolated using the reciprocity relationship with the w_E increasing linearly with the scattering angle. Additional nonreciprocal pattern changes are associated with aberration and astigmatism, which may be partially corrected by tuning the transverse optics as well as the alignment of the beam across the cavity. Meanwhile, the dose is identified from the detector signal counting and the illuminated area, easily obtained from various imaging modalities (including shadow imaging in projection geometry).

The phase-space characterization is demonstrated for N_e of 10^6 and 10^7 in Figure 11.19. The diagnosis builds on reaching the two "extrema" of pulse compression, given by minimizing the spread of the pulse in energy (ΔE), directly taken from the width of θ_{hkl}, and the pulse width, given by the Δt of the response function measured by UEC; see ref. 143 for details. Under these two scenarios, the phase space structure has zero chirp, which means that the emittance from a given N_e can be determined directly from $\Delta t\Delta E$. Alternatively, the crystallographic approach also provides the transverse emittance; as described in eqn (11.36)–(11.40), these phase-space densities are directly connected with the beam brightness discussed in Section 11.5 (see Table 11.1) for the multimodality consideration of the high-throughput UEM system.

11.6.2 Optimization of the Source Brightness

The electron pulses used in the UEC/UEM systems are typically created by applying the fs ultraviolet (266 nm) laser pulses on a flat photocathode housed within the Pierce gun compartment, proximity-coupled with a C0 lens and an aperture to maintain central extraction of the beam. The emission characterized by the bunch charge N_e, ranges typically from 10^4 to 10^7 with the adjustable energy range of 40–100 keV. The high temporal and spatial resolutions achievable by the high-throughput UEC and UEM systems introduced in Sections 11.4 and 11.5 evidence not only the ability to generate but also to maintain a high bunch phase-space density from the photocathode to the detector. The most important metric B_{CW} discussed in Section 11.5 that underscores this superiority of the photo-emitter compared to the conventional thermionic emitter (operating at one-electron limit) builds on the favourable scaling in filling the phase space as the emission current increases,[144] leading to an increase in ρ_{4D}. One also found that this brightness enhancement in the multiparticle bunch beam extraction shall continue to increase until the virtual cathode limit – presented when the field from the increased surface counter ions density starts to supress the emission efficiency.[144,180–182] Figure 11.20 underscores the physics for this

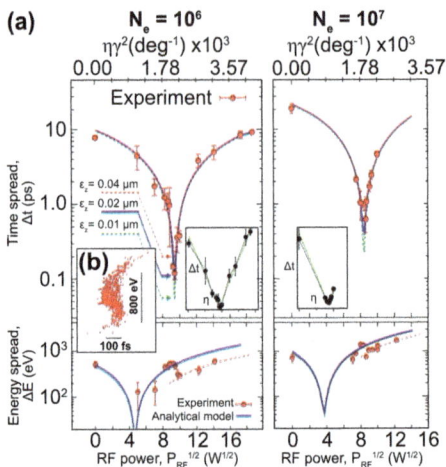

Figure 11.19   Characterization of high-intensity bunch phase-space evolution
under RF optics. (a) The RMS time (Δt) and RMS energy (ΔE) spreads
of the electron bunches at N_e of 10^6 and 10^7. The experimental data
are depicted in solid symbols, whereas the lines are simulation
results from the analytical model. The solid line represents the
best fit to the experimental data. The logarithmic scale in the vertical
axis highlights the values near the compression point, which is the
most sensitive range for determining the emittance e_z. The insets
show the data plotted on linear scales along with the best fits
obtained using the analytical model (green). A visible deviation
from the analytical model prediction is identified in the energy
compression experiments presented in the lower part of the panels
where the data after the time compression points collectively shift
downward. This deviation can be attributed to the transformation of
the phase space structure induced by the space charge effects at the
temporal focal plane. (b) The phase space structure at the temporal
focal plane obtained for $N_e = 10^6$ using the MLFMM approach on an
exaggerated scale along the time-axis is shown. Reproduced from
ref. 143, https://doi.org/10.1063/1.4999456, under the terms of the
CC BY 4.0 license https://creativecommons.org/licenses/by/4.0.

brightness enhancement effect using the multi-level fast multiple method
(ML-FMM) calculations designed to track the stochastic scattering effect
inherent in the short-pulse-driven dense beam formation.[183]

   We consider a locally flat Pierce-photocathode geometry, where the initial
emission current is driven by the linear extraction field F_a perpendicular to
the surface. The initial phase space is occupied by the stochastic particles
emitted from the cathode surface with a thermal energy profile[78,181] depicted
by the three-step-model[184] with the Schottky effect,[185] as presented in the
short-time map (Figure 11.20) of the particle flows along the transverse (x)
and longitudinal (z) directions, with $N_e^{emit} = 2 \times 10^7$. The extracted pulse is
energy-boosted under $F_a = 10$ MV m^{-1} to reach the terminal stage of
extraction depicted in the right panel with the bunch collective dynamics
developing into a laminar flow as viewed in the rest frame of the bunch.

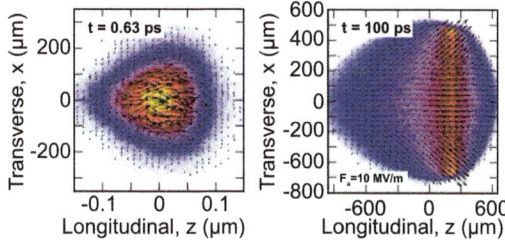

Figure 11.20 Multi-level fast-multiple model (ML-FMM) simulation of the photo-electron bunch extraction. The panels depict particle trajectories for electron bunch extraction under a field $F_a = 10$ MV m^{-1} from a flat photocathode at two stages (630 fs and 100 ps) along the transverse (x) and the longitudinal (z) directions. The particle momenta (p_x and p_z) depicted by the arrows give a certain spreading but largely are correlated with the position (x and z) influenced by the field structures extended within the multi-particle beams. Reproduced from ref. 144 with permission from AIP Publishing, Copyright 2013.

From the individual particle's transverse and longitudinal trajectories, we tracked the emittance growth from the initial thermal slices as a function of time and particle number in more comprehensive studies.[144]

Results emerging from joint experimental and numerical investigations confirm not only the strong influences of the space-charge effects on bunch profiles,[6,181,186–191] but also the essential early-stage emittance growth sensitively depending on the seeded thermal distribution and the cathode geometry. In particular, a bright beam generation is ensured by achieving and maintaining the laminar flow.[144] If such a condition is met, one finds the transverse emittance grows sub-linearly, whereas the longitudinal emittance grows nearly linearly, with respect to N_e until the virtual cathode limit that severely disrupts the laminar flow condition. Furthermore, once the bunch is fully extracted, one also finds that a quasi-conservation rule applies in the collective phase-space evolution along both transverse and longitudinal directions. This leaves the rest-frame phase-space density as a conserved quantity (see Section 11.5) that may be regarded as a mean-field equivalence of the Liouville theorem. These conservation rules largely apply, irrespective of the strength of the internal Coulombic forces,[144,164] providing a theoretical baseline for designing the high bunch charge UEC/UEM systems.

11.6.3 Precision RF Optical Control and Beam Delivery

While the core optical technologies combining magnetic and RF condenser lenses promise to deliver a bright beam with sub-100 fs temporal resolution limited only by the emittance (*e.g.*, the simulation presented in Figure 11.17 promises 30 fs RMS delivered at the focal plane), translating such short bunch delivery to performance requires high stability and precision in the control system. To quantitatively resolve the phase stability requirement, we resort again to the atomic grating measurements. The inset of Figure 11.18a shows a phase offset leading to a linear change in the observed beam velocity that is translated linearly to the arrival time difference. The arrival time jitter here is given under η_t, and from eqn (11.50), this condition is set by the focal length $L_{\text{lens-sample}}$ and the pre-cavity chirp – the instability of the RF field only couples to the defocusing effect as a second-order expansion, the phase jitter is expected to play a bigger role in determining the overall precision of UEC/UEM in a relatively amplitude-stable system. We will focus on reducing the RF phase jitter to meet the fundamental limit presented by emittance. To see how phase instabilities impact the energy spread, we further arrive at

$$\Delta V_0 = \frac{-\eta V_0 \gamma^2}{1 + \gamma^2 \beta^2} \Delta \phi, \tag{11.52}$$

based on the formalism in Section 11.6.1. Since the arrival time $t_{\text{arr}} = L_{\text{lens-sample}}/V_0$, we calculate the jitter in the arrival time based on the jitter present in the RF optics, measured in $\Delta \phi_{\text{rms}}$:

$$\Delta t_{\text{arr}} = \frac{\eta L_{\text{lens-sample}} \gamma^2}{V_0 \left(1 + \gamma^2 \beta^2\right)} \Delta \phi_{\text{rms}}. \tag{11.53}$$

From eqn (11.53), the time-to-phase ratio, $k_{t\phi} = \dfrac{\eta L_{\text{lens-sample}} \gamma^2}{V_0 \left(1 + \gamma^2 \beta^2\right)}$. In a typical arrangement, with $L_{\text{lens-sample}}$ being 0.5–1 metre, and V_0 in the range of 60–100 keV, a $k_{t\phi}$ value of 3.5–4.5 is common. This means that an RMS phase jitter at the level of 0.015° or better (at f_0 of 1.013 GHz) is required.

The basic UEC/UEM phase-locked loop (PLL) design is based on comparing the output signal with a reference signal derived from the high-harmonic signal where $f_{\text{ref}} = N \cdot f_{\text{osc}}$, with $N = 13$ in our case to lock the RF cavity to the laser oscillator frequency f_{osc}. Converting the frequency in the optical domain to the RF domain is accomplished by utilizing a fast-rectifying photodiode to re-create the frequency comb of the laser oscillator system in the RF signal domain. The PLL then generates a master RF control signal by locking at the f_{ref} to maintain the phase stability. In our system, this is typically better than 30 fs RMS or 0.015° RMS. As the pump/probe pulses and the RF system derive their signals from the same timing base, the synchronization between the probe/pump and the RF optical control thus is ensured under a reasonable beam pointing stability; here, the pump–probe overlap determines the time-zero for the entire experiment. However, one frequently finds that phase variation is invariably introduced into the RF signal loop with the main source of noise (jitter and phase shift) created in the RF amplification process and the cavity itself. Several additional sources of noise will couple to the more extended PLL system when the laser system is not proximity-coupled to the RF optics and beam delivery systems. Since every system configuration might be different, the key to success is to identify these spurious noise sources and correcting them through proportional–integral–derivative (PID) control loops over the appropriate timescales of the experiments.

Several types of noise reduction schemes based on feedback-control have been implemented for the RF systems used in keV-scale UED[192,193] and UEC/UEM[111,143] systems, and have been proven successful in quelling the integrated low-frequency (≤10 Hz) drift to ≤100 fs timescales. To deal with higher frequency noise to achieve the sub-30 fs stability, the new MSU phase locked loop (PLL) design employed two PID systems. One is FPGA-based and directly couples to the low-level RF (LLRF) system, operating at a frequency higher than 4 MHz.[194] The high-frequency PID can easily reject any phase noise spectrum within the cavity bandwidth. The precision of the LLRF determines the maximum correction time for the high-frequency PID loop. We use a second digitized PID loop that operates at the 10 Hz level, but with a sampling bandwidth from 100 Hz to several kHz based on averaging from a 4 kHz digitizer. The two PID loops mesh over the entire RF circuit responses to correct the phase drift from milliseconds to a very long period (>1 day), with intervening RF filters to reject biases introduced into the global RF system.

Acknowledgements

The work described in this chapter is based on many efforts at Michigan State University. In particular, I would like to acknowledge Ramani Raman, Ryan Murdick, and Yoshie Murooka in establishing the UEnC system and conducting the nanoparticle diffraction and photovoltammetry measurements, and Kiseok Chang, Fei Yuan and Terry Han for assistance in further developing the photovoltammetry protocols. I am grateful for Terry Han, Zhengsheng Tao and Faran Zhou for carrying out the ultrafast electron crystallography measurements and implementing the new analysis protocols. In addition, I also acknowledge Ming Zhang and Xiaoyi Sun for their efforts in upgrading the UEC system. The herculean efforts of Zhensheng Tao, Kiseok Chang, Tianyin Sun, Joe Williams. Shuaishuai Sun, and Xiaoyi Sun in setting up the UEM systems are thanked wholeheartedly. Particularly, I would like to thank Phil Duxbury, Bhanu Mahanti, Martin Berz, Kyoko Makino, He Zhang, Jenni Portmann, Brandon Zerbe, Simon Billinge, Mercouri Kanatzidis, Christos Malliakas, Nelson Sepulveda, Ti Ruan, Shen Zhao and Shriraj Kunjir for good collaborations. Finally, the work would not have been possible without the significant financial support of several research grants, including those from the US National Science Foundation (DMR 1126343 and 1625181), US Department of Energy (DE-FG0206ER46309 and SC0018529), and the MSU Foundation, for which I am most grateful.

References

1. J. C. H. Spence and J. M. Zuo, *Electron Microdiffraction*, Springer US, Boston, MA, 1992.
2. P. J. Brown *et al.*, *International Tables for Crystallography*, 2004, vol. C.
3. L. Mino, *et al.*, *Rev. Mod. Phys.*, 2018, **90**, 025007.

4. D. L. Dorset, *Structural Electron Crystallography*, Springer US, Boston, MA, 1995.
5. A. H. Zewail, *J. Phys. Chem. A*, 2000, **104**, 5660–5694.
6. R. Srinivasan, *et al.*, *Helv. Chim. Acta*, 2003, **86**, 1761–1799.
7. H. Mark and R. Wierl, *Naturwissenschaften*, 1930, **18**, 205–205.
8. P. Debye, L. Bewilogua and F. Ehrhardt, *Phys. Z.*, 1929, **30**, 84.
9. A. H. Zewail, *Phys. Today*, 1980, **33**, 27–33.
10. J. C. Williamson and A. H. Zewail, *Proc. Natl. Acad. Sci. U. S. A.*, 1991, **88**, 5021–5025.
11. H. E. Elsayed-Ali and G. A. Mourou, *Appl. Phys. Lett.*, 1988, **52**, 103–104.
12. S. Williamson, G. Mourou and J. C. M. Li, *Phys. Rev. Lett.*, 1984, **52**, 2364–2367.
13. A. A. Ischenko, J. D. Ewbank and L. Schafer, *J. Phys. Chem.*, 1995, **99**, 15790–15797.
14. A. A. Ischenko, P. M. Weber and R. J. D. Miller, *Chem. Rev.*, 2017, **117**, 11066–11124.
15. J. C. Williamson, *et al.*, *Nature*, 1997, **386**, 159–162.
16. H. Ihee, *et al.*, *Science*, 2001, **291**, 458–462.
17. R. Srinivasan, *et al.*, *Science*, 2005, **307**, 558–563.
18. V. A. Lobastov, *et al.*, *J. Phys. Chem. A*, 2001, **105**, 11159–11164.
19. H. Ihee, *et al.*, *J. Phys. Chem. A*, 2002, **106**, 4087–4103.
20. J. Cao, H. Ihee and A. H. Zewail, *Proc. Natl. Acad. Sci. U. S. A.*, 1999, **96**, 338–342.
21. H. Ihee, J. Cao and A. H. Zewail, *Chem. Phys. Lett.*, 1997, **281**, 10–19.
22. J. Cao, H. Ihee and A. H. Zewail, *Chem. Phys. Lett.*, 1998, **290**, 1–8.
23. C.-Y. Ruan, *et al.*, *Proc. Natl. Acad. Sci. U. S. A.*, 2001, **98**, 7117–7122.
24. B. M. Goodson, *et al.*, *Chem. Phys. Lett.*, 2003, **374**, 417–424.
25. J. C. Williamson and A. H. Zewail, *Chem. Phys. Lett.*, 1993, **209**, 10–16.
26. I. Hargittai and M. Hargittai, *Stereochemical Applications of Gas-Phase Electron Diffraction, Part A*, John Wiley & Sons, 1988.
27. A. H. Zewail, *Berichte der Bunsengesellschaft für physikalische Chemie*, 1985, **89**, 264–270.
28. A. I. Anatoli and A. A. Sergei, *Advances in Imaging and Electron Physics*, Elsevier, 1st edn, 2014, vol. 184.
29. M. Ben-Nun, J. Quenneville and T. J. Martínez, *J. Phys. Chem. A*, 2000, **104**, 5161–5175.
30. T. J. A. Wolf, *et al.*, *Nat. Chem.*, 2019, **11**, 504–509.
31. S. Ino and S. Ogawa, *J. Phys. Soc. Jpn.*, 1967, **22**, 1365–1374.
32. D. J. Wales, *Science*, 1996, **271**, 925–929.
33. W. Eberhardt, *Surf. Sci.*, 2002, **500**, 242–270.
34. W. D. Luedtke and U. Landman, *J. Phys. Chem.*, 1996, **100**, 13323–13329.
35. R. L. Whetten, *et al.*, *Adv. Mater.*, 1996, **8**, 428–433.
36. D. Zanchet, *et al.*, *Chem. Phys. Lett.*, 2000, **323**, 167–172.
37. L. D. Marks, *Rep. Prog. Phys.*, 1994, **57**, 603.
38. D. L. Klein, *et al.*, *Nature*, 1997, **389**, 699–701.
39. Y. Y. Illarionov, *et al.*, *Nat. Commun.*, 2020, **11**, 3385.

40. A. N. Shipway, E. Katz and I. Willner, *Chem. Phys. Chem.*, 2000, **1**, 18–52.
41. A. P. Alivisatos, *Science*, 1996, **271**, 933–937.
42. D. D. Awschalom, D. P. DiVincenzo and J. F. Smyth, *Science*, 1992, **258**, 414–421.
43. M. Haruta, *Catal. Today*, 1997, **36**, 153–166.
44. A. Hanisch-Blicharski, *et al.*, *Ultramicroscopy*, 2013, **127**, 2–8.
45. C.-Y. Ruan, D.-S. Yang and A. H. Zewail, *J. Am. Chem. Soc.*, 2004, **126**, 12797–12799.
46. C.-Y. Ruan, *et al.*, *Science*, 2004, **304**, 80–84.
47. C.-Y. Ruan, *et al.*, *Nano Lett.*, 2007, 7, 1290–1296.
48. P. Baum, D.-S. Yang and A. H. Zewail, *Science*, 2007, **318**, 788–792.
49. R. K. Raman, *et al.*, *Phys. Rev. Lett.*, 2008, **101**, 077401.
50. A. Plech, *et al.*, *Nat. Phys.*, 2006, **2**, 44–47.
51. C.-Y. Ruan, *et al.*, *Microsc. Microanal.*, 2009, **15**, 323–337.
52. R. K. Raman, *et al.*, *Phys. Rev. Lett.*, 2010, **104**, 123401.
53. S. Schäfer, W. Liang and A. H. Zewail, *Chem. Phys. Lett.*, 2011, **515**, 278–282.
54. F. Ercolessi, W. Andreoni and E. Tosatti, *Phys. Rev. Lett.*, 1991, **66**, 911–914.
55. N. Wang, S. I. Rokhlin and D. F. Farson, *Nanotechnology*, 2008, **19**, 415701.
56. R. L. McGreevy, *J. Phys.: Condens. Matter*, 2001, **13**, R877.
57. A. P. Pogany and P. S. Turner, *Acta Crystallogr., Sect. A: Cryst. Phys., Diffr., Theor. Gen. Crystallogr.*, 1968, **24**, 103–109.
58. L. M. Peng, S. L. Dudarev and and M. J. Whelan, *High Energy Electron Diffraction and Microscopy*, Oxford University Press, Oxford, New York, 2011.
59. H. Park and J. M. Zuo, *Appl. Phys. Lett.*, 2009, **94**, 251103.
60. S. Schäfer, W. Liang and A. H. Zewail, *Chem. Phys. Lett.*, 2010, **493**, 11–18.
61. H. Rose, *Geometrical Charged-Particle Optics (Springer Series in Optical Sciences, 142)*, Springer, Berlin, 2010.
62. R. A. Murdick, *et al.*, *Phys. Rev. B: Condens. Matter Mater. Phys.*, 2008, 77, 245329.
63. K. Chang, *et al.*, *Mod. Phys. Lett. B*, 2011, **25**, 2099–2129.
64. T. M. Brenner, *et al.*, *Nat. Rev. Mater.*, 2016, **1**, 1–16.
65. Y. Yin and A. P. Alivisatos, *Nature*, 2005, **437**, 664–670.
66. M. Grätzel, *J. Photochem. Photobiol., C*, 2003, **4**, 145–153.
67. V. I. Klimov, *et al.*, *Science*, 2000, **290**, 314–317.
68. H. Utzat, *et al.*, *Science*, 2019, **363**, 1068–1072.
69. X. Duan, *et al.*, *Nature*, 2003, **421**, 241–245.
70. J. Yu, *et al.*, *Sci. Adv.*, 2019, 5, eaav3140.
71. Y. Xia, *et al.*, *Adv. Mater.*, 2003, **15**, 353–389.
72. N. A. Anderson and T. Lian, *Annu. Rev. Phys. Chem.*, 2005, **56**, 491–519.
73. A. Kubo, *et al.*, *Nano Lett.*, 2005, 5, 1123–1127.
74. S. Chen, *et al.*, *Light: Sci. Appl.*, 2019, **8**, 17.

75. O. A. Aktsipetrov, *et al.*, *Phys. Rev. B: Condens. Matter Mater. Phys.*, 1999, **60**, 8924–8938.
76. T. Sato, *et al.*, *J. Appl. Phys.*, 1997, **82**, 696–701.
77. F. Vigliotti, *et al.*, *Angew. Chem., Int. Ed.*, 2004, **43**, 2585.
78. S. Sun, *et al.*, *Struct. Dyn.*, 2020, 7, 064301.
79. G. Sciaini and R. D. Miller, *Rep. Prog. Phys.*, 2011, **74**, 096101.
80. M. Gao, *et al.*, *Nature*, 2013, **496**, 343–346.
81. B. J. Siwick, J. R. Dwyer, R. E. Jordan and R. J. D. Miller, *Science*, 2003, **302**, 1382–1385.
82. A. de la Torre, D. M. Kennes, M. Claassen, S. Gerber, J. W. McIver and M. A. Sentef, *Rev. Mod. Phys.*, 2021, **93**, 041002.
83. Y. Tokura, M. Kawasaki and N. Nagaosa, *Nat. Phys.*, 2017, **13**, 1056–1068.
84. B. Keimer and J. E. Moore, *Nat. Phys.*, 2017, **13**, 1045–1055.
85. Z. Sun and A. J. Millis, *Phys. Rev. X*, 2020, **10**, 021028.
86. W. Fu, L.-Y. Hung and S. Sachdev, *Phys. Rev. B: Condens. Matter Mater. Phys.*, 2014, **90**, 024506.
87. B. Damski and W. H. Zurek, *Phys. Rev. A: At., Mol., Opt. Phys.*, 2006, **73**, 063405.
88. J. Berges, A. Rothkopf and J. Schmidt, *Phys. Rev. Lett.*, 2008, **101**, 041603.
89. T. W. B. Kibble, *J. Phys. Math. Gen.*, 1976, **9**, 1387.
90. T. W. B. Kibble, in *Patterns of Symmetry Breaking*, ed. H. Arodz, J. Dziarmaga and W. H. Zurek, Springer Netherlands, Dordrecht, 2003, pp. 3–36.
91. P. Calabrese and J. Cardy, *Phys. Rev. Lett.*, 2006, **96**, 136801.
92. C. Giannetti, *et al.*, *Adv. Phys.*, 2016, **65**, 58–238.
93. A. Mitra, *Annu. Rev. Condens. Matter Phys.*, 2018, **9**, 245–259.
94. M. Eckstein, M. Kollar and P. Werner, *Phys. Rev. Lett.*, 2009, **103**, 056403.
95. P. M. Chaikin, *Principles of Condensed Matter Physics*, Cambridge University Press, Cambridge, Revised edn, 2000.
96. W. L. McMillan, *Phys. Rev. B: Solid State*, 1977, **16**, 643–650.
97. L. D. Landau, *Zh. Eksp. Teor. Fiz.*, 1937, **11**, 19.
98. L. Landau, *Zh. Eksp. Teor. Fiz.*, 1937, 7, 1232.
99. P. A. Lee, T. M. Rice and P. W. Anderson, *Phys. Rev. Lett.*, 1973, **31**, 462–465.
100. X. Sun, S. Sun and C.-Y. Ruan, *C. R. Phys.*, 2021, **22**, 15–73.
101. Y. Feng, *et al.*, *Proc. Natl. Acad. Sci. U. S. A.*, 2012, **109**, 7224–7229.
102. U. C. Täuber, *Critical Dynamics: A Field Theory Approach to Equilibrium and Non-Equilibrium Scaling Behavior*, Cambridge University Press, Cambridge, 2014.
103. P. C. Hohenberg and A. P. Krekhov, *Phys. Rep.*, 2015, **572**, 1–42.
104. U. C. Täuber, in *Ageing and the Glass Transition*, ed. M. Henkel, M. Pleimling and R. Sanctuary, Springer, Berlin, Heidelberg, 2007, pp. 295–348.
105. R. Yusupov, *et al.*, *Nat. Phys.*, 2010, **6**, 681–684.

106. C. D. Malliakas and M. G. Kanatzidis, *J. Am. Chem. Soc.*, 2006, **128**, 12612–12613.
107. N. Ru, *et al.*, *Phys. Rev. B: Condens. Matter Mater. Phys.*, 2008, 77, 035114.
108. V. Brouet, *et al.*, *Phys. Rev. B: Condens. Matter Mater. Phys.*, 2008, 77, 235104.
109. H. Yao, *et al.*, *Phys. Rev. B: Condens. Matter Mater. Phys.*, 2006, **74**, 245126.
110. A. Kogar, *et al.*, *Nat. Phys.*, 2020, **16**, 159–163.
111. F. Zhou, *et al.*, *Nat. Commun.*, 2021, **12**, 566.
112. M. Maschek, *et al.*, *Phys. Rev. B: Condens. Matter Mater. Phys.*, 2018, **98**, 094304.
113. H.-M. Eiter, *et al.*, *Proc. Natl. Acad. Sci. U. S. A.*, 2013, **110**, 64–69.
114. J. P. Hinton, *et al.*, *Phys. Rev. B: Condens. Matter Mater. Phys.*, 2013, **88**, 060508.
115. J. Chang, *et al.*, *Nat. Phys.*, 2012, **8**, 871–876.
116. H.-H. Kim, *et al.*, *Science*, 2018, **362**, 1040–1044.
117. E. H. da Silva Neto, *et al.*, *Science*, 2014, **343**, 393–396.
118. E. A. Nowadnick, *et al.*, *Phys. Rev. Lett.*, 2012, **109**, 246404.
119. A. W. Overhauser, *Phys. Rev.*, 1962, **128**, 1437–1452.
120. A. W. Overhauser, *Phys. Rev. B: Solid State*, 1971, **3**, 3173–3182.
121. Y. Wang, *et al.*, *Nature*, 2022, **606**, 896–901.
122. E. Hall *et al.*, *Future of Electron Scattering and Diffraction*, US Department of Energy, Washington, DC (United States), 2014.
123. W. E. King, *et al.*, *J. Appl. Phys.*, 2005, **97**, 111101.
124. A. H. Zewail, *Science*, 2010, **328**, 187–193.
125. V. A. Lobastov, R. Srinivasan and A. H. Zewail, *Proc. Natl. Acad. Sci. U. S. A.*, 2005, **102**, 7069–7073.
126. B. Barwick and A. H. Zewail, *ACS Photonics*, 2015, **2**, 1391–1402.
127. T. LaGrange, *et al.*, *Ultramicroscopy*, 2008, **108**, 1441–1449.
128. J. S. Kim, *et al.*, *Science*, 2008, **321**, 1472–1475.
129. T. Voisin, *et al.*, *Mater. Today*, 2020, **33**, 10–16.
130. S. V. Kutsaev, *Eur. Phys. J. Plus*, 2021, **136**, 446.
131. F. Zhou, J. Williams and C.-Y. Ruan, *Chem. Phys. Lett.*, 2017, **683**, 488–494.
132. J. M. Cowley, *Diffraction Physics*, North Holland, Amsterdam, New York, 3rd edn, 1995.
133. M. Born and E. Wolf, *Principles of Optics: 60th Anniversary Edition*, Cambridge University Press, Cambridge, 7th edn, 2019.
134. L. Reimer and H. Kohl, *Transmission Electron Microscopy: Physics of Image Formation: 36*, Springer, New York, NY, 5th edn, 2008.
135. *High-Resolution Transmission Electron Microscopy: and Associated Techniques*, ed. P. Buseck, J. Cowley and L. Eyring, Oxford University Press, Oxford, New York, 1989.
136. R. F. Egerton, *Rep. Prog. Phys.*, 2008, 72, 016502.
137. N. W. Bigelow, *et al.*, *ACS Nano*, 2012, **6**, 7497–7504.
138. E. Pomarico, *et al.*, *MRS Bull.*, 2018, **43**, 497–503.

139. *Scanning Transmission Electron Microscopy: Imaging and Analysis*, ed. S. J. Pennycook and P. D. Nellist, Springer, New York, NY Heidelberg, 2011th edn, 2011.
140. *Science of Microscopy*, ed. P. W. Hawkes and J. C. H. Spence, Springer, 2008.
141. F. Zernike, *Physica*, 1938, **5**, 785–795.
142. J. Frank, *Optik*, 1973, **38**, 519.
143. J. Williams, *et al.*, *Struct. Dyn.*, 2017, **4**, 044035.
144. J. Portman, *et al.*, *Appl. Phys. Lett.*, 2013, **103**, 253115.
145. H. Bethe, *Ann. Phys.*, 1930, **397**, 325–400.
146. L. D. Landau and E. M. Lifshitz, *Quantum Mechanics: Non-Relativistic Theory: 3*, Butterworth-Heinemann, Singapore, 3rd edn, 1981.
147. P. Schattschneider and W. S. M. Werner, *J. Electron Spectrosc. Relat. Phenom.*, 2005, **143**, 81–95.
148. H. Raether, *Excitation of plasmons and interband transitions by electrons*, Springer-Verlag, 1980, vol. 88.
149. B. Barwick, D. J. Flannigan and A. H. Zewail, *Nature*, 2009, **462**, 902–906.
150. H. Liu, J. S. Baskin and A. H. Zewail, *Proc. Natl. Acad. Sci. U. S. A.*, 2016, **113**, 2041–2046.
151. T. T. A. Lummen, *et al.*, *Nat. Commun.*, 2016, 7, 13156.
152. Z. Su, *et al.*, *J. Am. Chem. Soc.*, 2017, **139**, 4916–4922.
153. X. Fu, *et al.*, *Nat. Commun.*, 2020, **11**, 5770.
154. K. Nakanishi, *et al.*, *J. Phys. Soc. Jpn.*, 1977, **43**, 1509–1517.
155. K. Nakanishi and H. Shiba, *J. Phys. Soc. Jpn.*, 1984, **53**, 1103–1113.
156. A. Suzuki, M. Koizumi and M. Doyama, *Solid State Commun.*, 1985, **53**, 201–203.
157. A. Spijkerman, *et al.*, *Phys. Rev. B: Condens. Matter Mater. Phys.*, 1997, **56**, 13757–13767.
158. T. Ishiguro and H. Sato, *Phys. Rev. B: Condens. Matter Mater. Phys.*, 1991, **44**, 2046–2060.
159. B. Sipos, *et al.*, *Nat. Mater.*, 2008, 7, 960–965.
160. R. E. Thomson, *et al.*, *Phys. Rev. B: Condens. Matter Mater. Phys.*, 1988, **38**, 10734–10743.
161. R. E. Thomson, *et al.*, *Phys. Rev. B: Condens. Matter Mater. Phys.*, 1994, **49**, 16899–16916.
162. X. L. Wu and C. M. Lieber, *Science*, 1989, **243**, 1703–1705.
163. K. Nakanishi and H. Shiba, *J. Phys. Soc. Jpn.*, 1977, **43**, 1839–1847.
164. M. Berz *et al.*, in *Advances in Imaging and Electron Physics*, ed. P. W. Hawkes, Elsevier, 2015, vol. 191, pp. 1–133.
165. T. van Oudheusden, *et al.*, *Phys. Rev. Lett.*, 2010, **105**, 264801.
166. M. Gao, *et al.*, *Appl. Phys. Lett.*, 2013, **103**, 033503.
167. R. P. Chatelain, *et al.*, *Appl. Phys. Lett.*, 2012, **101**, 081901.
168. O. Zandi, *et al.*, *Struct. Dyn.*, 2017, **4**, 044022.
169. S. Weathersby, *et al.*, *Rev. Sci. Instrum.*, 2015, **86**, 073702.
170. P. Zhu, *et al.*, *New J. Phys.*, 2015, **17**, 063004.

171. M. Hada *et al.*, in *Research in Optical Sciences*, Optica Publishing Group, 2012, p. JT2A.47.

172. J. Wu, *et al.*, *Proc. Natl. Acad. Sci. U. S. A.*, 2022, **119**, e2111949119.

173. L. Serafini, *IEEE Trans. Plasma Sci.*, 1996, **24**, 421–427.

174. S. Thorin *et al.*, in *International Free Electron Laser Conference, 2010*, 2010.

175. S. G. Anderson, *et al.*, *Phys. Rev. Lett.*, 2003, **91**, 074803.

176. D. H. Dowell, *et al.*, *Nucl. Instrum. Methods Phys. Res., Sect. A*, 1997, **393**, 184–187.

177. F. Qi, *et al.*, *Phys. Rev. Lett.*, 2020, **124**, 134803.

178. E. C. Snively, *et al.*, *Phys. Rev. Lett.*, 2020, **124**, 054801.

179. D. Ehberger, *et al.*, *Phys. Rev. Appl.*, 2019, **11**, 024034.

180. Á. Valfells, *et al.*, *Phys. Plasmas*, 2002, **9**, 2377–2382.

181. Z. Tao, *et al.*, *J. Appl. Phys.*, 2012, **111**, 044316.

182. P. Zhang, *et al.*, *Appl. Phys. Rev.*, 2017, **4**, 011304.

183. H. Zhang and M. Berz, *Nucl. Instrum. Methods Phys. Res., Sect. A*, 2011, **645**, 338–344.

184. W. F. Krolikowski and W. E. Spicer, *Phys. Rev. B: Solid State*, 1970, **1**, 478–487.

185. D. H. Dowell and J. F. Schmerge, *Phys. Rev. Spec. Top.–Accel. Beams*, 2009, **12**, 074201.

186. B. J. Siwick, *et al.*, *J. Appl. Phys.*, 2002, **92**, 1643–1648.

187. A. Gahlmann, S. T. Park and A. H. Zewail, *Phys. Chem. Chem. Phys.*, 2008, **10**, 2894–2909.

188. B. W. Reed, *J. Appl. Phys.*, 2006, **100**, 034916.

189. B. S. Zerbe, *et al.*, *Phys. Rev. Accel. Beams*, 2018, **21**, 064201.

190. N. Bach, *et al.*, *Struct. Dyn.*, 2019, **6**, 014301.

191. B.-L. Qian and H. E. Elsayed-Ali, *J. Appl. Phys.*, 2002, **91**, 462–468.

192. M. R. Otto, *et al.*, *Struct. Dyn.*, 2017, **4**, 051101.

193. L. Zhao, *et al.*, *Struct. Dyn.*, 2021, **8**, 044303.

194. S. Kunjir *et al.*, in *NAPAC2022 - Proceedings*, 2022.

195. K. Chang *et al.*, in *Quantum Dot Solar Cells*, ed. J. Wu and Z. M. Wang, Springer, New York, NY, 2014, pp. 311–347.

Imaging Ultrafast Structural Dynamics with Megaelectronvolt Ultrafast Electron Diffraction

M.-F. LIN,[a] A. H. REID,[a] X. SHEN[a] AND T. J. A. WOLF*[a,b]

[a] Linac Coherent Light Source, SLAC National Accelerator Laboratory, 2575 Sand Hill Road, Menlo Park, CA 94025, USA; [b] Stanford PULSE Institute, SLAC National Accelerator Laboratory, 2575 Sand Hill Road, Menlo Park, CA 94025, USA
*Email: thomas.wolf@slac.stanford.edu

12.1 Introduction

The observation of structural changes in real time with atomic precision (*i.e.*, on the femtosecond (fs) timescale and the sub-Ångström (Å) length scale) has been a long-held dream of physicists and chemists. The realization of this dream started to come into reach with the development of ultrafast lasers. Today, powerful time-resolved diffraction methods based on X-ray free electron lasers (see Chapters 8 and 9), laser-based X-ray sources (see Chapter 4), and ultrafast electron sources (see Chapters 11, 13, 14, and 16) enable detailed studies of femtosecond time-resolved structural dynamics across the three phases of matter: in the gas phase, the liquid phase, and in solid state samples. This chapter focuses on recent advancements in using ultrashort electron pulses with megaelectronvolt (MeV) kinetic energies for ultrafast electron diffraction (UED) experiments. It highlights the

Theoretical and Computational Chemistry Series No. 25
Structural Dynamics with X-ray and Electron Scattering
Edited by Kasra Amini, Arnaud Rouzée and Marc J. J. Vrakking

transformative opportunities of this capability as demonstrated at the MeV-UED facility at the SLAC National Accelerator Laboratory (SLAC-UED). Since 2019, SLAC-UED has operated as part of the Linac Coherent Light Source (LCLS) to serve users from ultrafast science communities worldwide.[1]

In the following, we will briefly highlight the differences between the electron and X-ray diffraction observables and review the historical background of MeV-UED (Section 12.1.1). We will discuss the specific advantages of using MeV electrons for UED experiments and describe the SLAC-UED facility and its capabilities in detail in Section 12.2. We will highlight recent results from structural dynamics investigations in the gas phase, liquid phase, and solid state in Sections 12.3–12.5. We conclude the chapter with an outlook on future developments for MeV-UED in Section 12.6.

Experimental investigations of structural dynamics by diffraction methods with sub-Å resolution are typically performed using either ultrafast X-ray or electron sources. X-rays and electrons can provide complementary information for atomic structure determination because of the different mechanisms through which X-rays and electrons interact with matter. A typical X-ray diffraction experiment is based on the Thomson scattering process, whereby X-rays exert a Lorentz force on the atomic electrons, driving them to oscillate and re-radiate. In a typical electron diffraction experiment, electrons follow the Mott scattering process, where incoming electrons are scattered by the atomic electrons and nuclei *via* Coulomb interaction. Therefore, X-ray diffraction provides information about the electron density distribution, which largely follows the positions of the atomic nuclei. In contrast, electron diffraction is directly sensitive to both the electron density distribution and the atomic nuclei. Combining information from X-rays and electrons could help achieve a more comprehensive picture of atomic structure evolution, specifically in cases where the electron density does not necessarily follow the atomic motion. A prominent example is proton-coupled electron transfer, where electron and proton motion can occur simultaneously between different donors and acceptors.[2]

On the other hand, UED experiments can use several unique features of electrons and the electron diffraction observable. First, the Mott scattering cross-section for electrons is four-to-six orders of magnitude larger than that for the X-ray Thomson scattering.[3] The strong nature of the Coulomb interaction in Mott scattering makes electrons a highly sensitive probe, which is particularly well-suited for studies of specimens with a small number of scattering centres, such as nanoscale thin films or gas-phase molecules. Second, the amount of energy deposited per elastic scattering event is three orders of magnitude lower for electrons than for X-rays.[3] Therefore, electrons can be considered a gentle probe that causes much less sample damage compared to X-rays, where in the latter, the damage processes must be outrun in a diffraction-before-destruction scheme.[4–6]

Ultrafast electron scattering instrumentation has enabled various powerful capabilities, including UED, ultrafast electron microscopy (UEM, see Chapters 14 and 15) for real-space imaging, and ultrafast electron energy loss spectroscopy (EELS), and achieved significant breakthroughs in condensed matter physics and chemical science.[7] Most existing UED instruments implement non-relativistic electron beams with energies between 20–200 keV, taking advantage of commercially available direct current (DC) electron sources with an acceleration gradient of $\sim$10 MV m^{-1}.[8] The advent of radio-frequency (RF) photoelectron guns with >100 MV m^{-1} gradient enabled access to ultrashort relativistic electron beams at a few MeV energy.[9] MeV-UED has opened a new era for ultrafast electron scattering instrumentation and enabled vast opportunities for ultrafast science.

12.1.1 History of the MeV-UED Development

The first experimental demonstration of UED can be traced back to the pioneering work of Mourou and Williamson in 1982.[10] They captured atomic structural changes during the melting process of a 15 nm-thick aluminium (Al) film using 20 keV electron beams with a time resolution of $\sim$20 picoseconds (ps). Since then, the field of UED has grown rapidly. In the past three decades, significant advancements in UED instrumentation have been achieved, primarily driven by the Zewail group at the California Institute of Technology[11] and the Miller group at the University of Toronto,[12] making UED a routine and reliable instrument for ultrafast science with sub-ps time resolution. Further developments were conducted by the Baum,[13] Ropers,[14] Ernstorfer,[15] Siwick,[16,17] Centurion,[18] and Ruan groups.[19,20] For a more detailed review of these developments, the reader is referred to Chapter 11.

All these developments were achieved using non-relativistic electrons in an energy range of 20–200 keV. A big challenge in keV UED is the strong space-charge effect. The Coulomb repulsion force between electrons can cause severe degradation of beam quality after the birth of an electron beam from the cathode, including pulse duration lengthening as the electron beam propagates to the sample. The effect of the space-charge repulsion force scales as $1/(\beta^2\gamma^3)$,[21] where $\beta = v/c$ and $\gamma = \sqrt{1 - v^2/c^2}$ are the Lorentz factors. Therefore, an efficient approach for suppressing the space-charge effect in UED is to implement relativistic electron beams. This is one of the strong motivations to develop MeV-UED, as it allows achieving a shorter electron pulse duration for UED experiments with higher time resolution and packing more electrons in a short bunch to enable high signal-to-noise-ratio (SNR) single-shot diffraction for irreversible dynamics.

The Wang group pioneered the development of MeV-UED. As early as 1996, Wang *et al.* proposed an S-band SLAC/UCLA/BNL RF photoinjector as the electron source for MeV-UED.[22] The electron beam produced by the RF photoinjector was experimentally characterized and demonstrated a strong potential for application in UED. Later, the concept of MeV-UED was further developed with considerations of a dedicated experimental setup[23] and other practical details.[24] In 2015, a complete experimental setup for MeV-UED was demonstrated by Wang *et al.* at the Brookhaven National Laboratory, achieving a time resolution of 305 fs full width at half maximum (FWHM).[25] This work laid a solid foundation for the success of the MeV-UED facility developed by the Wang group at the SLAC National Accelerator Laboratory.[26]

In the meantime, MeV-UED attracted attention and efforts from various other groups. In 2006, Hastings *et al.* demonstrated the acquisition of high SNR single-shot diffraction patterns from a 160 nm Al foil using a 5.4 MeV electron beam generated by an RF photocathode from the Gun Test Facility at the Stanford Linear Accelerator Center (now SLAC National Accelerator Laboratory).[27] Groups at the University of California Los Angeles and Tsinghua University carried out significant efforts in MeV electron beam characterization and optimization. They successfully acquired high SNR single-shot diffraction patterns with 3 MeV electron beams,[28–30] and for the first time, demonstrated the capability of MeV electrons to resolve structural dynamics of laser-induced melting of a gold thin film in time with single-shot electron diffraction patterns.[31–33] Efforts at Osaka University quickly followed up and demonstrated a 3 MeV setup producing high SNR diffraction patterns,[34] and high-quality atomic structural dynamics of laser-excited solids from MeV-UED.[35] More recently, research and development (R&D) efforts have been undertaken in multiple MeV-UED projects, including at Shanghai Jiaotong University,[36] the Relativistic Electron Gun for Atomic Exploration (REGAE) project at the Deutsches Elektronen-Synchrotron (DESY),[37] the Korea Atomic Energy Research Institute,[38] and the High Repetition-rate Electron Scattering (HiRES) facility at the Lawrence Berkeley National Laboratory.[39]

Started through SLAC's UED/UEM initiative in 2014, the SLAC-UED facility has become the most successful MeV-UED program of the past decade. Benefiting greatly from SLAC's expertise in electron beam physics and RF technology, as well as the ultrafast science community, SLAC-UED has developed a multi-functional platform for ultrafast science, serving as a complementary tool to the diffraction capabilities at the LCLS XFEL at SLAC. The multi-functional platform of SLAC-UED (see Figure 12.1) is capable of studying solid state thin film specimens for condensed matter physics,[40] liquid-phase[41] and gas-phase[42] molecules for chemical science, irreversible processes for high energy density science,[43] as well as application device samples for *in situ/operando* characterizations.[44]

12.2 The MeV-UED Technique

Megaelectronvolt electron beams have demonstrated numerous unique features for applications in ultrafast electron scattering instrumentation. This section will highlight the significant advantages of MeV electron beams. As an example of general technical considerations for MeV-UED, the technical details of the SLAC-UED beamline will be discussed.

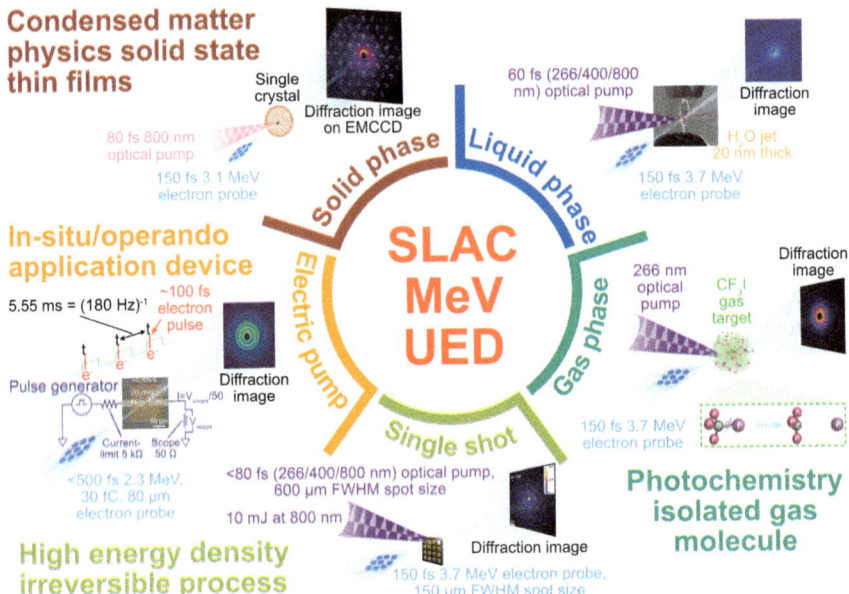

Figure 12.1 The SLAC-UED multi-functional platform for ultrafast science. The platform provides versatile access for ultrafast science studies over solid-phase, gas-phase and liquid-phase specimens, as well as for single-shot diffractive imaging of irreversible processes and *in situ* application devices.

12.2.1 Advantages of MeV Electron Beams for Ultrafast Electron Diffraction

A significant advantage of MeV electron beams is the suppression of space-charge forces. This can be illustrated using a simple model (see Figure 12.2) where two point-charges with charge q move in the same direction with velocity $\vec{v}$. When both point-charges are at rest (*i.e.*, $\vec{v} = 0$), the space-charge force $\vec{F}_{sc}$ experienced by one point charge due to the other is purely the Coulomb (electric) repulsion force $\vec{F}_E$. When $\vec{v} \neq 0$, there will be a magnetic attraction force $\vec{F}_B$ to partially compensate for the electric repulsion force $\vec{F}_E$. As a result, the repulsion from the space-charge force $\vec{F}_{sc} = \vec{F}_E + \vec{F}_B$ becomes weaker. When the point charges are travelling at the speed of light, the magnetic attraction force $\vec{F}_B$ will completely cancel the electric repulsion force $\vec{F}_E$.

Using rigorous derivations, one can show that the effect of space-charge forces on electron beam dynamics scales as $1/\beta^2\gamma^3$,[21] where $\beta = v/c$ and $\gamma = \sqrt{1 - v^2/c^2}$ are the Lorentz factors, and c is the speed of light in vacuum. Therefore, the space-charge effect for a 4 MeV relativistic electron beam is 1322 times weaker than that for a 100 keV nonrelativistic electron beam. This allows MeV electron beams to achieve a shorter pulse duration than nonrelativistic electron beams for a fixed electron bunch charge. On the

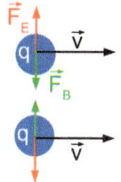

Figure 12.2 Schematic of space-charge forces between two point-charges q co-propagating with velocity $\vec{v}$. The electric repulsion forces $\vec{F}_E$ are indicated by the red arrows, while the magnetic attraction forces $\vec{F}_B$ are indicated by the green arrows. For relativistic electrons, the magnetic attraction forces $\vec{F}_B$ would largely cancel the electric repulsion forces $\vec{F}_E$, such that the total space-charge forces are significantly suppressed.

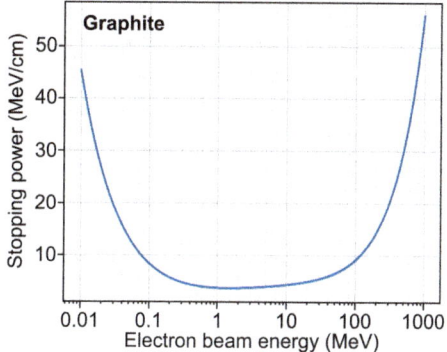

Figure 12.3 Electron stopping power for graphite as a function of electron beam energy.[158]

other hand, an electron beam with a fixed pulse duration can contain higher bunch charges using MeV electrons compared to nonrelativistic electrons. Therefore, MeV electrons provide access to high-intensity ultrashort beams, which are highly desirable for the studies of ultrafast dynamics. In particular, MeV electrons can enable a single-shot acquisition mode to capture ultrafast irreversible processes.

Given the $1/\beta^2\gamma^3$ scaling law of space-charge effects, it is tempting to consider implementing electron beams with energies even higher than a few MeV to achieve further space-charge suppression. However, electrons with energies at a few MeV exhibit another unique feature which makes them well-suited for UED applications. The stopping power of electrons as a function of electron beam energy is illustrated with graphite (see Figure 12.3). The minimum stopping power is found for electrons with energies around 1–4 MeV. At energies lower than 1 MeV, the stopping power is dominated by energy loss to atomic electrons and nuclei in collision interactions. For energies higher than ~ 10 MeV, energy loss due to bremsstrahlung radiation becomes the main contribution to the stopping power.[45]

Note that this form of electron stopping power as a function of electron energy is generally the same for other materials. Therefore, electron beams with energies of a few MeV can provide clean UED signals with the least amount of inelastic scattering effects.

Another important advantage of MeV electron beams compared to non-relativistic electron beams is the larger penetration depth. The calculated elastic scattering mean free path of electron beams in gold as a function of electron energy (see Figure 12.4) shows that electrons at 4 MeV have a three times longer mean free path for elastic scattering compared to those at 100 keV.[46] A short elastic scattering mean free path can cause multiple scattering effects, resulting in undesired events such as the appearance of forbidden Bragg peaks, which pose complications for data interpretation. Therefore, MeV electron beams can provide a clean kinematic scattering signal for a solid-state sample with a given thickness.

This feature is further illustrated through pump–probe UED data from a 30 nm-thick gold film acquired at SLAC-UED using a 4 MeV electron beam (see Figure 12.5). The solid data points show the value of the quantity $-\log(I_\infty/I_0)$ as a function of s^2, where I_0 is the intensity of a Bragg peak before the absorption of the pump laser energy into the gold film, I_∞ is the intensity of the same Bragg peak ~ 20 ps after the incidence of the pump laser, and s is the amplitude of the momentum transfer for corresponding Bragg peaks. At ~ 20 ps after the incidence of the pump laser onto the gold film, the atomic lattice of the gold film reached a thermal equilibrium at an elevated temperature. According to the Debye–Waller effect,[47] the intensities of the Bragg peaks will decrease due to the thermal vibration of the atoms. The logarithm of intensity change, $-\log(I_\infty/I_0)$, at a given s satisfies $-\log(I_\infty/I_0) = \alpha s^2$, where α is a

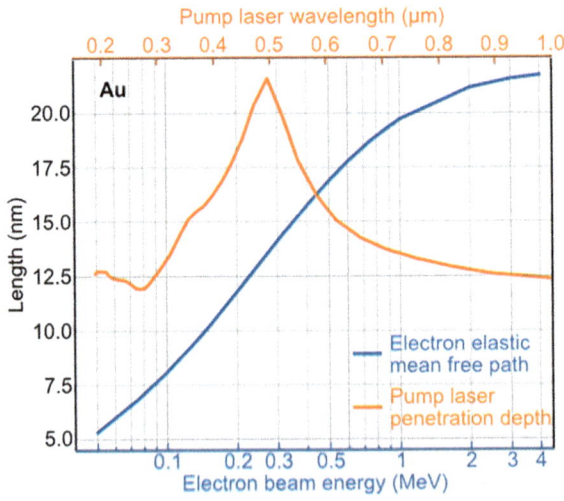

Figure 12.4   Electron elastic scattering mean free path in gold as a function of electron energy[46] (blue) and laser penetration depth for gold as a function of laser wavelength[159] (orange).

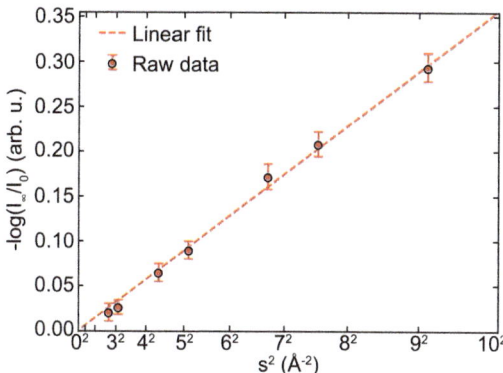

Figure 12.5   UED data for a 30 nm thick gold thin film sample measured using 4 MeV electron beams acquired at SLAC-UED. Here, I_0 denotes the intensity of a Bragg peak before the incidence of the pump laser and I_∞ denotes the intensity for the same Bragg peak ~ 20 ps after the incidence of the pump laser. The dashed lines show a linear fit of $-\log(I_\infty/I_0)$ as a function of s^2 to the experimental data. For kinematic scattering signals, the offset term for the linear fit is zero.

scalar factor. This enabled a linear fit of the experimental data to the Debye–Waller model to be applied (see dashed lines in Figure 12.5). The excellent agreement between the experimental data and the Debye–Waller model demonstrates the capability of MeV electron beams to capture kinematic scattering signals, which is critical for the quantitative investigation of the underlying ultrafast dynamics.

Furthermore, the pump laser penetration depth as a function of laser wavelength for gold (see orange lines in Figure 12.4) shows that the elastic scattering mean free paths of electron beams at a few MeV are larger than the pump laser penetration depths for the UV-to-NIR range (wavelengths from 266 nm to 1 μm) which is a range of common interest in typical pump–probe experiments for solid state samples. Thus, the sample thickness can be matched to the pump laser penetration depth, resulting in an efficiently pumped sample in combination with a purely elastic diffraction signal. On the other hand, when the electron elastic mean free path is shorter than the pump laser penetration depth, a thin sample matched to the electron elastic mean free path should be chosen to avoid multiple scattering effects.

Megaelectronvolt electron beams travel at relativistic speeds, resulting in a significantly smaller group velocity mismatch with the pump laser than nonrelativistic electron beams. This is particularly important for achieving high temporal resolution in UED of gas phase samples. For example, a 100 keV electron beam travels at 55% of the speed of light. It will take ~ 800 fs longer for a 100 keV electron beam to travel through a gas jet with a typical thickness of 300 μm compared to a pump laser travelling collinearly with the electron beam, strongly limiting the achievable temporal resolution. On the other hand, the group velocity mismatch for a 4 MeV electron beam for the same gas jet is only 6 fs, which is negligibly short compared to typical timescales of structural dynamics in gas molecules. The velocity mismatch can be overcome by tilting the pulse front of the optical pulses.[48] However, such an approach adds a non-trivial level of difficulty to the experimental setup.

Another vital advantage of MeV electron beams for practical feasibility is the high flexibility of the experimental setup. The reduced space-charge effects allow for distances in the metre range between the photocathode and the interaction point with the sample without substantial impact on the pulse duration. The relatively large distance permits a high gas density (up to 10^{-4} torr level) at the interaction point while maintaining an ultra-high vacuum level (10^{-10} torr level) at the photocathode.

Similar flexibility is afforded by the larger separation between the interaction point and the detection region of the diffraction signal: the de Broglie wavelength of MeV electrons is on the order of 0.001 Å, which is >10 times shorter than that of nonrelativistic electrons. According to Bragg's law, the diffraction angle for MeV electrons is also >10 times smaller than that of nonrelativistic electrons. Therefore, MeV electrons allow for a much longer separation distance between the sample location and the electron detector for capturing the electron diffraction pattern.

12.2.2   Technical Details of SLAC MeV-UED

The SLAC-UED facility was commissioned in 2014 to take full advantage of the abovementioned unique features of MeV electron beams. Since 2014, SLAC-UED has evolved into a state-of-the-art MeV UED beamline, enabling numerous ground-breaking scientific opportunities in ultrafast science. The SLAC-UED beamline (see Figure 12.6) consists of an electron gun, a space for exchangeable in-vacuum instruments, and a high-efficiency electron detector. The beamline is compact with an overall footprint of ~4.5 metres.

It can be operated in a single-shot[49] or pulsed mode up to a 360 Hz repetition rate (see Table 12.1 for typical operational parameters of SLAC-UED). It routinely delivers electron beams with 2–4 MeV energies and 10^4–10^6 electrons

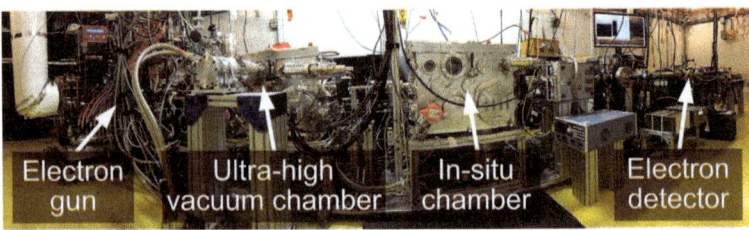

Figure 12.6   Layout of SLAC-UED beamline with the installed MeV-SUED instrument. The positions of the electron gun, the two MeV-SUED instruments, and the electron detector are marked. See the main text for details about each instrument component.

Table 12.1 Typical operational parameters for SLAC-UED.

Parameter	Value
Repetition rate	Single-shot to 360 Hz
Electron beam energy	2–4 MeV
Electron beam charge	2–200 fC
	(10^4–10^6 electrons per pulse)
Electron beam emittance	2–20 nm rad
Electron beam spot size (FWHM) at sample/detector	100–200 µm/300 µm
Momentum transfer s range/resolution	12 Å^{-1}/0.17 Å^{-1}
Temporal resolution (FWHM)	<150–300 fs[a]
Operation stability	Non-stop operation for >5 days

[a] Dependent on electron beam charge.

per pulse. Generally speaking, a minimum of 10^6 electrons is needed to obtain a single diffraction pattern with a high SNR.[8] It provides access to a large momentum transfer range of 12 Å^{-1}, with a reciprocal-space resolution of 0.17 Å^{-1}. The electron spot size at the sample (detector) location is 100–200 µm (300 µm) FWHM. The achievable temporal resolution is <150 fs FWHM.[50] The beamline can continuously operate for >5 days with high stability.[50] Such a stable long-term operation is a key factor in enabling excellent scientific outcomes for solid, gas, and liquid phase MeV-UED, which will be discussed in the following sections.

The electron gun is a SLAC LCLS-type photocathode RF gun operated at 2.856 GHz.[51,52] The implementation of the RF power source enables an acceleration gradient of 100 MV m^{-1} in the photocathode RF gun,[26] which is much higher than the 10–20 MV m^{-1} level in DC photoinjectors implemented by nonrelativistic UED instruments, and allows the production of high-intensity and high-brightness MeV electron beams. The photocathode RF gun was originally developed for the LCLS-I photoinjector, which was operated at a 120 Hz repetition rate. The system was later optimized for operation at 360 Hz for MeV-UED applications.

The electron beam is collimated using a combination of a solenoid lens and a collimator directly downstream of the electron gun.[26] A second solenoid lens is placed upstream of the experimental instruments to adjust the electron beam focus and optimize for diffraction quality.[53]

The SLAC-UED beamline is equipped with a Ti:Sapphire laser system producing 14.5 mJ, 50 fs laser pulses with a central wavelength of 800 nm at a repetition rate of 360 Hz. The laser is split into two parts, a 0.8 mJ branch is frequency-tripled to 266 nm with µJ-level energy for electron generation at the RF gun photocathode, while the remaining portion (13.7 mJ) in the second branch is reserved for the generation of pump pulses. SLAC-UED offers an extensive range of pump wavelengths, including up to the 4th harmonic of the laser, fully tunable wavelengths between 240 nm–2 µm and 3–15 µm, and wavelengths down to the single THz range.[54]

A low-level RF (LLRF) control system based on the advanced tele-communication computing architecture (ATCA) was built in-house for high-repetition-rate ($\geq$360 Hz) RF operation,[55] providing a highly stable RF performance which has a 0.06% root mean square (RMS) RF gun amplitude jitter and 0.04° RMS phase jitter. Taking advantage of the compact and modular design of the ATCA technology, a laser-LLRF timing synchroniza-tion system was also built in-house using the same ATCA platform, achieving a 10 fs RMS laser-LLRF timing jitter.[54] Active monitors and feedback systems are also implemented to enable reliable operation of all hardware, providing a continuous stable operation of the SLAC-UED beamline.

The SLAC-UED beamline is equipped with a high-efficiency electron detector composed of a 4 cm diameter phosphor screen (P43) positioned perpendicular to the electron beam path in a back-illuminated geometry with the aluminium substrate facing the electron gun. The phosphor screen is im-aged through a high-reflectivity 45° mirror and a vacuum viewport onto an Andor iXon Ultra 888 electron-multiplying charge-coupled device (EMCCD) camera using an out-of-vacuum 40 mm f/0.85 lens. 2.9 mm diameter holes both at the centre of the phosphor screen and the 45° mirror allow the passage of the un-diffracted electron beam to a downstream beam dump.[49] The detector is typi-cally used in an integrating mode with a typical exposure time of 1–10 seconds. The vacuum in the electron detector area is maintained at a 10^{-7} torr level.

Three interchangeable instruments providing tailored environments for investigations of gas phase, liquid phase, and solid-state samples can be in-stalled in the SLAC-UED beamline. They are described in more detail below.

12.2.2.1   *The Liquid Phase and Gas Phase UED Instruments*

The current liquid and gas phase SLAC-UED instruments are products of a common evolutionary process of continuous development and improvement of capabilities. The first gas phase MeV-UED (MeV-GUED) experiments were performed in a multipurpose instrument described in ref. 50. The lessons learned from this original instrument design were incorporated into the current liquid phase UED (LUED) instrument (see Figure 12.7).[56] The cur-rently used, dedicated MeV-GUED instrument represents a further, small design iteration. Due to their similarity, we will discuss the design and capabilities of the LUED and MeV-GUED instruments together.

The challenge of extremely low (sub-μm) mean free paths of electrons in liquids (see Sections 12.2.1 and 12.4) is overcome at SLAC-UED by imple-menting sub-μm thin liquid sheet jets which were developed within the sample delivery group at LCLS.[57,58] All LCLS sheet jets make use of com-parably cheap glass chip nozzles. These apply transverse momenta to the liquid jets on two opposing sides, leading to the consecutive formation of several flattened leaf-shaped sheets. The first sheet typically spans ~1 mm × 4 mm (width × length). Its spatial extent can be adjusted by controlling the flow rate of a HPLC pump located outside of the vacuum chamber. The transverse momenta can be generated by directing helium

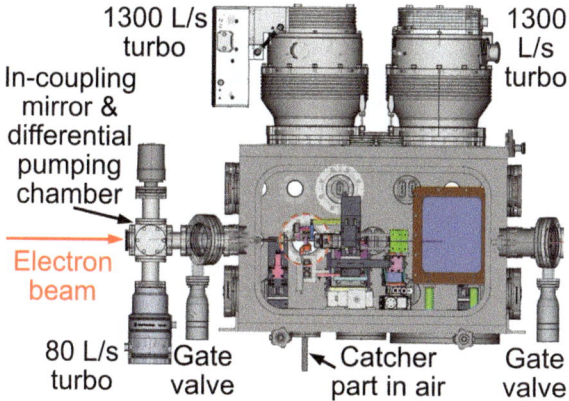

Figure 12.7 Overview drawing of the LUED instrument. The system contains one differential pumping stage and a main chamber. The electron beam propagates from left to right. Two $1300 \, L \, s^{-1}$ turbo molecular pumps are installed on the main LUED chamber. A small $80 \, L \, s^{-1}$ turbo pump keeps the vacuum level at 10^{-6} torr in the differential pumping chamber, which also hosts the laser in-coupling mirror. The interaction point between the laser, electrons, and the liquid sample is highlighted with a red circle. Its details are shown in Figure 12.8. The interaction point can be accessed through a large O-ring sealed front door.

gas onto both sides of the jet or through the geometry of the nozzle. Both approaches have been used for LUED experiments. Flow rates of ~ 0.5 to $3 \, mL \, min^{-1}$ can provide liquid sheets of sub-μm thickness, which are stable for up to 12 hours.[56–58]

The liquid is collected using a heated catcher (Innovative Research Solutions GmbH) at a distance of a few mm vertically downstream of the nozzle exit to reduce evaporation into the vacuum chamber (see Figure 12.8). The pressure in the LUED instrument is kept at $\leq 10^{-3}$ torr during jet operations using two $1300 \, L \, s^{-1}$ turbo molecular pumps. The liquid collected using the catcher is removed through a feedthrough into an external container for recycling. To efficiently reduce the backflow of gas from the container to the chamber through the catcher hole, the container is cooled down and is further evacuated using a chemical-resistant vacuum pump to keep the container pressure below ~ 7.5 torr.

The position of the liquid sheet jet relative to the catcher can be controlled in three dimensions to maximize the collection of the liquid sample during the experiments. Additionally, the optimized jet–catcher assembly can be independently controlled in three dimensions to align it to the electron beam. Spatial and temporal overlap is obtained using solid state samples on a sample card positioned in the same plane as the liquid sheet, perpendicular to the electron beam direction. The sample card is equipped with a Ce:YAG screen to adjust the spatial overlap of optical and electron beams, knife edges to measure their spot sizes, and bismuth and silicon samples to determine temporal overlap *via* the Debye–Waller effect. The position of the sample card

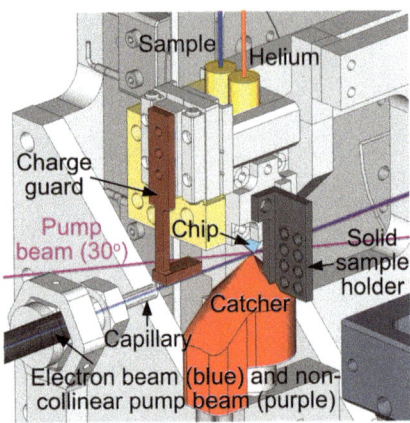

Figure 12.8 Detailed view of the interaction point of the LUED instrument. The electron probe beam propagates from left to right through the metal capillary when entering the LUED chamber. A charge guard upstream of the liquid jet protects the nozzle from damage and charging, which causes jet instability. A sample card next to the liquid jet containing Bi and Si crystals is used for the optimization of spatiotemporal overlap before and during the experiment. Reproduced from ref. 56, https://doi.org/10.1063/1.5144518, under the terms of the CC BY 4.0 license http://creativecommons.org/licenses/by/4.0.

with respect to the sheet jet allows to directly transfer the optimized spatial and temporal pump–probe overlap to the liquid target by moving the whole assembly of the liquid jet and sample card. The spatiotemporal overlap is typically optimized within ∼30 minutes of data accumulation so that rare timing drifts of 100s of femtoseconds over hours can be readily corrected.

Two differential pumping sections isolate the LUED instrument vacuum conditions from the upstream electron gun region. The downstream pressure differential consists of a capillary with an inner diameter of ∼2 mm and a length of ∼30 mm connected to the inner wall of the LUED instrument. It reduces the pressure to the 10^{-6} torr range. The capillary position can be adjusted along the plane perpendicular to the incoming pump and probe beams.

The pressure differential is also instrumental in preventing contamination of the laser in-coupling mirror surface (placed directly upstream from it) with sample molecules. Such contamination would otherwise lead to rapid degradation of the mirror reflectivity, especially in the case of laser wavelengths in the ultraviolet (UV). The laser in-coupling mirror (1″ diameter) has a central hole (2 mm diameter) to transmit the electron beam and to enable quasi-collinear (<1° angle) pump–probe geometries. To avoid clipping the laser beam by the capillary walls of the pressure differential, the laser focus is placed inside the capillary.

In addition to the quasi-collinear pump laser in-coupling, the LUED instrument is equipped with an extra in-coupling port at an angle of 30°

relative to the probe beam direction. In this geometry, the focus of the pump beam can be placed at the liquid jet position so that the jet is exposed to the maximum available laser intensity for strong-laser field ionization and warm dense matter experiments.[56]

The temporal resolution of the LUED system was measured using strong-field ionization of liquid water with an 800 nm laser pulse.[59] The temporal overlap between laser and electrons is characterized by an intense, short-lived transient signature at $s < 1$ Å^{-1}. A fit of this signature revealed an instrument response function of 140 fs FWHM. The signature was also intense enough (~10% change) so that single time scans averaging several seconds per delay step could be post-corrected for laser-electron timing drifts on the timescale of hours.

Compared to the LUED instrument, the liquid sheet jet and catcher assembly is replaced with a mounting point for several different gas phase sample delivery options in the MeV-GUED instrument. As for the LUED instrument, the design of the mounting point places the gas sample source in the same plane perpendicular to the beam direction with the solid-state sample card.

The SLAC-UED facility has an active gas phase sample delivery R&D program. It currently offers two sample sources for user experiments, a pulsed valve (Parker) and a flow cell. The pulsed valve (100 µm orifice) can be operated at repetition rates up to 180 Hz and temperatures up to 180 °C. It is the standard gas delivery option for chemicals with vapour pressures considerably below 1 torr at ambient temperature. In this case, samples must be heated to temperatures exhibiting vapour pressures of tens of torr for sufficient diffraction signal levels. The flow cell can deliver sample chemicals with vapour pressures of ≥1 torr at room temperature. It consists of a tube, which is closed at one end and contains entrance and exit holes for laser and electron beams perpendicular to the tube axis with typical openings of 550 µm. The inner tube diameter determines the interaction length of ~3 mm. Both gas sources can be connected to inside and outside vacuum sample compartments. In the case of the pulsed valve, the sample can be seeded with helium.

12.2.2.2 The Solid State UED Instrument

The MeV solid state UED (MeV-SUED) instrument at SLAC-UED has evolved from simple sample manipulation into a system allowing a variety of *in situ* environmental controls. The current MeV-SUED instrument has two separate chambers (see Figure 12.6) to allow a diversity of sample environments. The upstream chamber is configured for low-temperature experiments, while the downstream chamber allows for the reconfiguration of the sample environment (high temperature, in operando device structure, *etc.*).

The upstream MeV-SUED low-temperature sample chamber (see Figure 12.9) has samples mounted on a six-axis in-vacuum manipulator (see Figure 12.9c) and manoeuvred using a motor stack with linear XYZ manipulators together with pitch, roll and yaw rotational motors. Samples are held on interchangeable

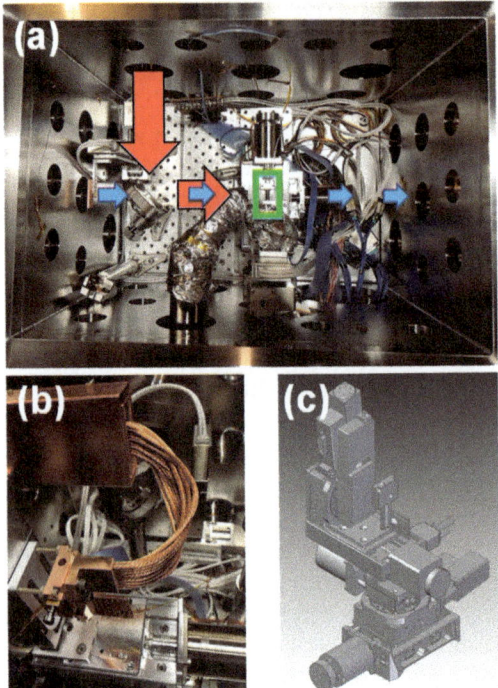

Figure 12.9   Views of the low-temperature solid-state chamber at the SLAC MeV-SUED instrument. (a) A top-view image of the SUED chamber. Blue and red arrows indicate the path of the electron and optical beams, respectively, to the sample interaction point (indicated by a green box). (b) Shows the cryogenic system that cools the copper sample card at the interaction point. Cooling is provided from a closed cycle cryocooler using a flexible copper braid. (c) Six-axis sample positioning, shown in (c), is provided by in-vacuum stages, which are thermally isolated from the sample card by glass-filled nylon screws. (image credits Alexander Reid/SLAC).

copper cards designed to fit standard 3 mm TEM-style grids. Up to 42 grids can be loaded onto a single card. An Advanced Research Systems DE-215S closed cycle cryocooler cools the card. The base temperature of the system is 20–30 K depending on the specifics of the configuration. Vibrations are decoupled from the cold head using a flexible copper braid (see Figure 12.9b). At the same time, the sample card is thermally isolated from the motion stages, which remain at room temperature, using low-thermal-conductivity screws. The low-temperature chamber is pumped *via* a 500 l s^{-1} turbomolecular pump and generally operates in a pressure range below 10^{-7} Torr. Near-collinear laser pumping is usually achieved using a mirror with a central hole for the electron beam (see Figure 12.9a).

   The downstream MeV-SUED sample chamber is a reconfigurable solid-state chamber. It is physically identical to the low-temperature chamber, and laser

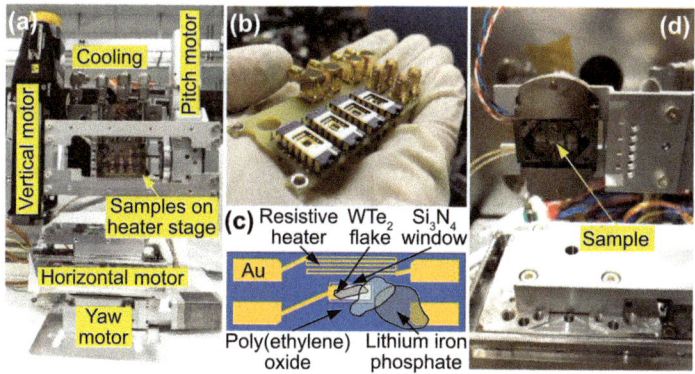

Figure 12.10   Various interchangeable sample environment systems at the MeV-SUED multi-purpose chamber. (a) A high-temperature sample system was used for sample temperatures from room temperature to 800 K (image credit Alexander Reid/SLAC). (b) Electrical pump sample holder (image credit Aditya Sood/Stanford). (c) Intercalation platform developed by Muscher *et al.*[60] (d) Uniaxial Razorbill strain cell incorporated into the UED system (image credit Alexander Reid/SLAC).

in-coupling is achieved similarly to the cryogenic chamber. The chamber is used for interchangeable sample environments, including high-temperature measurements, electrically driven dynamics, intercalation, and strain. Many of these environments limit the base vacuum pressure for the operation of the chamber.

To allow a greater range of sample temperatures, a custom heating system was developed by Instec for SLAC-UED. The system allows up to nine samples to be heated from room temperature to 800 K, has a temperature stability of ± 0.1 K over 1 hour, and has a five-axis motion system (see Figure 12.10a). The reconfigurable (downstream) MeV-SUED chamber has also been used to develop novel electrical pump methods, as utilized by Sood *et al.* in their study of the electrical switching of the metal-to-insulator transition in VO_2.[44] Here, in this interchangeable sample card with integrated electrical connections (see Figure 12.10b), electrical pulses are supplied by vacuum-compatible coax cables from pulse generators synchronized to the photocathode radio frequency. Other sample environments deployed include *in situ* electrochemical cells developed by Muscher *et al.* for use at the MeV-SUED (see Figure 12.10c),[60,61] and a commercial RazorBill uniaxial strain cell (see Figure 12.10d).

12.3   MeV-UED of Gas Phase Molecules

Gas phase ultrafast electron diffraction (GUED) has a long and impactful history far beyond the first demonstrations with MeV electrons.[11,18,62] Specifically, the pioneering work of the Zewail group with picosecond time resolution starting in the 1990s had a tremendous impact on our understanding of structural evolution in gas phase photochemistry (see, for example, ref. 11 and 62). A picosecond time resolution allows for the observation of the evolution of the temporal average of atomic distances in, for example, a reactant species into the time average of the atomic distances in a product species (*i.e.*, for the observation of picosecond reaction kinetics). However, the natural timescale of interatomic motion is in the femtosecond regime. Thus, experimental access to reaction dynamics and the atomic motion (*i.e.*, the detailed dynamic trajectory of the structural changes leading from a reactant to a product equilibrium structure, often referred to as a "molecular movie") requires a femtosecond temporal resolution. The first (nonrelativistic) GUED experiments with sub-picosecond time resolution were demonstrated by the Centurion group in the last decade.[18,63-66] The first gas phase diffraction experiments with sub-200 fs temporal resolution were demonstrated using hard X-rays (see Chapter 9).[67] The time resolution of these gas-phase ultrafast X-ray diffraction (GUXD) experiments was higher than at the SLAC-UED, albeit at a smaller accessible momentum transfer range. The combination of <150 fs temporal and sub-Å spatial resolution demonstrated at SLAC-UED has opened transformative opportunities for research into ultrafast photochemical dynamics in isolated molecules.[42,68-73] In the following, we will review methods for extracting structural information from MeV-GUED data and highlight recent results.

12.3.1 Extraction of Structural Information from Gas Phase MeV-GUED Data

This section briefly reviews strategies to extract time-dependent structural information, specifically focusing on MeV-GUED data. For detailed information about the electron diffraction observable, the reader is referred to Chapter 3 of this book. Several review articles also provide comprehensive descriptions of the theory of gas-phase electron diffraction.[62,74]

Gas phase electron diffraction patterns are typically centrosymmetric due to the random orientation of molecules in gas samples. Within the independent atom model (IAM, see Chapter 3), the s-dependent diffraction intensity $I_{tot}(s)$ represents the sum of atomic scattering contributions I_{at} as well as the interference terms between atomic scatterers containing the molecular structure information I_{mol}. The exact separation of experimental diffraction into atomic and molecular contributions is non-trivial. However, reliable methods have been developed during the long history of static gas-phase electron diffraction.[62,75] It is common for gas-phase electron diffraction to plot I_{mol} rescaled by I_{at} and s to compensate for the strong s-dependence in the scattering intensity (see Chapter 3) as the so-called modified molecular diffraction intensity (similar to the so-called molecular contrast factor, MCF, employed in Chapter 13)

$$sM(s) = \frac{I_{mol}}{I_{at}} s. \tag{12.1}$$

The focus of GUED is on molecular structure *changes*, which are conveniently expressed as molecular difference intensity $\Delta sM(s, t)$

$$\Delta sM(s, t) = sM(s, t) - sM(s, t_{ref}), \tag{12.2}$$

which is the difference between a diffraction signal recorded at a pump-probe delay time t and a diffraction signal taken at a reference time t_{ref}. At time t_{ref}, diffraction is measured with an electron probe pulse in the absence (or much before the arrival) of a photon pump pulse at the sample. At typical pump–probe delays on the femtosecond and picosecond timescale, the number of atoms in the interaction volume is approximately constant as sample diffusion in and out of the interaction volume with laser and electron pulses is negligible. Therefore, the atomic contributions are subtracted from time-dependent difference signals.

Analogous to GUXD (see Chapter 9), if linearly polarized light is used in the optical excitation, the directionality of the transition dipole moment leads to an anisotropic orientational distribution of the photoexcited molecules and, hence, to anisotropy in the pump-induced changes of the diffraction patterns. This effect can be used to differentiate between simultaneous dynamics being triggered by excitations into different states with different transition dipole moment orientations, as demonstrated in a study of the photodissociation dynamics of trifluoromethyl iodide (CF_3I).[42]

Several different approaches for extracting and interpreting structural information from GUXD and GUED experiments have been employed in the literature. It is instructive to discuss them together with the respective limits of the two methods. Elastic (coherent) scattering cross-sections of MeV electrons are in a regime of 10^{-19} cm^2 for light elements like carbon and not strongly dependent on the electron kinetic energy (see Figure 12.4).[3,76] In contrast, X-ray elastic (coherent) scattering cross-sections in the suitable and available photon energy regime for GUXD experiments at current and planned XFELs (10–25 keV) are several orders of magnitude smaller and more strongly dependent on the X-ray photon energy (in the regime from 10^{-24} to 10^{-25} cm^2 for carbon[77]). In the same photon energy range, cross-sections of inelastic (incoherent) X-ray scattering (on the order of 10^{-24} cm^2), which do not contain structural information, are much less photon energy dependent and overtake the coherent scattering cross-sections at high X-ray photon energies.[77] Inelastic (incoherent) electron scattering cross-sections are higher than coherent scattering cross-sections in the MeV range, but their energy dependence is also weak.[3] Additionally, incoherently scattered electrons are scattered into small s-ranges allowing for the separation of their contributions from coherently scattered electrons (see below). In contrast, incoherently scattered X-rays are scattered into a large Q range,[77] which makes it more difficult to separate their contributions from coherent scattering.

The disadvantage in the coherent scattering cross-sections of X-rays with respect to electrons can be compensated by orders of magnitude larger available flux of XFELs. However, the choice of the energy of the scattering particle has a significantly larger impact on the achievable signal quality for GUXD than for MeV-UED, also due to a decreasing detection efficiency of X-rays with increasing photon energy. The choice of X-ray photon energy also impacts the achievable signals in a different aspect: due to the large scattering angles of X-rays, the accessible momentum-transfer range and, thus, the spatial resolution, is determined by the geometry of the experiment (typically 5.8 Å^{-1} at 10 keV, 70° solid angle[78]). In contrast, the accessible momentum-transfer range in MeV-UED is determined by the available signal-to-noise levels due to much smaller scattering angles (<0.3° solid angle for 10 Å^{-1} at 4 MeV) and a steeper s-dependence of the differential scattering cross-sections.

Thus, the methods to extract structural information from GUXD, which have been established so far, target datasets obtained with X-ray photon energies in the range of 10 keV, which exhibit superb signal-to-noise levels, albeit at limited momentum transfer range. A good example of such a method is the process of extracting structural information by fitting experimental diffraction with simulations based on a pool of candidate structures, which is described in detail in Chapter 9.[79] In contrast, the established methods to extract structural information from MeV-UED target datasets with lower signal-to-noise levels but a substantially higher momentum transfer range. Approaches using genetic algorithms to extract structural information by fitting MeV diffraction have been successfully applied to MeV-GUED data.[68,80] However, the most common approach for extracting structural information from GUED data uses a sine transformation of the reciprocal-space $\Delta sM(s,t)$ to create time-dependent real-space difference pair distribution functions $\Delta\text{PDF}(r,t)$

$$\Delta\text{PDF}(r,t) = \int_{s=0}^{s_{\text{max}}} \Delta sM(s,t)\sin(sr)e^{-ks^2}\,\mathrm{d}s, \qquad (12.3)$$

where s_{max} is the maximum of the experimental momentum transfer range, r the atomic pair distance, and k a damping factor (see below).

For the real-space transformation of experimental data, care must be taken to avoid artefacts in the $\Delta\text{PDF}(r,t)$ from edge-effects. Edge effects from the cutoff at $sM(s_{\text{max}})$ are mitigated in eqn (12.3) by the e^{-ks^2} factor effectively dampening the diffraction at high s values. The $sM(s)$ signal is not measured down to $s=0$ due to the presence of the main beam of undiffracted electrons. The reader is referred to the supplemental materials of ref. 42 and 69–71 for detailed information on mitigating edge effects from missing signals at small s values.

Due to their differential nature, each pump-induced change in the molecular structure appears as a combination of contributions with positive and negative amplitude in $\Delta\text{PDF}(r,t)$. The contributions with negative amplitudes are static and localized at the initial values of atomic distances in the molecules, which experience pump-induced changes. The contributions

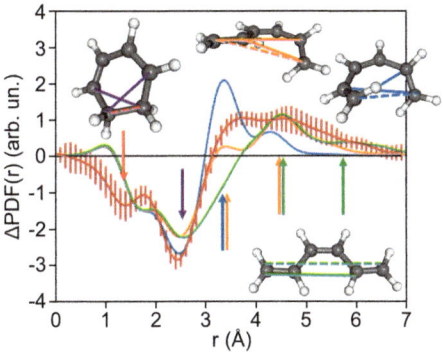

Figure 12.11 Experimental ΔPDF (red) of 1,3-cyclohexadiene 550 fs after photo-excitation. The experimental signal is compared to simulations based on the optimized reactant and product geometries of the ring-opening reaction. The reactant and product geometries are visualized, and the carbon–carbon distances showing the most significant changes during the reaction are highlighted with the colour of the corresponding ΔPDF. The carbon–carbon distance contributions to the ΔPDFs (negative for the reactant and positive for the products) are marked by colour-coded arrows. Reproduced from ref. 69 with permission from Springer Nature, Copyright 2019.

with positive amplitude appear at the atomic distances of the structure which the molecule has evolved into at a specific pump–probe delay t.

A good example is the signatures in the $\Delta PDF(r, t)$ of the photochemical ring-opening reaction of 1,3-cyclohexadiene (CHD, see Figure 12.11).[69] The ring-opening changes two sets of carbon–carbon distances in the molecule: the distance between two carbons involved in the bond which dissociates during the ring-opening (red, 1.4 Å), and two additional carbon–carbon distances across the CHD ring (purple, 2.4 Å). During the ring-opening, both sets of distances increase beyond 3 Å with values depending on the generated photoproduct conformer (blue, green, and orange), which leads to a combination of negative signatures at their initial positions in CHD, 1.4 Å and 2.4 Å, and a broad distribution of positive signatures at distances >3 Å in the ΔPDF (see Figure 12.11).

Interpretation of signatures from photochemical dynamics in time-resolved diffraction in real space as $\Delta PDF(r, t)$ has the advantage of giving intuitive access to the evolution of the nuclear wavepacket during the photochemical dynamics. Additionally, its model dependence is minimal. Moreover, the experimental observable can be easily simulated based on *ab initio* excited state wavepacket simulations.[69,70,81,82] Recently, a new method for the real-space transformation based on super-resolution approaches, which have been applied for biological imaging and optical microscopy, has shown significantly higher spatial resolution.[83]

Similar to observations in GUXD, as described in Chapter 9, signatures beyond the IAM were recently observed in MeV-GUED. In contrast to GUXD, the non-IAM signatures are primarily caused by inelastic rather than elastic electron scattering.[68,74,84] In line with previous observations of static electron scattering beyond the IAM,[85] these time-dependent inelastic scattering signatures are confined to a region of $s \leq 2$ Å^{-1}.

12.3.2 Science Opportunities from Femtosecond Time-resolved GUED

The significant advancement which can be achieved for a mechanistic understanding of fundamental photochemistry from direct imaging of nuclear motion in space and time is best exemplified by comparing studies of the same photochemical reaction performed with picosecond time resolution in the 1990s and 2000s with recent studies using MeV-GUED: the trifluoromethyl iodide (CF_3I) molecule is a standard system for the study of fundamental photodissociation reactions. Photoexcitation in the UV range promotes an electron from an iodine-centred lone pair orbital to the C–I σ^* orbital. The molecule subsequently photodissociates on a sub-100 fs timescale *via* parallel adiabatic and nonadiabatic pathways.[86,87] Previous diffraction experiments with picosecond resolution were able to image the average structure of the CF_3 reaction product.[88] A more recent MeV-GUED study with <150 fs temporal resolution was able to image the large amplitude internal motion of the CF_3 fragment triggered by the kinetic energy release from the photodissociation.[42] The dissociation initially pushes the carbon atom of the trigonal pyramidal CF_3 fragment towards the plane defined by the three fluorine substituents. This initial motion transforms into coherent wavepacket motions in the CF_3 umbrella mode and C-F_3 symmetric stretching mode.

In addition, photochemical electrocyclic ring-opening reactions further exemplify the capability of MeV-GUED to unravel intricate details of more complex photochemical dynamics. Electrocyclic reactions are characterized by the concerted formation and cleavage of multiple single and double bonds. The concertedness of these processes results in an essential property of electrocyclic reactions, namely, their stereospecificity (the specificity of reaction products to the reactant stereoconfiguration). Furthermore, electrocyclic reactions can proceed through both thermal pathways on the potential energy surface of the electronic ground state as well as photochemical pathways involving the potential energy surfaces of multiple electronic states, which are connected through conical intersections. The thermal and photochemical pathways show inverted stereospecificity. This behaviour is predicted by applying the famous Woodward–Hoffmann rules.[89]

A textbook example of electrocyclic reactions and the role of Woodward–Hoffmann rules is the photochemical ring-opening of CHD. The reaction is extremely well-studied using time-resolved spectroscopic methods in the gas phase and is known to proceed within less than 200 fs through a conical intersection between the first excited state and the ground state.[90,91] The reaction was investigated with ps time-resolved GUED by the Zewail and

Figure 12.12 Visualization of the conrotatory ring-opening motion triggered by photoexcitation of 1,3-cyclohexadiene.

Weber groups.[62,92–94] Some of the investigations revealed that the substantial excess energy, which is injected into the vibrational degrees of freedom of the photoproduct during the nonadiabatic reaction dynamics, shows a surprisingly long-lived non-statistical distribution. A recent CHD investigation using MeV-GUED followed the carbon–carbon distance changes during the ring-opening reaction in real-time *via* ΔPDFs (see above).[69] Additionally, the likely source of the previously observed non-statistical vibrational energy distribution could be identified. The Woodward–Hoffmann rules predict the ring-opening reaction to proceed in a conrotatory fashion, *i.e.*, by the ends of the emerging open carbon chain rotating away from each other in the same clockwise or counter-clockwise direction (see Figure 12.12). During the non-adiabatic dynamics, most of the absorbed photon energy is selectively redistributed into this rotational motion of the carbon chain ends, forming a nuclear wavepacket on the ground state potential energy surface of the 1,3,5-hexatriene photoproduct. The corresponding degrees of freedom in the photoproduct are sufficiently decoupled from the other vibrational degrees of freedom that this wavepacket could be followed during the entire investigated time window of 1 ps. It is, therefore, likely that the decoupling is strong enough to lead to the non-statistical energy distributions, which were previously observed by GUED over hundreds of ps.[94]

The MeV-GUED studies of electrocyclic photochemistry were recently extended to derivatives of CHD.[70,82] In the ring-opening dynamics of the derivative α-phellandrene, the stereospecificity of electrocyclic reactions predicted by the Woodward–Hoffmann rules could, for the first time, be directly followed in real space and time.[70] Here, α-phellandrene exhibits an isopropyl substituent (shown as "R" in Figure 12.13a) at one of the two sp^3 hybridized carbon ring atoms directly involved in the bond, which is cleaved during the ring-opening (see Figure 12.13a).

Due to the hybridization of the ring carbon, α-phellandrene has two possible conformer geometries, one where the isopropyl group is in axial (perpendicular) orientation with respect to the ring plane and the second where it is in an equatorial orientation with respect to the ring plane (see Figure 12.13a). The static diffraction of α-phellandrene revealed that equatorial conformers dominate the gas phase sample. The signatures in the time-resolved ΔPDFs are in quantitative agreement with simulations of the reaction dynamics starting from equatorial conformer geometries and showing their stereospecific evolution into the photoproduct isomer predicted by the Woodward–Hoffmann rules (see Figure 12.13b and c) Corresponding

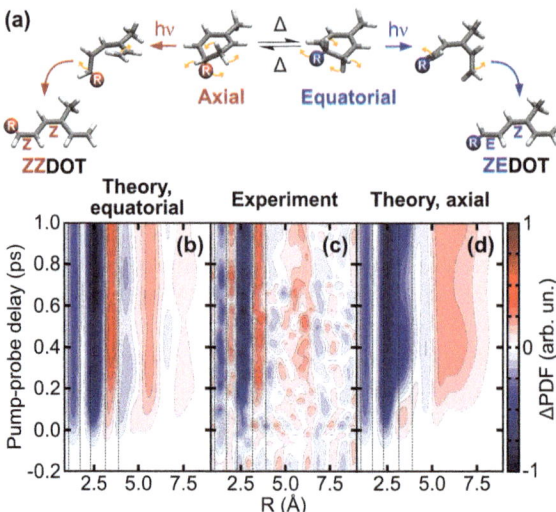

Figure 12.13 (a) Visualization of the conrotatory ring-opening motion of two conformers of α-phellandrene leading to different (ZZ and ZE) isomers of the photoproduct 3,7-dimethylocta-1,3,5-triene (DOT). The isopropyl substituent of the molecule is labelled as "R". (c) Time-resolved experimental and (b and d) simulated ΔPDFs of α-phellandrene. The experimental data are in good agreement with *ab initio* multiple spawning (AIMS) simulations based on the equatorial conformer of α-phellandrene. Analogous simulations based on the axial conformer of α-phellandrene show qualitative differences from the experiment. Adapted from ref. 70 with permission from The Authors, Copyright 2021.

simulations starting from axial conformer geometries qualitatively disagree with the experimental signatures (see Figure 12.13c and d).

The opportunity to follow both electronic and nuclear structure changes during nonadiabatic dynamics through conical intersection has recently been demonstrated with the molecule pyridine.[68] Directly after photo-excitation, an electronic state with nπ* character is populated in pyridine, which exhibits a strong and characteristic inelastic scattering signature. The confinement of the excited state signature to small *s*-values (see Section 12.3.1) allows for the simultaneous and independent observation of the response from the nuclei to population relaxation from the nπ* state back to the ground state.

12.4   Liquid Phase MeV UED Studies

While a detailed mechanistic understanding of fundamental photochemical processes can be gained from time-resolved investigations of isolated model systems in the gas phase, some of the most important ultrafast photochemical reactions, such as natural and artificial photosynthesis or human vision, take place in solution or soft tissue. Therefore, probes of ultrafast photochemistry in the liquid phase with structural sensitivity are desirable. Two diffraction-based techniques with atomic resolution are now available for ultrafast liquid phase dynamics investigations, liquid phase ultrafast X-ray scattering (LUXS) and liquid phase ultrafast electron diffraction (LUED). Since the advent of suitable and powerful sources for femtosecond time-resolved LUXS with the development of the first hard X-ray free electron laser LCLS, LUXS has matured into a reliable and sensitive method (see Chapter 8). Since diffraction methods are *a priori* equally sensitive to solute and solvent molecules, experiments are currently only feasible if the solute–solvent contrast is enhanced in the scattering signal through the presence of heavy elements in the solute. Therefore, LUXS studies have mainly focused on the photochemistry of transition metal complexes with application to photocatalysis and artificial photosynthesis.[95–97]

Three components of time-resolved diffraction signals in the liquid phase can be distinguished: signatures from (i) the dynamics of the solute, (ii) the interactions between the solute and the solvent cage, and (iii) the dynamics of the solvent. The available hard X-ray photon energies at XFELs have so far restricted the accessible X-ray scattering momentum-transfer space to a region ($s < 6 \ \text{Å}^{-1}$) where the contributions of all three components strongly overlap. Therefore, the interpretation of LUXS results relies strongly on theoretical modelling.

Compared to LUXS, much larger momentum-transfer ranges are, in principle, accessible with LUED, including regions where the solute component dominates. Due to the sub-μm attenuation length of electrons in liquids resulting from their high scattering cross-section, only a limited number of (static) electron diffraction studies have been published. Liquid phase electron diffraction has been performed using thin liquid cells from silicon nitride membranes.[98–100] The Zewail group pioneered utilising this type of cell to study the rotational dynamics of gold nanoparticles with picosecond resolved keV electron diffraction.[98] A combination of longer attenuation lengths of MeV electrons and the liquid sheet jet development at SLAC (see Section 12.2.2.1) enables more routine experiments in free-standing liquid sheets.

The number of scattering events per molecule in LUXS and LUED are on a similar level due to the compensation of the lower scattering cross-section in LUXS by the higher available X-ray photon flux (see Section 12.3.1). However, the requirement of sub-μm column lengths in LUED experiments and the associated drastically lower amount of scattering molecules compared to typical jet thicknesses employed in LUXS experiments ($\sim 50 \ \mu$m) put LUED at a disadvantage with respect to LUXS. However, the larger sensitivity of electron diffraction to, for example, proton dynamics makes it an attractive observable for studying such structural motion. Therefore, the focus of the initial LUED studies was on the dynamics of bulk water (see below).[41,59] However, LUED has been successfully used to study solute dissociation dynamics.[101]

12.4.1 Extraction of Structural Information from Liquids

The extraction of structural information from LUED data is similar to the procedures for obtaining MeV-GUED data, as shown in Section 12.3.1. The main difference to gas phase UED is the I_{mol} signal being no longer limited to intramolecular contributions but also containing contributions from intermolecular diffraction within the coherent volume of the electron beam. Static structural information can be extracted using the recently established charge-pair distribution function (CPDF) formalism.[102]

In contrast to static diffraction measurements of a solvent or solution, the data analysis of a pump–probe experiment does not require separation of the atomic part from the total diffraction signal since any time-dependent diffraction changes are solely due to changes in I_{mol}. The conversion of the diffraction signal to real-space is given in Section 12.3.1. The temporal evolution of the sample structure is commonly presented as a difference signal in real space, $\Delta PDF(r, t)$.[41,59] As described above, care must be taken to accurately separate solute, solute–solvent, and solvent dynamics.[103,104]

12.4.2 Overview of the First LUED Results

As mentioned above, two types of experiments in the liquid phase have so far been published from SLAC-UED, experiments on pure water and solutes in water. Specifically, investigations of dynamics in pure water use the unique ability of UED to image proton dynamics. Structural dynamics were investigated following (i) strong-field ionization with an intense near-infrared pulse (IR) and (ii) vibrational excitation with a weaker IR pulse.[41,59]

The reaction dynamics following the ionization of water have particular relevance for processes in biochemistry, laser surgery, and radiolysis mechanisms in nuclear power plants. Dynamics following strong-field ionization have received much attention from experimental and theoretical studies due to the formation of solvated electrons.[105,106] Moreover, H_2O^+ cations, initially generated by the photoionization process, undergo proton transfer to nearby water molecules and experience further chain reactions on the picosecond-to-microsecond timescales.[107]

LUED provides a novel observable to investigate these processes with direct sensitivity to the elementary and microscopic proton transfer reaction coordinate. Liquid water was strong-field ionized in a nine-photon process using an 800 nm laser pulse with a peak intensity of 2×10^{13} W cm^{-2}. The proton transfer was observed to occur within 140 fs producing a short-lived $H_3O^+\cdots{}^{\bullet}OH$ complex before its dissociation to a separated H_3O^+ ion and an ${}^{\bullet}OH$ radical through a secondary proton transfer to the solvation shell within 250 fs. The process could be followed by the signatures of new atomic distances at 1.4 Å and 2.4 Å, which originate from the hydrogen bond and the oxygen–oxygen pair in the short-lived complex, respectively.

The second LUED experiment targeted hydrogen bond dynamics in a neutral water network. The early stage of energy transfer and proton dynamics in water has been previously studied using spectroscopic methods with merely indirect sensitivity to proton motion.[108] LUED provides the ability to directly follow proton motion during the dissipation of vibrational energy from a specific vibrational degree of freedom. The OH stretch mode was selectively excited with an ultrashort pulse at a wavelength of 3 μm.[41] The LUED probe showed that the water network responded to the deposition of vibrational energy by strengthening the hydrogen bond, which in turn led to a contraction of the O···O distance within ~80 fs after the OH stretch excitation. The observation could only be explained by considering the quantum nature of the hydrogen bond. The hydrogen bond network relaxes further through a Fermi resonance mechanism (*i.e.*, energy transfer between OH stretching and the overtone of the water bending mode) by dissipating the absorbed energy into the bulk liquid water resulting in a temperature rise of ~32 K within ~1 ps, consistent with other experimental measurements using IR absorption measurement and molecular dynamics simulations.[41,109]

The observation of photochemical reaction dynamics in a solute using LUED, going beyond experiments in bulk liquids, has been recently demonstrated in a study of the dissociation of I_3^- ions in water.[101] From previous experiments, it is well-known that I_3^- can be photodissociated at 400 nm.[110,111] The dissociation fragments are produced in a common solvent cage and can, therefore, undergo geminate recombination. In this LUED experiment, the dissociation dynamics were directly measured on the femtosecond timescale. Furthermore, it revealed the speed of separation between the I_2^- and the neutral I fragments to be 5.8 ± 0.3 Å ps^{-1}. Additionally, the signature of geminate recombination within 0.6 ps was observed.

12.5 Time-resolved MeV Electron Diffraction on Solid-state Materials

While dilute solutes present many challenges in terms of scattering cross-sections, solid-state materials often represent the opposite end of the signal continuum. There is a long history of using electron diffraction to explore structural dynamics in solid-state materials. In this section, we will discuss the benefits and challenges related to solid-state megaelectronvolt ultrafast electron diffraction (MeV-SUED), particularly the preparation of appropriate samples for measurement, and several examples of the scientific insights gained with this technique.

The measurement of structural dynamics of thin film materials is typically the most straightforward experimental geometry to realize with MeV-SUED. The materials of interest are measured in a transmission geometry with a collinear or near-collinear optical pump to trigger the dynamics of interest. With MeV electrons, the source–sample and source–detector distances are longer than for keV instruments — the SLAC MeV UED instrument has a total beam length of 4.5 metres, with a typical photocathode-to-sample distance of 1.5 metres followed by a sample-to-detector distance of 3 metres.

This relaxes the engineering challenges related to vacuum isolation, sample motions and laser pump integration, which allows a great deal of flexibility with the sample environment.

12.5.1 Sample Preparation and Requirements

Electron diffraction is most versatile in a transmission geometry. In the MeV range, there is a relatively small velocity mismatch between the relativistic MeV electrons and optical pump beams, and generally a good match between the penetration depth of the optical laser pump and the mean scattering depth for MeV electrons (see Figure 12.4). To minimize dynamical scattering effects, it is also advantageous to keep the sample thickness below this mean scattering depth, typically in the range 5–50 nm for most materials of interest.

Fabricating high-quality samples of a few nanometres in thicknesses and sufficient lateral areas for probing (>0.1 mm^2) is challenging for solid-state UED experiments. Here, we will discuss several approaches for sample preparation together with their merits and limitations. Ultrathin commercially available Si_3N_4 membranes provide excellent support for ultrathin film growth or as a support for exfoliated samples. Membranes with a thickness ≤ 50 nm have a relatively low cross-section, and the highly amorphous Si_3N_4 structure with blurred diffraction signals can be separated from the sharp diffraction features of crystalline materials. In addition, the large optical bandgap of Si_3N_4 (≈ 5 eV) helps minimize the substrate's thermal heating due to the absorption of the optical pump laser pulse.

The primary constraint with Si_3N_4 is that it is a very challenging platform for single crystal sample growth. While success has been achieved with advanced techniques such as ion-beam-assisted deposition (IBAD),[112] this remains a technically difficult and complex method of growth for many samples.

An alternative approach for single-crystal sample growth is the growth of samples on sacrificial crystalline substrates to produce single-crystal freestanding films. For chemically inert materials such as FePt, this can be achieved by directly etching the MgO single-crystal substrates.[113] In most cases, it is preferential to use water-soluble substrates due to the sensitivity of the sample of interest to the etch. Single-crystal NaCl is a popular choice for a water-soluble substrate,[114,115] but growing high-quality crystals with this method has been challenging. Hwang and co-workers recently pioneered an alternative approach using sacrificial hygroscopic oxide thin films such as $Sr_3Al_2O_6$ that allow *in situ* growth and lift-off of many types of high-quality oxide films.[116,117]

More traditional thinning techniques, such as focused ion beam milling, can also create thin single crystals from bulk.[118] While these methods are the mainstay of TEM sample preparation, it is a challenge to produce samples that span the square-millimetre-sized areas required for MeV-UED experiments. However, such large-area films with excellent quality have been produced by combinations of mechanical polishing and low-angle ion milling.[119]

Finally, van der Waals materials are a natural fit for MeV-UED experiments. Scotch tape exfoliation techniques can generally be used to realize measurable-sized samples with relative ease. The fabrication techniques for van der Waals solids are advancing quickly with methods such as gold-tape exfoliation, which are now used to realize monolayers with centimetre dimensions.[120] Additional methods for creating complex heterostructures and twisted layer assemblies have been demonstrated, such as the robotic assembly of N-layer systems.[121] These methods promise excellent compatibility with UED measurements in the future.

12.5.2 MeV-SUED Applications to Quantum Materials

A key topic where MeV-SUED has demonstrated its versatility is in understanding and controlling the dynamics of emergent quantum order. For example, in charge density wave (CDW) materials, electrons are diffracted from the periodic density distribution of the valance electrons and the often-accompanying periodic lattice distortion.[122] These appear as a superstructure order in the diffraction pattern, allowing the dynamics of the CDW order to be separated from other structural dynamics in momentum space. The relatively flat Ewald sphere afforded by MeV electrons enables multiple reflections to be accessed at once, allowing studies of not only the magnitude of the order parameter but also access to the dynamical rearrangement of the order (see Figure 12.14). Le Guyader *et al.* followed

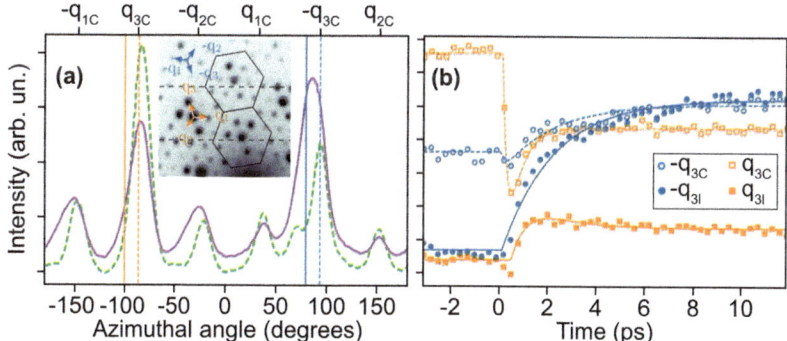

Figure 12.14 Dynamic intensity transfer between superlattice reflections of commensurate and incommensurate CDW stacking order in 1T-TaS$_2$. (a) Intensity profiles of the six-fold CDW superlattice reflection observed around the 200 Bragg reflection in 1T-TaS$_2$ (diffraction pattern shown as the inset) as a function of the azimuthal angle. Intensities are shown for the commensurate phase 1 ps before pumping with an optical laser pulse (dashed green) and 10 ps after excitation (solid purple). (b) The intensity–time evolution of the commensurate (open symbols) and incommensurate (filled symbols) $\pm q_3$ CDW superlattice reflection. Reproduced from ref. 123, https://doi.org/10.1063/1.4982918, under the terms of the CC BY 4.0 license http://creativecommons.org/licenses/by/4.0.

the dynamics of the stacking order in 1T-TaS$_2$ by monitoring the dynamic intensity transfer between superlattice reflections when the sample is oriented at five degrees to normal incidence.[123] The six superlattice peaks appearing around each Bragg reflection are labelled $\pm q_1$, $\pm q_2$, and $\pm q_3$ (see the inset of Figure 12.14a). The azimuthal angular position and intensity of the super-lattice reflect whether the stacking order is that of the commensurate phase or incommensurate phase. The azimuthal intensity profile of the superlattice reflections around the 200 Bragg peak (see Figure 12.14a) for 1 ps before (green dashed line) and 10 ps after (solid purple line) pump excitation shows a shift from the commensurate positions (dashed vertical lines) towards the incommensurate positions (solid vertical lines). The time evolution of the $\pm q_{3C}$ peak and $\pm q_{3I}$ peak intensities following excitation is depicted in Figure 12.14b. This is a clear signature of the dynamic rearrangement of the CDW stacking order following laser excitation.

In further work on the same material, Zong *et al.* used MeV-UED to explore how laser illumination could inject domain walls into the CDW condensate of 1T-TaS$_2$.[124] This was again visualized through the intensity transfer between CDW reflections on passing through the domain wall. Observing multiple superlattice reflections simultaneously is vital for understanding complex heterogenous dynamics in such systems.

An additional benefit of MeV-SUED for studying quantum materials is that it is a gentle probe. Electrons at energies below 10 MeV impart relatively little energy to the sample per elastic scattering event. This limits the thermal load associated with accessing the dynamics in quantum materials using electron beams as a probe. These properties were used to great effect by Kogar and collaborators to study a novel CDW order in LaTe$_3$.[125] Kogar *et al.* found that light pulses triggered the emergence of a second non-equilibrium CDW phase along the *a*-axis in LaTe$_3$ (see Figure 12.15). Careful consideration of the LaTe$_3$ electron diffraction pattern 0.3 ps before laser excitation and 1.8 ps after laser excitation (see Figure 12.15a) shows that two additional superlattice peaks along the vertical direction are clearly observed in the electron diffraction measurement post-excitation. These peaks correspond to the nonequilibrium *a*-axis CDW. The time evolution of the equilibrium *c*-axis CDW (in blue) and the nonequilibrium *a*-axis CDW (in red) following laser excitation was measured (see Figure 12.15b). In a follow-up work by Zong *et al.*, the light-induced state was found to be the result of fluctuations in the CDW order parameter.[126]

12.5.3   MeV-SUED Applications to Nanoscale Thermal Transport

Understanding the nanoscale dynamics of thermal transport is another area where MeV-SUED proves to be highly versatile. At sub-μm length scales, the transport of energy becomes complex. Often both phonon and electronic transport of energy need to be considered, and furthermore, both transport

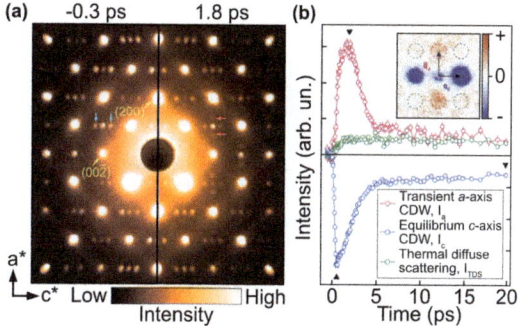

Figure 12.15   MeV-UED measurements of LaTe$_3$ showing the emergence of CDW superstructure peaks along the a^*-axis after photo-illumination.[125] (a) The LaTe$_3$ electron diffraction pattern 0.3 ps before excitation (left half of the image) and 1.8 ps after laser excitation (right half of the image). A new CDW superstructure peak appears in the laser-pumped diffraction pattern oriented along the a^*-axis. (b) The time evolution of the equilibrium c^*-axis CDW peak (blue symbols) and the non-equilibrium a^*-axis CDW peak (red symbols). Note that the y-axis for the upper half of this graph is multiplied by 2.8 times. Reproduced from ref. 125 with permission from Springer Nature, Copyright 2020.

mechanisms transition between a diffusive transport regime and super-diffusive and ballistic regimes as dimensionality is reduced. This is further complicated by heterogeneity in the systems of interest.

MeV-SUED experiments offer significant insights into the microscopic processes involved. The lower cross-section of MeV electrons compared to keV electrons enables quantitative measurements of the Deybe–Waller effect for a much wider range of systems. In a series of investigations, Sokolowski-Tinten and co-workers made use of this property of MeV electrons to study electron–phonon coupling and energy transport in thin films[127] and heterostructures.[128,129] Using a similar method, Sood *et al.* were recently able to explore energy transport between layers in WSe$_2$/WS$_2$ monolayer heterostructures.[130]

Another powerful method available to MeV-SUED in the kinematic scattering regime is accessing momentum-resolved phonon populations *via* diffuse scattering. This allowed contruction of a momentum-dependent picture of electron–phonon coupling. In the kinematic regime, the intensity of the diffuse scattering can be directly related to the phonon mode population at the corresponding momentum point through:[131,132]

$$I(S) = \sum_{j} \left(n_j(q) + \frac{1}{2} \right) \frac{|F_j(S)|^2}{\omega_j(q)}, \tag{12.4}$$

where $I(S)$ is the scattering intensity, S and q are the scattering and phonon momenta respectively, $n_j(q)$ is the mode occupation and $\omega_j(q)$ the mode

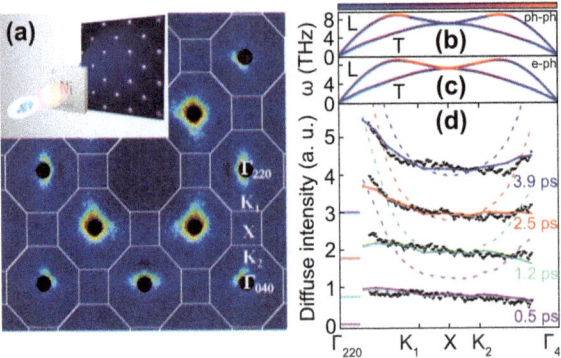

Figure 12.16 MeV-UED measurement of anisotropic electron–phonon coupling in nickel using diffuse scattering.[134] (a) The positive differential changes in scattering intensity following optical laser excitation. The inset shows the experimental geometry and electron diffraction pattern. (b and c) The calculated phonon energy dispersions along the Γ_{220}–X–Γ_{400} direction. Phonon linewidth contributions are denoted by colour for (a) phonon–phonon interactions and (b) phonon–electron interactions. (d) Measured diffusive scattering intensities (black symbols) along the Γ_{220}–X–Γ_{400} direction for different time delays (offset in the vertical direction) are compared to theoretical calculations for ferromagnetic nickel (dashed lines) and paramagnetic nickel (solid lines). Reproduced from ref. 134, https://doi.org/ 10.1103/ PhysRevB.101.100302, under the terms of the CC BY 4.0 license http:// creativecommons.org/licenses/by/4.0.

frequency, and the sum over is taken over all phonon branches j. Several experiments have used this relationship to study the anisotropy in electron–phonon coupling.[133,134] Maldonado *et al.* measured the time- and momentum-resolved dynamics of phonons in laser-heated nickel (see Figure 12.16).[134] The laser-induced change in diffuse scattering intensity is directly observable from the electron diffraction pattern (see Figure 12.16a). The diffuse scattering intensity along the Γ_{220}–X–Γ_{400} high-symmetry line is compared to calculations of the diffraction intensity based on nickel phonon-mode populations (see Figure 12.16d). These quantitative results allow benchmarking band-resolved models of the electron–phonon and phonon–phonon coupling (see Figure 12.16b and c, respectively). However, the relatively low cross-section of the phonons means that these experiments remain challenging for low atomic number elements and monolayer systems due to the low fluxes available at present MeV-UED sources. Future developments in photocathode brightness and higher source repetition rates will significantly improve the utility of this method.

12.5.4 MeV-SUED Applications to Structural Dynamics

MeV-SUED measurements of structural dynamics are exceptionally versatile because of the relatively flat Ewald sphere, which can simultaneously monitor multiple reflections. This allows understanding of complex motions with a limited number of measurements. Reid *et al.* utilized this to understand dynamic stresses in FePt nanoparticles allowing them to dissect the stresses into their microscopic contributions.[135] Dynamic strain is a powerful method for tuning a material's functional properties. For example, Sie *et al.* used THz laser excitation to drive a coherent interlayer shear strain in the Weyl semi-metal WTe_2.[40] Using MeV-SUED, they were able to quantify the amplitude of the induced strain and observe a longer-lived shear displacement associated with a distinct topological phase. The work demonstrates the utility of MeV-SUED as a method to quantify dynamic strain without strong perturbation of the system.

12.5.5 Single-shot MeV-SUED Applications to Stochastic Dynamics

Single-shot time-resolved MeV-SUED is also a powerful tool for measuring stochastic dynamics. The high electron energy enables a significant amount of charge to be present in a single pulse while maintaining temporal and momentum resolution. Mo *et al.* showed that a 20 fC electron bunch (~ 125k electrons) was sufficient to record single-shot diffraction patterns of crystalline gold with a time resolution of approximately 350 fs.[49] This enabled a first-of-its-kind single-shot study of the melting dynamics of gold with sub-picosecond time resolution.[43] The relative amounts of crystalline and liquid gold could be tracked in each single-shot measurement allowing a phase-diagram-like characterization of time-energy melting regimes. The relatively low energy deposited per elastic scattering event involving MeV electrons may make single-shot UED a particularly versatile method for time-resolved studies of non-destructive stochastic dynamics in materials.

12.6 Outlook

This chapter highlighted the various opportunities arising from existing MeV-UED experimental capabilities. However, these capabilities are currently far from the physical limits in terms of sensitivity and temporal resolution. In the following, we briefly highlight ongoing and future developments which will move future MeV-UED beamlines far beyond the current state-of-the-art.

The sensitivity of MeV-UED instrumentation can be improved through a higher electron flux by an increase in the repetition rate at a constant electron pulse charge. An ongoing upgrade effort will increase the repetition rate of SLAC-UED from the current 360 Hz to 1.08 kHz. However, the repetition rate of the current SLAC-UED photocathode gun is ultimately limited by heat deposition from the RF power. By lowering the acceleration field gradient to 20 MV m^{-1}, the APEX RF gun can be operated at 1 MHz and has demonstrated its feasibility for UED applications, albeit at considerably lower electron kinetic energies (820 keV).[39,136] In addition, superconducting RF gun technology can realize operation with very low heat dissipation. Therefore, it can enable >1 MHz repetition rates with higher acceleration field gradients than high-repetition-rate normal conducting RF gun designs such as the APEX gun. There are ongoing research efforts directed at such superconducting gun designs.[137,138]

Another important factor influencing the sensitivity and temporal resolution of MeV-UED instrumentation is the electron beam quality, which is critically affected by the acceleration field gradient of the electron source. When electrons are born from the cathode of an electron source, they are most vulnerable to space-charge effects because their energies are highly nonrelativistic. A higher acceleration field gradient can boost the as-born electrons to relativistic energies more rapidly, preventing space-charge-induced beam quality degradation in longitudinal (affecting the temporal resolution) and transverse directions (affecting the sensitivity) more efficiently. Dedicated designs to modify state-of-the-art photocathode RF guns for acceleration field gradient enhancements have been proposed and predicted to show significant performance improvements through computer simulations.[139,140] Novel schemes of cryogenic photocathode RF guns to push the field gradient limit have been conceived and are actively being explored.[141,142] Advanced accelerator techniques, such as laser wakefield acceleration (LWFA), can achieve extremely high acceleration field gradients on the order of $GV\,m^{-1}$ and potentially be employed as a future high-performance UED electron source.[143,144]

A further critical factor influencing the sensitivity of MeV-UED instrumentation is the detection of the diffracted electrons. The detection of individual electrons would increase the detection sensitivity of MeV-UED to the theoretical limit. However, single electron detection requires the detector to be read out at the repetition rate of the electron source to prevent a pile-up of electron hits. The TEAM (transmission electron aberration-corrected microscope) detector, which provides both shot-by-shot and single electron detection capabilities, has demonstrated its feasibility for MeV electrons as well at SLAC-UED.[145] It is based on a complementary metal–oxide–semiconductor active-pixel sensor technique and was originally designed as a prototype for high-resolution imaging for transmission electron microscopy using keV energy electrons. This type of direct electron detector could potentially enable high spatial resolution detection in UED and UEM. Hybrid pixel detectors are another promising type of shot-by-shot direct electron detector, which provide a very high dynamic range. One example is the ePix detector family, originally developed to detect X-ray photons at LCLS.[146] Preliminary studies at SLAC-UED have shown the applicability of ePix detectors for MeV electrons.

The development of high-sensitivity single-shot detectors for MeV-UED would also help significantly improve the achievable temporal resolution. The currently achievable time resolution at SLAC-UED (<150 fs) results from a convolution of electron pulse duration, electron arrival time jitter, and the detection scheme integrating over many electron pulses. Even with the current SLAC-UED beamline, electron pulse durations of <60 fs have been demonstrated, albeit at dramatically increased arrival time jitter.[147] However, methods for determining the relative arrival time of pump and probe pulses on a shot-by-shot level are now routinely used in XFEL experiments to improve the temporal resolution during postprocessing of the data.[148–150] The required shot-by-shot arrival time measurement of electron bunches based on THz streaking has already been demonstrated.[147,151,152] Thus, combining a single-shot detector and arrival time monitor capabilities would improve the temporal resolution of SLAC-UED significantly below 100 fs.

Electron pulse compression schemes can achieve further improvements. Using a time-varying RF field, an RF compressor can compress the electron beam pulse duration to <24 fs FWHM.[153] However, such a compression scheme requires the correction of additional arrival time jitter contributions. An alternative compression scheme based on intense THz pulses, which does not introduce additional arrival time jitter, was recently demonstrated.[154,155] Moreover, a beamline module composed of a designed sequence of electron magnets can realize simultaneous electron bunch compression and time-of-arrival jitter reduction, which is another efficient approach for achieving higher temporal resolution.[156,157]

The abovementioned developments will enable exciting experimental opportunities. The expected increases in electron flux will allow for the investigation of structural dynamics in biomolecules with low vapour pressure or photochemical reactions of transition metal based photocatalysts in solution. In combination with the expected improvements in temporal resolution, proton transfer reaction dynamics investigations will come into reach.

Acknowledgements

The authors wish to acknowledge Xijie Wang for his foundation of the SLAC MeV-UED program and the entire SLAC-UED team for their continuing support. The authors also acknowledge the U.S. Department of Energy, Office of Science, Office of Basic Energy Sciences, and the SLAC National Accelerator Laboratory for their continuing support of the SLAC-UED. TJAW was supported by the AMOS program within the DOE Office of Science, Basic Energy Sciences, Chemical Sciences, Geosciences, and Biosciences Division. The SLAC-UED is operated as part of the Linac Coherent Light Source at the SLAC National Accelerator Laboratory, supported by the U.S. Department of Energy, Office of Science, Office of Basic Energy Sciences under contract no. DE-AC02-76SF00515.

References

1. SLAC Megaelectronvolt Ultrasfast Electron Diffraction Instrument, https://lcls.slac.stanford.edu/instruments/mev-ued (accessed 14 November 2022).

2. S. Hammes-Schiffer and A. A. Stuchebrukhov, *Chem. Rev.*, 2010, **110**, 6939–6960.

3. R. Henderson, *Q. Rev. Biophys.*, 1995, **28**, 171–193.

4. H. N. Chapman, P. Fromme, A. Barty, T. A. White, R. A. Kirian, A. Aquila, M. S. Hunter, J. Schulz, D. P. DePonte, U. Weierstall, R. B. Doak, F. R. N. C. Maia, A. V. Martin, I. Schlichting, L. Lomb, N. Coppola, R. L. Shoeman, S. W. Epp, R. Hartmann, D. Rolles, A. Rudenko, L. Foucar, N. Kimmel, G. Weidenspointner, P. Holl, M. Liang, M. Barthelmess, C. Caleman, S. Boutet, M. J. Bogan, J. Krzywinski, C. Bostedt, S. Bajt, L. Gumprecht, B. Rudek, B. Erk, C. Schmidt, A. Hömke, C. Reich, D. Pietschner, L. Strüder, G. Hauser, H. Gorke, J. Ullrich, S. Herrmann, G. Schaller, F. Schopper, H. Soltau, K.-U. Kühnel, M. Messerschmidt, J. D. Bozek, S. P. Hau-Riege, M. Frank, C. Y. Hampton, R. G. Sierra, D. Starodub, G. J. Williams, J. Hajdu, N. Timneanu, M. M. Seibert, J. Andreasson, A. Rocker, O. Jönsson, M. Svenda, S. Stern, K. Nass, R. Andritschke, C.-D. Schröter, F. Krasniqi, M. Bott, K. E. Schmidt, X. Wang, I. Grotjohann, J. M. Holton, T. R. M. Barends, R. Neutze, S. Marchesini, R. Fromme, S. Schorb, D. Rupp, M. Adolph, T. Gorkhover, I. Andersson, H. Hirsemann, G. Potdevin, H. Graafsma, B. Nilsson and J. C. H. Spence, *Nature*, 2011, **470**, 73–77.

5. M. M. Seibert, T. Ekeberg, F. R. N. C. Maia, M. Svenda, J. Andreasson, O. Joensson, D. Odic, B. Iwan, A. Rocker, D. Westphal, M. Hantke, D. P. DePonte, A. Barty, J. Schulz, L. Gumprecht, N. Coppola, A. Aquila, M. Liang, T. A. White, A. Martin, C. Caleman, S. Stern, C. Abergel, V. Seltzer, J.-M. Claverie, C. Bostedt, J. D. Bozek, S. Boutet, A. A. Miahnahri, M. Messerschmidt, J. Krzywinski, G. Williams, K. O. Hodgson, M. J. Bogan, C. Y. Hampton, R. G. Sierra, D. Starodub, I. Andersson, S. Bajt, M. Barthelmess, J. C. H. Spence, P. Fromme, U. Weierstall, R. Kirian, M. Hunter, R. B. Doak, S. Marchesini, S. P. Hau-Riege, M. Frank, R. L. Shoeman, L. Lomb, S. W. Epp, R. Hartmann, D. Rolles, A. Rudenko, C. Schmidt, L. Foucar, N. Kimmel, P. Holl, B. Rudek, B. Erk, A. Hoemke, C. Reich, D. Pietschner, G. Weidenspointner, L. Strueder, G. Hauser, H. Gorke, J. Ullrich, I. Schlichting, S. Herrmann, G. Schaller, F. Schopper, H. Soltau, K.-U. Kuehnel, R. Andritschke, C.-D. Schroeter, F. Krasniqi, M. Bott, S. Schorb, D. Rupp, M. Adolph, T. Gorkhover, H. Hirsemann, G. Potdevin, H. Graafsma, B. Nilsson, H. N. Chapman and J. Hajdu, *Nature*, 2011, **470**, 78–81.

6. I. Inoue, V. Tkachenko, K. J. Kapcia, V. Lipp, B. Ziaja, Y. Inubushi, T. Hara, M. Yabashi and E. Nishibori, *Phys. Rev. Lett.*, 2022, **128**, 223203.

7. A. H. Zewail, *Science*, 2010, **328**, 187–193.

8. X. Wang and P. Musumeci, *FUTURE OF ELECTRON SOURCES*, SLAC National Accelerator Laboratory, 2016.

9. C. Limborg, Z. Li, L. Xiao, J. Schmerge, D. Dowell, S. Gierman, E. Bong and S. Gilevich, RF design of the LCLS Gun, http://www-ssrl.slac.stanford.edu/lcls/photoinjector/internals/rfgun/.

10. G. Mourou and S. Williamson, *Appl. Phys. Lett.*, 1982, **41**, 44–45.

11. A. H. Zewail, *Annu. Rev. Phys. Chem.*, 2006, **57**, 65–103.
12. G. Sciaini and R. J. D. Miller, *Rep. Prog. Phys.*, 2011, **74**, 096101.
13. C. Kealhofer, W. Schneider, D. Ehberger, A. Ryabov, F. Krausz and P. Baum, *Science*, 2016, **352**, 429–433.
14. A. Feist, N. Bach, N. Rubiano da Silva, T. Danz, M. Möller, K. E. Priebe, T. Domröse, J. G. Gatzmann, S. Rost, J. Schauss, S. Strauch, R. Bormann, M. Sivis, S. Schäfer and C. Ropers, *Ultramicroscopy*, 2017, **176**, 63–73.
15. L. Waldecker, R. Bertoni and R. Ernstorfer, *J. Appl. Phys.*, 2015, **117**, 044903.
16. R. P. Chatelain, V. R. Morrison, C. Godbout and B. J. Siwick, *Appl. Phys. Lett.*, 2012, **101**, 081901.
17. M. R. Otto, L. P. René de Cotret, M. J. Stern and B. J. Siwick, *Struct. Dyn.*, 2017, **4**, 051101.
18. Y. Xiong, K. J. Wilkin and M. Centurion, *Phys. Rev. Res.*, 2020, **2**, 043064.
19. J. Williams, F. Zhou, T. Sun, Z. Tao, K. Chang, K. Makino, M. Berz, P. M. Duxbury and C.-Y. Ruan, *Struct. Dyn.*, 2017, **4**, 044035.
20. S. Sun, X. Sun, D. Bartles, E. Wozniak, J. Williams, P. Zhang and C.-Y. Ruan, *Struct. Dyn.*, 2020, **7**, 064301.
21. D. A. Edwards and M. J. Syphers, *An Introduction to the Physics of High Energy Accelerators*, Wiley, 1st edn, 1993.
22. X. J. Wang, X. Qiu and I. Ben-Zvi, *Phys. Rev. E*, 1996, **54**, R3121–R3124.
23. X. J. Wang, Z. Wu and H. H. Ihee, in *Proceedings of the 2003 Bipolar/ BiCMOS Circuits and Technology Meeting (IEEE Cat. No. 03CH37440)*, IEEE, Portland, OR, USA, 2003, pp. 420–422.
24. X. Wang, D. Xiang, T. Kim and H. Ihee, *J. Korean Phys. Soc.*, 2006, **48**, 390–396.
25. P. Zhu, Y. Zhu, Y. Hidaka, L. Wu, J. Cao, H. Berger, J. Geck, R. Kraus, S. Pjerov, Y. Shen, R. I. Tobey, J. P. Hill and X. J. Wang, *New J. Phys.*, 2015, **17**, 063004.
26. S. P. Weathersby, G. Brown, M. Centurion, T. F. Chase, R. Coffee, J. Corbett, J. P. Eichner, J. C. Frisch, A. R. Fry, M. Gühr, N. Hartmann, C. Hast, R. Hettel, R. K. Jobe, E. N. Jongewaard, J. R. Lewandowski, R. K. Li, A. M. Lindenberg, I. Makasyuk, J. E. May, D. McCormick, M. N. Nguyen, A. H. Reid, X. Shen, K. Sokolowski-Tinten, T. Vecchione, S. L. Vetter, J. Wu, J. Yang, H. A. Dürr and X. J. Wang, *Rev. Sci. Instrum.*, 2015, **86**, 073702.
27. J. B. Hastings, F. M. Rudakov, D. H. Dowell, J. F. Schmerge, J. D. Cardoza, J. M. Castro, S. M. Gierman, H. Loos and P. M. Weber, *Appl. Phys. Lett.*, 2006, **89**, 184109.
28. P. Musumeci, J. T. Moody and C. M. Scoby, *Ultramicroscopy*, 2008, **108**, 1450–1453.
29. P. Musumeci, J. T. Moody, C. M. Scoby, M. S. Gutierrez, H. A. Bender and N. S. Wilcox, *Rev. Sci. Instrum.*, 2010, **81**, 013306.
30. R. Li, C. Tang, Y. Du, W. Huang, Q. Du, J. Shi, L. Yan and X. Wang, *Rev. Sci. Instrum.*, 2009, **80**, 083303.

31. P. Musumeci, J. T. Moody, C. M. Scoby, M. S. Gutierrez and M. Westfall, *Appl. Phys. Lett.*, 2010, **97**, 063502.

32. P. Musumeci, J. T. Moody, C. M. Scoby, M. S. Gutierrez, M. Westfall and R. K. Li, *J. Appl. Phys.*, 2010, **108**, 114513.

33. R. Li, W. Huang, Y. Du, L. Yan, Q. Du, J. Shi, J. Hua, H. Chen, T. Du, H. Xu and C. Tang, *Rev. Sci. Instrum.*, 2010, **81**, 036110.

34. Y. Murooka, N. Naruse, S. Sakakihara, M. Ishimaru, J. Yang and K. Tanimura, *Appl. Phys. Lett.*, 2011, **98**, 251903.

35. Y. Giret, N. Naruse, S. L. Daraszewicz, Y. Murooka, J. Yang, D. M. Duffy, A. L. Shluger and K. Tanimura, *Appl. Phys. Lett.*, 2013, **103**, 253107.

36. F. Fu, S. Liu, P. Zhu, D. Xiang, J. Zhang and J. Cao, *Rev. Sci. Instrum.*, 2014, **85**, 083701.

37. S. Manz, A. Casandruc, D. Zhang, Y. Zhong, R. A. Loch, A. Marx, T. Hasegawa, L. C. Liu, S. Bayesteh, H. Delsim-Hashemi, M. Hoffmann, M. Felber, M. Hachmann, F. Mayet, J. Hirscht, S. Keskin, M. Hada, S. W. Epp, K. Flöttmann and R. J. D. Miller, *Faraday Discuss.*, 2015, **177**, 467–491.

38. S. Setiniyaz, H. W. Kim, I.-H. Baek, J. Nam, M. Chae, B.-H. Han, B. Gudkov, K. H. Jang, S. Park, Y. U. Jeong, S. Miginsky and N. Vinokurov, *J. Korean Phys. Soc.*, 2016, **69**, 1019–1024.

39. D. Filippetto and H. Qian, *J. Phys. B: At., Mol. Opt. Phys.*, 2016, **49**, 104003.

40. E. J. Sie, C. M. Nyby, C. D. Pemmaraju, S. J. Park, X. Shen, J. Yang, M. C. Hoffmann, B. K. Ofori-Okai, R. Li, A. H. Reid, S. Weathersby, E. Mannebach, N. Finney, D. Rhodes, D. Chenet, A. Antony, L. Balicas, J. Hone, T. P. Devereaux, T. F. Heinz, X. Wang and A. M. Lindenberg, *Nature*, 2019, **565**, 61–66.

41. J. Yang, R. Dettori, J. P. F. Nunes, N. H. List, E. Biasin, M. Centurion, Z. Chen, A. A. Cordones, D. P. Deponte, T. F. Heinz, M. E. Kozina, K. Ledbetter, M.-F. Lin, A. M. Lindenberg, M. Mo, A. Nilsson, X. Shen, T. J. A. Wolf, D. Donadio, K. J. Gaffney, T. J. Martinez and X. Wang, *Nature*, 2021, **596**, 531–535.

42. J. Yang, X. Zhu, T. J. A. Wolf, Z. Li, J. P. F. Nunes, R. Coffee, J. P. Cryan, M. Gühr, K. Hegazy, T. F. Heinz, K. Jobe, R. Li, X. Shen, T. Veccione, S. Weathersby, K. J. Wilkin, C. Yoneda, Q. Zheng, T. J. Martinez, M. Centurion and X. Wang, *Science*, 2018, **361**, 64–67.

43. M. Z. Mo, Z. Chen, R. K. Li, M. Dunning, B. B. L. Witte, J. K. Baldwin, L. B. Fletcher, J. B. Kim, A. Ng, R. Redmer, A. H. Reid, P. Shekhar, X. Z. Shen, M. Shen, K. Sokolowski-Tinten, Y. Y. Tsui, Y. Q. Wang, Q. Zheng, X. J. Wang and S. H. Glenzer, *Science*, 2018, **360**, 1451–1455.

44. A. Sood, X. Shen, Y. Shi, S. Kumar, S. J. Park, M. Zajac, Y. Sun, L.-Q. Chen, S. Ramanathan, X. Wang, W. C. Chueh and A. M. Lindenberg, *Science*, 2021, **373**, 352–355.

45. J. A. Meli, in *The Physics of Radiation Dosimetry*, AIP Publishing LLC, 2020, pp. 1–4.

46. F. Salvat, J. D. Martnez, R. Mayol and J. Parellada, *Phys. Rev. A*, 1987, **36**, 467–474.

47. B. E. Warren, *X-ray Diffraction*, Courier Corporation, 1990.
48. P. Zhang, J. Yang and M. Centurion, *New J. Phys.*, 2014, **16**, 083008.
49. M. Z. Mo, X. Shen, Z. Chen, R. K. Li, M. Dunning, K. Sokolowski-Tinten, Q. Zheng, S. P. Weathersby, A. H. Reid, R. Coffee, I. Makasyuk, S. Edstrom, D. McCormick, K. Jobe, C. Hast, S. H. Glenzer and X. Wang, *Rev. Sci. Instrum.*, 2016, **87**, 11D810.
50. X. Shen, J. P. F. Nunes, J. Yang, R. K. Jobe, R. K. Li, M.-F. Lin, B. Moore, M. Niebuhr, S. P. Weathersby, T. J. A. Wolf, C. Yoneda, M. Guehr, M. Centurion and X. J. Wang, *Struct. Dyn.*, 2019, **6**, 054305.
51. R. Akre, D. Dowell, P. Emma, J. Frisch, S. Gilevich, G. Hays, Ph. Hering, R. Iverson, C. Limborg-Deprey, H. Loos, A. Miahnahri, J. Schmerge, J. Turner, J. Welch, W. White and J. Wu, *Phys. Rev. Spec. Top. - Accel. Beams*, 2008, **11**, 030703.
52. D. H. Dowell, E. Jongewaard, J. Lewandowski, C. Limborg-Deprey, Z. Li, J. Schmerge, A. Vlieks, J. Wang and L. Xiao, *ICFA Beam Dyn. Newslett.*, 2008, **46**, 162–192.
53. X. Shen, R. K. Li, U. Lundström, T. J. Lane, A. H. Reid, S. P. Weathersby and X. J. Wang, *Ultramicroscopy*, 2018, **184**, 172–176.
54. B. K. Ofori-Okai, M. C. Hoffmann, A. H. Reid, S. Edstrom, R. K. Jobe, R. K. Li, E. M. Mannebach, S. J. Park, W. Polzin, X. Shen, S. P. Weathersby, J. Yang, Q. Zheng, M. Zajac, A. M. Lindenberg, S. H. Glenzer and X. J. Wang, *J. Instrum.*, 2018, **13**, P06014.
55. L. Ma, X. Shen, K. Kim, D. Brown, M. D'Ewart, B. Hong, J. Olsen, S. Smith, D. Van Winkle, E. Williams, S. Weathersby, X. Wang, A. Young and J. Frisch, *arXiv*, 2019, DOI: 10.48550/arXiv.1910.02296.
56. J. P. F. Nunes, K. Ledbetter, M. Lin, M. Kozina, D. P. DePonte, E. Biasin, M. Centurion, C. J. Crissman, M. Dunning, S. Guillet, K. Jobe, Y. Liu, M. Mo, X. Shen, R. Sublett, S. Weathersby, C. Yoneda, T. J. A. Wolf, J. Yang, A. A. Cordones and X. J. Wang, *Struct. Dyn.*, 2020, 7, 024301.
57. J. D. Koralek, J. B. Kim, P. Brůža, C. B. Curry, Z. Chen, H. A. Bechtel, A. A. Cordones, P. Sperling, S. Toleikis, J. F. Kern, S. P. Moeller, S. H. Glenzer and D. P. DePonte, *Nat. Commun.*, 2018, **9**, 1353.
58. C. J. Crissman, M. Mo, Z. Chen, J. Yang, D. A. Huyke, S. H. Glenzer, K. Ledbetter, J. P. F. Nunes, M. L. Ng, H. Wang, X. Shen, X. Wang and D. P. DePonte, *Lab Chip*, 2022, **22**, 1365–1373.
59. M.-F. Lin, N. Singh, S. Liang, M. Mo, J. P. F. Nunes, K. Ledbetter, J. Yang, M. Kozina, S. Weathersby, X. Shen, A. A. Cordones, T. J. A. Wolf, C. D. Pemmaraju, M. Ihme and X. J. Wang, *Science*, 2021, **374**, 92–95.
60. P. Muscher, A. Sood, F. Lu, D. Rehn, D. Luo, X. Shen, E. Reed, X. Wang, W. Chueh and A. Lindenberg, *Microsc. Microanal.*, 2020, **26**, 1152–1154.
61. P. K. Muscher, D. A. Rehn, A. Sood, K. Lim, D. Luo, X. Shen, M. Zajac, F. Lu, A. Mehta, Y. Li, X. Wang, E. J. Reed, W. C. Chueh and A. M. Lindenberg, *Adv. Mater.*, 2021, **33**, 2101875.
62. R. Srinivasan, V. A. Lobastov, C.-Y. Ruan and A. H. Zewail, *Helv. Chim. Acta*, 2003, **86**, 1761–1799.

63. J. Yang, J. Beck, C. J. Uiterwaal and M. Centurion, *Nat. Commun.*, 2015, **6**, 8172.

64. O. Zandi, K. J. Wilkin, Y. Xiong and M. Centurion, *Struct. Dyn.*, 2017, **4**, 044022.

65. Y. Xiong, K. Borne, A. M. Carrascosa, S. K. Saha, K. J. Wilkin, M. Yang, S. Bhattacharyya, K. Chen, W. Du, L. Ma, N. Marshall, J. P. F. Nunes, S. Pathak, Z. Phelps, X. Xu, H. Yong, K. Lopata, P. M. Weber, A. Rudenko, D. Rolles and M. Centurion, *Faraday Discuss.*, 2021, **228**, 39–59.

66. C. J. Hensley, J. Yang and M. Centurion, *Phys. Rev. Lett.*, 2012, **109**, 133202.

67. M. P. Minitti, J. M. Budarz, A. Kirrander, J. S. Robinson, D. Ratner, T. J. Lane, D. Zhu, J. M. Glownia, M. Kozina, H. T. Lemke, M. Sikorski, Y. Feng, S. Nelson, K. Saita, B. Stankus, T. Northey, J. B. Hastings and P. M. Weber, *Phys. Rev. Lett.*, 2015, **114**, 255501.

68. J. Yang, X. Zhu, J. P. F. Nunes, J. K. Yu, R. M. Parrish, T. J. A. Wolf, M. Centurion, M. Gühr, R. Li, Y. Liu, B. Moore, M. Niebuhr, S. Park, X. Shen, S. Weathersby, T. Weinacht, T. J. Martinez and X. Wang, *Science*, 2020, **368**, 885–889.

69. T. J. A. Wolf, D. M. Sanchez, J. Yang, R. M. Parrish, J. P. F. Nunes, M. Centurion, R. Coffee, J. P. Cryan, M. Gühr, K. Hegazy, A. Kirrander, R. K. Li, J. Ruddock, X. Shen, T. Vecchione, S. P. Weathersby, P. M. Weber, K. Wilkin, H. Yong, Q. Zheng, X. J. Wang, M. P. Minitti and T. J. Martínez, *Nat. Chem.*, 2019, **11**, 504–509.

70. E. G. Champenois, D. M. Sanchez, J. Yang, J. P. Figueira Nunes, A. Attar, M. Centurion, R. Forbes, M. Gühr, K. Hegazy, F. Ji, S. K. Saha, Y. Liu, M.-F. Lin, D. Luo, B. Moore, X. Shen, M. R. Ware, X. J. Wang, T. J. Martínez and T. J. A. Wolf, *Science*, 2021, **374**, 178–182.

71. Y. Liu, S. L. Horton, J. Yang, J. P. F. Nunes, X. Shen, T. J. A. Wolf, R. Forbes, C. Cheng, B. Moore, M. Centurion, K. Hegazy, R. Li, M.-F. Lin, A. Stolow, P. Hockett, T. Rozgonyi, P. Marquetand, X. Wang and T. Weinacht, *Phys. Rev. X*, 2020, **10**, 021016.

72. J. Yang, M. Guehr, X. Shen, R. Li, T. Vecchione, R. Coffee, J. Corbett, A. Fry, N. Hartmann, C. Hast, K. Hegazy, K. Jobe, I. Makasyuk, J. Robinson, M. S. Robinson, S. Vetter, S. Weathersby, C. Yoneda, X. Wang and M. Centurion, *Phys. Rev. Lett.*, 2016, **117**, 153002.

73. J. Yang, M. Guehr, T. Vecchione, M. S. Robinson, R. Li, N. Hartmann, X. Shen, R. Coffee, J. Corbett, A. Fry, K. Gaffney, T. Gorkhover, C. Hast, K. Jobe, I. Makasyuk, A. Reid, J. Robinson, S. Vetter, F. Wang, S. Weathersby, C. Yoneda, M. Centurion and X. Wang, *Nat. Commun.*, 2016, **7**, 11232.

74. M. Centurion, T. J. A. Wolf and J. Yang, *Annu. Rev. Phys. Chem.*, 2022, **73**, 21.

75. I. Hargittai and M. Hargittai, *Stereochemical Applications of Gas-Phase Electron Diffraction, Part A*, John Wiley & Sons, 1988.

76. I. Plante and F. A. Cucinotta, *New J. Phys.*, 2009, **11**, 063047.

77. J. H. Hubbell, Wm. J. Veigele, E. A. Briggs, R. T. Brown, D. T. Cromer and R. J. Howerton, *J. Phys. Chem. Ref. Data*, 1975, **4**, 471–538.

78. J. M. Budarz, M. P. Minitti, D. V. Cofer-Shabica, B. Stankus, A. Kirrander, J. B. Hastings and P. M. Weber, *J. Phys. B: At., Mol. Opt. Phys.*, 2016, **49**, 034001.

79. B. Stankus, H. Yong, N. Zotev, J. M. Ruddock, D. Bellshaw, T. J. Lane, M. Liang, S. Boutet, S. Carbajo, J. S. Robinson, W. Du, N. Goff, Y. Chang, J. E. Koglin, M. P. Minitti, A. Kirrander and P. M. Weber, *Nat. Chem.*, 2019, **11**, 716–721.

80. J. P. F. Nunes, M. Williams, J. Yang, T. J. A. Wolf, C. D. Rankine, R. M. Parrish, B. Moore, K. Wilkin, X. Shen, M.-F. Lin, K. Hegazy, R. Li, S. P. Weathersby, T. J. Martinez, X. Wang and M. Centurion, *Res. Sq.*, DOI: 10.21203/rs.3.rs-1837872/v1.

81. R. M. Parrish and T. J. Martínez, *J. Chem. Theory Comput.*, 2019, 1523–1537.

82. Y. Liu, D. M. Sanchez, M. R. Ware, E. G. Champenois, J. Yang, J. P. F. Nunes, A. Attar, M. Centurion, J. P. Cryan, R. G. Forbes, K. Hegazy, M. C. Hoffmann, F. Ji, M.-F. Lin, D. Luo, S. K. Saha, X. Shen, X. Wang, T. J. Martínez and T. J. A. Wolf, *arXiv*, *Nat. Commun.*, 2023, **14**, 2795

83. A. Natan, *Phys. Rev. A*, 2023, **107**, 023105.

84. E. G. Champenois, N. H. List, M. Ware, M. Britton, P. H. Bucksbaum, X. Cheng, M. Centurion, J. P. Cryan, R. Forbes, I. Gabalski, K. Hegazy, M. C. Hoffmann, A. J. Howard, F. Ji, M.-F. Lin, J. P. Nunes, X. Shen, J. Yang, X. Wang, T. J. Martinez and T. J. A. Wolf, *Phys. Rev. Lett.* 2023, **131**, 143001.

85. M. Breitenstein, H. Meyer and A. Schweig, *Chem. Phys.*, 1987, **112**, 199–203.

86. Y. S. Kim, W. K. Kang and K.-H. Jung, *J. Chem. Phys.*, 1996, **105**, 551–557.

87. L. Inhester, Z. Li, X. Zhu, N. Medvedev and T. J. A. Wolf, *J. Phys. Chem. Lett.*, 2019, **10**, 6536–6544.

88. M. Dantus, S. B. Kim, J. C. Williamson and A. H. Zewail, *J. Phys. Chem.*, 1994, **98**, 2782–2796.

89. R. B. Woodward and R. Hoffmann, *Angew. Chem. Int. Ed. Engl.*, 1969, **8**, 781–853.

90. B. C. Arruda and R. J. Sension, *Phys. Chem. Chem. Phys.*, 2014, **16**, 4439–4455.

91. S. Deb and P. M. Weber, *Annu. Rev. Phys. Chem.*, 2011, **62**, 19–39.

92. R. C. Dudek and P. M. Weber, *J. Phys. Chem. A*, 2001, **105**, 4167–4171.

93. H. Ihee, V. A. Lobastov, U. M. Gomez, B. M. Goodson, R. Srinivasan, C.-Y. Ruan and A. H. Zewail, *Science*, 2001, **291**, 458–462.

94. C.-Y. Ruan, V. A. Lobastov, R. Srinivasan, B. M. Goodson, H. Ihee and A. H. Zewail, *Proc. Natl. Acad. Sci.*, 2001, **98**, 7117–7122.

95. E. Biasin, Z. W. Fox, A. Andersen, K. Ledbetter, K. S. Kjær, R. Alonso-Mori, J. M. Carlstad, M. Chollet, J. D. Gaynor, J. M. Glownia, K. Hong, T. Kroll, J. H. Lee, C. Liekhus-Schmaltz, M. Reinhard, D. Sokaras, Y. Zhang, G. Doumy, A. M. March, S. H. Southworth, S. Mukamel, K. J. Gaffney, R. W. Schoenlein, N. Govind, A. A. Cordones and M. Khalil, *Nat. Chem.*, 2021, **13**, 343–349.

96. K. S. Kjær, T. B. V. Driel, T. C. B. Harlang, K. Kunnus, E. Biasin, K. Ledbetter, R. W. Hartsock, M. E. Reinhard, S. Koroidov, L. Li,

M. G. Laursen, F. B. Hansen, P. Vester, M. Christensen, K. Haldrup, M. M. Nielsen, A. O. Dohn, M. I. Pápai, K. B. Møller, P. Chabera, Y. Liu, H. Tatsuno, C. Timm, M. Jarenmark, J. Uhlig, V. Sundstöm, K. Wärnmark, P. Persson, Z. Németh, D. S. Szemes, É. Bajnóczi, G. Vankó, R. Alonso-Mori, J. M. Glownia, S. Nelson, M. Sikorski, D. Sokaras, S. E. Canton, H. T. Lemke and K. J. Gaffney, *Chem. Sci.*, 2019, **10**, 5749–5760.

97. K. Kunnus, M. Vacher, T. C. B. Harlang, K. S. Kjær, K. Haldrup, E. Biasin, T. B. van Driel, M. Pápai, P. Chabera, Y. Liu, H. Tatsuno, C. Timm, E. Källman, M. Delcey, R. W. Hartsock, M. E. Reinhard, S. Koroidov, M. G. Laursen, F. B. Hansen, P. Vester, M. Christensen, L. Sandberg, Z. Németh, D. S. Szemes, É. Bajnóczi, R. Alonso-Mori, J. M. Glownia, S. Nelson, M. Sikorski, D. Sokaras, H. T. Lemke, S. E. Canton, K. B. Møller, M. M. Nielsen, G. Vankó, K. Wärnmark, V. Sundström, P. Persson, M. Lundberg, J. Uhlig and K. J. Gaffney, *Nat. Commun.*, 2020, **11**, 634.

98. X. Fu, B. Chen, J. Tang, M. Th. Hassan and A. H. Zewail, *Science*, 2017, **355**, 494–498.

99. S. Azim, L. A. Bultema, M. B. de Kock, E. R. Osorio-Blanco, M. Calderón, J. Gonschior, J.-P. Leimkohl, F. Tellkamp, R. Bücker, E. C. Schulz, S. Keskin, N. de Jonge, G. H. Kassier and R. J. D. Miller, *Microsc. Microanal.*, 2021, **27**, 44–53.

100. M. B. de Kock, S. Azim, G. H. Kassier and R. J. D. Miller, *J. Chem. Phys.*, 2020, **153**, 194504.

101. K. Ledbetter, E. Biasin, J. P. F. Nunes, M. Centurion, K. J. Gaffney, M. Kozina, M.-F. Lin, X. Shen, J. Yang, X. J. Wang, T. J. A. Wolf and A. A. Cordones, *Struct. Dyn.*, 2020, 7, 064901.

102. J. Yang, J. P. F. Nunes, K. Ledbetter, E. Biasin, M. Centurion, Z. Chen, A. A. Cordones, C. Crissman, D. P. Deponte, S. H. Glenzer, M.-F. Lin, M. Mo, C. D. Rankine, X. Shen, T. J. A. Wolf and X. Wang, *Phys. Chem. Chem. Phys.*, 2021, **23**, 1308–1316.

103. H. Ihee, M. Lorenc, T. K. Kim, Q. Y. Kong, M. Cammarata, J. H. Lee, S. Bratos and M. Wulff, *Science*, 2005, **309**, 1223–1227.

104. H. Ihee, *Acc. Chem. Res.*, 2009, **42**, 356–366.

105. O. Marsalek, C. G. Elles, P. A. Pieniazek, E. Pluhařová, J. VandeVondele, S. E. Bradforth and P. Jungwirth, *J. Chem. Phys.*, 2011, **135**, 224510.

106. Y. Kimura, J. C. Alfano, P. K. Walhout and P. F. Barbara, *J. Phys. Chem.*, 1994, **98**, 3450–3458.

107. S. Le Caër, *Water*, 2011, 3, 235–253.

108. F. Perakis, L. De Marco, A. Shalit, F. Tang, Z. R. Kann, T. D. Kühne, R. Torre, M. Bonn and Y. Nagata, *Chem. Rev.*, 2016, **116**, 7590–7607.

109. R. Yuan, J. A. Napoli, C. Yan, O. Marsalek, T. E. Markland and M. D. Fayer, *ACS Cent. Sci.*, 2019, 5, 1269–1277.

110. K. H. Kim, J. H. Lee, J. Kim, S. Nozawa, T. Sato, A. Tomita, K. Ichiyanagi, H. Ki, J. Kim, S. Adachi and H. Ihee, *Phys. Rev. Lett.*, 2013, **110**, 165505.

111. P. Vester, K. Kubicek, T. A. Assefa, E. Biasin, M. Christensen, A. O Dohn, T. B. van Driel, A. Galler, W. Gawelda, C. H. Tobias, N. E. Henriksen, K. S.

Kjaer, T. S. Kuhlmann, Z. Németh, Z. Nurekeyev, M. Pápai, G. Vankó, H. Yavas, D. B. Zederkof, U. Bergmann, M. M. Nielsen, K. B. Møller, K. Haldrup and C. Bressler, *J. Chem. Phys.*, 2022, **157**, 224201.

112. C. P. Wang, K. B. Do, M. R. Beasley, T. H. Geballe and R. H. Hammond, *Appl. Phys. Lett.*, 1997, **71**, 2955–2957.

113. P. W. Granitzka, E. Jal, L. Le Guyader, M. Savoini, D. J. Higley, T. Liu, Z. Chen, T. Chase, H. Ohldag, G. L. Dakovski, W. F. Schlotter, S. Carron, M. C. Hoffman, A. X. Gray, P. Shafer, E. Arenholz, O. Hellwig, V. Mehta, Y. K. Takahashi, J. Wang, E. E. Fullerton, J. Stöhr, A. H. Reid and H. A. Dürr, *Nano Lett.*, 2017, **17**, 2426–2432.

114. J. W. Matthews, *J. Vac. Sci. Technol.*, 1966, **3**, 133–145.

115. S. Nie, X. Wang, H. Park, R. Clinite and J. Cao, *Phys. Rev. Lett.*, 2006, **96**, 025901.

116. D. Lu, D. J. Baek, S. S. Hong, L. F. Kourkoutis, Y. Hikita and H. Y. Hwang, *Nat. Mater.*, 2016, **15**, 1255–1260.

117. D. Ji, S. Cai, T. R. Paudel, H. Sun, C. Zhang, L. Han, Y. Wei, Y. Zang, M. Gu, Y. Zhang, W. Gao, H. Huyan, W. Guo, D. Wu, Z. Gu, E. Y. Tsymbal, P. Wang, Y. Nie and X. Pan, *Nature*, 2019, **570**, 87–90.

118. J. Mayer, L. A. Giannuzzi, T. Kamino and J. Michael, *MRS Bull.*, 2007, **32**, 400–407.

119. C. Træholt, J. G. Wen, V. Svetchnikov, A. Delsing and H. W. Zandbergen, *Phys. C: Supercond.*, 1993, **206**, 318–328.

120. F. Liu, W. Wu, Y. Bai, S. H. Chae, Q. Li, J. Wang, J. Hone and X.-Y. Zhu, *Science*, 2020, **367**, 903–906.

121. A. J. Mannix, A. Ye, S. H. Sung, A. Ray, F. Mujid, C. Park, M. Lee, J.-H. Kang, R. Shreiner, A. A. High, D. A. Muller, R. Hovden and J. Park, *Nat. Nanotechnol.*, 2022, **17**, 361–366.

122. J. Li, J. Li, K. Sun, L. Wu, R. Li, J. Yang, X. Shen, X. Wang, H. Luo, R. J. Cava, I. K. Robinson, X. Jin, W. Yin, Y. Zhu and J. Tao, *Phys. Rev. B*, 2020, **101**, 100304.

123. L. Le Guyader, T. Chase, A. H. Reid, R. K. Li, D. Svetin, X. Shen, T. Vecchione, X. J. Wang, D. Mihailovic and H. A. Dürr, *Struct. Dyn.*, 2017, **4**, 044020.

124. A. Zong, X. Shen, A. Kogar, L. Ye, C. Marks, D. Chowdhury, T. Rohwer, B. Freelon, S. Weathersby, R. Li, J. Yang, J. Checkelsky, X. Wang and N. Gedik, *Sci. Adv.*, 2018, **4**, eaau5501.

125. A. Kogar, A. Zong, P. E. Dolgirev, X. Shen, J. Straquadine, Y.-Q. Bie, X. Wang, T. Rohwer, I.-C. Tung, Y. Yang, R. Li, J. Yang, S. Weathersby, S. Park, M. E. Kozina, E. J. Sie, H. Wen, P. Jarillo-Herrero, I. R. Fisher, X. Wang and N. Gedik, *Nat. Phys.*, 2020, **16**, 159–163.

126. A. Zong, P. E. Dolgirev, A. Kogar, Y. Su, X. Shen, J. A. W. Straquadine, X. Wang, D. Luo, M. E. Kozina, A. H. Reid, R. Li, J. Yang, S. P. Weathersby, S. Park, E. J. Sie, P. Jarillo-Herrero, I. R. Fisher, X. Wang, E. Demler and N. Gedik, *Phys. Rev. Lett.*, 2021, **127**, 227401.

127. K. Sokolowski-Tinten, R. K. Li, A. H. Reid, S. P. Weathersby, F. Quirin, T. Chase, R. Coffee, J. Corbett, A. Fry, N. Hartmann, C. Hast, R. Hettel,

M. H. von Hoegen, D. Janoschka, J. R. Lewandowski, M. Ligges, F. M. zu Heringdorf, X. Shen, T. Vecchione, C. Witt, J. Wu, H. A. Dürr and X. J. Wang, *New J. Phys.*, 2015, **17**, 113047.

128. K. Sokolowski-Tinten, X. Shen, Q. Zheng, T. Chase, R. Coffee, M. Jerman, R. K. Li, M. Ligges, I. Makasyuk, M. Mo, A. H. Reid, B. Rethfeld, T. Vecchione, S. P. Weathersby, H. A. Dürr and X. J. Wang, *Struct. Dyn.*, 2017, **4**, 054501.

129. N. Rothenbach, M. E. Gruner, K. Ollefs, C. Schmitz-Antoniak, S. Salamon, P. Zhou, R. Li, M. Mo, S. Park, X. Shen, S. Weathersby, J. Yang, X. J. Wang, R. Pentcheva, H. Wende, U. Bovensiepen, K. Sokolowski-Tinten and A. Eschenlohr, *Phys. Rev. B*, 2019, **100**, 174301.

130. A. Sood, J. B. Haber, J. Carlstrom, E. A. Peterson, E. Barre, J. D. Georgaras, A. H. M. Reid, X. Shen, M. Zajac, E. C. Regan, J. Yang, T. Taniguchi, K. Watanabe, F. Wang, X. Wang, J. B. Neaton, T. F. Heinz, A. M. Lindenberg, F. H. da Jornada and A. Raja, *Nat. Nanotechnol.*, 2022, **18**, 29.

131. M. Holt, Z. Wu, H. Hong, P. Zschack, P. Jemian, J. Tischler, H. Chen and T.-C. Chiang, *Phys. Rev. Lett.*, 1999, **83**, 3317–3319.

132. R. Xu and T. C. Chiang, *Z. Kristallogr. - Cryst. Mater.*, 2005, **220**, 1009–1016.

133. T. Chase, M. Trigo, A. H. Reid, R. Li, T. Vecchione, X. Shen, S. Weathersby, R. Coffee, N. Hartmann, D. A. Reis, X. J. Wang and H. A. Dürr, *Appl. Phys. Lett.*, 2016, **108**, 041909.

134. P. Maldonado, T. Chase, A. H. Reid, X. Shen, R. K. Li, K. Carva, T. Payer, M. Horn von Hoegen, K. Sokolowski-Tinten, X. J. Wang, P. M. Oppeneer and H. A. Dürr, *Phys. Rev. B*, 2020, **101**, 100302.

135. A. H. Reid, X. Shen, P. Maldonado, T. Chase, E. Jal, P. W. Granitzka, K. Carva, R. K. Li, J. Li, L. Wu, T. Vecchione, T. Liu, Z. Chen, D. J. Higley, N. Hartmann, R. Coffee, J. Wu, G. L. Dakovski, W. F. Schlotter, H. Ohldag, Y. K. Takahashi, V. Mehta, O. Hellwig, A. Fry, Y. Zhu, J. Cao, E. E. Fullerton, J. Stöhr, P. M. Oppeneer, X. J. Wang and H. A. Dürr, *Nat. Commun.*, 2018, **9**, 388.

136. F. Ji, D. B. Durham, A. M. Minor, P. Musumeci, J. G. Navarro and D. Filippetto, *Commun. Phys.*, 2019, **2**, 1–10.

137. L. W. Feng, L. Lin, S. L. Huang, S. W. Quan, T. Jiang, P. F. Zhu, J. K. Hao, F. Zhu, F. Wang, F. Fu, R. Wang, L. Zhao, D. Xiang and K. X. Liu, *Appl. Phys. Lett.*, 2015, **107**, 224101.

138. B. Alberdi Esuain, J.-G. Hwang, A. Neumann and T. Kamps, *Sci. Rep.*, 2022, **12**, 13365.

139. E. Pirez, P. Musumeci, J. Maxson and D. Alesini, *Nucl. Instrum. Methods Phys. Res., Sect. A*, 2017, **865**, 109–113.

140. Y. Song, J. Yang, J. Wang, J. Urakawa, T. Takatomi and K. Fan, *Nucl. Instrum. Methods Phys. Res., Sect. A*, 2022, **1031**, 166602.

141. A. D. Cahill, A. Fukasawa, R. Pakter, J. B. Rosenzweig, V. A. Dolgashev, C. Limborg-Deprey, S. Tantawi, B. Spataro and G. Castorina, *Nucl. Instrum. Methods Phys. Res., Sect. A*, 2017, **865**, 105–108.

142. R. R. Robles, O. Camacho, A. Fukasawa, N. Majernik and J. B. Rosenzweig, *Phys. Rev. Accel. Beams*, 2021, **24**, 063401.

143. Z.-H. He, A. G. R. Thomas, B. Beaurepaire, J. A. Nees, B. Hou, B. Malka, K. Krushelnick and J. Faure, *Appl. Phys. Lett.*, 2013, **102**, 064104.

144. Y. Fang, J. Hua, F. Li, B. Guo, L. Zhou, B. Zhou, Z. Zhou, Y. Wu, Y. Du, R. Li and W. Lu, *arXiv*, 2022, DOI: 10.48550/arXiv.2210.12093.

145. T. Vecchione, P. Denes, R. K. Jobe, I. J. Johnson, J. M. Joseph, R. K. Li, A. Perazzo, X. Shen, X. J. Wang, S. P. Weathersby, J. Yang and D. Zhang, *Rev. Sci. Instrum.*, 2017, **88**, 033702.

146. T. B. van Driel, S. Nelson, R. Armenta, G. Blaj, S. Boo, S. Boutet, D. Doering, A. Dragone, P. Hart, G. Haller, C. Kenney, M. Kwaitowski, L. Manger, M. McKelvey, K. Nakahara, M. Oriunno, T. Sato and M. Weaver, *J. Synchrotron Radiat.*, 2020, **27**, 608–615.

147. R. K. Li, M. C. Hoffmann, E. A. Nanni, S. H. Glenzer, M. E. Kozina, A. M. Lindenberg, B. K. Ofori-Okai, A. H. Reid, X. Shen, S. P. Weathersby, J. Yang, M. Zajac and X. J. Wang, *Phys. Rev. Accel. Beams*, 2019, **22**, 012803.

148. M. R. Bionta, N. Hartmann, M. Weaver, D. French, D. J. Nicholson, J. P. Cryan, J. M. Glownia, K. Baker, C. Bostedt, M. Chollet, Y. Ding, D. M. Fritz, A. R. Fry, D. J. Kane, J. Krzywinski, H. T. Lemke, M. Messerschmidt, S. Schorb, D. Zhu, W. E. White and R. N. Coffee, *Rev. Sci. Instrum.*, 2014, **85**, 083116.

149. M. Harmand, R. Coffee, M. R. Bionta, M. Chollet, D. French, D. Zhu, D. M. Fritz, H. T. Lemke, N. Medvedev, B. Ziaja, S. Toleikis and M. Cammarata, *Nat. Photonics*, 2013, **7**, 215–218.

150. S. Droste, L. Shen, V. E. White, E. Diaz-Jacobo, R. Coffee, S. Zohar, A. H. Reid, F. Tavella, M. P. Minitti, J. J. Turner, K. L. Gumerlock, A. R. Fry and G. Coslovich, in *2019 Conference on Lasers and Electro-Optics (CLEO)*, 2019, pp. 1–2.

151. L. Zhao, Z. Wang, C. Lu, R. Wang, C. Hu, P. Wang, J. Qi, T. Jiang, S. Liu, Z. Ma, F. Qi, P. Zhu, Y. Cheng, Z. Shi, Y. Shi, W. Song, X. Zhu, J. Shi, Y. Wang, L. Yan, L. Zhu, D. Xiang and J. Zhang, *Phys. Rev. X*, 2018, **8**, 021061.

152. M. Gao, Y. Jiang, G. H. Kassier and R. J. Dwayne Miller, *Appl. Phys. Lett.*, 2013, **103**, 033503.

153. J. Maxson, D. Cesar, G. Calmasini, A. Ody, P. Musumeci and D. Alesini, *Phys. Rev. Lett.*, 2017, **118**, 154802.

154. E. C. Snively, M. A. K. Othman, M. Kozina, B. K. Ofori-Okai, S. P. Weathersby, S. Park, X. Shen, X. J. Wang, M. C. Hoffmann, R. K. Li and E. A. Nanni, *Phys. Rev. Lett.*, 2020, **124**, 054801.

155. L. Zhao, H. Tang, C. Lu, T. Jiang, P. Zhu, L. Hu, W. Song, H. Wang, J. Qiu, C. Jing, S. Antipov, D. Xiang and J. Zhang, *Phys. Rev. Lett.*, 2020, **124**, 054802.

156. H. W. Kim, N. A. Vinokurov, I. H. Baek, K. Y. Oang, M. H. Kim, Y. C. Kim, K.-H. Jang, K. Lee, S. H. Park, S. Park, J. Shin, J. Kim, F. Rotermund, S. Cho, T. Feurer and Y. U. Jeong, *Nat. Photonics*, 2020, **14**, 245–249.

157. F. Qi, Z. Ma, L. Zhao, Y. Cheng, W. Jiang, C. Lu, T. Jiang, D. Qian, Z. Wang, W. Zhang, P. Zhu, X. Zou, W. Wan, D. Xiang and J. Zhang, *Phys. Rev. Lett.*, 2020, **124**, 134803.
158. M. J. Berger, J. S. Coursey and M. A. Zucker, ESTAR, PSTAR, and ASTAR, https://www.nist.gov/publications/estar-pstar-and-astar-computer-programs-calculating-stopping-power-and-range-tables-0 (accessed 12 November 2022).
159. D. I. Yakubovsky, A. V. Arsenin, Y. V. Stebunov, D. Y. Fedyanin and V. S. Volkov, *Opt. Express*, 2017, **25**, 25574–25587.

CHAPTER 13

Laser Induced Electron Diffraction

L. F. DIMAURO[a] AND C. I. BLAGA[*,b]

[a] The Ohio State University, Columbus, OH, USA; [b] James R. Macdonald Laboratory, Kansas State University, Manhattan, KS, USA
*Email: blaga@ksu.edu

13.1 Introduction: Diffraction and Interference as Tools for Determining Molecular Structure

In classical physics, interference is the process of addition of two or more waves – or disturbances – at a particular location and at a particular moment in time. Diffraction on the other hand, is the process by which waves bend around obstacles or pass through apertures whose spatial extent D is comparable to the wavelength λ of the incident wave. The exact nature of the wave is not important, as one can observe interference and diffraction with sound waves, water ripples, light, *etc.* Parts I and II of this chapter are written in the format and spirit of an upper undergraduate or graduate textbook, our goal being to provide a solid foundation to laser-induced electron diffraction (LIED) for senior undergraduates and graduate students aiming to carry out LIED experiments. We will not derive the equations provided, preferring instead to provide an expanded list of references that the reader can use as needed. We will however do our best to illustrate the concepts of LIED, its limitation, practical implementation and of course, potential.

Theoretical and Computational Chemistry Series No. 25
Structural Dynamics with X-ray and Electron Scattering
Edited by Kasra Amini, Arnaud Rouzée and Marc J. J. Vrakking
© The Royal Society of Chemistry 2024
Published by the Royal Society of Chemistry, www.rsc.org

13.1.1   From Diffraction of Light to Electron Diffraction

Diffraction and interference are powerful imaging tools. To illustrate this claim, we start our discussion with the well-known illustration of Fraunhofer diffraction from optics. When a laser impinges on an opaque small spherical obstacle, one observes the well-known Airy disk pattern on a far-field screen (see Figure 13.1a).

For a given laser wavelength, as a function of deflection angle θ, the light intensity $I(\theta)$ measured radially on the screen from the centre maximum is given by:[1]

$$\frac{I(\theta)}{I(\theta=0)} = \left[\frac{2J_1(kD\sin(\theta))}{kD\sin(\theta)}\right]^2 \tag{13.1}$$

where $k = 2\pi/\lambda$ is the photon wavenumber, D is the diameter of the aperture and J_1 is the Bessel function of order 1. The normalized radial intensity distribution is plotted in Figure 13.1b for three aperture sizes: small $(D=1\times)$, medium $(D=1.5\times)$ and large $(D=2\times)$. The effect of the aperture size is clear: as the aperture size increases, more minima and maxima are observed. The locations of the minima are determined by the zeroes of the Bessel function, themselves uniquely determined by the size of the aperture. Thus, knowing the wavelength of light and measuring the angles at which the intensity vanishes, allow us to precisely determine the diameter of the aperture. If the laser impinges on two obstacles separated by a distance R, the airy pattern becomes more complex, as the diffracted electromagnetic fields from each aperture interfere on the screen, producing interference fringes whose separation depends on R. Thus, if the wavelength of the light

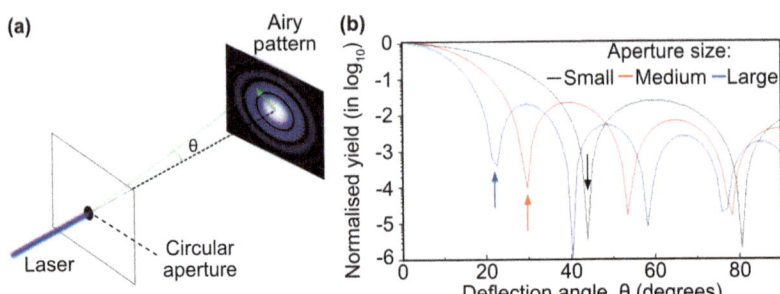

Figure 13.1   Diffraction as a structural tool. (a) Typical arrangement for observing diffraction on a circular aperture: a laser shining on a circular aperture generates the well-known Airy disk pattern of concentric rings on an observation screen behind the aperture. (b) Simulated intensity profiles as a function of deflection angle θ, plotted for three different aperture sizes: small, medium, and large (shown in blue, red and black, respectively). The arrows indicate the angular position of the first diffraction minimum. Clearly, the location of the minimum occurs at higher values as the aperture size decreases.

is known, analysing the diffraction and interference pattern permits the determination of the obstacle diameter D and separation R.

Clearly, a small diatomic molecule can be considered as two small obstacles separated by an interatomic distance R. As the optical method presented above works well for measuring aperture sizes from 10^{-4} m down to 10^{-6} m when visible light is used ($\lambda \sim 10^{-6}$ m), one can readily estimate that using X-rays would allow us to measure the size of atoms, whose spatial extent is on the order of 10^{-10} m. Unfortunately, the interaction of X-rays with matter is rather weak, and until the advent of high-brightness synchrotrons and free electron X-ray sources, X-rays were not used for structural studies of isolated molecules in the gas phase.

The solution to this issue came in 1924, when French physicist Louis De Broglie postulated that all matter exhibits a wave-like behaviour. For example, an electron of mass $m = 9.31 \times 10^{-31}$ kg possessing $E = 100$ eV of kinetic energy (1 eV $= 1.6 \times 10^{-19}$ J) has an associated de Broglie wavelength $\lambda_{dB} = h/(2mE)^{0.5} = 1.23 \times 10^{-10}$ m, where $h = 6.626 \times 10^{-34}$ J s^{-1} is Planck's constant. As electrons possess charge, their scattering cross-section is 5–6 orders of magnitude higher than that of X-rays.[2] Consequently, electron diffraction seemed better suited for studying small molecules in the gas phase. The first successful demonstration of elastic electron scattering as a molecular structural probe was carried out in 1930 by Mark and Wierl who investigated carbon tetrachloride (CCl_4) in the gas phase, thus marking the beginning of conventional gas-phase electron diffraction (CED).[3]

Before we begin our discussion about molecules, it is worth discussing the simplest targets: atoms. Figure 13.2 presents the elastic differential cross-sections for a 100 eV quasi monoenergetic electron beam scattering on argon (a), krypton (b), and xenon (c), as curated by Salvat.[4] Clearly, the observations we made above for light scattering on a circular aperture apply: the larger the atom, the higher the number of minima and maxima observed. In each

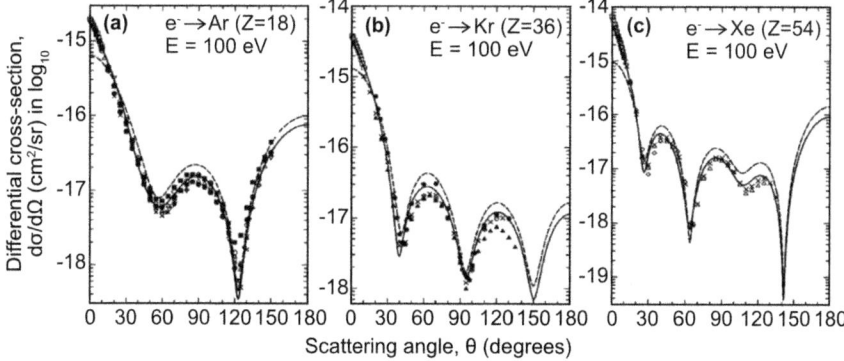

Figure 13.2 Electron diffraction of noble gases using a 100 eV electron beam shown for (a) argon, (b) krypton, and (c) xenon. Adapted from ref. 4 with permission from the American Physical Society, Copyright 2003.

panel, various symbols represent experimental data, whereas various lines represent numerical estimates (see ref. 4 and references within).

Unfortunately, this is where the comparison ends. For light, the theoretical expression listed in eqn (13.1) was obtained treating the incoming light as a classical field, while for electrons, the treatment is eminently quantum mechanical, starting with the inhomogeneous Schrödinger equation[2] given by

$$\Delta\psi + k_0{}^2\psi + \frac{2me}{\hbar^2}\varphi\psi = 0 \tag{13.2}$$

where the total wavefunction $\psi = \psi_0 + \psi'$, with ψ_0 being the wavefunction of the incoming wave satisfying the homogeneous Schrödinger equation $\Delta\psi_0 + k_0{}^2\psi_0 = 0$ and ψ' being the spherical scattered wave on potential φ. Given the cylindrical symmetry, the scattered wave at a distance R far from the atom can be expressed as

$$\psi' = \frac{A}{R}\, f(\theta)e^{ik_0 R} \tag{13.3}$$

where the factor A is a normalization constant, and the complex variable $f(\theta)$ is the all-important complex scattering amplitude. Employing a partial wave expansion, the scattering amplitude can be expressed as

$$f(\theta) = \frac{1}{2ik_0}\sum_{j=0}^{\infty}(2l+1)\left(e^{2i\delta_l} - 1\right)P_l(\cos(\theta)) \tag{13.4}$$

where $P_l(\cos(\theta))$ is the Legendre polynomial of order l. The partial phases δ_l depend on the electron energy and the potential φ and have the physical interpretation of a delay experienced by the incoming electron wave as it passes by the atom. Finally, the expression for the scattered electron intensity, I_s, can be found by calculating the probability current, yielding

$$I_s(\theta) = \frac{I_0}{R^2}\,|f(\theta)|^2 \tag{13.5}$$

where I_0 is the incoming electron current. If the R^{-2} dependence is the well-known inverse square law for spherical waves, we note here that the entire angular dependence is contained in the scattering amplitude.

Calculating scattering amplitudes is generally difficult and beyond the scope of this chapter. It is worth mentioning that at very large electron energies (say above 5 keV), the potential φ can be treated in the static-field approximation, the atom being modelled as a static charge distribution.[4] Unfortunately, as we will see in part 2, the electron energies in LIED are typically in the 50–300 eV range. In this case, the static-field and static-exchange approximations are not valid and more complicated models must be used to obtain accurate scattering amplitudes.[4] Nonetheless, as seen in Figure 13.2, even for 100 eV, modern theoretical models are in perfect agreement with the experimental data, and therefore do not preclude LIED from being a valuable and accurate ultrafast molecular imaging tool.

Obtaining a theoretical expression for the scattering signal from molecules is even more complicated. However, since the incoming electron energy is usually significantly larger than the binding energies of valence electrons, the atomic core is predominantly responsible for electron scattering at large angles, whereas the binding orbitals are essentially transparent. Consequently, electron diffraction experiments are typically agnostic to the chemical nature of the molecule. Furthermore, if one can ignore multiple scattering (*i.e.*, the scattering of an already scattered wave), the diffracted waves from each atom do not depend on the existence of its neighbours. In this so-called independent atom model (IAM; see Chapter 3), the scattering wave from a rigid molecule composed of N atoms fixed in space can be expressed as

$$\psi(\vec{R}) = \frac{KA}{R} e^{ik_0 R} \sum_{j=1}^{N} f_j(\theta) e^{i\vec{q}\vec{r}_j} \tag{13.6}$$

where $|\vec{q}| = |\vec{k} - \vec{k}_0| = 2k_0 \sin(\theta/2)$ is the momentum transfer and $\vec{r}_j$ are the coordinates of atom j. As gas molecules are randomly oriented in space, the measured scattering signal is an incoherent sum over all orientations, each occurring with equal probability. Introducing the internuclear separations $\vec{r}_{ij} = \vec{r}_i - \vec{r}_j$, a spatial average of eqn (13.5) can be performed to yield a scattering signal $I_s(\theta)$:

$$I_s(\theta) = \frac{K^2 I_0}{R^2} \left(\sum_{j=1}^{N} |f_j(q)|^2 + \sum_{i=1}^{N} \sum_{j=1, i \neq j}^{N} f_i(q) f_j^*(q) \frac{\sin(qr_{ij})}{qr_{ij}} \right) \tag{13.7}$$

The first term on the right-hand side of the above equation is a purely atomic scattering factor that contains no structural information. Nonetheless, it usually has the largest overall contribution to the differential cross-section (DCS). The second term is the so-called molecular term. Structural information can be extracted through r_{ij}. However, the contribution of the molecular term to the total measured signal is only 5–10% compared to the atomic one. Thus, if one is to measure a DCS, then very good signal-to-noise ratios (SNRs) with good statistics are required to determine various r_{ij} values accurately. Moreover, as the internuclear distances only appear in the $\sin(qr_{ij})/qr_{ij}$ terms, it is obvious that the momentum transfer, not the magnitude of the incoming electron momentum k_0, determines the spatial resolution when extracting r_{ij} from measured DCSs. Ideally, large momentum transfer ranges are desired, so that the several extrema of $\sin(qr_{ij})$ are observed. For example, for a typical internuclear distance of 1 Å, a DCS spanning a momentum transfer range of 7 Å^{-1} will exhibit two extrema, sufficient for attaining spatial resolutions in the few pm range or better. Unfortunately, increasing q further will produce diminishing returns given that q appears in the denominator of the molecular contrast term. Thus, DCSs measured for very large q ranges require low noise measurements with exceptional statistics to extract the small molecular contrast factor from the large atomic term.

13.2 The Nuts and Bolts of Laser Induced Electron Diffraction

13.2.1 The Origin of Laser-induced Electron Diffraction

For decades, CED has been the method of choice to determine internuclear separations of isolated molecules at equilibrium, with modern implementations reaching spatial resolutions on the femtometre scale. Unfortunately, CED's reliance on an electron gun precluded the method from attaining the temporal resolutions necessary to see atomic motion during chemical transformations, since the electrostatic repulsion in the electron bunch and the electronic devices used to condition the beam limit the electron pulse duration to ~ 1 picosecond. In recent years, an ultrafast variant dubbed ultrafast electron diffraction (UED) using femtosecond UV pulses to generate sub-picosecond electron bunches has emerged as an alternative, although its temporal resolution currently limits its applicability to the motion of slower, heavier atoms. Scientifically, diffraction methods that can follow the motion of smaller atoms such as hydrogen, carbon or oxygen inside organic molecules would be the most fruitful. For example, the C–H stretch mode in organic molecules is approximately 10 fs, whereas the C–C stretch mode is approximately 20 fs. Thus, to observe bonds breaking and forming, it is necessary to employ imaging methods with comparable temporal resolutions.

In 1994, Paulus and colleagues[5] reported the existence of a rescattering plateau, confirming the so-called three step model of photoionization in strong field physics.[6-8] In step one, an atom or molecule is tunnel-ionized by the intense, yet slowly varying linearly polarized laser field, releasing an electron wavepacket into the continuum. In the second step, the released wavepacket undergoes a quiver motion along the laser polarization, driven by the laser field, while the electrostatic effect of the parent ion is neglected. Finally, in the third step, a part of the quivering wavepacket revisits and collides with the parent ion before moving away towards the detector. Two years later, Zuo and colleagues[9] demonstrated theoretically that the recollision event encodes structural information, giving birth to laser-induced electron diffraction (LIED). As the entire process takes place during the ultrafast pulse, if successfully implemented, the new tool could become an imaging method with temporal resolutions no longer than the pulse duration (*i.e.*, a few tens of femtoseconds or better as we will see below).

The implementation of LIED seems straightforward enough: illuminate the target with an intense, linearly polarized ultrafast laser pulse and obtain the DCS from the measured 3D photoelectron spectra. Unfortunately, several complications arise. First, the momentum of the electron and the angle at which it scatters are not the same as the detected momentum and angle. Unlike CED/UED, in LIED, diffraction takes place in the presence of a strong laser field, which acts and modifies the electron momentum post-collision, an effect that needs to be addressed when experimental photoelectron angular distributions (PADs) are analysed. Second, interaction effects must be understood and accounted for. These include the dependence of the ionization rate as a function of molecular alignment, laser intensity, and wavelength. Third, the nonlinear nature of the laser–matter interaction mandates that focal averaging must be accounted for since molecules experience different intensities as a function of their location in the laser focus. Finally, as we will see below, even when accurate DCSs are extracted from experimental data, theoretical modelling could be challenging in LIED, since collision energies are significantly lower than in CED/UED, thus making the validity of IAM somewhat problematic.

More than a decade of research, both theoretical and experimental advances, was required to address these issues. In the remainder of this section, we will systematically illustrate these developments. First, let us exemplify how well LIED works when properly implemented. In Figure 13.3, we showcase the electron-argon ion LIED DCS at 100 eV return energy extracted from the PAD recorded under irradiation of 200 TW cm^{-2}, 2 μm, 50 fs pulses (green squares) overlaid on the CED result for electron-neutral argon as reported by Rudd and DuBois[10] in 1976 (red symbols). Clearly, the location of diffraction extrema (minima at 60° and 125° and maxima at 90° and 180° degrees) are well reproduced. Furthermore, as the recollision electron momentum is $k_0 = 5.1$ Å^{-1}, the momentum transfer span covered by the 20–180° angular range is 1.8–10.2 Å^{-1}. Looking at the phase term in eqn (13.7), such a momentum transfer range is more than sufficient to

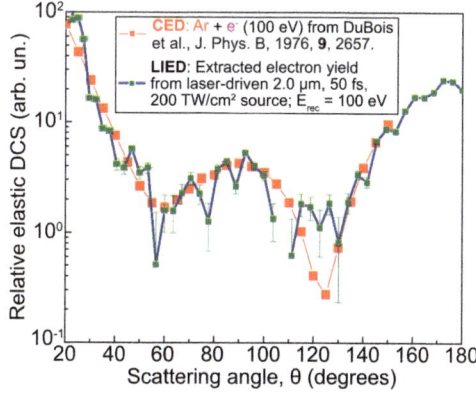

Figure 13.3 Comparison between conventional electron diffraction (red symbols) and laser-induced electron diffraction (green symbols) for the case of argon atoms at 100 eV electron (re)scattering energy.

observe internuclear distances with 0.1 Å resolutions, as we will see shortly. We also notice that for CED, the target was a neutral atom, whereas in LIED, the target was a cation, since in LIED the electron rescatters from the parent ion. Finally, we observe that the error bars present in the LIED data of Figure 13.3 are only due to limited data collection statistics, which can easily be improved.

13.2.2 Experimental Implementation

The LIED DCS (see Figure 13.3) was extracted from the angular distribution of Figure 13.4 below. The data were collected in a standard time-of-flight (TOF) ultrahigh vacuum electron spectrometer with a background pressure of few 10^{-10} mbar. The spectrometer length was ~25 cm and had an energy resolution $\Delta E/E = 1.5\%$. The electrons were recorded with a multichannel plate detector in front of which a small pinhole was placed to augment the angular resolution of the apparatus to ~2° by limiting the collection angle. The target gas – argon in this case – was introduced effusively *via* a precision leak valve. Normal operating pressures were adjusted to assure the highest data collection rates but without space-charge effects. The linearly polarized laser was focused with a relatively strong positive lens ($f = 100$ mm) to minimize the focal volume and hence the background gas contribution. The angle φ between the laser polarization and the spectrometer axis was controlled using a zero-order half waveplate. To obtain the 2D PAD, spectra for φ from 0° to 90° in steps of 2° were collected randomly to eliminate systematic errors due to changes in the experimental conditions.

A close inspection of Figure 13.4 reveals that the centre of the image (warm colours) contains most of the electrons. These low momentum electrons are direct electrons that have not scattered and therefore are not useful in LIED. As we will see later, they are the reason why the DCS shown in Figure 13.3 starts at $\theta = 20°$, not at $\theta = 0°$. However, the LIED relevant region are the electrons in the green and blue regions. These higher momentum electrons are those that have rescattered on their parent ion and thus, their distribution

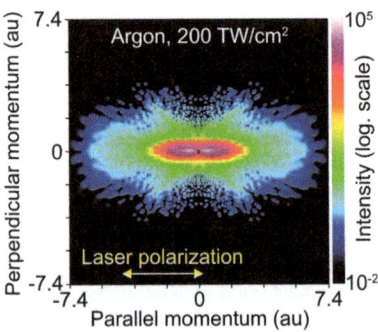

Figure 13.4 Experimental photoelectron angular distribution for argon irradiated with 2 μm, 50 fs pulses with a peak intensity of 200 TW cm^{-2}.

encodes the spatial information. They are significantly fewer compared to the direct electrons. Consequently, to obtain a good DCS, it is essential to record data with good statistics in this region. For the data shown in Figure 13.4, this is accomplished using a high repetition laser (1 kHz in this case) and long acquisition times (20 minutes for each value of angle φ). Consequently, we obtained a PAD with a seven orders-of-magnitude dynamic range.

An alternative experimental implementation for collecting PADs is based on velocity map imaging (VMI) spectrometers. The main advantages of VMI are superior counting rates and better angular resolution compared to a typical TOF, but usually the dynamic range and the energy resolution are poorer. The selection of the experimental apparatus is often application-dependent, mostly dictated by counting rates and/or angular or energy resolution requirements. However, in all cases, higher laser repetition rates are always preferred.[11]

13.2.3 Theoretical Underpinnings

13.2.3.1 The Ionization Step

Let us begin our discussion of a photoelectron's journey from birth to detection with the first step: ionization. Historically, ultrafast photoionization in an intense laser pulse interacting with matter is thought to happen in two regimes: multiphoton and tunnelling. In 1964, Keldysh introduced an adiabacity parameter γ defined as[12]

$$\gamma = \sqrt{\frac{I_P}{2U_P}} \qquad (13.8)$$

where I_P is the ionization potential of the target and U_P is the ponderomotive potential, defined as the cycle-averaged kinetic energy of a free electron oscillating in the laser field. If $\gamma < 1$, the photoelectron leaves the atom *via* tunnel ionization, whereas if $\gamma > 1$, the ionization is multiphoton. The ponderomotive potential is a function of laser intensity I and laser angular frequency ω. For linear polarization in atomic units $(m_e = q_e = \hbar = 1)$,

$$U_P = \frac{I}{4\omega^2} \qquad (13.9)$$

In the next subsection when we discuss rescattering, we will see that the energy of the recolliding electron has an upper bound of $\sim 3U_P$. As LIED requires relatively energetic recollisions to attain satisfactory spatial resolutions, U_P should be greater than 30 eV. As most molecules of interest have ionization potentials on the order of 10–15 eV, we can clearly see that the ionization in all LIED experiments is eminently tunnelling in nature. An exhaustive treatment of tunnel ionization is beyond the scope of this chapter. For the interested reader, the modern treatment by Bisgaard and Madsen[13] and the original work of Keldysh,[12] Ammosov, Delone and

Krainov[14] are excellent starting points in the case of atoms. For molecules, the seminal paper of Tong, Zhao and Lin is a must-read.[15] Here, we will only summarize the key findings of these important manuscripts.

First, at a given time t during the laser pulse, the tunnel ionization rate depends exponentially on the instantaneous laser field strength $F(t)$ and the ionization potential I_P. Additionally, the rate depends on the nuclear charge Z and all quantum numbers (n,l,m), but in general $m=0$ is dominant. Furthermore, the rate is the highest for the photoelectron possessing a null momentum on the outer turning point of the barrier. For the simple case of the hydrogen atom, we have

$$W_{\text{static}}(F,\vec{p}=0)=A\left(\frac{2(2I_P)^{1.5}}{F}\right)^{\frac{2}{\sqrt{2I_P}}-1}\exp\left(-\frac{(2I_P)^{1.5}}{3F}\right), \qquad (13.10)$$

where the constant A depends on the quantum number Z. Integrating eqn (13.10) over a laser cycle with F_0 being the maximum electric field strength, we obtain the total ionization rate per laser cycle W_{avg}:

$$W_{\text{avg}}=\left(\frac{3F_0}{\pi(2I_P)^{1.5}}\right)^{0.5}\cdot W_{\text{static}}(F_0,\vec{p}=0). \qquad (13.11)$$

The ionization probabilities for photoelectrons born with nonzero momenta were obtained by Nikisov and Ritus[16] and Delone and Krainov[17] and given by

$$W_{\text{static}}(p_{\parallel},F)=W_{\text{static}}(F,\vec{p}=0)\exp\left(-\frac{p_{\parallel}^2\omega^2(2I_P)^{1.5}}{3F^3}\right), \qquad (13.12)$$

and respectively

$$W_{\text{static}}(p_{\perp},F)=W_{\text{static}}(F,\vec{p}=0)\exp\left(-\frac{p_{\perp}^2(2I_P)^{0.5}}{F}\right). \qquad (13.13)$$

Eqn (13.10)–(13.13) are the heart of short-range atomic tunnelling approximation. More recently, Ristić et al. included a coulombic correction in the calculation of the outer turning point, essentially extending the validity of the tunnel approximation for long range potentials.[18]

For molecules, the tunnel ionization rate also depends on the orientation of the molecules with respect to the laser polarization axis. Analytical formulas similar to eqn (13.11) were obtained by Tong, Zhao, and Lin.[15] While the dominant exponential dependence seen for atoms remained the same, the constant A now acquires an orientation dependence. The net effect of the DCS is that even for a randomly oriented molecular gas, in LIED, the sample looks partially aligned, as certain orientations are more likely to ionize – and hence be imaged – compared to others. This effect is considered when performing the spatial average of eqn (13.6). In principle, this makes DCS calculations *via* the IAM slightly more complex. However, the amount of averaging is less and therefore the diffraction fringes obtained with LIED are more defined compared to CED. This effect is more pronounced for molecules with a greater asymmetric shape and becomes smaller for medium- and large-sized molecules, especially if they are relatively close to a spherical shape.

13.2.3.2 Focal Averaging Effects

Due to diffraction, it is impossible to focus a laser beam down to a point. For a laser beam with a Gaussian temporal and spatial profile propagating along the z axis, the focal intensity $I(r, z)$ in cylindrical coordinates is given by

$$I(r,z) = I_0 \frac{\omega_0{}^2}{\omega(z)^2} \exp\left(-\frac{2r^2}{\omega(z)^2}\right), \tag{13.14}$$

where I_0 is the peak intensity at $z = r = 0$ and $\omega(z)$ is the beam waist radius defined as the radial distance where $I(\omega_0, z)$ drops by a factor of e^2, with $\omega_0 = \omega(z = 0)$. The beam waist can be expressed as a function of z, the laser wavelength λ and ω_0 as $\omega(z) = \sqrt{\omega_0{}^2 + \dfrac{z^2 \lambda^2}{\pi^2 \omega_0{}^2}}$.

Clearly, in any experiment, molecules will experience a laser intensity depending upon where they are located inside the focal volume. Thus, any experiment essentially samples a continuous intensity distribution from I_0 to 0 and therefore all observables – including our photoelectron yields – are so-called focal averaged. Fortunately, in the case of tunnel ionization, the exponential dependence of the ionization rate limits the contribution of intensities not too much less than I_0. To illustrate this effect, we can adopt the onion peel model for modelling the focal intensity distribution.[19] Performing a volume integral over an isointensity surface I_s, we can determine that the volume $V(I_0, I_s)$ encased is:

$$V(I_0, I_s) = \frac{\pi^2 \omega_0{}^4}{\lambda} \left\{ \frac{4}{3} \left(\frac{I_0}{I_s} - 1\right)^{0.5} + \frac{2}{9} \left(\frac{I_0}{I_s} - 1\right)^{1.5} - \frac{4}{3} \text{atan}\left[\left(\frac{I_0}{I_s} - 1\right)^{0.5}\right] \right\} \tag{13.15}$$

Taking the difference $V(I_0, I_s + dI_s) - V(I_0, I_s)$, one can estimate the volume of the peel of intensity I_s. Finally, multiplying the peel volume with the ionization rate given using eqn (13.11) with F_0 determined from I_s, it can easily be determined how much each peel contributes to the ionization yield. Figure 13.5 illustrates this calculation for the case of hydrogen atoms ($I_P = 13.6$ eV), although the result is generally appliable.

Clearly, the most "visible" isointensity is 68 TW cm^{-2}, approximately 83% from $I_0 = 82$ TW cm^{-2}. The distribution is somewhat broad, getting even

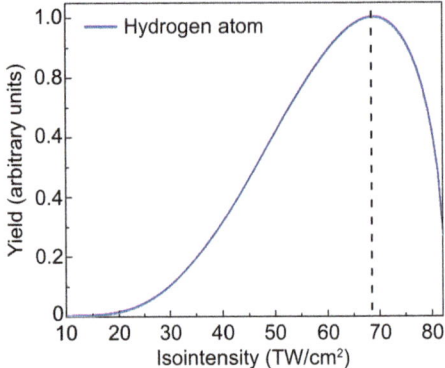

Figure 13.5   Contribution of isointensity peels to the ionization rate in the case of the hydrogen atom for a peak intensity of 82 TW cm^{-2}.

broader as one approaches the over-the-barrier intensity $I_{OTB} = \left(\frac{I_P}{2}\right)^4$ in atomic units. Therefore, if possible, in LIED experiments, it is always desired to keep the peak intensity I_0 well below I_{OTB}. A value of 50% is ideal, but higher ones could be used if data acquisition rates become prohibitively low.

13.2.3.3   *Photoelectron Propagation in the Continuum*

In the previous sections, the photoionization was treated quantum mechanically. However, to treat the photoelectron in the continuum during and after the pulse, we will consider the electron as a classical point particle with charge e and mass m_e. To a sensible physicist, this approach should be rather unpalatable and indeed, this treatment has been (and maybe still is) contentious with some colleagues. Nonetheless, the deterministic nature of the classical treatment is necessary to connect an electron detected with momentum p at an angle φ with respect to the laser polarization to the recollision event that took place when the photoelectron of momentum k_0 moving along the laser polarization scattered off its parent at angle θ. Although one can simply argue that the classical treatment "just works" as exemplified in Figure 13.3, it was a purely quantum result that convinced us more than a decade ago to pursue LIED. In Tate *et al.*, we calculated, using the numerical integration of the time-dependent Schrödinger equation (TDSE) in the single active electron approximation, at which time during the laser cycle XUV photons are emitted.[20] In the tunnelling regime, the quantum result overlapped perfectly with the classical result. This result (see Figure 13.6) was a direct confirmation that indeed, the returning photoelectron wavepacket, albeit a quantum entity, behaved as a classical electron emitted in the field at time t_b with zero momentum as predicted by the tunnelling theory and returning at time t_r. Furthermore, the quantum result demonstrated that the classical approximation that the ionic potential of the parent could be ignored was valid. Consequently, ignoring again the ionic potential post collision, one could connect the

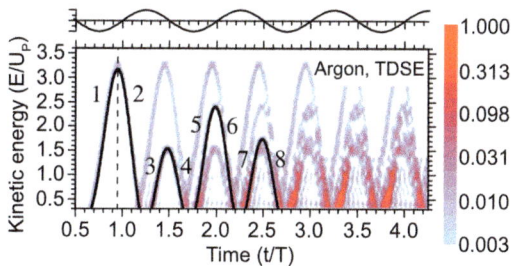

Figure 13.6 High-harmonic generation emission as a photoelectron wavepacket revisits its parent calculated *via* TDSE for argon with 2 μm pulses with a peak intensity of 160 TW cm^{-2}. Adapted from ref. 20 with permission from the American Physical Society, Copyright 2007.

detected electron to the recollision event, allowing us to extract accurate DCSs.

Hamiltonian dynamics is the most convenient approach to describe a classical electron in the laser field when the ionic potential $V(\vec{r})$ is neglected. Starting with the classical Hamiltonian of an electron oscillating in the laser field expressed in atomic units given by

$$H = \frac{\left(\vec{k}(t) - \vec{A}(t) \right)^2}{2},$$

(13.16)

where $\vec{k}(t)$ and $\vec{A}(t)$ are the mechanical momentum and the vector potential at time t, respectively; it can be readily shown using the Hamilton–Jacobi equations that the generalized momentum $\vec{P}(t) = \vec{k}(t) - \vec{A}(t)$ is a constant of motion from birth to rescattering and from rescattering to detection. At birth, the mechanical momentum is approximated as zero as in accordance with the tunnelling theory, whereas at detection the vector potential vanishes as the laser pulse is not present. Hence, the mechanical momentum at instant t prior to rescattering is simply:

$$\vec{k}(t) = \vec{A}(t) - \vec{A}(t_{\text{birth}})$$

(13.17)

This equation can be integrated between t_{birth} and t to obtain the electron position $x(t)$. Finally, finding the roots of the transcendent equation $x(t) = 0$, the return times of all classical returns are determined. Once return times t_{r} are known, using eqn (13.17), the return momenta k_0 and the return kinetic energies $(E_{\text{r}} = k_0^2/2)$ are calculated. The return energies of classical trajectories are shown in Figure 13.6 as solid black lines as a function of t_{r}, only assuming ionization during the first half cycle. The first classical returns labelled 1 and 2 are the so-called short and long trajectories, respectively. Clearly, they are the most energetic, with the two trajectories merging into one at their maximum return energy of 3.17U_{p}. The subsequent returns all have their birth times when the long trajectory was launched, but they revisit the parent at subsequent times. The second return labelled 3 and 4 lead to the softest collisions and plays the key role in the Coulomb focusing of the returning wavepacket, giving rise to the so-called low energy structure (LES).[21] The maximum return energy for the second return is $1.5U_P$. Slightly more energetic is the third return, labelled 5 and 6. The fourth return labelled 7 and 8 is more energetic than the second, but less energetic than the third. In general, the contribution of multiple returns diminishes with increasing order as the spectral density of the retuning electron wavepacket decreases due to its expansion perpendicular to the laser polarization direction, as seen from eqn (13.13).

At the time of rescattering, the electron wavepacket diffracts on its parent ion. Classically, the event is interpreted as the incoming electron with momentum $\vec{k}_0$ rescattering elastically and becoming scattered with momentum $\vec{k}_r$, the probability being the DCS, itself a function of k_0 and scattering angle θ. After rescattering, the electron leaves the vicinity of the parent ion and once again, its motion can be described using the ionic potential-free Hamiltonian. Therefore, while the generalized momentum immediately after rescattering is conserved up until detection, we obtain the electron momentum at detection $\vec{p}$:

$$\vec{p} = \vec{k}_r - \vec{A}_r \tag{13.18}$$

The last two equations allow us to attribute, for given detected electron momentum $\vec{p}$, a rescattering momentum $\vec{k}_r$ and the vector potential at rescattering $\vec{A}_r$. Unfortunately, this attribution is not unique, as several classical trajectories might contribute. However, as we will see below, at the highest rescattering energies, only one classical trajectory contributes, allowing us to attribute a unique $(\vec{A}_r, \vec{k}_r)$ with the highest detected momentum $\vec{p}$. As seen in Figure 13.6, above $2.5U_P$ rescattering energies, only two classical trajectories exist. We examine these trajectories further by considering the returning electron momenta (see solid lines, left vertical axis of Figure 13.7) and the corresponding detected electron energy

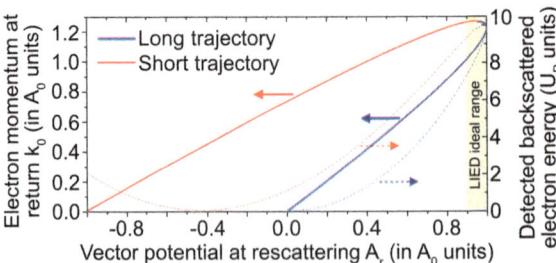

Figure 13.7   The returning electron momenta (solid lines, left axis) and the corresponding detected energies (dash lines, right axis) for the first classical returns calculated for zero initial electron momentum while ignoring the ionic potential during propagation.

(dashed lines, right vertical axis of Figure 13.7) for the long and short trajectories in the backscattering geometry $(\theta = 180°)$. Several important conclusions can be drawn. First, we notice that the highest energy collisions produce the highest energy electrons at detection, reaching up to $10U_P$. This is not only a consequence of high recollision momenta, but also because the recollision takes place near the maximum value of the vector potential, or near the zero crossing of the laser's electric field. Second, the long and short trajectories are near-degenerate close to the maximum return energy, lifting the ambiguity of associating $(\vec{A}_r, \vec{k}_r)$ to the maximum detected electron energies. Finally, we notice that for intermediate electron return momenta, the vector potential at return for the short trajectory is smaller compared to the long trajectory. Based on these observations, we conclude that the highest detected electrons in a measured photoelectron angular distribution are ideal for LIED. They not only correspond to the highest electron momentum transfers, but also, they are not contaminated with lower energy electrons resulting from (i) multiple returns and (ii) ionization events at the lower intensities present in the laser focal volume.

13.2.3.4   The Quantitative Rescattering Theory and DCS Extraction

The retrieval of the relative elastic electron–ion differential cross-section from experimental angular distributions is relatively straightforward. However, the physical justification of why the detected electron yield is directly linked to the DCS is not. In the earliest LIED experimental works of Okunishi *et al.*[22] and Meckel *et al.*,[23] the extracted electron yields from PADs were compared with calculated DCS, but the authors stopped short of directly equating the two. Shortly thereafter, the seminal work of Chen *et al.* has put LIED on a firm theoretical foundation with the introduction of the quantitative rescattering theory (QRS).[24] A full description of QRS is beyond the scope of this chapter. However, a summary is necessary. Briefly, QRS postulates that the detected electron yield $D(p, \varphi)$ in a high energy momentum distribution can be factorized into a product of the form:

$$D(p, \varphi) = W(k_r)\sigma(k_r, \theta), \tag{13.19}$$

where $W(k_r)$ is the returning electron wavepacket (RWP) and $\sigma(k_r, \theta)$ is the field-free DCS. Benchmarking the numerically integrated time-dependent Schrödinger equation (TDSE) and the strong field approximation with rescattering (so-called SFA2), Chen and co-workers demonstrated that RWP can be accurately evaluated from SFA2. Furthermore, RWP only depends on the laser parameters and has a negligible dependence on the nature of the molecule under study. Therefore, structural information is contained only in the DCS. Finally, for a given k_r, in eqn (13.19), the RWP is a simple numerical

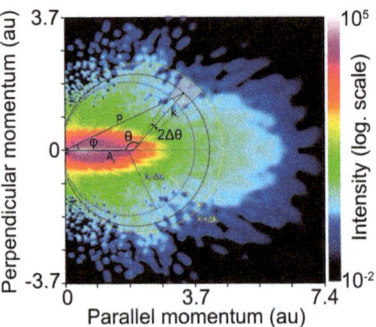

Figure 13.8   Photoelectron angular distribution and the electron–ion differential cross-section extraction for electrons scattering elastically with momentum k_r and detected with momentum p. The experimental data are the same as Figure 13.4.

scaling factor. Therefore, the electron yield measured along a circle of radius k_r and offset A_r from origin as determined using eqn (13.18) is the desired DCS. In practice, yield integration must be performed carefully since a change of variables from (p, φ) or $(p_\parallel, p_\perp)$ to (k_r, θ) is performed and the proper Jacobian applied for norm conservation. The yield is integrated between $k_r \pm \Delta k_r$ and $\theta \pm \Delta \theta$, with Δk_r and $\Delta \theta$ kept as small as possible given the statistics and the signal to noise of the data. The graphical representation of this procedure is presented in Figure 13.8. It is worth pointing out that the vector potential offsets for the circles of radii $k_r - \Delta k_r, k_r$, and $k_r + \Delta k_r$ are not the same, so the integration area varies with angle θ, as indicated. Following this procedure, accurate DCSs can be extracted, as illustrated in Figure 13.3.

13.3   Laser Induced Electron Diffraction *Versus* Ultrafast Electron Diffraction

Ultrafast electron diffraction (UED) and laser-induced electron diffraction (LIED) are ultrafast imaging methods capable of determining molecular structures on femtosecond timescales.[25] Of the two, UED is the direct descendent of the venerable conventional gas phase electron diffraction (CED). UED owes its ultrafast capabilities to its state-of-the-art electron beam generation (see Chapters 11 and 12), using techniques developed for X-ray free electron lasers (see Chapter 8). Currently, the best known and most successful MeV-UED instrument is located at the Linac Coherent Light Source (LCLS) at the SLAC National Accelerator Laboratory[26] (see Chapter 12). It operates with a 2–4 MeV electron beam at a 360 Hz repetition rate and achieves a temporal resolution of 150 fs while offering a momentum transfer range of 0.5–12 Å^{-1} in a small angle deflection geometry. Under these conditions, the UED apparatus allows users to measure DCS that are essentially identical with CED, rendering analysis straightforward using the independent atom model (IAM). The electron beam is typically paired with various ultraviolet-visible (UV-VIS) femtosecond laser pulses for pump–probe experiments investigating ultrafast molecular dynamics in various gas-phase samples.

Laser-induced electron diffraction's unique approach to deliver the electron on target brings unique opportunities and challenges. First, it offers better temporal resolutions compared to UED. In 2012, using a tunable infrared laser source, we demonstrated a temporal resolution of 5 femtoseconds while maintaining a spatial resolution of 5 picometres.[27] In the worst case, LIED's temporal resolution equals the duration of the laser pulse itself. In practice, the temporal resolution is better than this since the removal of the electron is performed at the peak of the laser field when the laser is the most intense. As modern lasers can generate mid-infrared pulses with few tens of femtoseconds and even shorter, typical LIED experiments can achieve 15–50 fs resolutions with relative ease, allowing LIED to "follow" low-Z nuclei in ultrafast molecular studies. Nonetheless, advances in mitigating the electrostatic repulsion inherently present in the UED electron beam had led to continuous improvements in UED's temporal resolutions and recently two state-of-the-art UED implementations in solid state samples have reached the 50 fs barrier[28,29] (see Chapter 11). The second significant difference between UED and LIED is that in the latter, the probing electron has significantly less energy. In fact, if the classical treatment for the electron propagation in the continuum is valid, there is no inherent lower limit for the electron energy. Of course, the IAM will fail at low energies, but this is not an inherent limitation of LIED. It simply means one requires more complex DCS modelling. If one can extract accurate DCS at low electron return energies, LIED could in principle acquire chemical sensitivity, which is not the case for high energy UED. Finally, we point out that combining LIED with other ultrafast tools such as Coulomb explosion imaging and high harmonic generation is likely easier than in the case of UED, since the detection apparatus are similar and more compatible.

13.4 Recent Developments in Laser Induced Electron Diffraction

The preceding sections of this chapter introduced the reader to LIED, illustrating how a typical experiment is performed, how the DCS is extracted and finally, how the encoded structural information is retrieved. In the following paragraphs, we will summarize a series of impactful scientific findings based on LIED reported in the literature.[25] The list is not intended to be exhaustive. As ultrafast molecular imaging is a very active area or research, the list is expected to become outdated in a few years. Nonetheless, our aim is to provide the reader with a handful of successful LIED implementations that showcase the technique's ultrafast molecular imaging qualities and hopefully inspire future LIED applications.

The classical treatment of the electron wavepacket post ionization has been the most objectionable ingredient of LIED. In our view, this issue has been properly addressed and investigated in the literature, both theoretically and experimentally. Successful DCS extractions for atoms that are in perfect agreement with CED experiments (see Figure 13.3) are sufficiently compelling. In fact, we are not aware of any experimentally retrieved DCS in atomic targets that does not agree with the CED result, provided that the strong field limit of LIED is fulfilled. Nonetheless, we would like to highlight here other investigations that lend additional support for the classical treatment. First, we highlight the systematic investigation in the emergence of classicality in strong field ionization by Colosimo *et al.*, demonstrating that as the Keldysh parameter becomes subunitary, atomic photoelectron spectra along the laser polarization acquire the expected classical shape: a steep exponential decay from zero to $2U_P$ for direct electrons, followed by a long plateau extending up to $10U_P$.[30] The experimental results of Colosimo were subsequently used to demonstrate the validity of the quantitative rescattering theory by Chen and collaborators.[24] Second, in 2012, Xu *et al.* reported an extensive LIED investigation in noble atoms.[31] Accurate DCSs were extracted for Ar, Kr and Xe at electron scattering energies ranging from 50 to 200 eV. Furthermore, the authors used the factorization formula (eqn (13.19)) to extract the spectral density of the returning electron wavepacket from experimental data, obtaining the expected universal dependence in accordance with the quantitative rescattering theory. As an independent experimental support for the classical treatment, in 2012, DiChiara *et al.* demonstrated that the classically estimated returning wavepacket can be used in a QRS-like approach to model the measured ionic yields in multiple nonsequential ionization experiments.[32] Finally, we mention that QRS has been applied successfully to estimate high harmonic generation yields in strong field physics.[33]

The first experimental LIED demonstrations were reported independently in 2008 by Okunishi *et al.*[22] and Meckel *et al.*[23] In both cases, the driving laser was 800 nm, somewhat insufficient for reaching the strong field limit. Nonetheless, structural information in the extracted DCS was clearly identified, paving the way for future LIED investigations. The first successful, quantitative DCS extraction and confirmation that LIED is an ultrafast molecular imaging tool was reported by Blaga *et al.* in 2012.[27] Using tunable infrared pulses, it was demonstrated that the removal of the imaging electron from O_2 molecules leads to a shortening of $\sim 100 \pm 5$ pm of the O–O bond by the time the electron revisits the parent ion after approximately 5 fs. This experiment demonstrated that LIED can attain a spatial resolutions of 5 pm and a temporal resolutions of 5 fs. A few years later, Wolter and collaborators reported another significant breakthrough: the first direct observation of molecular bond breaking.[34] Using a high repetition mid-infrared laser, the deprotonation of acetylene 9 fs after strong field ionization was observed in the experiment. Furthermore, using an alignment laser pulse, the team demonstrated control over the dissociation process. In 2014,

Xu *et al.* demonstrated that the broadband nature of the returning electron wavepacket could be employed in LIED.[35] Measuring the photoelectron spectra only in the backscatter geometry, it was shown that the internuclear separation in N_2 molecules can be accurately retrieved. The dubbed fixed-angle broadband laser-driven electron scattering (FABLES; also called the Fourier-transform LIED, FT-LIED) method has the advantages that it is easier to implement experimentally and that a simple Fourier transform of the data can retrieve bond lengths. Shortly thereafter, Pullen and colleagues further developed FABLES, expanding it to O_2 and polyatomic molecules (C_2H_2).[36,37] In 2017, Walt *et al.* successfully combined LIED with photoelectron holography to investigate the electron and nuclei dynamics in NO molecules.[38] In recent years, LIED experiments studying targets of increased complexity followed: ammonia,[39] benzene,[40] carbonyl sulfide,[41,42] carbon disulfide,[43] 1,3-butadiene[44] and C_{60} fullerenes.[45] Moreover, the influence of the electronic structure and molecular orbitals on the LIED imaging process has been studied.[44,46] Clearly, LIED is a powerful emerging molecular imaging method that is quickly being adopted by the ultrafast community. On the theoretical side, significant advances also took place, primarily on DCS retrieval and analysis. We only mention here two developments. First, it has been shown by Yu *et al.* and Wang *et al.* that more accurate structural information can be extracted from experimental data by a comprehensive analysis of the entire 2D photoelectron angular resolution.[47,48] Second, we highlight the first demonstration of machine learning in LIED DCS analysis as reported by Liu *et al.*[49] This approach can provide a significant advantage over earlier methods which is especially well suited for larger molecules, as demonstrated in the case of fenchone ($C_{10}H_{16}O$).

13.5 Advantages and Disadvantages of Laser Induced Electron Diffraction

The few-femtosecond temporal resolution constitutes the main advantage of LIED,[27] which, as of today, to the best of our knowledge, is the record for all electron diffraction methods. The main disadvantage of LIED, which somewhat counterintuitively stems from its advantage, is the strong field nature of the ultrashort laser pulse. As discussed in the previous section, exposing the target to an intense laser pulse could induce molecular dynamics during the ramp up of the pulse, altering the molecular dynamics before the probing electron is removed from the molecule in one of the laser cycles near or at the peak of the laser field. Normally, if the LIED pulse is non-resonant with molecular transitions of the target, this issue might not preclude the successful implementation of LIED. However, if the laser is resonant with a molecular transition, which at the typical midinfrared wavelengths typically used in LIED is likely to be a vibrational resonance, excitation of the target by the probing pulse could be a serious concern.

Obviously, if available, the solution to this issue is to tune the laser wavelength away form the resonance.

The dependence of the ionization rate with the molecular orientation relative to the laser polarization axis is another issue that is both an advantage and a disadvantage. It is an obvious advantage as the orientation-dependent tunnelling rate partly removes the spatial randomness that is inherent in conventional methods, but also in UED, unless molecular alignment is used.[50] This effect is most pronounced for molecules with large structural asymmetry,[51] but it can be neglected for molecules with large structural symmetry, as it was shown in the case of C_{60} fullerenes.[45] Nonetheless, when the effect is non-negligible, the LIED DCS will differ from the CED DCS even for the same electron return energy, since probing electrons will be produced preferentially by molecules whose orientation maximize the ionization rate. Although the LIED DCS has better fringe visibility, molecular ionization rates must be calculated and incorporated during structural retrieval from DCSs, making the analysis more cumbersome.

Similar with conventional and ultrafast electron diffraction, the maximum momentum transfers achievable with LIED are, in principle, unlimited. As discussed in the preceding sections, the maximum kinetic energy attainable during rescattering is $\sim 3U_P$. As U_P scales linearly with intensity and quadratically with the laser wavelength, using laser drivers with longer wavelengths will result in more energetic recollisions and thus a higher momentum transfer range. However, in practice, one should choose laser parameters producing the smallest acceptable momentum transfers for a given situation, as longer wavelengths reduce the spectral density of the returning electron wavepacket due to its transverse spread during propagation. This effect is proportional to λ^2 as shown by Colosimo *et al.*[30] Nonetheless, mid-infrared pulses with wavelengths in the 3–5 μm range should be more than sufficient in most cases. For example, at intensities reaching 50 TW cm^{-2}, a range suitable for medium-sized molecules with ionization potentials around 10 eV, a 5 μm laser pulse will produce returning electrons reaching 350 eV, leading to a maximum momentum transfer of ~ 10 Å^{-1}. Such a range is generally sufficient for determining the motion of nuclei over distances of several Å from equilibrium. Adjusting the intensity of the laser to obtain large momentum transfers is not possible past the over-the-barrier intensity, as discussed in Section 13.2.3.2.

The last LIED advantage that we mention here is the broadband nature of the returning electron wavepacket. If in typical CED and UED experiments, the electron beam is monochromatic, in LIED the returning electron energy range is 0–$3.17U_P$ (this is an estimate based on classical trajectories, but as seen in Figure 13.6, the classical result is in excellent agreement with the quantum estimation). Therefore, although one can use only the highest energy electrons collected in the angular distributions to extract only the DCS of electrons scattering with $3U_P$, we demonstrated that a large portion of the PAD could be used for analysis.[47,48] In essence, the PAD contains redundant information in DCSs at energies below the classical cut-off.

This leads to more accurate structural retrieval, as more data are used in the analysis.

13.6 Outlook

Since the first demonstration of conventional gas-phase electron diffraction (CED) experiments by Mark and Weirl nearly a century ago,[3] electron diffraction has remained one of the most important molecular imaging tools in atomic, molecular, and optical physics. Over the years, sustained experimental, theoretical, and technological advances and refinements have made CED a robust and mature indispensable tool. For ultrafast molecular imaging, modern CED variants possessing temporal resolutions, allowing us to record the motion of nuclei during chemical reactions like LIED and UED, will very likely follow a similar development and refinement path. For LIED specifically, we estimate that the method will continue to be applied to more and more complex molecular targets undergoing chemical reactions. One of the most significant technological advancements in recent years with transformative potential for LIED is the emergence of Yb-based very high repetition, high average power driving lasers. With operating rates exceeding 100 kHz, these lasers are ideal for producing the mid-infrared laser pulses required in LIED *via* frequency down conversion. Photoelectron angular distributions like that presented in Figure 13.3 that took several hours to acquire will be recorded in mere minutes provided that all other experimental conditions like gas pressure, focusing geometry, *etc.* remain the same. This will make VIS/UV pump, LIED probe experiments that can record molecular movies rather routine. One such apparatus is currently under construction at the NeXUS laser facility at The Ohio State University in Columbus, OH, USA. On the theoretical front, refinements of DCS extraction and internuclear bond length fitting methods will lead to improved LIED applications. LIED's unique feature of using large angle scattering at intermediate energies will lead to further improvements in DCS modelling beyond the independent atom model. Finally, we remark that all ultrafast molecular imaging techniques will continue to push advancements in theoretical and numerical quantum chemical dynamics, as accurate nuclear wavepacket propagation following excitation by the pump pulse remains a significant challenge. This issue is not unique to LIED, but given LIED's exceptional temporal resolution, it is reasonable to assume that LIED could provide valuable benchmarks for theoretical refinements.

Acknowledgements

LFD acknowledges support from the U.S. Department of Energy, Office of Science, Basic Energy Sciences, under award # DE-FG02-04ER15614. CIB acknowledges support from the U.S. Department of Energy, Office of Science, Basic Energy Sciences, under award # DE-FG-02-86ER134910054.

References

1. E. Hecht, *Optics*, Pearson, London, UK, 5th edn, 2015.
2. I. Hargittai and M. Hargittai, *Stereochemical Applications of Gas-Phase Electron Diffraction*, VCH Publishers Inc., New York, NY, USA, 1st edn, 1988.
3. H. Mark and R. Weirl, *Z. Phys.*, 1930, **60**, 741.
4. F. Salvat, *Phys. Rev. A: At., Mol., Opt. Phys.*, 2003, **68**, 012708.
5. G. G. Paulus, W. Nicklich, H. Xu, P. Lambropoulos and H. Walther, *Phys. Rev. Lett.*, 1994, **72**, 2851.
6. H. B. van Linden van den Heuvell and H. G. Muller, *Multiphoton Processes*, Cambridge University Press, Cambridge, UK, 1st edn, 1987.
7. K. J. Schafer, Baorui Yang, L. F. DiMauro and K. C. Kulander, *Phys. Rev. Lett.*, 1993, **70**, 1599.
8. P. B. Corkum, *Phys. Rev. Lett.*, 1993, **71**, 1994.
9. T. Zuo, A. D. Bandrauk and P. B. Corkum, *Chem. Phys. Lett.*, 1996, **259**(3–4), 313.
10. R. D. DuBois and M. E. Rudd, *J. Phys. B: At. Mol. Phys.*, 1976, **9**, 2657.
11. B. Wolter, M. G. Pullen, M. Baudisch, M. Sclafani, M. Hemmer, A. Senftleben, C. D. Schröter, J. Ullrich, R. Moshammer and J. Biegert, *Phys. Rev. X*, 2015, **5**, 021034.
12. L. V. Keldysh, *Sov. Phys. JETP*, 1964, **20**, 1945.
13. C. Z. Bisgaard and L. B. Madsen, *Am. J. Phys.*, 2004, **72**, 249.
14. M. V. Ammosov, N. B. Delone and V. P. Krainov, *Sov. Phys. JETP*, 1986, **64**, 1191.
15. X. M. Tong, Z. X. Zhao and C. D. Lin, *Phys. Rev. A: At., Mol., Opt. Phys.*, 2002, **66**, 033402.
16. A. I. Nikishov and V. I. Ritus, *Sov. Phys. JETP*, 1966, **23**, 168.
17. N. B. Delone and V. P. Krainov, *J. Opt. Soc. Am. B*, 1991, **8**, 1207.
18. V. M. Ristić, J. M. Stevanović and M. M. Radulović, *Laser Phys. Lett.*, 2006, **3**, 298.
19. P. Wang, A. M. Sayler, K. D. Carnes, B. D. Esry and I. Ben-Itzhak, *Opt. Lett.*, 2005, **30**(6), 664.
20. J. Tate, T. Auguste, H. G. Muller, P. Salières, P. Agostini and L. F. DiMauro, *Phys. Rev. Lett.*, 2007, **98**, 013901.
21. C. I. Blaga, F. Catoire, P. Colosimo, G. G. Paulus, H. G. Muller, P. Agostini and L. F. DiMauro, *Nat. Phys.*, 2008, **5**, 335.
22. M. Okunishi, T. Morishita, G. Prümper, K. Shimada, C. D. Lin, S. Watanabe and K. Ueda, *Phys. Rev. Lett.*, 2008, **100**, 143001.
23. M. Meckel, D. Comtois, D. Zeidler, A. Staudte, D. Pavičić, H. C. Bandulet, H. Pépin, J. C. Kieffer, R. Dörner, D. M. Villeneuve and P. B. Corkum, *Science*, 2008, **320**, 1478.
24. Z. Chen, A.-T. Le, T. Morishita and C. D. Lin, *Phys. Rev. A: At., Mol., Opt. Phys.*, 2009, **79**, 033409.
25. K. Amini and J. Biegert, *Adv. At., Mol., Opt. Phys.*, 2020, **69**, 163.
26. S. P. Weathersby, G. Brown, M. Centurion, T. F. Chase, R. Coffee, J. Corbett, J. P. Eichner, J. C. Frisch, A. R. Fry, M. Gühr, N. Hartmann,

C. Hast, R. Hettel, R. K. Jobe, E. N. Jongewaard, J. R. Lewandowski, R. K. Li, A. M. Lindenberg, I. Makasyuk, J. E. May, D. McCormick, M. N. Nguyen, A. H. Reid, X. Shen, K. Sokolowski-Tinten, T. Vecchione, S. L. Vetter, J. Wu, J. Yang, H. A. Dürr and X. J. Wang, *Rev. Sci. Instrum.*, 2015, **86**, 073702.

27. C. I. Blaga, J. Xu, A. D. DiChiara, E. Sistrunk, K. Zhang, P. Agostini, T. A. Miller, L. F. DiMauro and C. D. Lin, *Nature*, 2012, **483**, 194.

28. F. Qi, Z. Ma, L. Zhao, Y. Cheng, W. Jiang, C. Lu, T. Jiang, D. Qian, Z. Wang, W. Zhang, P. Zhu, X. Zou, W. Wan, D. Xiang and J. Zhang, *Phys. Rev. Lett.*, 2020, **124**, 134803.

29. Y. Morimoto and P. Baum, *Nat. Phys.*, 2018, **14**, 252.

30. P. Colosimo, G. Doumy, C. I. Blaga, J. Wheeler, C. Hauri, F. Catoire, J. Tate, R. Chirla, A. M. March, G. G. Paulus, H. G. Muller, P. Agostini and L. F. DiMauro, *Nat. Phys.*, 2008, **4**, 386.

31. J. Xu, C. I. Blaga, A. D. DiChiara, E. Sistrunk, K. Zhang, Z. Chen, A.-T. Le, T. Morishita, C. D. Lin, P. Agostini and L. F. DiMauro, *Phys. Rev. Lett.*, 2012, **109**, 233002.

32. A. D. DiChiara, E. Sistrunk, C. I. Blaga, U. B. Szafruga, P. Agostini and L. F. DiMauro, *Phys. Rev. Lett.*, 2012, **108**, 033002.

33. A.-T. Le, R. R. Lucchese, S. Tonzani, T. Morishita and C. D. Lin, *Phys. Rev. A: At., Mol., Opt. Phys.*, 2009, **80**, 013401.

34. B. Wolter, M. G. Pullen, A.-T. Le, M. Baudisch, K. Doblhoff-Dier, A. Senftleben, M. Hemmer, C. D. Schröter, J. Ullrich, T. Pfeifer, R. Moshammer, S. Gräfe, O. Vendrell, C. D. Lin and J. Biegert, *Science*, 2016, **354**, 308.

35. J. Xu, C. I. Blaga, K. Zhang, Y. H. Lai, C. D. Lin, T. A. Miller, P. Agostini and L. F. DiMauro, *Nat. Commun.*, 2014, **5**, 4635.

36. M. G. Pullen, B. Wolter, A.-T. Le, M. Baudisch, M. Hemmer, A. Senftleben, C. Dieter Schröter, J. Ullrich, R. Moshammer, C. D. Lin and J. Biegert, *Nat. Commun.*, 2015, **6**, 7262.

37. M. G. Pullen, B. Wolter, A.-T. Le, M. Baudisch, M. Sclafani, H. Pires, C. D. Schröter, J. Ullrich, R. Moshammer, T. Pfeifer, C. D. Lin and J. Biegert, *Nat. Commun.*, 2016, 7, 11922.

38. S. G. Walt, N. Bhargava Ram, M. Atala, N. I. Shvetsov-Shilovski, A. von Conta, D. Baykusheva, M. Lein and H. J. Wörner, *Nat. Commun.*, 2017, **8**, 15651.

39. B. Belsa, K. Amini, X. Liu, A. Sanchez, T. Steinle, J. Steinmetzer, A. T. Le, R. Moshammer, T. Pfeifer, J. Ullrich, R. Moszynski, C. D. Lin, S. Gräfe and J. Biegert, *Struct. Dyn.*, 2021, **8**, 014301.

40. Y. Ito, C. Wang, A.-T. Le, M. Okunishi, D. Ding, C. D. Lin and K. Ueda, *Struct. Dyn.*, 2016, **3**, 034303.

41. E. T. Karamatskos, G. Goldsztejn, S. Raabe, P. Stammer, T. Mullins, A. Trabattoni, R. R. Johansen, H. Stapelfeldt, S. Trippel, M. J. J. Vrakking, J. Küpper and A. Rouzée, *J. Chem. Phys.*, 2019, **150**, 244301.

42. A. Sanchez, K. Amini, S.-J. Wang, T. Steinle, B. Belsa, J. Danek, A. T. Le, X. Liu, R. Moshammer, T. Pfeifer, M. Richter, J. Ullrich, S. Gräfe, C. D. Lin and J. Biegert, *Nat. Commun.*, 2021, **12**, 1520.

43. K. Amini, M. Sclafani, T. Steinle, A.-T. Le, A. Sanchez, C. Müller, J. Steinmetzer, L. Yue, J. R. Martínez Saavedra, M. Hemmer, M. Lewenstein, R. Moshammer, T. Pfeifer, M. G. Pullen, J. Ullrich, B. Wolter, R. Moszynski, F. J. García de Abajo, C. D. Lin, S. Gräfe and J. Biegert, *Proc. Natl. Acad. Sci. U. S. A.*, 2019, **116**, 8173.
44. F. Schell, T. Bredtmann, C. Peter Schulz, S. Patchkovskii, M. J. J. Vrakking and J. Mikosch, *Sci. Adv.*, 2018, **4**, eaap8148.
45. H. Fuest, Y. H. Lai, C. I. Blaga, K. Suzuki, J. Xu, P. Rupp, H. Li, P. Wnuk, P. Agostini, K. Yamazaki, M. Kanno, H. Kono, M. F. Kling and L. F. DiMauro, *Phys. Rev. Lett.*, 2019, **122**, 053002.
46. F. Brausse, F. Bach, F. Krečinić, M. J. J. Vrakking and A. Rouzée, *Phys. Rev. Lett.*, 2020, **125**, 123001.
47. C. Yu, H. Wei, X. Wang, A.-T. Le, R. Lu and C. D. Lin, *Sci. Rep.*, 2015, **5**, 15753.
48. S.-J. Wang, J. Daněk, C. I. Blaga, L. F. DiMauro, J. Biegert and C. D. Lin, *J. Chem. Phys.*, 2021, **155**, 164104.
49. X. Liu, K. Amini, A. Sanchez, B. Belsa, T. Steinle and J. Biegert, *Commun. Chem.*, 2021, **4**, 154.
50. C. J. Hensley, J. Yang and M. Centurion, *Phys. Rev. Lett.*, 2012, **109**, 133202.
51. S.-F. Zhao, J. Xu, C. Jin, A.-T. Le and C. D. Lin, *J. Phys. B: At., Mol. Opt. Phys.*, 2011, **44**, 035601.

CHAPTER 14

Electron Imaging in Action: Attosecond Electron Diffraction and Microscopy

M. TH. HASSAN

Department of Physics, University of Arizona, Tucson, AZ 85721, USA
Email: mohammedhassan@email.arizona.edu

14.1 Introduction

The discovery of X-rays, electrons, and their diffraction properties allowed for resolving the three-dimensional (3D) structures of matter.[1-3] Because electrons have a shorter wavelength and a larger scattering cross-section, electron diffraction can characterise thin and low-density samples (two-dimensional materials and gases). Moreover, the energy deposition from electron beams to the sample is a thousand times less than that from X-rays, inducing less radiation damage per scattering event. Accordingly, electron diffraction is more powerful when studying chemical and biochemical systems. Importantly, electron diffraction is more sensitive to valence electron distributions than X-ray diffraction at small scattering vectors, providing the capability to resolve the electronic structural information of matter to trace electron motion in atoms and molecules.

Over many decades, electron diffraction imaging has enabled the resolution of static spatial arrangements of atomic structures with picometer resolution, and this technique has different applications in many fields of science.[1] Introducing the time dimension to electron diffraction was a revolutionary advancement that provided access to structural dynamics in both time and space domains. The earliest efforts at developing time-resolved analogues of

Theoretical and Computational Chemistry Series No. 25
Structural Dynamics with X-ray and Electron Scattering
Edited by Kasra Amini, Arnaud Rouzée and Marc J. J. Vrakking
© The Royal Society of Chemistry 2024
Published by the Royal Society of Chemistry, www.rsc.org

conventional electron diffraction were demonstrated by studying gas-phase radical dynamics on sub-millisecond timescales.[4,5] Then, the temporal resolution of electron diffraction imaging was enhanced to the sub-nanosecond scale by generating electron bursts using a pulsed laser beam, recording images of thin aluminium films in a phase-transmission mode with a 100 picosecond (ps) resolution.[6–8]

In the last decade, the establishment of ultrafast electron diffraction (UED), based on generating ultrafast electron pulses by photoemission from a photocathode, has enabled the imaging of atomic motion in real space and time. This essential tabletop tool paved the way for a variety of applications in fields of science spanning chemistry, physics, materials science, and biology.[1] In the UED technique, ultrafast dynamics is triggered by a laser pulse, and the ensuing structural dynamics is then tracked in real-time by recording electron diffraction patterns using a charge-coupled device (CCD) camera. An electron diffraction pattern is formed by the interaction between an ultrafast electron "probe" pulse and the sample under study. A video of the atomic motion can be obtained by acquiring a sequence of diffraction-pattern snapshots at different instances of time delay between the excitation pump "laser" pulse and the probe "electron" pulse. Then, the 3D structural dynamics can be retrieved from the measured diffraction patterns as the transient structure evolves in time. Specifically, the change in interatomic distances can be extracted from the recorded diffraction movie using the Fourier transform relationship between reciprocal-space and real-space. The extracted molecular movie of transient changes in the position of atoms provides intricate details about the reaction mechanism(s) that the sample under study proceeds through as it evolves from the ground state to its final state(s), which is beyond the reach of time-resolved optical spectroscopy measurements.

The UED technique has several advantages over time-resolved X-ray and high-harmonic generation XUV diffraction in resolving structural dynamics in real time with a high spatial resolution.[9–18] First, the cross-section for electron scattering is approximately six orders of magnitude larger than that of X-ray and XUV scattering.[19,20] Second, the quantum efficiencies for generating electron pulses based on photoemission are on the order of 10^{-4} and 10^{-1} for metals and semiconductor photocathodes, respectively,[21] whereas the conversion efficiency is typically 10^{-6} for the generation of X-rays and XUV light.[22–24] This efficiency is further reduced by approximately 10^{-2} at the position of the sample because of a limited solid angle of collection as well as the use of X-ray and XUV optics. Third, the electron–matter interaction is stronger, and, therefore, electron diffraction can reveal the transient structures of gases, surfaces, and (thin) crystals. Fourth, electrons are less damaging to specimens per elastic scattering event,[25] which is vital in real-life applications (*e.g.*, studying the ultrafast dynamics of chemical and biochemical samples). Moreover, UED has a crucial advantage over other electron imaging tools such as ultrafast electron microscopy (UEM), namely, setup flexibility, which allows for compressing the probe

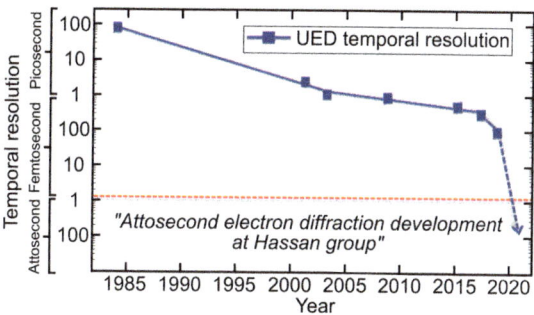

Figure 14.1   The development of the temporal resolution in electron diffraction imaging over time. The significant enhancement of breaking the sub-picosecond temporal resolution barrier in 2000 was followed up by the development of electron pulse compression techniques allowed to reach 100 fs imaging resolution. The Hassan group is developing attosecond electron diffraction (AED) based on the optical gating and taming of electrons in UED by attosecond laser pulses. This AED resolution will open the door to simultaneously image electron motion and structural dynamics in matter.[1-3]

pulse to shorter durations[26-28] and studying ultrafast dynamics in the gas phase.[29]

In time-resolved UED measurements, the temporal resolution of electron diffraction imaging, typically called the instrument response function, is defined by (i) the pulse duration of the ultrashort electron probe beam; (ii) the pulse duration of the excitation pump pulse (which is a crucial parameter at higher time resolutions); and (iii) the synchronisation stability between the electron probe and optical pump pulses. Despite the tremendous efforts that have been dedicated to improving UED temporal resolution over the last decade, from 100 ps to the sub-ps regime, the total resolution is limited to >100 femtoseconds (fs)[1] (see Figure 14.1). The main factors limiting the temporal resolution of UED are the pulse duration of the ultrashort electron probe beam and its synchronisation stability with the optical pump beam, given the fact that commercial laser systems can deliver pump optical pulses as short as 5 fs.

The electron pulse suffers from temporal broadening due to the space-charge effect and energy dispersion during its propagation from the cathode to the sample, as an external field does not correct the energy dispersion. Therefore, several strategies and techniques have been proposed and employed to generate ultrashort electron pulses, such as suppressing the space-charge effect by compressing the electron pulse with radiofrequency (RF) cavities and terahertz (THz) waveguides, compensating for energy dispersion with reverse bending magnets combined with an achromatic compressor, increasing the electron energy to the relativistic regime, and shortening the electron propagation length with very compact experimental setup designs.[30-36] Among these approaches, electron pulse compression using an RF cavity is a very effective technique in the sub-relativistic regime while simultaneously keeping a high brightness or electron flux in a single electron pulse required for single-shot measurements (although the shortest electron pulse duration reported is >28 fs[37] while most UED measurements employ compressed electron pulses of duration >100 fs). Alternatively, other techniques, such as THz radiation compression, have been used to generate a few tens of femtosecond electron pulses.[38] Nevertheless, RF-based compression intrinsically causes an additional timing jitter between the reference optical pump and the electron probe because it changes the time-of-flight (TOF) of electron pulses due to the jitter in the RF phase and amplitude. A time-stamping technique based on THz deflectors is being developed to measure and correct this timing jitter, thus improving the temporal resolution of electron diffraction imaging *via* UED. However, this is a shot-to-shot correction technique, which only applies to single-shot or multi-shot experiments using electron pulses containing many electrons so that an electron diffraction pattern can be analysed in a single shot. More-over, employing such a time-stamping method is challenging in high-repetition-rate experiments as a very fast detector response time is also required. Alternatively, a more practical solution is using an active synchronisation system that has corrected the RF-laser timing jitter to <20 fs (full width at half maximum, FWHM) in ultrafast electron diffraction and imaging experiments.[39] In this approach, an RF signal is generated from the laser oscillator train pulse which is then compared to the RF signal generated by the electron compression cavity. The phase mismatch between these two RF signals is corrected on a shot-to-shot basis.

So far, UED has been exploited for imaging transient dynamics related to phase transitions, charge density waves, lattice vibration and oscillation, and chemical bond forming and breaking, which occur on timescales of picosecond-to-100 fs and length scales of sub-picometer.[40–45] Nevertheless, to resolve and image dynamic processes taking place in atoms, molecules, and solids on atomic and electronic timescales (*i.e.*, a few femtoseconds to hundreds of attoseconds, as), a significant enhancement (two orders of magnitude) in the total temporal resolution of UED imaging is required. Thus, new electron diffraction tools that provide a sub-femtosecond imaging resolution sufficient to image electron motion in action must be developed. We are currently developing this tool, which we call "Attosecond Electron Diffraction" (AED), based on the concept of taming ultrafast electron pulses using attosecond laser pulses to achieve the desired attosecond resolution in electron diffraction imaging. In this chapter, we explain in detail the development stages of UED and its reported applications, as well as advances in AED and its potential imaging applications.

14.2   Ultrafast Electron Diffraction

In a typical UED experimental setup (see Figure 14.2), the electron probe pulse is generated by illuminating a photocathode with an ultraviolet laser (UV) pulse *via* the photoemission process in an electron gun. The generated

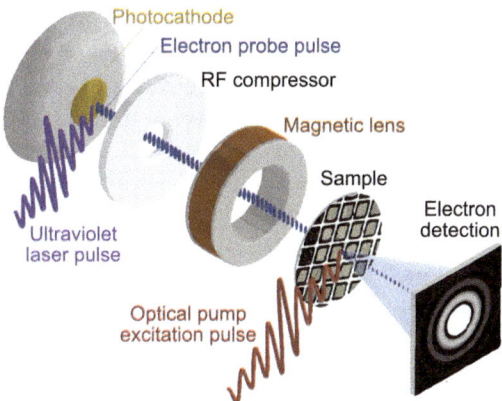

Figure 14.2   Schematic of a typical UED setup. Here, an ultraviolet optical pulse is focused on the photocathode to generate picosecond-to-femtosecond electron pulses. The electron pulses are accelerated by an external DC/RF field, and focused in space and time by magnetic coils and external RF and THz fields, respectively, onto the sample. The electron pulse probes the transient structural dynamics of matter triggered by another optical pump excitation pulse. A full electron diffraction movie is obtained by recording electron diffraction patterns at various pump–probe delays.

electron pulse is first accelerated and then compressed using an RF compressor to a few hundred fs duration at the sample position (depending on the number of electrons per pulse). The compressed electron beam is then steered and spatially focussed onto the sample using magnetic fields generated by electrostatic magnets. The transmitted electrons from the sample are detected using a photoemission/direct electron detection camera, enabling a static diffraction pattern of the sample under study to be recorded. A pump laser pulse illuminates the sample to trigger the ultrafast dynamics. The wavelength and intensity of the pump pulse are predetermined based on the target dynamics to be studied. In time-resolved electron diffraction measurements, the delay between the electron and laser pulses is controlled precisely by a linear delay stage, which controls the beam path difference between the pump and UV (and consequently the electron) pulses, called the pump–probe delay. The diffraction pattern of the sample is acquired at different pump–probe delay times between the pump laser and the probe electron pulses. These diffraction images are analysed to obtain a movie of the ultrafast dynamics of the sample in real time and space.

   Extracting useful structural information from a sample of 10–100 nm thickness through 1 kHz UED measurements requires employing electron pulses containing at least a thousand to ten thousand electrons to resolve diffraction peaks and obtain clear diffraction images.[35,41] Measuring UED data with a higher signal-to-noise ratio (SNR) is achievable with higher brightness electron sources, which have been developed with pulse durations on the order of a few hundreds of femtoseconds,[32,46,47] and higher repetition-rate laser systems employed in high repetition-rate rescattering-based UED measurements.[48,49] More details about the developments of keV-UED are explained in Chapter 11. Moreover, the second generation of UED, with mega-electron-volt (MeV) electron sources, has been developed recently at SLAC[50,51] (see Chapter 12). This MeV-UED high-brightness source tool has the unique advantage of a high SNR within a reasonable exposure time, enabling single-shot time-resolved measurements in biological and biochemical systems.[52] Exploiting MeV-UED in real-life applications will significantly impact science and humankind—a game-changer, elevating the entire electron diffraction imaging field to the next level.

14.3 Applications of Ultrafast Electron Diffraction (UED)

Over the last two decades, UED has found a vast range of applications, especially in chemistry and material sciences. The UED technique was first demonstrated in 2001,[40] where the ultrafast dynamics of the 1,2-diiodo-tetrafluoroethane $(C_2F_4I_2)$ chemical reaction to produce tetrafluoroethene and iodine were studied following the photodissociation of a C–I bond. In the same work, the real-time ring-opening dynamics of 1,3-cyclohexadiene (CHD) to form 1,3,5-hexatriene were also studied with a spatiotemporal resolution of 0.01 Å and 1 ps.[40] Time-resolved diffraction measurements of the first reaction reveal that the fragmentation of $C_2F_4I_2$ proceeds *via* two steps at timescales of 5 ps and 31 ± 4 ps (see Figure 14.3a). In contrast, the ring-opening of CHD, which involves structural rearrangement rather than fragmentation, occurs on the tens-to-hundreds of picosecond timescale (see Figure 14.3b). The reaction dynamics and structural rearrangements in such molecules remained never before seen until the establishment of UED.

In subsequent work, the UED temporal resolution was further improved using RF compression, breaking the temporal resolution barrier below 1 ps.[41]

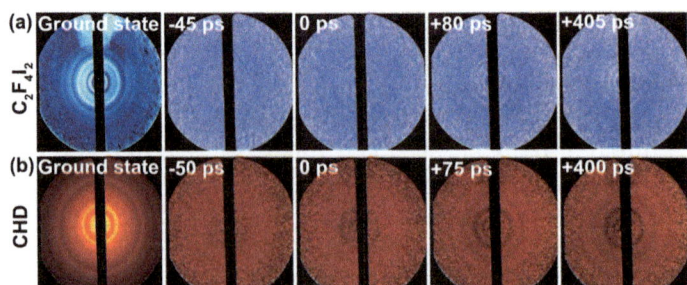

Figure 14.3   UED studies of ultrafast dynamics in gas-phase molecules. Electron diffraction patterns of (a) 1,2-diiodotetrafluoroethane $(C_2F_4I_2)$ and (b) 1,3-cyclohexadiene (CHD) at different pump–probe delay times.[40] See the main text for more details. Reproduced from ref. 1 with permission from IOP Publishing Ltd, Copyright 2018.

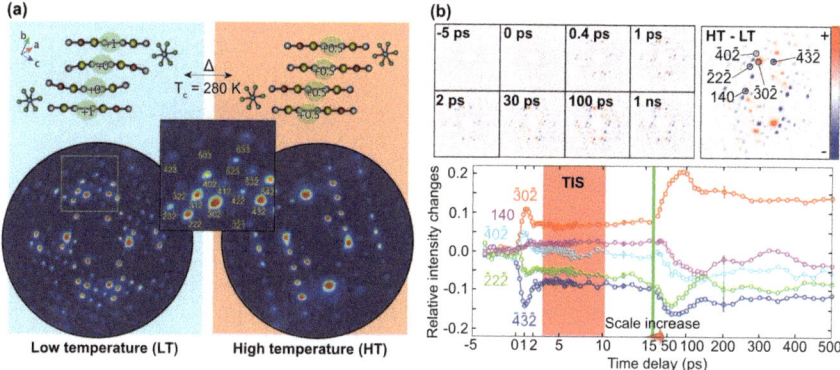

Figure 14.4   UED study of insulator-to-metal phase transition and the ensuing ultrafast lattice dynamics in the molecular organic salt (EDO-TTF)$_2$ PF$_6$.[53] See the main text for more details. Reproduced from ref. 1 with permission from IOP Publishing Ltd, Copyright 2018.

This crucial advancement in reversing space-charge dispersion with RF fields allowed a single-shot electron imaging experiment to study the insulator-to-metal phase transition and charge delocalisation driven by molecular motions in an organic salt (EDO-TTF)$_2$PF$_6$, which has a weak scattering center.[53] In this salt, in the low temperature (LT) phase, new diffraction peaks appear that are not present in the high temperature (HT) phase and are indicative of a symmetry-breaking "cell doubling" effect (see Figure 14.4a). The clear difference between the diffraction patterns between the two phases confirmed that a non-negligible bending of the neutral crystal orientation promotes a Peierls-like mechanism (*e.g.*, see Chapter 2) that leads to the replication of the unit cell. Snapshots of the structural dynamics evolving in time were mapped (see Figure 14.4b), which indicated that the system undergoes phase transition in two steps. First, a transient intermediate structure (TIS) is generated within <5 ps during the early stage of charge delocalisation. Then, the system converts to a metallic structure over hundreds of picoseconds following an insulator-to-metal phase transition driven by uncorrelated motions as the system thermally relaxes to a metallic HT state. In general, the relative intensity changes of various diffraction peaks (*i.e.*, ($\bar{3}0\bar{2}$), ($\bar{4}0\bar{2}$), ($\bar{2}2\bar{2}$), and ($\bar{4}3\bar{2}$)) between the two temperature phases occur because of the contrasting changes in the unit cell dimension of each phase.

Furthermore, the bright MeV ultrafast electron pulses from the UED facility at SLAC, as explained earlier, provide a significant enhancement of the SNR when imaging the photoexcited coherent motion of a vibrational wave packet in iodine molecules (I$_2$) in the gas phase.[54] The images of this motion were recorded with sub-angstrom spatial resolution and a 230 fs temporal resolution. This demonstration opened the door for different applications of this MeV UED instrument in real-time imaging of more complex gas-phase molecular structures undergoing chemical reactions.

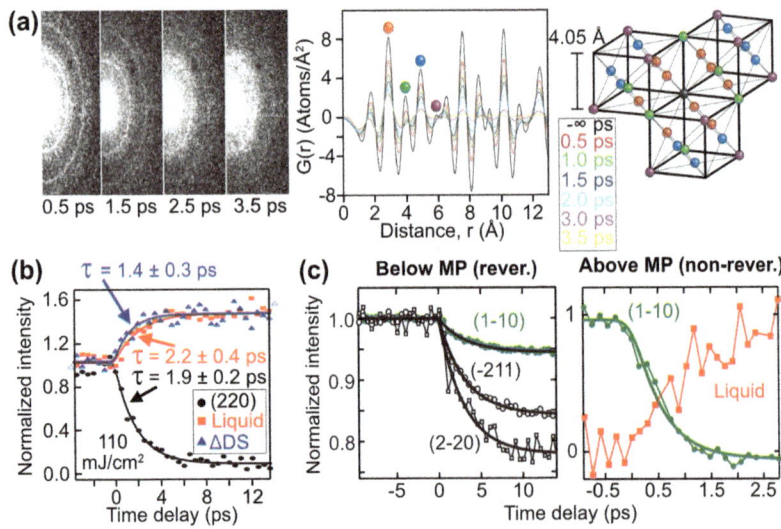

Figure 14.5   Imaging the ultrafast phase transition dynamics in solid-state by UED. (a) Study of the ultrafast melting phase transition process in polycrystalline aluminium nanofilms with the single-shot UED technique.[42] (b) Single-shot UED imaging of the polycrystalline gold nanofilm phase transition.[43] (c) Electron diffraction imaging of the structural changes in the photo-induced melting process of crystalline bismuth.[46]

Moreover, the MeV UED setup's capabilities were further extended to liquid phase time-resolved UED measurements, such as in the mapping of structural dynamics during water dissociation.[55,56]

   In this chapter, the main focus is placed on phase transitions in solid-state samples studied with nonrelativistic keV UED sources. The generation of compressed, high-brightness, ultrafast electron pulses in keV UED is a critical enabling factor for studying phase transitions in solids. For instance, nonrelativistic UED has been used to image the melting process in aluminium films in real-time, which proceeds on a picosecond timescale as a direct consequence of the strong electron–phonon coupling in aluminium (see Figure 14.5a).[41,42] Moreover, the structural dynamics of melting in polycrystalline gold films have also been reported.[43] The gold nanofilm undergoes a phase transition from a solid to a high-density plasma on a sub-picosecond timescale, also referred to as warm dense matter (see Figure 14.5b). Such a phase transition caused a reduction in the measured (220) Bragg intensity peak which occurred simultaneously with the appearance of a liquid phase diffraction peak accompanied by a background thermal diffuse scattering (ΔDS) signature. The moderate electron–phonon coupling in gold enables the few-ps observed changes under such high fluence conditions (110 mJ cm^{-2}). In addition, phase transitions in crystalline bismuth were imaged with femtosecond UED under two fluence conditions.[46] The observed lattice dynamics proceed on two timescales of several hundred femtoseconds and several picoseconds for a sample that is excited above and below the melting point (MP), respectively (see Figure 14.5c). In the below MP case, the decrease in the $(1\bar{1}0)$, (-211) and $(2\bar{2}0)$ diffraction peaks is due to the Debye–Waller effect. While in the above MP case, the observed transient is a non-reversible solid-to-liquid phase transition measured with single-shot measurements. Moreover, UED has been used to study the structural dynamics of different organic molecular systems such as pyridine, 2-methylpyridine, 2,6-dimethylpyridine, benzaldehyde,[57] and interfacial water following photo-induced substrate heating.[58]

One of the exciting materials science applications of femtosecond UED is the measurement of atomic motion dynamics in quasi-two-dimensional charge-density-wave (CDW) materials.[30] For example, it was determined that the phase transition of the 2D 1T-TaS_2 system from a periodic lattice distortion phase to a CDW phase was due to changes in its electronic spatial distribution.[30] In another study, the CDW ultrafast dynamics of $LaTe_3$ were investigated, revealing the transient dynamics of a non-equilibrium phase of matter with no equilibrium counterpart.[59,60] The ultrafast phase transition dynamics of cuprates, which have superconductivity properties below the critical temperature (T_c) and metallic properties at room temperature, were determined *via* UED.[61] The phase transition of polycrystalline vanadium dioxide, VO_2, to a semimetal-like state was imaged with UED.[45] The direct connection between the temporal evolution of a phase transition and the subsequent transient spatial and electronic rearrangements revealed that a phase transition is driven by electron–lattice and electron–electron interactions.

The temporal-resolution limitation of nonrelativistic UED, due to the technical challenge of pulse compression and timing jitter corrections, constrains the imaging of atomic motions that span several hundred femtoseconds or longer, and keeps the imaging of electron motion beyond reach. Therefore, a revolutionary temporal resolution enhancement of UED to attosecond timescales is highly desired. This extreme attosecond temporal resolution, in addition to the sub-picometer spatial resolution in UED, will enable the simultaneous measurement of a complete map of atomic dynamics and their corresponding electronic motion in real time and space, opening new prospects for femtosecond and attosecond UED applications and promising to break new ground in ultrafast science.

14.4 Attosecond Electron Diffraction (AED)

An enhancement in the temporal resolution of UED systems to between hundreds-of-attoseconds and a few-fs is a prerequisite to image electron motion. Recently, we exploited a free-electron optical gating and control approach, which we demonstrated for the first time in 2015,[62] to generate the shortest single-isolated (30 fs) free-electron pulse to date inside an ultrafast electron microscope.[63,64] This optical gating of electron pulses can be explained as follows: the free electrons and laser pulses interact, and inelastic scattering occurs between both pulses on a nanostructure. This interaction leads to the gain/loss of photon quanta by electrons. Therefore, due to photon–electron coupling, some of the electrons (within the electron pulse) gain or lose one or multiple photon quanta in the presence of the optical laser pulse. The optical pulse acts as a "temporal optical gate" for these electrons. The gated electrons have a temporal profile identical to the gating window (i.e., the optical laser pulse duration). These electrons can be filtered out to obtain ultrashort electron pulses, thus significantly enhancing the temporal resolution of UED for exploring the ultrafast dynamics of matter triggered by another ultrashort optical laser pulse. This optical-gating approach has several advantages over conventional electron-pulse compression techniques for generating shorter electron pulses. First, the generated electron pulse duration is limited only by the gating laser pulse duration, which could span the attosecond timescale. Second, active phase-locking between the two optical (pump and gating) pulses can achieve extreme phase and synchronisation stability. Other electron-pulse-compression techniques cannot accomplish this. Therefore, this approach is considered the best candidate for achieving attosecond resolution in UED and other electron imaging tools.

The optical gating approach has been proposed to generate a train of attosecond electron pulses.[65,66] Later, the generation of an attosecond electron pulse train was demonstrated experimentally from a single picosecond electron pulse accelerated at 70 keV.[67] Also, a train of attosecond electron pulses was generated based on the dielectric acceleration on a single photonic chip,[68] and inverse free-electron-laser process.[69] Moreover, these pulse trains were generated inside the transmission electron microscope using a continuous-wave laser.[70,71] Notably, in a realistic time-resolved electron imaging experiment, the electron pulse train provides the temporal resolution of the primary electron beam, which remains on the order of 50–100 fs.

In the Hassan group, we are utilising this optical gating approach to generate single-isolated attosecond electron pulses (AEPs) by gating the free electrons of ultrashort laser pulses in UED using an optical attosecond pulse (OAP) to establish AED.[62,63,72] Attosecond electron imaging will provide further insights into electron dynamics in real-time and space for attosecond applications in physics, chemistry, biology, and materials science.

14.5   Attosecond Electron Diffraction Beamline

The development of AED, based on attosecond electron gating in a UED system (see Figure 14.6), can be explained as follows: first, the main laser beam is generated from the optical parametric chirped-pulse amplification (OPCPA) laser system (the oscillator runs at a repetition rate of 41.6 MHz). The output, amplified laser pulse has a central wavelength of 800 nm, duration of 15 fs, maximum power of 20 W, and repetition rate of 20 kHz. The output laser pulse is split into two parts. The first portion (1 W) is frequency-tripled to generate a UV laser (266 nm, 4.65 eV). The UV pulse illuminates a

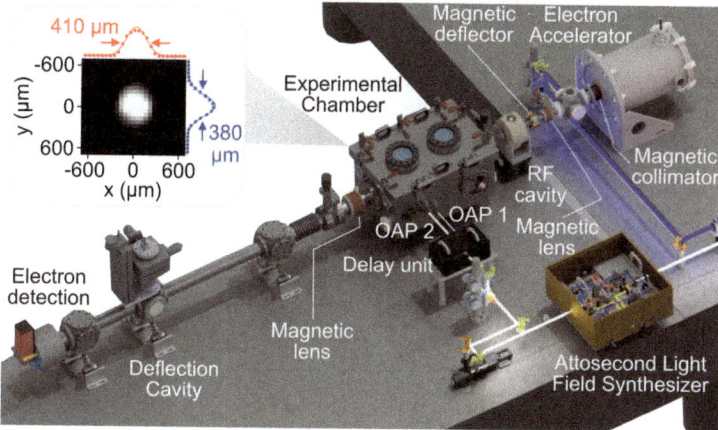

Figure 14.6 Schematic of the attosecond electron diffraction (AED) beamline in the Hassan lab. See the main text for more details. The inset shows the measured electron pulse beam profile at the detector with the FWHM value indicated by arrows.

bulk copper photocathode (4.6 eV working function) to generate a free-electron pulse. Then, the generated electron pulse is accelerated to a kinetic energy of 100 kV in a DC electric field of strength 10 MV m^{-1} across a ~10 mm distance between the −100 kV cathode and grounded anode. The second portion (19 W) is used in the optical gating approach to generate OAPs, which will be explained later in more detail. The generated electron flux is estimated to be a few million electrons per second. The accelerated electrons are temporally compressed using an RF compression cavity to compensate for space-charge broadening. They are spatially confined and manipulated using magnetic focusing coils and steering coils. A collimator aperture (100 μm) is introduced before the samples to filter out stray electrons to provide a symmetric, Gaussian beam profile with a FWHM diameter size of ~400 μm (see the inset of Figure 14.6).

Inside the experimental chamber, the electron beam illuminates the sample under study and is mounted on a sample holder (containing nine slots to easily exchange one sample with another). The sample holder is mounted on a compact five-dimensional stage, which provides complete control of the sample direction and incident angle of the electron beam. A beam blocker is introduced after the sample to block the transmitted unscattered primary electron beam before the detector to prevent image saturation. After the sample, there is another magnetic focusing lens to collimate electrons and to be able to change the magnification to collect higher-order diffraction peaks. A deflection cavity is mounted 2.3 m away from the cathode for time-of-flight (TOF) electron energy loss spectroscopy (EELS) measurements. At the end station of the beamline (3 m away from the cathode), a direct electron detection camera records electron diffraction patterns.

The second part of the AED beamline (the pump laser arm) starts by sending the second portion (19 W) of the main laser beam through a hollow-core fibre (HCF) to generate OAPs (see Figure 14.7a). The nonlinear propagation generates a supercontinuum inside the HCF, filled with ∼2 bar Ne gas.[73]

This broadband spectrum (see Figure 14.7b) enters the attosecond light field synthesiser (LFS; see Figure 14.8a) which is designed to synthesise the

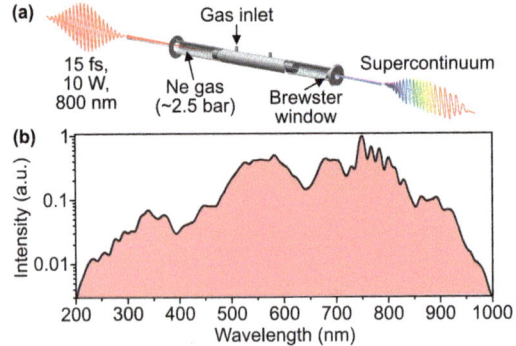

Figure 14.7   Supercontinuum light generation spanning more than two octaves. (a) Schematic of the hollow core fibre (HCF) setup used in generating the two-octave supercontinuum. The output laser pulse is focused into the HCF entrance (with a 250 μm diameter), and the laser beam is nonlinearly propagating inside the fibre to generate a supercontinuum extending over the visible and neighbouring ranges (200–1000 nm). (b) The measured supercontinuum spectrum. Reproduced from ref. 73, https://doi.org/10.1063/5.0082958, under the terms of the CC BY 4.0 license https://creativecommons.org/licenses/by/4.0/.

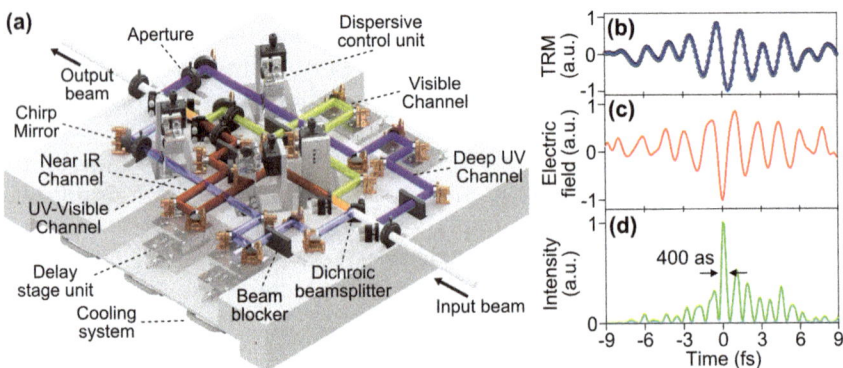

Figure 14.8   Optical attosecond light field synthesiser (LFS). (a) Schematic of the prototypical four-channel light field synthesiser in the Hassan lab. (b) The measured total reflectivity modulation (TRM). The (c) electric field and (d) instantaneous intensity of OAPs were retrieved from the measured TRM data. The temporal duration of the main crest in panel (c) is ∼400 as FWHM. Reproduced from ref. 73, https://doi.org/10.1063/5.0082958, under the terms of the CC BY 4.0 license https://creativecommons.org/licenses/by/4.0/.

light field with attosecond resolution. Using dichroic beam splitters (see Figure 14.8a), the supercontinuum is divided into four (nearly equal) spectral bandwidth constituent channels. The four constituent spectral channels comprise (i) near-infrared (NIR; 700–1300 nm), (ii) visible (500–700 nm), (iii) visible-UV (350–500 nm), and (iv) deep ultraviolet (270–350 nm). Dividing the ultra-broadband spectrum inside the synthesiser is the key to compressing this two-octave light using chirp mirror compressors to an extreme limit. At the exit of the apparatus, pulses from the constituent channels are spatio-temporally superimposed with the same type of beam-splitter to generate OAPs, which are sampled by the attosecond all-optical reflectivity modulation (Atto-ARM) methodology reported previously.[74] The total reflectivity modulation (TRM) of an OAP has been measured (see Figure 14.8b), and the electric field and temporal profile have been retrieved (see Figure 14.8c and d), the latter showing that an OAP has been generated with a duration of 400 as FWHM.

Notably, it takes a few iterations to adjust the relative delays and intensities between the LFS channels to generate the OAPs. After exiting the LFS, the OPA enters a delay unit (see Figure 14.9). The OAP is split into two identical beams in this delay unit using a mask with two holes. One of these beams is reflected by two mirrors mounted on top of the high-precision delay stage for measurements with attosecond resolution. The second beam reflects directly to the exit (see Figure 14.9). The two beams at the exit of the delay unit are directed to the sample inside the experimental chamber. The first beam acts as a gating OAP to generate attosecond electron pulses (AEPs), while the second beam is a pumping OAP that photoexcites a target system to trigger the dynamics of interest in time-resolved AED measurements.

In the AED setup (see Figure 14.10), the isolated AEP generation occurs as follows: the ultrafast electron pulse, which is generated by illuminating the photocathode in the electron accelerator with a few-fs UV pulse, interacts

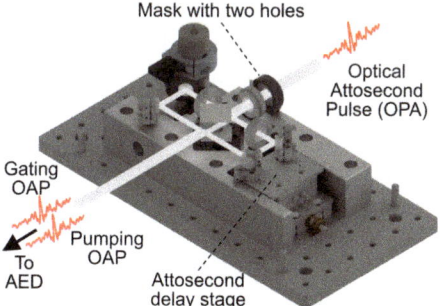

Figure 14.9 Schematic of the attosecond delay unit setup. In configuration A, the optical attosecond pulse (OAP) from the LFS is divided into two identical OAP pulses with the same polarization, while in configuration B, the two output beams have different polarization. Then, the two output beams are directed to the AED setup. Reproduced from ref. 73, https://doi.org/10.1063/5.0082958, under the terms of the CC BY 4.0 license https://creativecommons.org/licenses/by/4.0/.

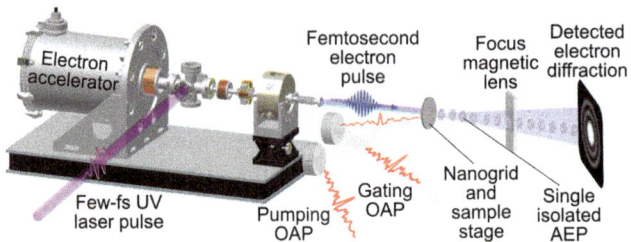

Figure 14.10   Attosecond electron diffraction (AED) setup. See the main text for more details. This AED setup can be modified for TOF EELS measurements by placing a TM_{110} RF cavity between the focussing magnetic lens and the electron detector.

with the OPA in our AED setup. The interaction takes place on an aluminium film fixed on a transmission electron microscopy (TEM) grid tilted at 30° to achieve the maximum coupling efficiency between the OAP and the electron pulse. The OAP acts as a temporal "gating" filter for the electron wavepacket in the electron pulse. The electrons generated by optical gating have a sub-fs temporal profile.[62,63] The gated electrons are spatially separated from ungated electrons and filtered out using a dielectric deflection cavity 0.5 m before the CCD (see Figure 14.6).

The separation mechanism can be explained as follows. After photo-emission at the photocathode, the electron pulse is generated with an appreciable spread in longitudinal and transverse momenta due to internal space-charge forces, leading to a significant spread in the initial energy of the electron beam as it travels to the sample. Optical gating of the electron pulse occurs on the sample so that the electron beam will have a specific gain or loss in its kinetic energy distribution. After interaction with the target sample, the electron beam propagates a considerable distance to separate electrons in the TOF that have gained or lost energy relative to a zero-loss primary electron beam energy of ~ 100 keV. The cavity generates an RF field (3 GHz) and operates in the TM_{110} mode. In this mode, a transverse magnetic field is generated, which deflects the electrons in the same propagation direction. The dispersed electron beam enters the cavity at different times, so electrons with different kinetic energy will obtain different transverse momenta such that the longitudinally distributed pulses will be deflected or projected onto different transverse positions on the detector. Therefore, the EELS distribution can be extracted from the measured deflected spatial distribution on the detector. This is the basis of the TOF EELS process.

Moreover, the temporal profile of the gated electrons can be measured by cross-correlation means. Another OAP from the delay unit illuminates the sample to interact with the gated electrons in this measurement. In this case, a lower number of gated electrons will be detected in the presence of the second OAP. The measured change in the detected number of gated electrons, as a function of the delay between the two OAPs, contains the cross-correction temporal profile of the gated electron. Achieving a gated isolated AEP enables the ability to perform AED measurements with attosecond and nanometre spatiotemporal imaging resolution. The isolated "gated" AEP can be used as a "probe," with another OAP used as a "pump" to ultimately image electron motion in real-time with an AED setup (see Figure 14.10). Notably, the number of gated electrons in the single isolated attosecond pulses depends on the optical gating efficiency and coupling between the optical and electron pulses on the gating medium. This efficiency will be different based on the type of the gating medium and its structure (*i.e.*, nanofilms, nanostructures, nanoparticles, nanowire networks). Also, the number of gated electrons will depend on the repetition rate of the laser pulses.

In our AED setup, we employ a 20 kHz laser system that generates an electron pulse containing several 10^6 electrons per second. We expect to produce an isolated AEP with an electron flux of 500–1000 electrons per seconds (*i.e.*, 0.05% of the primary electron beam before gating), which would be sufficient to perform attosecond electron imaging experiments.

14.6 From Attosecond Electron Diffraction to Attosecond Electron Microscopy: "Attomicroscopy"

Optical gating can be exploited to enhance the temporal resolution in various electron imaging tools, such as UEM, to establish the field of "attomicroscopy". Recently, we demonstrated the enhancement of temporal resolution in UEM by more than an order of magnitude (16 times) by generating the shortest-to-date, intense, isolated, ~30 fs electron pulses accelerated to 200 keV. These short electron pulses were generated by the temporal optical gating of the "original" electron pulse, which was generated by the photoemission process of the photocathode inside the microscope, using a 30 fs visible laser pulse. The temporal profile of the "original" electron pulse was characterised by cross-correlation two-pulse measurements between the "gating" visible laser pulse and the electron pulse. Since the "gating" pulse duration is much shorter than the electron pulse duration, the cross-correlation directly reflects the temporal profile of the electron pulse.[65] The retrieved pulse duration of the "original" electron pulse is on the order of 500 fs (see Figure 14.11). Furthermore, to generate an isolated, ultrashort, "gated" electron pulse with maximum counts, the "gating" pulse is kept at the optimal temporal overlap with a 500 fs electron pulse.

The gated electron pulse, with the same temporal profile as that of the gating pulse, is characterised by using another laser pulse and measuring its cross-correlation temporal profile with the gated electrons. The principle of this three-pulse measurement will be briefly discussed (see Figure 14.12a). The cross-correlation spectrogram (see Figure 14.12b) was recorded first. Then, the temporal profile of this cross-correlation was retrieved from the

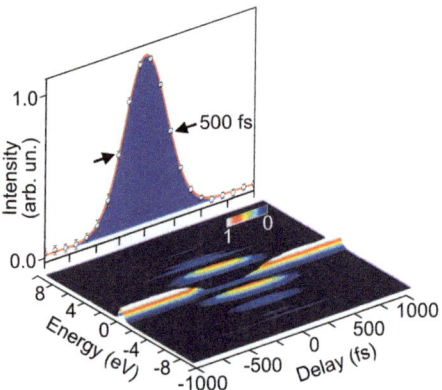

Figure 14.11   Characterizing an ultrafast electron pulse in UEM. The measured spectrogram of the coupling between the visible gating optical pulse and the original ultrafast electron pulse. The two-pulse cross-correlation temporal profile shows that the duration of the original electron pulse from the photocathode is on the order of 500 fs. Reproduced from ref. 63 with permission from Springer Nature, Copyright 2017.

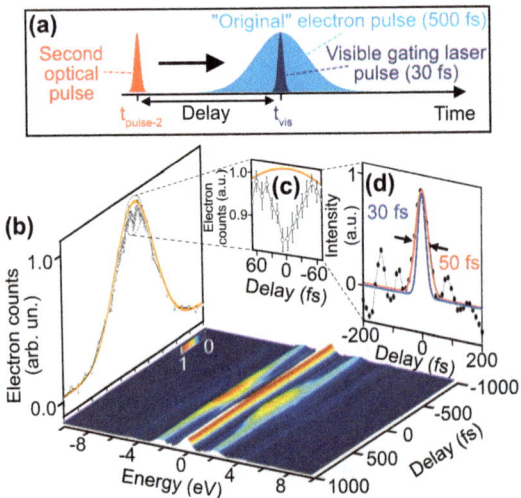

Figure 14.12   Temporal characterization of the gated isolated electron pulse. (a) Illustration of the gated electron pulse temporal characterization principle based on a three-pulse cross-correlation measurement scheme. (b) Electron energy spectrogram of the coupling between the "original" electron pulse and the two "gating" visible pulses. (c) Cross-correlation temporal profile (50 fs) extracted from the spectrogram. (d) Retrieved temporal profile of the gated electron from (c) with a 30 fs duration FWHM. Reproduced from ref. 63 with permission from Springer Nature, Copyright 2017.

spectrogram, which carries the signature of the coupling between the original electron pulse and the first visible laser pulse (which generates the gated electron pulse) and that between the gated electron pulse and the second, additional laser pulse (see Figure 14.12c). The cross-correlation temporal profile of the latter is 50 fs FWHM (see Figure 14.12d). Since the pulse duration of the second optical pulse was 30 fs, the gated electron pulse duration must be on the order of 30 fs, which is similar to the temporal profile of the (first optical) gating pulse. The generated "gated" electron pulse had sufficient electron counts ($\sim 8\%$ of the total electron counts, or <1 electron per pulse) for probing the ultrafast electron dynamics of matter. The achieved temporal resolution (30 fs) allowed us to image electron dynamics over a few tens of femtoseconds previously out of reach, such as those associated with the electron–electron scattering and electron–phonon coupling in semiconductors,[75] and those of surface plasmons.[76]

14.7 Attosecond Electron Microscopy, "Attomicroscopy"

In our group, we also exploited the attosecond optical gating to generate a single-isolated AEP inside the microscope to establish the field of attomicroscopy (see Figure 14.13); the OAP will be generated by the LFS and divided in the delay unit, as explained earlier. The first gating OAP output beam, together with the short electron pulse generated from the photocathode of the microscope, illuminates the sample. The gating OAP pulse is kept at a low power to avoid saturation. At the spatiotemporal overlap between the two pulses, gating takes place. In this coupling process, the inelastic interaction between the electron pulse and the gating OAP leads to the gain/loss of photon quanta by some electrons in the electron pulse, which can be energy-resolved by measuring an electron energy spectrum using an energy analyzer.

The availability of an electron energy spectrometer in the microscope would allow the measurement of the electron energy spectrum, consisting of discrete peaks that are spectrally separated by n multiples of photon energy ($n\hbar\omega$), where ω is the center frequency of the OAP on both sides of the zero-loss peak (see kinetic energy spectrum illustration in Figure 14.13). These discrete peaks correspond to electrons that have gained or lost energy only in the presence of the laser pulse (see the grey shaded area in the spectrum of Figure 14.13), while the energy analyzer inside the microscope filters out the gated electrons to obtain single-isolated AEPs. Such pulses will significantly improve the temporal resolution in UEM measurements, enabling the study of electron dynamics in matter triggered by another laser pulse[63] in different UEM modes (*i.e.*, diffraction, electron spectroscopy, or direct imaging). The gating medium should have a broad frequency response for such a broadband OAP. For example, a 30 nm standard aluminium thin film is an

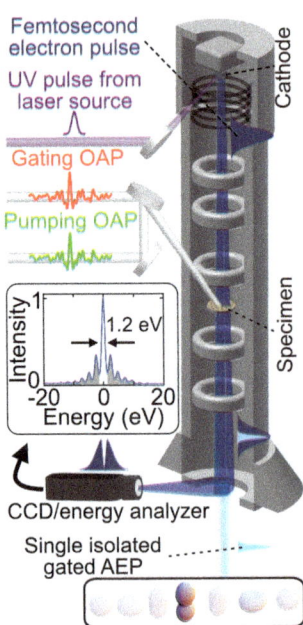

Figure 14.13 Illustration of the attomicroscopy principle and the attosecond electron imaging experiment. The optical attosecond pulse (OAP) provides the sub-femtosecond optical gating window to generate an isolated attosecond electron pulse from the original electron pulse inside the microscope. The optical gating and generation occur on the sample surface. A pumping OAP also photoexcites the sample to initiate reaction dynamics of interest. The ungated electron pulse travels towards an energy analyzer that can measure a TOF EELS spectrum (see the left side inset) with a potential energy resolution of <1.2 eV. While the energy analyzer can filter the gated electron pulse, providing a single isolated gated attosecond electron pulse (AEP) suitable for capturing electron motion in real-time with attosecond resolution (*e.g.*, such as changes in the electron orbital densities shown in the bottom inset). Reproduced from ref. 1 with permission from IOP Publishing Ltd, Copyright 2018.

excellent candidate. The potential sample to be studied will be located underneath the thin aluminium film. In this way, the generated AEP will not suffer from temporal broadening due to the space-charge effect caused by electron pulse propagation (based on calculations, a 200 keV generated AEP can propagate a few millimetres before it loses its attosecond resolution).

In a potential attomicroscopy electron imaging experiment, the OAP gating beam is kept at the spatiotemporal overlap to gate the electrons and generate the AEPs, as explained above. Another OAP beam (the pumping OAP) from the delay unit will be directed to the microscope, as shown in (see Figure 14.13). The pumping OAP will trigger the electron dynamics of the sample under study. Then, a set of images (electron spectra) at different time delays between the gating OAP and pumping OAP is recorded. This delay between the two OAPs can be controlled by the high-precision piezo stage implemented in the delay unit. This piezo stage provides a delay step size on the order of 100 as. The recorded images at different pump–probe delays can be combined to generate a movie of the ensuing electron dynamics in real time and space.

14.8   Outlook

The gating and taming of free-electron pulses by ultrashort laser pulses open the door to significant enhancement in the temporal resolution of electron imaging (diffraction and microscopy) tools. Utilising attosecond laser pulses for gating electrons can provide attosecond temporal resolution in UED and UEM imaging to establish the fields of AED and attomicroscopy. These pivotal tools will allow imaging and recording movies of electron motion in action, providing greater insight into the electron and quantum world in real-time and space. Optical gating using synthesised OAPs based on LFS[73,77] is challenging as it requires unique and complex knowledge and setups. Therefore, other approaches that can generate a gating laser pulse with more straightforward implementation are sought after for use in tabletop and accelerator facilities, which the Hassan group is currently developing. Therefore, in a few years, we expect these advancements to be utilised on larger scales, helping to establish the attosecond electron imaging field. Moreover, adopting this optical gating approach at electron accelerator facilities will provide powerful tools to explore structural dynamics at the smallest and fastest scales, promising to break new ground in different fields of science and technology.

Acknowledgements

The author acknowledges the financial support by the Gordon and Betty Moore Foundation grant (grant# GBMF7938) and Air Force Office of Scientific Research (grant# FA9550-19-1-0025). Moreover, the author is grateful for support from the W.M. Keck Foundation.

References

1. M. T. Hassan, *J. Phys. B: At., Mol. Opt. Phys.*, 2018, **51**, 032005.
2. A. Zewail, *The Chemical bond: structure and dynamics*, Elsevier, 1992.
3. J. Deisenhofer, O. Epp, K. Miki, R. Huber and H. Michel, *Nature*, 1985, **318**, 618–624.
4. A. H. Zewail, *Annu. Rev. Phys. Chem.*, 2006, **57**, 65–103.
5. A. P. Rood and J. Milledge, *J. Chem. Soc., Faraday Trans. 2*, 1984, **80**, 1145–1153.
6. J. D. Ewbank, W. L. Faust, J. Y. Luo, J. T. English, D. L. Monts, D. W. Paul, Q. Dou and L. Schäfer, *Rev. Sci. Instrum.*, 1992, **63**, 3352–3358.

7. V. A. Lobastov, J. D. Ewbank, L. Schäfer and A. A. Ischenko, *Rev. Sci. Instrum.*, 1998, **69**, 2633–2643.
8. S. Williamson, G. Mourou and J. C. M. Li, *MRS Online Proc. Libr.*, 1984, **35**, 87–96.
9. J. R. Rouxel, D. Keefer and S. Mukamel, *Struct. Dyn.*, 2021, **8**, 014101.
10. R. J. D. Miller, *Science*, 2014, **343**, 1108–1116.
11. C. Bressler and M. Chergui, *Chem. Rev.*, 2004, **104**, 1781–1812.
12. A. Rousse, C. Rischel, S. Fourmaux, I. Uschmann, S. Sebban, G. Grillon, P. Balcou, E. Förster, J.-P. Geindre and P. Audebert, *Nature*, 2001, **410**, 65–68.
13. A. Chin, R. Schoenlein, T. Glover, P. Balling, W. Leemans and C. Shank, *Phys. Rev. Lett.*, 1999, **83**, 336.
14. A. Lindenberg, I. Kang, S. L. Johnson, T. Missalla, P. Heimann, Z. Chang, J. Larsson, P. Bucksbaum, H. Kapteyn and H. Padmore, *Phys. Rev. Lett.*, 2000, **84**, 111.
15. D. von der Linde, *Science*, 2003, **302**, 1345–1346.
16. K. Sokolowski-Tinten, C. Blome, J. Blums, A. Cavalleri, C. Dietrich, A. Tarasevitch, I. Uschmann, E. Förster, M. Kammler and M. Horn-von-Hoegen, *Nature*, 2003, **422**, 287–289.
17. D. Oulianov, I. Tomov, A. Dvornikov and P. Rentzepis, *Proc. Natl. Acad. Sci. U. S. A.*, 2002, **99**, 12556–12561.
18. F. Raksi, K. R. Wilson, Z. Jiang, A. Ikhlef, C. Y. Côté and J. C. Kieffer, *J. Chem. Phys.*, 1996, **104**, 6066–6069.
19. J. R. Dwyer, C. T. Hebeisen, R. Ernstorfer, M. Harb, V. B. Deyirmenjian, R. E. Jordan and R. D. Miller, *Philos. Trans. R. Soc. London, Ser. A*, 2006, **364**, 741–778.
20. R. D. Miller, R. Ernstorfer, M. Harb, M. Gao, C. T. Hebeisen, H. Jean-Ruel, C. Lu, G. Moriena and G. Sciaini, *Acta Crystallogr., Sect. A: Found. Crystallogr.*, 2010, **66**, 137–156.
21. W. E. Spicer and A. Herrera-Gomez, Proceedings of SPIE–the International Society for Optical Engineering, 1993.
22. A. Egbert, B. Mader and B. Chichkov, *Laser Phys.*, 2002, **12**, 403–408.
23. M. Hagedorn, J. Kutzner, G. Tsilimis and H. Zacharias, *Appl. Phys. B: Lasers Opt.*, 2003, **77**, 49–57.
24. F. Zamponi, Z. Ansari, C. V. Korff Schmising, P. Rothhardt, N. Zhavoronkov, M. Woerner, T. Elsaesser, M. Bargheer, T. Trobitzsch-Ryll and M. Haschke, *Appl. Phys. A: Mater. Sci. Process.*, 2009, **96**, 51–58.
25. R. Henderson, *Q. Rev. Biophys.*, 1995, **28**, 171–193.
26. F. Qi, Z. Ma, L. Zhao, Y. Cheng, W. Jiang, C. Lu, T. Jiang, D. Qian, Z. Wang and W. Zhang, *Phys. Rev. Lett.*, 2020, **124**, 134803.
27. H. W. Kim, I. H. Baek, J. Shin, S. Park, H. S. Bark, K. Y. Oang, K.-H. Jang, K. Lee, N. Vinokurov and Y. U. Jeong, *Struct. Dyn.*, 2020, 7, 034301.
28. J. Yang, X. Zhu, T. J. Wolf, Z. Li, J. P. F. Nunes, R. Coffee, J. P. Cryan, M. Gühr, K. Hegazy and T. F. Heinz, *Science*, 2018, **361**, 64–67.
29. Y. Xiong, K. J. Wilkin and M. Centurion, *Phys. Rev. Res.*, 2020, **2**, 043064.
30. M. Eichberger, H. Schafer, M. Krumova, M. Beyer, J. Demsar, H. Berger, G. Moriena, G. Sciaini and R. J. D. Miller, *Nature*, 2010, **468**, 799–802.

31. B. W. Reed, *J. Appl. Phys.*, 2006, **100**, 034916.
32. T. van Oudheusden, P. Pasmans, S. van der Geer, M. de Loos, M. van der Wiel and O. Luiten, *Phys. Rev. Lett.*, 2010, **105**, 264801.
33. A. Michalik, E. Y. Sherman and J. Sipe, *J. Appl. Phys.*, 2008, **104**, 054905.
34. A. Michalik and J. Sipe, *J. Appl. Phys.*, 2009, **105**, 084913.
35. G. Sciaini and R. D. Miller, *Rep. Prog. Phys.*, 2011, **74**, 096101.
36. C. Kealhofer, W. Schneider, D. Ehberger, A. Ryabov, F. Krausz and P. Baum, *Science*, 2016, **352**, 429–433.
37. A. Gliserin, M. Walbran, F. Krausz and P. Baum, *Nat. Commun.*, 2015, **6**, 8723.
38. D. Ehberger, K. J. Mohler, T. Vasileiadis, R. Ernstorfer, L. Waldecker and P. Baum, *Phys. Rev. Appl.*, 2019, **11**, 024034.
39. M. R. Otto, L. P. René de Cotret, M. J. Stern and B. J. Siwick, *Struct. Dyn.*, 2017, **4**, 051101.
40. H. Ihee, V. A. Lobastov, U. M. Gomez, B. M. Goodson, R. Srinivasan, C.-Y. Ruan and A. H. Zewail, *Science*, 2001, **291**, 458–462.
41. B. J. Siwick, J. R. Dwyer, R. E. Jordan and R. D. Miller, *Science*, 2003, **302**, 1382–1385.
42. B. J. Siwick, J. R. Dwyer, R. E. Jordan and R. J. D. Miller, *Chem. Phys.*, 2004, **299**, 285–305.
43. R. Ernstorfer, M. Harb, C. T. Hebeisen, G. Sciaini, T. Dartigalongue and R. D. Miller, *Science*, 2009, **323**, 1033–1037.
44. M. Gao, C. Lu, H. Jean-Ruel, L. C. Liu, A. Marx, K. Onda, S.-Y. Koshihara, Y. Nakano, X. Shao and T. Hiramatsu, *Nature*, 2013, **496**, 343–346.
45. V. R. Morrison, R. P. Chatelain, K. L. Tiwari, A. Hendaoui, A. Bruhács, M. Chaker and B. J. Siwick, *Science*, 2014, **346**, 445–448.
46. G. Sciaini, M. Harb, S. G. Kruglik, T. Payer, C. T. Hebeisen, F.-J. M. Z. Heringdorf, M. Yamaguchi, M. H.-V. Hoegen, R. Ernstorfer and R. J. D. Miller, *Nature*, 2009, **458**, 56–59.
47. T. van Oudheusden, E. F. de Jong, S. B. van der Geer, W. P. E. M. Op 't Root, O. J. Luiten and B. J. Siwick, *J. Appl. Phys.*, 2007, **102**, 093501.
48. K. Amini and J. Biegert, in *Advances In Atomic, Molecular, and Optical Physics*, ed. L. F. Dimauro, H. Perrin and S. F. Yelin, Academic Press, 2020, vol. 69, pp. 163–231.
49. B. Wolter, M. G. Pullen, A.-T. Le, M. Baudisch, K. Doblhoff-Dier, A. Senftleben, M. Hemmer, C. D. Schröter, J. Ullrich, T. Pfeifer, R. Moshammer, S. Gräfe, O. Vendrell, C. D. Lin and J. Biegert, *Science*, 2016, **354**, 308–312.
50. S. Weathersby, G. Brown, M. Centurion, T. Chase, R. Coffee, J. Corbett, J. Eichner, J. Frisch, A. Fry and M. Gühr, *Rev. Sci. Instrum.*, 2015, **86**, 073702.
51. X. Shen, F. Ji, K. Michael, D. Luo, M.-F. Lin, A. Reid, S. Weathersby, J. Yang and X. Wang, *Microsc. Microanal.*, 2020, **26**, 208–209.
52. F. Qi, Z. Ma, L. Zhao, Y. Cheng, W. Jiang, C. Lu, T. Jiang, D. Qian, Z. Wang, W. Zhang, P. Zhu, X. Zou, W. Wan, D. Xiang and J. Zhang, *Phys. Rev. Lett.*, 2020, **124**, 134803.

53. M. Gao, C. Lu, H. Jean-Ruel, L. C. Liu, A. Marx, K. Onda, S.-Y. Koshihara, Y. Nakano, X. Shao, T. Hiramatsu, G. Saito, H. Yamochi, R. R. Cooney, G. Moriena, G. Sciaini and R. J. D. Miller, *Nature*, 2013, **496**, 343–346.

54. J. Yang, M. Guehr, X. Shen, R. Li, T. Vecchione, R. Coffee, J. Corbett, A. Fry, N. Hartmann and C. Hast, *Phys. Rev. Lett.*, 2016, **117**, 153002.

55. J. Cao, X. Wang and D. Zhong, *Science*, 2021, **374**, 34–35.

56. M.-F. Lin, N. Singh, S. Liang, M. Mo, J. P. F. Nunes, K. Ledbetter, J. Yang, M. Kozina, S. Weathersby, X. Shen, A. A. Cordones, T. J. A. Wolf, C. D. Pemmaraju, M. Ihme and X. J. Wang, *Science*, 2021, **374**, 92–95.

57. R. Srinivasan, J. S. Feenstra, S. T. Park, S. Xu and A. H. Zewail, *Science*, 2005, **307**, 558–563.

58. C.-Y. Ruan, V. A. Lobastov, F. Vigliotti, S. Chen and A. H. Zewail, *Science*, 2004, **304**, 80–84.

59. A. Kogar, A. Zong, P. E. Dolgirev, X. Shen, J. Straquadine, Y.-Q. Bie, X. Wang, T. Rohwer, I.-C. Tung and Y. Yang, *Nat. Phys.*, 2020, **16**, 159–163.

60. A. Zong, P. E. Dolgirev, A. Kogar, Y. Su, X. Shen, J. A. Straquadine, X. Wang, D. Luo, M. E. Kozina and A. H. Reid, *Phys. Rev. Lett.*, 2021, **127**, 227401.

61. N. Gedik, D.-S. Yang, G. Logvenov, I. Bozovic and A. H. Zewail, *Science*, 2007, **316**, 425–429.

62. M. T. Hassan, H. Liu, J. S. Baskin and A. H. Zewail, *Proc. Natl. Acad. Sci. U. S. A.*, 2015, **112**, 12944–12949.

63. M. T. Hassan, J. S. Baskin, B. Liao and A. H. Zewail, *Nat. Photonics*, 2017, **11**, 425–430.

64. M. T. Hassan, *Attomicroscopy: attosecond electron microscopy*, SPIE, 2018.

65. A. Feist, K. E. Echternkamp, J. Schauss, S. V. Yalunin, S. Schäfer and C. Ropers, *Nature*, 2015, **521**, 200–203.

66. K. E. Priebe, C. Rathje, S. V. Yalunin, T. Hohage, A. Feist, S. Schäfer and C. Ropers, *Nat. Photonics*, 2017, **11**, 793–797.

67. Y. Morimoto and P. Baum, *Nat. Phys.*, 2018, **14**, 252–256.

68. N. Schönenberger, A. Mittelbach, P. Yousefi, J. McNeur, U. Niedermayer and P. Hommelhoff, *Phys. Rev. Lett.*, 2019, **123**, 264803.

69. C. M. S. Sears, E. Colby, R. Ischebeck, C. McGuinness, J. Nelson, R. Noble, R. H. Siemann, J. Spencer, D. Walz, T. Plettner and R. L. Byer, *Phys. Rev. Spec. Top.–Accel. Beams*, 2008, **11**, 061301.

70. A. Ryabov, J. W. Thurner, D. Nabben, M. V. Tsarev and P. Baum, *Sci. Adv.*, 2020, **6**, eabb1393.

71. Y. Morimoto and P. Baum, *Phys. Rev. Lett.*, 2020, **125**, 193202.

72. M. T. Hassan, *Ultrafast Nonlinear Imaging and Spectroscopy VI*, 2018.

73. H. Alqattan, D. Hui, V. Pervak and M. T. Hassan, *APL Photonics*, 2022, **7**, 041301.

74. D. Hui, H. Alqattan, S. Yamada, V. Pervak, K. Yabana and M. T. Hassan, *Nat. Photonics*, 2022, **16**(1), 33–37.

75. F. Rossi and T. Kuhn, *Rev. Mod. Phys.*, 2002, **74**, 895.

76. M. I. Stockman, *Opt. Express*, 2011, **19**, 22029–22106.

77. M. T. Hassan, T. T. Luu, A. Moulet, O. Raskazovskaya, P. Zhokhov, M. Garg, N. Karpowicz, A. M. Zheltikov, V. Pervak, F. Krausz and E. Goulielmakis, *Nature*, 2016, **530**, 66–70.

CHAPTER 15

RF Cavity-based Ultrafast Transmission Electron Microscopy

S. BORRELLI, S. T. KEMPERS, P. H. A. MUTSAERS
AND O. J. LUITEN*

Eindhoven University of Technology, PO Box 513, 5600 MB, Eindhoven,
The Netherlands
*Email: o.j.luiten@tue.nl

15.1 Ultrafast Electron Microscopy: The Reason Why

Through the centuries, microscopy has arguably been the primary tool to probe the spatial structure of matter, helping to unveil nature's intricate mysteries. Conventional microscopy is an imaging technique that can capture static images of any material structure with an achievable spatial resolution depending on the probe used. Matter can be imaged on length scales in the micrometre regime using optical microscopes in which light acts as a probe. Electron microscopes possess a greater resolving power and can probe matter on the nanometre and even sub-nanometre scales. State-of-the-art electron microscopes can achieve a spatial resolution better than 50 pm,[1] which is less than one-tenth of the size of the hydrogen atom. It is worth noting that this resolution is approximately only an order of magnitude away from the fundamental diffraction limit associated with the quantum mechanical de Broglie wavelength of electrons (2 pm at 300 keV kinetic energy). Although electron microscopes are capable of impressively high spatial resolution, they can only provide snapshots of the microscopic

Theoretical and Computational Chemistry Series No. 25
Structural Dynamics with X-ray and Electron Scattering
Edited by Kasra Amini, Arnaud Rouzée and Marc J. J. Vrakking
Published by the Royal Society of Chemistry, www.rsc.org

world frozen in time. The required exposure time, typically milliseconds or longer, restricts conventional electron imaging techniques almost exclusively to equilibrium systems.

The goal of generating a "movie" of the process under investigation by adding the temporal dimension to images has been a long sought-after milestone. The human eye cannot resolve an action evolving on a timescale faster than a few tens of milliseconds.[2] The desire to study motion on shorter timescales prompted the development of chronophotography, and an exemplary case study is the first motion picture, The Horse in Motion (1878), by Muybridge. Before this work, capturing details of the animal's locomotion with the human eye was impossible. Through chronophotography with 1 ms resolution, Muybridge revealed that a galloping horse had all four hooves off the ground, which gathered beneath the body at regular intervals. Marey's chronophotographic gun (1882), an instrument capable of taking 12 consecutive frames per second and recording all frames on the same picture, can be considered the earliest device capable of capturing and displaying moving images at their own dynamic time. Nowadays, the corresponding scientific challenge consists of shooting movies manifesting the motion of atoms in molecules, the structural changes of nano-materials, the dynamics underlying fast processes in biological systems, chemical reactions, or phase transitions in condensed matter. All these processes proceed on a picosecond (ps; $1 \text{ ps} = 10^{-12}$ s) or sub-ps time scale, which is many orders of magnitude smaller than the time resolution of conventional electron microscopy. By contrast, optical microscopy can provide the required temporal resolution;[3–5] however, it has a limited spatial resolution which is constrained by the wavelength of the employed light, typically in the range of 200–800 nm. The ideal technique should exhibit electron microscopy's sub-atomic picometre spatial resolution and atomic-scale femtosecond (fs; $1 \text{ fs} = 10^{-15}$ s) temporal resolution of optical imaging.

Electron microscopy was revolutionised in 2005 when Zewail and co-workers at the California Institute of Technology reported experiments using a modified transmission electron microscope (TEM) capable of directly observing the structural dynamics of matter on sub-picosecond timescales.[6] The development of Zewail's 4D imaging technique marked the birth of ultrafast electron microscopy (UEM), a method capable of adding the temporal dimension to sub-atomic-resolution electron imaging. It is also worth noting that the pioneering works of Bostanjoglo at Technical University Berlin have been essential in achieving time-resolved transmission electron microscopy.[7–9]

To image material structures whose properties vary over time, short electron pulses are required to probe the specimen synchronously with the structural changes. At the arrival instants of the electron pulses, the measured electron image of the sample provides a snapshot of the material's structure under investigation. Caltech scientists realised time-resolved electron microscopy (see Figure 15.1) by incorporating a femtosecond

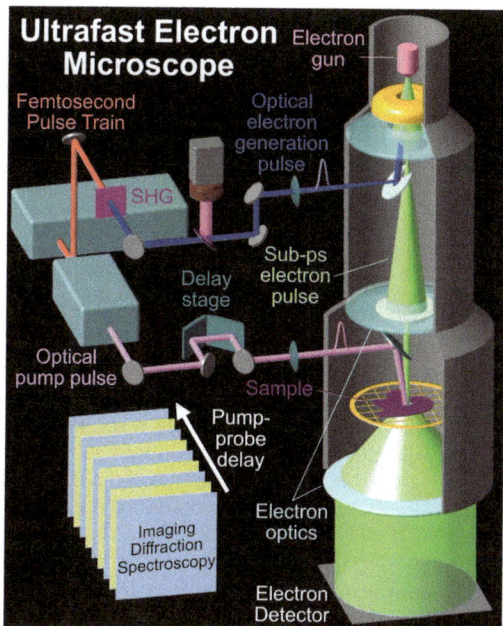

Figure 15.1 Schematic of the first ultrafast TEM developed at the California Institute of Technology by Zewail and co-workers. Picture shows the integration of a femtosecond laser pulse into a conventional TEM's electron gun and column. Part of the fs-laser beam generates on-axis ultrashort electron pulses by photoemission. A second part of the optical beam excites a specimen in the TEM's sample plane.[6] Reproduced from ref. 6 with permission from The National Academy of Sciences of the USA, Copyright 2005.

laser in the electron gun of a TEM to generate sub-ps electron pulses by photoemission. Part of the same pulsed laser radiation was used to excite a specimen inside the electron microscope column and induce changes in its structure occurring at the laser frequency. The photoelectron pulses probed the spatial–temporal structure of the sample with high resolution. A delay stage enabled adjusting the relative arrival time of the optical-pump and electron-probe pulses at the specimen. The concept behind the stroboscopic pump–probe experiments pioneered by Zewail forms the basis of all current ultra-high-resolution experimental studies on matter dynamics.

One of the first sequences of snapshots with ultrashort temporal resolution, obtained using UEM,[10] provided a movie of a laser-induced phase transition in vanadium dioxide (see Figure 15.2). The UEM real-space images of the metal-to-insulator phase transition (see Figure 15.2a) with nanometre resolution were captured before (left) and after (right) the phase transition. The corresponding diffraction images with atomic-scale resolution (see Figure 15.2b) were also obtained before (M) and after (R) the phase transition. It is worth noting that the timescale of the measured phase

(a)   **(b)**

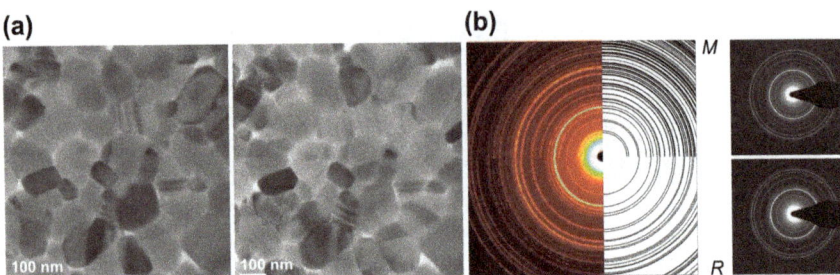

Figure 15.2   Ultrafast transmission electron microscopy images of laser-induced metal-to-insulator (MIT) phase transition in vanadium dioxide. (a) Real-space images with nanometre resolution captured before (left) and after (right) the MIT phase transition. (b) On the right, diffraction patterns with atomic-scale resolution obtained before (M) and after (R) the MIT phase transition. On the left, the composite diffraction pattern of the two phases experimentally observed (red) and theoretically evaluated (black).[10] Reproduced from ref. 10 with permission from The National Academy of Sciences of the USA, Copyright 2006.

transition spanned approximately 3 ps, a timescale that could not be probed before the development of UEM.

15.1.1   State-of-the-art in Ultrafast Electron Microscopy

Following the seminal works of Zewail and Bostanjoglo, the groups at Caltech,[11] Technical University Berlin,[12] University of Toronto[13] and Lawrence Livermore National Laboratory[14–16] firmly established the feasibility of laser-pump electron-probe studies in electron microscopy. In the last fifteen years (2006–2021), the scientific community has made enormous progress in advancing the performances of ultrafast TEMs. In this context, particular attention has been devoted to improving the electron beam's figure-of-merit and space-time-energy limits towards Ångström-fs-meV resolutions.[17–23] As a result of this development, transient structures that appear during the ultrafast dynamics of materials can currently be probed on a micrometre, nanometre, and even Ångström (Å) scale together with up to femtosecond temporal resolution in a growing number of laboratories. Two notable examples of UEM applications are the first UEM application in plasma lensing[24] and the femtosecond imaging of single-phonon transport in crystalline Ge and WSe$_2$ by fs-electron microscopy.[25]

The generation and propagation of ultrashort electron pulses are at the core of time-resolved electron microscopy. The ability to achieve high spatial and temporal resolutions requires the generation of electron pulses with extremely high brightness and ultrashort duration while preserving these properties up to the interaction point with the sample. Furthermore, to fully harness the imaging and spectroscopy capabilities of ultrafast TEMs, it is advisable to generate electron pulses with low energy spread at a high repetition rate in combination with a number of particles per pulse that satisfy the requirements of each specific experiment.

There are two leading modes of operation in time-resolved UEM, both based on pump–probe schemes but addressing different spatial–temporal scales: (i) the fs-stroboscopic mode, which is directly derived from the original approach of Zewail and uses stroboscopic illumination with electron pulses containing on average one electron and (ii) the single shot operation mode using a single pulse containing more than 10^8 electrons to acquire an image of the sample in a single exposure. The stroboscopic mode enables accessing a timescale as low as a few hundred femtoseconds.[26] However, it is only applicable to reproducible processes induced in samples that recover during the time interval between two successive laser pulses (typically nanoseconds to milliseconds). The single shot mode can only reach nanosecond (ns) resolution, despite efforts to push this limit down to ~ 10 ps without degrading the spatial resolution.[27,28] Thus, the single-mode operation is well suited for probing the dynamics of irreproducible processes on a material that does not recover on this timescale.

In the fs-stroboscopic single-electron mode, the spatial resolution limit is not different from that in a conventional TEM, and this limit is predominantly determined by the brightness of the electron source and lens quality.[29] At the same time, the temporal resolution is determined by the electron pulse duration, which is mainly established by the pulses' generation process. Presently, fs-stroboscopic mode UEM has a few-100 fs temporal resolution, a spatial resolution as low as 2 Å in real-space imaging, and an energy resolution of ~ 1 eV, which is the intrinsic TEM energy resolution.[30–34]

Even though the stroboscopic single-electron pulse mode allows for resolving 100 fs- and Å-scale dynamics, it suffers from a low number of electrons per pulse. On the other hand, in multi-electron pulses, space-charge effects (*i.e.*, electron–electron repulsion occurring at the photoelectron source and during pulse propagation from the source to the specimen) induce a broadening of the electron beam in the longitudinal (temporal) and transverse (imaging) directions. These effects must be circumvented to avoid limiting the spatial and temporal resolutions, which could occur *via* the degradation of coherence lengths.[35–38] For this reason, temporal and spatial resolutions better than ~ 10 ns and sub-10 nm have not yet been experimentally shown in the single-shot mode, even though it is theoretically possible.[26] Many approaches have been proposed to overcome the limitations on the number of electrons per pulse imposed by space-charge effects. Among these, it is worth referencing photoelectron-pulse compression schemes, for example, using synchronised electromagnetic fields of RF cavities,[27,39,40] and operation at ultra-relativistic electron energies (*i.e.*, megavolt accelerating voltages instead of kilovolt).[41,42] In particular, already in 2010, the Coherence and Quantum Technology (CQT) group at Eindhoven University of Technology (TU/e) demonstrated electron pulse compression of more than two orders of magnitude down to 100 fs pulse duration employing a miniaturised 3 GHz RF cavity operating in the

010 transverse magnetic (TM$_{010}$) mode (see Section 15.2.1 for a detailed discussion).[43] Integrating such a cavity into an ultrafast TEM allows operating the instrument with a higher electron pulse intensity without compromising the temporal resolution.

The UEM community is also working on techniques enabling trade-offs between energy and time resolutions on one side and beam intensity on the other, tailored to each specific experiment requirement. The ultimate goal is to establish an ultrafast TEM equipped with a gun whose parameters can be changed to switch from highly populated electron pulses and short exposure times to nearly single-electron pulses and ultrashort temporal resolution but longer exposure times. Different research groups have demonstrated that this is possible in ultrafast electron microscopes based on thermionic guns equipped with a Wehnelt electrode. They proved that by simply changing the gun parameters (*e.g.*, cathode–Wehnelt gap and the bias applied to the Wehnelt), it is possible to generate electron pulses with a population spanning over many orders of magnitude, from 1 to about 10^5 electrons, while still achieving high time-energy resolution. The attained electron pulse duration ranges from ~ 100 fs in the single electron regime to 1–10 ps in the highly populated pulse regime.[17,44,45]

Currently, the ultrafast electron microscopes employed in most laboratories worldwide rely on the photoelectric generation of electron pulses from a metal or ceramic photocathode, typically with UV femtosecond laser pulses. This generation process enables achieving extremely high but intrinsically limited temporal resolution. The latter is ultimately constrained by the combination of the laser pulse length, the photoemission process duration, the velocity dispersion associated with the finite energy spread, and the Coulomb forces between the generated electrons. Chong-Yu Ruan *et al.* at Michigan State University incorporated an RF cavity into a photoemission-based ultrafast electron microscope to reduce the electron pulse duration *via* phase-space manipulation. The proposed setup enables temporal resolution improvement without sacrificing the pulse population. Ruan and co-workers achieved a spatial resolution of ~ 10 nm and a temporal resolution of about 100 fs with up to 10^5 electrons per pulse at a 1 kHz repetition rate.[46] Using (RF or THz) compression techniques, the fundamental limitation is the longitudinal emittance and the allowed energy spread.[40,47] Furthermore, an RF cavity requires synchronising the compression field to the photo-injection laser, as discussed in the following sections. More generally, simply using shorter fs-laser pulses is insufficient to significantly improve the UEM temporal resolution in the single-electron regime. Electron generation by photoemission also suffers from other limitations. It requires expensive laser systems and intrusive modification of the electron gun. Moreover, the average current inside the microscope is limited by the repetition rate of the fs-laser employed (typically in the 1 kHz to the few-MHz range). Finally, the peak brightness of the generated electron pulse and the ensuing achievable spatial resolution are limited by the relatively large area of a flat photocathode.

The groups at Stanford University, the University of Göttingen, and the Max Planck Institute in Erlangen have devised a novel photoemission approach based on sideways illumination of a nanoscale field emitter cathode.[22,48,49] The Ropers' group at the University of Göttingen was the first to show that this approach enables the generation of ultrashort electron pulses with a peak brightness comparable to a state-of-the-art continuous field emission gun (FEG), thus overcoming the spatial resolution limitations mentioned above. It also allows for continuous tuning of the temporal structure of the electron beam from a continuous to a fs pulsed mode. In pulsed mode operation, they achieved a duration of 200 fs for a beam diameter of 9 Å and an energy spread of 0.6 eV.[50]

An alternative to the photoelectric generation of electron pulses is incorporating a beam blanker (also known as a beam chopper) into the TEM column to realise pulsed operation.[51] Blanking requires deflection of a continuous beam across an aperture, and it has the essential advantages that (i) it does not necessarily require high-power, ultrafast photoemission lasers; (ii) it can be combined with state-of-the-art high-brightness continuous electron sources; and (iii) it allows to quickly change from pulsed to continuous operation and *vice versa*. Beam blanking is typically achieved by fast switching of $\sim$MV m^{-1} transverse electric fields. Electronic switching limits the switching time to $\sim$ns. Ultrafast laser-triggered photoconductive switches, the so-called Auston switches, are employed to realise the high slew rates required for sub-ps temporal resolution.[52] Ultrafast electron streak cameras for ultrafast electron diffraction have been developed using this technology.[53] An interesting approach is being pursued by the TU Delft group, which employs MEMS technology to implement a miniaturised Auston-switch-based beam blanker into an existing electron microscope.[54] Although theoretically promising, ps temporal resolution has not yet been achieved. Moreover, the technique is still dependent on ultrafast lasers.

15.1.2 Ultrafast Electron Microscopy with Microwave Cavities

To overcome the limitations imposed by the photoelectric generation of pulsed electron beams, the TU/e group proposed the generation of ultrashort electron pulses by chopping a continuous electron beam using an RF cavity-based fast blanker. In collaboration with Thermo Fisher Scientific, the group designed and developed an RF cavity-based ultrafast TEM, which produces electron pulses of $\sim$100 fs duration at a high repetition rate (3 GHz or 75 MHz).[55] This chapter focuses on RF cavity-based ultrafast transmission electron microscopy. Section 15.2 discusses the use of microwave cavities for UEM and, in particular, describes the operating principle of the TU/e RF cavity-based ultrafast electron microscope. Section 15.3 gives an overview of the applications enabled by the TU/e microscope, highlighting the main advantages of the method and the entailed demands. Section 15.4 depicts the challenges and limitations of the technique and the possible future advancements and applications.

15.2 RF Cavity-based Ultrafast Transmission Electron Microscopy

In 1973, it was already suggested to employ the strong and rapidly oscillating electromagnetic fields in RF cavities to realise ultrafast electron pulses.[56,57] Such pulses were subsequently applied for stroboscopic measurements of fast processes in an electronic circuit.[58] The CQT group at TU/e was the first to recognise the possibility of using resonant microwave cavities synchronised to a fs-laser for ultrafast electron beam manipulation in UEM. The group demonstrated that a 3 GHz RF cavity operating in the TM_{110} mode could be used as an ultrafast beam blanker to chop the continuous electron beam of a conventional TEM into pulses with a few-100 fs duration.[55] They showed that this ultrafast blanking could be performed without affecting the peak brightness and the energy spread of the continuous beam produced by a Schottky FEG.[59] In 2018, the TU/e group successfully developed and commissioned an RF cavity-based ultrafast transmission electron microscope (UTEM) by equipping a commercial FEI Tecnai TF20 system with a 3 GHz deflection cavity operating in the TM_{110} mode and a small aperture. Synchronising the RF phase of the electromagnetic fields inside the cavity to a fs-laser enables pump–probe experiments while simultaneously avoiding the need for powerful photoemission lasers to generate the electron-probe pulses.[60]

Such a design overcomes the limitations imposed by the photoemission process on the duration of the electron pulses. In principle, the pulse length can be made almost arbitrarily short since it is only limited by the combination of the chopping field's strength, the beam diameter, and the chopping aperture size (see Section 15.2.1). Additionally, chopping is independent of the continuous source, so further development in high-brightness source technology will immediately be available to UEM. Furthermore, the TU/e UTEM provides a GHz repetition rate of the pulses, two-to-three orders of magnitude higher than that achievable by photoemission, while at the same time guaranteeing the same high source brightness and average current. However, such a high repetition rate is not desirable for typical pump–probe experiments. It would require a pump laser with the same high pulse repetition rate, which is not trivial, if possible. Moreover, studying dynamic processes with a stroboscopic method requires a period between probing electron pulses longer than the relaxation time of the process under study. For these reasons, the RF chopping cavity can also be operated in the so-called dual-mode. As explained in more detail in Section 15.2.1.2, a dual-mode TM_{110} cavity supports two orthogonal resonant modes with slightly different resonance frequencies and a frequency ratio that is a rational number. As a result, the RF cavity deflects the electron beam in a Lissajous pattern (see Figure 15.4), which passes the chopping aperture at a reduced frequency. This technique allows producing a pulsed electron beam running at the same repetition rate as the laser pulse train generated by the mode-locked laser of choice, typically in the range of a few tens-to-hundreds MHz.[23] In the TU/e lab, the dual-mode TM_{110} cavity runs at frequencies of 3.000 GHz and 3.075 GHz, respectively, with the RF phases locked to a 75 MHz Ti:Sapphire laser oscillator. Interestingly, the interchangeability of the cavity's operation modes between 3 GHz and 75 MHz presents a compelling opportunity to modify the period of the electron pulses from 333.3 ps to 13.3 ns. This feature allows for investigating phenomena occurring on different timescales using a single instrument.

15.2.1 TM_{110} RF Cavities for Ultrafast Electron Microscopy

Over the years, scientists extensively used microwave cavities for electron beam manipulation due to their capability of generating and storing high-amplitude electromagnetic fields with relatively low input power by resonant enhancement. Different electromagnetic field (modes) configurations can be excited inside a microwave cavity. The TM_{010} and TM_{110} cavity modes are the most relevant for electron microscopy applications. A TM_{010} mode sustains an oscillating electric field pointing along the cavity axis, which can induce a time-dependent acceleration or deceleration of the electrons. RF cavities in the TM_{010} mode are used in both ultrafast electron diffraction[43] and ultrafast electron microscopy[46] to compress electron pulses, thus improving the temporal resolution. In contrast, a TM_{110} mode supports an on-axis transverse magnetic field, which can deflect the electron beam off-axis periodically. Hence, cavities operating in this mode are primarily used as streak cameras for pulse length measurements,[43] time-resolved diffraction studies,[61] and time-of-flight electron energy loss spectroscopy.[62]

The TM_{110} microwave cavity incorporated in the TU/e RF cavity-based UTEM is a pillbox cavity with cylindrical geometry. In what follows, the z-axis of the Cartesian coordinate system indicates the direction of the central axis of the cavity, while the x- and y-direction match the transverse axes. In the proximity of the cavity's geometrical axis (*i.e.*, for $k\rho \ll 1$, with $k = \omega/c$ denoting the wavenumber, ω representing the cavity's resonance frequency, and $\rho = \sqrt{x^2 + y^2}$ denoting the radial distance from the cavity's axis), the electromagnetic field distribution inside the TM_{110} pillbox cavity is given as

$$E_z = \omega B_0 x \cos(\omega t + \phi_0), \tag{15.1}$$

$$B_x = 0, \tag{15.2}$$

$$B_y = B_0 \sin(\omega t + \phi_0), \tag{15.3}$$

where $B_0 = \dfrac{E_0}{2c}$ is the amplitude of the on-axis magnetic field and ϕ_0 is the phase of the electromagnetic field experienced by the electron when it enters the cavity at $t = 0$. Hence, an electron traveling along the cavity's geometrical axis $(x = y = 0)$ experiences only a transverse magnetic field oscillating at the cavity's resonance frequency.[†] The corresponding Lorentz force $F_x = -q_e\nu_0 B_{y,0}(z)\sin(\omega t + \phi_0)$ periodically diverts the particle in the transverse direction. Here, ν_0 is the longitudinal velocity of the electron at the entrance of the cavity. Consequently, when an electron beam with a spot size much smaller than the cavity's cross-section moves along the cavity's geometrical axis, it is transversely deflected at the resonance frequency ω. The continuous electron beam diverted by the cavity can be chopped into ultrashort pulses using a small aperture of diameter s located at a distance l from the cavity's centre and aligned with its geometrical axis (see Figure 15.3).

Assuming a top-hat profile of the magnetic field amplitude as a function of z and the absence of fringe fields around the cavity's apertures, the deflection angle at which an electron exits the cavity is given as

$$\alpha = \frac{\omega_c}{\omega}(\sin(\phi_0) - \sin(\phi_0 + \Lambda)). \tag{15.4}$$

Here, ω_c is defined as $\omega_c \equiv \dfrac{q_e B_0}{\gamma_0 m_e}$, where γ_0, m_e, q_e are the electron's Lorentz factor, mass, and charge, respectively. The parameter Λ is the dimensionless cavity length parameter that is defined as $\Lambda \equiv \dfrac{\omega L}{\nu_0}$, with L equal to the cavity length. Under the assumption of a negligible transverse offset, electrons pass through the aperture only if they experience an RF-phase ϕ_0 such that the deflection angle is in the range $|\alpha(\phi_0)| < s/2l$. As a result, an ideal continuous electron beam is thus chopped into temporally uniform pulses at a repetition rate equal to twice the cavity's resonance frequency $f_r = \dfrac{\omega}{2\pi}$. The duration τ_e of each electron pulse is determined from the specific range of phases $\Delta\phi_0$ defining the aperture's acceptance, as expressed by the formula

$$\tau_e = \frac{\Delta\phi_0}{\omega} = \frac{\gamma_0 m_e s}{2l q_e B_0 \sin(\Lambda/2)}. \tag{15.5}$$

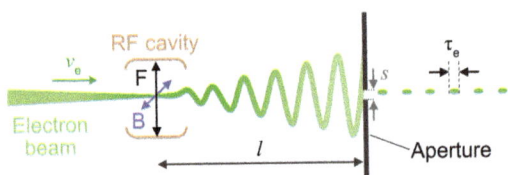

Figure 15.3   Principle of beam chopping. An RF cavity periodically deflects the continuous electron beam towards an aperture. Consequently, ultrashort electron pulses are created at twice the cavity's resonance frequency.[23]

[†]The longitudinal electric field acting on electrons traveling along the cavity axis is zero. However, electrons slightly off-axis will also experience a longitudinal accelerating force due to the longitudinal electric field. The effects due to the transverse broadening of the electron pulses will be neglected here. Details are provided in ref. 59.

According to this expression, choosing the cavity length parameter as $\Lambda = \pi$, or equivalently the cavity's length as $L = \dfrac{\pi v_0}{\omega}$, guarantees a minimum pulse duration. Correspondingly, the transit time of the electrons inside the cavity indeed equals half an RF-field oscillation period.

Preserving the high brightness and small energy spread of the conventional Schottky FEG of a TEM is one of the main advantages of cavity-based generation of electron pulses. Scientists at TU/e demonstrated that the beam quality is preserved for a wide range of ϕ_0 and z_0 values by focusing the electron beam to a cross-over at $z = z_0$ inside the cavity. The TU/e group has been the first to apply this scheme, first developed in light optics and called conjugate blanking, to beam quality preservation in RF cavity-based ultrafast electron microscopy. In the conjugate blanking mode, the normalised beam emittance ε_n just behind the cavity matches its initial value. Under the specific conditions providing the shortest electron pulses (*i.e.*, RF-phase equal to $\phi_0 = \dfrac{1}{2}(\pi - \Lambda)$ and cavity length equal to $L = \dfrac{\pi v_0}{\omega}$), the transverse normalised emittance is preserved if the beam is focused precisely at the centre of the cavity at $z_0 = \dfrac{L}{2}$. The energy spread σ_u is another crucial parameter that determines the quality of the beam. The electron beam in reality has a non-negligible transverse spread in spot size. Therefore, while moving through the cavity, off-axis electrons experience the longitudinal electric field of the TM_{110} mode (see eqn (15.1)–(15.3)). This field causes a variation in the total energy of the beam and an increase in the energy spread. The TU/e group has shown that the conjugate blanking scheme also minimises the increase in energy spread. Hence, the TEM's field-emission gun performance is fully preserved in the pulsed mode, provided that the conjugate blanking scheme is applied. Details are provided in ref. 59 and 63.

Focusing the electron beam to the centre of the cavity leads to an increase in its spot size at the aperture location. Consequently, the pulse length cannot be determined from eqn (15.5), which is valid under the assumption that beam spot size is negligible compared to the aperture cross-section. More generally, the temporal profile of the electron pulses is proportional to the convolution of the aperture's profile and the approximately top-hat distribution of the electron beam at the aperture location. The correct expression for the pulse length (FWHM) is given by

$$\tau = \frac{\sqrt{2}\gamma_0 m_e \sigma_{x'}}{q_e B_0 \sin(\Lambda/2)} = \frac{\gamma_0 m_e (s+w)}{4 q_e l B_0 \sin(\Lambda/2)}, \tag{15.6}$$

where $\sigma_{x'}$ is the root mean square (RMS) divergence angle of the beam, which is related to the beam width w at the location of the chopping aperture, the aperture width s, and the distance l between the cavity and the aperture.[64] Generating short pulses while focusing into the middle of the cavity thus requires the beam's divergence angle to be as small as possible, which can be achieved by selecting a chopping aperture with a small diameter.[‡] From the previous expression, it is clear that the duration of the generated pulses can be almost arbitrarily small since the combination of the strength of the magnetic field inside the cavity and the ratio between the chopping aperture and the beam diameter only limits it.

15.2.1.1 Design of a TM_{110} Cavity for UEM

In principle, when designing a cavity with the sole purpose of employing it as an ultrafast beam blanker to chop a continuous electron beam, no constraints apply to the cavity size, shape, and resonance frequency. However, incorporating a cavity as an electron-optical element inside a TEM requires two main constraints to be overcome. The cavity must be compact and the power consumption must be modest. The TU/e group selected an S-band pillbox cavity with a resonance frequency of 2.9985 GHz (~3 GHz) operating in the TM_{110} mode as a starting point for the design. The radius of a TM_{110} cavity at a given resonance frequency ω_0 is given by

$$a = \frac{\chi_{11}}{k} = \frac{3.8317}{\sqrt{\varepsilon\mu\omega_0}}, \qquad (15.7)$$

where k is the wavenumber of the electromagnetic waves inside the cavity, ε and μ are the absolute permittivity and permeability of the medium filling the cavity, respectively, and χ_{11} is the first zero of the cylindrical Bessel function of the first type.[65] Based on this expression, a vacuum cavity operating at 3 GHz has a radius of 60.9 mm, which is too big to fit into the column of a conventional TEM. However, filling the cavity with a high-permittivity dielectric material reduces the cavity radius by a factor of $\sqrt{\varepsilon_r}$ at the same resonance frequency. A systematic study by the TU/e group proved that the ceramic material $ZrTiO_4$ doped with less than 20% $SnTiO_4$ is a good candidate for developing a compact dielectric-filled cavity. Its dielectric permittivity is $\varepsilon_r = 36.5$–38 in the 0.3–24 GHz frequency regime,[66] thus resulting in a cavity radius a of ~10 mm at the target resonance frequency of 3 GHz, which is a factor six lower than the radius of an empty cavity.[65] Therefore, the dielectric material guarantees the reduction of the cavity size required to fit into the confined space of a transmission electron microscope.

Filling a cavity with a dielectric material also leads to a decrease in the power consumption. The power loss for a dielectric-filled cavity consists of two contributions. The first is the ohmic dissipation in the non-perfectly conducting cavity walls, common to all cavities. The second is the power absorption into the dielectric material itself. Even if the $SnTiO_4$ doped

---

[‡] The electron beam's divergence angle is determined by the strength of the second lens of the TEM's condenser system (the C2 lens) and by the diameter of the associated C2-aperture. Eqn (15.6) holds for a cavity located between the C2-aperture and the chopping aperture, and equal C2- and chopping aperture diameters.

$ZrTiO_4$ dielectric yields an additional power loss, reducing the cavity radius due to the dielectric itself implies a reduction in the ohmic losses. The sum of the ohmic and dielectric power losses in a dielectric-filled cavity is thus significantly lower than the sole ohmic losses in a vacuum cavity at the same resonance frequency.[65] Verhoeven *et al.* optimised the design to reduce the power consumption in the partially-filled cavity even more. As a comparison, the optimised partially-filled cavity requires an input power of 14.2 ± 0.2 W to generate a field strength of 2.84 ± 0.07 mT. In contrast, a vacuum pillbox equivalent would require about 393 W.[65]

To conclude, due to the low input powers and small dimensions, partially-filled cavities with $ZrTiO_4$ dielectric materials are well-suited as electron-optical elements for transmission electron microscopy.

15.2.1.2   Design of a TM_{110} Cavity for UEM

TU/e scientists proved that a 3 GHz RF cavity operating in the TM_{110} mode combined with a small aperture (diameter of a few 10 µm) enables the generation of electron pulses with ~ 100 fs duration at a repetition rate of twice the cavity's resonance frequency.[59] Although such a high repetition rate offers unique imagining possibilities (see Section 15.3.1), it is not advisable for typical pump–probe experiments, as discussed earlier in the introduction of this section. A repetition rate in the 100 MHz regime would be best-suited. However, because the cavity's resonance frequency is inversely proportional to its radius, a ~ 100 MHz cavity would be far too large to fit in the confined space of an electron microscope column. To overcome this limitation, Lassise *et al.* at TU/e proposed to drive the cavity with two orthogonal TM_{110} modes of different resonance frequencies.[67]

Transverse magnetic modes with different orientations are degenerate in a circular cylindrical cavity. However, a slightly elliptical geometry allows to break the rotational symmetry and excite two orthogonal modes at different but well-defined frequencies. The electromagnetic field configuration of a so-called dual-mode cavity induces a bi-dimensional deflection of the continuous electron beam following a Lissajous pattern. When the two modes are driven at two different higher harmonics of a fundamental frequency ν_0, the Lissajous pattern is traced at a frequency equal to the difference between the two harmonics. In the TU/e dual-mode cavity, the microwave signals that drive the two orthogonal modes are generated from the 40th and 41st harmonics of the same 75 MHz Ti:Sapphire laser oscillator. The two modes with resonance frequencies of $\nu_1 = 3$ GHz and $\nu_2 = 3.075$ GHz are thereby accurately phase-locked to the laser beam and each other. As a result, the continuous electron beam is transversely deflected into a Lissajous pattern, traced at the same frequency $(\nu_2 - \nu_1 = 75$ MHz$)$ as the laser oscillator. Figure 15.4 shows an image of the Lissajous pattern focused on the fluorescent screen located downstream of the dual-mode cavity in the TU/e UTEM. It has been generated by independently driving the two phase-locked signals while feeding both modes with a low input power of about 1 W. The pattern features

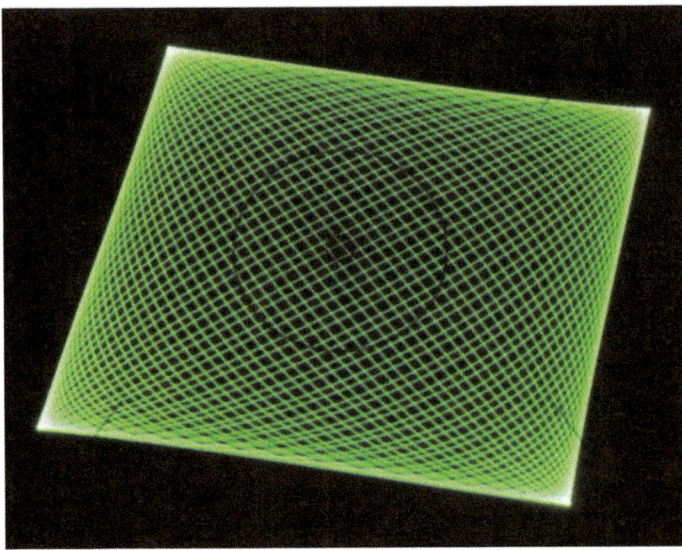

Figure 15.4   Photograph of the Lissajous figure on the fluorescent screen of the TU/e RF cavity-based UTEM. The figure has been generated by driving the dual-mode cavity in the microscope with two phase-locked signals of frequencies $\nu_1 = 3$ GHz and $\nu_2 = 3.075$ GHz.[64] Reproduced from ref. 23 with permission from AIP Publishing, Copyright 2018.

distinguishable lines, thus proving that both modes of the dual-mode cavity can be phase-locked correctly.

An aperture downstream of the dual-mode cavity and centred on a single line of the Lissajous pattern allows for generating ultrashort (~ 100 fs) electron pulses at a 75 MHz repetition rate. Dual-mode cavity-generated electron trains thus enable the temporary resolution of dynamic processes with a relaxation time of up to 13 ns with ~ 100 fs resolution, corresponding to a dynamic range greater than 10^4. The brightness and energy spread of the TEM conventional source can be preserved by operating the cavity in the conjugate blanking mode (see Section 15.2.1) also in dual-mode.

The manufactured TM_{110} dual-mode cavity (see Figure 15.5) is based on the same optimised design explained in the previous subsection (see Section 15.2.1.1). It is a pillbox cavity with a cylindrical copper housing partially filled with a dielectric material ($ZrTiO_4$ doped with <20% of $SnTiO_4$) featuring an elliptical cross-section. A vacuum 'moat' surrounds the dielectric stone and hosts two linear antennas and two tuning stubs. An antenna and a tuning stub are located along the short axis of the dielectric elliptical cross-section, while the others are positioned along the long axis. The antennas enable loading the power into each one of the two cavity modes. The metal tuning stubs are used to tune the resonance frequency of each mode over a range of several MHz. This configuration gives nearly independent

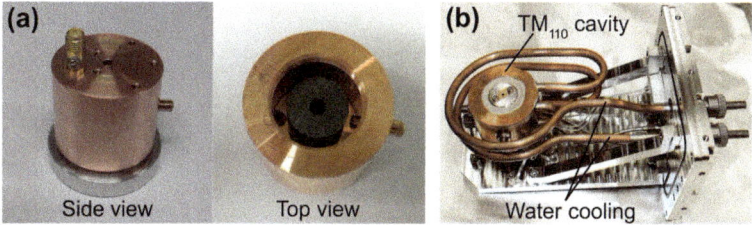

Figure 15.5 (a) Pictures of the manufactured TU/e TM$_{110}$ dual-mode cavity. A side view of the cavity and a top view of the cavity with the lid removed and the partially filled dielectric material exposed.[64] (b) Photograph of the dual-mode streaking cavity (side view) on the UTEM mount.

control over each mode. The cavity design also provides control of the mechanical stress and temperature fluctuations to maintain the cavity's resonance frequencies stable within 1 MHz.

15.2.2 The TU/e RF Cavity-based Ultrafast Transmission Electron Microscope

The TU/e RF cavity-based ultrafast transmission electron microscope (see Figure 15.6) is based on a 200 keV FEI Tecnai TF20 TEM comprising an additional 203 mm vacuum section between the C2-aperture and the mini-condenser lens. This section accommodates the TM$_{110}$ dual-mode cavity described in Section 15.2.1.2. An extra set of beam deflectors has been added above the cavity in the same section. Below the cavity, an additional aperture holder enables inserting a chopping aperture at a distance $l = 122.2$ mm from its centre and aligning it its optical axis. In the conjugate blanking operation mode, a beam cross-over is created in the middle of the cavity, precisely at a distance l from the chopping aperture, to prevent emittance and energy spread growth. The aperture holder design allows selection between three different chopping aperture diameters $d = 50, 30, 10$ µm according to the experimental needs. The cavity can be independently operated in a single- or dual-mode. As explained in Section 15.2.1, in a single mode it creates electron pulses at a 3 GHz repletion rate. In a dual-mode instead, the microscope produces electron pulses at the same repetition rate as a 75 MHz fs-oscillator, ensuring synchronisation with the laser pulse train. Details on the microwave setup and synchronisation system used at TU/e are provided in ref. 23. The presented setup reduces the timing jitter between the laser pulses and the microwave signals to below 100 fs.[60] Variable attenuators and a combination of two 3 GHz phase shifters offer independent control of the amplitudes and the phase of each mode in the cavity.

The laser oscillator integrated into the TU/e UTEM setup is a Ti:Sapphire femtosecond mode-locked oscillator (Femtosource Pro Seed) customised to

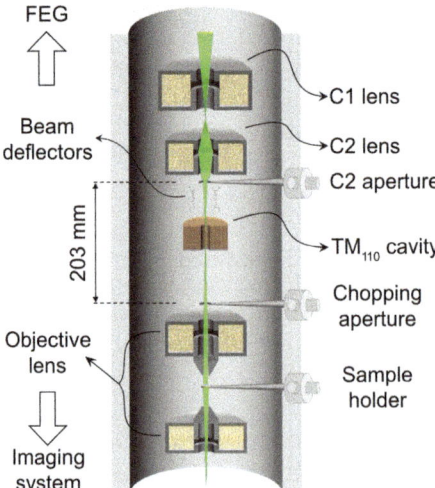

Figure 15.6   Schematics of the TU/e RF cavity-based ultrafast electron microscope. An additional 203 mm vacuum section has been added to a 200 keV FEI Tecnai TF20 TEM below the C2-aperture. This section hosts a TM_{110} dual-mode cavity, an additional set of beam deflectors, and an aperture-holder that allows the insertion of different-sized chopping apertures.

produce pulses at a repetition rate of 74.9625 MHz (~ 75 MHz). This frequency has been selected so that its 40th harmonic matches the European standard-ised frequency of 2.9985 GHz of an S-band cavity. A customised control system guarantees frequency stability within a few Hz. The central wavelength of the laser beam is 790 nm with a (FWHM) bandwidth of 116 nm (corresponding to a Fourier transform limit of ~ 8 fs). The laser pulse (FWHM) duration at the instrument is ~ 30 fs. The mode-locked oscillator produces an output power of about 420 mW, a fraction of which is split off to drive the synchronisation system. An objective module is installed in one of the side ports of the TEM's octagon, only a few metres away from the fs-oscillator source. It guides and focuses the laser beam inside the microscope up to the sample plane. An active laser beam stabilisation system by MRC[68] stabilises the position of the laser beam in this plane within a few micrometres.

To prove the performances of the TU/e UTEM, van Rens *et al.* fully char-acterised the electron pulses generated by beam chopping with a 10 μm aperture while operating the cavity in the dual-mode.[23] First, the independent characterisation of both cavity modes proved that the pulse length changes as a function of the magnetic field inside the cavity according to the expected $\tau \propto 1/B_0 \propto 1/\sqrt{P}$ behaviour. Measurements demonstrated that a pulse length of $\tau = 630 \pm 10$ fs can be achieved with a modest input power of $P = 3.7 \pm 0.1$ W, while a power $P = 14$ W would guarantee a pulse duration of ~ 100 fs. Second, TU/e researchers generated sub-picosecond electron pulses at a 75 MHz repetition rate by driving both modes simultaneously with phase-locked microwave signals synchronised to the fs-oscillator.

During the measurements, they operated the cavity in the conjugate blanking mode by using the C2-lens to create a beam cross-over at the cavity centre and the minicondenser to focus back the beam on the fluorescent screen or the camera. Electron pulses have been obtained by driving the 40th and 41st harmonics at input powers of 1.3 ± 0.1 W and 1.1 ± 0.1 W, respectively. The resulting pulse length is $\tau = 750 \pm 10$ fs. This has been determined by measuring the continuous and pulsed beam currents using a Faraday cup. During the measurements, the beam peak current (*i.e.*, the time-averaged current of the unchopped beam) was $I_p = 0.4 \pm 0.1$ nA. The measured beam (RMS) transverse sizes are $\sigma_x = 1.3 \pm 0.1$ nm and $\sigma_y = 0.8 \pm 0.1$ nm, where the asymmetry is due to lens aberrations inside the microscope. The measured beam (RMS) divergence angles are $\sigma_x' = \sigma_y' = 1.7 \pm 0.1$ mrad. The resulting beam horizontal and vertical (RMS) normalised emittances are thereby $\varepsilon_{n,x} = 2.1 \pm 0.2$ pm rad and $\varepsilon_{n,y} = 1.3 \pm 0.2$ pm rad, respectively. The (RMS) normalised peak brightness of the beam is defined as

$$B_{RMS} = \frac{q_e}{m_e c^2} \frac{I_p}{4\pi^2 \varepsilon_{n,x} \varepsilon_{n,y}}. \tag{15.8}$$

Accordingly, the measured pulsed beam has a (RMS) normalised peak brightness of $B_{RMS} = (7 \pm 1) \times 10^6$ A m^{-1} sr^{-1} V^{-1}. This value is in good agreement with the peak brightness of the continuous electron beam, $B_{RMS} = (7 \pm 1) \times 10^6$ A m^{-1} sr^{-1}, and the peak brightness of the pulsed electron beam generated by single-mode operation, $B_{RMS} = (6.6 \pm 1) \times 10^6$ A m^{-1} sr^{-1}, previously measured at the TU/e cavity-based UTEM.[63] Finally, the energy spectrum of the 75 MHz pulsed electron has been measured using a Gatan ENFINA spectrometer, with a dispersion of 0.05 eV per channel. The full width at half maximum energy spread extrapolated from a pseudo-Voigt fit is $\Delta U = 0.90 \pm 0.05$ eV,[23] and it matches the energy spectrum of the continuous electron beam.[63]

Results prove that the generation of a pulsed electron beam by leveraging a dual-mode cavity preserves the high brightness and low energy spread of the field emission gun of conventional TEMs. This highlights one of the greatest assets of the microwave cavity generation of pulsed electron beams in UEM. Furthermore, since the peak brightness of the original beam is conserved after chopping with an RF cavity, any future improvements in the continuous source will directly apply to the UTEM. It is sobering to note that although the 3 GHz repetition rate of an RF cavity-based UEM is several orders of magnitude higher than in conventional UEMs based on photo-emission, the low charge per pulse limits the average current. LaB$_6$-emitters promise a brightness up to $B_r = 10^{10}$ A m^{-2} sr^{-1} eV^{-1}.[69,70] Provided that they can be operated at sufficient current, combining such a thermionic source with an RF cavity could result in ultrafast electron microscopy with unprecedented spatial and temporal resolution at the average current of present-day continuous electron microscopes. The combination of high brightness, low energy spread, and fast switching between 333.3 ps to 13.3 ns time delay between electron pulses results in unique imaging opportunities offered by the TU/e UTEM.

15.3 Applications

An RF cavity-based ultrafast electron microscope represents an appealing alternative to the most commonly used time-resolved microscopes and significantly contributes to the field of UEM. Interestingly, integrating a fs-oscillator and microwave cavities in the column of a TEM also opens up the possibility of coherently manipulating the electron beam using time-dependent electromagnetic fields. Several applications highlight this broad potential of the TU/e ultrafast microscope. For example, manipulating the longitudinal phase space of electron beams *via* electromagnetic fields in microwave cavities provides advanced instrument performances and electron-probe pulse characteristics (*e.g.*, repetition rate and pulse length). These in turn guarantee remarkable imaging capabilities by offering remarkable radiation damage suppression and the unique possibility of performing pump–probe experiments at an extremely high repetition rate (75 MHz–3 GHz). Furthermore, the wavefunction of a single electron can be manipulated and shaped by employing time-dependent fields of the focused high-intensity femtosecond laser pulses inside the TEM column. Details about the applications emerging from the TU/e UTEM are provided in the following subsections.

15.3.1 Pump–Probe Experiments at High Repetition Rate

In most ultrafast electron microscopes, electron pulses are generated from a cathode through the photoelectric effect. This electron generation process requires high-energy laser pulses with a wavelength of a few hundred nm. To accomplish high pulse energies, it is necessary to employ chirped pulse amplification (CPA) laser systems. These CPA systems can create laser pulses with an energy of 10 µJ–10 mJ but at a relatively low repetition rate of 1 kHz–1 MHz. In the stroboscopic single-electron mode of UEM, a laser pulse initiates a process of interest inside a sample material (providing the starting time of the reaction, t_0), which is then probed by an electron pulse. Varying the time delay between the pump excitation pulse and the electron-probe pulse, the time evolution of the process can be investigated. However, in the stroboscopic mode, each probe-pulse contains, on average, a single electron. In contrast, about 10^8 electrons must illuminate the sample and reach a detector to obtain an image of the specimen with a sufficient signal-to-noise ratio. For this reason, the image signal is accumulated over many laser cycles for each fixed time delay (*i.e.*, at a fixed state of the investigated process). The laser system's 1 kHz–1 MHz repetition rate limits the average current of an ultrafast TEM based on photoemission. However, long relaxation times of dynamical processes can also limit the maximal repetition times used. Due to this low repetition rate, acquiring an image requires longer exposure times.

In the TU/e setup, laser pulses from a 75 MHz mode-locked fs-oscillator are used to trigger dynamical processes on a sample inserted in the UTEM column. A 3 GHz microwave TM_{110} cavity operated in the dual-mode produces electron-probe pulses at the same repetition rate of the fs-oscillator (75 MHz). The microwave cavity's phase must be accurately synchronised to a pump laser pulse to perform pump–probe experiments. The time jitter between the electron and laser pulses is suppressed to levels well below 100 fs using state-of-the-art synchronisation schemes.[60,71] Thus, the TU/e UTEM enables pump–probe experiments at a high repetition rate of 75 MHz without requiring an amplified laser. The higher repetition rate translates into a higher average beam current in the TEM that offers an order of magnitude faster signal acquisition, thus enabling the investigation of samples with relaxation times shorter than 13 ns. A fast beam blanker can be inserted below the cavity to pick specific pulses to enable the study of samples with longer relaxation time, which require repetition rates lower than tens of MHz.

On the other hand, performing pump–probe experiments at a repetition rate higher than tens of MHz would be advisable when investigating samples with relaxation time faster than ~13 ns. It has been shown that a microwave signal can be successfully used as a pump pulse on electronic or semiconductor devices to perform stroboscopic imaging of the induced dynamics with a few GHz repetition rates.[72] The 3 GHz microwave signal of the TM_{110} cavity operated in a single-mode can thus be used as a pump pulse. This approach guarantees laser-free stroboscopic imaging with repetition rates in the GHz regime.

Therefore, the TU/e UEM allows optimising the setup's repetition rate according to each experiment's requirements. Last but not the least, it is important to realise that a laser fluence of about 100 J m^{-2} is sufficient to trigger nearly all of the most interesting effects in pump–probe experiments. The laser beam from a Ti:Sapphire oscillator with an energy of ~10 nJ per pulse can guarantee this required laser fluence when focused to a spot size of about 100 μm.[73] However, mode-locked oscillators have the advantages of being more compact, cheap, and commercially available in turn-key systems that are easy to operate and maintain. Besides, they provide higher beam stability and shorter pulses.

As an exemplary application of the TU/e UTEM in pump–probe microscopy, we envisage the real-time imaging of single-phonon dynamics in two-dimensional quantum materials with atomic space and time resolution. The imaging of the emergence and evolution of acoustic phonons at individual defects in crystalline Ge and WSe_2 by fs-electron microscopy[25] has been one of the first applications of ultrafast transmission electron microscopy. The high brightness and average beam current of the RF cavity-based UTEM will enable similar experiments with enhanced spatial and temporal resolution and a much higher repetition rate, which results in an improved signal-to-noise ratio.

15.3.2 Radiation Damage Suppression

One of the most challenging applications in UEM is the investigation of radiation-sensitive samples, such as biological materials and some types of synthetic catalyst materials. The ultimate performance of modern ultrafast transmission electron microscopes has improved to a level where they can achieve a 2 Å spatial resolution in real-space imaging on suitable samples. However, as a rule of thumb, a higher spatial resolution requires a higher dose, D, on the specimen (*i.e.*, more electrons per square Ångström). Radiation-sensitive materials cannot endure a high dose release without permanent damage occurring. Therefore, radiation sensitivity restricts the achievable spatial resolution in many cases. In addition, this class of materials typically exhibits low image contrast. They would thus require a more intense electron beam to be imaged with a sufficient signal-to-noise ratio using amplitude contrast.

Biological samples, for example, are prevalently composed of light elements (*e.g.*, carbon and hydrogen) characterised by a low critical dose.[§] Microscopists measured that the maximum total dose releasable to biological specimens is limited to $D = 20$ electrons per $Å^2$.[74] However, to image a single carbon atom in graphene, a 500 times higher dose of $D = 10^4$ electrons per $Å^2$ would be required.[75,76] In addition to biological samples, other types of material are highly sensitive to electron radiation, some of which are vital in industrial processes.

The development of cryo-EM marked a stride in radiation-sensitive material imaging.[77] This technique involves cooling the specimen to cryogenic temperatures and embedding it in an environment of vitreous water. Even though this approach does not directly prevent radiation damage, it allows for higher spatial resolutions by keeping the atoms in place while imaging with the probe-electron pulses. Recently, cryo-EM imaging of many different biomolecules has been carried out with resolutions as low as 2.2–2.8 Å.[78,79] However, cryo-EM images are often extremely noisy. Moreover, while providing high-resolution static images of biomolecules, the technique does not allow for studying dynamical processes in biological samples at the atomic level.

The recent discovery of a non-linear dependence of radiation damage at the applied dose rate represents a breakthrough in radiation-sensitive material imaging. For a long time, the material science community believed that beam-induced damage to a sample depended only on the total accumulated dose and acceleration voltages applied to the electron probe beam. After a long-standing debate, different research groups instead showed a dependence of the radiation damages not only on the total dose released but also on the dose rate,[75,80,81] and observed damage reduction by employing pulsed illumination of the sample.[82,83] This observation opened up a

---

[§] The critical dose is the maximum radiation dose appliable to a specimen. Administering a total dose higher than the critical dose would irreversibly alter the material.

pathway to a new and unexpected application of photoemission- and RF cavity-based UEM. In the following, two recent works on the topic will be discussed.

Van den Bussche and Flannigan at the University of Minnesota[83] performed a systematic study of the effect of photoemission-based pulsed irradiation and conventional irradiation methods on hexatriacontane ($C_{36}H_{74}$) microcrystals. They showed that delivering single-electron pulses with a well-defined and uniform electron arrival time reduces irreversible structural damage compared to conventional ultra-low dose methods for the same total accumulated dose and dose rate. Furthermore, they proved that the damage increases by decreasing the arrival time between each electron, depending on the number of electrons per pulse. More specifically, there is a certain threshold number of electrons per pulse above which the benefits of the pulsed illumination compared to conventional methods disappear. These observations suggest that the pulsed-beam irradiation mitigates the effects of sequential inelastic scattering events occurring within the characteristic duration of a given damage mechanism, in contrast to what happens during the temporally random irradiation characteristic of conventional low-dose methods.

Kisielowski's group at the Lawrence Berkeley National Laboratory, in collaboration with the TU/e group, performed a study of the Ziegler–Natta catalyst $MgCl_2$-supported $TiCl_4$, using both a continuous electron beam and a pulsed beam generated in an RF cavity-based UTEM.[30] The $MgCl_2$ supporting material is extremely sensitive to electron beam radiation and ambient conditions.[77] The low critical dose of this material has hindered the possibility of thoroughly characterising its atomic and defect structure by conventional electron microscopy. Kisielowski *et al.*[30] demonstrated a clear dependence of the beam-induced damage on the dose rate, which the standard model of radiation damage cannot explain. They show that the critical dose of this radiation-sensitive material increases by two orders of magnitude when probing it with ultrashort, high repetition rate electron pulses produced by a streaking cavity. Authors also show that limiting the dose rate makes it possible to identify a radiation-damage-free regime where they acquired the first atomic-resolution phase-contrast images of pristine $MgCl_2$ supporting material.

15.3.3 Coherent Manipulation of the Electron Wavefunction with Light

Beyond time-resolved investigation of specimens with an atomic-scale spatio-temporal resolution, RF cavity-based UEM established new free electron quantum optics based on the interaction of ultrashort electron pulses with the time-dependent electromagnetic fields of light. Integrating a 75 MHz mode-locked oscillator in the TU/e UTEM setup creates the unique opportunity of inducing and observing the synchronised interaction between free electrons and fs-laser pulses inside the microscope column. Moreover, the beam coherence in state-of-the-art TEMs approaches the quantum mechanical limit imposed by Heisenberg's uncertainty principle.

Nowadays, researchers aim at uncovering new techniques to control and shape both the amplitude and the phase distribution of the electron wavefunction. The ultimate goals are establishing coherent detection techniques such as holography, interferometry, and interaction-free measurements, developing perfect aberration correction schemes, and ascertaining distinct phase-contrast imaging. Over the years, relevant results have been reported;[84–87] however, all these techniques leverage the use of matter to shape the electron wavefunction, which inevitably induces unintended electron scattering and image distortion due to the potential generated by volume or surface charges. Furthermore, matter-based methods for phase manipulation can only provide a fixed phase variation, always accompanied by undesired losses of a beam fraction. For all the above reasons, controlling the quantum-mechanical phase using the time-dependent electromagnetic field of light represents an appealing alternative.

The following subsections will provide details of two kinds of coherent light-electron interaction investigated at TU/e: the photon-induced near-field electron microscopy (PINEM) and the light-based Zernike phase plate.

15.3.3.1   *Photon-induced Near-field Electron Microscopy (PINEM)*

The widespread use of ultrafast electron microscopes based on pulsed electron beams has given rise to new techniques that are well-established among the electron microscopy community. A well-known example is PINEM, which detects modulations of the electrons' energy loss spectrum (EELS) due to the interaction with electromagnetic near-fields.[88,89]

Even though PINEM was initially motivated by the desire to map the optical near-field induced changes in nanostructures, it is nowadays successfully exploited to provide *in situ* characterisation of the electron pulses. It has been proven that the temporal cross-sections of high-order peaks in the energy spectrum measure the electron pulse duration.[90] Accurately measuring the duration of the electron pulses is paramount to determining the microscope's temporal resolution and imaging capabilities, particularly in the case of imaging dynamics occurring near the temporal resolution limits. Likewise, the electron pulse length is a crucial parameter in all the experiments involving coherent manipulation of a quantum system with light, particularly in quantum-mechanical phase control (see Section 15.3.3.2).

For all the reasons above, the TU/e group harnesses the PINEM effect to characterise the duration of the electron pulses produced in the RF cavity-based UTEM. In addition, PINEM represents a benchmark for electron wavefunction manipulation with light in the microscope column. The experimental setup used to observe PINEM in the TU/e RF cavity-based UTEM setup (see Figure 15.7) employs p-polarized laser pulses generated by a

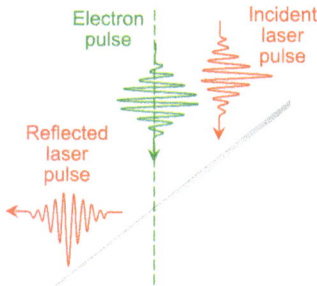

Figure 15.7 Schematics of a PINEM experiment in the TU/E RF cavity based UTEM. An aluminium mirror reflects an incoming laser pulse, while a 200 keV electron pulse interacts with the light field resulting from the superposition of the incident and reflected light beams.

790 nm fs-oscillator (see Section 15.2), impinging at an angle $\alpha = 45°$ on an aluminum mirror and reflected at the same angle. The aluminium mirror with reflectance $R = 0.82$ at the laser wavelength[91] abruptly interrupts the light field in free space. Electron pulses generated by operating the cavity in the dual-mode are synchronised to the laser pulse train within less than a few ~ 100 fs. While traveling at an angle $\theta = \alpha = 45°$ to the mirror, the electron pulses interact with the incident and reflected light fields.

The interaction gives rise to the absorption or emission of photons. As a result, side-band peaks on the right (energy gain region) and left (energy loss region) of the zero-loss peak appear in the measured EEL spectrum of the electron pulses. The spacing between the side-band peaks is an integral multiple of the laser photon energy. This is exemplified by the PINEM signals in Figure 15.8a (orange dots) measured with the above-described TU/e experimental set-up showing first-order side peaks, proving the emission and absorption of single photons by an electron.

Here, the measurement was performed with a laser and an electron pulse both of duration $\tau_1 \sim \tau_e \sim 1$ ps, and a laser power of ~ 120 mW focused to a waist of ~ 20 μm at the interaction point. The spectrum was acquired using a Gatan Enfina spectrometer of the modified FEI Tecnai TF20 with a dispersion of 0.05 eV per channel and an exposure time of 2 seconds. Fitting the measured EELS with the sum of three Gaussian functions provides estimations of the electrons' energy at the peak locations and the amplitude of the ZLP and the side-band peaks. The cross-correlation between the laser and the electron pulses determines the temporal duration of the PINEM event. More precisely, the spectrum in Figure 15.8a was measured in the case of perfect synchronisation between laser and electron pulses. However, the probability of observing absorption/emission events decreases when the laser pulse arrives before or after the electron pulse. As a result, the amplitude of the side-band peaks decreases as the absolute value of the temporal distance between the laser and the electron pulses increases. The evolution of the measured relative amplitude of the first-order absorption/emission peak as a function of the laser

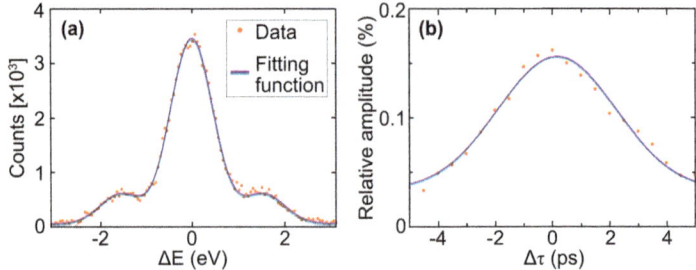

Figure 15.8 (a) PINEM signal measured in the TU/e RF cavity-based UTEM. The measured EEL spectrum (orange dots) displays quanta of photon absorption and emission on either side of the zero-loss peak (ZLP). The sum of three Gaussian functions (solid purple line) has been fitted to the measured EELS. (b) Evolution of the first-order peak relative amplitude (orange dots) as a function of the laser delay with respect to the electron pulses. The first-order peak amplitude is expressed as a percentage of the ZLP intensity for each time-delay. The measured curve results from the cross-correlation between laser and electron pulse duration. A Gaussian function has been fitted to the data (solid purple line).

delay with respect to the electron pulse $\Delta\tau$ is shown in Figure 15.8 (orange dots). This temporal evolution results from the cross-correlation between the laser and the electron pulses. The figure also shows the result of fitting a Gaussian to the measured profile (see solid purple lines in Figure 15.8b). The electron pulse length τ_e can be determined by precise measurement of the (RMS) laser pulse duration τ_l and the standard deviation of the fitted temporal profile τ, using the relation $\tau = \sqrt{\tau_l^2 + \tau_e^2}$. Since electron–photon coupling occurs only in evanescent near-fields, the measured pulse properties occur precisely at the specimen location.

15.3.3.2 Pulsed-laser Zernike Phase Plate

Scientists have been tackling the challenge of manipulating, shaping, and controlling the quantum-mechanical phase of free electrons for decades. The first demonstration of quantum optical control of free electrons dates back to 2001 when Freimund *et al.*[92] observed the Kapitza–Dirac effect almost seventy years after its formulation.[93] Only in 2018, Schwartz *et al.* at the Lawrence Berkeley National Laboratory (LBNL) were able to reproduce the effect in a conventional transmission electron microscope using continuous (CW) laser and electron beams.[94] Due to the short electron-light interaction time in a micron-scale laser focus, establishing the diffraction effect in a TEM requires an intensity of tens of GW cm^{-2}. To boost the laser intensity to this order of magnitude, the Berkeley scientists incorporated a Fabry–Pérot cavity in the TEM that magnified the optical power by a factor of 4000. The achieved power of 43 GW cm^{-2} is a world-record in intra-cavity CW-laser intensity.[95] The diffraction process induces a variation of the electron wavefunction's phase. The maximum phase shift achieved in the experiment mentioned above is $\pi/2$. In the interaction with a tightly focused and high-intensity CW-laser beam, electrons experience the same phase shift when traversing a material known as a Zernike phase plate. Therefore, Müller's group at LBNL realised the first light-based Zernike phase plate, which was successfully used for imaging graphene by phase-contrast.[94]

Zernike phase plates are essential tools in phase-contrast imaging, a microscopy technique based on converting changes in the phase of electrons passing through a specimen to amplitude variations in the image. This technique is well suited for investigating weakly scattering and absorbing materials that produce a too low amplitude contrast, such as many biological samples. Even though these specimens generate low amplitude contrast, they typically imprint a weak position-dependent phase shift on the electron wavefunction, which can be conveniently exploited to image their structure using a Zernike phase plate.[96] This approach relies on producing a relative shift of $\pi/2$ between the phases of the electron waves scattered and unscattered by the specimen. Such a phase difference enables conversion of the phase modulation induced by the material into detectable amplitude modulation. Zernike phase-contrast in transmission electron microscopy has been demonstrated using carbon thin films located in the back focal plane of the objective lens, where electrons are spatially separated according to their diffraction order. This method, however, is affected by limitations inherently related to the presence of matter since interaction with matter inevitably leads to loss of electrons and decoherence. The result therefore limits the spectroscopic capability of this method. Moreover, material phase plates produce a non-tunable and time-varying phase shift.[97]

In contrast, light-based phase plates provide a stable, tunable phase shift,[95,98] thus providing a potentially powerful alternative tool for phase-contrast imaging in TEM.[99,100] Furthermore, they could be used in conjunction with spatial light modulators, which can module the structure of the light field interacting with the electrons to enable more complex types of phase modulation of the electron wave. This would, in turn, open up new approaches for aberration correction. Despite providing outstanding results, manipulating the electron wavefunction's phase with light appears cumbersome. Developing an extremely high-finesse Fabry–Pérot cavity and implementing it inside a TEM is a significant challenge. The sophistication of this setup might represent a significant hurdle to the wide adoption of light-based phase plates.

Additionally, it is non-trivial to integrate spatial modulators with a CW-laser beam inside an optical cavity. A breakthrough would be using a high repetition rate (70–90 MHz) femtosecond mode-locked laser oscillator, which routinely provides the necessary peak power required to induce a substantial phase shift in the wavefunction of a free electron. However, this would require electron pulses generated at the same repetition rate as the laser. It is then evident that ultrafast TEMs offer the best-suited environment for developing phase shaping devices based on light, provided that a sufficient average electron current is achievable.

One of the main ongoing research activities of the TU/e group is the development of a pulsed-laser-based Zernike phase plate for phase-contrast imaging. The ultimate aim is to induce a variation of the electron wavefunction's phase equal to $\Delta\phi = \pi/2$ during the interaction with a time-dependent light field. Conventional Ti:Sapphire oscillators can produce laser pulses with an energy greater than 10 nJ and a duration lower than 100 fs. Thus, a peak power higher than 100 kW can be easily achieved. The TU/e group has developed a complete theory of the interaction between an electron pulse and a tightly focused laser pulse orthogonally propagating to each other, based on the relativistic ponderomotive potential.[101] van Leeuwen *et al.* developed a flexible calculation method where a quantal phase has been derived from the classical action integral along the relativistic classical path. Results extend to realistic configurations of the electromagnetic fields of a pulsed laser beam. The model proves that a phase shift of $\pi/2$ can be achieved by focusing the 800 nm laser pulses produced by a fs-oscillator to a waist size of 2λ. A pulse energy of only ~ 53 nJ can guarantee a phase shift in the target when using electron and laser pulses with a duration of 300 fs and 100 fs, respectively. The claim has been supported by accurate self-consistent simulations, which proved the feasibility of a pulsed-laser Zernike phase plate in a cavity-based UTEM.[101] It is worth mentioning that the high repetition rate of the synchronised electron and laser pulses in the RF cavity-based UTEM guarantees a sufficiently high average current for practical achievement.

More generally, the widespread adoption of ultrafast electron microscopes interfaced with dedicated laser systems makes it necessary to develop robust and practical free-electron quantum optical elements for dynamic phase-shaping of the electron wavefunction. For example, Thomas Juffmann's group at the University of Vienna has recently demonstrated programmable transverse electron beam shaping in free space based on ponderomotive interaction between electrons and intense short laser pulses. They realised convex and concave electron lenses with a focal length of a few millimetres, comparable to the state-of-the-art in EM.[102]

15.4   Challenges and Outlooks

The main challenge in operating a UTEM setup at the moment is the improvement of the synchronisation between the phase of the microwave field in the cavity and the fs pump laser, which is presently at the level of ~ 100 fs. State-of-the-art synchronisation systems allow for a reduction of timing jitter below ten femtoseconds, opening up new possibilities for coherent manipulation of the electron wavefunction with light (see Section 15.3.3). Future systems could even provide synchronisation at the level of about one femtosecond. An upgrade of the TU/e microwave and synchronisation setup is underway, enabling a temporal overlap between laser and electron pulses significantly below 10 fs. Moreover, performing pump–probe measurements with a cavity in a single-mode at a high repetition rate of 3 GHz requires a 3 GHz fs laser of sufficiently high average power, which is not yet available. Thus, employing a cavity in a dual-mode (75 MHz) is only possible with existing laser technology. Current research in UTEM also includes the development of next-generation electron sources (see Chapter 16), detecting Coulomb-correlated two-electron pulses with high energy resolution and high detection efficiency,[103,104] and using isolated attosecond optical pulses (see Chapter 14).

Regarding the temporal resolution of the TU/e UTEM, the electron pulse duration can be made, in principle, arbitrarily small by increasing the microwave input power and/or selecting the C2 and chopping apertures with a smaller cross-section. Further development of dielectric cavities would allow for higher achievable field strength without damaging the dielectric material. Alternatively, a miniaturised RF cavity operating in the TM_{010} mode (see Section 15.2.1) can be incorporated into the setup to temporally compress the chopped electron pulses of the TU/e UTEM in shorter pulses or increase the charge per pulse at a given pulse length, thus guaranteeing a higher peak current. In this advanced UTEM, the TM_{110} chopping cavity produces 100 fs electron pulses with 1 eV energy spread. These pulses are then compressed to 10 fs duration in a downstream TM_{010} module while preserving a reasonable energy spread of 10 eV.

It should be noted that since the longitudinal emittance is conserved, compression always increases the energy spread. A compression cavity can create, in principle, electron pulses even shorter than 10 fs if a more extensive energy spread is allowed. In the envisioned chopping-compressing setup of a UTEM, the achievable temporal resolution in pump–probe experiments will be, in practice, limited only by the level of synchronisation between the fs-laser pulse and the microwave phase, which can be well below 10 fs with available technology.

Manipulating the longitudinal phase space of pulsed electron beams is one of the most exciting applications of microwave cavities in ultrafast electron microscopy. In this perspective, another appealing outlook is using a suitable combination of microwave cavities for advanced longitudinal manipulation techniques enabling a novel time-of-flight electron energy loss spectroscopy (ToF EELS) with high energy and temporal resolutions.[62] Verhoeven *et al.* devised an advanced method using a specific configuration of two TM_{110} and two TM_{010} cavities.[105] In the proposed setup, the first TM_{110} cavity chops a high-quality continuous beam into ultrashort pulses. A subsequent TM_{010} cavity stretches these pulses, thus lowering the uncorrelated energy spread. The pulses then interact with a sample, generating an energy loss distribution. A second TM_{010} cavity downstream of the sample compresses the pulses, while a second TM_{110} cavity finally streaks these pulses on a detector screen. This method enables measurements of the energy loss with an energy resolution of a few-10 meV at a few ps pulse length, only limited by the 'longitudinal aberrations' of the compression cavity. At the same time, it provides ~10 fs temporal resolution at eV energy spread. This advanced UTEM would allow for switching rapidly from conventional continuous operation to ultrafast operation with ~ 10 fs time resolution and ultrafast spectroscopic operation with ~ 10 meV energy resolution.

Acknowledgements

This publication is part of the project Dynamic Phase Space Shaping for Ultrafast Transmission Electron Microscopy (DPSS, project number 741.018.302) of the research programme ENW PPS-fonds (NWO-IPP) which is (partly) financed by the Dutch Research Council (NWO).

References

1. P. W. Hawkes and J. C. Spence, *Springer handbook of microscopy*, Springer, 2019.
2. M. C. Potter, B. Wyble, C. E. Hagmann and E. S. McCourt, *Atten. Percept. Psychophys.*, 2014, **76**, 270–279.
3. M. Seo, H. Yamaguchi, A. D. Mohite, S. Boubanga-Tombet, J.-C. Blancon, S. Najmaei, P. M. Ajayan, J. Lou, A. J. Taylor and R. P. Prasankumar, *Sci. Rep.*, 2016, **6**, 1–7.
4. M. M. Gabriel, J. R. Kirschbrown, J. D. Christesen, C. W. Pinion, D. F. Zigler, E. M. Grumstrup, B. P. Mehl, E. E. Cating, J. F. Cahoon and J. M. Papanikolas, *Nano Lett.*, 2013, **13**, 1336–1340.
5. D. Gerlich and J. Ellenberg, *Nat. Cell Biol.*, 2003, S14–S19.
6. V. A. Lobastov, R. Srinivasan and A. H. Zewail, *Proc. Natl. Acad. Sci. U. S. A.*, 2005, **102**, 7069–7073.
7. O. Bostanjoglo, *Phys. Status Solidi A*, 1982, **70**, 473–481.
8. O. Bostanjoglo, R. Elschner, Z. Mao, T. Nink and M. Weingärtner, *Ultramicroscopy*, 2000, **81**, 141–147.
9. H. Dömer and O. Bostanjoglo, *J. Appl. Phys.*, 2002, **91**, 5462–5467.
10. M. S. Grinolds, V. A. Lobastov, J. Weissenrieder and A. H. Zewail, *Proc. Natl. Acad. Sci. U. S. A.*, 2006, **103**, 18427–18431.
11. A. H. Zewail, *Science*, 2010, **328**, 187–193.
12. H. Dömer and O. Bostanjoglo, *Rev. Sci. Instrum.*, 2003, **74**, 4369–4372.
13. B. J. Siwick, J. R. Dwyer, R. E. Jordan and R. D. Miller, *Science*, 2003, **302**, 1382–1385.
14. N. Browning, M. Bonds, G. Campbell, J. Evans, T. LaGrange, K. Jungjohann, D. Masiel, J. McKeown, S. Mehraeen and B. Reed, *et al.*, *Curr. Opin. Solid State Mater. Sci.*, 2012, **16**, 23–30.
15. T. LaGrange, M. Armstrong, K. Boyden, C. Brown, G. Campbell, J. Colvin, W. DeHope, A. Frank, D. Gibson and F. Hartemann, *et al.*, *Appl. Phys. Lett.*, 2006, **89**, 044105.
16. T. LaGrange, G. H. Campbell, B. Reed, M. Taheri, J. B. Pesavento, J. S. Kim and N. D. Browning, *Ultramicroscopy*, 2008, **108**, 1441–1449.
17. L. Piazza, D. Masiel, T. LaGrange, B. Reed, B. Barwick and F. Carbone, *Chem. Phys.*, 2013, **423**, 79–84.

18. E. Kieft, K. B. Schliep, P. K. Suri and D. J. Flannigan, *Struct. Dyn.*, 2015, **2**, 051101.

19. G. Cao, S. Sun, Z. Li, H. Tian, H. Yang and J. Li, *Sci. Rep.*, 2015, **5**, 1–7.

20. M. Kuwahara, Y. Nambo, K. Aoki, K. Sameshima, X. Jin, T. Ujihara, H. Asano, K. Saitoh, Y. Takeda and N. Tanaka, *Appl. Phys. Lett.*, 2016, **109**, 013108.

21. W. A. Curtis and D. J. Flannigan, *Phys. Chem. Chem. Phys.*, 2021, **23**, 23544–23553.

22. P. Hommelhoff, Y. Sortais, A. Aghajani-Talesh and M. A. Kasevich, *Phys. Rev. Lett.*, 2006, **96**, 077401.

23. J. Van Rens, W. Verhoeven, E. Kieft, P. Mutsaers and O. Luiten, *Appl. Phys. Lett.*, 2018, **113**, 163104.

24. O. Zandi, A. E. Sykes, R. D. Cornelius, F. M. Alcorn, B. S. Zerbe, P. M. Duxbury, B. W. Reed and R. M. van der Veen, *Nat. Commun.*, 2020, **11**, 1–11.

25. D. R. Cremons, D. A. Plemmons and D. J. Flannigan, *Nat. Commun.*, 2016, **7**, 1–8.

26. D. A. Plemmons, P. K. Suri and D. J. Flannigan, *Chem. Mater.*, 2015, **27**, 3178–3192.

27. R. Li and P. Musumeci, *Phys. Rev. Appl.*, 2014, **2**, 024003.

28. D. Xiang, F. Fu, J. Zhang, X. Huang, L. Wang, X. Wang and W. Wan, *Nucl. Instrum. Methods Phys. Res., Sect. A*, 2014, **759**, 74–82.

29. H. S. Park, J. S. Baskin, O.-H. Kwon and A. H. Zewail, *Nano Lett.*, 2007, **7**, 2545–2551.

30. C. Kisielowski, P. Specht, B. Freitag, E. R. Kieft, W. Verhoeven, J. F. van Rens, P. Mutsaers, J. Luiten, S. Rozeveld and J. Kang, *et al.*, *Adv. Funct. Mater.*, 2019, **29**, 1807818.

31. B. Barwick, H. S. Park, O.-H. Kwon, J. S. Baskin and A. H. Zewail, *Science*, 2008, **322**, 1227–1231.

32. K. Bikker, M. Picher, O. Cregut, T. Lagrange, B. W. Reed, S. T. Park, D. J. Masiel and F. Banhart, *Ultramicroscopy*, 2016, **171**, 8–18.

33. A. Gahlmann, S. T. Park and A. H. Zewail, *Phys. Chem. Chem. Phys.*, 2008, **10**, 2894–2909.

34. A. Janzen, B. Krenzer, O. Heinz, P. Zhou, D. Thien, A. Hanisch, F.-J. Meyer Zu Heringdorf, D. Von Der Linde and M. Horn von Hoegen, *Rev. Sci. Instrum.*, 2007, **78**, 013906.

35. B. J. Siwick, J. R. Dwyer, R. E. Jordan and R. J. D. Miller, *J. Appl. Phys.*, 2002, **92**, 1643–1648.

36. M. R. Armstrong, K. Boyden, N. D. Browning, G. H. Campbell, J. D. Colvin, W. J. DeHope, A. M. Frank, D. J. Gibson, F. Hartemann and J. S. Kim, *et al.*, *Ultramicroscopy*, 2007, **107**, 356–367.

37. B. W. Reed, *J. Appl. Phys.*, 2006, **100**, 034916.

38. Z. Tao, H. Zhang, P. Duxbury, M. Berz and C.-Y. Ruan, *J. Appl. Phys.*, 2012, **111**, 044316.

39. R. P. Chatelain, V. R. Morrison, C. Godbout and B. J. Siwick, *Appl. Phys. Lett.*, 2012, **101**, 081901.

40. C. Kealhofer, W. Schneider, D. Ehberger, A. Ryabov, F. Krausz and P. Baum, *Science*, 2016, **352**, 429–433.
41. P. Zhu, Y. Zhu, Y. Hidaka, L. Wu, J. Cao, H. Berger, J. Geck, R. Kraus, S. Pjerov and Y. Shen, *et al.*, *New J. Phys.*, 2015, **17**, 063004.
42. S. Weathersby, G. Brown, M. Centurion, T. Chase, R. Coffee, J. Corbett, J. Eichner, J. Frisch, A. Fry and M. Gühr, *Rev. Sci. Instrum.*, 2015, **86**, 073702.
43. T. van Oudheusden, P. L. E. M. Pasmans, S. B. van der Geer, M. J. de Loos, M. J. van der Wiel and O. J. Luiten, *Phys. Rev. Lett.*, 2010, **105**, 264801.
44. D. A. Plemmons and D. J. Flannigan, *Chem. Phys. Lett.*, 2017, **683**, 186–192.
45. K. Bücker, M. Picher, O. Crégut, T. LaGrange, B. Reed, S. Park, D. Masiel and F. Banhart, *Ultramicroscopy*, 2016, **171**, 8–18.
46. S. Sun, X. Sun, J. Williams and C.-Y. Ruan, *Microsc. Microanal.*, 2020, **26**, 430–433.
47. J. Maxson, D. Cesar, G. Calmasini, A. Ody, P. Musumeci and D. Alesini, *Phys. Rev. Lett.*, 2017, **118**, 154802.
48. M. Krüger, M. Schenk and P. Hommelhoff, *Nature*, 2011, **475**, 78–81.
49. A. Feist, K. E. Echternkamp, J. Schauss, S. V. Yalunin, S. Schafer and C. Ropers, *Microsc. Microanal.*, 2015, **21**, 1203–1204.
50. A. Feist, N. Bach, N. R. da Silva, T. Danz, M. Möller, K. E. Priebe, T. Domröse, J. G. Gatzmann, S. Rost and J. Schauss, *et al.*, *Ultramicroscopy*, 2017, **176**, 63–73.
51. L. Zhang, J. P. Hoogenboom, B. Cook and P. Kruit, *Struct. Dyn.*, 2019, **6**, 051501.
52. D. H. Auston, *Appl. Phys. Lett.*, 1975, **26**, 101–103.
53. M. Gao, Y. Jiang, G. H. Kassier and R. Dwayne Miller, *Appl. Phys. Lett.*, 2013, **103**, 033503.
54. I. Weppelman, R. Moerland, L. Zhang, E. Kieft, P. Kruit and J. Hoogenboom, *Struct. Dyn.*, 2019, **6**, 024102.
55. W. Verhoeven, J. Van Rens, E. Kieft, P. Mutsaers and O. Luiten, *Ultramicroscopy*, 2018, **188**, 85–89.
56. H. Fujioka and K. Ura, *Scanning*, 1983, **5**, 3–13.
57. J. Thong, *Meas. Sci. Technol.*, 1991, **2**, 207.
58. D. Winkler, R. Schmitt, M. Brunner and B. Lischke, *IBM J. Res. Dev.*, 1990, **34**, 189–203.
59. J. Van Rens, W. Verhoeven, J. Franssen, A. Lassise, X. Stragier, E. R. Kieft, P. Mutsaers and O. Luiten, *Ultramicroscopy*, 2018, **184**, 77–89.
60. G. Brussaard, A. Lassise, P. Pasmans, P. Mutsaers, M. Van Der Wiel and O. Luiten, *Appl. Phys. Lett.*, 2013, **103**, 141105.
61. P. Musumeci, J. Moody, C. Scoby, M. Gutierrez, M. Westfall and R. Li, *J. Appl. Phys.*, 2010, **108**, 114513.
62. W. Verhoeven, J. F. M. van Rens, M. A. W. van Ninhuijs, W. F. Toonen, E. R. Kieft, P. H. A. Mutsaers and O. J. Luiten, *Struct. Dyn.*, 2016, **3**, 054303.

63. W. Verhoeven, PhD thesis, Eindhoven University of Technology, 2018.
64. J. F. M. van Rens, PhD thesis, Eindhoven University of Technology, 2019.
65. W. Verhoeven, J. van Rens, A. Kemper, E. Rietman, H. van Doorn, I. Koole, E. Kieft, P. Mutsaers and O. Luiten, *Rev. Sci. Instrum.*, 2019, **90**, 083703.
66. T-Ceram website, http://www.t-ceram.com/ceramic-materials.htm.
67. A. C. Lassise, PhD thesis, Eindhoven University of Technology, 2012.
68. MRC website, https://www.mrc-systems.de/en/products/laser-beam-stabilization.
69. H. Zhang, J. Tang, J. Yuan, Y. Yamauchi, T. T. Suzuki, N. Shinya, K. Nakajima and L.-C. Qin, *Nat. Nanotechnol.*, 2016, **11**, 273–279.
70. W. Toonen, A. Rajabi, R. van den Berg, X. Stragier, P. Mutsaers, P. Smorenburg and O. Luiten, *Nucl. Instrum. Methods Phys. Res., Sect. A*, 2021, **1013**, 165678.
71. M. Walbran, A. Gliserin, K. Jung, J. Kim and P. Baum, *Phys. Rev. Appl.*, 2015, **4**, 044013.
72. J. Qiu, G. Ha, C. Jing, S. V. Baryshev, B. W. Reed, J. W. Lau and Y. Zhu, *Ultramicroscopy*, 2016, **161**, 130–136.
73. P. Pasmans, D. Van Vugt, J. Van Lieshout, G. Brussaard and O. Luiten, *Phys. Rev. Accel. Beams*, 2016, **19**, 103403.
74. J. Hemminger, G. Fleming and M. Ratner, *Directing matter and energy: Five challenges for science and the imagination*, DOESC (USDOE Office of Science (SC)), 2007.
75. J. Z. Chen, C. Sachse, C. Xu, T. Mielke, C. M. Spahn and N. Grigorieff, *J. Struct. Biol.*, 2008, **161**, 92–100.
76. N. Uyeda, T. Kobayashi, K. Ishizuka and Y. Fujiyoshi, *Nature*, 1980, **285**, 95–97.
77. R. Fernandez-Leiro and S. H. Scheres, *Nature*, 2016, **537**, 339–346.
78. A. Bartesaghi, A. Merk, S. Banerjee, D. Matthies, X. Wu, J. L. Milne and S. Subramaniam, *Science*, 2015, **348**, 1147–1151.
79. M. G. Campbell, D. Veesler, A. Cheng, C. S. Potter and B. Carragher, *eLife*, 2015, **4**, e06380.
80. R. Egerton, *Ultramicroscopy*, 2013, **127**, 100–108.
81. C. Kisielowski, *Adv. Mater.*, 2015, **27**, 5838–5844.
82. C. Kisielowski, L.-W. Wang, P. Specht, H. A. Calderon, B. Barton, B. Jiang, J. H. Kang and R. Cieslinski, *Phys. Rev. B: Condens. Matter Mater. Phys.*, 2013, **88**, 024305.
83. E. J. VandenBussche and D. J. Flannigan, *Nano Lett.*, 2019, **19**, 6687–6694.
84. H. Lichte and M. Lehmann, *Rep. Prog. Phys.*, 2007, **71**, 016102.
85. G. Möllenstedt and H. Düker, *J. phys.*, 1956, **145**, 377–397.
86. R. Danev and K. Nagayama, *Ultramicroscopy*, 2001, **88**, 243–252.
87. J. Verbeeck, A. Béché, K. Müller-Caspary, G. Guzzinati, M. A. Luong and M. Den Hertog, *Ultramicroscopy*, 2018, **190**, 58–65.
88. S. T. Park, M. Lin and A. H. Zewail, *New J. Phys.*, 2010, **12**, 123028.
89. B. Barwick, D. J. Flannigan and A. H. Zewail, *Nature*, 2009, **462**, 902–906.

90. D. A. Plemmons, S. T. Park, A. H. Zewail and D. J. Flannigan, *Ultramicroscopy*, 2014, **146**, 97–102.

91. P. B. Johnson and R.-W. Christy, *Phys. Rev. B: Condens. Matter Mater. Phys.*, 1972, **6**, 4370.

92. D. L. Freimund, K. Aflatooni and H. Batelaan, *Nature*, 2001, **413**, 142–143.

93. P. Kapitza and P. Dirac, in *Mathematical Proceedings of the Cambridge Philosophical Society*, Cambridge University Press, 1933, vol. 29, pp. 297–300.

94. O. Schwartz, J. Axelrod, S. Campbell, C. Turnbaugh, R. Glaeser and H. Müller, 2018, arXiv preprint, arXiv:1812.04596.

95. O. Schwartz, J. Axelrod, D. Tuthill, P. Haslinger, C. Ophus, R. Glaeser and H. Müller, *Opt. Express*, 2017, **25**, 14453–14462.

96. F. Zernike, *Physica*, 1942, **9**, 686–698.

97. R. Danev, D. Tegunov and W. Baumeister, *eLife*, 2017, **6**, e23006.

98. H. Mueller, J. Jin, R. Danev, J. Spence, H. Padmore and R. M. Glaeser, *New J. Phys.*, 2010, **12**, 073011.

99. R. Danev and W. Baumeister, *Curr. Opin. Struct. Biol.*, 2017, **46**, 87–94.

100. A. Merk, A. Bartesaghi, S. Banerjee, V. Falconieri, P. Rao, M. I. Davis, R. Pragani, M. B. Boxer, L. A. Earl and J. L. Milne, *et al.*, *Cell*, 2016, **165**, 1698–1707.

101. K. van Leeuwen, W. Schaap, B. Buijsse, S. Borelli, S. Kempers, W. Verhoeven and O. Luiten, *New J. Phys.*, 2023, **25**(2), 023031.

102. M. C. C. Mihaila, P. Weber, M. Schneller, L. Grandits, S. Nimmrichter and T. Juffmann, *Phys. Rev. X*, 2022, **12**(3), 031043.

103. R. Haindl, A. Feist, T. Domröse, M. Möller, S. V. Yalunin and C. Ropers, *Nat. Phys.*, 2023, **19**, 1410–1417.

104. S. Meier, J. Heimerl and P. Hommelhoff, *Nat. Phys.*, 2023, **19**, 1402–1409.

105. W. Verhoeven, J. F. M. van Rens, W. F. Toonen, E. R. Kieft, P. H. A. Mutsaers and O. J. Luiten, *Struct. Dyn.*, 2018, **5**, 051101.

CHAPTER 16

Next-generation Electron Sources

M. KRÜGER*[a] AND P. HOMMELHOFF[b]

[a] Department of Physics and Solid State Institute, Technion – Israel Institute of Technology, 32000 Haifa, Israel; [b] Department of Physics, Friedrich-Alexander-Universität Erlangen-Nürnberg (FAU), Staudtstraße 1, Erlangen 91058, Germany
*Email: krueger@technion.ac.il

16.1 Introduction

With the limited space available, this chapter cannot give a complete overview of all aspects of currently used or envisioned electron sources. We will hence select recent highlights in the understanding and development of new ultrafast and highly coherent electron sources. Various excellent reviews have recently appeared that give a more comprehensive overview of these fields, which we will only briefly discuss here. We will point the interested reader to these reviews wherever helpful.

This chapter focuses on needle tip-based electron sources for a simple reason: the smaller the size of the source, the better the transverse beam properties in principle (see Figure 16.1a and b for an illustration). This improvement in transverse beam properties comes at a price: the smaller the physical source size for a given bunch charge, the more Coulomb repulsion the electrons in the bunch will experience at the source. For this reason, the ultrafast electron sources we discuss here are limited to small bunch charges, which, depending on the beam parameters, can range from less than 1 electron per pulse (*i.e.*, <0.16 aC) to roughly 1000 electrons per pulse (*i.e.*, 16 fC). The small yield can be mitigated by

Theoretical and Computational Chemistry Series No. 25
Structural Dynamics with X-ray and Electron Scattering
Edited by Kasra Amini, Arnaud Rouzée and Marc J. J. Vrakking
© The Royal Society of Chemistry 2024
Published by the Royal Society of Chemistry, www.rsc.org

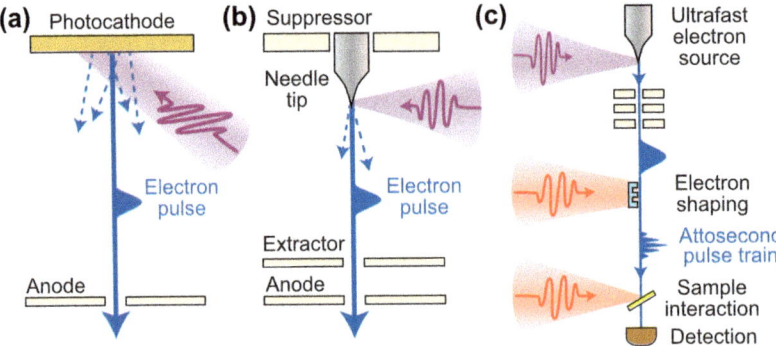

Figure 16.1 Ultrafast electron pulse generation. (a) Laser-triggered flat photocathode electron sources suffer from a large source size (micrometre
range), essentially given by the laser spot size. A multitude of
electron trajectories emerge at various angles from a large area,
leading to a spatially incoherent electron beam. (b) In contrast,
needle tip-based sources are almost point-like with excellent
coherence properties. Here, geometric source sizes are tied to the
size of the tip apex and can reach 100 nm and below. More
importantly, the virtual or effective source size can easily reach
the sub-nanometre range. (c) Electron pulse shaping can be
achieved downstream from the source by bringing the electron
beam into the optical near-field of a laser-illuminated photonic
nanostructure. Most prominently, attosecond electron pulses can
emerge, enabling ultrafast probing of the dynamics of a sample. The
photonic nanostructure is required for efficient electron-light coupling. Alternatively, a two-photon process or ponderomotive scheme
can be used for efficient free-space coupling (not shown here but
discussed in Section 16.3.3).

operating the source at high repetition rates, quickly leading to average
currents of nanoamperes. For many applications, this average beam
current is more than sufficient.

   In addition to Coulomb repulsion of the electrons within a pulse,
vacuum dispersion of electrons also has detrimental effects on electron
pulses. Because of the finite energy width of any electron source, extremely short electron pulse durations created by ultrafast laser triggering cannot be straightforwardly transported to the sample. However,
elegant techniques have recently been demonstrated in shaping electron beams away from the source (see Figure 16.1c). In the second part of
this chapter, we will review highlights around dielectric laser acceleration (DLA), photon-induced near-field electron microscopy (PINEM)
physics, and ponderomotive methods to shape and control electron
pulses. We conclude our chapter with a review of applications of laser-
triggered needle tips and electron beam shaping in ultrafast science
and take a glimpse into the future, focusing on quantum optics with
electrons.

16.2 Laser-triggered Electron Emission from Needle Tips

In this section, we will introduce needle tip emitters and the properties of the resulting electron beams when operated in an ultrafast mode. We will discuss the laser-triggered emission mechanisms that enable femtosecond pulsed operation of the tips and discuss several selected sources.

16.2.1 Needle Tip Electron Emitters and Their Properties

Needle tip emitters have the advantage that they can be used in both DC operation and pulsed operation. Here, we will introduce basic concepts and fundamental properties of needle tip emitters pertaining to both operation modes.

16.2.1.1 *Materials and Geometries*

Needle tips are manufactured from metals (*e.g.*, tungsten, gold, silver, tantalum, iridium, platinum or aluminium) or n-doped semiconductors (*e.g.*, silicon). To obtain nano-scale metal needle tips, electrochemical etching of thin metallic wires is commonly employed (see, for example, ref. 1). This can lead to the production of a typical tungsten tip with a conical shank (see Figure 16.2a). The sharp apex of such a needle tip is approximately hemispherical (see Figure 16.2b), with a radius of curvature down to 5 nm. The resulting emission area is tiny compared to flat photocathodes and micrometre-sized flat-top tips ("microtips"). To obtain larger emission currents, needle tip arrays[2,3] have also been explored (see Figure 16.2c). However, it is challenging to focus the resulting electron beam on a small spot;[4] hence, such arrays are not used in narrow-beam applications such as electron microscopy. Adding a surface adsorbate to metallic needle tips is another important pathway for optimizing emission efficiency in both DC and pulsed operations. This way, typical metallic work functions in the 4–5 eV range can be decreased to below 3 eV. Prominent examples are ZrO_x-coated tungsten W(100) tips with a resulting work function of $\Phi \sim 2.8$ eV[5,6] (see Figure 16.2d–f) and diamond-coated tungsten needle tips ($\Phi \sim 2.8$ eV).[7] This is comparable to the work function of conventional flat-top LaB_6 emitters ($\Phi \sim 2.7$ eV).[8] Heating the tip can create a thermally excited electron distribution at energies above the Fermi level, effectively decreasing the energy required to overcome the work function and further improving laser-driven emission processes.

16.2.1.2 *Static Field Enhancement and Schottky Barrier Lowering*

Compared to flat photocathodes, needle tips exhibit two important effects when DC voltages are applied. First, the sharp apex geometry of the tip leads to DC field enhancement *via* the lightning rod effect, resulting in extremely high electric field strengths on the order of $V\,nm^{-1}$ with modest voltages

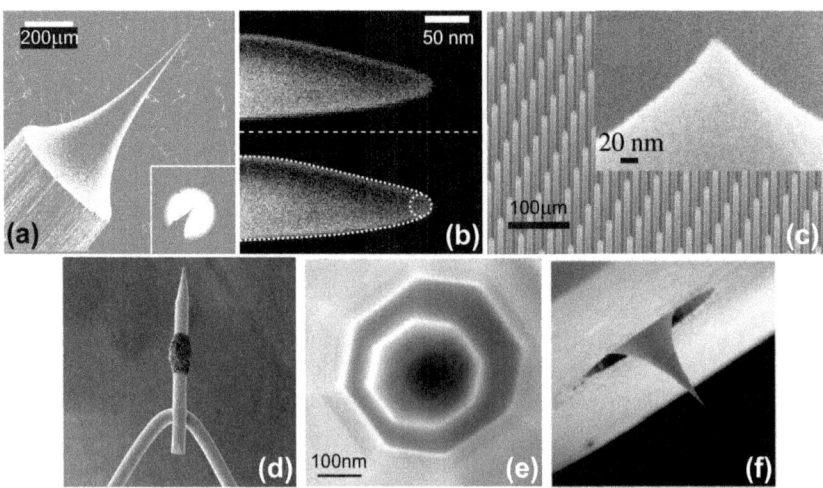

Figure 16.2   Needle tip electron emitters. (a) Scanning electron microscope (SEM) image of a tungsten needle tip (inset: optical microscope image). (b) SEM micrograph of the tip apex. In the lower copy of the image, the lines mark the tip shank and the tip radius (~12 nm). (c) Silicon tip array (inset: zoom-in on the tip). Reproduced from ref. 3 with permission from American Chemical Society, Copyright 2014. (d) ZrO$_x$/W(100) needle tip assembly used for ultrafast transmission electron microscopy. The ZrO reservoir is located in the centre of the image. (e) Top view of the needle tip apex. (f) The needle tip is situated in a suppressor electrode. Panels (d–f) reproduced from ref. 6 with permission from Elsevier, Copyright 2017.

below 100 V, for example. The field strength is given by $F_{DC} = U/(kR)$, where U is the applied DC voltage, R is the tip radius and k is a geometry-dependent field reduction factor[9] (typically $k \sim 5...15$). Second, the use of large field strengths leads to a pronounced Schottky effect, causing an effective reduction in the barrier height due to the combined influence of a negative DC voltage and the image charge potential. The resulting effective barrier height including the Schottky effect is given by

$$\Phi_{\text{eff}} = \Phi - \sqrt{\frac{e^3 F_{DC}}{4\pi\varepsilon_0}} \qquad (16.1)$$

where e is the elementary charge and ε_0 is the vacuum permittivity. Typically, the Schottky effect can reduce Φ by up to 2 eV. Both effects (Schottky and cold field emission) lead to strong spatial localization of DC electron emission to the tip apex.

16.2.1.3   Optical Near-field Enhancement

Like DC field enhancement, optical near-field enhancement occurs at the apex of needle tips, leading to strongly localized AC fields.[10–13] Optical field enhancement factors are typically in the range of 3–8, meaning that the required laser intensity is reduced by up to two orders of magnitude. In special settings, field enhancement factors can be even more prominent. This effect relaxes the requirements on the laser system so that high repetition rate sources such as oscillators can be employed. The field driving the photoemission is no longer the incoming field but the strongly enhanced optical near-field. Like its static counterpart, it enables the spatial confinement of the ultrafast electron emission to the tip apex. For needle tip experiments, the field component parallel to the tip's pointing direction is enhanced strongly, whereas the enhancement of perpendicular field components is insignificant. Beyond free-space irradiation of the tip apex, there are two additional pathways to obtain a strongly enhanced field at the apex. First, in plasmonic nanofocusing,[14,15] a near-infrared field irradiates a discontinuous structure at the tip's shank, such as a grating, groove, or grain boundary. A travelling surface plasmon is generated, which propagates down the tapered shank and localizes at the tip apex (see Figure 16.3). Large field strengths can be reached, leading to efficient nonlinear photoemission.[16–18] Second, the needle tip can be made from an optical fibre and coated with a thin metal layer. This enables fibre coupling of the light to the apex, resulting in photoemission.[19] Both pathways enable the placement of a sample close to the tip while avoiding uncontrolled irradiation of the sample with the laser light that is supposed to trigger photoemission only.

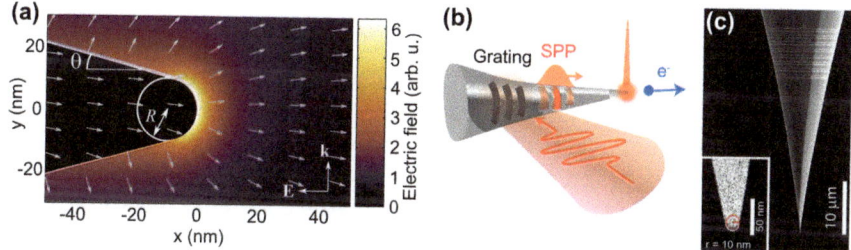

Figure 16.3   Coupling light to the apex of a metallic needle tip. (a) Near-field enhancement due to free-space irradiation of the apex. The colour plot shows the local electric field strength calculated for a sharp tungsten tip, 800 nm light (radius $R = 10$ nm, shank opening angle $\theta = 15°$). The grey arrows indicate the direction of the field in space. Reproduced from ref. 10, https://doi.org/10.1088/1367-2630/17/6/063010, under the terms of the CC BY 3.0 license https://creativecommons.org/licenses/by/3.0/. (b) Principle of plasmonic nanofocusing. A laser pulse excites a surface plasmon polariton (SPP) at the shank of the tip with the help of a grating coupler. The SPP travels down the shank and localizes at the apex, triggering photoemission. (c) Electron microscope image of a gold needle tip with a grating coupler (inset: apex region). Reproduced from ref. 15, https://doi.org/10.1080/23746149.2016.1177469, under the terms of the CC BY 4.0 license https://creativecommons.org/licenses/by/4.0/.

16.2.1.4 Repetition Rate, Peak Current and Coulomb Repulsion

In the ultrafast operation of a needle tip electron source, the average beam current is best represented as $J_{avg} = Nef_{rep}$, where N is the number of electrons per electron bunch, Ne is the total electron bunch charge, and f_{rep} is the laser repetition rate. For characterizing an ultrafast emitter, the definition of a peak current is useful, $J_{peak} = Ne/\tau$, where τ is the electron pulse duration. Coulomb repulsion between individual electrons within an electron pulse becomes important, which is a major limiting factor for these sources. This effect must be carefully considered near the emitter where it is strongest and where a multitude of electrons might be present, strongly confined in space (due to the size of the emitter) and time (due to the duration of the laser pulse that drives the photoemission). Many fundamental science experiments are hence carried out in the single electron bunch regime to avoid Coulomb repulsion effects. Several experimental studies found that Coulomb repulsion can significantly increase the electron pulse duration, broaden the energy spectrum and enlarge the size of the electron beam spot size downstream at the sample.[20–22] Such effects are clearly observed in an electron holography experiment with a tungsten needle tip source (400 fs electron pulse duration at the sample). However, the quality of the hologram appears to be unaffected by electron repulsion effects for up to about 10 electrons per pulse,[23] which is encouraging for applications requiring high peak currents. As an alternative to increasing the peak current, f_{rep} can be increased for optimizing the signal in downstream measurements. Similarly, quantum-path interference with multi-colour fields may be used to increase the yield.[24]

16.2.1.5 Spatial Coherence and Brightness

Due to their sharp apex geometry, needle tips are essentially point-like electron sources and possess excellent spatial coherence properties. If a sample of a certain size is illuminated coherently, effects like electron diffraction and holography yield excellent interference patterns with a high degree of visibility. Here, the quantity of interest, the transverse spatial coherence length $\xi_\perp$ of the electron beam at the sample, must be larger than the characteristic length scale of the sample (*e.g.*, the unit cell dimensions of a crystalline material). The van Cittert–Zernike theorem[25] relates $\xi_\perp$ to the virtual source radius $r_\perp$ of the emitter through $\xi_\perp \sim 1/r_\perp$. The virtual source radius is typically smaller by a factor of ten compared to the geometrical source radius so that needle tip sources can possess virtual source sizes on the order of 1 nm or less even for considerably large apex radii.[26–28] Nearly perfect coherence is achieved by single atoms tips (*i.e.*, nanotips with a single protruding atom located at the apex), only limited by the fundamental Heisenberg uncertainty principle.[29–31] The geometric beam, current and coherence properties are captured by the reduced source brightness B_r, which is conserved along the beam path regardless of acceleration, assuming the absence of aberrations. The source brightness is given by

$$B_r = \frac{J_{avg}}{\pi r_\perp^2 \Omega U},$$ (16.2)

where Ω is the solid angle of emission and U is the acceleration voltage. Apertures in the beam path do not affect B_r because Ω and J_{avg} decrease together, but a decrease in the current should be avoided as much as possible for practical purposes. For ultrafast sources, it is useful to replace J_{avg} with J_{peak}, hence yielding a reduced peak brightness. In our discussion of photoemission mechanisms, we will highlight a few state-of-the-art sources and their brightness properties. For a recent overview of reduced peak brightness values for laser-driven tip sources, we refer the reader to ref. 32.

16.2.2 Photoemission Physics of Needle Tips

This part will introduce the various electron photoemission mechanisms from needle tip emitters. We will also highlight a few sources and the properties of the resulting electron pulses.

16.2.2.1 Linear Photoemission

The photoelectric effect provides the most straightforward mechanism for ultrafast electron emission. Through the absorption of a photon of energy $\hbar\omega$, an electron can overcome the work function Φ_{eff} (see Figure 16.4a). Assuming that the electron is promptly released from the surface,[†] the electron pulse duration corresponds to the duration of the intensity envelope of the laser pulse (see Figure 16.4b). Ultrafast linear photoemission uses ultraviolet (UV) or visible light, and has been observed from a wide range of tip emitters. Needle tip cold field emitters and Schottky emitters are employed for linear photoemission, such as tungsten needle tips,[27] $ZrO_x/$W(100) needle tips[5,6] and diamond-coated tungsten needle tips.[7] A special case of linear photoemission is photofield emission,[34,45] where electrons excited by single-photon absorption are subsequently emitted through DC field emission. Ultrafast photofield emission has been demonstrated from tungsten needle tips.[35,36]

An example for linear photoemission is shown in Figure 16.4c, where a $ZrO_x/$W(100) needle tip is triggered by UV light inside of an ultrafast SEM.[5] In this pioneering study, decreasing the wavelength leads to higher source stability (see Figure 16.4d). A similar source[37] driven at 2–8 MHz produces a pulsed electron beam with $J_{avg} \sim 3$ nA, $N \sim 840$ and $\tau \sim 650$ fs. For the latter source, ref. 32 estimates an upper bound of $B_r \sim 7.9 \times 10^8$ $A\,m^{-2}\,sr^{-1}\,V^{-1}$. The spectral bandwidth for typical applications in ultrafast transmission electron microscopes (TEMs) is usually below 1 eV for needle tips employed in an ultrafast TEM, such as the one described by Feist *et al.*[6] (see Figure 16.4e). In this study, currents up to hundreds of fA have been achieved at $f_{ref} = 800$ kHz. An optimization of this system for the highest coherence and smallest

---

[†] The emission delay on a metal surface amounts up to hundreds of attoseconds, depending on the material and wavelength, and has been well studied for extreme-ultraviolet light pulses.[33]

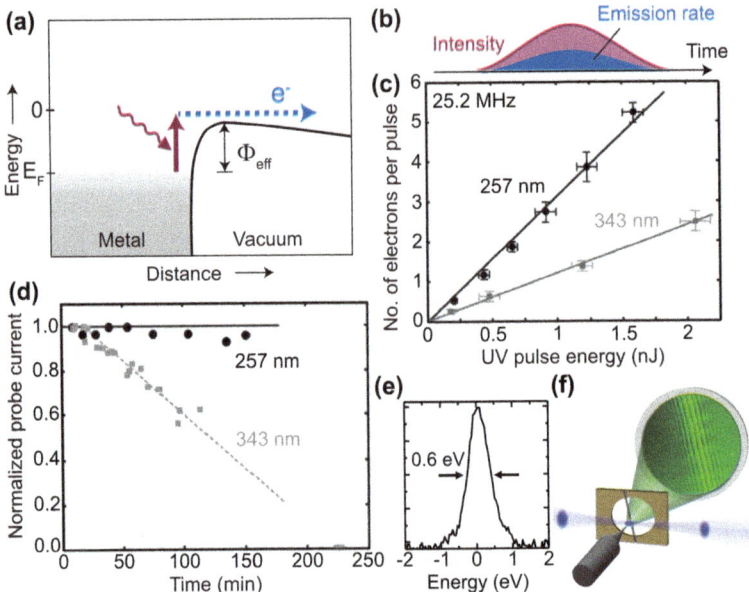

Figure 16.4   Linear photoemission regime. (a) An electron (blue; Fermi energy E_F, Schottky-reduced effective work function Φ_{eff}) is emitted after absorbing a photon (violet). (b) The emission rate (blue) follows the intensity envelope of the pulse (violet). (c) Dependence of the number of electrons per pulse on the pulse energy for a ZrO$_x$/W(100) needle tip triggered by two UV wavelengths. (d) Emission current stability when triggered by two UV wavelengths. Panels (c and d) reproduced from ref. 5 with permission from the National Academy of Sciences, Copyright 2010. (e) Typical energy spectrum for a ZrO$_x$/W(100) needle tip. Reproduced from ref. 6 with permission from Elsevier, Copyright 2017. (f) Spatial coherence study of a UV-driven tungsten needle tip emitter. Reproduced from ref. 27 with permission from the American Physical Society, Copyright 2015.

spot size below 1 nm yields $J_{avg} \sim 0.3$ fA, $N \sim 2.3 \times 10^{-3}$ (at the sample), $\tau \sim 200$ fs and $B_r \sim 1 \times 10^8$ A m^{-2} sr^{-1} V^{-1}. The corresponding effective source radius is estimated to be smaller than 100 nm. A coherence measurement performed on a laser-driven W(310) needle tip emitter[27] (see Figure 16.4f) demonstrated a transverse coherence length of below 1 nm, on par with the cold field emission operation of the emitter, and the coherence length was determined from the visibility of an interference pattern created by a nearby free-standing carbon nanotube (CNT).

16.2.2.2   Multiphoton Photoemission

In multiphoton photoemission, the emission of an electron involves the absorption of at least two photons (see Figure 16.5a). The photocurrent

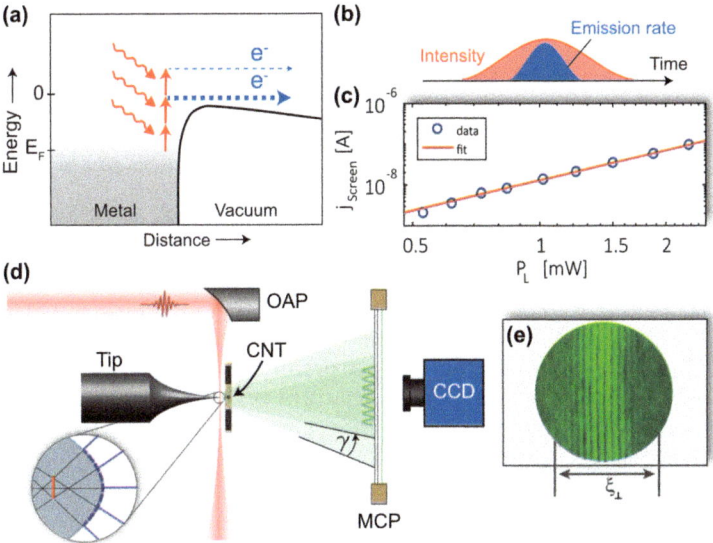

Figure 16.5 Multiphoton photoemission regime. (a) Multiple photons need to be absorbed to lift an electron over the barrier. At higher intensities, more emission channels beyond the lowest multiphoton order are observed (above-threshold photoemission), indicating the generation of higher kinetic energy electrons. (b) Due to the nonlinear emission process, the emission rate (blue) is confined to a shorter time than the envelope of the light pulse (red). (c) Typical intensity scaling, directly revealing the leading multiphoton order (here: ∼2). (d) Setup for measuring the transverse spatial coherence of multiphoton-emitted electrons with the help of a carbon nanotube (CNT) electron beamsplitter. OAP: off-axis parabolic focusing mirror and MCP: micro-channel plate. The inset shows the geometric source area (blue dashed surface) and the virtual or effective source area (red). (e) Camera image of the phosphor screen behind the MCP electron detector shown in (d). Clear electron interference fringes are observed. The transverse coherence length, $\xi_\perp$ at the MCP is indicated, here roughly 6 mm. Panels (c–e) reproduced from ref. 28 with permission from AIP Publishing, Copyright 2018.

scales according to the power law $J \propto I^n$, where I is the light intensity and n is the minimum number of photons required to overcome Φ_{eff}. The electron pulse duration at the source is shorter than the laser pulse duration (see Figure 16.5b) due to the nonlinear scaling, which can be advantageous for ultrafast applications. At larger intensities, above-threshold photoemission occurs (*i.e.*, the generation of photoelectrons with energies corresponding to the absorption of more photons than necessary[38,39]). This process marks the beginning of the transition into the strong-field regime of photoemission.[40] Sub-picosecond multiphoton photoemission has been observed in a wide range of systems, including flat metal surfaces[41,42] and needle tips.[28,43]

Figure 16.5c shows a typical multiphoton current-intensity scaling of a W(310) needle tip source.[28] The source has excellent coherence properties (see Figure 16.5d and e). The effective source radius is determined to be 0.65 nm, which is significantly smaller than the geometrical tip radius of about 6 nm. A multiphoton needle tip source based on a W(310) tip has recently been integrated into a TEM by Houdellier *et al.*[44] At a repetition rate of $f_{rep} = 1$ MHz, an average current of $J_{avg} \sim 80$ fA (or $N \sim 0.5$) is generated. At the sample, the pulse duration is $\tau \sim 500$ fs and the reduced peak brightness is estimated to be $B_r \sim 1 \times 10^9$ A M^{-2} sr^{-1} V^{-1}. The energy width of the electron pulse is about 1 eV when operated at $f_{rep} = 2$ MHz in the single electron regime.[23] An improved average current of $J_{avg} \sim 4$ pA (or $N \sim 16$) can be achieved, but with energy width and pulse duration broadening.[23,45] Another multiphoton-regime tungsten needle tip source used for ultrafast low-energy electron diffraction (LEED) is reported in ref. 46, underlining the suitability of tip sources for diffraction experiments.

16.2.2.3   *Strong-field Tunnelling Photoemission*

In the strong-field regime, the optical field is sufficiently strong that the surface potential barrier is reduced at its extrema, and a significant number of electrons can tunnel out when the barrier is sufficiently small (see ref. 11–13 for reviews). Hence, electron emission here is confined to bursts of sub-optical-cycle duration, enabling the generation of attosecond electron wavepackets.[39] Optical near-field enhancement at tips in conjunction with ultrashort laser pulses enables strong-field physics on metallic surfaces and helps avoid the destruction of the emitter. An extreme case of strong-field tunnelling photoemission is THz-driven field emission,[47–49] nicely representing the transition to the DC case. To the best of our knowledge, tunnelling photoemission has not been used as a source for downstream spatiotemporal studies because of the large energy spread and the concomitant loss of temporal resolution upon long propagation distances of the electron beam.

16.3   Laser-driven Electron Acceleration and Modulation

For many applications, important electron pulse characteristics cannot be obtained solely from a bare electron source. Examples include attosecond electron pulses, spectrally very narrow pulses, or high energy pulses. For this, we now expand the notion of the electron source to include a bare source element and additional components or interaction zones that help shape the electron beam to the desired parameter range. In the following, we highlight three highly successful means recently introduced and explored, all relying on electron manipulation using laser light.

16.3.1 Dielectric Laser Acceleration

In dielectric laser acceleration (DLA), a photonic nanostructure generates an optical near-field mode with a phase velocity matched to the velocities of electrons to be accelerated (or bunched or shaped). A near-field mode is required because otherwise an efficient interaction with electrons and light cannot be sustained over extended distances much longer than a wavelength of the driving light. For an electron travelling with a speed $v = \beta c$, where c is the speed of light, the fields decay away from the nanostructure's surface with a decay length given by $\beta\gamma\lambda/(2\pi)$; here, γ is the Lorentz factor and λ is the driving wavelength. The decay of this near-field is due to the evanescent nature of the near-field accelerating mode; this evanescent nature is well known to appear, for instance, on the outside of an entirely reflective surface. Hence, for efficient electron–light interaction, the electrons should be less than a wavelength away from the surface, putting stringent requirements on the electron beam quality. To excite the optical mode co-propagating with the fast electron, a synchronicity condition has to be met: $\lambda_p = n\beta\lambda$, where λ_p represents the period of the nanostructure, and n is an integer, referring to higher spatial harmonics if $n > 1$. Most often, $n = 1$ is chosen, but up to $n = 5$ has been demonstrated[50] (see Figure 16.6 for some examples of DLA nanostructures). We note in passing that the nanostructure does not have to be periodic; a flat surface suffices, but in this case, the illumination has to be well chosen.[51]

The initial ideas for dielectric laser acceleration date back to the invention of the laser.[57,58] The initial experimental demonstration of laser acceleration at a periodic metal structure was performed using terahertz radiation in the 1980s,[59] representing the demonstration of the inverse Smith–Purcell effect. Energy modulation at a single interface was shown in 2005.[60] Shortly thereafter, in a series of proposal papers, Plettner, Byer and colleagues have developed various devices arising from nanostructured dielectrics, including compact on-chip light sources (see, for example, ref. 61–63). In 2013, dielectric laser acceleration was demonstrated with both relativistic electrons[52] and non-relativistic electrons.[53,64]

Given that transparent optical materials (dielectrics) have damage thresholds reaching 10 GV m^{-1}, dielectric laser acceleration is poised to accelerate electrons much more efficiently than conventional radio-frequency (RF) accelerators, which are limited to roughly 100 MV m^{-1}. The two orders of magnitude larger damage threshold of dielectrics at optical frequencies should allow reaching two orders of magnitude larger acceleration gradients. Since the acceleration gradient determines the size of an accelerator, a 1 m long hospital accelerator might be shrunk to a 1 cm long device, which, if appropriately powered *via* fibres, might lead to new endoscopic proximity radiation tools for clinicians, or compact on-chip light sources, for example.[65] The largest acceleration gradient reported to date is 1.8 GeV m^{-1} for relativistic electrons,[66] and 376, 210 and 178 MeV m^{-1} for non-relativistic electrons in various settings.[67–69]

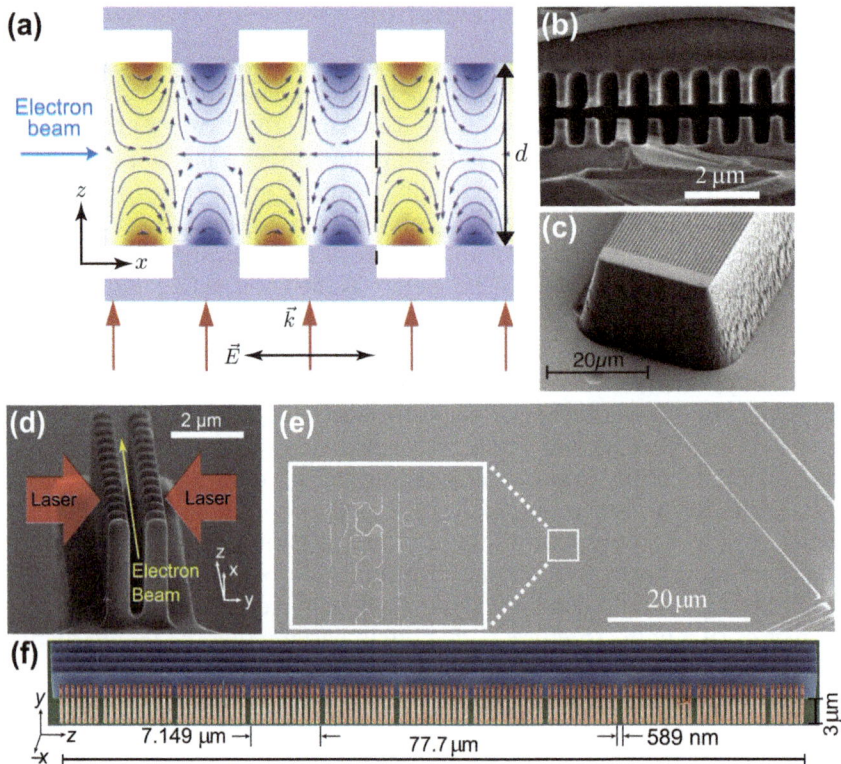

Figure 16.6   Structures for dielectric laser acceleration. (a) The concept: an optical mode co-propagating with the electrons in the *x*-direction is generated with a pulsed laser beam impinging in the *z*-direction. The colour scale shows the field of the co-propagating mode. The channel width *d* should be smaller than the driving wavelength. (b) A dielectric laser acceleration structure made from two etched fused silica grating bonded together, enabling a proof-of-concept experiment with relativistic electrons. Reproduced from ref. 52 with permission from Springer Nature, Copyright 2013. (c) The single-sided fused silica grating structure used for the proof-of-concept of the non-relativistic DLA experiment. Reproduced from ref. 53 with permission from American Physical Society, Copyright 2013. (d) Symmetrically driven silicon dual pillar gratings for electron acceleration. Reproduced from ref. 54 with permission from The Optical Society, Copyright 2018. (e) Inversely designed nanophotonic structure for highly efficient DLA. Reproduced from ref. 55 with permission from the American Association for the Advancement of Science, Copyright 2020. (f) Silicon photonic nanostructure for electron beam phase-space control. Reproduced from ref. 56 with permission from Springer Nature, Copyright 2021.

Since dielectric laser acceleration is based on the very same principles as conventional RF acceleration, it allows performing many of the same features well known from classical accelerator physics, but on much smaller length scales and much faster time scales – both scaled by the driving frequency ratio of microwaves to the optical drive (*i.e.*, roughly a factor of 10 000). The following applications have been demonstrated so far in DLA:

- *Deflection and streaking.* In addition to energy modulation, providing transverse forces to keep electrons on track in a DLA structure is mandatory. With two-sided driving fields, transverse and longitudinal forces can be coupled and controlled elegantly by shifting the relative phase between the two pulsed laser beams[54] (see Figure 16.7a).

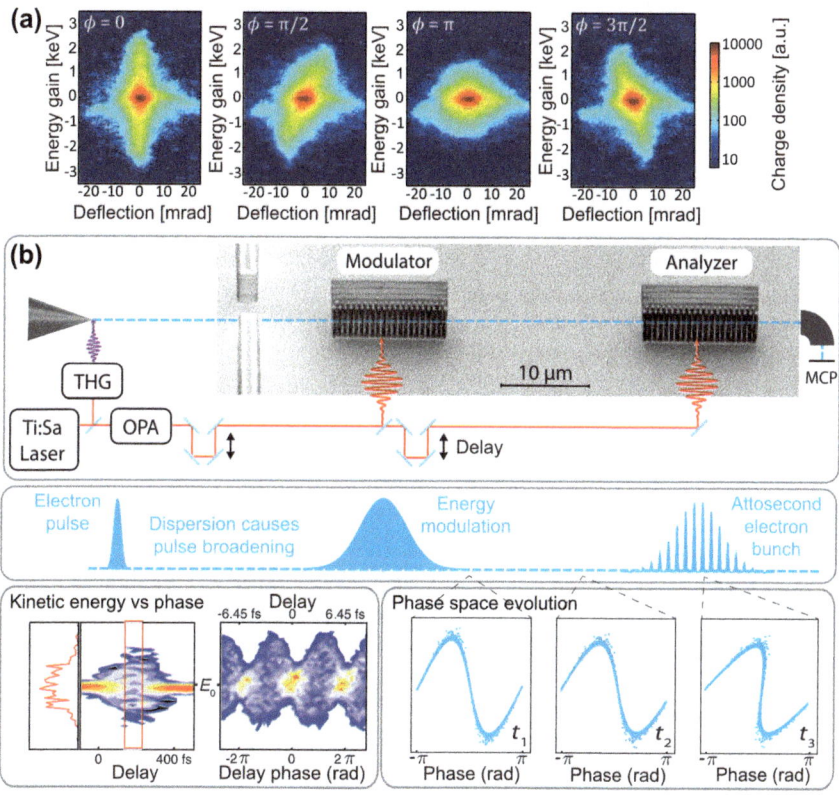

Figure 16.7 Acceleration, deflection and attosecond bunch generation in a DLA. (a) Electron energy spectra (vertical axis) and transverse position (horizontal axis) for four different relative phases between two pulsed laser beams impinging from opposite directions on a structure, as shown in Figure 16.6d. The electron cannot only experience acceleration but can also be deflected left and right. Reproduced from ref. 54 with permission from The Optical Society, Copyright 2018. (b) Electrons traverse the first DLA interaction structure, the modulator, where they experience a periodic energy modulation. After ballistic propagation, this energy modulation translates into a density modulation, which can be probed in a second DLA interaction region, the analyzer. Recording spectra as a function of the phase between the two interaction regions allows reconstruction of the electron (phase space) density. Reproduced from ref. 70 with permission from the American Physical Society, Copyright 2019.

- *Sub-cycle electron bunch train generation: attosecond pulse generation.* Given the optical period length of 6 fs (for 2 μm central wavelength), it is nearly impossible to inject electron pulses into a DLA structure with a duration shorter than an optical period. This means that the electron pulse experiences accelerating and decelerating parts of the synchronous mode repeatedly (see Figure 16.7b). Because of the non-relativistic nature of the electrons, a drift space after an energy modulation region is sufficient to realize a periodic density modulation so that the faster (accelerated) electrons catch up with the slower (decelerated) ones, which occurs periodically. This results in an attosecond electron pulse train. In a second (analyzer) interaction region, a second laser pulse, which is phase-stable with respect to the first (buncher) laser pulse in the interaction region, can probe the electron density distribution. Here, spectrograms are measured as a function of phase delay between the two pulsed laser beams, or transverse streaking in the analyzer structure can be used to probe the bunched electron pulse. The shortest duration of a bunchlet (*i.e.*, of a single pulse within the pulse train) measured this way was 270 ± 80 as (FWHM).[70,71] For bunching with a narrow energy spread, a more intricate design has been proposed.[72]

- *Net energy gain.* Once the electrons (or the periodically spaced electron bunchlets) are shorter than the driving period, they can be injected into the purely accelerating phases of the phase-synchronous mode. While an important achievement in itself, a demonstration of this net acceleration is part of the attosecond bunching demonstration.[70,71] Similarly, a distribution of the incoming electron energy spectrum to acceleration and deceleration, and the resulting double-horn spectra, can be found in ref. 73.

- *Complex electron phase space manipulation by alternating phase focusing.* In any accelerator, electron loss needs to be avoided. With purely accelerating forces present, electrons can inevitably be lost by crashing into the walls of the accelerating structure. The alternating phase focusing scheme, well-known in accelerator physics for decades, is well suited for keeping electrons bunched together while accelerating them, also in the DLA realm.[74,75] In the first step, pure (active) transport has been demonstrated in a structure (see Figure 16.6f). The gaps in the colonnade structure can lead to phase jumps between the propagating electron and the optical near-field mode. The phase jumps can be chosen such that, as a net effect, electrons are kept together both longitudinally and transversely.[56]

- *Optical waveguide-fed DLA.* For future applications, it will be highly beneficial if the driving laser light could be directed into the DLA channel through on-chip waveguides instead of relying on free space propagation and focusing (see the structure shown in Figure 16.6e). Here, the acceleration channel is powered by optical waveguides from either side.[55] A part of the grating coupler is used to couple the laser light into the waveguide (see the lower right part of Figure 16.6e).

- *Inverse nanophotonics design for high-efficiency DLA coupling.* High-efficiency DLA coupling can be achieved with a periodic structure but with much more complex features (see Figure 16.6e) than in all other DLA structure examples shown elsewhere (see Figure 16.6). These features result from a numerical optimization process called inverse design. Inverse design determines an optimal solution for a given set of boundary conditions. For DLA, excitation of the accelerating mode with the highest efficiency is most important and can be achieved this way.[55] Similarly, easier to handle geometries, such as the top illumination, are currently being explored.[76]
- *Material optimization.* So far, most studies have been done with standard materials (*e.g.*, fused silica and silicon). Special coatings[77] or materials such as diamond[78] may enable even larger driving field strengths.

16.3.2 Photon-induced Near-field Electron Microscopy

Combining electron beams and optical near-fields has also been conceived with imaging in mind. Zewail and co-workers invented photon-induced near-field electron microscopy (PINEM[79]). Here the electron beam from a TEM interacts with a near-field laser pulse and enables the imaging of the optical near-field with the high spatial resolution capabilities of a (scanning) TEM. The near-field can be generated, for example, by a nanostructure (see ref. 79 and 80). The near-field modifies the energy of the electron beam like discussed before, leading to sidebands in the energy spectrum (see Figure 16.8a–d). We can easily measure PINEM spectra with a standard electron spectrometer attached to the TEM. In another seminal contribution, Ropers and co-workers have shown the quantum coherent nature of the PINEM scheme,[80] leading to attosecond electron bunch train generation in a similar fashion as discussed above (see Figure 16.8d). Also, two interaction regions could be addressed independently here, enabling Ramsey-type experiments on the electron energy ladder.[81] In later work, the Wigner function of the electron was reconstructed,[82] showing clear evidence for attosecond pulses (see Figure 16.8e).

Also, other setups and structures can be used to achieve PINEM-type energy modulation and attosecond electron pulse trains. For instance, an electron beam can be phase-matched to an evanescent field at a prism,[51] resulting in strong PINEM-type spectra[83] (see Figure 16.8b). Moreover, an electron beam passing through a thin metallic foil illuminated with laser light from one side experiences the same kind of energy modulation as in PINEM, resulting in sidebands[85] and attosecond electron pulse train generation,[82,86,87] with applications in electron diffraction.[86] Recently, thin foils were used to modulate continuous wave (CW) electron beams,[88] hence lifting the requirements to operate the source in an ultrafast mode; however, only one sideband order could be generated. As another example of achieving PINEM-type energy modulation, DLA and PINEM have been brought closer together by introducing periodic DLA structures into PINEM-type setups. The long

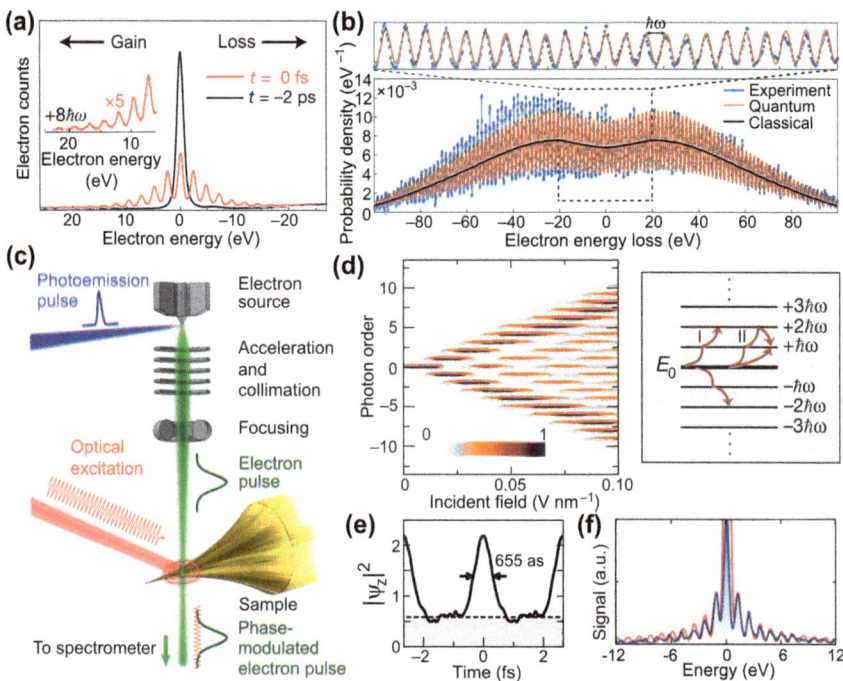

Figure 16.8   Photon-induced near-field electron microscopy (PINEM). (a) An electron pulse passes a nanostructure excited by a laser pulse. When the two pulses overlap in time and space, energetic sidebands (red) appear around the initial energy peak (black) position. Reproduced from ref. 79 with permission from Springer Nature, Copyright 2009. (b) Increasing the interaction length enables extremely broad PINEM-type spectra. Reproduced from ref. 83 with permission from Springer Nature, Copyright 2020. (c) Typical experimental setup for PINEM in a TEM. (d) Electron spectrum as a function of the excitation field. Interference structures are visible due to quantum path interference of Rabi transitions between energy ladder states. (c and d) Reproduced from ref. 80 with permission from Springer Nature, Copyright 2015. (e) The electron wavefunction probability reconstructed from a PINEM-type experiment, exhibiting a train of attosecond pulses. Reproduced from ref. 82 with permission from Springer Nature, Copyright 2017. (f) PINEM observed in an SEM at 17 keV. Reproduced from ref. 84 with permission from the American Physical Society, Copyright 2022.

interaction distance facilitated by the phase-matched DLA structure in a TEM allowed observation of pronounced PINEM sidebands with a CW electron beam.[89] Two recent works employed photonic cavities to achieve pronounced PINEM spectra in pulsed operation[90] and CW operation;[91] similarly, the DLA structure used in ref. 89 was a low-Q resonator around the electron beam, enabling CW operation even with non-coherent light. Although not surprising from the fundamental physics perspective, it is interesting to note that PINEM physics has now also been observed in a scanning electron microscope (SEM) with electron energies down to 10 keV (see Figure 16.8f). These microscopes are much better suited for extended light-electron interaction regions, and might enable reaching electron-based multi-interaction zone schemes more easily than what is possible in the small sample chambers of a TEM.[84] Also, low-energy electrons below 1 keV can be subjected to PINEM, as proposed in ref. 92.

16.3.3   Ponderomotive Acceleration and Modulation

DLA and PINEM represent processes in which the optical field directly interacts with the electrons. For this to work over extended distances and to observe a net electron momentum change, a third body is needed to fulfil energy and momentum conservation. Free space or true vacuum schemes, not requiring a third body, can be realized by utilizing a two-photon process or, in other words, a ponderomotive scheme. Naturally, such a scheme requires the presence of two light beams, analogous to atomic physics experiments.[93]

If only forward momentum modulation is achieved again, a stationary optical wave can be generated in the rest frame of the electrons (*i.e.*, an optical travelling wave co-moving with the electrons; see Figure 16.9a and c). In remarkable similarity to what we discussed above, the electron energy is also modulated here depending on which phase of the travelling wave an electron experiences.[94,96] Intriguingly, because this is a vacuum scheme, extremely high acceleration gradients may be reached, only limited by the Schwinger limit of electron pair production (and dephasing, which can be compensated in principle). So far, 2.2 GeV m^{-1} has been demonstrated with modest laser pulse parameters.[94] Again, this may lead to attosecond pulse trains, with the shortest bunchlet duration within the train reported to be 260 as.[97] Interestingly, this scheme can be extended to single attosecond pulse generation with the help of a third, correctly timed laser pulse,[98] and may even generate electron vortex beams all-optically.[99]

Based on the same effect but used in a somewhat different geometry, optically generated phase masks have also been demonstrated. Electron-optical phase masks are extremely important for various electron-based imaging modalities, and an optical vacuum scheme is highly attractive for its immaterial character, meaning that the electron beam cannot damage the phase mask, nor can the phase mask change its properties because of deposition of atoms and molecules from the rest gas. Such a phase mask was recently demonstrated based on an optical cavity with a CW intracavity power sufficient for significant electron phase modulation[95] (see Figure 16.9d–g).

16.4   Applications and Future Perspectives

High-brightness ultrafast needle tip sources and coherent downstream manipulation of free electrons are promising for many applications and new fundamental insights.

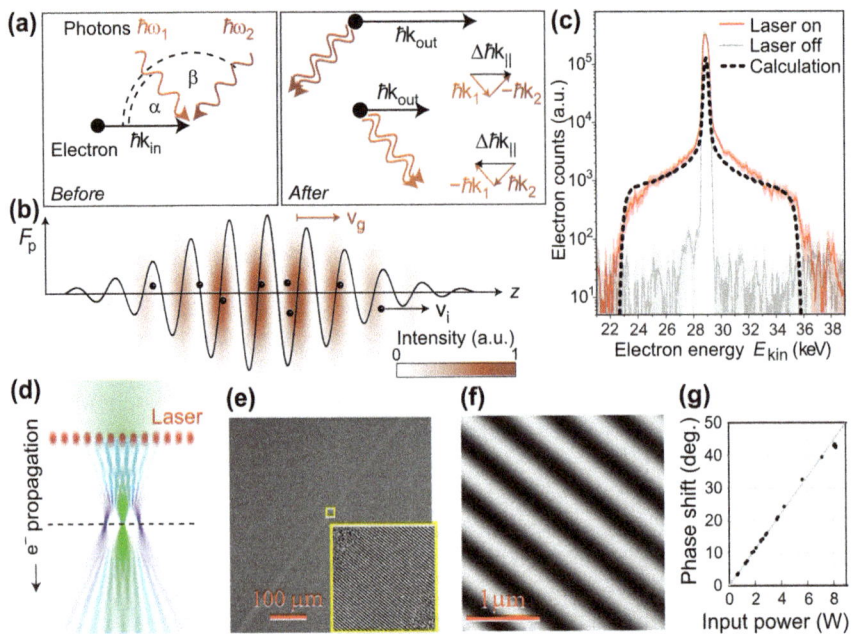

Figure 16.9   Ponderomotive electron acceleration and modulation. (a) Illustration of two-colour, two-photon driven electron acceleration and deceleration. Energy and momentum conservation can be fulfilled in these processes. (b) Time-domain picture: an optical travelling wave with velocity V_g is co-moving with the electron beam. (c) Strong energy gain and loss is imprinted on the electron spectrum. (a–c) Reproduced from ref. 94 with permission from Springer Nature, Copyright 2018. (d) A TEM beam passes through the standing wave of a high-intensity optical cavity, resulting in an electron phase mask. (e) Electron micrograph of the standing wave. (f) High-resolution micrograph. (g) The phase shift scales linearly with intensity, confirming the ponderomotive nature of the process. (d–g) Reproduced from ref. 95 with permission from Springer Nature, Copyright 2019.

16.4.1   Applications

Most importantly, ultrafast needle tip sources (see Section 16.2) led to the development of ultrafast TEM (UTEM), described in detail in Chapter 15 of this book, along with ultrafast SEM. As introduced in Section 16.3, downstream laser acceleration and modulation can help turn an otherwise "long" electron pulse into an attosecond pulse train or even a single pulse. In the following, we will describe a few selected applications of ultrashort electron pulses generated along the lines described in this chapter. Like most ultrafast measurements, these applications are based on the pump–probe paradigm, where a laser pulse induces structural dynamics in a sample, and a synchronized electron pulse probes the subsequent temporal and spatial evolution.

Starting with the attosecond timescale, ultrafast electron pulses allow for the microscopy of electromagnetic waveforms. While the first experiment demonstrated time-resolved imaging of THz domain waveforms in a ring resonator,[100] the downstream generation of attosecond electron pulse trains enables imaging in other spectral domains, such as in the infrared.[86] Different snapshots of the field at a foil induced by a laser pulse (see Figure 16.10a) reveal that it is possible to reach a temporal accuracy of 10 as and micrometre spatial resolution. Here, the laser pulse must possess the same periodicity as the electron pulse modulation, and this requirement can be lifted by employing a single isolated electron pulse.[87]

Moving to picosecond timescales, Kurman *et al.* demonstrated the imaging of phonon polariton propagation with femtosecond pulses in a UTEM.[101]

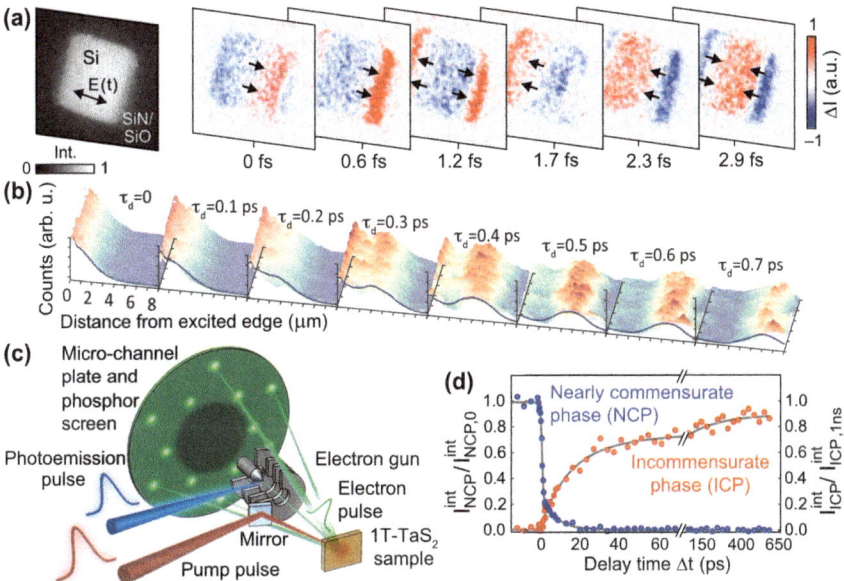

Figure 16.10 Applications in microscopy. (a) Electromagnetic waveform imaging. Left: Shadow image of a silicon membrane (grey) irradiated with an excitation field $E(t)$ (black arrow). Right: Images of intensity differences as a function of time delay reveal the electromagnetic waveform in space and time. Reproduced from ref. 86 with permission from Springer Nature, Copyright 2018. (b) Phonon polariton imaging. Measurement of the electrons that gained energy for different time delays between the laser pulse and the electron probe pulse, revealing movie frames of the propagating phonon polariton wave. Reproduced from ref. 101 with permission from the American Association for the Advancement of Science, Copyright 2021. (c) Experimental setup for ultrafast low-energy electron diffraction. (d) Watching the intensity of Bragg spots as a function of the delay between the pump pulse and the electron probe pulse allows for the observation of a transition between two charge density wave phases in 1T-TaS$_2$. (c and d) Reproduced from ref. 105 with permission from Springer Nature, Copyright 2018.

Here, the interaction of the electron pulse with a mid-infrared-range phonon wave leads to PINEM sidebands. Filtering only electrons that have experienced energy gain reveals the time-dependent intensity of the wave, creating a movie of the dynamics (see Figure 16.10b).

Ultrafast electron diffraction directly tracks structural dynamics, primarily rapid changes in the lattice structure, through observing Bragg spots as a function of pump–probe delay.[102,103] The use of needle tip emitters enables relatively compact experiments with low-energy electrons in transmission[104] and reflection.[105,106] Vogelgesang *et al.* used an ultrafast low-energy electron diffraction (ULEED) setup (see Figure 16.10c) to track pump-laser-induced transitions between charge-density wave phases in a 1T-TaS$_2$ sample[105] (see Figure 16.10d). Also, downstream manipulation enhances the capabilities of ultrafast electron diffraction. For example, Tauchert *et al.* imaged the ultrafast demagnetization of a Ni film with sub-ps time resolution and pm lattice movement resolution.[107] The authors found that pump-laser-induced demagnetization leads to an intermediate excitation of polarized phonons. Demagnetization can also be observed by microscopy in a UTEM. Da Silva *et al.* invented femtosecond Lorentz microscopy and imaged magnetic vortices with sub-ps and 100 nm resolution.[108]

16.4.2   Future Perspectives

Here, we focus on a research direction exploring electron–light interactions at the quantum level, which has only started in recent years. Numerous theoretical studies have shed light on new possibilities, such as coherent control of quantum systems by shaped electron bunches,[109–113] electron–photon entanglement,[90] non-classical light emission driven by electrons[114,115] and quantum coherence measurements in materials,[116] to only name a few. As an example of an experimental study involving dielectric laser acceleration, a recent measurement revealed the quantum-classical correspondence principle, with the aid of electron acceleration, through the observation of quantized peaks in the electron energy spectra.[117] Likewise, and with an inverse photonics-designed high-efficiency coupler, even the quantum statistics of the driving light could be revealed from the shape of the energy sideband spectra: thermal light could be directly distinguished from coherent light as the quantum coherence on the electron energy ladder was either gone (thermal) or present (coherent), in direct similarity of a classical Galton board as compared to its quantum version.[89] Quantum electron optics, as this new field may be named, goes hand in hand with current efforts to further increase the light-electron-matter coupling strength,[83,89,91,118] which enables continuous wave operation and single-photon generation heralded by a single electron.[119]

Recently observed electron correlations in the photoemission current from a needle tip have led to another new contribution to the field of quantum electron optics. Two independent studies[120,121] found, with the help of energy-time measurements of the emitted electrons and thorough analysis, that two-electron states are characterized by post-emission Coulomb repulsion, accompanied by a corresponding energy gap between 2 and 4 eV (see Figure 16.11a and b). Cross-correlation experiments reveal that the gap exists only around the temporal overlap between two laser pulses (see Figure 16.11c). Exciting possibilities are opened up by the fact that two-electron states can be separated from single-electron states by spatial and spectral filtering with satisfactory fidelity (see Figure 16.11d). This enables electron heralding, similar to photon heralding,[122] which is a standard approach in quantum optics. Here, one electron serves as a herald, which allows for a deterministic knowledge of the presence of the second electron and its energy. The second electron can be then used for a measurement, providing a viable scheme for quantum imaging with electrons.

A further perspective is opened up by the fact that needle tips emit highly coherent electron pulses with low kinetic energies on the order of 100 eV,

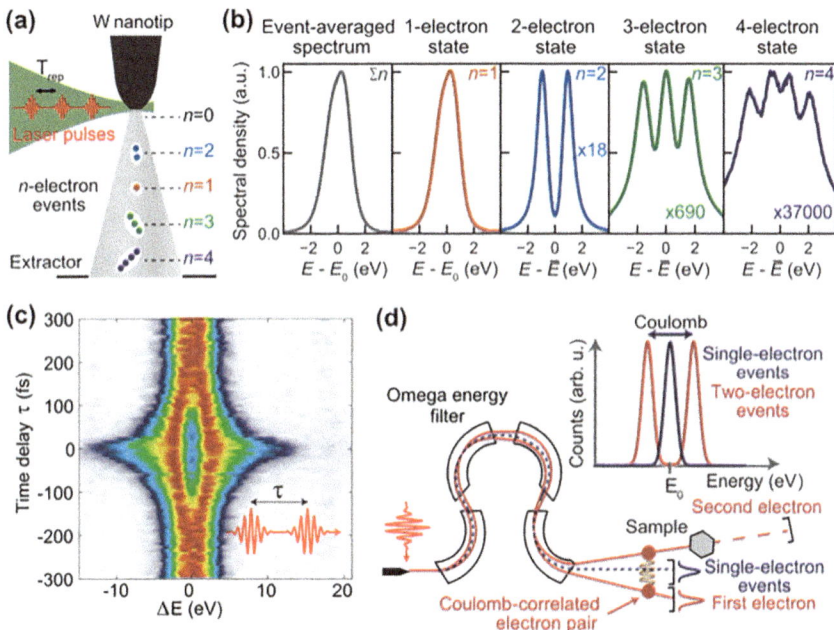

Figure 16.11 Multi-electron effects. (a) Laser-triggered emission from a needle tip in an UTEM leads to single-electron emission and multiple electron states. (b) Coulomb repulsion leads to energy gaps between the individual electrons in multiple electron states. Panels (a and b) reproduced from ref. 121 with permission from the Authors. (c) Temporal cross-correlation measurement of two-electron states emitted from a needle tip. (d) Proposed scheme for heralded electron quantum imaging utilizing the two-electron states. An omega filter enables a high-fidelity spatial separation of the two-electron state and suppression of single-electron states. Panels (c and d) reproduced from ref. 120 with permission from the Authors.

far lower than those of standard TEMs or SEMs, without the need for complex electron optics. In this way, low-energy electron diffraction,[46,105,106] point-projection microscopy[123,124] and holography[125] in an ultrafast mode are becoming possible. Finally, ultrafast electron pulses from needle tips have been produced with spin polarization, enabling the observation of magnetic phenomena.[126]

References

1. M. Klein and G. Schwitzgebel, *Rev. Sci. Instrum.*, 1997, **68**, 3099.
2. A. Mustonen, P. Beaud, E. Kirk, T. Feurer and S. Tsujino, *Appl. Phys. Lett.*, 2011, **99**, 103504.
3. M. E. Swanwick, P. D. Keathley, A. Fallahi, P. R. Krogen, G. Laurent, J. Moses, F. X. Kärtner and L. F. Velásquez-García, *Nano Lett.*, 2014, **14**, 5035.
4. P. Helfenstein, A. Mustonen, T. Feurer and S. Tsujino, *Appl. Phys. Express*, 2013, **6**, 114301.
5. D.-S. Yang, O. F. Mohammed and A. H. Zewail, *Proc. Natl. Acad. Sci. U. S. A.*, 2010, **107**, 14993.
6. A. Feist, N. Bach, N. R. da Silva, T. Danz, M. Möller, K. E. Priebe, T. Domröse, J. G. Gatzmann, S. Rost, J. Schauss, S. Strauch, R. Bormann, M. Sivis, S. Schäfer and C. Ropers, *Ultramicroscopy*, 2017, **176**, 63.
7. A. Tafel, S. Meier, J. Ristein and P. Hommelhoff, *Phys. Rev. Lett.*, 2019, **123**, 146802.
8. L. Piazza, D. Masiel, T. LaGrange, B. Reed, B. Barwick and F. Carbone, *Chem. Phys.*, 2013, **423**, 79.
9. T. T. Tsong, *Atom-probe field ion microscopy*, Cambridge University Press, Cambridge, 1990.
10. S. Thomas, G. Wachter, C. Lemell, J. Burgdörfer and P. Hommelhoff, *New J. Phys.*, 2015, **17**, 063010.
11. P. Hommelhoff and M. F. Kling, *Attosecond Nanophysics: From Basic Science to Applications*, Wiley-VCH, Weinheim, 2015.
12. M. Krüger, C. Lemell, G. Wachter, J. Burgdörfer and P. Hommelhoff, *J. Phys. B: At., Mol. Opt. Phys.*, 2018, **51**, 172001.
13. P. Dombi, Z. Pápa, J. Vogelsang, S. V. Yalunin, M. Sivis, G. Herink, S. Schäfer, P. Groß, C. Ropers and C. Lienau, *Rev. Mod. Phys.*, 2020, **92**, 025003.
14. M. I. Stockman, *Phys. Rev. Lett.*, 2004, **93**, 137404.
15. P. Groß, M. Esmann, S. F. Becker, J. Vogelsang, N. Talebi and C. Lienau, *Adv. Phys.: X*, 2016, **1**, 297–330.
16. J. Vogelsang, J. Robin, B. J. Nagy, P. Dombi, D. Rosenkranz, M. Schiek, P. Groß and C. Lienau, *Nano Lett.*, 2015, **15**, 4685–4691.
17. B. Schröder, M. Sivis, R. Bormann, S. Schäfer and C. Ropers, *Appl. Phys. Lett.*, 2015, **107**, 231105.
18. M. Müller, V. Kravtsov, A. Paarmann, M. B. Raschke and R. Ernstorfer, *ACS Photonics*, 2016, **3**, 611–619.

19. S. Keramati, A. Passian, V. Khullar and H. Batelaan, *Appl. Phys. Lett.*, 2020, **117**, 061102.

20. M. Kozák, J. McNeur, N. Schönenberger, J. Illmer, A. Li, A. Tafel, P. Yousefi, T. Eckstein and P. Hommelhoff, *J. Appl. Phys.*, 2018, **124**, 023104.

21. N. Bach, T. Domröse, A. Feist, T. Rittmann, S. Strauch, C. Ropers and S. Schäfer, *Struct. Dyn.*, 2019, **6**, 014301.

22. S. Meier and P. Hommelhoff, *ACS Photonics*, 2021, **9**, 3083.

23. G. M. Caruso, F. Houdellier, S. Weber, M. Kociak and A. Arbouet, *Advances in Physics: X*, 2019, **4**, 1660214.

24. A. Li, Y. Pan, P. Dienstbier and P. Hommelhoff, *Phys. Rev. Lett.*, 2021, **126**, 137403.

25. J. C. H. Spence, W. Qian and M. P. Silverman, *J. Vac. Sci. Technol.*, 1994, **12**, 542–547.

26. J. Spence, *High-Resolution Electron Microscopy*, Oxford University Press, 2013.

27. D. Ehberger, J. Hammer, M. Eisele, M. Krüger, J. Noe, A. Högele and P. Hommelhoff, *Phys. Rev. Lett.*, 2015, **114**, 227601.

28. S. Meier, T. Higuchi, M. Nutz, A. Högele and P. Hommelhoff, *Appl. Phys. Lett.*, 2018, **113**, 143101.

29. T.-Y. Fu, L.-C. Cheng, C.-H. Nien and T. T. Tsong, *Phys. Rev. B*, 2001, **64**, 113401.

30. C.-C. Chang, H.-S. Kuo, I.-S. Hwang and T. T. Tsong, *Nanotechnology*, 2009, **20**, 115401.

31. P. Hommelhoff, C. Kealhofer, A. Aghajani-Talesh, Y. R. Sortais, S. M. Foreman and M. A. Kasevich, *Ultramicroscopy*, 2009, **109**, 423.

32. L. Zhang, J. P. Hoogenboom, B. Cook and P. Kruit, *Struct. Dyn.*, 2019, **6**, 051501.

33. A. L. Cavalieri, N. Müller, T. Uphues, V. S. Yakovlev, A. Baltuška, B. Horvath, B. Schmidt, L. Blümel, R. Holzwarth, S. Hendel, M. Drescher, U. Kleineberg, P. M. Echenique, R. Kienberger, F. Krausz and U. Heinzmann, *Nature*, 2007, **449**, 1029.

34. M. J. G. Lee, *Phys. Rev. Lett.*, 1973, **30**, 1193.

35. P. Hommelhoff, Y. Sortais, A. Aghajani-Talesh and M. A. Kasevich, *Phys. Rev. Lett.*, 2006, **96**, 077401.

36. H. Yanagisawa, M. Hengsberger, D. Leuenberger, M. Klöckner, C. Hafner, T. Greber and J. Osterwalder, *Phys. Rev. Lett.*, 2011, **107**, 087601.

37. J. Sun, A. Adhikari, B. S. Shaheen, H. Yang and O. F. Mohammed, *J. Phys. Chem. Lett.*, 2016, 7, 985.

38. M. Schenk, M. Krüger and P. Hommelhoff, *Phys. Rev. Lett.*, 2010, **105**, 257601.

39. M. Krüger, M. Schenk and P. Hommelhoff, *Nature*, 2011, **475**, 78.

40. R. Bormann, M. Gulde, A. Weismann, S. V. Yalunin and C. Ropers, *Phys. Rev. Lett.*, 2010, **105**, 147601.

41. W. S. Fann, R. Storz and J. Bokor, *Phys. Rev. B*, 1991, **44**, 10980.

42. M. Reutzel, A. Li and H. Petek, *Phys. Rev. B*, 2020, **101**, 075409.

43. C. Ropers, D. R. Solli, C. P. Schulz, C. Lienau and T. Elsaesser, *Phys. Rev. Lett.*, 2007, **98**, 043907.

44. F. Houdellier, G. Caruso, S. Weber, M. Kociak and A. Arbouet, *Ultramicroscopy*, 2018, **186**, 128.

45. F. Houdellier, G. Caruso, S. Weber, M. Hýtch, C. Gatel and A. Arbouet, *Ultramicroscopy*, 2019, **202**, 26.

46. G. Storeck, S. Vogelgesang, M. Sivis, S. Schäfer and C. Ropers, *Struct. Dyn.*, 2017, **4**, 044024.

47. G. Herink, L. Wimmer and C. Ropers, *New J. Phys.*, 2014, **16**, 123005.

48. H. Li, B. Mignolet, G. Wachter, S. Skruszewicz, S. Zherebtsov, F. Süßmann, A. Kessel, S. A. Trushin, N. G. Kling, M. Kübel, B. Ahn, D. Kim, I. Ben-Itzhak, C. L. Cocke, T. Fennel, J. Tiggesbäumker, K.-H. Meiwes-Broer, C. Lemell, J. Burgdörfer, R. D. Levine, F. Remacle and M. F. Kling, *Phys. Rev. Lett.*, 2015, **114**, 123004.

49. D. Matte, N. Chamanara, L. Gingras, L. P. R. de Cotret, T. L. Britt, B. J. Siwick and D. G. Cooke, *Phys. Rev. Res.*, 2021, **3**, 013137.

50. J. McNeur, M. Kozak, D. Ehberger, N. Schönenberger, A. Tafel, A. Li and P. Hommelhoff, *J. Phys. B: At., Mol. Opt. Phys.*, 2016, **49**, 034006.

51. M. Kozák, P. Beck, H. Deng, J. McNeur, N. Schönenberger, C. Gaida, F. Stutzki, M. Gebhardt, J. Limpert, A. Ruehl, I. Hartl, O. Solgaard, J. S. Harris, R. L. Byer and P. Hommelhoff, *Opt. Express*, 2017, **25**, 19195.

52. E. Peralta, K. Soong, E. R. England, R. J.and Colby, Z. Wu, B. Montazeri, C. McGuinness, J. McNeur, K. J. Leedle, D. Walz, E. Sozer, B. Cowan, B. Schwartz, G. Travish and R. L. Byer, *Nature*, 2013, **503**, 91.

53. J. Breuer and P. Hommelhoff, *Phys. Rev. Lett.*, 2013, **111**, 134803.

54. K. J. Leedle, D. S. Black, Y. Miao, K. E. Urbanek, A. Ceballos, H. Deng, J. S. Harris, O. Solgaard and R. L. Byer, *Opt. Lett.*, 2018, **43**, 21814.

55. N. V. Sapra, K. Y. Yang, D. Vercruysse, K. J. Leedle, D. S. Black, R. J. England, L. Su, R. Trivedi, Y. Miao, O. Solgaard, R. L. Byer and J. Vučković, *Science*, 2020, **367**, 79.

56. R. Shiloh, J. Illmer, T. Chlouba, P. Yousefi, N. Schönenberger, U. Niedermayer, A. Mittelbach and P. Hommelhoff, *Nature*, 2021, **597**, 498.

57. K. Shimoda, *Appl. Opt.*, 1962, **1**, 33.

58. A. Lohmann, *IBM Tech. Note*, 1962, **5**, 169.

59. K. Mizuno, J. Pae, T. Nozokido and K. Furuya, *Nature*, 1987, **328**, 45.

60. T. Plettner, R. L. Byer, E. Colby, B. Cowan, C. M. S. Sears, J. E. Spencer and R. H. Siemann, *Phys. Rev. Lett.*, 2005, **95**, 134801.

61. T. Plettner, P. P. Lu and R. L. Byer, *Phys. Rev. Spec. Top.-Accel. Beams*, 2006, **9**, 111301.

62. T. Plettner and R. L. Byer, *Phys. Rev. Spec. Top.-Accel. Beams*, 2008, **11**, 030704.

63. T. Plettner, R. L. Byer, C. McGuinness and P. Hommelhoff, *Phys. Rev. Spec. Top.-Accel. Beams*, 2009, **12**, 101302.

64. J. Breuer, R. Graf, A. Apolonski and P. Hommelhoff, *Phys. Rev. Spec. Top.-Accel. Beams*, 2014, **17**, 021301.

65. R. J. England, R. J. Noble, K. Bane, D. H. Dowell, C.-K. Ng, J. E. Spencer, S. Tantawi, Z. Wu, R. L. Byer, E. Peralta, K. Soong, C.-M. Chang, B. Montazeri, S. J. Wolf, B. Cowan, J. Dawson, W. Gai, P. Hommelhoff, Y.-C. Huang, C. Jing, C. McGuinness, R. B. Palmer, B. Naranjo, J. Rosenzweig, G. Travish, A. Mizrahi, L. Schachter, C. Sears, G. R. Werner and R. B. Yoder, *Rev. Mod. Phys.*, 2014, **86**, 1337.

66. D. Cesar, S. Custodio, J. Maxson, P. Musumeci, X. Shen, E. Threlkeld, R. J. England, A. Hanuka, I. V. Makasyuk, E. A. Peralta, K. P. Wootton and Z. Wu, *Commun. Phys.*, 2018, **1**, 46.

67. K. J. Leedle, A. Ceballos, H. Deng, O. Solgaard, R. F. Pease, R. L. Byer and J. S. Harris, *Opt. Lett.*, 2015, **40**, 4344.

68. M. Kozák, M. Förster, J. McNeur, N. Schönenberger, K. Leedle, H. Deng, J. Harris, R. Byer and P. Hommelhoff, *Nucl. Instrum. Methods Phys. Res., Sect. A*, 2017, **865**, 84.

69. Y. Miao, D. S. Black, K. J. Leedle, Z. Zhao, H. Deng, A. Ceballos, R. L. Byer, J. S. Harris and O. Solgaard, *Opt. Lett.*, 2020, **45**, 391.

70. N. Schönenberger, A. Mittelbach, P. Yousefi, J. McNeur, U. Niedermayer and P. Hommelhoff, *Phys. Rev. Lett.*, 2019, **123**, 264803.

71. D. S. Black, U. Niedermayer, Y. Miao, Z. Zhao, O. Solgaard, R. L. Byer and K. J. Leedle, *Phys. Rev. Lett.*, 2019, **123**, 264802.

72. U. Niedermayer, D. S. Black, K. J. Leedle, Y. Miao, R. L. Byer and O. Solgaard, *Phys. Rev. Appl.*, 2021, **15**, L021002.

73. P. Yousefi, N. Schönenberger, J. Mcneur, M. Kozák, U. Niedermayer and P. Hommelhoff, *Opt. Lett.*, 2019, **44**, 1520.

74. U. Niedermayer, T. Egenolf, O. Boine-Frankenheim and P. Hommelhoff, *Phys. Rev. Lett.*, 2018, **121**, 214801.

75. U. Niedermayer, T. Egenolf and O. Boine-Frankenheim, *Phys. Rev. Lett.*, 2020, **125**, 164801.

76. R. Shiloh, T. Chlouba, P. Yousefi and P. Hommelhoff, *Opt. Express*, 2021, **29**, 14403.

77. H. Deng, K. J. Leedle, Y. Miao, D. S. Black, K. E. Urbanek, J. McNeur, M. Kozák, A. Ceballos, P. Hommelhoff, O. Solgaard, R. L. Byer and J. S. Harris, *Adv. Opt. Mater.*, 2020, **8**, 1901522.

78. T. Chlouba, R. Shiloh, P. Forsberg, M. Hamberg, M. Karlsson, M. Kozák and P. Hommelhoff, *Opt. Express*, 2022, **30**, 505.

79. B. Barwick, D. J. Flannigan and A. H. Zewail, *Nature*, 2009, **462**, 902.

80. A. Feist, K. E. Echternkamp, J. Schauss, S. V. Yalunin, S. Schäfer and C. Ropers, *Nature*, 2015, **521**, 200.

81. K. E. Echternkamp, A. Feist, S. Schäfer and C. Ropers, *Nat. Phys.*, 2016, **12**, 1000.

82. K. E. Priebe, C. Rathje, S. V. Yalunin, T. Hohage, A. Feist, S. Schäfer and C. Ropers, *Nat. Phot.*, 2017, **11**, 793.

83. R. Dahan, S. Nehemia, M. Shentcis, O. Reinhardt, Y. Adiv, X. Shi, O. Be'er, M. H. Lynch, Y. Kurman, K. Wang and I. Kaminer, *Nat. Phys.*, 2020, **16**, 1123.

84. R. Shiloh, T. Chlouba and P. Hommelhoff, *Phys. Rev. Lett.*, 2022, **128**, 235301.

85. F. O. Kirchner, A. Gliserin, F. Krausz and P. Baum, *Nat. Photonics*, 2014, **8**, 52.

86. Y. Morimoto and P. Baum, *Nat. Phys.*, 2018, **14**, 252.

87. Y. Morimoto and P. Baum, *Phys. Rev. Lett.*, 2020, **125**, 193202.

88. A. Ryabov, J. W. Thurner, D. Nabben, M. V. Tsarev and P. Baum, *Sci. Adv.*, 2020, **6**, eabb1393.

89. R. Dahan, A. Gorlach, U. Haeusler, A. Karnieli, O. Eyal, P. Yousefi, M. Segev, A. Arie, G. Eisenstein, P. Hommelhoff and I. Kaminer, *Science*, 2021, **373**, eabj7128.

90. O. Kfir, H. Lourenço-Martins, G. Storeck, M. Sivis, T. R. Harvey, T. J. Kippenberg, A. Feist and C. Ropers, *Nature*, 2020, **582**, 46.

91. J.-W. Henke, A. S. Raja, A. Feist, G. Huang, G. Arend, Y. Yang, F. J. Kappert, R. N. Wang, M. Möller, J. Pan, J. Liu, O. Kfir, C. Ropers and T. J. Kippenberg, *Nature*, 2021, **600**, 653.

92. N. Talebi, *Phys. Rev. Lett.*, 2020, **125**, 080401.

93. P. Radcliffe, M. Arbeiter, W. B. Li, S. Düsterer, H. Redlin, P. Hayden, P. Hough, V. Richardson, J. T. Costello, T. Fennel and M. Meyer, *New J. Phys.*, 2012, **14**, 043008.

94. M. Kozák, T. Eckstein, N. Schönenberger and P. Hommelhoff, *Nat. Phys.*, 2018, **14**, 121.

95. O. Schwartz, J. J. Axelrod, S. L. Campbell, C. Turnbaugh, R. M. Glaeser and H. Müller, *Nat. Methods*, 2019, **16**, 1016.

96. P. Baum and A. H. Zewail, *Proc. Nat. Acad. Sci. U. S. A.*, 2007, **104**, 18409.

97. M. Kozák, N. Schönenberger and P. Hommelhoff, *Phys. Rev. Lett.*, 2018, **120**, 103203.

98. M. Kozák, *Phys. Rev. Lett.*, 2019, **123**, 203202.

99. M. Kozák, *ACS Photonics*, 2021, **8**, 431.

100. A. Ryabov and P. Baum, *Science*, 2016, **353**, 374.

101. Y. Kurman, R. Dahan, H. H. Sheinfux, K. Wang, M. Yannai, Y. Adiv, O. Reinhardt, L. H. G. Tizei, S. Y. Woo, J. Li, J. H. Edgar, M. Kociak, F. H. L. Koppens and I. Kaminer, *Science*, 2021, **372**, 1181.

102. B. J. Siwick, J. R. Dwyer, R. E. Jordan and R. J. D. Miller, *Science*, 2003, **302**, 1382.

103. P. Baum, D.-S. Yang and A. H. Zewail, *Science*, 2007, **318**, 788.

104. M. Gulde, S. Schweda, G. Storeck, M. Maiti, H. K. Yu, A. M. Wodtke, S. Schäfer and C. Ropers, *Science*, 2014, **345**, 200.

105. S. Vogelgesang, G. Storeck, J. G. Horstmann, T. Diekmann, M. Sivis, S. Schramm, K. Rossnagel, S. Schäfer and C. Ropers, *Nat. Phys.*, 2018, **14**, 184.

106. G. Storeck, J. G. Horstmann, T. Diekmann, S. Vogelgesang, G. von Witte, S. V. Yalunin, K. Rossnagel and C. Ropers, *Struct. Dyn.*, 2020, 7, 034304.

107. S. R. Tauchert, M. Volkov, D. Ehberger, D. Kazenwadel, M. Evers, H. Lange, A. Donges, A. Book, W. Kreuzpaintner, U. Nowak and P. Baum, *Nature*, 2022, **602**, 73.

108. N. Rubiano da Silva, M. Möller, A. Feist, H. Ulrichs, C. Ropers and S. Schäfer, *Phys. Rev. X*, 2018, **8**, 031052.
109. A. Gover and A. Yariv, *Phys. Rev. Lett.*, 2020, **124**, 064801.
110. F. J. García de Abajo, *Phys. Rev. Lett.*, 2021, **126**, 019501.
111. A. Gover and A. Yariv, *Phys. Rev. Lett.*, 2021, **126**, 019502.
112. Y. Morimoto and P. Hommelhoff, *Phys. Rev. Res.*, 2020, **2**, 043089.
113. Y. Morimoto, P. Hommelhoff and L. B. Madsen, *Phys. Rev. A*, 2021, **103**, 043110.
114. A. Karnieli, N. Rivera, A. Arie and I. Kaminer, *Phys. Rev. Lett.*, 2021, **127**, 060403.
115. O. Kfir, V. D. Giulio, F. J. G. de Abajo and C. Ropers, *Sci. Adv.*, 2021, 7, eabf6380.
116. R. Ruimy, A. Gorlach, C. Mechel, N. Rivera and I. Kaminer, *Phys. Rev. Lett.*, 2021, **126**, 233403.
117. Y. Adiv, K. Wang, R. Dahan, P. Broaddus, Y. Miao, D. Black, K. Leedle, R. L. Byer, O. Solgaard, R. J. England and I. Kaminer, *Phys. Rev. X*, 2021, **11**, 041042.
118. O. Kfir, *Phys. Rev. Lett.*, 2019, **123**, 103602.
119. A. Feist, G. Huang, G. Arend, Y. Yang, J.-W. Henke, A. S. Raja, F. J. Kappert, R. N. Wang, H. Lourenço-Martins, Z. Qiu, J. Liu, O. Kfir, T. J. Kippenberg and C. Ropers, *Science*, 2022, 377, 777.
120. (a) S. Meier, J. Heimerl and P. Hommelhoff, ArXiv, 2022, preprint, arXiv:2209.11806; (b) *Nat. Phys.*, DOI: 10.1038/s41567-023-02059-7.
121. (a) R. Haindl, A. Feist, T. Domröse, M. Möller, S. V. Yalunin and C. Ropers, ArXiv, 2022, preprint, arXiv:2209.12300; (b) *Nat. Phys.*, DOI: 10.1038/s41567-023-02067-7.
122. P. J. Mosley, J. S. Lundeen, B. J. Smith, P. Wasylczyk, A. B. U'Ren, C. Silberhorn and I. A. Walmsley, *Phys. Rev. Lett.*, 2008, **100**, 133601.
123. E. Quinonez, J. Handali and B. Barwick, *Rev. Sci. Instrum.*, 2013, **84**, 103710.
124. M. Müller, A. Paarmann and R. Ernstorfer, *Nat. Commun.*, 2015, 5, 5292.
125. J. Vogelsang, N. Talebi, G. Hergert, A. Wöste, P. Groß, A. Hartschuh and C. Lienau, *ACS Photonics*, 2018, 5, 3584.
126. E. Brunkow, E. R. Jones, H. Batelaan and T. J. Gay, *Appl. Phys. Lett.*, 2019, **114**, 073502.

Subject Index